ENCYCLOPAEDIA OF
ENVIRONMENTAL SOIL AND MARINE POLLUTION

ENCYCLOPAEDIA OF ENVIRONMENTAL SOIL AND MARINE POLLUTION

[Vol. 1]

Edited by

G. R. CHHATWAL

ANMOL PUBLICATIONS PVT LTD

NEW DELHI- 110 002 (INDIA)

ANMOL PUBLICATIONS PVT LTD
4374/4B, Ansari Road, Daryaganj
New Delhi- 110 002

Encyclopaedia of Environmental Soil and Marine Pollution

First Edition 1997

ISBN 81-7488-434-3

PRINTED IN INDIA

Published by J. L. Kumar for Anmol Publications Pvt Ltd, New Delhi and Printed at Mehra Offset Press, Delhi.

Contents

Preface

Encyclopaedia of Environmental Soil and Marine Pollution, organised in two volumes, is a veritable mine of authoritative information on the menacing soil and marine pollution conditions endangering plant and marine lives.

Pollution is "the wrong amount of the wrong things in the wrong place at the wrong time".

Soil connects the biosphere to a source of mineral elements in the lithosphere, the earth's rock mantle, to a source of gaseous elements in the atmosphere, and to a source of water from the hydrosphere. It plays critical roles in the earth's carbon, nitrogen, sulfur, and hydrologic cycles.

Introduction by man, directly or indirectly, of substances or energy into the marine environment (including estuaries) resulting in such deleterious effects as harm to living resources, hazards to human health, hindrance to marine activities including fishing, impairment of quality for use of sea water and reduction of amenities.

The present encyclopaedia will prove useful and informative to researchers and teachers besides the general readers, policy planners and activists in the field of environment.

— Editor

1

Soil: Physical Condition and Plant Growth

I. Introduction

Crops, the source of food and fiber for bulk of the world's billions of people, derive many benefits from the soil on which they grow. No single feature of a soil is more important than the characteristics of its pore system. Pore system characteristics are a function of soil structure, the arrangement of mineral fragments into larger secondary units. Soil structure imparts tilth to a soil and determines its behaviour in the biosphere, the carbon cycle, the nitrogen cycle, the sulfur cycle, and the hydrologic cycle. Maintenance of soil structure requires energy inputs and, if these are not made, desertification or salinization — destruction of the soil resource — is the consequence.

Soil is a body in nature composed of organized, unconsolidated mineral and organic material on the immediate surface of the earth that serves as a natural medium for the growth of land plants. The soil is the consequence of the parent material from which it is formed and the environmental factors climate, micro- and macroorganisms, topography, and relief acting over a period of time. Soils differ from parent material in many physical, chemical, and biological features. The very thin layer of soil on the land surface of the earth has special characteristics because of its formation processes, and these special properties allow it to perform for the terrestrial biosphere in a fashion similar to the function of the placenta. These special characteristics are largely a function of the pore system produced by gravity, swelling and shrinking forces, seismic noise, forces arising from the growth and activities of soil biota, capillary and drag forces arising from the properties and motion of water, and sundry other naturally induced forces. This system of forces arranges and connects the unconsolidated mineral and organic material in structures which are described here.

Soil connects the biosphere to a source of mineral elements in the lithosphere, the earth's rock mantle, to a source of gaseous elements in the atmosphere, and to a source of water from the hydrosphere. It plays critical roles in the earth's carbon, nitrogen, sulfur, and hydrologic cycles. In the carbon cycle, plants absorb carbon dioxide from the atmosphere and fix carbon in organic form through the process of photosynthesis. Plants can do this because during the growing season, or in the case of perennials for several sequential growing seasons and intervening times, many of their needs are provided by the soil.

In addition, as part of the carbon cycle, carbon is stored in the soil in the form of organic matter, a collection of relatively stable compounds which decompose slowly. The amount of organic carbon stored in the soil per unit area of land surface is frequently equivalent to 5-10 times the amount of carbon fixed annually by photosynthesizers occupying the same area. It is especially high in forest soils, wetlands, and native grasslands. The physical condition of soil is important in the soil's role in the carbon cycle because (1) it influences the amount of photosynthesis and plant growth that occurs at a given site during a given season and (2) it influences the nature of the organic residue decomposition processes that occur at a given location over a time interval. Critical processes which produce these impacts are described below.

The role of soils in the nitrogen cycle is manifold. Soils provide suitable sites for nitrogen fixation by free-living and symbiotic microorganisms and for organic nitrogen mineralization by decomposers. They furnish favourable locations for oxidation of reduced nitrogen compounds to nitrite and nitrate and for absorption of nitrogen in the ammonium and nitrate form by microorganisms and higher plants. Soils provide potential sites for denitrification of nitrate to nitrous oxide and nitrogen gases. Thus, nitrogen is biologically commuted from the atmosphere to the biosphere, where it plays a critical role, and returned to the atmosphere in the completion of the cycle. The physical condition of the soil — the characteristics of the substratum in concert with local and regional topography — determine the rate of removal of excess (drainable) moisture from the soil. This in turn controls the nature of the nitrogen transformations in the medium during various seasons because it regulates soil temperature and oxygen supplies. [See Nitrification; Nitrogen in Soil.]

Soils play a similar role in the sulfur cycle. Hydrogen sulfide escaping from wetlands, swamps, and marshes is oxidized in the atmosphere to sulfur dioxide and dissolved in rainwater. It is further oxidized to sulfuric acid and returned to the land or sea surface. Plants absorb sulfate from soils and incorporate sulfur into protein, which returns to the soil and in well-aerated soils is reoxidized to sulfate. Thus soil mediates a mini oxidation - reduction cycle of sulfur in the biosphere which involves microorganisms and frequently higher plants. Reduced sulfur escapes from lower portions of soil profiles in the form of hydrogen sulfide. It can be oxidized to sulfuric acid in the aerated upper portion of the soil. The sulfuric acid participates in dissolving minerals, thus affecting soil formation and release of nutrients from unweathered rock and mineral fragments. [See Soil Biochemistry.]

Again, as in the case of nitrogen, the nature, magnitude, and consequences of the transformations of sulfur depend on soil physical conditions and their impact on flow of gases and aqueous solutions into and out of soils.

Another highly significant process in which soils engage is a buffering action in the hydrologic cycle. To survive and grow plants need water every day. In most terrestrial climatic zones water accumulated during storms must be stored in the root zones and carried over into dry periods. Soils have the capacity to store water in a condition in which it is available to plants. Even in humid tropical ecoregions plants transpire water from that stored in the soil. this water becomes part of atmospheric water vapour which contributes to downwind precipitation.

In addition, soil absorbs moisture from rains and snowmelts, which flows slowly through the soil toward groundwater and downhill toward surface streams, rivers, and lakes. This slow release of water is a critical part of the hydrologica cycle. It reduces peak stream flows and increases flows between precipitation events. [See Water Cycle, Global.]

What features of soils enable them to be so instrumental in the biosphere and hydrosphere? What characteristics contribute to their roles in the carbon, nitrogen, sulfur, and hydrologic cycles? The single most important feature of soils in these respects is an internal porosity, a pore system with just the right dimensions to interact expeditiously with water and air. The dimensions of the pore system are such that drainage of excess water is rapid, but that drainage virtually ceases while there still remains a fair proportion of water in the soil. In the ideal pore system,. a large proportion of this remaining water will be available for use by plants. The flow rates and the partitioning of water into drainage water, available water, and unavailable water depend on the pore size distribution, that is, the dimensions of the pore system. A mechanical model of such a system may be represented in thought by a leaky barrel filled with sponges of varying types, tubes of varying lengths with diameters ranging from micrometers to millimeters, larger pipes, dead-end cavities, springs for strength and elasticity, and dashpots to regulate the response of the springs to forces. Keeping this model in mind, we may proceed to examine the functions and characteristics of soils in their role in the growth and development of crops for food and fiber for people the world over.

II. Function of Soil

In relation to plant growth, under most natural horticultural and agronomic circumstances, soils have the following functions:

1. Provide anchorage and support to enable plants to stand erect and absorb solar radiation. Provide support for the foot and vehicle traffic in connection with husbandry of crops.
2. Provide water to plant root systems.
3. Furnish plant nutrients at adequate rates and levels.
4. Permit the flow of gases to provide oxygen for respiring root systems.
5. Expedite the flow of heat to subsurface layers.
6. Provide a physical chemical regime for regulation of toxins.

Soils must have sufficient strength to provide the anchorage necessary for plants to stand erect and hold their leaves in the sun against the force of gravity and the drag forces of winds. Simultaneously, soils must be weal enough or have regions of weakness so that they can be adequately penetrated by roots. Roots can create pressures of several atmospheres, yet can use these pressures only if they have short thrust work to do. Over greater lengths, roots bend, go around obstacles, and follow paths of least resistance in directions dependent on reaction strengths to thrust against. Thus, roots need pores which are greater in effective diameter than the relatively blunt root tip. Soils, then, must be strong (hard) and weak (soft) at the same time to meet the needs of plants.

Soils used in people's food and fiber systems must also have sufficient bearing capacity to support the weight of walking men and women, draft animals, grazing animals and vehicular traffic. soils must also have adequate shear strength to provide traction for wheeled and track-type vehicles.

Mineral soils are composed of finely comminuted mineral particles of irregular shape and sizes ranging from about 0.0001 to 2 mm effective diameter. (Particles larger than 2 mm effective diameter are present in soils but considered to be gravel or stone admixtures rather than soil.) Such mixtures of sizes and shapes would rarely have appropriate strength for the purposes of crop pants under both wet and dry conditions if they were only mixtures of solid mineral particles without some special structural features such as those described in this article.

Plants use a great deal of water while standing in sunlitfields. Most plants transpire between 500 and 1000 lb of water for each pound of dry matter produced. The ;water passes through the plant and into the atmosphere because the plant must expose a wet surface to a dry atmosphere in order to absorb carbon dioxide,. In this exposure a great deal of water evaporates from the moist tissue surfaces even though these are somewhat protected. The evaporation of water from leaf tissue cools the tissue and prevents the attainment of temperatures harmful to the living system. Water is also the carrier for chemicals which enter the plant from the soil. The transpiration stream moves essential chemicals to all portions of plant tissue, especially to growing portions. In addition, water is the medium in the plant cells in which the various biophysical and biochemical reactions occur. Water contributes hydrogen for the synthesis of organic compounds produced in plant cells. Plant tissue is ordinarily stiff and moderately rigid because of turgor pressure, the wall pressure caused by water drawn into cells by osmotic potential and pressing against cell membranes to impart stiffness. [See Evaporation, Transpiration, and Water Uptake of Plants.]

Seventeen chemical elements are necessary for plant growth. These are known as plant nutrients. Most are supplied to the plant via the soil and most of these are in the water in the soil (the soil solution) before they are absorbed by the plant. Part of the process of absorption of nutrients from the soil is performed with the plant acting as a nutrient pump, that is, moving nutrients from places where their concentration is low to places where their concentration is high. This process requires energy, which is obtained from biological processes that require oxygen. The availability of nutrients to plants depends on the abundance of the nutrients in the soil and their chemical form. Soils which contain no air will provide reducing conditions for chemical reactions, and this can dramatically reduce the availability of some nutrients to plants. Examples of nutrients thus affected are nitrogen, sulfur, and several heavy metals. Metals may precipitate as relatively insoluble sulfides under anaerobic conditions.

For water and nutrients for upland plants, and oxygen supply in the soil, below the ground surface, is absolutely essential. Oxygen is essential for releasing the metabolic energy to do the work of overcoming mechanical impedance of hard (strong) soils to root elongation and growth, the work of moving chemical ions against concentration gradients, and the pressurization work against the strength and elasticity of cell walls and membranes. Aeration or oxygenation of the soil affects the availability of several nutrients essential to the growth, vigour, and health of upland types of plants. Soil aeration is also essential for the decomposition of organic matter and the release of nutrients contained therein.

These systems cannot perform if they are excessively cold. Heat must be provided to warm the root system — to warm the soil. Heat must be conducted downward from the soil surface, and this heat conduction is an important aspect of the physical behavior of the soil with respect to the welfare of the plant. Heat released to a cold atmosphere can reduce frost damage suffered by tender new tissue during early morning hours if the air temperature is near 32° F (0° C).

Finally, the soil is a repository of many dead and dying organic materials, as well as a plethora of inorganic substances, many of which would be toxic if they were not quickly decomposed or rendered ineffective by some means, such as precipitation or sorption on the surfaces of the soil particles. One of the functions performed by soil for crops is the inactivation of hazardous chemicals in the environment of the plant root system.

How can all of these functions be performed simultaneously in the root zone for the welfare of our crops? The physical features of the system which are important are not the particles of solids, but the surfaces of these particles and the pore spaces between them. The surfaces of particles and the pores are where the action is. Good physical condition is a matter of an appropriate distribution of pore sizes and particle surfaces to achieve an adequate balance of the various physical and chemical activities which must occur to meet the needs of the crop plant.

III. The Pore System

It is convenient to introduce the concept of capillarity, that is, the rise of fluid such as in a wick or in a fine tube against the force of gravity. Capillarity plays an important role in the functioning of a soil to meet the needs of crops. The laws of capillarity dictate that after a soil has been thoroughly wetted and time is allowed for drainage, the very large pores, spaces between soil particles, will drain free of water but the very smallest pores are likely to remain filled. They remain filled because capillary forces cause the water to be suspended by providing an upward force opposing gravity. If the soil is to be well drained, it must have pores which are sufficiently large that the capillary forces will be overcome by gravity and the water will flow out. With this draining out of the water, air is sucked into the soil, the system becomes aerated, and the necessary processes referred to earlier can occur. [See Soil Physics.]

Consider tow boxcars of equal dimensions sitting on a railroad track. One boxcar is filled with basketballs, each basketball contained within a cubic box just large enough to hold it. The other boxcar is filled with BBs similarly contained in tiny cubic boxes. In both cars the boxes containing the spheres are arranged in the same fashion. The boxes are stacked in orthogonal straight lines funning the length, the width, and the height of the interior of each car.

Obviously, each of the eight corners of each small box is empty, contains only air. That is, it is outside the sphere contained within the box. this is equally true for the BBs and the basketballs. In each box, approximately 52.4% of the vólume is within the sphere (the basketball or the BB) and the remaining volume, about 47.6% of the total, is outside the contained sphere; it is pore space. If the boxes were magically removed, and the stacking or the packing remained the same, this pore space would be seen as openings running in straight lines traversing the length, the width, and the height of each boxcar.

The pore volume — the porosity, or fraction of the total volume occupied by pores — would be the same in the two boxcars (47.6%) but obviously the pore systems would be vastly different. The pores between BBs would be much smaller in average cross-sectional area and average diameter but, being far closer to each other, would be smaller than those in the basketball system, any water filling them would be closer to a solid surface and therefore have far greater viscous resistance to movement that the water in the basketball system. This is very similar to the behavior of water in soils. Water flows with ease rapidly through coarse sands (effective diameter of 0.5-1.0 mm) and very, very slowly through clays (effective diameter of 0.0001-0.002 mm). [See Geophysical Fluid Mechanics.]

If the basketballs and BBs were removed from the cubic containers and poured and packed into the boxcars, basketballs into one and BBs into the other, it is virtually impossible that they would end up in any uniform, consistent, or systematic packing or arrangement. The packing would be random, a mixture of possible arrangements varible from place to place and, on the average, more dense, or less porous, than the regular system described above. The cubic system, with 47.6% porosity, is the loosest, most porous of six possible regular packing arrangements. When uniform spheres of glass, all the same size, settle out of a suspension in water, they usually pack in an arrangement close to the tightest theoretical packing, the tetrahedral packing with a porosity of approximately 26%. In such a packing, the pores are not straight but are rather tortuous, as well as being smaller than pores in cubic packing.

Thus far our consideration of porosity has dealt with systematic packs of solid rigid spheres with porosity ranging from 26 to 48%. The range of porosities found in soils is similar; the bulk of soil porosities would all in or near this range. However, there are a few soils with porosities below 26% and many with porosities greater than 48%. These departures deserve explanation. Two other observations about soils are also worthy of our consideration at this time: (1) fine-textured soils usually have more porosity than coarse-textured soils, and (2) in many fine-textured soils a large portion of the porosity is due to pores whose effective diameter is very large compared to the size of the particles making up the walls of the pores. In the theoretical systems described above, this could not occur. In those systems the pores have maximum effective diameters approximately equal to the effective diameters of the constituent particles and minimum diameters of approximately one-sixth to two-fifths of the effective diameters of the constituent particles.

IV. Aggregations of Fines

In Fig. 1.1, P represents a fine particle of diameter D. Imagine these particles arranged in close or tetrahedral packing as shown diagrammatically in Fig. 1.1. In this packing the volume of voids (pores) would be 0.26 of the the total volume of the pack and the volume of solids would be 0.74. If there were a means by which these primary particles P could be cemented together in large secondary units A, as shown in Fig. 1.2, and the particles between aggregates A moved to new locations to agglomerate to additional aggregates, the volume of the system would increase by occupation of new locations and the porosity of the system

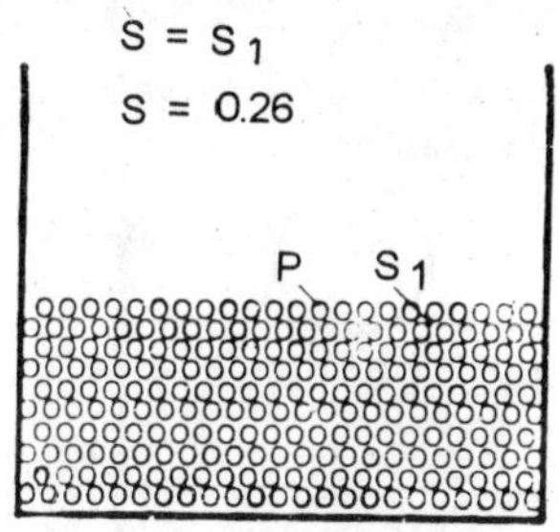

Fig. 1.1 Schematic representation of interparticle pores in a system of solid rigid spheres in tetrahedral (close) packing. The pores, S_1, are shown at their minimum cross-sectional area, where their "diameter" is approximately one-sixth the diameter of the solid spheres P. Pores in tetrahedral packing vary in both size and shape along their length and are tortuous, not straight. The diagram is slightly expanded to improve clarity.

would increase by the generation of pore system S_2 as in Fig. 1.2. Pores of the system S_2 would be much coarser, larger pores than those of system S_1. The ratio of diameters of the pore systems would be related to the ratio of diameters of the pore systems would be related to the ratio of the diameter of the particle, D, to the diameter of the aggregates, A, By this modeling of the generation of a secondary pore system, the pores between aggregates, one can visualize the generation of systems of high total porosity, higher than attainable by "open" packing, and a system of coarse pores, in which there are pores larger than the solid particles making up the wall system of the pores. As indicated, this is known to occur in soils.

Fine-textured soils, low in sand and high in clay, have more porosity than coarse-textured sandy soils because more large aggregates are formed and survive wetting in these fine-textured materials. Survival of wetting is critical for the effectiveness of aggregation. Soils are subject to great changes in moisture content. They are wetted by rains and irrigations and dried by drainage, water use by plants, and evaporation. If the aggregation were not water stable and could not withstand such cycling, all fine-textured soils would upon wetting consolidate into dense, low-porosity systems because of the collapse of the aggregate structure under the forces of gravity and negative water pressures pulling particles closer together, just as wet paper sticks to a mirror or to one's face. If such were the conditions of the world's soils, people's diet would be much different because most fine soils would be too compact and poorly drained to produce most upland crops.

A few soils have total porosities below 26% because lack of aggregation into water-stable units allows small particles to fill pores between larger particles. Silt and fine sand, for example, may fill pores in coarse sand and gravel. When cementing agents are deficient, when water-stable aggregates are absent or nearly so, and when the natural particle size distribution allows fine particles to fill the pores between coarse particles, the porosity of the system can fall to below the theoretical porosity of close-packed systems. Such systems can arise because of the features of certain particle size distributions with no cementing agents or because excessive disruptive forces on the same size distributions containing adequate cements overcome their stabilizing capability.

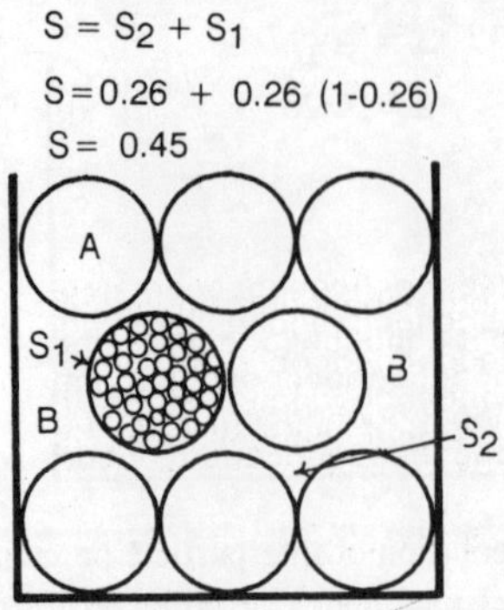

Fig. 1.2. Schematic representation of a cross section of interparticle (S_1) and interaggregate (S_2) pores in a tetrachedral pack of aggregates A which are very large compared to the primary particles (labeled P in Fig. 1). As in Fig. 1.1, the pores are shown on an intersection iwth a plane where their effective "diameter" is minimum. Note the calculation of the pore volume S, which is the sum of two components, S_1 and S_2. Areas B represent an anomaly due to the container walls and would not exist in a large volume of porous medium. The diagram is expanded slightly to enhance clarity.

A. Aggregation Process

Water-stable aggregates are formed by a two-step process. First, soil particle systems, mixture of fragments of rocks, minerals, and organic compounds of an infinite variety of shapes and ranging in effective size from about 0.0001 to about 1.0 mm (see earlier footnote), are subjected to forces such as seismic noise, root pressures, swelling and shrinking, pore water pressure, wetting and drying, and the attraction and /or repulsion of electrically charged surfaces and dipoles. This results in quasistable, temporary orientations of particles relative to each other. Such quasistable orientations are easily disrupted by moving water, raindrop impact, seismic vibration, gravity, and so forth, and as existing "weak" orientations are disrupted and new ones are established, a larger proportion of more stable configurations evolves. Certain agents in soils can act as cements and stabilize particle configurations against weak disruptive forces. Ordinarily the cementation is not strong and can easily be overcome, but it is of sufficient strength to impart stability against a variety of naturally occurring processes. Such cementation produces water stability and allows aggregate systems to be wetted and dried without disintegrating. Cementing agents will be identified and discussed in detail later.

B. Pore Size Distribution

If aggregation of primary particles creates a secondary pore system of larger pores, it follows that aggregation of aggregates would create a tertiary pore system containing still larger pores. This stepwise process could continue until aggregates were hundreds to thousands of times larger than the primaries of which they were composed. A pore system would be created in which the coarse pores were thousands of times larger than the finest pores. Such a phenomenon does occur, and pore systems of very wide size distribution occur in soils. These pore systems make soils into three-phase systems simultaneously containing solids, liquids, and gases under the influence of sorption, surface tension forces, and gravity.

V. Pore System Stability

A model based on characteristics of clean, uniform-bore capillary tubes is used to describe soil pore systems. Let us suppose that glass tubes of a variety of bores ranging from 0.0001 mm to 10 cm are bundled together. Each tube is 20 cm long. After being fixed together, the bundle of tubes is submersed in water. Appropriate action is taken to remove all air from inside the tubes and accomplish complete filling with water. Next, the bundle is lifted from the water, held with tubes in a vertical position, and observed. The large tubes, having bore diameters of 1 cm or 5 mm, drain completely. Midsize tubes lose some water, but the fine tubes suffer virtually no drainage. We have created a three-phase system containing solids, liquids, and gases under the influence of gravity and at equilibrium with the atmosphere. This system can do what a soil must do for plants: have rigidity, provide water, and provide for flow of oxygen to root systems and of carbon dioxide from root respiration to the atmostphere. If we make our bundle of tubes longer, more water will drain from the system before equilibrium is attained. If we apply suction to the lower end of the tubes, we can extract more water. If we apply small positive air pressures at the top of the bundle, we can force more water out. As we continue pressure application, we can measure the size distribution of the tubes as we observe the effects of changes in applied pressure on the volume of water remaining in the tubes (or the volume drained out). We have a capillary model which can be used to characterize the pore size distribution in a soil and evaluate the capacity of the soil to provide the air and water a plant needs.

Soil scientists and engineers have made thousands of these measurements. This is one possible way to describe the soil physical condition as it affects plant growth, and it provides an important tool for understanding tilth. We can measure the coarse pores which must provide the avenues for rapid drainage. Upon drainage, these pores are free of water and can serve as conductors for gases to provide oxygen for root respiration and permit outflow of carbon dioxide, preventing its accumulation to suffocative levels. We can measure the drainability and thereby estimate the soil warming rats in the spring. A wet soil is a cold soil because of the large amounts of hear needed to elevate water temperature, Obviously, the characteristics of the pore system (voids) influence soil strength and soil hardness at various moisture contents. They influence tilth. [See Soil Water: Liquid, Vapor, and Ice.]

For many years measurement of the water content at certain free energy levels, i.e., certain negative water pressures, has been used to characterize the capacity of a soil to store water which will be available to plants. Fir very many soils, the laboratory-determined water content judged to be "in equilibrium" with an applied pressure of - 0.033 MPa corresponds well to a relative wetness called field capacity. Field capacity is defined as the moisture content of a soil when drainage virtually ceases, usually measured 48-72 h after a rain or irrigation. This could be taken as the upper limit of plant available moisture. Although useful, this concept is vague, and field capacity is not a soil property. There is no physical reason for field capacity to correspond to - 0.033 MPa. It does so because at this level of wetness, the hydraulic conductivity of a soil is very low and falling rapidly with decreasing pore water pressure (decreasing soil water potential). Thus, further changes by drainage are very slow. In the case of weekly or more frequent irrigation or substantial rainfall, additional drainage losses are negligible, and thus field capacity is a useful concept.

However, in dryland agriculture with an extended dry period between the rainy season and water use by crops, the concept is far less useful and may even be misleading. For example, in the Columbia Plateau dryland wheat area of the Pacific Northwest (Oregon, Washington, and Idaho) a soil with a laboratory-determined volumetric field capacity of 25% may drain to a volumetric water content of 18% by late June if rainfall virtually ceases at mid-April. In a soil depth of 60 cm, that much change represents 42 mm of water. This is a water supply for wheat for 1 week and empirically for that area translates into a decreased yield of 400 kg/ha.

In the case of drip or trickle irrigation systems, and more modern, energy-and capital-intensive systems, the concept of field capacity is also meaningless. Here, soil moisture is relatively invariant in time but varies spatially relative to the location of emitters and the plant root system. Under drip irrigation, some part of the root system may be in soil which is too dry, some where water is available, and some where the soil is too wet, that is, saturated. Only a trivial fraction of the root zone is likely to be at a moisture content equal to "field capacity" at any particular time. The concept has no meaning under these conditions.

At the drier end of the available range of soil water contents, the science is also not exact. Empirically, pore water pressure of - 1.5 MPa applied to a moist soil until outflow ceases produces a soil moisture content corresponding to that in the soil in which container-grown sunflower plants suffer "permanent wilting." Thus, this energy level is known as the wilting coefficient potential, and the moisture content of a soil at this level is called the permanent wilting percentage. Again, there is no rational physical argument explaining the wilting percentage pore water pressure. Soil scientists offer explanations based on the unsaturated hydraulic conductivity of soils being of extremely low magnitude, but this parameter has already been assigned a value of virtually zero at "field capacity", which is at an energy level 50 times higher.

The Clausius-Clapeyron equation relates the free energy status of a liquid and its vapor pressure. When this treatment is used in the case of soil water, the relative humidity of soil air is predicted to be very close to 100% at high soil water energy levels but to drop rapidly as the energy levels (expressed as pressure) of the liquid state reach - 1 to - 3 MPa. This may be at least a partial explanation of the wilting coefficient. The soil atmosphere becomes desiccative to the root system.

Water held at pore water pressures between - 0.03 and - 1.5 MPa would be considered available to plants according to this scheme. This concept, although not completely sound scientifically, has proved useful in agriculture for land selection for cropping and in irrigation and drainage design.

A. Particle Statics

As indicated earlier, before cementation can (should) occur, particles are (should be) oriented in loose conglomeration relative to each other to provide good tilth. Otherwise, the pore system will not be sufficiently continuous; it will be too tortuous, too fine grained with inadequate coarse pore, or too small in total volume (per cent pore space). It is not difficult to imagine how this could happen, since most soil primary particles are far from being solid rigid spheres. Many are tiny, flat, plate-shaped objects, relatively small in one dimension and very irregular in the other two. Statics dictates that under pressure these will orient fact to face, exposing maximum available area in directions perpendicular to the major effective stress. Yes this rarely occurs except under heavy loading.

Much of the plate-shaped fine material is in the clay size range (less than 0.002 mm effective diameter) and is colloidal in character. Many of the minerals in this class fraction are secondary and not found in rocks or parent material; they are formed in the soil-forming process. Many of the particles have a net residual charge, usually a negative charge balanced by sorbed cations. These particles can react electrostatically with one another, with active radicals on organic molecules, or with charge sites on larger primary mineral fragments. They can produce weak dipole-dipole or dipole-induced dipole interactions, especially because their shape provides ample opportunity for charge concentration and separation and their composition makes them internally excellent insulators.

Considering these properties, it is likely that edge-to-face orientations are favored in the clay fraction, which leads to an open house-of-cards orientation with opposing rotational moments providing stability. Such conditions can occur only given an appropriate charge density and salt solution environment. Hence flocculated clays are good cementing agents and dispersed clays are not.

Flocculation and dispersion are processes that occur in a dilute suspension of colloid-sized particles in a dispersing medium. A dilute suspension of a few grams of clay in a liter of water can be such a system. As water is removed by evaporation, absorption, or drainage and the system thickens, ordinary colloid mechanics no longer apply. Precisely what happens between floccules (or domains) remains to be investigated and described. Soil scientists disagree about whether or how flocculated clay acts as a aggregate-stabilizing cement. It is known that soil containing clay which would be flocculated in a disperse system tends to have aggregates with higher water stability than soil containing clay which would be dispersed in a dilute suspension in water. For example, clay with calcium, magnesium, hydrogen, and virtually no sodium in its cation exchange complex would be flocculated in dilute suspension, and such clay is found in water-stable aggregates. As the level of exchangeable sodium increases, the clay observed in dilute suspension may register more dispersion and less flocculation, and the corresponding state in the drier condition would be a lower degree of aggregate stability under mild manipulation in water.

B. Cementation

As indicated earlier, orientation and cementation are the two steps in creation of structural stability, that is, aggregate stability, which in turn leads to pore system stability in soils.

Flocculated clay is a very important cement for stabilizing structure to accept and permit the movement of water. Clay particles bind to other clay, to larger particles, to organic molecules or tissue fragments, to microorganisms, and to root systems. Clay binding cannot be stable if excessive shrinking and swelling occurs. Organic skins and edge-to-face binding can reduce swelling upon wetting and shrinkage upon drying. [See Clay Sedimentology.]

Another important cement for stabilizing aggregates against disruption by water-related forces is organic matter. The most effective material is a mixture of large, usually ploymeric molecules with numerous well-spaced active radicals to react with charged mineral surfaces or other organic compounds. Undecomposed plant or animal tissue simply dilutes soils and acts as a spacer between mineral fragments; it is not a stabilizing cementing agent. Effective organic molecules include polysaccharides from partially decomposed cellulose, gums, resins, and organix acids produced by microorganisms acting on incorporated organic residues, dead root systems, animal manure, or other incorporated dead or living plant or animal matter.

Soils contain amounts of organic matter ranging from near zero in recently converted dune sands to over 90% by weight in clean peats and mucks. However, the mean organic matter content of most mineral soils is of the order of 2-6% by weight. Usually, fine-textured soils such as clays or clay loams have higher organic matter contents than sands. Clay protects organic matter from attack by microorganisms by absorbing it and occupying active radicals which microorganisms would attack; it does this by blocking physical access of microorganisms to organic matter sorbed between clay layers and by holding additional water, thus reducing aeration in the microenvirons of organic molecules.

A third type of cementing agents for stabilizing aggregates against disintegration in water consists of the sesquioxides of gelatinous oxides of iron and aluminum. In soils of appropriate chemistry, moderate acidity, well leached of bases and containing rapidly decomposing organic matter because of elevated soil temperatures and humidity, iron (and aluminum) oxides dissolve and move with water. They can be deposited as cements at particle contact points, thus stabilizing particle configurations against the impact and viscous forces of moving water. This cementation leads to stable aggregation and stable pore size distributions.

Lime and dolomite represent a fourth type of stabilizing agent. These materials are common in many young immature soils in semiarid to arid climatic zones where the soil has not been subjected to adequate weathering and leaching to remove the lime. The action is similar to that of iron cementation. Dissolved calcium and /or magnesium carbonates precipitate out of the soil solution at particle contact points as the soil dries. Prolonged and continued action goes far beyond cementation of aggregates. Indurated and cemented "hard-pans" can be formed in this fashion.

Flocculated clay, active organix matter, sesquioxides, and carbonates can act as stabilizing agents or cements. But in and of themselves they can only stabilize existing configurations. Suitable configurations must be produced first by particle movement. Forces causing relative movement can originate from gravity, seismic noise, swelling pressure, capillary pressure, surface tension, root growth pressures, small animal activity, earthworms, and other sources. Changes in particle positions relative to each other continue until table configurations are achieved. Obviously, this takes time and energy. It may take years or decades to achieve stability in soil structure. The source of the energy is the sun. Solar energy in the form of heat, organic molecules, or living organisms provides the driving force to create soil aggregation, the crux of soil structure, which in turn is the major parameter of soil tilth.

VI. Soil Tilth and Tillage

Soils in ideal physical condition to nurture and enhance plant growth gave a total porosity that is roughly 35-60% of total soil volume and is made up of a relatively wide range of pore sizes. They have a pore size distribution such that approximately one-half of the pore volume is drained free of water by gravity after a saturating rainfall or irrigation. This means that 15-30% of the soil volume will be occupied by air when drainage virtually ceases after wetting. Soils in ideal physical condition are crumbly, have a granular feel and appearance at a range of water contents, and drain free of excess moisture rapidly, within 24-36 h after rain or irrigation ceases. Such soils are said to have good tilth. They are friable, soft but firm when moist, and easily crumbled when dry.

Tilth is manifestation of the fact that strong interparticle forces exist within aggregates or granules, but only weak forces can be developed, even on drying, between aggregates. Particles within aggregates are in close proximity, whereas contacts between aggregates are small, few and far between. Therefore failure occurs along planes between aggregates with relative ease. In a soil with ideal tilth, aggregates (peds or clods) resulting from tillage are roughly similar in size to crop plant seeds.

With good tilth, drainage of excess water out of the soil, given appropriate outlet conditions, is relatively rapid. Water leaves the largest pores very rapidly, and then the drainage rate progressively diminishes as water flows from finer and finer pores and through thinner and thinner water films on particle surface. Aggregates tend to remain filled with water because the pores are fine and the water is tightly held. However, larger interaggregate pores drain free of water and fill with air. A ring of water remains around the contact point because there solid surfaces are close enough together and hold water by capillary forces. Water flows through the aggregates and circumcontact wager bodies to continue drainage more slowly after the large pores are drained.

As indicated, soils will contain 15-30% air by volume when drainage is virtually complete if the tilth (structure) is ideal. Most upland plants appear to perform at maximum levels at 15% or more soil air space. At 10-20% soil air space, plants begin to show effects of adverse soil physical condition. Below 10% air space, plant growth rates are reduced because of inadequate oxygen supply for roots, reduced availability of water to the root system, or excessive mechanical impedance to elongation and elaboration of branch roots and root hairs. Several research efforts have demonstrated that when soils have 10% or less of their volume occupied by gases their ability to conduct or transmit gases is nil. That is, the 10% volume which contains gases their ability to conduct or tansmit gases is nil. That is, the 10% volume which contains gases is probably blocked, discontinuous, or dead-end air space making virtually no contribution to mass flow or diffusion of gases through the soil and water mass. Such soil have poor tilth which can be improved by tillage, land resting, rotations, and treatments designed to increase soil organic matter or otherwise increase water-stable aggregation of soil particles into secondary units.

A. Tillage

In modern agriculture, with access to power units and sophisticated soil-working equipment, tillage is performed for many reasons: to loosen firm or compacted soil, to bury crop residues, to bury weeds and weed seeds, to create a dust mulch to conserve moisture, to produce a seedbed with crumb size compatible with the size of the seeds to be planted, to bury crop disease organisms, and/or to improve soil tilth.

If a soil has good tilth, tillage may not be necessary or even appropriate. Since tillage can damage tilth by compacting, puddling, or smearing soils if improperly performed or performed on soils at excessive wetness, one should be certain of the need for tillage, the reason for each operation, before undertaking it. Most crops can be successfully produced in soils with good tilth with a minimum of tillage, just enough to plant the seed. No-till planting can leave land surfaces covered with residues and litter from a previous crop to protect the soil surface from erosion by raindrop impact and runoff water.

On the other hand, in cases where tilth is naturally limited, where soil abuse has damaged tilth and destroyed structure, or where compaction has diminished the proportion of coarse pores, attempts to produce crops without tillage are ordinarily not successful. Also, no-till planting fails to achieve adequate results where drainage is poor and limited water outflow rates prevent completion of need drainage in a timely fashion.

Strip tillage is frequently adequate to meet the seedbed and rootbed needs of crops if soil tilth is at satisfactory level. In this process narrow bands are tilled just for planting seeds and strips between the rows are left unbroken. The tilled strips may be 10-20 cm wide and the intervening undisturbed strips may be 50-90 cm wide, depending on the crop plant size.

B. Rotations

Tillage is but one of a variety of parameters affecting soil tilth. As indicated earlier, particle orientations and stabilization of aggregates by cements are the two primary processes of tilth development as soils form or otherwise respond to imposed forces. Tilth is affected by cropping, crop harvest, and grazing by animals, by the way people use or abuse the land, and by changes in soil organic matter, soil acidity, soil salinity, and crop rotations.

To build and protect soil tilth, rotations should include crops, such as cereal or maize, which have dense, deeply penetrating fibrous root systems. As these decay, they put considerable organic matter in intimate contact with soil particles, more or less uniformly throughout a large volume of soil. Alfalfa or other deep-rooted legumes with a heavy vigorous taproot are also a desirable choice for penetrating intractable soil layers and improving their tilth.

Rotations also provide opportunities to rest the land, permitting forces of shrinking and swelling, freezing and thawing, and consolidation due to drainage and consequent subatmospheric water pressures to reorganize and reorder primary particles into more stable configurations. These natural forces are directed on a fine (primary particle-sized) scale and are primary movers for generation and maintenance of ideal soil physical condition.

VI. Tilth and Soil Conservation

Maintenance and protection of soil tilth is an *a priori* and essential step in soil conservation.

If land is overworked, if too much of its product is removed for use by people and their livestock, its organic matter content will decrease and infiltration and drainage of water will slow down. This has two detrimental impacts on the characteristics of the land. First, less water is available for production of crop or cover and for increasing soil organic matter. Thus, a decrease in the tilth level provides a negative feedback which induces additional damage to the tilth. Second, decreased infiltration means increased runoff, which, coupled with decreased cover, accelerates soil erosion. The surface soil is the portion most vulnerable to erosive action fo raindrops and running water, so it si lost rapidly. As surface soil is lost, soil organic matter and soil tilth, highest in surface soil, are lost with it. Subsequent erosion events are more severe because infiltration and water storage in the soil are reduced and morc watcr is availablc for thc crosion proccss. [Scc Erosion and Weathering.]

These are the steps of the desertification process which has destroyed millions of hectares of arable land in northwest China, the eastern Mediterranean, the African Sahal, and other places. According to some authors, feedback in the desertification process is even more severe than the steps reported here; it is thought that desertification in some areas can decrease rainfall in downwind areas and further intensify desertification by depriving vegetation of necessary moisture. [See Desertification.]

In irrigated agriculture, even where drainage is adequate, an equivalent of desertification is salinization, which also destroys land productivity. As organic matter is lost and soil compaction increases due to intensification and land overloading, percolation of water decreases as the coarse pore system disappears. With this comes an accumulation of salts in the profile. Depending on the physics and chemistry of the particular case, the salts which accumulate may be neutral salts (salinization) or basic salts (sodification). In irrigated agriculture where there is not initially adequate drainage, failure due to salinization occurs quickly because leaching requirements cannot be met. [See Soil Salinity.]

Preventing these two destructive processes requires tilth maintenance. Tilth maintenance requires energy input. This energy input is usually solar, that is, organic matter from crop residues, cover crops, green manures, animal manure, or human waste. However, if ancillary energy supplies are abundant, other kinds of energy inputs such as irrigation, fertilizers, composts, dried and sterilized sludges, or organic industrial wastes can be used along with land resting and land regenerative rotations.

This returns us to the beginning of this article. A resource has relatively reduced entropy because an energy input has been involved in its evolution. soil is such a resource. To conserve this resource, energy inputs must be continued. Otherwise, the resource will degrade to its parent material, a mass if lifeless mineral fragments cycling in the plate tectonics of the lithosphere.

The finiteness of petroleum reserves aggravates the problem of resource protection and conservation. Ancillary energy systems such as fertilization, irrigation, mechanized tillage and synthetic soil amendments will not be the ideal pathway in the long run unless they can be based on solar energy. They may be used temporarily to boost productivity and improve tilth if other options are available for substitution in the long run as the availability of petroleum-based fuels diminishes.

In the long run, human population control to decrease the loading on resources, which is already excessive in many parts of he world, is the only sure way to conserve and protect soil physical condition, soil fertility, plant growth, and land productivity.

Glossary

Aggregation. Organization and arrangement of primary particles in soils, mineral fragments, into much larger secondary units which are sufficiently stable to persist through wetting and drying as well as moderately vigorous soil manipulation such as tillage.

Denitrification. Microbiologically driven process in which the form of nitrogen is changed from the water-soluble nitrate ion, a part of a water-soluble salt, to a gas which escapes to the atmosphere.

Desertification. Transformation of farm, forest, or pasture land from resource status to desert-an area virtually devoid of plants and animals, supporting only simple ecosystems and unable to support more without very large ancillary energy inputs.

Flocculation. Agglomeration of colloid-sized matter into loosely structured clusters (floccules) where it is held together by weak forces. The clusters behave as physical entities in interactions with the environs.

Photosynthesizer. Organisms, both macro and micro, which possess the chemical compounds and the physical chemical systems necessary to absorb solar energy and use it to remove carbon dioxide from the atmosphere and incorporate it into organic molecules such as carbohydrates, fats, and proteins.

Soil structure. Combination of soil primary units into secondary units or peds. The secondary units are characterized and classified with respect to size, shape, and degree of distinctness. The secondary units classified as structural units may or may not be stable through wetting and drying cycles.

Soil tilth. Attribute a soil posses wherein its strength and consistency at a range of moisture conditions can be described as favourable for agricultural use. This is related to relative ease of tillage, fitness as a seedbed, and low but adequate level of impedance to seeding emergence and root penetration.

Transpiration. Loss of internal water from special openings on plant tissue external surfaces. The water moves as a vapour into the atmosphere.

2

Soil Salinity

I. Introduction

Soil salinity refers to dissolved salts present in the liquid phase of the soil and those salts not presently in solution but that are readily solubilized. The term saline is used to describe a soil with sufficiently high salinity to impair productivity. For agricultural use these soils require special remedial measures and management practices. An arbitrary value of 4 dS m^{-1} for the electrical conductivity of a saturation extract is often used to define a saline soil. However, a specific value is not possible since the salt content at which an adverse effect on plant growth occurs is dependent on the plant selected and various environmental factors.

Salinity problems affect the agricultural productivity of large areas (about 25% of all the irrigated land) of the world and occur in almost all irrigated regions. The total acreage of salt-affected soils in the world (arable and nonarable) is estimated at 1.0 billion hectares, a value equal to 60% of the cultivated land in the world. Soils are considered salt-affected whenever their soluble salt content is sufficiently high to reduce their agricultural productivity. These soluble salts consist primarily of the cations sodium, magnesium, and calcium and the anions chloride, sulfate, bicarbonate, and nitrate. The cation potassium and the anions carbonate and borate are usually, but not always, present in lesser quantities. Excessive quantities of these soluble salts cause either a reduction in plant growth or complete crop failure-depending on the quantity of salt present and the plant sensitivity to salts. Salt-affected soils occur in every continent and under a variety of environmental conditions, although the majority are found in arid and semiarid regions. Poor agricultural productivity resulting from increasing salinity in major irrigated areas threatens the economy and well-being of many arid land countries. [See Arid and Semiarid Ecosystems].

II. Development of Salt-Affected Soils

A. Geologic Origins

The processes of mineral weathering commenced with the release of water and volcanic gases from the developing planet, early in the history of the earth. Soluble salts resulted from the release of the acidic gases CO_2, HCl, and SO_2 and the weathering of silicate rocks. Most scientists who have attempted to reconstruct the early composition and development of the earth consider that the ocean quickly reached the approximate salinity of present seawater (about 35 parts per thousand salt by mass). Early geologists through that the age of the earth could be estimated by summing the mass of salts entering the ocean at the present time and calculating the years necessary to reach the present mass of salts in the ocean, considering that the volume of the ocean was constant. Other geologists subsequently advanced the concepts of the hydrologic cycle and of salt cycling—that is, water and salts enter the ocean via rivers, water returns to the land via ocean evaporation and subsequent condensation and rainfall, while salts is removed by precipitation into ocean sediments and to a lesser extent by transport to the continents via dry particle deposition and rainfall. These processes are considered to be in balance, thus the composition of seawater is quite constant in concentrations of the major dissolved ions. Residence time in the ocean equals the total mass of the ion in the ocean divided by its rate of input or rate of removal. Residence times vary depending on the solubility of the minerals formed; thus they range from approximately 100 years for iron and aluminum to about 8 million years for calcium and 260 million years for sodium. Salts are returned to the land both in rain and snow and by uplift of marine deposits. At times in the geologic past, large shallow inland seas developed along with the deposition of large quantities of salts in various mineral phases. Subsequent uplift and erosion of these areas releases these salts for a renewed cycle of dissolution, transport, and redeposition. [See Water Cycle, Global.]

Leaching of soluble salts from saline formations is the major immediate source of salt in arid landscapes. These salts do not necessarily leave the region but may instead be relocated to low-lying areas, in inland seas (such as the Great Salt Lake), or in lakes that contain water only intermittently (called playas). Transport of salts also occurs via transport in particulate form by wind action. For example, salts are transported from playa regions via dust storms during the dry season.

B. Evaporation and Transpiration as Concentrating Processes

The degree of salt occurrence at a particular location is the result of a complex interaction of geologic makeup, climate, topography, and land use. In regions with high rainfall and well-developed drainage, the soluble salts are continuously transported into rivers, underlying sediments, or aquifers. As a result, soils containing high salt concentrations are not common in humid regions. Compared to humid areas the chemical weathering of rock is lower in arid areas, the removal of salts by rainfall is greatly reduced, and salinity levels are almost always greater. Evaporation and plant transpiration (evapotranspiration) that exceed rainfall in such arid regions serve to further concentrate the salt mass. As a result, salt-affected soils are commonly found in such regions. [See Evaporation and Evapotranspiration].

The concept of potential evapotranspiration is utilized when estimating crop water requirements. The potential evapotranspiration is increased with decreasing humidity and with increasing temperature and wind speed. The high water evapotranspiration demands of hot arid areas, combined with the low water inputs, concentrates the salts into a small volume of water. Plants do not take up appreciable dissolved salts with the water they consume. Thus transpiration is not distinctly different from evaporation; both remove water from the liquid phase and leave essentially all the salts behind. [See Evaporation, Transpiration, and Water Uptake of Plants].

Typically between 50 and 95% of rainwater in arid regions is consumed in evapotranspiration. This implies a concentration factor of 2-to 20-fold for the salts of the area. In specific locales the values can be much greater, since much rainwater escapes as runoff during high-intensity rainstorms, causing the remaining water to be even more highly consumed and thus concentrated. Often the soluble salts are most concentrated near or at the soil surface. The evaporation process causes an upward movement of water, along with dissolved salts.

Weathering of silicate minerals, which often provide the major salt input in humid areas, is greatly reduced in arid regions. However in some instances it can have a very important effect on water composition. The western slope of the Sierra Nevada in California provides a good example. The very dilute recharging waters that result from snowmelt, pick up relatively minor amounts of Na, Mg, and Ca bicarbonate from weathering of silicate rocks as they move downstream at relatively low pH. But after evaporation and calcite precipitation of these waters in the valley, they are transformed into relatively saline, magnesium, sodium, carbonate waters of high pH. Dissolved HCO_3 and CO_3 can be derived from either dissolution of carbonates or silicate weathering. For silicate weathering the sum of HCO_3 and CO_3 is greater than CA + Mg (when expressed in equivalents); for carbonate dissolution Ca + Mg equals HCO_3 + CO_3. For most saline waters the much higher concentrations of the anions Cl and SO_4 over HCO_3 and CO_3 indicate the predominance of cycling of soluble salts and the lesser role of silicate weathering.

C. Chemical Processes Operative in the Soil that Affect Salinity

The major chemical processes in saline soils are dissolution and precipitation of relatively soluble salts and cation exchange reactions at the clay mineral surface.

Although there are a wide range of salt compositions in saline soils, these chemical processes limit the possible combinations. This is most evident for mineral precipitation processes where specific ions are removed in fixed proportions. Calcite and gypsum are the two most important minerals whose chemical precipitation impacts saline-water chemistry. In waters where Ca is initially greater than one-half that of bicarbonate, the process of evapotranspiration causes calcite precipitation according to the following stoichiometry:

$$Ca^{2+} + 2HCO^-_3 - C_aCO_3 + CO_2 + H_2O$$

With increasing evaporation and precipitation, the Ca/HCO_3 ratio in the remaining water increases; thus calcium concentrations increase and bicarbonate decreases, and the water evolves into a calcium-rich and bicarbonate-depleted water. Similarly, if $Ca^{2+} < 1/2HCO^-_3$, the water will eventually evolve into a calcium-deficient, bicarbonate-rich water. As a result, saline waters tend to evolve into distinctive compositions, whereas nonsaline waters exhibit continuous distribution in compositional makeup. For example, a water cannot be high in both calcium and bicarbonate, for as the salinity becomes concentrated one or the other of these ions will predominate. Similarly, precipitation of the mineral gypsum ($CaSO_4$. $2H_2O$) results in essentially a 1:1 removal of calcium and sulfate. Unless calcium initially exceeds that of one-half bicarbonate plus sulfate, the resulting water will become relatively low in calcium and will tend toward high sulfate and bicarbonate concentrations. The limited solubility of calcite and gypsum coupled with the presence of sulfate, bicarbonate, and carbonate generally restricts the Ca concentration. Under very saline conditions Na and Mg sulfates and chlorides precipitate. These general occur as salt crusts at or near the soil surface and are readily dissolved by rainfall. In some landscapes salts are deposited in a pattern similar to that of a chromatographic separation. In other words, the least soluble calcite an gypsum are found in the regions subject to the weak concentrating influences (such as in the upper regions of a river), and the more soluble salts of sodium, magnesium sulfate, and carbonate and chloride are found in the regions of higher salinity (such as in the lower regions of a river). [See Soil Water: Liquid, Vapour, and Ice].

The potential exists for apparently inert rock formations to release large amounts of slats when influenced by irrigation practices. The Mancos shale of Colorado, when ground up and reacted in water, produces a solution of low salt concentration. However given sufficient time, the following reactions occur in this soil type: Organic matter is oxidized with the subsequent release of H_3O^+ and sulfate ions; these ions react with calcite and dolomite resulting in the release of Ca^+ ions, which in turn exchange with Na and Mg present in exchange form on the clay mineral illite, to produce Na, Mg, SO_4 waters with lesser amounts of calcium and the secondary precipitation of gypsum.

D. Influences of Rainfall

As mentioned earlier humid regions tend not to be saline, since the existing drainage systems are well developed and precipitation is greater than potential evaportranspiration. The salts exposed by erosion or generated by weathering are transported out of the basin rapidly in surface runoff at relatively dilute concentration, although the total mass of salts transported per unit area may be in excess of that from arid regions. Under arid or semiarid conditions the regional salt load is not so rapidly flushed away and the soils typically develop saline conditions. Some salt transport by water can still occur in very dry regions. Recharge to groundwater systems have been documented in the most arid regions of the world. Despite rainfall of less than 20 mm per year, infrequent rain events can occasionally cause recharge.

E. Drainage and Lateral Movement of Groundwater

Most of the salinity problems associated with irrigated agriculture are related to over irrigation and/or inadequate drainage. Lack of sufficient drainage results in a rising water table or perched water condition. Once the saturated zone is within approximately 1-2 m of the soil surface, the potential for the upward movement of water and salt deposition within the soil by transpiration increases markedly. Unfortunately, many large scale irrigation projects have been adversely affected by inadequate drainage. Many alluvial, high clay content soils do not have the natural permeability to allow excess irrigation water or drainage water to percolate below the root zone. The problems of over irrigation are often difficult to solve because the excess water usually moves downslope laterally. Since soils in irrigated basins are usually alluvial in nature, they are layered. As a result the horizontal permeability is often many times greater than the vertical permeability. Lateral movement of water often results in development of secondary salinity problems in the topographically lower areas. Artificial drainage systems have been utilized to collect perched water and to lower the depth of the water table. [See Ground Water].

In dry-land areas native vegetation such as prairie grasses have been replaced with crops, such as wheat. Often there is not sufficient rainfall or water stored in the soil to sustain yearly crop production or crop production during the summer months. Leaving the land fallow decreases the amount of infiltrated water depleted by evapotranspiration and more water percolates through the root zone. This percolating water picks up salts from the subsurface (salts that were concentrated over thousands of years) and moves it along an impermeable layer until it intersects the land surface in a lower-lying region. Evaporation then results in the concentration of salts in relatively small area often called a saline seep. These seeps are characterized by very high salinities, little or no plant growth, and wet soil conditions; often salt crusts are evident.

III. Effects of Salt on Plants

A. Visual Symptoms and Yield Loss

Plants adversely affected by salinity often do not show any characteristic injury symptoms. Affected annual plants generally are simply smaller with shorter and smaller stems and fewer and smaller leaves. Leaves may be darker in colour, either dark green (due to more chlorophyll) or a blue to grayish green (due to an increased wax layer). Leaves of perennial woody species and some field crops may exhibit marginal or tip necrosis due to excessive sodium or chloride concentrations. Under severe salinity conditions injury symptoms are usually similar to those of plants suffering from severe water stress. Most species are most sensitive to salinity during the seedling stage; thus areas of high salinity in a field will be bare, with stunted plants in the marginal areas. Plants exhibit a wide range of salt tolerance with differences among cultivated crops exceeding an order of magnitude. Salt sensitive plants typically begin to suffer adverse effects at soilwater electrical conductivities of about 4 dS m^1. [See Environmental Injury to Plants].

B. Osmotic Potential

The adverse effect of salinity on plants is considered to result from the osmotic potential of the soil solution, expressed as quantity of work per unit mass of pure water (J/kg). The osmotic potential thus represents the work required by a plant to extract relatively pure water from the soil solution. This provides an analogy to matric potential in which the plant has to exert work to remove water from the soil pores. Based on limited date, there is support for the concept that the matric and osmotic stresses are additive and thus plant response can be related to total potential (or water stress); thus where ψ_g, ψ_o, and ψ_p represent gravitational,

$$\psi_t = \psi_g + \psi_o + \psi_p,$$

osmotic, and pressure potential, respectively. This expenditure of energy to acquire water reduces that available to produce growth—hence the smaller plant under saline conditions. [See Soil Physics].

C. Yield Reduction, Ion Toxicities, and Nutritional Imbalance

It appears insufficient to simply classify plants based on relative salt tolerance, since the actual yield response depends on the specific salinity level chosen. Some plants, such as rice, suffer little yield loss at moderate salinity and then only lose yield gradually with further increases in salinity. However, other plants, such as peanut, tolerate moderate salinity levels with little yield loss but then lose yield rapidly with further increases in salinity. For example, at 6.5 dS m^{-1} rice would experience a 25% yield loss and peanut a 60% yield loss.

Specific ion toxicities and nutritional imbalances may modify crop response to salinity. With increasing salinity, the Ca/Mg ratio generally decreases; this can cause a Ca deficiency in some crops starting at ratios as high as 1:1. Conditions of relatively high soil pH (>8-8.5) restrict mineral solubilities and can cause deficiencies of P, Zn, Mn, Fe, and other minerals.

The born concentration of saline waters is often sufficiently high to be toxic to plants. Although born is an essential element for plant growth, the optimal boron concentration range is rather narrow. Plants vary in their tolerance to boron but typically visual toxicity symptoms are apparent at levels of about 1.5 to 2.0 times the optimal. Yield losses occur only at boron concentrations considerably greater than those associated with the onset of visual symptoms. Large differences within a species are attribution to difference in boron uptake, which is dependent on root characteristics. Similarly, various woody plants are sensitive to Na or Cl ions. Stone fruits and grapes are particularly sensitive. These specific toxicities are also highly variable depending on root characteristics.

D. Climatic Factors

Since salt stress can be represented by the osmotic potential of the soil solution it is not surprising that climatic factors affect plant response to salinity. Most crops are more sensitive to salinity under conditions of high evaporative demand (high temperature and low relative humidity). Under cool, humid conditions there is reduced plant uptake of water and plants expend less energy in extracting saline water from the soil. Wind speed under dry conditions would also be expected to increase salt sensitivity, although this has not been demonstrated. Air pollutants that cause plant stress increase the apparent salt tolerance.

E. Morphological and Biochemical Factors

Because plant response to salinity is primarily osmotic, it is reasonable to consider that any factors that reduce the quantity of water passing through the plant, relative to its mass, will help the plant be more salt tolerant. As a result plants with low leaf-area/plant-mass ratios and plants with thick waxy layers, which reduce transpiration, will be expected to be more salt tolerant than plants with high leaf-area/mass and finer leaf type. Salt tolerant plants (halophytes) are slower growing and transpire less water; thus, the energetics of water uptake under saline conditions is apt to be less limiting for such plants compared to other plants.

IV . Effect of Salt on Soil Properties

A. Type of Salts

In addition to the direct effects that salinity has on plant growth, it may also reduce the suitability of soil as a plant medium. The suitability of a soil for plant growth depends on its ability to both hold and transport water, its air permeability, and structural properties that relate to soil tilth and friability. [See Soil Physical Condition and Plant Growth].

Saline soils generally have relatively good physical properties. Sodic soils, with low to moderate levels of salinity, have reduced permeability to air and water, poor soil structure, and poor soil tilth. These characteristics are related to the effects of exchangeable sodium, electrolyte concentration, and pH on the physical condition of the soil clays.

Smectite (usually montmorillonite) and illite are the most prevalent soil clays in arid and semi-arid regions. A net deficiency of + charge in the mineral structure results in a negatively charged surface. This negative charge is balanced by cations that are present in the solution adjacent to the surface. The distance out into the solution at which the surface charge is neutralized by excess solution cations is controlled by competing processes. Cations are electrostatically attracted to the negatively charged surface. The higher cation concentration near the surface sets up a concentration gradient with cation diffusion away from the surface and toward the more dilute bulk solution. The distance of charge neutralization is very important to the structural characteristics of the soil. Divalent cations have a greater attraction for the surface than do monovalent cations. In addition, the hydrated Na ion occupies a larger volume per charge than do Ca or Mg ions. As a result Ca^{2+} and Mg^{2+} ions are able to neutralize the charge at relatively shorter distances, in contrast to Na^{+} ions. In all instances an increase in concentration of the ions in the bulk solution compresses this diffuse double layer toward the clay surface. If this layer is sufficiently compressed, clay platelets can approach each other sufficiently close for van der Waals forces of attraction to overcome the electrostatic repulsion, and the clay platelets can aggregate.

The potential for aggregation is controlled by the surface charge, ion composition, electrolyte content (salinity), and pH. The ion composition of concern is not that of the bulk solution but rather that of the exchangeable ions. The exchangeable sodium percentage (ESP) is related to the solution composition via an ion exchange relation. For these purposes the ESP can usually be estimated by the expression $Na/(Ca+Mg)^{1/2}$ where concentrations are expressed in mmol $liter^{-1}$. At high pH, variable charge oxides and clay edges become negatively charged and contribute to the repulsion of soil particles.

B. Soil Structure, Permeability and Crusting

Under saline conditions, high-exchangeable sodium is not deleterious to soil structure and permeability. Infiltration problems may develop in such soils either during rainfall or irrigation with high-quality waters. In this instance surface salinity levels decrease rapidly with leaching, but the ESP may remain high. Surface crusting develops as a result of the combined effects of energy inputs (rainfall, sprinkler irrigation, and to a lesser extent furrow irrigation) and the chemical factors of clay dispersion.

The repulsive forces between particles, coupled with the energy inputs, result in particle migration and clogging of soil pores. The decrease in porosity results in poor soil aeration and a severe reduction in the ability to transmit water.

The soil surface is particularly susceptible to dispersion, since it is subjected to more physical forces for disaggregation and dispersion than is subsurface soil. Surface crusting due to the combined physical and chemical forces of dispersion is thus a major problem with sodic soils, especially after rainfall. Development of surface crusts causes reduced infiltration and an increase in surface runoff during storm events. Sodic soils are thus susceptible to erosion despite the fact that soils naturally high in Na are generally found in areas with little slope. Although surface crusts can be easily broken up by surface tillage, loss of soil structure and clay plugging of pores can be an almost irreversible process when it occurs at depth.

V. Distribution and Agricultural Impact of Soil Salinity

A. Geographical Areas Affected

Naturally occurring salt-affected soils are generally found in poorly drained low-lying lands in arid and semiarid regions. In these regions large quantities of salts have accumulated as a result of transport from higher elevations and from shallow water tables. There is usually a good relationship between the depth to the water table, groundwater salinity, and the salinity build-up in the soil. In addition to these regions the development of large-scale irrigation projects with inadequate drainage have also resulted in salinity problems over extensive areas.

On a percentage as well as absolute area basis, Australia has the greatest area of saline soils at 350 million hectares, followed by North and Central Asia at 210 million hectares. Serious salt-related problems occur within the boundaries of the following countries:

- Europe : Austria, Bulgaria, Cyprus, Czechoslovakia, France, Greece, Hungary, Italy, Portugal, Romania, the Soviet Union, Spain, Yugoslavia.
- North and Central America : Canada, Cuba, Mexico, the United States.
- South America : Argentina, Bolivia, Brazil, Chile, Colombia, Ecuador, Peru, Uruguay, Venezuela.
- Middle East and South Asia : Afghanistan, Bangladesh, Burma, India, Iran, Iraq, Jordan, Lebanon, Kuwait, Oman, Pakistan, Qatar, Saudi Arabia, Sri Lanka, Syria, Turkey, United Arab Emirates, Yemen (Aden), Yemen (San'a').
- North and East Asia : China, Mongolia, Soviet Union.
- Southeast Asia and Australia : Australia, Indonesia, Kampuchea, Malaysia, Thailand, and Vietnam.
- Africa : Algeria, Angola, Botswana, Cameroon, Chad, Djibouti, Egypt, Ethiopia, Gambia, Guinea, Guinea-Bissau, Kenya, Liberia, Libya, Mali, Mauritania, Morocco, Niger, Senegal, Sierra Leone, Somalia, Namibia, The Sudan, Tanzania, Tunisia, Zaire, Zambia.

Thus, salt-affected soils occur under widely varying conditions of climate, geology, agriculture, and, of course, social and cultural systems.

The world total of salt-affected soils is likely in excess of 1 billion hectares. Although salt-affected soil occur in almost all climatic and geologic conditions, the adverse effects are most serious in arid zones where the population is most dependent on irrigated agriculture.

The United Nations Food and Agriculture Organization estimates that the expansion of saline areas causes the abandonment of 10 million hectares of existing irrigated land every year.

B. Impact of Productivity

The economic impacts of salinity on agricultural productivity are very large but almost impossible to quantify on a global basis. Maps of salt-affected areas do not have a detailed breakdown of the salinity at each location; thus, yield losses cannot be easily estimated. In addition to direct losses, high salinity limits crop options and results in the cultivation of lower value crops. In the United States it is estimated that about 30-35% of all irrigated land experiences crop yield reductions resulting from salinity. In the lower Colorado River basin regions irrigated with river water, the estimated salinity damage to agricultural production was estimated at an average of 115 million dollars a year for the years 1976-1985. Total damage (including municipal and industrial water use) in the lower Colorado River basin totaled an estimated 830 million dollars per year.

VI. Controlling Soil Salinity

A. Drainage Requirements

Reclamation of saline soils requires leaching with better quality water. This generally means the installation of artificial drainage systems to remove the water used for leaching the soil. With modern installations, plastic drain tubes surrounded with synthetic filter material are installed to precise depth and grade. Typically, a self-propelled tractor unit installs continuous sections of perforated plastic tubing using a laser guidance system. Artificial drainage costs vary according to depth, spacing, and materials used; however, typical 1987 costs in the lower Colorado River basin were $850 per acre. In addition, soil profile modification is often needed to either invert soil profiles or to disrupt impermeable layers down to 2.5 m. The extremely high cost of artificial drainage makes it difficult to utilize in many regions of the world.

Often salinity problems are caused by over irrigating, which causes shallow or perched water tables and salt build-up in the soils of low-lying regions. Soils must have a minimum capability for draining water to the subsurface to avoid salinity development. Since plants transpire water and concentrate the salts in the soils, irrigation constantly adds salts to the soil and some leaching is essential. Typically, irrigation water contains between 0.05 and 3.5 tons of salt per 1000 cubic meters. For typical irrigations of 6000 to 9500 cubic meters of water per year per hectare, approximately 0.3 to 33 tons of salt are added to the soil per hectare annually. Unfortunately, most irrigation projects throughout the world have been developed without consideration of the potential need for drainage. In many instances the subsequent drainage requirements exceed the capability of the water districts to finance the costs. In most cases better irrigation design and distribution systems along with improved farm management can reduce excessive groundwater recharge and greatly minimize the need for artificial drainage. This is generally not possible once salinity problems develop. If the hydraulic properties of the subsurface allow for good lateral water movement, open drains can be constructed. Use of open drains was once the only method of artificial drainage, but these systems have become less prevalent because of high maintenance costs (siltation) and loss of irrigable land occupied by these drains.

B. Farm and Delivery System Management

Salinity control commensurate with water conservation and environment protection requires a net downward movement of soil water while the quantity of water percolated is minimized. The concerns of salt build-up have led to excessively liberal recommendations for required leaching and over irrigation, which generally aggravate the salinity problems. Construction of concrete canals or earthen canals with plastic liners greatly reduces conveyance losses. It is not unusual for up to 50% of a total project's water to be lost due to leakage in the conveyance systems when unlined conveyance systems are used. Such losses not only make less water available but contribute no benefit to soil leaching while placing greater demands on the drainage system. Lined canals and farm laterals can reduce conveyance losses to 10% or less of the diverted water.

Relatively recent improvements in farm management practices allow for more uniform application of water. Poor water distribution, in traditional irrigation systems and practices, required over irrigation in many in a field in order to assure that all areas would receive adequate water. With the use of closed irrigation systems, such as sprinkler or trickler for field crops and bubbler or drip for trees and vegetables, water can be more uniformly applied and the application rate can be more closely matched to the intake rate of that soil. In a similar fashion, with proper tillage and laser leveling, basin irrigation can achieve sufficient irrigation uniformity to also restrict water application. In this system, level basins of 2-15 hectares are constructed with large-flow water delivery systems that allow for rapid application of water to floor the basin with 5-10 cm of water. Automated gravity irrigation systems are also available for sloping areas that regulate water flow to specific furrows and achieve more uniform irrigation.

Efficient irrigation systems also require a more flexible water delivery system. Typically, farmers are provided water on a regular schedule, so that their lateral systems will have water only on specific days. The nature of the gravity systems dictates that water in the canals must either be diverted on farm or spilled at the end of the systems.

Efficient irrigation practices require water on demand as needed for each crop, particular soil, and climatic condition this requires a closed-pipe delivery system, which is still not common.

C. Minimized Leaching

Improved delivery systems and irrigation practices enable farmers to control salinity yet reduce the amount of deep percolation. The leaching requirement depends on the crops grown, climatic factors, and the salinity of the irrigation water. For irrigation with water below an electrical conductivity of 1 dS m^{-1}, leaching requirements are not likely to exceed 10% of the applied water. Typically, leaching fractions are 30-50% under conventional irrigation practices. Improved management practices can thus reduce the volume of on-farm deep percolation by 60-80%. Coupled with lining of delivery systems, these improvements can greatly reduce or eliminate artificial drainage requirements.

D. Drainage Disposal

Drainage water is typically returned to the river from which the irrigation water was obtained. Rivers may increase in salinity downstream; irrigation practices greatly accentuate this trend. Downstream water has reduced utility because of increased salinity and the potential presence of contaminants such as pesticides, boron, and selenium contributed in the drainage water. Changes in irrigation management have direct effects on water quality. Irrigated agriculture is a major source of salinity for many rivers and groundwater systems. Although it is not possible to maintain irrigated agriculture without leaching and drainage, its adverse impacts can be reduced. In regions where dissolution and displacement of native salts occur, there is a high correlation between salt load and volume of water drained. In some instances the concentration of salts in the drainage water if practically constant, independent of the volume of water leached. In these regions it is clearly beneficial to reduce leaching to the minimum necessary to maintain a downward movement of water and salts. In situations where the primary salt source is the irrigation water, reducing the amount of leaching also reduces the salt load returned to rivers or groundwater. However, in these instances there is an increase in salinity in the unsaturated zone. These water-quality benefits combined with the reduced need for drainage water make reduced leaching management a generally beneficial alternative to typical irrigation practices.

Ideally, drainage water from irrigated lands would be reduced to the minimum necessary to maintain crop yield and then would be kept separate from the high-quality water. In instances where efficient irrigation practices are not feasible, drainage water may be reused for irrigation of salt-tolerant crops. In some instances drainage water is disposed of to a low-lying evaporation basin, such as Dead Sea in the Middle East or the Salton Sea in Southern California. In these situations reuse of drainage water before disposal allows for increased water utilization and reduces disposal costs. Desalination of drainage water has been implemented in a few instances. At present it is not economically justified in terms of the value of the water and desalination costs.

Erosion and Weathering

1. Physical and Chemical Weathering

Erosion and weathering are the processes by which geological materials are modified and transported at the earth's surface. Weathering involves physical disintegration and chemical dissolution of rock materials in situ in response to the environmental conditions prevailing at the earth's surface. Erosion is the transport of the materials from the site of weathering to a site of deposition.

The term weathering originally referred to the effects of physical and chemical processes in causing the disintegration of rock exposed to the weather on land. Physical weathering takes place as a result of temperature changes. It is caused by differential expansion and contraction of mineral grains of different composition or by the expansion of ice in cracks and crevices. Chemical weathering involves attack by active agents at the exposed rock surface or at the soil-rock interface. Physical weathering breaks down rock by causing disaggregation of the constituent mineral grains; chemical weathering dissolves susceptible grains and the cement that holds them together. Biological agents contribute to both physical and chemical weathering. The expansion of roots as plants grow may widen cracks, and the humic acids released into the soil by plants are an important agent in dissolution of rock materials. weathering releases material into solution and leaves a residue of detrital particles that can be transported by water, ice, and wind to a new site of deposition.

The major processes involved in chemical weathering on land are hydrolysis, oxidation, and carbonation and, to a lesser extent, reaction with sulfuric, nitric, and humic acids. These processes affect all rocks and sediments, but the rates at which the weathering reactions proceed are strongly dependent on composition of the material being weathered.

Recently, the term weathering has been expanded to include analogous processes that occur subaquatically, referring particularly to the interaction of seawater and ocean crustal rocks in hydrothermal systems. [See Hydrothermal Systems].

A. Weathering of Silicate Minerals

Igneous and many metamorphic rocks consist chiefly of silicate minerals. The products of weathering of igneous and metamorphic rocks are sediments that, on compaction and induration, become sedimentary rocks. It is the chemical weathering of silicate minerals that alters minerals that are stable at the high temperatures and pressures prevailing in the interior of the earth to minerals stable under the conditions existing on the surface of the earth. Weathering of silicate minerals involves breakdown of the parent mineral to release cations and silica. The cations and silica may migrate into new minerals or be removed in solution. [See Chemical Weathering Controls on Sand Composition; Silicates].

The major weathering reactions are hydrolysis, oxidation, and carbonation, shown in the following examples:

$$\text{Hydrolysis: } \underset{\text{Orthoclase}}{2KAlSi_3O_8} + 11H_2O \rightarrow$$

$$\underset{\text{Kaolinite}}{Al_2Si_3O_8} + 4H_4SiO_4 + 2K^+ + 2OH^-$$

$$\text{Oxidation: } \underset{\text{Fayalite}}{2Fe_2SiO_4} + 4H_2O + O_2 \rightarrow$$

$$\underset{\text{Hematite}}{2Fe_2O_3} + 2H_4SiO_4$$

$$\text{Carbonation: } \underset{\text{Forsterite}}{Mg_2SiO_4} + 4H_2O + 4CO_2 \rightarrow$$

$$2Mg^{2+} + 4HCO_3 + H_4SiO_4$$

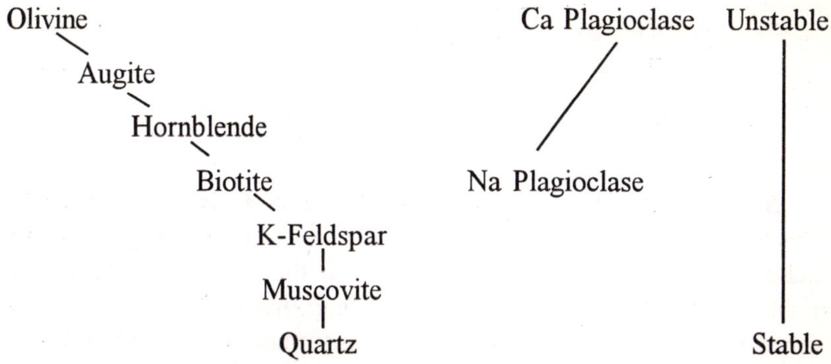

Fig. 2.1. Stability of minerals found in igneous rocks under conditions prevailing at the earth's surface. The order of increasing stability is the inverse of their order of crystallization (Bowen's reaction series)and corresponds to decreasing temperature of formation.

Orthoclase is the common feldspar of granitic rocks. Kaolinite is a clay mineral. Fayalite and forsterite are olivines, found in ultrabasic igneous rocks and very susceptible to weathering at the earth's surface. Hematite is the rust-colored mineral that often stains the surface of ultrabasic rocks. H_4SiO_4 is silicic acid, the soluble form in which silica is transported by surface-and groundwater. All of these reactions can affect all silicate minerals; rates depend on temperature, presence of water, pH, and Eh.

The increasing resistance of silicate minerals to weathering is shown in Fig. 2.1. The order is the same as that of Bowen's reaction series describing the order of crystallization of minerals from a silicate melt and corresponds to the increasing number of Si-O-Si bonds, from zero in olivine to four in quartz.

Mechanical weathering, the result of differential expansion and contraction in response to temperature changes, freezing and thawing of intergranular water, growth of plant roots, etc. causes the mineral grains to separate. This both greatly increases the surface area for chemical reaction and enhances the likelihood that the grains will be moved by transporting media.

B. Alteration and Dissolution of Sediments

Sedimentary minerals are derived from silicate minerals by alteration to become more stable under conditions at the earth's surface and from other minerals produced during the weathering process. Igneous silicate minerals become clay minerals that are progressively leached of all soluble cations. The clay mineral products of silicate weathering depend on the climate. Thus albite, the Na-plagioclase feldspar, weathers to kaolinite under wet tropical conditions but to montmorillonite under wet tropical conditions but to montmorillonite under less humid, cooler subtropical conditions.

$$\underset{\text{Albite}}{2NaAlSi_3O_8} + 2CO_2 + 11H_2O \rightarrow$$

$$\underset{\text{Kaolinite}}{Al_2Si_2O_5(OH)_4} + 2Na^+ + 4HCO_3 + 4H_4SiO_4$$

$$2NaAlSi_3O_8 + 2CO_2 + 6H_2O \rightarrow$$
Albite

$$Al_2Si_4O_{10}(OH)_2 + 2Na^+ + 2HCO_3^- + 2H_4SiO_4$$
Montmorillonite

Some sedimentary minerals are the ultimate residual products of weathering [e.g., gibbsite, $Al_2(OH)_6$] and are virtually insoluble and immune to further decomposition. The leached materials are carried in solution to the sea or to playas where, under evaporation conditions, they are precipitated as salts [e.g., halite (NaCl)]. The sedimentary minerals thus formed may be highly soluble. Sedimentary minerals are affected by hydrolysis, oxidation, and carbonation reactions, but exhibit an even greater range of sensitivity than do igneous silicates.

C. Development of Residuum

Weathering of rock results in solution of some components and their removal by percolating water. Disaggregated rock that is little altered chemically is termed "gruss". The smaller particles formed as the rock disintegrates and disaggregates increase the surface area available for chemical reaction. As weathering proceeds, soluble material goes into solution and is removed. The insoluble material that remains behind is termed "residuum" and forms a mantle over the parent material. The nature and thickness of this weathering mantle varies with climate; in high latitudes and the arid zone the absence of liquid water inhibits chemical weathering, so that it is largely rubble resulting from physical weathering. In the humid belts chemical weathering dominates, and the residuum consists of those minerals that are most stable and resistant to solution. In the tropics the warmth makes the chemical reactions proceed rapidly, so that the weathering mantle is often very thick and leaching is extreme. Fig. 2.2 shows the changes in thickness of the weathering mantle with latitude.

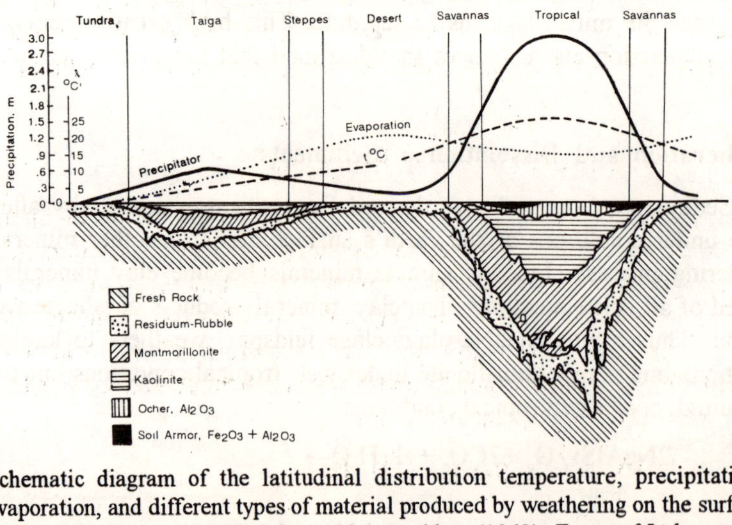

Fig. 2.2. Schematic diagram of the latitudinal distribution temperature, precipitation, evaporation, and different types of material produced by weathering on the surface of the earth today. [Modified from N.M. Strakhov (1963). *Types of Lithogenesis and Their Evolution in the History of the Earth* (in Russian). Gasgeoltekhizdat, Moscow].

D. Soil Formation

Soil is a complex material containing not only the residues of rock weathering but also moisture, air, decomposing organic matter and the associated organic compounds, along with living bacteria, plants and animals. In the upper part of the soil, the zone of eluviation, chemical reactions result in leaching and mobilization of material that is carried downward by percolating waters. In the lower part of the soil, or zone of illuviation, some of the material dissolved from the upper part of the soil is deposited. Other soluble materials is flushed from the soil, infiltrating the underlying rock or sediment, and is carried away in solution in groundwater.

Soil formation is strongly dependent on the nature of the parent materials, slope of the land, climate, and the vegetation cover. Weathering processes take place within the soil, as well as within the weathering mantle beneath the soil and at the rock interface. Fig. 2 shows the variability of weathering mantle and soil types with latitude and climate.

II. Environmental Factors Influencing Weathering Rates

A. Climate

Climatic factors are significant in influencing weathering rates. Mechanical erosion depends on rapid temperature changes. Temperature variations are greatest at the rock- air interface but decrease rapidly downward. The daily cycle of temperature change penetrates only a few centimeters, and the annual cycle is usually damped out below a meter depth. Thus the insulating effect of a thickness of soil or weathering mantle of more than a few centimeters greatly reduces physical weathering. Chemical reactions proceed more rapidly at higher temperature and virtually stop at the freezing point of water. Precipitation-evaporation balance is important in determining the moisture content of the weathering mantle. If precipitation exceeds evaporation, soluble material will be flushed away with water infiltration to the groundwater system; if evaporation exceeds precipitation, solutes are drawn with the evaporating water to the soil surface and form a crust. Seasonal rainfall may result in an alternation of hydrolysis and carbonation with the oxidation reaction.

B. Atmospheric and Soil-Atmospheric Compositions

Although soil contains atmospheric gases, the soil-atmosphere composition differs from that of the atmosphere itself. Soil atmosphere is enriched in CO_2 supplied from decomposing organic matter. Rain water always contains some carbonic acid (H_2CO_3), from solution of atmospheric CO_2 in rain drops, and is hence at least weakly acidic. However, the water percolating through the soil into the deeper weathering mantle becomes enriched in carbonic acid due to the higher CO_2 pressure in the soil and is more strongly acidic. Oxygen is also present in soil atmosphere, but its partial pressure is less than that of the atmosphere because of oxygen consumption in the decomposition of organic matter. Oxygen is also consumed in the weathering of sulfide minerals at the rock interface, producing sulfuric acid as one of the weathering products.

$$\underset{\text{Pyrite}}{4FeS} + 4H_2O + 9O_2 \rightarrow \underset{\text{Hematite}}{2Fe_2O_3} + 4H_2SO_4$$

The atmosphere contains some SO_2 from natural volcanic as well as anthropogenic sources; hence rainwater already contains small amounts of H_2SO_4, sulfuric acid. However, the sulfuric acid content of waters percolating through weathering mantle or rock containing sulfides increases greatly. Sulfides occur in most mineral deposits, hence the waters draining from mine tailings are often highly acid. Sulfides also occur in some sediments, notably black shales, making the waters percolating through these sediments acid. [See Acids and Bases in Watersheds; Land-Atmosphere Interactions].

C. Rainwater-Soil-Groundwater Compositions

Average rainwater contains salts in proportions that resemble very dilute (0.02%) seawater. Groundwater differs markedly from rainwater because it contains the solutes derived from the weathering process. River water a mixture of about two—thirds direct runoff, water that has run over or through the soil, and one-third groundwater. The average groundwater has a content of dissolved solids about 20 times that of rainwater and about 0.4% that of seawater; average river water is slightly more dilute. The carbonation reaction dominates the composition of groundwater, so most groundwater and river waters are bicarbonate solutions, with Ca^{2+} the major cation. Limestone is readily dissolved by carbonic acid and transported to be redeposited in the sea mostly through the intermediate agency of organisms that precipitate $CaCO_3$ as either calcite or aragonite skeletal material.

$$\underset{\text{Limestone}}{CaCO_3} + H_2CO_3 \rightarrow Ca^{2+} + 2HCO_3^{2-}$$

$$\underset{\text{Shells}}{CaCO_3} + H_2O + CO_3$$

The CO_2 for the carbonic acid came from the atmosphere either directly or indirectly through the intermediate agency of fixation of carbon by plants and subsequent decomposition of the organic matter. It returns to seawater and the atmosphere when calcium carbonate is deposited in the sea. The weathering of limestone (and dolomite) uses atmospheric CO_2 as a catalyst to transport $CaCO_3$ from one site to another; it does not consume atmospheric CO_2. Atmospheric proceeds much more slowly and uses bicarbonate for transport in solution.

The composition of groundwater is usually dominated by Ca^{2+} and $HCO^{2-}{}_3$, but a variety of other cations and anions are present, reflecting particularly the abundance of soluble sedimentary rocks in the drainage basin. Na^+ and Cl^- are present because of their introduction as salt recycled from the sea through rainwater, but their concentrations become significant where circulating groundwaters bathe halite deposits. Significantly, human activities, such as bringing brine from great depths to the surface in pertroleum extraction and in direct mining of salt, has increased the Na^+ Cl^- content of many rivers. Similarly, the SO^2 content of rivers is high if the river drains areas underlain by gypsum or anhydrite ($CaSO_4$). Weathering of dolomite, leaching of clay minerals, and weathering of igneous rocks contributes Mg^{2+}; and K^+ is derived from weathering of igneous rocks and leaching of clay minerals. Because of the differences in solubility of these ions in silicates and sedimentary rocks, the composition of rivers is largely controlled by the availability of soluble sedimentary rocks in the drainage basins.

D. Elevation and Slope

The regional elevation and local slope of the ground both significantly affect the rates of weathering. Slope is a major factor contributing to the removal of mechanically separated detrital particles. Steep slopes may not develop a residuum adequate to retain moisture or a soil that can supply acids for chemical weathering. Also, one in transport, chemical decomposition is curtailed, so that steep slopes promote physical disintegration not chemical decay. Higher elevations have lower temperatures, reducing the rates of chemical weathering but are drained by rivers steep enough to remove mechanically derived particles.

E. Vegetation

Vegetation plays a major role in determining the nature of the soil, and hence the composition of waters percolating down to the horizons where active chemical weathering proceeds. Coniferous forests develop soils rich in organic acids. Vegetation also modifies the way in which rainfall reaches the soil surface. Forests that develop canopies intercept much of the rain and both slow the rate of rainfall onto the forest floor and prolong the time of throughfall as the canopy continues to drop after the rain has stopped. In many trees, the rainfall intercepted by the canopy is directed to the forest floor by flowing along the branches and trunk, where the water may become enriched in organic acids. Finally, vegetation plays a major role in binding the soil and weathering mantle with its roots, retarding mechanical erosion.

III. Weathering as the Rate-Limiting Process in Denudation of the Continents

Weathering is the rate-limiting process in denudation of the continents. Chemical weathering dissolves the cement that holds the particles of rocks together. Chemical and mechanical weathering prepare the rocks for transport by water, wind, and ice. The surface of the land cannot be lowered by erosion and transport of the material until weathering has reduced the detrital particles to a size small enough to be moved by ice, water, or wind.

IV. Global Change and Weathering Rates

Natural changes, such as mountain building and climatic change, will affect rates of weathering. Weathering processes go to completion only after they have operated for very long periods of time, probably millions of years. During the past few million years most of the earth has been subjected to repeated rapid climate change, and as a consequence there are very few areas where the weathering process has developed stable soil profiles. Australia is one of the few continents where the climate has remained nearly constant, and soils are quite ancient. In the regions of North America and Europe covered with debris left by the glaciers, soil formation has just begun, and the soils are especially fertile.

As an agent for exposing rock and sediment to weathering, human activity exceeds natural processes. Farming and ranching have altered the natural vegetation cover an exposed bare soil to erosion. Plowing churns the soil and exposes it to the oxygen-rich atmosphere much more rapidly than the natural activity of burrowing animals. Construction activity pulverizes and exposes rock that would otherwise remain buried. Mining activity has brought to the surface, pulverized, and spread out as tailings minerals containing rare metals that would otherwise reach the weathering zone at a much slower rate.

Global warming will change weathering processes in a variety of ways, Mechanical weathering would be reduced by lessened seasonal temperature variation in high latitudes, but it may be increased if climate change involves a greater frequency of extreme events. chemical weathering will increase with increased temperature. Increased atmospheric CO_2 may have some effect on increasing the rates of chemical weathering, but this is moderated by the already high CO_2 content of soil atmosphere. Increased SO_2 and NO_2 in the atmosphere, causing "acid rain" may become more important as chemical weathering factors. [See Acid Rain].

V. Weathering of Rocks in the Ocean

Submarine weathering is the result of the chemical reaction of seawater with igneous rocks forming the ocean floor. Near the mid-ocean ridges where new sea floor forms by volcanic activity, seawater percolates into the hot rock through cracks and clefts, and the reactions lead to hydrothermal alteration of the rock. The basalt of the sea floor consists mostly of plagioclase feldspars, amphiboles, and pyroxenes-minerals that are rich in Ca and Fe. In reaction with seawater, Ca is replaced by Mg, Na by K, less abundant metals are mobilized, and silica goes into solution. The result is conversion of the minerals typical of basalt into clay minerals. Some of the cations carried in solution in the hot waters precipitate on contact with cold seawater, but some other cations,such as Ca^{2+}, are introduced into the ocean in quantities that are of the same order of magnitude as the inputs from land. This large submarine input acts to stabilize the ocean chemical system, so that perturtabtions from the delivery of material from land may be less important in changing ocean composition than was previously thought.

Reverse weathering is a term used to describe the silicate mineral-forming processes that occur in the sea, removing solutes from seawater to form clay and other minerals that would be unstable if exposed to the atmosphere and rain water.

VI. Erosion

Erosion is removal of weathered material or sediment and its transport to a new site of residence or deposition. The major processes of erosion on land involve transport by ice, water, or air. Erosional processes are usually measured in terms of transport of material to the sea. At present the total amount of material eroded and delivered to the sea is about 250 X 10^{14} g/yr, equivalent to lowering the land surface by about 0.07 mm/yr. Of this, the particulate load of rivers accounts for about 73% of the total, that of glaciers 8% the dissolved load of rivers 16%, the dissolved load of groundwater 2%, erosion on coasts 1%, and atmospheric dust only about 0.2%. The comparison of processes in terms of transfer from land to sea is misleading because some transport processes are more effective in moving material from erosional to depositional sites on land. Also, erosion, transport, and redeposition of sediment may occur entirely subaqueously; erosion and redeposition are important in both shallow seas and the deep ocean basin. [See Sediment Transport Processes; Sedimentary Geofluxes, Global].

A. Glacial Erosion

Glacial erosion is the result of moving ice. Because the temperatures in glaciated regions are usually below the freezing point of water, chemical weathering is reduced, and physical weathering dominates. The expansion of water as it freezes to ice is also an important factor in producing debris that can be moved by glaciers. Glacial transport can accommodate large boulders, but the effectiveness of glaciers as erosive agents is dependent on the slopes down which they flow and whether the flow is confined. Mountain glaciers carry large amounts of debris plucked from the valley walls and floors. Continental ice sheets may be much less effective because the flow of the ice takes place above its floor. Although glacial transport today accounts for about 8% of the total of material delivered to the ocean, glaciers play a larger role in moving material from higher to lower elevations in mountainous regions. Glaciers are effective in pulverizing rock, reducing it to grain sizes more readily transported by rivers. glaciers often oversupply sediment to their meltwater; so that the streams emerging from them are usually braided, unable to fix a channel because of the deposition of excess sediment. Glaciers can move detritus uphill as well as downhill. The Pleistocene ice cap scooped the basins of the Great Lakes from the soft sediments surrounding the crystalline rocks of the Canadian Shield, depositing the excavated sediment as the ground moraine of the midwestern states of the United States and the prairie provinces of Canada.

B. Fluvial Erosion

Fluvial erosion dominates transport of material from land to sea. The chemical load of rivers is dependent on the area drained and on the rocks, sediments, and soils exposed to the percolating and moving waters. The concentration of solutes is inversely proportional to the runoff; rivers draining wet areas are dilute, those draining arid regions have higher concentrations of dissolved soils, reflecting the rate-limiting nature of weathering processes. The dissolved load of rivers is easily measured and well known, but analysis of the solutes does not reveal all of the cations being transported. Significant cation transport may take place by their absorption on the surface of clay minerals with a high exchange capacity.

These adsorbed cations may be exchanged and and released where the river enters the sea but there are very few data on these processes. Most of the load of rivers is transported as fine detrital particles, mostly clay minerals and quartz, in suspension. Kaolinite, which has a low exchange capacity, is the dominate clay mineral in tropical rivers; at intermediate and higher latitude, where weathering proceeds under cooler conditions, it is replaced by montmorillonite, which has a higher exchange capacity. [See Chemical Mass Balance Between Rivers and Oceans].

The bed load of rivers consists of coarser sand and gravel moved by traction, averaging 10% of the suspended load; but the proportion of bed load varies greatly among rivers. The particulate loads of rivers are highly dependent on water volume and flow rate. If these are adequate to move all of the sediment supplied, the river is said to be competent. Evaporation from the surface of the river may reduce the volume flow so that the river is not longer able to transport the sediment introduced in its supper reaches; the river becomes incompetent, and deposition occurs. Global warming may result in major changes of the flow regimes of rivers and change the competence of rivers to carry loads. Human modifications of rivers have greatly altered their sediment transport characteristics. The construction of dams, from those used for farm ponds to major barriers, has completely changed the way in which detrital sediment is transported. Although agricultural activity has increased erosion rates, dams trap sediment, so that less sediment reaches the sea than under natural conditions. Dams change the competence along the length of the river. sediment-free water released below a dam has a high erosive capacity, and the still waters of the lakes have low transport competence. [See Rivers].

C. Aeolian Erosion

Under natural conditions, aeolian erosion is limited to the subtropical desert regions and to periglacial areas. In the desert regions, lack of moisture result in sparse or nonexistent vegetative cover to bind the weathering mantle and soil. In periglacial regions, glacial outwash plains are devoid of vegetation, and when dry the fine sediment is readily transported by the wind. Human agricultural practices seasonally destroy vegetation cover and expose soil to erosion by the winds in regions where aeolian erosion would not normally occur. In times of drought, when the vegetative cover is not regenerated wind erosion may become severe. [See Aeolian Dust and Dust Deposits].

D. Coastal Erosion

Coastal erosion is the result of wave action that mobilizes particles and currents that transport them to a new site of deposition. On rocky coasts, wave energy is concentrated on promontories and material removed from them is deposited in the adjacent embayments. On coasts bordered by a coastal plain the effects of coastal erosion may be more pronounced. Erosion occurs where wave energy is concentrated, and the mobilized sediment is then transported, usually longshore, to a deposition site. In many regions ;the balance between sediment supply by rivers and longshore transport determines the position of the coast. Human activities often upset this balance by changing the river supply or longshore transport. If the longshore transport is interrupted by an artificial feature, such as a dredged harbour entrance, the result is downstream erosion of the coast.

VII. Environmental Factors Affecting Erosion Rates

A. Climate

Cold temperatures affect erosion rates because frozen ground cannot be readily eroded and frozen rivers do not transport sediment. The sediment loads of Arctic rivers are only a few per cent of the loads of rivers in warmer parts of the world, because significant parts of their drainage basins are underlain by permafrost and their flow is limited to a few months of the year. As global warming melts permafrost and allows rivers to flow for longer periods of time, sediment transport budgets in high-latitude regions can be expected to increase markedly.

The precipitation-evaporation balance is very important in sediment transport. High erosion rates occur at the margin between savannas and desert areas, where rainfall occurs but is infrequent, and vegetation cover is sparse. Temporal distribution of rainfall is also important. Exceptional storms are responsible for mobilizing weathered material by slope wash and transporting it downslope into local streams, where it is usually redeposited but remains available for transport by lesser flood events.

B. Atmospheric Composition

The short-term effects of changes in atmospheric composition on erosion are largely indirect, through climate change. If increases in atmospheric greenhouse gases induce global warming, the precipitation-evaporation balance will favor evaporation, expanding regions of sparse vegetation and ;high fluvial and erosion rates.

C. Elevation and Slopes

On the long term, increases in atmospheric CO_2 must in increased weathering of silicate minerals of igneous and metamorphic rocks. The formation of mountain ranges and their erosion modulate atmospheric CO_2 content. Volcanic activity during mountain building releases CO_2 into the atmospheres, and the later weathering and erosion of silicate exposed in the mountain ranges consumes CO_2 from the atmosphere.

D. Vegetation

Detrital erosion and denudation rates vary with the nature of the vegetation cover that determines how rainfall reaches the ground. Where vegetation is sparse and the plant cover incomplete, soil or weathering mantle is directly exposed to the rain. At the other extreme, rain forest and jungle canopies prevent most rainfall from reaching the ground directly, reducing its erosive effect. Chemical erosion and denudation rates also vary with the nature of the vegetation cover. The rates of production of humic acids and CO_2 in the soil are ultimately controlled by the vegetation.

VIII. Transport, Storage, Delivery to Endoric Regions, Delivery to the Sea

Most of the global information on loads of rivers, atmospheric dust transport, etc., refers to the flux from land to sea. The land-sea flux does not represent the global flux nor does it correspond to the denudation rate of specific land areas. Measurements are also made in other parts of the transport systems and show that eroded materials is temporarily stored within drainage basins as a result of decreasing competence of rivers. Changes in river competence- their ability to transport sediment-is a function of the gradient and amount of water in the river.

Some rivers do not reach the sea but. drain into closed basins in the interior of the continents, termed "endoric regions". These basins have no external drainage because evaporation rates within them exceed both precipitation and supply of water from rivers and the groundwater system. The increased evaporation expected with global warming may expand existing endoric regions and create new ones.

IX. Subaquatic Erosion, Transport, and Redeposition of Sediment

Erosion, transport and redeposition of sediment within lakes and the sea are not as well known as for land areas. Of the material brought to the sea in solution, much of the calcium and carbon are removed from the seawater and deposited by organisms as $CaCO_3$ shell material. The remaining carbon that is not recycled back to the atmosphere as CO_2 or other gases is buried as organic carbon. Most of the silicic acid brought to the sea is precipitated by organisms as opal skeletal material. Other salts are removed in playas, marine evaporite basins, and along the shore in very arid regions.

Detrital material brought to lakes and the sea is sorted by waves, with coarser material forming a moving band on and near the shore, kept in motion by longshore currents. Finer detrital material settles to the bottom further offshore. Deposition on the continental shelves may be only temporary, and during storms material is mobilized into dense flows of turbid water that flows down submarine canyons onto the continental rise and abyssal plains. Mass movements, slumps, and slides also transport large quantities of material to the deep sea floor. Very little is known about the relative importance of transport processes in the sea.

X. Global Change and Erosion

Natural global change affects erosion rates by changing elevations through mountain building and the distribution and amounts of rainfall, and vegetative cover is changed through climatic change. All of these changing conditions alter rates of slope erosion and the competence of rivers. Permafrost is extensively developed around the Arctic, and it effectively shields weathered material from erosion. If the permafrost melts, it frees large amounts of particulate debris for erosion and transport.

Human agricultural activity has destabilized the natural balance; farming practices have increased soil erosion rates, but dams trap the eroded material in reservoirs. Although denudation rates have generally increased, sediment delivery to the sea has decreased; and becaue of the lower sediment supply, coastal erosion has increased. Construction activities expose natural materials to erosion, and paving and buildings increase runoff. As a result of construction requirement, man has become a major agent in moving sand and gravel from one place to another on the surface of the earth.

Global warming will result in higher evaporation off reservoirs and rivers, reducing the amount of water and the competence of rivers, increasing storage of sediment in the drainage basins.

Human activity has also been significant in altering the budget of material near the shore; obstructions slow longshore currents and cause sediments to be deposited and down current the lost sediment supply results in coastal erosion. Warming of the ocean might cause destabilization of the continental slopes. Clathrates, ices formed by water and gases such as methane at high pressures and temperatures above the freezing point of water, and present under extensive areas of the continental slopes. They act as cement within the sediment, holding it in place. Rising oceanic temperatures may cause the clathrates to decompose to water and gas, and the resulting loss of cement may cause large areas of the continental slopes to fail catastrophically as massive subsea slides.

XI. Summary

Weathering involves mechanical and chemical processes that transform rocks and minerals to materials more stable under conditions prevailing at the surface of the earth. Weathering produces a mantle of residuum over the parent rock. Organisms modify the residuum, adding organic compounds and facilitating penetration of gases and moisture to produce soil. Erosion is the movement of weathered material to a new site of deposition. Detrital material is supplied to the sea from land. Solutes having mass of about one-fourth that of the detrital material are also supplied from land, but there are additional supplies of solutes from submarine weathering in hydrothermal systems and from waters being squeezed out of thick sediment wedges. The magnitude of these subsea sources of solutes is not known, but for some ions may be as large as or larger than the flux from rivers and groundwater. Human activity modifies the weathering and erosion system in many ways, affecting rates of weathering, transport and deposition.

3

Soil Biochemistry

I. Introduction

Soil biochemistry is the science that deals with the chemistry of living systems and their components if soils. All biochemical action is dependent on, or related to, the presence of enzymes produced by the living soils component (i.e., microorganisms, animals, and roots of higher plants). Important biochemical processes that occur in soils can be divided into several categories including (1) liter decomposition, (2) formation and degradation of soils humic substances, (3) mineralization and immobilization of nutrient elements, (4) degradation of pesticides added to the soil, and (5) oxidation and reduction of mineral compounds in soil.

The cycling of nutrients in soils involves biochemical, chemical, and physicochemical reactions. The biochemical reactions are mediated by microorganisms, plant roots, and soil animals and are catalyzed by enzymes. As a result of biochemical action, biochemicals accumulate in the soil. These include a wide range of substances such as organic acids, carbohydrates, amino acids, amino sugars, phosphate and sulfate esters, and humic substances.

During decomposition of plant and animal detritus, carbon (C) recirculates to the atmosphere as CO_2 and nutrients are released in plant-available mineral forms. During decomposition, some of the C and nutrients are assimilated by the microbial biomass (immobilized) and some are converted into soil organic forms. Indeed, the vast majority of soil nitrogen (N) and sulfur (S) and a sizable portion of the phosphorus (P) are present in organic forms. Most of this N, S, and P has undergone biochemical transformations; and much of it is bonded to, or associated with, the complex polymeric humic substances that are synthesized during decomposition. At the same time that new humus is being formed, native humus is being degraded (and organic N, S, and P is being mineralized) so that the total soil organic matter content can remain constant. The biochemical pathways involved in both humus formation and its degradation are extremely complex.

Biochemical reactions are important in the degradation of pesticides (such as herbicides, insecticides, and fungicides) applied to the soil by humans. Biochemical reactions involving oxidation-reduction of mineral components in soils are also important. Significant oxidation reactions include oxidation of ammonium-N to nitrate-N (nitrification) and oxidation of elemental S to sulfate-S. Important reduction reactions include denitrification (the reduction of nitrate to nitrous oxide and dinitrogen gases) and reduction of manganese oxide, iron hydroxide, and sulfate-S under waterlogged soil conditions. [See Nitrification].

Two groups of biochemicals that accumulate in soils, the carbohydrates and humic substances, are particularly important in binding soil aggregates together and thus maintaining soil structure. Constituents of soil organic matter can form stable complexes with metal ions. When such complexes form with insoluble humic polymers the availability of metals to plants is reduced. In contrast, complexation with soluble organic material can maintain metals in soil solution when insoluble precipitates would otherwise occur. Plants can excrete soluble complexing agents (e.g., organic acids) into their rhizosphere to increase the availability of some nutrients.

II. Enzyme Activity

A list of some of the biochemical reactions commonly occurring in soils and the enzymes that catalyze them is shown in Table I. Soil enzymes are not easily recovered from soils and only on rare occasions have they been isolated in pure form. As a consequence, soil enzymes are not determined by direct analysis but, rather, indirectly by the ability of a soil extract to transform an added substrate into known products.

Table I. Some Important Enzymes Present in Soils

Enzyme or enzyme system	Reaction catalyzed
α- and β-Amylase	Hydrolysis of 1, 4-glucosidic bonds
Arylsulfatases	$R\text{-}SO_3^- + H_2O \rightarrow ROH + H^+ + SO_4^{2-}$
Asparaginase	Asparagine + $H_2O \rightarrow$ aspartate + NH_3
Cellulase	Hydrolysis of β-1, 4-glucan bonds
Damidase	Carboxylic acid amide + $H_2O \rightarrow$ carboxylic acid + NH_3
Dehydrogenases	XH_2 + acceptor $\rightarrow$ X + acceptor H_2
α- and β-Galactosidase	Galactoside + $H_2O \rightarrow$ ROH + galactose
α- and β-Glucosidase	Glucoside + $H_2O \rightarrow$ ROH + Glucose
Lichenase	Hydrolysis of β-1, 3-cellotriose bonds
Lipase	Triglyceride + $3H_2O \rightarrow$ glycerol + 3 fatty acids
Nucleotidases	Dephosphorylation of nucleotides
Phenoloxidases	Diphenol + $\frac{1}{2}O_2 \rightarrow$ quinone + H_2O
Phosphatase	Phosphate ester + $H_2O \rightarrow$ ROH + PO_4^{3-}
Phytases	Inositol hexaphosphate + $6H_2O \rightarrow$ inositol + $6PO_4^{3-}$
Proteases	Proteins $\rightarrow$ peptides and amino acids
Pyrophosphatase	Pyrophosphate + $H_2O \rightarrow 2PO_4^{3-}$
Urease	Urea $\rightarrow 2NH_3 + CO_2$

Source: Stevenson (1986).

Each soil has its own characteristic pattern of enzyme activity. Enzyme activity is affected by many factors such as cropping history, vegetation cover, soil amendments, and environmental factors (e.g., soil moisture, temperature, and aeration). Marked seasonal fluctuations in enzyme activity typically occur. Enzyme activities decrease with increasing soil depth, and the activity in surface soils is closely correlated with soil organic matter content. Activity also tends to be positively correlated with soil clay contents.

Enzymes can be subdivided as intracellular and extracellular. Intracellular enzymes are those that function within the cytoplasm of proliferating microbial, plant, and animal cells. Many of these enzymes are involved with the metabolic activity of the cells. Often they cannot function outside the cell because their action depends on cofactors and/or other enzymes located inside the cell or some physical property of the cell. Some intracellular enzymes can, however, retain their activity outside the cell.

Extracellular enzymes are those secreted by living cells during normal growth and division. These are produced in abundance by bacteria, fungi, and plant roots. Extracellular enzymes are typically involved in the breakdown of high-molecular-weight compounds that are too large to enter cells directly. The cellulases, amylases, proteases, phosphatases, and sulfatases secreted by a wide range of bacterial and fungi are examples.

Because of the deposition of enzymes by lysed microbial cells as well as decaying roots and root debris, it is difficult to distinguish between truly extracellular enzyme activity in soils and that derived from death and lysis of cells and the release of cytoplasmic material. Enzymes may be present in the soil in a variety of locations. Enzymes that can function exocellularly (extracellular plus some intracellular enzymes) can be present free in soil solution, bound to inorganic or organic soil constituents or within particulate cell debris. Intracellular enzymes can be present in either dead or living cells. Thus, at any one time, the total activity of an enzyme in soil is composed of the activities associated with different soil constituents.

The contribution from enzymes of microbial origin is of great importance in soils, and enzyme activity is usually enhanced under conditions that favor microbial growth and activity. Nonetheless, soil enzyme activity is seldom consistently correlated with either microbial numbers in the soil or with soil respiratory activities. The major reason for this is the existence of stabilized exocellular enzymes in soils. Exocellular enzymes in soil solution are accessible to biological degradation by proteinase enzymes. Thus, such enzymes can turn over relatively rapidly. However, a portion of the active enzyme becomes protected from degradation through adsorption onto soil colloids, covalent bonding to resistant organic materials, or physical entrapment. These protected enzymes are generally less active than those in soil solution, particularly toward high molecular weight substances. In addition, many extracellular enzymes are induced in the presence of a specific substrate while the activity of others can be controlled by substrate or product concentrations. This further weakens any relationship between microbial activity and enzyme activity, as does the direct contribution of plant enzymes to the soil.

III. Decomposition

A. Process of Decomposition

Litter originating from above- and below-ground plant parts is the major pathway of supply of energy and nutrients to the soil in most terrestrial ecosystems. Other inputs include animal carcasses and excreta. Except for a small proportion of mineral constituents (1-8%), plant tissues are largely composed of complex organic components. These include lignin (5-30%); hemicelluloses (10-30%); cellulose (15-60%); protein (5-15%); fats, oils, and waxes (0.5-5%); and water soluble compounds (5-20%). Individual organic constituents decompose at different rates.

A number of stages can be identified during the decomposition of organic residues in soils. Soil fauna (such as earthworms, ants, mites, termites) play an important role in fragmenting litter components, thereby exposing a larger area for microbial attack. Further transformations are carried our through biochemical reactions catalyzed by enzymes principally of microbial origin. A wide range of microorganisms (bacteria, fungi, actinomycetes) are involved in decomposition processes.

The initial stages of microbial attack are characterized by rapid loss of readily decomposable organic constituents such as simple sugars, amino acids, most proteins, and some polysaccharides. The ensuing stages of decomposition involve enzymatic attack of organic intermediates and dead microbial tissue, with the production of CO_2 and new microbial bioinass. The latter stages of decomposition involve the slow decomposition of the more resistant plant constituents such as lignin.

The sugars and storage polysaccharides (e.g., starch) along with most of the essential mineral elements are located mainly in the cytoplasm and vacuoles of living cells. However, the main sources of C and energy are polysaccharides (cellulose, hemicellulose, and pectin) and lignin, which are present in the plant cell walls. These compounds have complex polymeric structures and must be broken down into smaller units by exocellular enzymes before they can be absorbed and assimilated by decomposer microorganisms. Hydrolytic depolymerization of polysaccharides in soils is catalyzed by a group of enzymes known as polysaccharidases. These enzymes catalyze the hydrolysis of glycosidic bonds; they are typically highly specific in terms of both constituent sugars of the polysaccharides and the type of glycosidic bond linking them. Thus, to depolymerize the complex array of polysaccharides in the plant cell wall a large number of enzymes are required.

Lignin is a complex heterogeneous polymeric substance. It is formed from phenylpropane (C_6-C_3) units linked by a number of different bonds in various three-dimensional forms. A restricted number of organisms, mainly basidiomycete fungi, have the capacity to metabolize lignin. It is thought that a number of different enzyme systems are involved and these may be differentially distributed among the lignolytic fungi. Enzymatic degradation of lignin results in the release of phenylpropane units. In turn, these units are broken down into simple phenolic compounds.

B. Anaerobic Conditions

The majority of fungi and actinomycetes are aerobes, as are many bacteria. Under anaerobic conditions, decomposition is dependent on anaerobic bacteria, which operate at a much lower energy level and are less efficient than aerobic organisms. Decomposition is therefore slower and less complete under anaerobic conditions. Under aerobic conditions carbon dioxide (CO_2) and water (H_2O) are the main decomposition products, but anaerobic metabolism results in the accumulation of intermediate products such as organic acids, methane (CH_4), hydrogen sulfide (H_2S), and a variety of secondary decomposition products such as ketones, aldehydes, amines, and mercaptans.

As already noted, fungi are the dominant microorganisms involved in lignin decomposition. The lack of fungi under anaerobic conditions results in lignin accumulating in the soil, albeit in a modified form. Other substances such as cellulose are decomposed by anaerobic bacteria, and the end products include CO_2, H_2O, ethanol, and various organic acids. The slow and incomplete decomposition of plant material under anaerobic conditions is reflected in the large accumulations of organic matter that occur in bog and marsh soils (e.g., peat bogs).

IV. Humus Formation and Degradation

A. Humus Formation

The residues of the decomposition process contribute to the formation of soil organic matter. Humified material represents a highly resistance component of soil organic matter and consists of a mixture of large, complex polymeric molecules that are synthesized by the decomposer microflora during the decomposition process.

The major pathway of humic polymer formation appears to be through condensation reactions involving the polymerization of quinones in the presence or absence of amino compounds (Fig. 3.1). The two major sources of quinones are polyphenols derived from the microbial degradation of plant lignins and phenolic polymers (melanins) synthesized by many bacteria, fungi, and actinomycetes from simple aliphatic compounds that arise from degradation

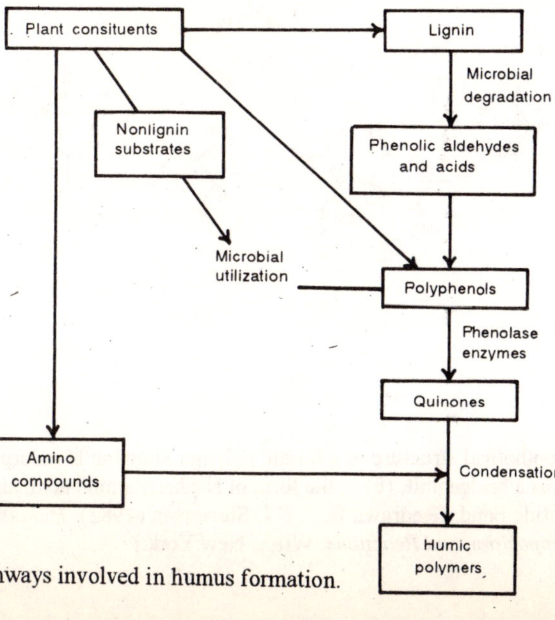

Fig. 3.1. Major pathways involved in humus formation.

of non-lignin C sources (e.g., cellulose). Polyphenols of plant origin may be of some importance, particularly in forest litter. Polyphenols are subject to decomposition but also can be oxidized to quinones through the actions of the phenolase and peroxidase enzymes that are synthesized by many soil microorganisms. Quinones may then undergo self-condensation or combine with amino compounds to form N-containing compounds. Condensation of quinones is greatly enhanced by the presence of amino compounds such as amino acids, peptides, and proteins.

Compounds and components of plant, animal, and microbial origin become bonded to the phenolic compounds. Thus, soil humic materials represent highly heterogeneous mixtures formed as a result of many chemical and biological synthesis reactions. Such mixtures consist of old and newly synthesized polymers, and the complex system of molecules have a wide range of generally high molecular weights. The average molecular weight range for humic acids (alkali-soluble, acid-insoluble humic fractions) and fulvic acids (alkali-soluble and acid-soluble fractions) is of the order of 50,000 to 100,000 and 500 to 2000, respectively. A hypothetical model of humic acid is presented in Fig. 2. Its basic structure is an aromatic ring, of di- or trihydroxyphenol type, bridged by -O-, -CH_2, -NH-, -N=, -S-, and other groups containing both free hydroxyl groups and double linkages of quinones. Carbohydrate and proteinaceous side-chains are attached to the phenolic core.

B. Humus Degradation

The complexity and variability of structure of humic substances means that a wide range of enzymatic activities are required to degrade them. The range of enzymes required includes those such as proteases, carbohydrases, and hydrolases, which degrade materials present on side chains (e.g., proteins and polysaccharides). Other enzyme systems such as monophenol oxidases, peroxidases, and etherases are involved in decomposition of the phenolic core of the humic substances.

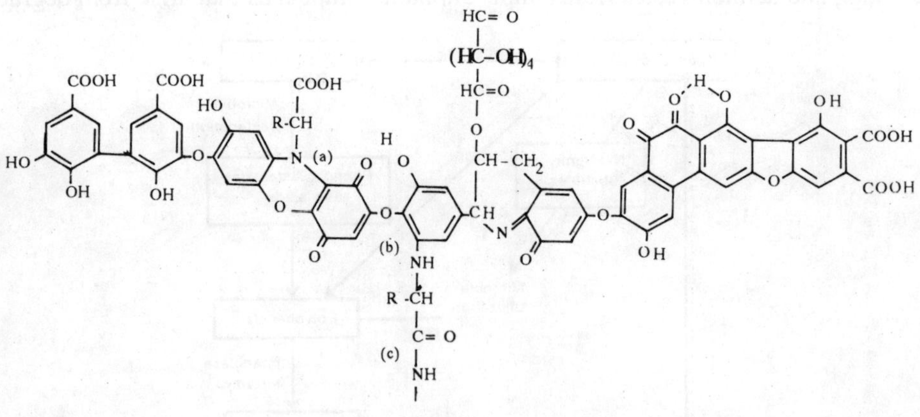

Fig. 3.2. Hypothetical structure of a humic polymer showing N incorporated in three ways: (a) as a bridge unit, (b) in the form of N-phenylamino acid, and (c) in the form of a peptide bond. [Redrawn from F.J. Stevenson (1982). *Humus Chemistry, Genesis, Composition and Reactions*. Wiley, New York.]

Organisms of most microbial groups have been shown to be able to degrade humic substances to some degree. In many of these cases, the microorganisms use humified proteins, peptides, and polysaccharides as C,N, and energy sources and their ability to degrade aromatic compounds varies considerably. Degradation can involve either oxidation of the individual aromatic rings on the humic molecule or depolymerization and subsequent oxidation of the aromatic compounds. In both cases, the reactions are slow and the most degradable fraction of humic material has decomposition times of hears or decades, and the most resistant forms have residence times of thousands of years.

V. Mineralization-Immobilization Turnover

Mineralization-immobilization turnover of N,S, and P occurs as follows:

$$\text{Organic N,S,P} \xrightarrow[\text{immobilization}]{\text{mineralization}} NH^+_4, SO_4^{2-}\ H_2PO_4^-$$

The organic N, S, and P may originate from plant and animal residues, dead microbial biomass, or soil humic substances. The biochemical processes of mineralization are catalyzed by a large group of enzymes and net mineralization occurs when there is a steady release of ammonium (NH_4^+), sulfate (SO_4^{2-}), or orthophosphate ($H_2PO_4^-$). At the same time as mineralization is occurring, some mineral N, S, and P is being immobilized by the microbial biomass. This involves a large number of intracellular enzyme systems. Where mineral forms of N, S, or P are in short supply and there is a plentiful supply of C for microbial growth, net immobilization is likely to occur. Net immobilization of N, S, or P usually occurs when added organic residues have C:N, C:P, and C:S ratios of over 30:1, 300:1, and 400:1, respectively.

A. Forms of Soil Organic N

Over 95% of N in surface soil is usually present in organic forms. The exact nature of this N is only partially understood. During the formation of humic substances nitrogenous materials such as amino acids and amino sugars, predominantly of microbial origin, become linked into the humic substances in many different ways. The N associated with humic materials may exist in the following linkages.

1. As free amino groups ($-NH_2$).
2. As open chain groups (-NH-, =N-)
3. As part of a heterocyclic ring (-NH- of indole or pyrrole or the -N=of pyridine)
4. As a bridge constituent linking quinone groups together.
5. As an amino acid group attached directly to aromatic rings.

About 30-45% of organic N is present as amino acids and another 5-10% is present as amino sugars. Much of the organic N that is as yet unidentified may occur as bridging constituents linking quinone groups together, as amino acids attached directly to aromatic rings, or as non a-amino N in amino acids such as arginine and lysine. [See Nitrogen in Soil.]

B. Mineralization of Organic N

During humus degradation there tends to be preferential loss of nitrogenous materials. The loss is thought to involve amino acids incorporated by peptide linkages being attacked and rapidly removed by hydrolytic enzymes. In contrast, amino acids bound to phenolic nuclei are much less susceptible to attack. Nonetheless, the oxidation of aromatic polymers and cleavage of aromatic rings during humus degradation eventually releases the N present as bridging units or as N-phenolamino acids. Thus, the arsenal of enzymes required to degrade humic substances will be needed to release much of the N from humic substances.

Some of the major biochemical reactions involved in mineralization of nitrogenous organic compounds are show in Table II. Important hydrolytic enzymes include the proteases and peptidases, which catalyze the hydrolysis of peptide bonds, and the enzymes involved in hydrolysis of amino sugars (aminoglycan hydrolases, amino sugar kinases, and deaminases). Other nitrogenous compounds in soils include nucleic acids, purines, and pyrimidines. A large number of enzymes are required to convert nucleic acid-N to NH^+_4.

C. Forms of Soil Organic S

Organic S usually accounts for over 95% of the total S content of soils. Soil organic S can be divided into two broad groups of compounds; (1) S bonded directly to C (C-bonded S) such as that in the S-containing amino acids cysteine and methionine and (2) sulfate esters (R-O-SO_3H). Although 90% or more of S in plant material is present in C-bonded form, approximately 30-70% of organic S in soils is in ester sulfate form. The ester sulfate is mainly synthesized by the microbial biomass and probably represents a reserve source of S for microbial growth.

Ester sulfates are thought to be mainly associated with the sidechains of humic substances. In contrast, much of the C-bonded S is believed to be associated with the aromatic core of humic substances. Complex condensation products may result from the reaction of S-containing compounds (e.g., cysteine) with quinones, similar to the way that amino compounds react with quinones.

D. Mineralization of Organic S

Under long-term cultivation where soil organic matter declines, more S is lost from C-bonded forms than from ester sulfate forms. However, ester sulfates are likely to be more transient and of greater significance in the short-term mineralization of organic S.

Table II. Some Major Biochemical Reactions involved in Nitrogen Mineralization

Substrate	Enzyme	Reaction	Major products
Proteins and peptides	Proteinase and peptidase	Hydrolysis of peptide bonds	Amino acids and peptides
Amides and amidines	Amidoydrolases and amidinohydrolases	Hydrolysis	Amino acids, ammonium, and urea
Amino acids	Amino acid dehydrogenase and amino acid oxidase	Oxidative deamination	a-oxo acids and ammonium
Aminopolysaccharides	Aminoglycanhydrolases	Hydrolysis	Aminomonosaccharides
Aminomonosaccharides	Amino sugar kinases plus amino sugar-6-phosphate isomerases	Hydrolysis followed by deamination	Ammonium plus sugar phosphates
Urea (from animal urine or fertilizer)	Urease	Hydrolysis	Carbon dioxide and ammonium

Sulfohydrolase enzymes present in soils are known to be capable of desulfonating ester sulfates of many different types. The enzymes are classified according to the type of ester they hydrolyze (e.g., arylsulfatases, alkylusulfatases, steroid sulfatases, glucosulfatases, chonodrosulfatases, and myrosulfatases). The most studied group is the arylsulfatases, which catalyze the hydrolysis of arylsulfate to release SO_4^{2-}

$$ROSO_3^{-}+H_2O \rightarrow ROH+H^{+}SO_4^{2-}$$

The hydrolysis occurs extracellularly. The activity of sulfohydrolases is generally inhibited by high concentrations of soil sulfate and the release of these enzymes by microorganisms is probably controlled by their need for S.

For C-bonded S that is bound to the aromatic core of humic substances, an array of enzymes is required to degradc humic substances before the S can be released. A number of pathways have been proposed for the degradation of S-containing amino acids such as cysteine and methionine. The diversity of microorganisms and enzyme systems present in the soil that are involved in this process suggests that several pathways are probably involved.

E. Forms of Soil Organic P

Organic P constitutes between 20 and 80% of the total P present in soils. Up to 70% of total organic P is thought to be present as monoester phosphate. Inositol phosphates make up the bulk of this P; they are derived from plant, animal, and most important, microbial residues. Diester phosphates make up some 1-80% of organic P and include phospholipids and nucleic acids. Traces of phospho-proteins, sugar phosphates, and pyrophosphate have also been measured. [See Phosphorus, Biogeochemistry].

Inositol phosphates are known to be associated with humic substances; they probably exist on the sidechains of these substances. They can also become stabilized through the formation of complexes with Fe and Al compounds in acid soils and Ca in calcareous soils.

The 20 to 50% of organic P that has not been accounted for in soils may be present in polymeric phosphate-containing compounds such as teichoic acids (constituents of bacterial walls), as phosphorylated polysaccharides, or as P linked into humic polymers.

F. Mineralization of Organic P

The phosphatase enzymes catalyze the final stage of the conversion of organic P to inorganic P in soils. Phosphatase is a general name used to describe a group of enzymes that catalyze the hydrolysis of esters and anhydrides of H_3PO_4. The enzymes include phosphoric monoester hydrolyses, phosphoric diester hydrolyses, triphosphoric monoester hydrolyses, as well as enzymes that act on phosphoryl-containing anhydrides and P-N bonds.

The phosphomonoesterases have been most studied. Two types exhibit different pH optima: acid phosphatase has an optimum in the range pH 4 to 6 (predominant in acid soils) and alkaline phosphatase has a pH optimum of 9 to 11 (predominant in alkaline soils). Both soil microorganisms and plant roots are sources of acid phosphatase, but plants appear devoid of alkaline phosphatase. Phosphomonerases catalyze the hydrolysis of monoesters with the release of phosphate.

$$R{-}O{-}\overset{\overset{O}{\|}}{\underset{\underset{O^-}{|}}{P}}{-}O + H_2O \xrightarrow[\text{enzyme}]{\text{Phosphatase}} HO{-}\overset{\overset{O}{\|}}{\underset{\underset{O^-}{|}}{P}}{-}O^- + ROH$$

Since the bulk of soil organic P exists as monoesters, phosphomonoesterase activity is central to mineralization of soil P. Its activity is inhibited by high concentrations of soil solution phosphate, and its release from soil microorganisms is thought to be controlled by their need for P.

Phosphodiesterases originate from plant, animal, and microbial residues. These enzymes catalyze the hydrolysis of nucleic acids and phospholipids with the release of phosphate. As a consequence, only small quantities of diester phosphates accumulate in the soil, even though they are being continually added to the soil through decomposition of plant, animal, and microbial remains.

VI. Pesticide Degradation

The use of synthetic organic compounds to control pests, diseases, and weeds (e.g., insecticides, fungicides, and herbicides) is commonplace on agricultural lands. Such compounds are often arbitrarily classified as biodegradable (rapidly degraded) and persistent (slowly degraded). A major factor affecting the persistence of pesticides is adsorption to clay and surfaces of organic matter. Such adsorption protects the pesticides from rapid degradation. The degree of adsorption is affected by factors such as chemical structure of the pesticide, soil clay, and organic matter content, soil moisture content, temperature, and pH.

Degradation of pesticides in soils is largely mediated by soil microorganisms, although it can occur nonbiologically. Microbially mediated, enzyme catalyzed, degradation reactions of pesticides include decarboxylation, deamination, hydroxylation, β-oxidation, hydrolysis of ester linkages, dehydrochlorination, and reductive dechlorination. Some typical reactions are shown in Table III. Abiotic degradation is mostly through hydrolysis and is influenced greatly by pH.

The persistence of insecticides in soils usually follows the order: organochlorines > carbamates > organophosphates and can range from 2 weeks to 20 years. For ring compounds (organochlorines) degradation of the side chain occurs first, mostly through β-oxidation. This is then followed by breaking of C-C bonds within the ring. The nature and number of side chains has a large effect on degradation rate, as does the number and position of Cl atoms in the ring.

VII. Oxidation and Reduction of Mineral Compounds

A. Oxidation

1. Nitrification

Nitrification involves the conversion of NH^+_4 to NO_3-N. In most soils nitrification is mediated by autotrophic bacteria belonging to the family Nitrobacteraceae. The initial conversion of NH^+_4 to NO^-_2 is mediated by five genera, Nitrosomonas, Nitrosococcus, Nitrosospira, Nitrosolobus, and Nitrosovibrio, while one genus, Nitrobacter, oxidizes NO^-_2 to NO^-_3.

During the conversion of NH_4^+ to NO_2^- there are at least two intermediates, hydroxylamine (NH_2OH) and probably nitroxyl (NOH).

$$NH_4^+ \rightarrow NH_2OH \rightarrow [NOH] \rightarrow NO_2^- \rightarrow NO_3^-$$

During the oxidation of NH_4^+ to NO_2^- two H^+ ions are produced per mole of NH_4^+ oxidized and nitrous oxide gas (N_2O) is evolved probably by reduction of NO_2^-.

Under acid soil conditions (e.g., pH < 5.0) heterotrophic microorganisms (bacteria, fungi, and actinomycetes) may be important in converting NH_4^+ and other reduced N-forms to NO_2^- and NO_3^-.

Nitrification is an important transformation in agricultural soils, since it converts the potentially phytotoxic NH_4^+ ion (often applied as fertilizer) to NO_3^-, which plants can tolerate at much great concentrations than that of NH_4^+. However, NO_3^- is highly mobile in soils and large leaching losses of NO_3^- can occur from agricultural systems. Continual use of NH_4^+ fertilizers on arable land and its subsequent nitrification is a major cause of soil acidification on such land. In many natural ecosystems nitrification does not occur, or occurs at a very slow rate, and plants use NH_4^+ as their major N-source.

2. Sulfur Oxidation

Numerous species of bacteria, fungi, and actinomycetes can oxidize reduced forms of S (elemental S, H_2S, FeS_2) in soils and related environments. In most soils some species of the genus Thiobacillus and a few heterotrophs are thought to be important. The oxidation process is of great importance within the S cycle because the primary source of S in soil is S-bearing minerals. In agricultural soils it is of increasing importance because of the increased use of elemental S as a fertilizer. The net result of S oxidation is the release of H^+ ions and acidification of the soil. Elemental S is sometimes used as an acidifying agent for soils of high pH. S oxidation is particularly important in acid sulfate soils. Such soils develop when swamps and tidal marshes are drained and reclaimed. Oxidation of FeS_2 (originating from surrounding rocks or sulfates in seawater) to H_2SO_4 (sulfuric acid) results in the development of extremely acid soils.

A general scheme for oxidation of reduced inorganic S compounds by thiobacilli is shown here.

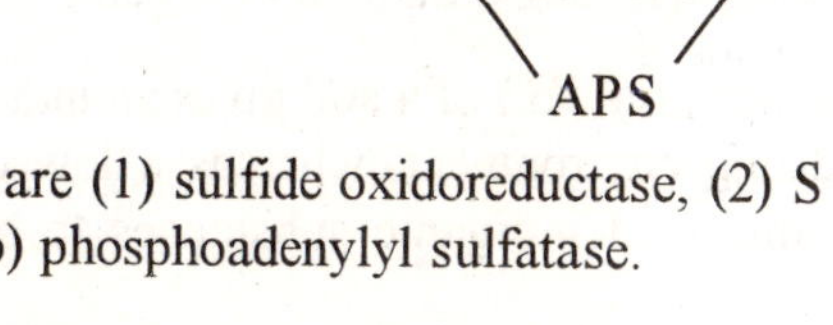

The enzymes involved are (1) sulfide oxidoreductase, (2) S oxidase, (3) sulfite oxidase, (4a) APS reductase, and (4b) phosphoadenylyl sulfatase.

B. Reduction

1. Reduction in Flooded Soils

In soils that are flooded the character of the microbial flora changes. Aerobes are

Table III Some Typical Reaction Types Involved in Transformations of Pesticides in Soils

Category	Reaction[a]	Example
Dehalogenation	$RCH_2CL \rightarrow RCH_2OH$	Propachlor
	$ArCL \rightarrow ArOH$	Nitrofen
	$ArF \rightarrow ArOH$	Flamprop-methyl
	$ArCl \rightarrow ArH$	Pentachlorophenol
	$Ar_2CHCH_2CL \rightarrow Ar_2C{=}CH_2$	DDT
	$Ar_2CHCHCl_2 \rightarrow Ar_2C{=}CHCl$	DDT
	$Ar_2CHCCl_3 \rightarrow Ar_2CHCHCl_2$	DDT
	$Ar_2CHCCl \rightarrow Ar_2C = CCl_2$	DDT
Deamination	$ArNH_2 \rightarrow ArOH$	Fluchloralin
Decarboxylation	$ArCOOH \rightarrow ArH$	Bifenox
	$Ar_2CHCOOH \rightarrow Ar_2CH_2$	DDT
	$RCH\ (CH_3)COOH \rightarrow RCH_2CH_3$	Dichlorofop-methyl
	$ArN(R)\ COOH \rightarrow ArN(R)H$	DDOD
Methyl oxidation	$RCH_3 \rightarrow RCH_2OH$ and/or $\rightarrow$ RCHO and/or $\rightarrow$ RCOOH	Bromacil, disopropylanphthalene pentachlorobenzyl alcohol
Hydroxylation	$ArH \rightarrow ArOH$	Benthiocarb, dicamba
β-oxidation	$ArO(CH_2)_NCH_2CH_2COOH \rightarrow ArO(CH_2)_NCOOH$	ω-(2, 4-Dichlorophenoxy)-alkanoic acids
Triple bond reduction	$RCCH \rightarrow RCH = CH_2$	Buturon
Double bond reduction	$Ar_2C{=}CH_2 \rightarrow Ar_2CHCH_3$	DDT
	$Ar_2C{=}CHCl \rightarrow Ar_2CHCH_2Cl$	DDT
Double bond hydration	$Ar_2C{=}CH_2 \rightarrow Ar_2CHCH_2OH$	DDT

Source: M. Alexander (1981). Biodegradation of chemicals of environmental concern. Science 211, 134.
a R = organic moiety; Ar = aromatic.

replaced by facultative anaerobes, which are then replaced by obligate anaerobes. Under anaerobic respiration, bacteria use inorganic compounds other than oxygen as the external electron acceptor. Substrates such as nitrate, manganese and ferric oxides, sulfate, and carbonate are reduced to molecular nitrogen, manganous and ferrous compounds, sulfide, methane, and other reduced compounds. Reduction of inorganic compounds during anaerobic respiration is coupled with energy-yielding oxidation of organic compounds during anaerobic respiration is coupled with energy-yielding oxidation of organic compounds.

The oxidation-reduction potential (E_h) of a soil gives an indication of reducing conditions. A high E_h indicates an oxidizing environment, whereas a low or negative value suggests a reducing environment. The ability of inorganic substances to be reduced follows the order shown in Table IV.

The changes in soil nutrient status under anaerobic conditions can sometimes result in nutritional disorders occurring in a rice paddy. For example, flooding results in increases in the concentrations of soluble ferrous ions, manganous ions, and H_2S, all of which can result in physiological disorders in rice plants. Methane and ethylene that are commonly released from anaerobic soils can also inhibit the growth of some plants.

2. Denitrification

Denitrification is a respiratory process that is present in a limited number of aerobic bacteria whereby they can grow in the absence of O_2 while reducing NO^-_3 or NO^-_2 to nitrous oxide (N_2O) and dinitrogen gas (N_2). The majority of such bacteria are heterotrophs; they obtain their energy and cellular C from organic substrates. In the absence of O_2, NO^-_3 and other oxides derived from it serve as terminal electron acceptors for respiratory electron transport during the oxidation of the organic substrate. This results in the product being a more reduced N oxide or N_2.

The pathway of N oxide reduction is shown here.

$$NO_3^- \rightarrow NO_2^- \rightarrow NO \rightarrow N_2O \rightarrow N_2$$

The enzymes responsible for the reductions are thought to be NO_3^- reductase, NO_2^- reductase, NO reductase, and N_2O reductase.

The denitrification process is promoted by anaerobic conditions, high levels of soil NO_3^-, and a readily available source of carbon. It is generally positively related to soil pH and temperature. Denitrification appears to occur even in well-structured aerobic soils due to the occurrence of anaerobic microsites within them.

Table IV. Oxidation-Reduction Potentials for Some Typical Soil Reactions

Reaction	Oxidation-reduction potential at 25°C and pH 7.0 (mV)
Disappearance of O_2	
$O_2 + 4H^+ 4e^- \rightarrow 2H_2O$	816
Disappearance of NO^-_3	
$NO^-_3 + 2H^+ 2e^- \rightarrow NO^-_2 + H_2O$	421
Formation of Mn^{2+}	
$MnO_2 + 4H^+ + 2e^- \rightarrow Mn^{2+} + 2H_2O$	396
Reduction of Fe^{3+} to Fe^{2+}	
$Fe(OH)_3 + 3H^+ + e^- \rightarrow Fe^{2+} + 3H_2O$	—182
Formation of H_2S	
$SO^{2-}_4 + 10H^+ + 8e^- \rightarrow H_2S + 4H_2O$	—215
Formation of CH_4	
$CO_2 + 8H^+ + 8e^- \rightarrow CH_4 + 2H_2O$	-244

Denitrification is of great significance in terms of the biogeochemical cycle of N in that it tends to balance N inputs to ecosystems through biological N_2 fixation (that is, N_2 is returned to the atmosphere). In agricultural systems, losses of fertilizer N through denitrification commonly average 10-15% of applied N. The loss of N_2O to the atmosphere through denitrification (and nitrification) has received much study in recent years because it may speed up the rate of reactions in the stratosphere that lead to destruction of the ozone (O_3) layer.

VIII. Soil Aggregation

Soil aggregates must be stable in order to maintain a soil structure conducive to the growth of plants. Two groups of soil biochemicals are involved in stabilizing soil aggregates: soil polysaccharides and humic substances. There are thought to be two stages to the formation of stable aggregates: an aggregating stage involving humic substances.

Many microorganisms produce exocellular mucilages and gums, which are also produced on root surfaces. These compounds are predominantly polysaccharides and act in the soil as glues. The rhizosphere, with its highly active microbial population, is a major site for the production of mucilages; but they are also produced in abundance by soil microflora wherever organic debris are being decomposed. The linear, flexible chain structure of polysaccharides makes them well suited to form bonds between particles or groups of particles that make up aggregates. In general, polysaccharides in soils are readily degraded by microorganisms; so their effect can be short-lived unless there is a continuous supply, such as in the rhizosphere.

The long-term stability of soil structure is thought to be conferred largely by humic substances. These relatively stable substances confer stability because they can form stable complexes with mineral fractions in the soil (e.g., layer silicates and Al and Fe hydrous oxides) through a variety of bonding mechanisms. It is thought that microaggregation occurs predominantly by the functional groups of organic matter (Fig. 3.2) being linked to polyvalent cations on layer silicate clay surfaces through a water bridge (clay-polyvalent-water-organic matter).

A breakdown of aggregate stability can result in aggregates breaking down into fine particles under the action of water (slaking). If a dry period follows slaking, a hard surface crust can develop over the soil surface. Such a crust can impede seedling emergence and limit entry of further rainfall or irrigation water, leading to ponding or runoff and water erosion. Erosion leads to loss of surface soil from sloping lands and its deposition on lowlands. It can also result in siltation of waterways. [See Soil Physical Condition and Plant Growth.]

IX. Solubilization and Precipitation

A. Formation of Metal Organic Complexes

Many soil biochemicals, including high-molecular-weight humic materials and other compounds as organic acids, amino acids, and polyphenols, can form stable complexes with metal ions. The stability of metal-organic matter complexes is determined by a variety of factors but usually follows the order:

$$Cu^{2+} > Ni^{2+} > Co^{2+} > Zn^{2+} > Fe^{2+}\ Mn^{2+}$$

Soluble complexes form with compounds such as organic acids, amino acids, polyphenols, and soluble humic material; these maintain metals in soil solution when insoluble precipitates would otherwise occur. These soluble complexes may enhance movement of metals toward the roots of higher plants.

Complexation of metal ions by insoluble humic polymers reduces their availability to plants. In some cases where potentially toxic levels of metals are added to soils in wastes (e.g., sewage sludges) their concentration may be reduced to nontoxic levels through complexation. The formation of insoluble complexes of Al^{3+} with organic matter can be particularly important in maintaining low concentrations of Al^{3+} in soil solution in some acid soils. High concentrations of Al^{3+} in solution inhibit plant root growth.

B. Root Exudates and Plant Nutrition

As already noted, plant roots release an appreciable amount of C into the rhizosphere. Root exudates include low-molecular-weight substances such as sugars, amino acids, organic acids, and phenolic compounds. Some of these biochemicals can significantly influence the nutrition of plants. For example, malic acid, which is in root exudates, can cause the increased solubilization of MnO_2. Solubilization involves reduction of MnO_2. Solubilization involves reduction of MnO_2 to Mn^{2+} (during the oxidation of the organic acid to CO_2); the subsequent chelation of Mn^{2+} by organic acids prevents its reoxidation and facilitates its movement in the rhizosphere, where it is readily available to the plant. Organic acids can also cause reduction of Fe^{3+} oxide and its solubilization and the solubilization of sparingly soluble phosphate compounds.

Phenolic compounds are of particular importance in relation to solubilization and uptake of Fe. In response to Fe deficiency, dicotyledonous plants release phenolics into the rhizosphere. These compounds promote reduction of Fe^{3+} (and to a lesser extent MnO_2), and subsequent chelation increases Fe mobility in the rhizosphere. In graminaceous plants, Fe deficiency results in the release of nonproteinogenic amino acids (known as phytosiderophores) from roots. These compounds chelate Fe^{3+} and thus enhance its solubility and uptake by the plant.

Root exudates may also be involved in detoxification of toxic nutrients in the rhizosphere. For example, some Al-tolerant plant species secrete organic acids into their rhizosphere. The organic acids can complex with Al^{3+} ions, thus rendering the Al less damaging to root growth and function.

complexation may occur. These soluble complexes may affect movement of metals towards roots of higher plants.

4

Orogenic Belts

1. Introduction and Historical Premise

In contemporary usages, orogenic belts are curvilinear segments of severely deformed, normally continental rocks, stretching from hundreds to thousands of kilometers. In many places parts of the belts are affected by regional metamorphism and intruded by large bodies of mainly graites and related rocks. The processes of deformation and transformation in such belts lasted from tens to a few hundreds of million years. Younger orogenic belts mark positions of major existing mountain chains, and it is inferred that the older belts in flatter terrains originally were also elevated.

A. Orogenesis

The process of mountain building is normally referred to as orogenesis, derived from the Greek words oros (mountain) and gennao (to produce), and the segment of the earth where the process takes place is known as an orogen, which on maps may be a partial or whole orogenic belt. The process of orogenesis involves not only the ;elevation of a part of the crust but also severe structural interpenetrative deformation. In this sense, major elevations of the earth's crust such as midoceanic rises, continental and oceanic plateaus, and volcanogenic edifices are not considered as mountain chains in an orogenic sense. Major orogenic belts usually represent a complex of orogens, each generated during a limited period of time and outcropping over a limited area. For instance, in the Appalachian orogenic belt of North America (Fig. 4.1) at least three orogens-Taconian, and Alleghenian-have been distinguished (Fig. 4.2). Each orogen is characterized by a particular stratigraphic package of sediments and associated volcanic rocks, intruded by granites and other mainly felsic (SiO_2-rich) rocks and affected by regional metamorphism. The sediments, volcanic formations, and some of the granites are widely affected by folding, Because these orogens, as shown in Fig. 4.2, tend to overlap each other in space and differ in the timing of orogenesis, their superposition leads to complexities that have not always been successfully deciphered. In the Appalachians each of the three orogens traditionally was deemed to have been affected by a specific, time-restricted episode of particularly severe orogenesis, generally referred to as orogeny. thus the Taconian orogen was the product of the mid-Ordovician Taconian or Taconian orogeny, the Acadian orogen was formed during the mid-Devonian to Early Carboniferous Acadian orogeny, and the Alleghenian orogen arose in the process of the Late Carboniferous to Early Permian Alleghenian orogeny.

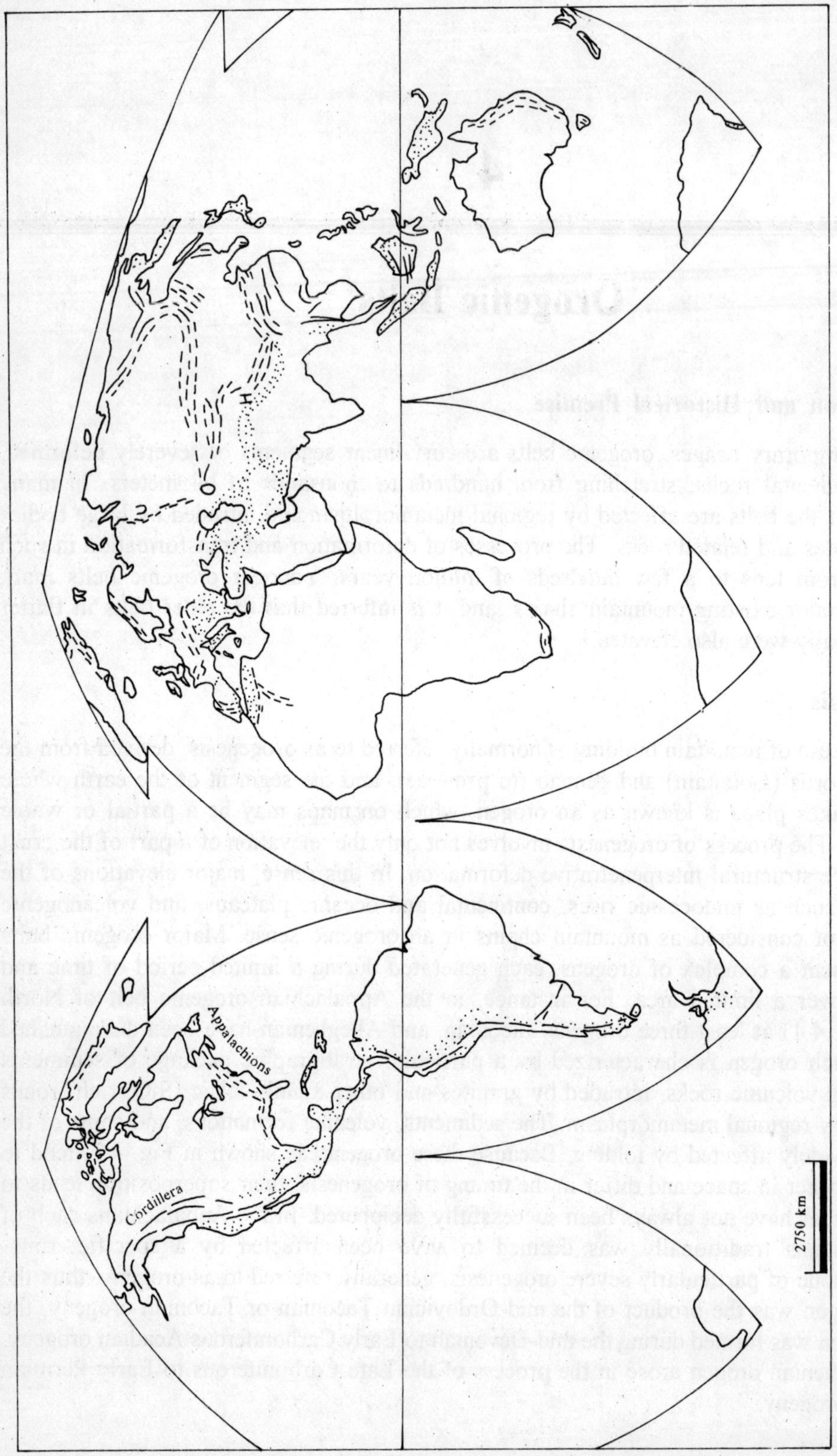

Fig. 4.1. Major Phanerozoic orogenic belts of the earth. Cenozoic belts, stippled; Mesozoic belts, long dashes; Paleozoic belts, short dashes. U, Urals; A, Alps, H, Himalayas. (Boggs Eumorphic Equal Area Projection.)

Initially these stratigraphic ages of orogenies were based on stratigraphic evidence, each orogeny being fitted into an interval of time separating the deposition of strata in an earlier orogen from that in the later orogen and represented by an angular unconformity. Each such unconformity was a surface of discordance between the deformed and eroded rocks of an early orogen and the often less deformed rocks of a later orogen, stratigraphically lying above the surface of discordance, as shown in Fig. 4.3. Later, isotopic dating was introduced and in general confirmed distinctions among individual orogens, their sediments, volcanic rocks, and periods of plutonism. In many places, as in the Appalachians,superposition of one orogen on the other led to mixed isotopic ages that were not amenable to simple analysis. In places, as in central Newfoundland, it appears that in terms of sediment deposition there is complete continuity from the Taconian into the Acadian orogen, without an intervening unconformity.

Because one of the earliest types of structure recognized in orogenic belts was the fold, affecting the originally subhorizontal strata, orogenic belts have been referred to as fold belts (foldbelts) and folded belts. The term mobile belt has also been employed to reflect movements

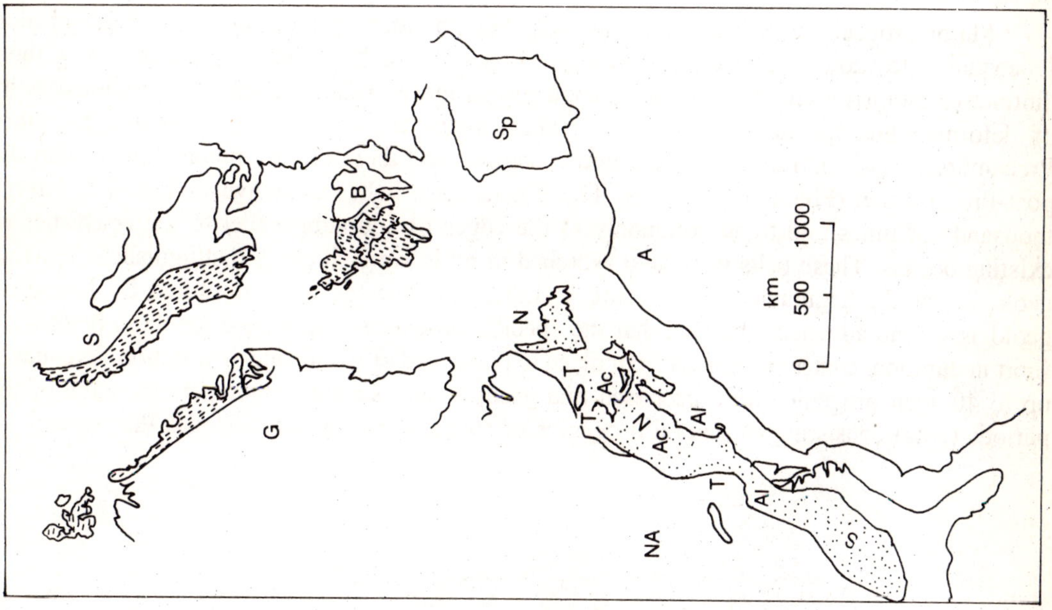

Fig. 4.2. Appalachian and Caledonian orogenic belts. Appalachian belt, open stipple: S, southern N, northern; Ac, Acadian, Al, Alleghenian (dotted lines separate these orogens); T, Taconian. Caledonian belt, dashed. Locations (large letters): A, Africa; B, Britain; G, Greenland; N, Newfoundland; NA, North America; S, Scandinavia; Sp, Spain.

throughout the period of deformation recorded within the belt. Even in individual orogens the period of time during which orogeny took place is debatable and prolonged, lasting up to tens of millions of years, especially if distal parts of the orogen are compared. As a result, orogeny, which was defined by G.K. Gilbert in 1890, was understood by many geologists in the past as a relatively short period of strong deformation. Thus earlier in the century many referred to such a period as a "revolution." It has since been recognized that orogenic periods are longer than formerly thought and their timing, as assessed in different parts of the orogen, varies. Those results led to the concept of time-transgressive orogeny on the one hand and pulselike subdivision of an orogeny into episodes or phases on the other. In the Acadian orogeny of North America, for instance, several phases have been recognized, some based on the interpretation of minor folds that have been refolded, others on successive regional metamorphic events,yet others on time relationship to granitic intrusions. Thus some granites are recognized to be preorogenic, others to be contemporaneous with the orogeny or synorogenic, and lastly others to succeed the orogenic deformation but precede the erosion of the mountains and thus be postorogenic. Detailed isotopic dating of granites allows an estimate of the age of the orogeny to which they are related. In the British Isles the extensively mapped and investigated Caledonian mountain chain has yielded very detailed data on regional metamorphic phases of the Caledonian orogeny and has permitted the cross-dating of such phases and intrusive granites.

Major orogenic belts have conventionally been divided by their ages into Precambrian, Paleozoic, Mesozoic, and Cenozoic-Recent. Precambrian belts have in general lost their dintinctive mountain edifices and are recognized as more or less elongated, curvilinear belts of deformed and metamorphosed rocks. These belts are commonly intersected by other Precambrian belts or truncated at oceanic coastlines and therefore are much shorter than the post-Precambrian (Phanerozoic) belts. The Phanerozoic belts, as shown in Fig. 4.1, stretch thousands of miles and trend commonly at the edges of and subparallel to the coastlines of existing oceans. These belts were first sketched in by Edward Suess, a distinguished Austrian geologist, at the beginning of this century. Later his schemes were amplified by European geologists, who advanced the idea that the specific orogenies responsible for these belts were short in duration, contemporaneous from place to place, and of worldwide distribution. Eventually up to 40 such orogenies were proposed and grouped into several major periods, each of the periods (eras) consisting of several orogenies or successive orogenic episodes that were

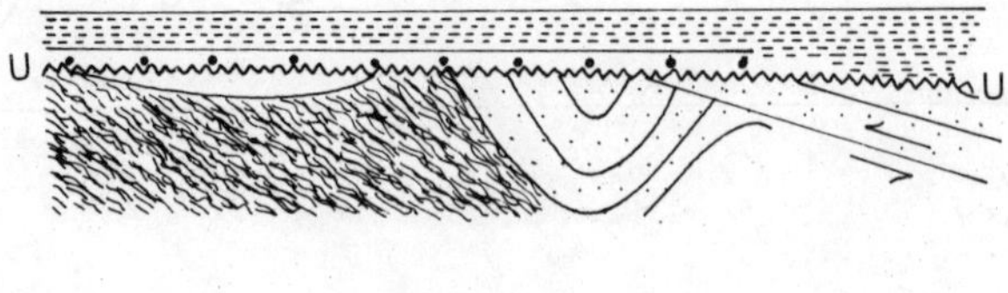

Fig. 4.3. Diagrammatic relationships at an unconformity separating orogenic events. U, Unconformity (zigzag line). Below unconformity succession: shales, anastomosing pattern; sandstones, open stipple. Over unconformity succession: conglomerate, O pattern; sandstone, stipple; shale, dash.

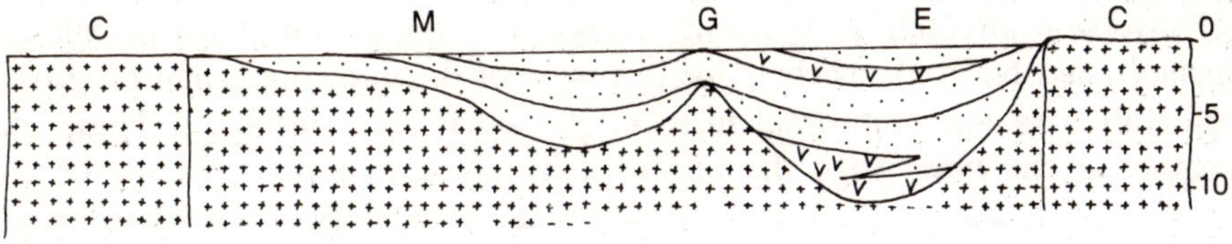

Fig. 4.4. Conceptual geosynclinal cross section. continental crust, crosses; sediments, stipple; V, volcanics; C, craton; C, symmetrical cratonic land; E, eugeosyncline, G, geanticline; M, miogeosyncline. Depths on the right of the section are in thousands of meters.

closely interrelated. This usage did not take root in North America. Thus, for instance, the Taconian and acadian orogenies of the Appalachians would in Europe be included in the Caledonian orogenic periods, as phases of the Caledonian orogeny. Moreover, the Early Paleozoic (Caledonian), Late Paleozoic (Variscan), and Meso-Cenozoic (Alpidic or Alpine) orogenic periods (eras) in Europe cannot be easily translated into the Taconian, Acadian, Alleghenian, and Laramide (Late Cretaceous) orogenies of North america.

A traditional European view has been that each orogeny is part of a geotectonic cycle, which with the deposition of sediments (depositional phase), succeeded by deformation and plutonism (orogenic phase), followed by uplift, crustal fracture, and erosion (the postorogenic morphological phase). The depositional phase as recognized in the orogens reflected the formation of a basin of sedimentation, known as a geosyncline. This basin was deduced to be marine, normally contained a great thickness of sediments deposited over a long period of time, and not uncommonly was conceived as structurally divided into two separate parallel subbasins. One was of shallow-water deposits without appreciable thickness of volcanic rocks and was referred to as a miogeosyncline and the other contained volcanic rocks and deep sediments and was referred to as a eugeosyncline, as shown in Fig. 4.4. The barrier between the two subbasins was called a geanticline, and over it the sediments and /or volcanic deposits were much thinner. Such asymmetric pairs of basins were frequently assumed to be marginal to continents, although hypothetical land was also placed on the other side of the pair, as in Fig. 4.4, which could have been a continental fragment or an island arc. The originally asymmetrical structure of such orogens was further developed in Europe to have two sets of paired orogens, disposed on either side of a median mass consisting of continental crust and margined on the far sides by major continents. Thus, after an orogeny and erosion those structures became intracontinental orogenic belts such as the Urals in Fig. 1. Most geologists found that eugeosynclinal belts were particularly affected by plutonism, with widespread zones of regional metamorphism affecting deformed and foliated rocks. These rocks were intruded by granitic bodies, of which most were syntectonic to posttectonic. These strongly deformed and metamorphosed belts became known as the internides, while the deformed, but much less plutonized, miogeosynclinal belts were called the externides and the comparatively undeformed continental crust adjacent to them the craton (Fig. 4). The process of conversion of the sediments and other rocks into a strongly deformed package, which at the completion of an orogeny adhered to the craton with the resulting disappearance of a marine geosynclinal domain, became known as inversion. It was maintained that each episode of inversion affecting geosynclines marginal to the continent led to enlargement of the continental mass (craton), whereas the inversion of the intracontinental orogens involved either preservation of the continental mass or its actual area shrinkage. These ideas were current mainly at the time when the prevalent hypotheses of the behavior of our planet through time envisioned it either in a steady state with static (permanent) positions of continents and oceans or in a state of relatively moderate contraction as a result of loss of heat. Modern investigations suggest that these "Fixist" premises were incorrect and that the earth is much more dynamic both crustally and subcrustally, and the new mobilistic considerations have led to different interpretations.

B. Forces Causing Orogenesis

The idea that the earth has evolved to the present state by the cooling of an originally molten mass was prevalent in the nineteenth century and naturally resulted in the contraction hypothesis. Its main geological difficulty was that most reasonable calculations suggested that contraction, assessed from mountain building, exceeded very significantly that expected from cooling of the molten mass. Moreover, the apparent episodicity of mountain building in time implied alternation of physical conditions, for which contraction did not provide a viable mechanism. On a theoretical level, the gradual cooling of the originally molten earth became an obsolete idea, because the discovery of radioactivity in the first decade of the twentieth century meant that the earth was not a passive cooling body but had an internal source of radioactive heat. In time it was proposed that contraction and expansion of the earth occur in successive pulses, the former generating mountains and the latter, new depressions of geosynclines. This oscillation hypothesis, which succeeded the contraction hypothesis and is still favored by some and particularly Soviet scientists, is based on the presupposition that vertical oscillatory movements cause orogenic effects. The movements were claimed to have been caused by forces produced by periodic melting and solidification of parts of the earth's mantle. Others sought recourse in periodic heat-activated expansion and contraction of the mantle as a whole, while yet another group of geologists disclaimed any notion of the mechanism of the apparent periodicity of event. This periodicity was very graphically named the "Pulse of the Earth" by the Dutch geologist J.H.F. Umbgrove in his book by that title. He, in particular, emphasized the observation that there is a connection between the growth and emergence of mountains and the resulting worldwide regressions of oceanic waters and between the denudation, collapse, and submergence of mountains and worldwide transgressions, thus establishing a tectonic explanation for worldwide (eustatic) oscillations of sea level. Yet he expressed agnosticism as to the real mechanisms causing these events, restricting himself to a simple statement that they must be internal to the earth.

All the ideas advance so far assumed that the continents and oceans retained their original positions with regard to each other. This basic approach was clearly fixist. Although they were first conceived of in the nineteenth century, the beginning of the twentieth century saw the emergence of a new set of hypotheses which claimed that the permanence of oceanic and continental positions is a myth, that continents broke up and the resulting pieces drifted from each other, and that in the intervening areas new oceans came into existence. The adherents of these hypotheses, of which the most developed one was proposed by A. Wegener, became known as mobilists and the basic theoretical tenet to which they adhered was the breakup and separation of continents called continental drift. One of the originally most significant items of evidence for it was the fact that South America and Africa had parallel coastlines, suggesting that the two were in the past a single continent and that the South Atlantic ocean came into existence in the aftermath of continental breakup. Later quantitative, computer-designed juxtaposition of America, Europe, and Africa resulted in a very convincing fit of these continents, shown in Fig 4.5. This inference was made in Cambridge, England, under the guidance of the late Sir Edward C. Bullard and is known as the Bullard's fit. The mechanism and indeed the forces responsible for the drifting apart of America, Europe, and Africa were not understood. Some claimed that the earth has suffered expansion. Others suggested that because the earth's mantle underlying the crust was viscous rather than solid, it developed internal motion as convection currents, rising under the central parts of oceans, diverging there and exerting forces to drive continents away from each other (Fig. 4.6) and thus cause continental drift. At the edges of continents the currents, which now lost heat, were deemed to plunge under the continents, exerting compression and generating mountain belts. Thus, as a result of a single process there would be oceans and accompanying mountain belts. The convection current hypothesis in modern literature is too often attributed to living or recently deceased scientists, yet it was formulated in the last century. One of the most elegant of such formulations was set forth in 1889 by Rev. Osmond Fisher, a mathematician, geologist, and Church of England vicar. Yet, the convection current hypothesis in its early form did not explain how the oceans form, how the intracontinental orogenic belts come into existence, and indeed why the drifting apart of,say, Africa and South america did not produce concomitant orogenic belts on either side of the South Atlantic ocean. The explanation of this dilemma had to await the post-World War II burst of oceanic and submarine scientific investigations by geophysicists and the development of the world seismic maps recording earthquake activity. [See Plate Tectonics, Crustal Deformation and Earthquakes].

The oceanographic investigations in the mid-1960s demonstrated the remarkable fact that the magnetic pattern characteristic of the oceanic crust is symmetrical across the midocean ridges, and isotopically dated, it was found that the symmetry reflected a more or less continuous aging of the crust. At the midocean ridge of the Atlantic Ocean (Fig. 4.1) the crust is young, whereas at the continental edges it is over 150 million years older, implying that the magnetic pattern recorded a gradual generation of the midocean ridge where the convection currents came up. This phenomenon was technically called ocean spreading (fig. 4.6). It was theoretically predicted by Harry Hess, formerly professor in Princeton, before the magnetic evidence was discovered. The opening of the Atlantic Ocean, some 2000 km wide, meant that in compensation for the formation of new oceanic crust, elsewhere on the globe the oceanic crust has to be consumed. Such consumption was inferred and consumption of oceanic crust was global rather than related to some specific ocean. Also, clearly the convection currents that came up in the central parts of the ocean did not necessarily plunge under the nearest continental margin. The basic observation for this was provided by seismologists, who first constructed a world map of all large earthquakes recorded by

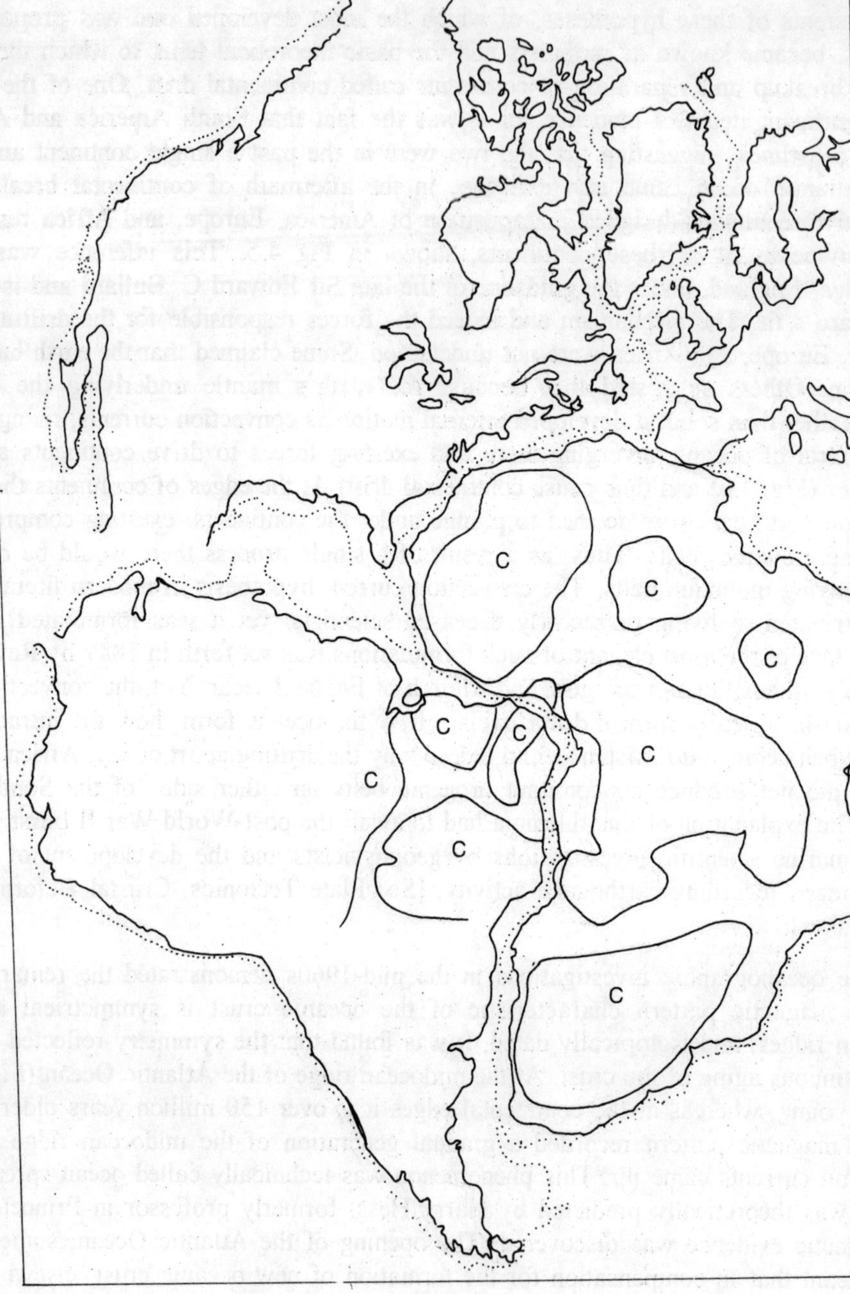

Fig. 4.5. Bullard's fit of continents prior to Jurassic rifting. The line of the continents against the oceans is solid black; the line of the - 1000-m contour is dotted; older cratons in Africa and South america are outlines and marked C; blank areas in between the cratons are the Pan-African domain. [Based on data from Sir Edward Bullard, H. Porada, and R. Caby].

seismometers. It was found that the oceanic ridges, where the convection currents came up, were shallow (less than 30 km), while at the edge of the Pacific Ocean, where the convection currents were going down and the oceanic crust was subducted, the zone of subduction was recognized by the fact that earthquakes occurred at progrssively greater depths and were situated along a rough planar zone inclined from the ocean under the continents. On the continental side of the zone and above it chains

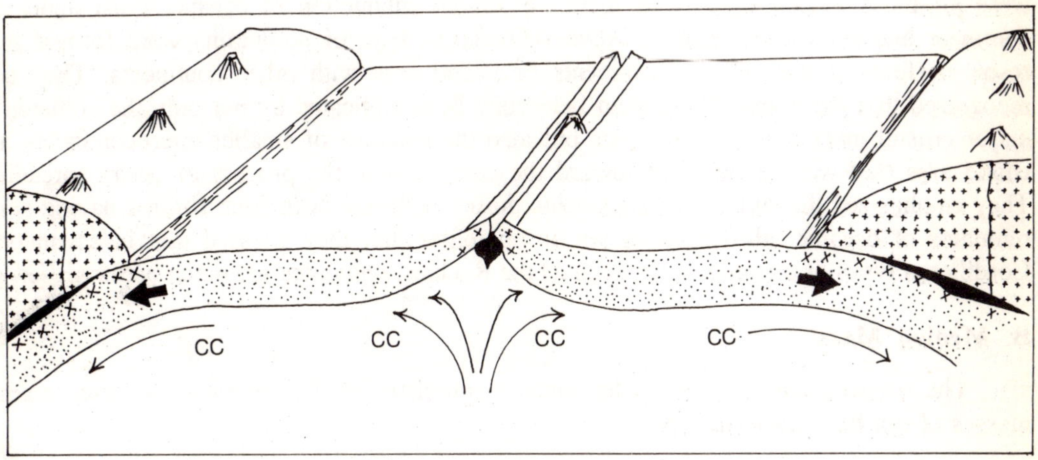

Fig. 4.6. The concept of convection currents with rifting in the middle of the ocean and compression and deformation at the edge of continents. Oceanic lithosphere, stipple; continental lithosphere, crosses; trenches at the edges of continents; volcanic cones of perioceanic volcanic arcs on the continents; magma chambers and conduits, black; earthquakes, heavy crosses; directions of movement of the oceanic lithosphere, heavy arrows; convection currents, light arrow marked CC.

of volcanic cones occur, often, as in South america, surmounting orogenic mountains. In between lines of earthquakes (fig.4.6), associated with midocean ridges and zones of subduction, large earthquakes are rare, and therefore it was proposed that the earthquakes represent mechanical interaction at the edges of plates, forming the upper rigid layer of the earth known as the lithosphere. The lithosphere was found to be much thicker than the earth's crust and involved also the upper part of the earth's mantle. The overall concept of the platelike mosaic forming the lithosphere is now known as plate tectonics. Its introduction into the geological sciences provided a global basis for the interpretation of the structure and evolution of orogenic belts. Although most plates have either oceanic or continental lithosphere, some are composite and therefore do not show an intervening subduction zone and the associated orogenic belt separating the two parts of the plate.

II. Types of Orogenic Belts

B. Bird - Dewey Hypothesis

The traditional theory of orogenesis recognized essentially two types of orogenic belts: (1) symmetrical, intracontinental and (2) asymmetrical, perioceanic. The symmetrical type implied a geosynclinal development within a continent with subsequent viselike deformation between two forelands and topographic inversion of the geosyncline into a mountain chain.

The asymmetrical type was supposed to have had a similar evolution, but the foreland on the ocean side somehow disappeared, or the geosyncline was an oceanic deep ;trench and the mountains were formed by queezing the sediments of the trench between the continent (island arc) and the ocean. The transition from the traditional to the plate tectonic interpretation took place in the late 1960s and early 1970s, in part as a result of the effort by J. Bird and J.F. Dewey, who suggested that the asymmetric (Cordilleran or Andean) mountain chains were produced in consequence of thermally driven subduction of oceanic crust under the continent and the symmetrical or Alpine-Himalayan type of mountains were formed as a result of forceful collision of continents or island arcs with other continents. They also recognized that the Alpine-Himalayan belts may be complicated by not only the collision of major crustal units such as continents but also the collision of smaller microcontinents and island arcs that were transported toward the continents in the process of ocean spreading. They emphasized the suturing of continents during collision, with such sutures having small remnants of oceanic lithosphere known as ophiolites, but they stressed that in asymmetric Cordilleran types of orogenic belts, such as the Rockies, ophiolites are much better preserved.

B. Median Mass

The intracratonic orogenic belts show a symmetry of deformation and often median masses of uplifted basement. The

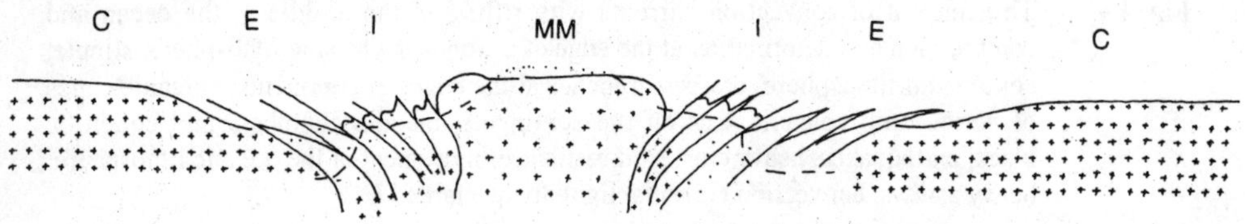

Fig. 4.7. Tectonic relationships in a symmetrical orogenic belt according to L.Kober Continental (cratonic) crust, crosses; geosynclinal sediments, open stipple; C, craton; E, externides showing thrusting; 1, internides showing thrusting and folding; MM, median mass, reconstituted and granitized crusts; boundary of high-grade metamorphism, dashed line.

Variscan orogenic belt of west-central Europe is commonly cited as an example. This median mass (Zwischengebirge in German) was first introduced by a well-known German tectonician, Leopold Kober, who conceived continents to have stable, immobile regions, which he called kratogens (cratons), and regions of high mobility that were the orogens with median masses (Fig. 4.7) that separated fold belts of divergent polarity of motion. These belts were divided into the outer zone (the externides) next to the craton, with few intrusive or extrusive igneous rocks, and the inner zone (the internides), with many intrusions and a high degree of metamorphism. It was held that orogenic deformation began in the median mass, produced as a result of melting and rising of he crust, and in time had spread toward the opposing cratons (Fig. 4.7). According to the plate tectonic hypothesis of Bird and Dewey, the so-called median mass was the site of collision of two continents (cratons) and was divisible by the suture into two separate parts. Instead of the median mass, they employed the term mobile core, accompanied by a broad zone of metamorphic effects. At present the preferable term is the metamorphic core. The terms internides and externides and still employed in geological literature but more or less in a purely descriptive sense.

The complex Pan-African late Precambrian-Cambrian orogenic belt does not easily fit the classifications applied to Phanerozoic belts. First, rocks affected by the Pan-African orogenesis, instead of normal curvilinear surface trends form a network of belts surrounding cratons of Archean to Middle Proterozoic ages. The Pan-African belt also extends into the Arabian Peninsula and possibly, as the so-called Cadomian belt, into Europe, as well as into South america, Antarctica, and South Australia. Second, the belt is so diffuse and widespread and its deformation and metamorphism are so long-lasting (500-900 million years) that it is referred to by many as having suffered a tectonothermal event or events, called the Pannotios cycle. Comparisons between such a widespread orogenic domain and Phanerozoic belts are at present premature, although the overall plate tectonic origin of the Pan African belts is difficult to doubt because in many places in Africa ophiolites representing remnants of the old oceanic lithosphere are preserved within the deformed belts. Third, despite this, certain of the belts have no ophiolites and presumably did not suffer ocean spreading prior to the onset of deformation. It is thus maintained that intracontinental orogenic belts were formed as a result of tensional rifting succeeded by compressional orogeny (Fig. 4.8).

III. Zonation of Orogenic Belts

The zonation of Alpine-type orogenic belts can be considered from tow standpoints: the zonation of sediments and other deposits prior to orogenesis, and the postdeformation zonation of the orogen following collision. As shown in Fig. 4.8, the plate tectonic interpretation of preorogenic sediments in a belt assumes that most have been deposited on the continental shelf of a rifted continent, where sedimentation involved the deposition oft a pile of deposits now reaching thicknesses of up to 18 km and overlying the collapsed continental crust, and that some sediments were transported farther afield into the oceanic crust, where they intermingled with volcanic rocks. The former pile, mainly shallow water and sometimes referred to as ensialic, consists mainly of clastic sediments and is called miogeoclinal. The latter pile, generally thinner, of deeper water intruded by minor igneous bodies and interbedded with lavas, is called eugeoclinal. Laterally toward the ocean the eugeoclinal sediments lose thickness and develop distinct oceanic types, such as radiolarian cherts, thin laminated shales, altered volcanic ash deposits, and impure graded sands.

The miogeoclinal deposits toward the continent range into more coastal sandstones, conglomerates, and reef limestones. In the present-day equatorial regions the continental edge separating the miogeocline from eugeocline is in some places occupied by a carbonate reef bank, such as the Great Barrier Reef off the northeast coast of Australia. A similar but much larger "fossil" Cambrian-Ordovician carbonate bank was recognized at the western edge of the Appalachian orogenic belt, suggesting that it also marks the former shelf edge of the Laurentian continent (craton). The transitional zone at the edge of the craton-shelf belt is the future marginal platform of the orogenic belt. The platformal sedimentation sensu stricto is thin, intermittent, and takes place in shallow-water epicontinental inland seas such as the present-day Baltic Sea in Europe. With regard to the Appalachians. Early Paleozoic rocks of central Kentucky can be cited as an example of this type of sedimentation.

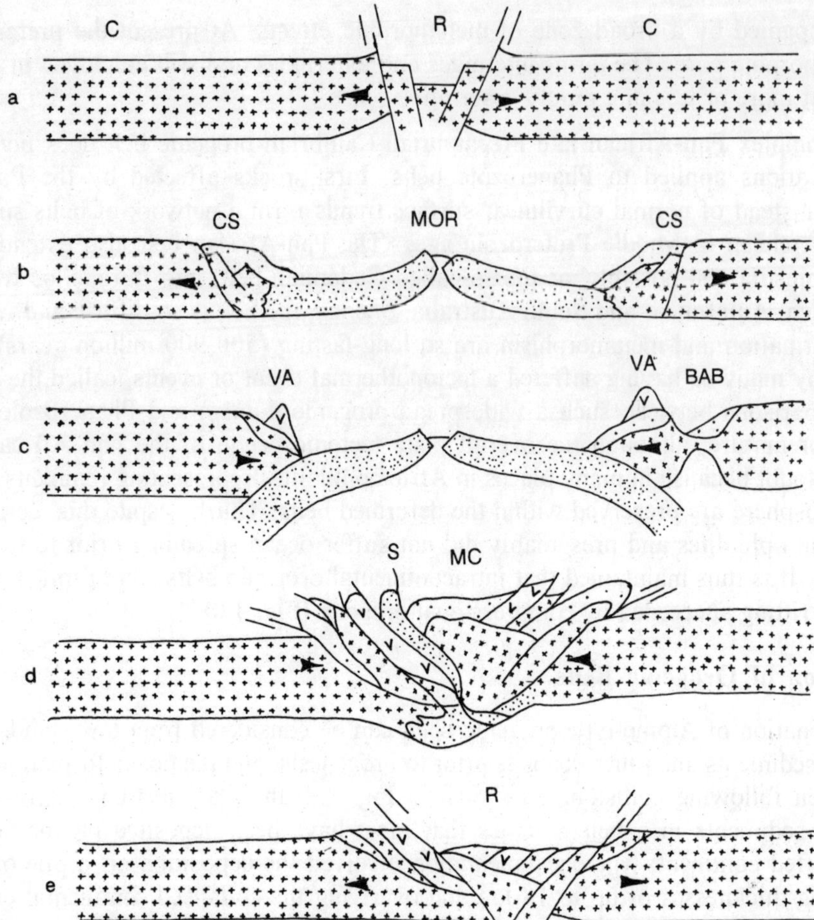

Fig. 4.8. Schematic development of a collisional orogenic belt and repetition of rifting. (a) Initial rifting; (b) ocean spreading; (c) initiation of subduction and convergence; (d) collision and orogenesis; (e) recurrent rifting. Continental lithosphere, crosses; oceanic lithosphere, close stipple; sediments, open stipple; volcanics, v; thrusts and rift faults, arrows; C, continental plate; R, rift; CS, continental shelf; MOR, midocean ridge; VA, volcanic arc; BAB, behind ;arc basin; S, suture; MC, metamorphic core. Arrowheads show movement of plates.

Thus, in terms of original, preorogenic sedimentation one can recognize the platformal zone, the marginal zone, the miogeoclinal zone, and the eugeoclinal zone. A rough parallelism exists between each of these zones and tectonic zones that originate during orogenic deformation. In the platformal zone there is little deformation, and tectonically this zone is referred as the foreland. In the marginal and miogeoclinal zones strong deformation, characterized by thrust faults and large overfolds with gentle axial planes as well as associated systems of high-angle reverse faults. is characteristic. The regional metamorphism is low grade (low temperature); few if any granitic intrusions penetrate these rocks. This zone of externides is sometimes associated with deformation that produces steep, subvertical fold structures accompanied by cleavage. This type of deformation for a short time enjoyed the title of paratectonic, although now this term is not extensively used. The eugeoclinal deposits are normally strongly deformed, folded and faulted, affected by high-grade regional metamorphism, and intruded by granitoids. This tectonic style occurs in the interior of the orogenic belt; the zone exhibiting it is known as the internides and is characterized by polyphase deformation and folded thrust faults.

The tectonic zonal divisions have been supplemented by others; Harold Williams from Newfoundland and introduced the term tectonostratigraphic, for combined structural-stratigraphic series of features, and applied it specifically to the Northern Appalachian orogenic belt. At first his zonation was applied only to the Taconian orogen, but progressively both Taconian and Acadian became included in single zones. In central Newfoundland recent work suggests that the distinction between the two may be blurred and that a continuous depositional sequence exists from the Ordovician to Silurian times.

Superposed on the sedimentary, structural, and tectonostratigraphic zones are zones of regional metamorphism. These generally are expressed in terms of single types or assemblages of metamorphic minerals that reflect the temperature and at times the pressure that the metamorphosed strata had experienced. An initial zonal scheme was developed in the Highlands of Scotland, where in the nineteenth century George Barrow recognized zones of growth of progressively higher-temperature metamorphic minerals including chlorite, biotite, garnet, staurolite, kyanite, and sillimanite.Since then, it has been discovered that kyanite is a pressure-induced mineral. Later pressure-dependent metamorphic rocks have been discovered, such as in blue schists containing as sodic amphibole (glaucophane), and found extensively developed around the Pacific Ocean and near the oceanic trenches, where the state of high compression was recognized from seismic evidence. The belts of high-pressure metamorphism were paralleled father inland by adjacent belts of high-teamperature metamorphism. Where erosion had cut deeply into an older orogenic belt, very high-pressure and high-temperature rocks, known as granulites, are found. Because early Precambrian orogenic belts are found in areas where several episodes of erosion have occurred and where it is still in process, in the past it was inferred that granulite-producing metamorphism is characteristic of older Precambrian rocks. It is now known that younger Phanerozoic rocks have also been affected by granulite metamorphism, and therefore the presence of granulites in an orogenic belt reflects temperature-pressure conditions referred to in terms of metamporphic facies rather than the age of metamorphism. Especially in strongly metamorphosed and polyphase-deformed parts of orogenic belts, there is often evidence for several episodes of metamporphism. Because metamorphic transformations of rocks are rarely absolutely complete, relicts of earlier metamorphic minerals are found among new mineral assemblages of the younger metamorphic episodes. In Vermont, for instance, remnant garnet crystals of earlier Taconian metamorphism were found to be overgrown by the garnets of Acadian metamorphism. This relationship was still earlier extensively described from the Caledonian orogenic belt of the Highlands of Scotland. Multiple metamorphisms proved amenable to age dating of episodes or regional metamorphism by isotopic methods involving the breakdown of such radioactive elements as potassium, rubidium, uranium, and samarium. Only under exceptional conditions was it possible to determine ages of distinct episodes unequivocally. In many instances "mixed" ages resulted from the overprinting of one episodes by another. In other cases prolonged cooling and diffusion of the by-products of radioactive decay of the metamorphosed pile of rocks, or alteration by the circulation of dccp ;fluids, produced ambiguous ages. Thus, although at some specific locality two or more episodes of mctamorphism can be recognized or even dated, elsewhere within the same orogenic belt this is currently impossible. Therefore, although in the northern Appalachians of New England and Maritime Provinces of Canada, Taconian and Acadian orogenic and metamorphic episodes can be distinguished, these cannot be properly traced throughout the region, which gives rise to extensive, often sterile controversies.

The common superposition of structural effects of successive orogenies leads to even more confusing situations. For instance, in the southern Appalachians, rocks involved in the Taconian and Acadian orogens were widely affected by Alleghenian (Late Carboniferous-Permian) thrusts, which transported large slices of those orogens toward the craton and piled them up on the platform. Such thrust slices, known as allochthons, there overlie either the original unmoved crust (the autochthon) or the underlying allochthonous parts of the orogen. Allochthonous relationships were first described in the Swiss Alps and northeast Highlands of Scotland and then recognized and described in the Appalachians. Recent deep seismic reflection investigations have confirmed the earlier geological inferences in this regard. Where the pile of thrusts in an orogen impinges on and climbs up the platform, it generates a substantial load, and the platform bends down, giving rise to a depression in front of the thrust pile. This downwarp becomes a basin of sedimentation known as a foreland basin or marginal downwarp. The basin, on the craton side, often has a somewhat discontinuous, broad, upwarped edge, referred to as the peripheral ;bulge. The recognition of these structures near the edge of the orogen in the southern. Appalachians permits separation of the Taconian and Alleghenian orogens, although in this region the edge of the Acadina orogen is currently detectable only in Alabama.

The zonation of intrusive rocks in Phanerozoic, Alpine-type orogenic belts is complex. In general diorite-granite types are most abundant. Compositionally, some of these rocks are calc-alkaline (equally rich in calcium and sodium plus postassium) and others are mainly alkaline (rich in sodium and potassium). The calc-alkaline rocks are held to the originated in island arcs, and the alkaline rocks along large tensile faults. The coexistence of both types in orogenic belts is attributed to either different types of contemporaneous deformation in a belt or different tectonic events at different times. In the Cordilleran types of orogenic belts, zones of different alkalinity have also been recognized.

IV. Theories of Origin in Time

The successive orogenic episodes that were inferred in many orogenic belts used to be attributed to spasmodic, regular episodes of deformation and plutonism accompanied by volcanicity. The episodes of orogeny were assumed to be of regular sequence, widespread distribution, and accompanied by periodic granite emplacement. The introduction of mobilistic theories, whether expansionist or plate tectonic, permits much less catastrophic interpretations and a considerable countinuity of events, especially where orogeny can be related to the interaction of large plates, such as the Pacific versus South American or North American. Nevertheless, the analysis of Phanerozoic plate motions indicates that in each large region there are plates. Oceanic plates are shown to have been entirely subducted away under continents, while continental plates grew as a result of collisions of smaller plates, of plates and adjacent island arcs, of smaller oceanic or continental plates with another major continent, and of midocean ridges with continental plates. The results of ;such complex happenings include not only orogenic deformation of the lithosphere, but also the removal of some lithosphere substance by scraping of plates (tectonic erosion) and by increasing the thickness of a plate by adherence of parts of another plate (underplating). In the latter case the plate may split along the crust-mantle interface; this is known as delamination. Underplating by material less dense than the plate above may lead to isostatic uplift of mountain chains. It has been suggested that the formation of particularly high mountains such as the Himalayas is due to the underplating of the Eurasian

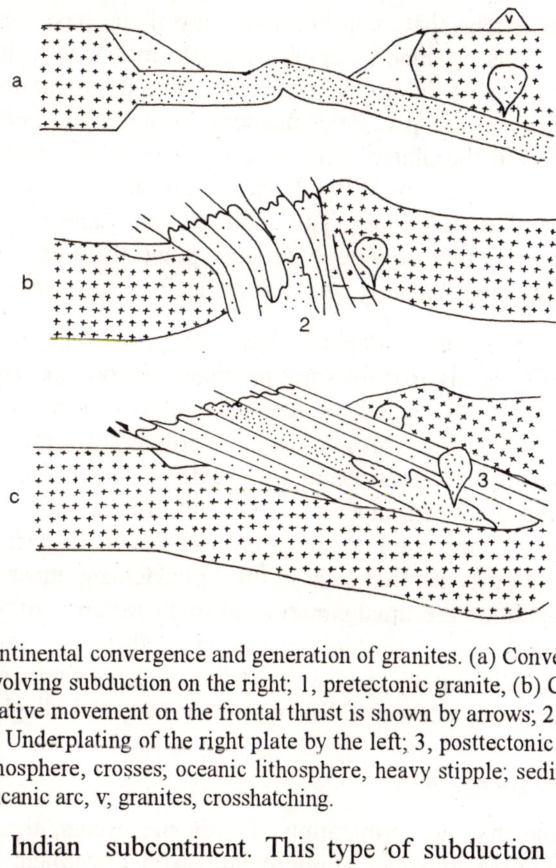

Fig. 4.9. Continental convergence and generation of granites. (a) Convergence of two plates involving subduction on the right; 1, pretectonic granite, (b) Continental collision; relative movement on the frontal thrust is shown by arrows; 2, syntectonic granite. (c) Underplating of the right plate by the left; 3, posttectonic granite. Continental lithosphere, crosses; oceanic lithosphere, heavy stipple; sediments, open stipple; volcanic arc, v; granites, crosshatching.

Tibet plateau by the Indian subcontinent. This type of subduction of oceanic lithosphere under the continent, as is the case with the Andean chain, is known as the B type. Thus it is proposed that prior to the collision of India with Eurasia, B-type subduction took place at a speed of 10 cm per year. about 40-60 million years ago the collision took place, and since then the subduction of India has been continuous at 5 cm per year. The suggested model is that collisional events generate the onset of orogeny in an orogenic belt, while B-type subduction causes the formation of island arcs and the emplacement of preorganic granitoids and associated volcanic rocks. Subsequent to the collision, A-typc subduction of continental plate under continental plate results in the generation of synorogenic granites that are commonly migmatitic, and the period of orogenic deformation lasts until the A-type subduction terminates. The ultimate relaxation of compression provides an opportunity for an upward rise of postorogenic granitoids and associated volcanic rocks. The model is shown schematically in fig. 4.9. The deformation associated with collisional orogenies develops structures indicative of strain such as folds and faults. In addition, a substantial part of bedded rocks is severely disturbed, converted into chaotic breccias known as melanges and affected by much pervasive shearing in zones of milled and partly recrstallized rocks, referred to as mylonites.

One of the results of collisional transition from type B to type A subduction is that parts of the oceanic lithosphere disintegrate into blocks and fragments. These becomes ophiolites, and their wedging between colliding plates is taken to represent sutures along which the adjacent plates adhere. The search for past geographic conditions is carried out by paleogeographic reconstructions that can be made only if the past position of a particular plate can be determined. This is done by employing paleomagnetic techniques, which enable an approximate determination of the position of a plate and its orientation with regard to the magnetic poles at any time in the pst. Paleomagnetic research has permitted reconstruction of the changes in position of the plates with respect to the earth's magnetic poles along the apparent polar wandering paths (APWPs) back as far as early Precambrian (Archean) times, more than 2500 million years ago. This indicates that plate tectonics probably was a mechanism of orogeny even then, although on this point there is still much controversy because patterns of fragmented Archean belts are difficult to reconstruct.

Cordilleran belts, which were thought to have been produced by a gradual process of subduction and continual orogeny that does not involve collision but only deformation at the oceanic trench, have in the past decade been reinterpreted. It is now thought that the multiple orogenic events encountered in the North American Cordillera are the results of either head-on or oblique collisions between continental fragments, volcanic arcs, and other relatively small oceanic objects and the North american continent. These microplates postcollisionally are referred to as terranes, and their original positions with respect to the continent are problematic. It is currently thought that even in the Appalachians most of the internides have been generated as a result of the amalgamation of such terranes of suspect origin, which accrued to the North American (Laurentian) craton at different times. These collisional accretions, when isotopically dated, yield a spectrum of dates that are difficult to fit with precision into the conventional scheme of Taconian, Acadian, and alleghenian orogenies.

V. Denudation and Rejuvenation

It is most probable that the termination of orogenic events, through either continent-continent collision or multiple collisions of microplates with a continent, results ultimately in a final elevation of mountains (morphogenesis) and the collapse of the orogen. These then become subject to erosion of supply sediments to successor basins overlying the stumps of collapsed and denuded mountains as well as new basins in the adjacent cratons. For instance, the termination of the Acadian and Caledonian orogenies in Devonian times resulted in a widespread development of clastic deposits known as the Old Red Sandstone. These conditions, however, were ephemeral, and the continental cratons of Laurentia (North America) and Baltica (northern Europe) that adhered in the Devonian time split again in the Mesozoic, resulting in the North Atlantic Ocean. J.F. Dewey claims that the extension following the generation of orogens occurs along them because here the lithosphere and the continental crust are thicker, and therefore isostatic rise is faster. Such events are likely to occur several times during the development of an orogenic belt. Thus one of the first results, in addition to erosion, would be tectonic denudation by gravitational collapse of the rapidly risen mountain chain. In the North american Rockies this is very well known, resulting in large-scale detachments of the uplifted surface sediments and exposure of deeper metamorphic cores, known as the metamorphic core complexes. The detachment resulted in the formation of large-scale, low-lying tensional faults marked by well-developed shear zones of mylonite. Such collapse events have also bee of late inferred in parts of the Appalachians, where the tensional normal faults approximately follow the previously formed compressional thrusts. Continental stresses set up at this stage commonly lead to the formation of large strike-slip faults (e.g., San Andreas fault of California) formed during the subduction phase of the oceans or the convergence of continents prior to collisions. Even the collapsed parts, however, are subject to subaerial erosion and denudation. Yet there is ample evidence that even the denuded orogens are often uplifted again, although the underlying reasons for such uplifts are at present controversial and inadequately documented.

5

Soil Pollution and Its Control

Introduction to Land Pollution

The land pollution has been mainly caused by solid wastes and chemicals. One of the major pollution problems of large cities has been the disposal of solid waste material including farm and animal manure, crop residues (agricultural wastes), industrial wastes like chemicals, fly ash and cinders which have been residues of combustion of solid fuels, garbage, paper, cardboard, plastics, rubber, cloth, leather, construction rubbish, brick, sand, metal and glass resulting from demolition of buildings, dead animals like cattles, dogs, cats, birds, containers, discarded manufactured products like old refrigerators, washing machines and autos. Huge quantities of unwanted material bring about serious disposal problems.

Further, many of the chemicals given out into the air like radioactive minerals, sulphur, and lead, eventually come to earth to pollute the soil. Many pesticides and herbicides have been applied by aerosol spray and enter into the soil. These chemicals may influence significant effects on plants and animals, bringing about a disruption in species composition of communities. In forest ecosystems, these chemical pollutants may be able to inhibit processes of soil formation, and reduce the capacity of the forest to maintain fertility of the soil.

MAIN LAND POLLUTANTS

(a) The major amount of solid rubbish has been provided by our households in the form of domestic wastes. Some common examples include groceries, food scraps, vegetable remains, packing materials, paper, remnants of used coal, ash, wood, metals, plastics, ceramics, glass, etc. Many of these have been non-reusable. All these form heaps of municipal refuse. It is not properly disposed off, this can prove perilous. Such places often become a dwelling place for rats, flies, bacteria, mosquitoes and a large number of other vectors, having the potential of causing many human diseases.

(b) Modern agriculture is mainly responsible for polluting soil through the non-judicious use of chemical fertilizers, herbicides, insecticides and fumignants. Most of these have been stable chemicals and remain in the soil for long period without degradation, and exert cumulative effect. Pollution caused by pesticides has been among the greatest causes of concern in the field of agricultural pollution. Pesticides residues occur in soil air and water as well as in living organisms. Besides killing the living

organism present on the surface of the soil, they reach even the deeper layers through tilling and irrigation on the land, killing still more living forms. With the continuous use the soil micro organisms lose their capacity of nitrogen fixation.

(c) Areas surrounding smelting and mining complexes are generally soiled by metals like cadmium, zinc, lead, copper, arsenic and nickel. These act as phytotoxic even in small quantities. They also make plants unsafe for human and animal consumption. Zinc, along with cadmium, has been released into the environment during the use or breakdown of lubricating oils, vehicle tyres, galvanized metals and fertilizers.

(d) The main sources of land pollution have been the industries like pulp and paper mills. oil refineries, power and heating plants chemicals and fertilizer manufactures, iron and steel plants, plastic and rubber producing complexes and so on. Thousands and thousands tons of solid wastes have been either dumped or burnt or emptied into rivers. Most industrial furnaces give rise to a grey, powder residue of unburnt material called fly ash, an important pollutant, besides huge mounds of solid wastes.

SOLID WASTE POLLUTION

Solid-waste pollution has necessarily always been more advanced in the largest cities. Disposal problems become difficult with increase of population density. Simultaneously there is a greater production of waste per unit area, and a decreased proportion of land available for its disposal. Therefore, the history of solid-waste pollution is largely connected with the histories of the largest cities.

From the earliest civilizations burial of wastes has always been comparatively easy in rural areas, and was specified by Moses as a required practice for Israel.

During the flourishing of the Minoan civilization on Crete from 3000 to 1000 BC, solid wastes in the capital of Knossos were placed in large pits with layers of earth at intervals; disposal and wastes accumulated in the streets and around the towns and villages (ibid). This practice was said to have persisted until the 19th century.

Ancient cities have yielded their secrets to archeologists partly through analysis of their refuse piles. Because so much waste material was refused in various ways the ultimate discards tended to be long-lived. Examples are ceramic pot shards and the dust and mud from floor and highway sweepings. Despite the small quantities of these materials which were produced, the steady accumulation year after year contributed in many cases to the eventual burying of the city and to its subsequent rebuilding either on the same site at a higher level or at another site.

The old Roman city of Bath in England is between 12 and 20 feet above the level of the Roman period. The rush-covered floors of houses in Tudor times contained debris up to two or three-feet thick and rich in nitrates. Until the first importation of nitrates in 1610 this debris was reputed to be the sole source of the saltpetre used in the manufacture of gun-powder. Even the right to "digge for saltpetre" was sold.

Some medieval German cities avoided the danger of being buried in their own wastes by requiring that departing wagons which had been used, perhaps, to bring farm produce into the city return with a load of wastes to be deposited in the countryside.

Few features of the environment have been more abrasive to the senses than the discarded objects of civilized life. Litter along the roadsides, trash on the beaches, heaps of refuse in the streets, the pungent piles of rotting garbage, and similar unwelcome debris are unpleasant reminders of untidy human habits. Thousands of alleys, backyards and vacant lots contain the remains of discarded refrigerators, defunct washing machines, and parts of shattered automobiles-rusting remnants of the chrome-plated glory of a technological society. Open dumps scar the landscape; and highway travelers view a country side spotted with the incongruity of beauty and blight.

The process of civilized living has been creating mountains of solid waste, the disposal of which has created a formidable problem. In countries where the population is high and land scarce, the conventional municipal refuse tip will soon become a thing of the past, unless we are prepared to sacrifice our valuable landscape and countryside. In many advanced countries, the actual amount of refuse which is produced is equal to 300 kg. occupying 1.5 m^3, per person per year. There is every sign that these quantities would go up in the near future. Two main factors are responsible for this:

1. In the past much domestic refuse was made to burn into open fires to burn. With the spread of central and electric heating this practice has been to some extent discontinued.
2. An increasing amount of packaging has been used for the goods we buy, more newspapers are being read, etc., all increasing the quantity of refuse generated by the average family.

The composition of refuse gets varied from area to area. Where open coal fires are still being used, a high proportion of the refuse would be having partly and fully burned clinker and ash; modern housing estates supply a less dense refuse which largely consist of packaging materials. A seasonal variation also occurs in the weight of refuse produced per household; it tends to be is at its maximum in winter and its minimum in summer.

Because of the changes of living standards and habits today, a rapid change in the nature and composition of refuse has taken place. The average composition of refuse in 1975 and the forecast for 1985 has been provided in the following table:

Component	*Percentage 1975*	*Percentage 1985*
Dust	24	12
Cinders	12	6
Vegetables and other rotting materials	17	13
Cardboard and paper	25	50
Metals	7	6
Textiles	3	3
Glass	8	6
Miscellaneous materials	6	4

It is expected that by 1990, the quantity of refuse per home will rise from 15 kg per home per week to 20 kg.

Plastic waste has been an important and increasing component of domestic refuse. About 100 g per week of waste plastics are given out per dwelling, but the nuisance created by waste plastics has been much greater than has been suggested by the moderate quantities involved. Plastics do not rot, like most other materials, although they can be burned. PVC (polyvinyl chloride) has been especially objectionable in that it forms highly corrosive hydrochloric acid when burned, which is an extreme nuisance in the operation of refuse incinerators. If we use an inadequately high temperature in burning waster PVC, it is even possible for highly poisonous phosgene ($POCI_3$) to get produced.

Modern man has been using an enormous quantity of basic raw materials, converting them at first into valuable products, and later, when the products get discarded, into waste. It has been estimated that the present annual rate of man-induced scatter of certain elements would be as follows:

Element	*Metric tonnes per annum*
Iron	345 $\times 10^6$
Manganese	2 $\times 10^6$
Copper	4.8 $\times 10^6$
Zinc	4.3 $\times 10^6$
Nickel	660 $\times 10^3$
Lead	2.5 $\times 10^6$
Molybdenum	70 $\times 10^3$
Chromium	7 $\times 10^3$
Silver	12 $\times 10^3$
Mercury	10 $\times 10^3$
Tin	168 $\times 10^3$

According to the Law of Conservation of Mass, it means that waste materials, theoretically, remain available for use on this planet. Availability in theory and usefulness in practice have been, however, very far apart, because the cost of recovering almost all of these elements invariably becomes much greater than the cost of obtaining them from out natural resources. Until quite recently, we have realised that they can be and will be severely depleted in the near future, and that we must now make greater efforts to salvage them from our own wastes.

This 'natural history' of the motor car has been a classic example of man's recklessness in the mater of essential raw materials. In one component part or another, most of the elements shown in the table above finds use in motor manufacture. These elements are first extracted from crude ore, then refined, alloyed, treated and formed into specific parts; once all the parts get assembled together, the completed motor car has a high monetary and utilitarian value. For several years it is usefully used, until it can no longer satisfy economically the needs of any owner. 'Planned obsolescence' ensures that this point would get reached after a given number of years or a given mileage travelled, whichever gets achieved the sooner. As often as not the vehicle is then abandoned to dis-figure the landscape, since its low scrap value is hardly enough to induce its owner to do anything else with it. In consequence, large quantities of valuable materials would get virtually lost to man for ever. Since a technological endeavour, salvaging these materials has been well within the ability of man; but economically, reclamation of specific elements from composite waste would not be a viable commercial proposition. It is widely accepted that the economic balance must be adjusted so that it is always economically extract useful raw materials from waste.

Not all of the components of solid waste are valuable. Some are useless and others are toxic. For example, selenium, a poisonous element that occurs naturally in the soil, is taken up by plants that are used for making paper and is present in newspaper in the amount of about 8.6 ppm. This element is found in incinerator stack gas, raw solid wastes, incinerator residue, and in water that has been used for removing fly ash. Many other toxic chemicals of various kinds and amounts are present in municipal and industrial waste.

The adverse economics of the disposal of solid wastes as commonly practiced are far-reaching. Mineral resources such as copper, zinc, lead, and tin are limited and the ores are nonrenewable. These metals are not destroyed by disposal but are often scattered beyond recovery. Even iron ore is in finite supply in economically desirable deposits. The squandering of these resources that occurs when wastes are dumped indiscriminately in spoil areas will inevitably produce a disruptive economic impact at some time in the future unless improvements are adopted. Development of salvage techniques will, therefore, have direct economic effects. Salvage and recycling will also alleviate other problems associated with the rapidly increasing density in population,s which results not only in the generation of wastes but also in wastes of greater complexity.

Today the disposal of garbage and other solid wastes appears to be much simpler: urban dwellers merely place them in bags or other containers, put these in garbage cans, and have them picked up by a public or private collection agency. Only when one passes an open dump or sees a landfill outside a big city just as outer Ring Road in New Delhi the real magnitude of the solid waste problem becomes apparent.

The agriculture and food processing industry is considered to be the largest single contributor to the total annual production of solid wastes, but mining operations and some industries are also important sources. Postconsumer wastes (those generated by end user, not by operations that make consumer products, and excluding junk vehicles) were responsible for solid wastes. Of this, 37% consisted of paper products alone, the largest postconsumer source of solid waste. Table 5.1 gives a breakdown of postconsumer wastes by material categories, together with quantities recycled and the net quantities disposed. Urban refuse annually includes about 60 million cans, 360 million bottles, 53 million tons of paper, 4 million tons of plastics, and about 1 million abandoned vehicles.

The environmentalists expect the quantities of total wastes to increase greatly by 1990, not only because of the increased population but also because an increased amount will be discarded per person. Even though larger quantities will be recycled. Information about solid wastes has been comparatively scanty for a number of reasons. The need for this information

Table 5.1
Postconsumer Solid Wastes, Residential and Commercial Sectors, (millions of tons)

Material Categories	Gross Discards	Material Recycled: Quantity	Material Recycled: Percent Recycled	Net Waste Disposal: Quantity	Net Waste Disposal: Percent Total Waste
Paper	53.0	8.7	16.5	44.2	32.8
Gass	13.5	0.3	2.1	13.2	9.9
Metals	12.7	0.2	1.6	12.5	9.3
Ferrous	11.2	0.2	1.4	11.0	8.2
Aluminum	1.0	0.04	4.0	1.0	0.7
Other non-ferrous metals	0.4	0.0	0.0	0.4	0.3
Plastics	5.0	0.0	0.0	5.0	3.7
Rubber	2.8	0.2	7.1	2.6	1.9
Leather	1.0	0.0	0.0	1.0	0.7
Textiles	1.9	0.0	0.0	1.9	1.4
Wood	4.9	0.0	0.0	4.9	3.6
Total non-food products	94.8	9.4	9.9	85.4	63.4
Food waste	22.4	0.0	0.0	22.4	16.6
Yard waste	25.0	0.0	0.0	25.0	18.5
Miscellaneous					
Inorganic waste	1.9	0.0	0.0	1.9	1.4
Total waste	144.0	9.4	6.5	134.8	100.0

could not be a number of years ago and many communities made no weight or other measurements. Measurements have been difficult to make due to the complexity of the material, and techniques for sampling and making laboratory measurements have not been standardized. The quantities and character of solid wastes also vary greatly with geographical location, food patterns, etc. Data collection has started to improve, however, with the holding of national conference and with the growth of the solid wastes program of the Central Government.

A typical classification of solid wastes is the following:

1. *Garbage*—putrescible (decomposable) wastes from food, slaughter-houses, canning and freezing industries, etc.
2. *Rubbish*—non-putrescible wastes, either combustible or non-combustible. Combustible wastes would include paper, wood, cloth, rubber, leather, and garden wastes. Non-combustibles would include metals glass, ceramics, stones, dirt, masonry and some chemicals.

3. *Ashes*—residues (such as cinders and fly ash) of the combustion of solid fuels for heating and cooking or the incineration of solid wastes by municipal, industrial, and apartment house incinerators.

4. *Large wastes*—demolition and construction rubble (pipes, lumber, masonry, brick, plastic, roofing and insulating materials) automobiles, furniture, refrigerators and other home appliances, trees, tires, etc.

5. *Dead animals*—household pets, birds rodents zoo animals, etc. There are also anatomical and pathological wastes from hospitals.

6. *Sewage treatment process solids*—screening settled solids, sludge.

7. *Industrial solid wastes*—chemicals, paints, sand, explosives etc.

8. *Mining wastes*—"tailings, slag heaps, culm piles at coal mines, etc.

9. *Agricultural wastes*—farm animal manures, crop residues, etc.

Some of these wastes have been the result of control techniques for air and water pollution, such as fly ash recovered from flue gases and sludge from sewage treatment plants. Others have been capable of producing air or water pollution quite readily, such as organic matter that can decay to produce odors or farm animal wastes that can enter water supplies. Much of the solid waste produced is collected or otherwise disposed of but much is permitted to litter the landscape-highway litter, dog wastes in cities, discarded oil drums etc.

The sources of solid wastes have been (1) municipal-street sweepings, sewage treatment plant wastes, wastes from schools and other institutions; (2) domestic-garbage, rubbish, and occasional large wastes from homes (and, in years past, large amounts of ashes from the use of coal); (3) commercial-from stores and offices; (4) industrial-from manufacturing plants; (5) mining-from coal mining, strip mining, etc.; (6) agricultural.

Solid wastes could be bound out in terms of their mass or their volume. Their densities generally vary over a wide range-typically 0.05 to 1.0g/cm^3- and mass (or weight) measurements are felt to be more useful. Solid wastes are rather low in density due to the large amounts of air present in them and it is usually not difficult to compact them to several times their original density, as is done in sanitary landfill operations or by home compactors.

The composition of refuse seems also to vary from area of area. Where open coal fires are still being used, a high proportion of the refuse consists of partly and fully burned clinker and ash; modern housing estates supply a less dense refuse which largely consists of packaging materials. There is also a seasonal variation in the weight of refuse produced per household; this is at its maximum in winter and its minimum in summer.

The composition of collected refuse gets changed over the years, with the quantity of ashes decreasing as coal heating of households get replaced by gas, oil, and electric heating and with the amount of paper increasing dramatically along with the growth in consumer packaging.

In its place the initial step in dealing step in dealing with refuse will have to be incineration. Against the capital cost of constructing incinerators on the scale needed there have been several advantages to be considered. Most organic waste gets oxidized by combustion to CO_2 and water. After incineration the volume of refuse substantially gets reduced. Valuable elements such as chromium, molybdenum and lead can be recovered from the residue with relative case, and the heat of combustion can be usefully employed. The ultimate residue has, of course, very little value, consisting mainly of compounds of silicon and aluminium, of which the earth's crust has a superabundance. Approximately 25.7 per cent of all mineral matter is having silicon and 7.4 per cent of aluminium. Wastage of some iron is not a particular tragedy, for at 4.7 per cent of the earth's crust iron has been the fourth most widely distributed element. A proportion of the ultimate residue can already be employed as an aggregate in building construction and as ballast in road or motorway construction, provided that the temperature of incineration has been sufficiently high. A further proportion can be employed in earth-filling works like dams, embankments and land reclamation. The rest amounting to less than 10 per cent of its original volume-can only be dumped; we have, therefore, to consider where best it may be placed.

Geophysical research has revealed that a number of 'tectonic trenches' exist on the ocean floor, where a kind of convection occurs within the earth's crust itself. In one of these, near the mouth of the River Fraser, Canada, the ocean floor has been found to be sinking continuously at the rate of some six metres per annum over an area of 12 km^2. If we deposit residual inorganic refuse in these tectonic trenches, it would rapidly sink and disappear. It is possible to defect tectonic positions with relative ease because they are associated with unusual fluctuations in gravity and with deep-seated seismic disturbances. Bostrom and Sherif, whose geophysical findings were published in Nature (10 October 1970), have concluded that there are sufficient areas exhibiting toctonic phenomena to cater for all the waste materials produced by man for many years to come. The initial costs of compressing and transporting residual waste to the tectonic trenches would be some what-but, it is thought, not unduly high once the practice became accepted and the techniques properly developed.

The simplest method has been crude tipping or open dumping, a common method which finds use is most Indian cities. More satisfactory has been controlled tipping or the sanitary landfill, which has been recently adopted in Delhi for solid waste disposal. In sanitary land fill, a layer of about 2 metres of refuse has been covered by at least 23 cm of earth, ash or other inert material, upto the level of the hole chosen. The surface can then be used for housing or sports fields. Before such filling, the wastes could be pulverized by machines to a uniform particle size: by this means the volume could be reduced and thus the life of the tip gets extended, and some of the refuge has been more quickly biodegraded.

Much of the discarded waste is swept under the rug, so to speak, in accordance with the anaesthetizing but naive philosophy, "out of sight, out of mind." But less visible parts of the iceberg like problem remain. Solid wastes are dumped in the ocean; mining wastes are produced at the rate of millions of tons a day; slag heaps and mill tailings accumulate near processing operations; and industrial refuse contaminates stream and lakes.

PROPERTIES OF SOLID WASTES MATERIALS

The first step in solving solid waste pollution problems is to discard the belief that "waste" is so heterogenous in its composition and so variable in its properties that problems cannot be defined, let alone solved. To be sure, waste streams exhibit great variability. The design of solid waste disposal systems must, therefore, provide for more operating flexibility, reserve capacity, and materials "stamina" than conventional process equipment. Nonetheless, characterizations of the average composition and properties of waste streams provide the starting point for design. An appreciation of the range of excursions from these averages has been also a necessary input. In addition, forecasts of likely changes in waste characteristics should be regarded in the design. Unlike manufacturing processes, the raw materials fed to solid waste disposal systems are usually specified by constraints and priorities beyond the influence of the waste waste manager.

Several properties of potential importance in waste characterization are shown in Table 5.2. Although it becomes seldom necessary to characterize wastes in all the areas in Table 5.2, such a checklist can be useful in alerting the waste manager to processing constraints and opportunities or safety hazards.

When designing waste-processing systems it will often be found that desired data have been lacking. Many industrial and municipal waste studies have shown the large errors possible from armachair estimates of the generation rate, composition or properties of waste. It has been strongly recommended, therefore, that the results of especially commissioned waste surveys and analyses be incorporated in the problem-definition phase of the design effort for waste management systems. For systems disposing of municipal waste, for example, the local mix of waste-generator types, climate, level of construction and demolition activity, and existing waste-management practices and expectations can greatly effect the characteristics of the wastes which have been collected. For industrial wastes, unique processing methods or integrated plant designs can result in net waste production rates which have been greatly from "norms" predicated on other plant configurations within the same industry.

Beyond the deviations in average properties resulting from local influences, the waste manager should also prepare for significant short-term or more modest long-term excursion from these averages. For instance, municipal waste may average 8% mental, but recently in Delaware Country, Pennsylvania, a full packer-ruck of magnesium alloy shavings was delivered and charged (almost intact) into a county incinerator furnace. The resulting damage to the equipment had been reported to be extensive. Although unexpected, this problem would have been moderated by an established regimen of refuse mixing. Long-term excursions could be typified by the large increase in the yard-waste content of municipal waste associated with leaf collection in the fall. Such changes in characteristics must be provided for in the design or the operating protocols of waste-processing systems. Even with such safeguards, however, the worst should get anticipated (live ammunitions, cans of flammable solvent containers of toxic chemicals, etc.).

SOLID WASTE PROPERTIES

Establishing the basic composition of mixed-waste streams has been regarded as the first step in analysis of solid-wastes-management systems. In order to design equipment and processes to cope with the problem, however, detailed understanding of the physical chemical

Table 5.2
General, Physical, and Chemical Parameters of Possible Significance in the Characterization of Solid Wastes.

General parameters	
Compositional weight fractions	C. Agricultural
A. Domestic, commercial and institutional	Field Processing
Paper (broken into subcategories)	Animal raising
Food waste	D. Industrial
Textiles	E. Mining/metallurgical
Glass and other ceramics	F. Special
Plastics	Radioactive
Rubber	Munitions, etc.
Leather	Pathogenic
Metals	Moisture
Wood (limbs, sawdust)	*Process weight factions*
Bricks, stones, dirt, ashes	Combustible
B. Other municipal	Compostable
Dead animals	Processable by landfill
Street sweepings	Salvageable
Catch-basin cleanings	Having intrinsic value

Physical Parameters	
Total Wastes	*Particle characteristics*
Size	Size distribution
Shape	"shape"
Volume	"surface"
Weight	"porosity"
Density	"sorption"
Density stratification	"density"
Surface area	"aggregation"
Compaction	*Liquid wastes*
Compactability	Turbidity
Temperature	Color
Color	Taste
Odor	
Age	Odor
Radioactivity	Temperature
Physical state	Viscosity date

(Contd.)

Total solids	Stratification
Liquid	Total solids (%)
Gas	Soluble (%)
Solid wastes	Suspended (%)
Soluble (%)	Settlable (%)
Suspendable (%)	Dissolved oxygen
Combustible (%)	Vapor pressure
Volatile (%)	Effect of shear
Ash (%)	rate
Soluble (%)	Effect of temperature
Suspendable (%)	Gel formation
Hardness	*Gaseous wastes*
	Temperature
	Pressure
	Volume Density
	Particulate (%)
	Liquid (%)
Chemical Parameters	
General	Heavy metals
pH	especially Mercury
Alkalinity	Lead
Hardness ($CaCo_3$)	Cadmium
MBAS (methylene-blue-active substances)	Copper Nickel
BOD (biochemical oxygen demand)	Toxic materials
COD (chemical oxygen demand)	Chromium
Rate of availability of nitrogen	especially Arsenic
Rate of availability of phosporus	Selenium
Crude fiber	Beryllium
Organic (%)	Asbestos
Cmbustion parameters	Eutrophic materials
Heat content	Nitrogen
Oxygen requirement	Potassium
Flame temperature	Phosphorus
Combustion products (including	*Organic*
ash)	Soluble (%)
Flash point	Protein nitrogen
Ash-fusion characterization	Phosphorus
Pyrolysis characterization	Lipids

(Contd.)

Toxicity	Starches
Corrosivity	Sugars
Explosivity	Hemicelluloses
Other safety factor	Lignins
Biological stability	Phenols
Attractiveness to vermin	Benzene oil
Inorganic and elemental	ASB (alkyl benzene sulfonate)
Moisture content	CCE (carbon chloroform extract)
Carbon	PCB (Polychlorinated biphenyls)
Hydrogen	PNH (Polynuclear hydrocarbons)
(P_2O_5 and phosphate)	Vitamins (e.g., B-12)
Sulfur content	Insecticides (e.g., Heptochlor,
Alkali metals	DDT, Dieldrin, etc.)
Alkaline-earth metals	
Precious metals	

and biological properties of the mixed refuse and its components has been needed. It has been the objective of this section to provide this understanding.

Recognizing that the properties of waste materials are not necessarily influenced by their source, it is evident that many of these data have direct applicability to non-municipal waste-management analysis.

Chemical Composition

Mixed Municiple Refuse. The chemical analysis of mixed municipal refuse gets varied throughout the year and from location to location. In an extensive study of solid-waste problems, average compositions for the various categories of waste were developed (Table 5.3).

Analysis for Specific Elements in Refuse. Due to toxicological hazard, enhancement of corrosion in refuse-processing equipment, air-or water-pollution potential, and similar problems, special attention has been drawn to the concentration of selected elements in refuse. The sections below concern several such elements.

Chlorine. Chlorine, usually as metal chlorides, such as sodium and calcium chloride, occurs in many refuse components. As such, it appears in the fly ash of municipal incinerators. These solid chlorides may take part in corrosion reactions with hot metal surfaces in incinerators or boilers. Also, the chlorides tend to accelerate metal corrosion in incinerator wet-scrubber systems which recirculate the scrubber water. In landfills, metallic chlorides (most of which are soluble) leach out into percolating rain water, thereby adding to the dissolved-salts content of the receiving waters.

Table 5.3
Estimated Ultimate Analysis of Refuse Categories (%). (Dry Basis)

Category	C	H	O	N	Ash	S	Ee	Al	Cu	Zn	Pb	Sn	P^b	Cl⁻	Se	% Fixed carbon (dry basis)
Metal	4.5	0.6	4.3	0.05	90.5	0.01	77.3	20.1	2.0		0.02	0.6	0.03			0.5
Paper	45.4	6.1	42.1	0.3	6.0	0.12									Trace	11.3
Plastics											0.01	6.0				5.1
Leather and	59.8	8.3	19.0	1.0	11.6	0.3										
Rubber									2.0							6.4
Textiles	46.2	6.4	41.8	2.2	3.2	0.2			2.0		0.03					3.9
Wood	48.3	6.0	42.4	0.3	2.9	0.11					0.05					14.1
Food wastes	41.7	5.8	27.6	2.8	21.9	0.25					0.24					5.3
Yard wastes	49.2	6.5	36.1	2.9	5.0	0.35					0.04					19.3
Glass	0.52	0.07	0.36	0.03	99.02											0.4
Misce-llaneous	1.30	2.0	12.0	3.0	70.0											7.5

In addition to the importance of the metallic chlorides, however, considerable attention has been assigned to the chlorine appearing in polyvinyl chloride (PVC) and vinylidene chloride (Saran etc.) These materials appear in refuse as housewares, packaging and film. toys, etc. Pure PVC resin (i.e., without plasticizers, fillers, pigments etc.) burns in air to quantitatively convert its chlorine content to hydrogen chloride (hydrochloric acid). Each kg of PVC thus yields about 0.59 kg of hydrogen chloride. Although some portion of the hydrogen chloride undoubtedly reacts with basic oxides in the fly ash, the majority remains in the gaswous phase. This gaseous hydrogen chloride attacks high-temperature metal surfaces in incinerator boilers like superheater tubes and is absorbed in scrubber waters to yield highly acidic, eorrosive scrubber liquor.

Estimates of the fraction PVC in plastics are shown in Table 5.4.

Table 5.4
Anticipated PVC Content of Municipal Refuse.

Year	*Estimated plastics content of refuse (weigh %)*	*Estimated % PVC resin in refuse plastics*
1968	1.1	7.8
1970	1.3	10.0
1975	1.8	10.6
1980	2.7	8.7
1990	3.5	10.9
2000	4.2	11.2

Selenium. As a large fraction of municipal refuse is paper and as selenium is known to be present in almost all types of paper, the possibility of exposure to selenium in waste-management activities has been significant. Selenium, a toxic and perhaps carcinogenic element, occurs, in trace quantities in refuse and in the by-products of refuse combustion.

Agricultural Wastes. Due to their entirely organic origins and water-pollution potential, anlyses of agricultural wastes have been reported in dimensional units quite different from the combustion process-oriented analyses of municipal and industrial refuse.

Density Characteristics

Probably no other property of solid waste has the engineering and economic importance of bulk density. The low initial density and poor compaction characteristics (at moderate pressures) of solid wastes has been found to limit carrying capacity and thus have been contributing to the high cost of collection and hauling. Indeed, few collection vehicles approach their maximum allowable payload weight when they must depart from their routes to be emptied. Incineration systems, build and operated at high expense, have been justified primarily as a volume-reduction process to extend landfill life and/or reduce long-haul transportation costs for refuse disposal.

Bulk Density. The bulk density of most solid waste has been appreciably lower than the ultimate density of its components.

Many of these wastes have been readily susceptible to two-or three-fold increases in density in conventional hydraulic-compactor refuse-collection vehicles or in stationary compactors. The effect of high-pressure compaction, as in baling equipment, has been described as follows:

Compaction Effects. With increasing solid-waste-generation rates and decreasing urban landfill capacity many municipal and regional agencies have been implementing transfer-station and railhaul operations to distant disposal areas. Also, attention has to be paid to the use of baling operations as a precursor to long-haul landfill or marine disposal. Each of these operations has been found to depend importantly on substantial refuse densification, usually by using hydraulic compaction equipment. Because of the importance of compaction, the various studies have been made about the refuse densification. The following is largely drawn from the results of that effort.

Experiments were run on raw, shredded and synthetic municipal refuse and oversized bulky waste. A small number of tests got incorporated water soluble adhesives which had been intended to improve the binding properties of wastes.

Although compaction data have been generally reported as a volume-reduction ratio, such "data" have been significantly less useful than documentation of ultimate density. It occurs due to the extreme variability in starting density occasioned by previous compaction and storage history and the mechanical/operational limitations of the compactor loading hopper and loading crew. For instance, reported compaction ratios at 3500 psi (maximum pressure) ranged from 7 : 1 to 23 : 1, a 300% variation. The same date reveal a range in ultimate density from 93 to 108 pounds per cubic foot lb/cu ft), a 15% variation. These considerations have been especially important in evaluating compaction equipment vendor-performance claims.

Ultimate density. As the density of different papers is some what between 43 and 71 lb/cu ft, it is clear that high-density waste components (glass, metals, dirt) contribute importantly to bale density. Prepackaging of refuse in paper bags provide almost identical results to those for unpackaged refuse.

Liquids and sludges expressed during compaction. Expected increases in bale density with wet refuse could not be realized due to the expulsion of moisture during compaction and the increased bulkiness of wetted papers. Weight losses of up to 37% had been recorded for wet refuse exposed to rain. Moderately wet refuse lost only 1 to 3% of its initial weight and no loss had been recorded for dry refuse.

The expressed liquids has been greenish in appearance. The settlable sludges were gray. Both had an offensive odor. Chemical analysis of the leachings yielded the data in Table 5.5. Review of these date reveal that no unusual pollutants will occur in the refuse pressing liquors.

Volume expansion after compaction. Compacted refuse under goes expansion upon release of the compaction pressure. The degree of expansion has been following the degree of pressure application: the greatest expansion occurs in the direction of ram 3 and the least in the direction of ram 1. Although expansion continues for several hours after the release of pressure, volume increases taking place in the initial 2 minutes were used as a benchmark measurement.

Expansion effects had been found to be less important if the force on the bales got held for several minutes. The bales compacted

Table 5.5
Chemical Analysis of Leachings

Sludge	*Sample 1*	*Sample 2*
Organic matter	38.9%	76.7%
Silica as SiO_2	30.8	6.9
Aluminum as Al_2O_3	15.2	5.8
Phosphates as P_2O_5	8.9	4.4
Calcium as CaO	2.0	1.3
Magnesium as MgO	1.4	0.0
Iron as Fe_2O_3	0.5	0.6
Sulfates as SO_3	Trace	Trace
Carbontes as CO_2	0.0	0.0
Liquid		
pH	7.8	4.5
Total dissolved solids	6090 ppm	6040 ppm
Organic matter	1720	4480

(Contd.)

Sludge	*Sample 1*	*Sample 2*
Bicarbonates	90	610
Sulphates	250	200
Chlorides	21	218
Heavy metals (suspension)	Large % amount	Small % amount

at about 1000 psi for 5 minutes had been as stable as those compacted at 1500 to 1750 psi without holding pressure. Similarly, 5 minutes at 1500 psi gave rise to product bales similar to those held only momentarily at 2000 psi. The dimensional stability of high-moisture refuse bales had been consistently poor.

Effects of adhesive addition on unsacked and sacked refuse. It is speculated that adhesives effective in bonding paper materials would be able to reduce the degree of post-compression bale expansion. The addition by spraying of several adhesives (including polysaccharide-, protein- and cellulose-base water-soluble compounds and a hydrocarbon base compound dissolved in mineral spirits) offered no improvement in compaction properties and only a slight reduction in short-term bale spring-back. the expansion of bales compacted at about 3400 psi had been about 8% less than that of the unsprayed bales. The main difference noted had been an improvement in the appearance and stability of the bales after standing. The sprayed bales got less distorted and more rugged than those not sprayed. The degree of improvement, however, appeared too little to justify the use of adhesives in commercial operation.

Some improvement in the stability of bales of plastic-sacked refuse had been noted with addition of adhesives to the outside of the sacks. It appeared importantly associated with the inadequate strength of the plastic-sacked refuse bales due to the lack of adhesion across the plastic film.

Effects of refuse shredding prior to campaction. Experimental compaction tests were undertaken with a single load of shredded refuse. The shredded refuse was obtained from the shredding facility in New Delhi. This limited test series do not show any improvement in ultimate bale density in the machine or after expansion had taken place. Date showing the volume increase of compacted, shredded refuse with time have been given in Table 5.6.

Gas formation in bales. In order to get information regarding the hazards of gas formation in compacted refuse bales, gas samples had been taken from six bales. The bales were stored at 85 + 5^0 F. Two bales had been wrapped in plastic immediately after leaving the press. After one day, they were warm to the touch and beads of moisture had been observed on the plastic cover. The unwrapped bales remained cool and dry to the touch. After three days of storage, the remaining four bales had been wrapped. Gas composition data had been recorded. Although the gas compositions suggest aerobic bacterial activity, the short duration of the test precludes definitive characterization of the long-term decay process and thus effluent gas composition.

Table 5.6
Volume Increase of Compacted Shredded Refuse with Time.

Sample	Time (min)	Volume increase (%)	Compaction pressure (psi)
1	1	44	3400
	5	49	
	30	58	
	1440	71	
2	1	60	1850
	5	59	
	30	82	
	1440	95	

Moisture effects on compaction and post-compaction expansion. Several sets of experiments had been made to run with mixed-paper and with newspapers of controlled moisture content. Beyond the addition of readily deformable mass with water addition, there appears to be a further increase in density, perhaps associated with enhanced pliability in the paper.

The degree of bale expansion following discharge from the compaction device was also shown to get affected by refuse-moisture content. Bales of more than 40% moisture by weight had been tended to disintegrate immediately upon release of moisture.

Compaction properties of selected refuse components. A series of compaction tests were made to run on a variety of homogeneous or mixed refuse components. The test results from experiments at compaction pressures of about 3500 psi have been reported in Table 1.7.

Combustion Properties

Combustion processes (which include on-site and municipal conventional incineration, incineration with heat recovery, combine firing of refuse and fossil fuels in utility boilers, and pyrolysis or incomplete oxidation) figure importantly in municipal and industrial waste-management activities.

Heat content. The most fundamental combustion property has been probably the heat content or heating value. In the tables which follow, the quantity reported is the higher heating value (HHV) expressed in Btu/lb. The HHV may be defined as the heat released from the complete combustion of the material in question starting with the reactants at 20°C and ending with all products at 20°C. Thus, with this definition, the latent heat of condensation of the water vapor formed by combustion has to be included in the heating value.

Recently a correlation based on thermochemical principles got developed to predict the heat of combustion (dry basis) from the ultimate analysis of the waste. The correlation neglects metal oxidation but the resulting error must be small in conventional incinerators. The correlation needs the knowledge of the proportion (lb per lb). of organic and inorganic (carbonate) carbon, oxygen, sulfur, nitrogen, and hydrogen (denoted C_{org}, C_{inorg}, O, S, N and H respectively), all on a dry basis, in the refuse sample. The final equation has been as follows:

Table 5.7
Baling Characteristics of Refuse Components (3500 psi Compaction Pressure).

1. *Rubber tyres*. Several samples of rubber tires were subjected to compaction. However, they could not be compacted. After release of pressure the tyres always scparated.
2. *Rubber tyres enclosed in wire mesh*. The compacted bundles of tyres enclosed in the wire mesh always expanded rapidly. The wire mesh, depending on its strength, usually burst open at a few or many welded spots.
3. *Shredded rubber and residential wastes*. Residential waste, containing about 25% of small rubber pieces, produced good bales. Shredded rubber on its own could not be compacted.
4. *Metal cans and oversized metallic wastes*. All metallic waste products, including refrigerators, washing machines, stoves and appliances, produced excellent bales.
5. *Oversized wastes*. Mixtures of refrigerators or other oversized metallic wastes, together with furniture and textiles produce very stable bales.
6. *Oversized wastes and household refuse*. These mixtures produced very good bales.
7. *Plastic scrap*. Compacted mixtures of plastics did not hold together after release of pressure.
8. *Very dry papers*. Such papers did not produce stable bales.
9. *Vegetables, fruits, meat scarps*. These wastes could not be compacted on their own or in combination.
10. *Vegetables, fruits and residential refuse*. An appreciable amount of food pulp was formed by compacting residential wastes which contained a large amount of fresh vegetables or fruits. Addition of about 25 weight % of vegetables to residential wastes produced a large amount of fine pulp which was partially released from press openings at about 900 psi. The pulp which adhered to the outside of the compacted bale fell off after removal of pressure.
11. *Cooked meats and residential refuse*. Similar results as those obtained with vegetables and fruits were obtained with meats containing a large amount of moisture.
12. *Vegetables and wood shavings*. The addition of vegetables to wood shavings produced unstable bales.
13. *Wood shavings*. Wood shavings, alone, produced relatively good bales.
14. *Wood chips and residential refuse*. Mixtures of residential wastes and wood chips produced good bales.

$$\text{Btu/lb (dry basis)} = 14{,}096C_{org}$$

$$+\,55{,}750\left(H-\frac{o}{8}\right) + 3982s - 6382C_{inorg}$$

$$+4274\left(\frac{o}{2}\right) + 1040N$$

Air requirement. the stoichiometric (theoretical) air requirement has been an important variable in many combustion calculations. The air requirement could be readily calculated from the ultimate analysis, taking proper account of the oxygen contained in the refuse. Oxidation of metals could be assumed to occur to the extent of about 10%. For computational purposes, it can be assumed that the refuse metal is iron. A sample calculation of the stoichiometric air requirement in presented in Table 5.8.

As an alternative, it can be noted that, for many fuels, the air requirement per million Btu has been approximately constant. For municipal refuse, an air requirement of 724 pounds per million Btu (HHV) has been an acceptable estimate. For industrial wastes, especially when a component other than paper has been present in high concentrations, it may be prudent to compute the air requirement from the ultimate analysis.

Residues from the combustion of municipal refuse. In the Incineration process, four solid residues have been produced generated: sifting, dust grate residue, and fly ash. Each of these residues has been a solid waste in its own right and needs disposal, usually via landfill. In the following, the chemical and physical properties of incinerator residues will be presented.

Siftings. In the operation of municipal incinerators which incorporate a grate, a fraction of the refuse and grate residue falls through the air openings in the grate. This material, called siftings, is characterized in tests on a four-section, Flynn and Emrich rocking grate.

The siftings got collected from beneath the first (feed) grate section and a composite sample got obtained from stoker sections 2, 3, and 4.

Table 5.8
Sample Computation of Stoichiometric Air Requirement from Ultimate Analysis

1. Using the refuse ultimate analysis the refuse composition and oxygen requirement on an as-fired basis are:

Component	*Pounds per 100 lb of refuse*	*Molecular weight*	*Moles*	*Required Moles O_2*
Moisture	28.16	18.0	1.564	0
Carbon	25.62	12.0	2.135	2.135
Hydrogen (bound)	2.65	2.0	1.326	1.326
Oxygen (bound)	21.21	32.0	1.326	1.326
Hydrogen	0.80	2.0	0.399	0.199
Sulfur	0.10	32.0	0.003	0.003

(Contd.)

Component	*Pounds per 100 lb of refuse*	*Molecular weight*	*Moles*	*Required Moles O_2*
Nitrogen	0.64	28.0	0.023	0.0
Metal (assumed iron)	8.70	55.8	0.156	0.010
Ash	12.12	—	—	0.0
	100.00			2.347

2. Stoichiometric air requirement

Moles oxygen = 2.35

Moles nitrogen = 8.83

11.18 moles air = 322 lb per 100 lb of refuse

Dust. In an incinerator operation large amounts of dust have been to be found. These dusts are characterized as to heat content as a consequence of concern regarding their explosive hazard. Using techniques developed by the Bureau of Mines, the Explosivity Index (relative to coal) of incinerator dusts was determined. The results are shown in Table 5.9. In general, the dusts were found not readily susceptible to initiation and propagation of an explosion.

Table 5.9
Calorific Value and Explosivity Index of Incinerator Dusts.

Dust source	Heating value (cal/g-dry)	Explosion concentration (mg/litre)	Explosivity index
Airborne	3390	—	—
Charging floor	2770	—	0.33
Walls	2030	—	<0.22
Platform by pit	2280	—	<0.25
Railing near pit	2614	—	<0.28
Ground refuse	3430	70	0.8
Pittsburgh coal	6667	55	1.0

Grate residue. The material discharged from the grate of a municipal incinerator has been comprised about 25-33% of the weight but only 5-15% of the volume of the incoming refuse. These residues have been generally landfilled although interest is increasingly being shown to processing these wastes for the recovery of ferrous and non-ferrous metals and glass.

In connection with studies of by-product recovery from incineration plants, analyses had been made of the ferrous-metal residues. The high copper content, (far higher than would be expected in the eron alloys analyzed) is presumed to result from the following sequence of reactions:

(i) Polyvinyl chloride in the refuse burns to hydrochloric acid.

(ii) Hydrochloric acid attacks hot copper to form cupric-chloride.

(iii) The cupric chloride dissolves in the quench water at the end of the grate.

(iv) The copper plates out on the iron by replacement.

The formation of clinkers in the refuse fire and the sintered and fused deposits of furnace enclosures has produced great interest in the fusion properties of the ash of refuse and refuse components.

Fly Ash. Fly ash, the fine particulate matter entrained in the gases leaving the burning refuse bed in an incinerator, has gained importance from two standpoints. Fly ash has been the material which gets accumulated as slag on the walls and in the flues of incinerator furnaces and has been the material which presents the most significant air-pollution control problem of incinerator systems.

The importance of the slag-forming nature of fly ash makes its chemical analysis important. The fly-ash composition data reveal the strong similarity between the compositions of refuse ash, incinerator slag and fly ash. Thus the ash-fusion data must be a useful guide in assessing the wall-temperature maximums which avoid fly-ash adherence and build-up.

The air-pollution problem to fly ash needs an understanding of the particles-size distribution of this material.

HEALTH AND INJURY HAZARDS OF SOLID-WASTE TREATMENT

As solid waste may potentially has any of the solid materials found in nature and in addition many of the man-made materials, they have been regarded as the most heterogeneous collection of substances possible. Laws and ordinances should prohibit certain materials from being put into the solid wastes stream. Domestic refuse has, with alarming frequency, been found to have, for instance, live ammunition or explosives; highly toxic substances; or diseased cultures.

Industrial wastes from any one location is generally having a lesser degree of heterogeneity than that of domestic refuse, but the variations of compositions and properties from one operation to another and one type of industry to another has been very large.

The scale of the potential dangers in solid wastes, has only recently been realized, and consequently data are extremely scanty.

Solid wastes could be hazardous as generated; they may subsequently combine to produce secondary hazards; or they could be hazardous by reason of their location or the method of processing used. These categories will find use in the following discussion.

Initially Hazardous Wastes

Solid wastes which have been dangerous at the point at which they have been discarded; such as explosive or incendiary devices, or toxic substances, require little discussion here because they constitute the most obvious category of hazard.

A related but distinct category has been including the broad range of wastes which, while not toxic, have the potential of causing injury. A plate-glass window becomes solid waste at the moment it has been broken, and at that point it has obvious injury potential for men and animals. Machine-shop turnings belong in the same category, with additional dangers because they have been generally hot enough to cause burns or alternatively may be contaminated with cutting oils which may bring about adverse physiological reactions if the skin has been punctured by such turnings.

A wide variety of substances can bring about harm or disfigurement either in general or to sensitive individuals. Acids and alkalis have been universal problems; though not strictly solid wastes, they appear along with many other liquid substances in discarded containers. Pesticide containers have caused intoxication and death after being played with by children. Minute quantities of foaming or hardening agents used for plastic resins bring about acute reactions in some individuals. In some cases, death may take place in otherwise fit individuals with in 30 minutes of exposure to, for instance, penicilin.

Although organisms which are able to transmit disease are organic and will live on organic refuse, the quantity of organic material necessary to maintain a culture could be an invisibly small coating on a plate or container. Or the culture might be maintained at the tip of a hypodermic needle. The careless discarding of hypodermic needles after use into wastes-baskets has unfortunately frequently resulted in janitors becoming infected when they employ their hands to clean out the residue from a waste container.

A German study on illnesses caused by refuse revealed that the practice of over 70% of medical laboratories and institutes of allowing their wastes to be collected long with regular domestic wastes has been particularly dangerous. All medical wastes should be kept in special, closed, plastic bags and collected and disposed of separately.

Half the outbreaks of salmonellosis were found to occur in hospitals. Regular fecal examination of food handlers in hospitals and institutions and action to eliminate salmonellae from animal feeds has been recommended.

Pathogenic bacteria can be and have been present in very high concentrations in hospital solid wastes. The highest cloriform counts were found by Trigg to occur in wastes from the intensive care, pediatric, and psychiatric wards.

Hospital refuse chutes were found by Armstrong to be potentially transmitters of airborne bacteria. The use of closed plastic bags in refuse chutes has been especially important, and an air-removal and purifying system to remove the smallest particles which can penetrate the respiratory tract is necessary.

The use of disposable needless, knives, swabs, diapers, examination gowns, thermometers, transfusion fittings, bed pads, sheets, plates and utensils, and so forth, in hospitals and institutions requires special handling if injuries and cross-infections have been to be avoided. A particular danger arises if a contaminated nonwoven item gets laundered alongwith washable linen. The nonwoven item will probably distintegrate and spread over all other clothing theory potentially spreading infection over the whole instition.

AIRBORNE PATHOGENS ASSOCIATED WITH SOLID WASTES HANDLING

Air samples taken from municipal incinerators and a municipal compost plant showed that the potential health hazard from the pathogens associated with the dust has been somewhat significant. The dust was found to carry a large number of microorganisms including pathogens of intestinal and respiratory tract origin. The degree of hazard is dependent on the concentration of the dust and on the type of organisms carried.

Table 5.1.10 show microbial characteristics of representative environments and of the ambient air in various locations in municipal incinerators. The dumping-floor values can be regarded to representative of dumping floors of other types of solid-wastes-handling plants; the microbial levels might be somewhat lower for incinerators because, in good designs, the combustion air gets extracted from the dumping area to reduce dust, smell and pathogen levels.

Fungus spores are also responsible for discase (mycosis) through inhalation and passage to the throat and lungs.

Table 5.1.10
Representative Environmental Microbial Levels.

Environment	Total microbial levels (colonies/cu ft air)
Country air	56
General offices and schools	95
City streets	72
Factories	113

Asbestos Dust

The high incidence of lung and other cancers among asbestos workers, including those in the demolition industry, has been recognized. Stringent regulations have been introduced in many countries to control one form of asbestos-the blue crocidolite-which has been found to be principally responsible.

Those engaged in the manufacture of asbestos products and those who both use and work with such products have been mainly aware of the health hazards in question and take every precaution to safeguard the health of employees. The industry is nevertheless vitally involved, because its members have so often to handle and process scarp materials which are contaminated with asbestos thermal lagging, e.g., pipes, boilers, tanks, etc., which are received into their years, and in addition often have to deal with such materials on demolition/dismantling sites, and the regulations quite specifically cover all operations which causes the liberation of absestos-containing dust to such as extent as may result in the emission of dangerous quantities of dust.

The health hazards involved have been primarily the risk of contracting asbestosis through the inhalation of heavy concentrations of asbestos dust over a long period. This sometimes gives rise to the development of lung cancer, particularly if the person concerned has been a heavy smoker. Another rare form of cancer, mesothelioma, affecting the chest or stomach wall, sometimes take place chiefly from exposure to certain types of 'blue' asbestos.

Types of asbestos. The main types of asbestos used in the past have been:

Amosite	—	Color — grey/brown
Anthophyllite	—	" — white
Chrysotile	—	" — white
Crocidolite	—	" —lavender blue.

Crocidolite has been the most dangerous types of asbestos and is likely to be found in acid fume areas such as chemical plants and gas works, and it is also found in power stations. Total emphasis upon the ability to recognise crocidolite by color alone cannot be relied upon because it has been sometimes found mixed with the other types of asbestos, and although the significant proportion of admixture of crocidolite may be as low as 1 to 2%, even in these proportions it could be highly hazardous if the dust gets inhaled and unless the original owners of asbestos lagged materials can certify that such lagging has been free from crocidolite, the only sure identification of its presence has been by x-ray diffraction of samples.

Asbestos Dust-Threshold limit value (TLV). The Asbestos Regulations have been based upon the concentrations of respirables asbestos dust in the atmosphere of the working area together with the duration of exposure to such dust, and threshold limit values have been set as follows:

Chrysotile (white) and Amosite (grey/brown) asbestos	—	2 fibers per cc average over 4-hr working period and not above 12 fibers per cc average in any 10 min. period.
Crocidolite (blue) asbestos	—	0.2 fibers per cc average over 10 min.

The Stripping of asbestos Lagging. The removal of old legging has been the cause for serious concern as this is generally an extremely dusty process and the following paragraphs attempt to deal with the problems associated with the stripping of such lagging in the varying circumstances likely to be encountered by the scrap processor.

Preparations Prior to Stripping Old Lagging. The first requirement is to establish the types of asbestos in the lagging that is to be stripped and where no record of this exists it will be necessary to determine if it contains 'blue' asbestos. If it is not apparent by inspections, samples should be taken from a number of different points for examination. Normally, positive-pressure respiratory equipment will be needed where 'blue' asbestos is present. In other cases a simpler from of approved dust respirator may suffice.

Protective Screens. Where stripping has been to be done occupied premises, screens or tents made from plastic sheets erected around the plant may be sufficient to reduce the asbestos dust level in adjacent working areas to an acceptable level.

Table 5.11
Identification of Possible Sources of Origin of Asbestos Materials

Industry	*Asbestos Usage*
Electricity generating	Thermal lagging and delagging (Special risk from 'blue' asbestos)
Steel	Thermal lagging and delagging.
Heavy engineering	Furnance insulation.
Locomotive building	Heat and sound insulation. (Special risk from 'blue' asbestos used in the past.)
Railroad-car building	Heat and sound insulation. (Special risk from 'blue' asbestos used in the past.)
Boiler making	Heat insulation.
Paper making	Manufacture of filter papers and grinding of rollers.
Motor-vehicle repaid	Repairs to brake and clutch parts.
Building trades	Asbestos cement sheets and insulation boards, asbestos spraying.
Electrical insulation	Insulation systems.
Chemical plants	Thermal insulation. Special risk from acid-resistant 'blue' absestos.)
Gas works	Thermal insulation. (Special risk from acid-resistant 'blue' asbestos.)
Other sources of Asbestos Lagged Materials	
Dockyards	
Shipbuilding	
Shiprepairing	
Installation of plant in heavy industries, e.g., steel.	
Large building projects-industrial and domestic-or insulation and repair of heating apparatus, e.g., schools.	

The wet method of stripping. Thorough saturation of asbestos lagging wherever possible prior to stripping can greatly reduce the dust generated. Each job has been likely to require its own technique, e.g., non-absorbent surfaces will need to be punctured to allow water to be introduced into the insulation. This operation could be by a hallow probe drilled along its length to permit water to peneterate into the insulation, such probes parallel being coupled to a suitable water supply. alternatively, a fine low-pressure water supply way be used in such a way that dust does not arise from the spray hitting the surface of the lagging. Spraying in this manner has been not as effective as total saturation in controlling the dust and the arrangements may require to be augmented by portable exhaust ventilation equipment and the use of respirators.

On removal, the lagging should not be permitted to fall but placed immediately in suitable plastic bags (200-gauge polyethylene double-sealed at bottom has been found satisfactory for removal from site and subsequent disposal by burying. The slurry created should not be allowed to dry out but should be removed from all resting places while still wet. Special additional precautions are needed when the waste contains 'blue' asbestos.

Table 5.12
Non-Hazardous Asbestos Wastes.

Linoleum	
Floor tiles	
Rubber Paints	Used as a filler and locked in with the other components.
Plastics	
Adhesives	
Roofing compounds	
Motor assembly	Grinding in assembly of brake and clutch parts.
Light engineering	Asbestors washers and gaskets.

Landfil disposal of asbestos wastes. The Asbestos Research Council has laid down a code of practice for landfill disposal, given below in part.

Asbestos waste should be disposed of in such a way that no dust in emitted into the air during transport, in the act of dumping, or after dumping. Where the waste has been derived from blue asbestos, the receptacles containing it should be clearly marked to show this. Waste which is adequately kept in impermeable bags or other suitably sealed receptacles does not require special transport.

Special containers can be obtained which, when filled, have been removed by specially fitted road vehicles direct to a landfill. Whatever type of container has been used, steps should be take to prevent emission of dust from containers when in use.

Wet waste should be transported to the disposal area in sludge tankers or similar vehicles. After use, all vehicles and containers should be cleared of loose fiber or dust by vacuum cleaning or other dustless method.

All waste deposited should be covered with at least 9 in. of consolidated earth. Covering should be done promptly (immediately in the case of waste derived from blue asbestos). No waste should be left uncovered at the end of a working day.

The disposal of wet waste on a dry waste landfill has been acceptable, provided that the quantities are no excessive. It should be remembered that, even if waste has been wet when dumped, it must still be covered so that dust cannot escape from the dried-out material.

RADIOACTIVE WASTES

These are materials with half lives of more than 100 days. U-236 has the longest half-life, over 23-million years.

Radioactive wastes cannot be "disposed of" but must be stored in perpetuity. Liquid wastes are converted to a solid or a greaselike from. Groundwater is being monitored.

While all agree upon the absolute requirement of extremely stringent standards, they believe that there is no safe way of handle radioactive wastes.

However, the minimum requirement for locations where radioactive wastes have been housed temporarily includes primary and secondary containment, with continuous monitoring of the air in the secondary containment (e.g., a building or a cave) to detect leakage, and the digging of observation wells to monitor groundwater quality continuously. Radioactive wastes should never be incinerated, nor accepted for any other form of conventional solid-waste disposal. .

HAZARDOUS INDUSTRIAL WASTES

The following notes on some particularly troublesome metals and organic wastes are meant to be illustrative rather than all inclusive of the health hazards that can result from incorrect handling of as-generated wastes.

Inorganic Materials

Lead. Lead poisoning causes brain damage, especially in children, and interferes with the ability of people of all ages to resist infection. Lead is dissolved out of lead water-supply pipes to enter the diet through the drinking water. Children take additional lead from eating paint chips in old lead-painted buildings. the disposal of metallic lead and demolition debris in landfills therefore has created a potential hazard if the leachate has been allowed to pass into ground-water feeding a water-supply system.

Mercury. Metallic mercury has been so valuable that it has been unlikely to appear as such in solid wastes to any extent. Organic and inorganic compounds of mercury have been widely used, however, and can appear in industrial and agricultural wastes. All mercury compounds have been found to be highly toxic to man and animals, giving severe mental and motor symptoms, neurological symptoms, kidney damage, gastrointestinal and pulmonary symptoms and genetic damage.

Beryllium. It is a low-density metal which finds use in many industries, principally in alloy form, and as a rocket fuel in the pure form. When ingested as atmospheric particles it causes severe disease involving respiratory-tract irritation, dermatitis and conjunctivitis and possibly involves other organs. Mortality has been 35% in its chronic form. Whether long-term, low-level exposure can lead to cancers is still uncertain.

Manganese. It finds use principally as an alloying element in steels. Inhalation of manganese oxides may cause chronic managanese poisoning or manganic pneumonia.

Organic Materials.

Benzopyrene and benzofluoranthene. 3, 4 benzopyrene and 3, 4-benzofluoranthene have been formed when organic substances have been heated above 500°C in incineration and in other processes. These hydrocarbons have found to be carcinogenic to animals. The substances occur in soils, incorporated from sediments from flue emissions, and in root vegetables (potatoes, carrots, radishes etc.) grown in such soils.

DEHP. Di-2-ethylhexyl phthalate finds use as a plasticizer for polyvinylchloride packaging. It has been found to be toxic to animals and man. It is suspected that deaths in patient s receiving blood transfusions from blood stored in PVC bags had been due to this plasticizer being transferred from the bag to the blood. Handling and disposal of quantities of waste DEHP must be done with stringent precautions.

PCB. Polychlorinated biphenyls had been previously used for no-carbon copy paper. A small proportion recycled into combination paper board used for packaging breakfast cereals allowed transfer of PCB to the foodstuffs and caused measureable levels in human fat. PCBs or an associated contaminant, chlorinated dibenzofuran, is responsible for liver malfunction and damage, severe skin eruptions and hydropericardium.

Fiber polymers and polyurethane foams. The open burning of some fiber polymers, such as Orlon, and foamed plastics, such as polyurethane, can give rise to the formation of potentially toxic concentrations of hydrogen cyanide.

HEALTH AND INJURY TO SANITATION WORKERS

Many organizations believed that there was no particular problem with hazards to sanitation workers. One reason for this generally false belief has been that people who work in solid-waste handling have been often undereducated, having little bargaining power, in largely transient employment, and they often leave the job when they are injured or become ill.

SECONDARY HAZARDS FROM SOLID WASTE

After disposal, solid wastes normally are allowed to pass first to a container, either immediately or via a refuse chute; then the wastes are generally transported by vehicle either in the original container or after transfer to a larger container; then there may occur a processing stage such as incineration, compaction, shredding, or reclamation; and so-called "final" disposal (if not by recycling) must by definition be back to the environment in solid, liquid, or gaseous form.

During this general sequence of events, non-toxic could become poisonous; sterile wastes may become capable of transmitting disease; nonexplosive wastes may cause explosions; and wastes which apparently have no injury-causing potential may in fact produce injury and even death. The mechanisms involved, and steps which can be taken to combat the problems, have been discussed below.

PUBLIC HEALTH HAZARDS

Within a matter of hours in a warm temperature, sterile organic matter, like cooked meat, can become a potentially lethal source of toxic or disease-producing organisms. The organisms do not have to be originally present in the host material as the environment is normally well provided with spores, bacteria, viruses, insects, vermin and other vectors awaiting a favourable site on which to multiply.

The mere presence of these cultures of potentially disease producing organisms in solid wastes has been not enough to cause a major health hazard. The blame for disease transmittal must be placed especially on the flies, mosquitos (table 5.13) and rodents (Table 5.14). It is estimated that 90% of the urban house-fly population breeds in the contents of open trash barrels. Rats also rely on readily available scraps of food such as those which are kept in open trash barrels. Refuse overflowing from or carelessly disposed around containers outside restaurants or supermarkets or the use of containers with ill-fitting or best closures have been equally responsible.

Alternative forms of treatment have been available which might be practicable in certain cases. One has been compaction. Reasonably high-density compaction (perhaps defined as that producing a density of about 40 lb/cu ft or 1000 lb/cu yd) eliminates most of the crevices and voids in which insects prefer to lay their eggs; greatly reduces the surface area available; and squeezes out many of the juices so leaving the surface dry, or perhaps at most uniformly moist without the local areas of high organic concentration which have been necessary for optimum hatching of fly larvae.

Table 5.13
Fly-Borne Diseases

Fly-Borne Diseases
Typhoid
Bacillary dysentery
Amoebic dysentery
Diarrheas
Asiatic cholera
Helminth/infections (worm)
Myiasis
Loiasis
Onchocerciasis
Ozzard's filariasis
Leishmaniasis
African sleeping sickness (trypanosomiasis)
Yaws
Tularemia
Bartonellosis
Cararrhal conjunctivitis
Sandfly fever

Table 5.14
Mosquito-Borne Diseases

Mosquito-Borne Diseases
Dengue
Encephalitis
Filariasis
Malaria
Yellow fever
Tularemia
Lymphocytic choriomeningitis
Melioidosis
Rift-Valley fever

A second approach which has been less desirable, has been to process refuse within two days of it having been discarded, so that fly larvae do not have time to emerge (2-3 days being required). Rates can still feed if containment is not needed and there is no presently available method of treatment other than denying rodents access to food.

Table 5.15

Echinostomiasis	Schistosomiasis
Hemorrhagic septicemia	Bilharziasis
Histoplasmosis	Swine erysipelas
Lymphocytic choriomeningitis	Trichinosis
Plague	Leptospirosis
Rat-bite fever	Leishmaniasis
Rat-mite dermatitis	Relapsing fever
Rat-tapeworm infection	Tularemoa
Rocky Mountain spotted fever	Rickettsial pox
Salivary-gland virus infection	Murine ryphus
Salmonellosis	

A third approach is some what contradictory to the first in that it involves shredding solid wastes. Shredding increases the surface area and decreases the density. In these circumstances an interesting transformation occurs in organic wastes. If shredded refuse is not placed in closed containers, and if the volume in any one place is not such as to give a depth of refuse greater than perhaps three feet, sufficient air is able to reach all parts of the mass so that aerobic rather than anaerobic bacteria grow vigorously. These bacteria have been found to work many times faster than the anaerobic variety and give off combustion heat as well as being thermophilic. Carefully controlled tests have revealed that insects are not able to hatch on aerobically decomposing refuse, and the material is also unattractive to rats and other vermin. The rate of air movement possible within such a mass has been small, and fires, if set, tend to smolder rather than to burn. Since most fire deaths have been due to smoke inhalation rather than to flame-produced injuries, the shredding of refuse could be considered as an acceptable way to reduce may of the secondary hazards of solid wastes so long as the shredded refuse has been not stored within a building or other enclosure in which people might be trapped.

Fire and Explosions

There has been an extreme danger of fire in most categories of untreated solid wastes simply since a large proportion of the material has been paper products, wood or plastics. The potential for these materials to be ignited by carelessly disposed of smoking materials has been high. However, whenever organic materials have been amassed together there has been the danger that these and less flammable materials may undergo spontaneous combustion.

Spontaneous Combustion

Spontaneous combustion takes place when organic materials decompose, usually initially by action of aerobic (oxygen-loving)bacteria and the heat released has been not all able to escape so that the temperature of the mass rises. Thermal insulation has been usually provided by the neighboring mass of non-reacting material, so that is an analogy to a nuclear reaction, a critical mass of organic material must be present before heat transfer outwards from the reaction zone has been sufficiently reduced for the temperature to rise and for high-temperature burning to commence.

Spontaneous combustion is a problem in agricultural materials like hay; in coal; and in other finely divided fuels like woodchips and saw-dust. A recent example emphasizing the need to avoid large masses of any materials which have combustion potential has been the fusing together of a pile of steel turnings which were awaiting reclamation in an open-air yard. The turnings had been wet with cutting. oil. The plant supervisor maintained that the inside of the pile got red hot and subsequently cooled down to form a solid mass. The resulting pyramidical block became too thick to be cut with torches and all other attempts to reduce it to moveable pieces failed.

There are not scientifically useful data to give guidance for avoiding spontaneous combustion in all types of solid wastes. Common-sense solutions have been to simply avoid storing solid wastes in warm moist conditions for several days when the smallest dimension of the waste mass becomes greater than 6 to 10 ft (2 to 3m). Smaller quantities, or very dry, or very wet conditions, have been the inhibitors of spontaneous combustion.

Explosions

While it has been obvious that explosive materials, like live ammunition, have been very liable to be detonated during handling and processing of mixed solid wastes, it has been less well known that combinations of, or large quantities of, apparently innocuous materials can also cause explosions in certain circumstances. Incinerator explosions, for example, have resulted from the charging of large masses of plastic wastes. It is assumed the combination of gaseous pyrolysis products with the available oxygen reached explosive proportions. The gas produced by anaerobic decomposition of organic waste can also give rise to certain circumstances to explosive mixtures, and this described as follows:

Gas Production

The main gaseous product of anaerobic decomposition of organic wastes has been methane, CH_4, although other gases like carbon dioxide, CO_2, nitrogen, N_2, and hydrogen sulfide, H_2S_2, are also present. Wastes buried in landfills undergo prefominantly anaerobic decomposition, and the gases normally take the shortest or easiest route to the surface. Sometimes the presence of fissures in or surrounding the filled land, possibly together with the presence of a surface barrier, like a roadway or parking area, will make the evolved gases to travel large distances horizontally. In certain cases these gases enter the basements of houses and buildings, and deaths and injuries have resulted from asphyxiation, poisoning, or from the detonation of explosions of air-methane mixtures.

Sanitary-landfill regulations need that ducts be incorporated into landfills to lead the evolved gases to be discharged at a safe location.

Leachate Production

During decomposition, organic wastes normally produce a liquid called a "leachate". In landfills, some of this liquid gets resulted from the pressure applied by the fill above squeezing of leachate has been rain water which, during its passage through the landfill, is bale to dissolve a large range of materials. Leachate may emerge from a landfill with a biological oxygen demand of over 20,000 mg/litre, which has been about 100 times stronger than raw sewage. Studies have revealed that the leachate concentration from a landfill might still be significant after 20 years.

DANGER POINTS IN SOLID-WASTE HANDLING

In this sections the progress of typical solid waste will be described through typical sequences of locations and operations and particular hazards would be identified.

Wastebaskets

For typical residential, institutional, and in some cases commercial refuse, the wastebasket has been the start of the journey to final disposal. Wastebaskets are generally dry and reasonably clean and are emptied every one or two days, so that problems of fly breeding and multiplication of undesirable bacteria becomes small. The principal hazards have been fires originating from wastebaskets and from personal injury to the personnel emptying wastebaskets.

Paper products predominate in wastebaskets. The contents have been therefore usually highly flammable. As wastebaskets are usually used as receptacles for ashes from cigarettes, cigars and pipes, and frequently for the butts and pipe residues, the contents of ash trays, and the matches and spent cigarette lighter cartridges, there exists only a very low incidence of any of these materials being in a smoldering condition for wastebasket fires to be comparatively frequent. Wastebaskets have been often placed half hidden behind curtains and drapes and by upholstered furniture so that the fire danger becomes typically extremely high.

Some obvious methods of control have been to limit smoking within buildings; to provide a large-capacity, floor-standing ash tray near each wastebasket; or to use one of the many types of selfclosing waste containers.

Danger to personnel emptying wastebaskets comes predominantly from thoughtlessly discarded dangerous materials.

Two avenues of correction suggest themselves in these cases. One has been public education to the dangers to personnel of dangerous substances and objects in the waste stream, along with education of the service personnel to empty wastebaskets directly into receiving containers without the use of hands. The second approach has been to promulgate regulations which make the person who discards prohibited materials into the regular waste stream legally liable for damages resulting from these actions. The second approach has been in fact a variant of the first in that education has been usually more rapidly assimilated when there have been strong incentives to learn.

Refuse Chutes

A frequently heard complaint against architects has been that the handling of solid wastes has been seldom taken into consideration in the design stage of buildings, large or small. Occasionally however, refuse chutes have been incorporated, and these undoubtedly greatly reduce handling costs and thereby reduce some of the dangers involved in solid-waste handling.

Some other dangers have been introduced, such as the extreme fire danger if a refuse chute has been designed without appropriate traps to disallow the chute acting s a chimney and if the materials and construction have been insufficiently fire-resistant. Building codes in most cities need that fire traps be used. A secondary danger, which they allow encourages people to dispose of all manner of prohibited, dangerous, or incriminating wastes.

Compactors

Compactors have been designed to withstand the high loads and pressures imposed on wastes. Accordingly, they have been able to withstand and have minor explosions from, for instance, vaporpropellant bottles. Fires are usually extinguished by compaction. However, fires have been comparatively frequent in collector compaction trucks after, for example, some hot ashes might be loaded toward the beginning of a collection run, before the vehicle contents have become compressed to any extent. If fire inside a truck body has been confirmed, standing instructions to the crew usually have been to dump the burning load in the middle of the road insuring first, of course, that other traffic gets alerted and rerouted. A fire department can more easily put out burning refuse than exploding gasoline.

Other dangers from compactors come from the loading process when loading is done by hand. In most stationary compactors there have been interlocks to insure that the compaction device, usually a ram and sometimes a screw, gets closed off before operation. Accidents to loaders of refuse-collection trucks have been fairly frequent due to imploding T.V. tubes, or the sudden movement of pieces of steel when these get pinched half in and half out of the mechanism. When workers have been on an incentive system, protective devices have been often removed in the interest of speed, and good management is required to educate what is usually a transient work force on the dangers involved.

Containers

The term "container" has been generally reserved for enclosures with volumes larger than those of wastebaskets or ash barrels: containers normally hold from 1 to 20 cu yd. However, containers have been suffering from many of the dangers of wastebaskets and refuse barrels: fire, dangerous materials, and the encouragement of insects and vermin are examples. In addition, the storage of large volumes of refuse in a compact mass tends to encourage biological degradation which can give rise to spontaneous combustion given the right combinations of moisture, warmth, insulation, and fuel availability.

Incinerators

Small incinerators as manufactured for apartment houses and commercial establishments have been banned in many areas due to the difficulty of reducing pollutant emissions from the stack. Larger incinerators (50 tons/day to 3000 tons/day) will normally be having a discharge or tipping floor; a refuse pit or horizontal refuse conveyors; generally an over-head crane; one or more furnaces; ashhandling equipment; gas cleaning equipment; and a stack. Such a plant is having all the dangerous characteristics of an industrial processing plant involving the mechanical handling of extremely varied materials, together with some danger of explosions resulting from intrinsically explosive materials or from a high concentration of plastics. Poisonous gases have been produced given off during the combustion of some plastics, especially from the urethanes and the vinyls, but because furnaces have been normally maintained under negative pressure the dangers to personnel should be small.

Shredders

Solid wastes have been increasingly comminuted, the most popular device being a hammermill. Single machines posses capacities from one-half ton per hour to fifty per hour. No incidence of danger from shredders has been reported, although fires and explosions have been occasionally experienced. The robust construction of shredders make them to contain all but the most violent explosion.

Transfer Station

In its irreducible minimum, a transfer station has been a location where solid wastes get transferred from, usually, the collection vehicles to long-haul vehicles without any additional processing. The long-haul vehicles have been usually large-capacity highway trucks, sometimes with trailers; sometimes railroad cars; and occassionally barges. Frequently, processing by compaction or comminution has been included as part of a transfer station operation. Danger points have been those with the unit operations as discussed above.

Reclamation Plants

Reclamation is done in a crude way at many town dumps by authorized and unauthorized personnel picking over refuse for possible valuable items. It is an extremely dangerous practice due to the intrinsically unsafe nature of the surface of a dump; because of the difficulty of extracting desired items from their resting position in the dump; and from the danger of being run down by vehicles while they discharge their loads. Dumps and landfills should always be adequately fenced and policed to disallow scavenging and the reasons for these restrictions should be clearly stated.

Some degree of mechanization got introduced into reclamation in the early decades of the present century by loading refuse onto traveling belts which then pass by so-called "picking stations" where individual personnel are having responsibility for removing items in particular categories (newspapers, corrugated cardboard, bottles and so forth). There have been obvious dangers involved, and in present day conditions it is found to be difficult to attract any but an extremely transient and poorly educated work force.

Modern automated reclamation plants have been developed which are generally marked by a lack of human contact with refuse and by the use of sophisticated control systems and protective devices. The dangers involved should therefore be small.

Landfills

Dangers from improper landfilling procedures have been described above, and fall into the categories of fires; gas production resulting in possible fires, asphyxiation and explosions in buildings; leachate production causing possible pollution of aquifers; rat and insect infestation, leading to spread of disease; and scavenging leading to a probability of injuries. Of these, the production of gases and leachates, and the spread of underground fires, which can persist for 20 years and more, have been the most troublesome. While easy to prevent during landfilling by following one of several approved techniques, subsequent prevention of these conditions after poorer landfilling methods have been used is extremely difficult, if not impossible. Even though, for instance, modern infrared measurements may locate with considerable accuracy the area where in underground fire has been smoldering, such as fire has been virtually impossible to extinguish with certainty. The cavern left by a smoldering fire may subsequently make a sudden collapse of the surface, with obvious dangers to people and to any buildings which are erected without sufficient design precautions.

POTENTIAL HEALTH HAZARDS OF SEWAGE-SLUDGE COMPOSTING AND LAND DISPOSAL

Recent findings at the University of Toronto indicate "alarming amounts of cadmium and lead occur in sewage sludge which is used as fertilizers for food crops. These metals could be assimilated by croups in amounts harmful to human and animal health. Medical evidence reveals a correlation between cadmium and hypertensive, cardiovascular, and respiratory problems. Lead can bring about brain damage or death."

Recent German research has also reported that there has been "no decrease in the number of pathogens in raw sewage over the last few years, and even increased incidence of hepatitis, salmonelloses and bovine cysticercoses has been noticed."

There has been much disagreement as to what extent high quality aerobic composting kills pathogenic bacteria present in the feed material (particularly in sewage sludge, if used, and household refuse). One view, given in a 1974 German report, has been that "after the temperature gets stayed for one day between 30^0 and 45^{0}C the temperature may further rise to 70^{0}C after one week all pathogenic germs are killed" Another 1974 German report states: "The temperature of 65^0 to 70^{0}C at which composting processes usually occur is not sufficiently high to kill all pathogenic germs, viruses, bacteria and bacilli found in garbage and sewage sludge. However, studies of the decomposition process in composting plants showed that microbial substances with antibiotic effects can form which kill even heat resistance forms of pathogenic germs." These antibiotic substances have been found to work effectively after the initial composting process, usually taking 3 to 7 days, during the so-called "curing" process. During this process, compost has to be stored in piles or windrows, often in the open, for perhaps 3 months. Formerly it was considered as a subsidiary process, but with this finding it takes on primary significance.

When sewage sludge has been used for field irrigation there has been no curing process. Enteroviruses (poliomyelitis, ECHO, Coxsackie) have been found to be taken up by the solid and to contaminate vegetables grown in them (see Tables 5.16 and 5.17).

Parasites could also be spread from sewage. Hookworm infection, prevalent over much of the world, has been reduced to a negligible point in most of Europe and the United States. The larvae of hookworm penetrate the skin, pass through the liver and lodge in the lungs. The pulmonary migration may bring about symptoms which have been ranging from mild, dry cough to severe dyspnoea, cyanosis, etc. Intestinal ascariasis tends to aggravate

Table 5.16
Diseases Associated with Human Fecal Waste

Bacterial infections	Protozoal infections
Typhoid fever	Entamoeba histolytica
Paratyphoid fevers	Helminthiasis
Cholera	Fish tapeworm
Shigellosis (bacillary dysentery)	Beef tapeworm
Viral infections	Pork Tapeworm
Poliomyelitis	Pinworm
Coxsackie infection	Roundworm
Infectious hepatitis	Whipworm
(very many other enteric viruses exist)	Hookworm

Table 5.18
Diseases Associated with Animal Fecal Waste

Salmonellae	Helminthiasis
Infection from pig intestinal protozoan (Balantidium coli)	Pig ascaris
	Animal tapeworms
	Hydatid worm

malnutrition and may cause dysentery. Hookworm infection brings about an anemia syndrome. Strongyloides infection is having a mild form characterized by catarrhal enteritis, with mucous secretions, intermediate forms characterized by edematous enteritis, with mucous secretions, intermediate forms characterized by edematous enteritis, and serious forms characterized by ulcerous enteritis. The VLM (visceral larva migrans) syndrome, like massive trichuriasis and ascariasis, could be seen in dirt-eating children aged one to four years. The principal symptoms have been eosinophilia, hepatomegaly, hyperglobulinemia, fever and pneumonitis. Larval toxocariasis may bring about damage to the eye.

HAZARDS FROM SEA DISPOSAL OF SOLID WASTES

The data about the harmful effects of dumping a vast range of solid wastes and sludges at sea has been not available, except for a few specific cases. These cases, involving DDT, oil, mercury, and bacteria from sewage, alarmed and alerted the public and regulatory agencies, so that the dumping of some hazardous or possibly hazardous materials has been outlawed.

Metals

The principal metals which may offer potential hazards have been arsenic, cadmium, chromium, cobalt, copper, iron, lead, manganese, mercury, nickel, silver and zinc.

Mercury discharging from a chemical plant in Japan had been taken up by the fish caught locally and extensively used for food. Deaths and serious disablement had been followed.

Sewage sludge contains metals in a quantity which can be toxic to marine life when the deposits have been allowed to build up rather than to disperse. Some of the metals, like lead and copper, dissolve out of water-supply systems; some, such as iron and lead, come from high way runoff into combined sewer systems; mercury may run into a sewer system after use as a garden fungicide; other metals like cadmium, silver and zinc get discharged, sometime illegally, into sewers by industry.

Inorganics

Inorganic materials strictly include the metals. However, the term has been more usually taken to include such materials as dredging spoils, sulfuric-acid sludges, glass from domestic refuse and demolition debris.

Organics

Wastes from the synthetic-chemical industry have been particularly complex, and new compounds are discovered and join the waste stream faster than they can be tested for long-term hazardpotential. It was the discovery in Berkeley, California that the decline of the brown pelican have been due to the effects of DDT on the shells of the eggs, and later that measurable concentrations of DDT could be detected in Antarctic penguins, which catalyzed public concern over pesticide use and the discharge of organic chemicals.

Biological Wastes

Three components of the sludge from human sewage have been of great concern. disease organisms, particularly infectious hepatitis, may contaminate shellfish which may reinfect people. Metals dissolved from plumbing systems and from other sources could be toxic to marine growth and may concentrate in shellfish and other diet fish (e.g., mercury in swordfish). The nutrients may bring about the growth of undesirable organisms.

This third concern, the possibility of excess nutrients being present, has been quite different from the problem which arises in inland waters. The ocean is starved of nutrients, and any excess is purely a local phenomenon. Dispersal has been therefore an acceptable solution.

AVAILABILITY OF WASTE IN INDIA

A detailed report on Utilisation and Recycling of Wastes (Research, Development and extension Requirements) was submitted in 1975 by the National Committee on Science and Technology of the Government of India (NCST, 1975). The NCST is the progenitor of the Scientific Advisory Committee to the Cabinet (SACC). A lot of effort has been done in the preparation of this Report but not much cognizance has been taken of it so far. Although over a decade old, this report along with a series of other papers published by O.P. Vimal and his associates (1975-85) constitute the only available information on the extent and nature of wastes available and the R & D effort in progress as also possible on the subject. India being a developing country can ill afford to waste any wastes from agriculture, forestry, animal husbandry, industry, minerals urban and rural community and building materials.

By proper utilisation of these materials it becomes possible to conserve the national resources, reduce the cost of production in many cases and solve the problem of popllution and generate some employment. The basic rationale on which the Report has been based is "What is waste to one industry may be raw material for another"

During the last decade, waste recycling and reuse has been a very major activity in most of the developed countries and considerable progress has been made. It has been fast becoming a hightech area. There is, therefore, need now for a very concerted national R and D and pilot level effort in this area. This should enable to conserve the resources as also energy and at the same time add to the national wealth. If a country like India does not take up serious work in this area, time will come the country will have to go round buying technology for waste recycling.

The Japanese example has been before us where work on waste recycling and utilisation has paid rich dividend. An illustrative list of types of wastes available in India is given below. The quantities wherever given are based on 1975 data.

Depending on their origin, the wastes could be grouped under seven heads namely; Agricultural, Fruit and Vegetable, Animal, Aquatic weeds, Community and Industrial and Construction materials.

I. *Agricultural Wastes*

— Eighty species of Non-edible oil seeds (1 Mt.). Sal seeds (40.000t), Groundnut cake after extraction of oil (2-2.5 Mt), Rice barn (2.4 Mt) and Rice husk, Rice straw; Tea waste (8-900t). Bagasse (5.25 Mt), Molasses (0.17 Mt), Press mud, Tobacco waste (62,000 t); Protein rich plants, Mahaua flowers: wood waste, saw dust, bark from industrial wood for venneer logs; and other agricultural wastes.

— Coconut waste by products (other than those used for construction materials), Coconut shell (0.88 Mt), Coconut husk (0.5 Mt), Coconut water.

— Lac waste.

— Arecanut waste/by-products (other than those used for construction), Arecanut husk (0.1 Mt). Leaf sheath (90,000 t), Chogaru liquid.

— Cashew waste/by-products (other than those used for construction needs), Cashew apple (30 Mt), Cashew testa (3-4000 t) with 61% tannin, Cashewnut shell liquid (10-14000t).

— Coffee waste.

— Rubber wastes/by-products, Rubber seed, Rubber wood.

— Tobacco waste, leaf scrap and stalks seed.

— Cotton wastes/by-products. Cotton stalks (2.5 Mt), Cotton leaves; Cotton linters.

— Jute wastes and by-products, Jute sticks (2.5 Mt), Jute mill waste (50,000 t).

II. *Fruit and Vegetable Processing Industries*

— Mango peels and kernels.

— Citrus peels, pomace, etc. (0.48 Mt).

— Apple pomace.

— Pineapple peels, ears and eyes.

— Banana pseudostems.

— Waste from vegetable canning industry.

III. *Animal Wastes*

— Cattle shed waste.

— Fisheries wastes.

— Prawn wastes (10,000 t).

— Squilla (0.1 Mt).

— Frog wastes (5,000 t).

— Carcass utilisation (12.5 M carcasses).

— Slaughter house waste (27,000 t intestines, 55,000 t blood, etc.).

— Wood wax.

IV. *Aquatic Weeds*

They are very prolific in growth and a menace by blocking water ways and interfering with hydroelectric production, wasting water in evapotranspiration, hindering traffic and fishing, as also help in increasing water borne diseases. The negative impacts of the aquatic weeds can be counter balanced by converting these into food (for fish, duck, donkeys, pigs, sheep and even processed animal feed for cattle), fertiliser, pulp and paper, fibre energy, (biogas and waste waster treatment (NAS, 1976).

Aquatic plants including algae from marine sources need to be utilised in a planned manner.

V. *Community Waste.*

— Night soil, sewage, sludge, cattle dung and urine.

— 7 Mt of dry matter containing 2 Mt of nitrogen, 0.5 Mt of each of P_2O_5 and K_2O.

— City refuse.

VI. *Industrial Wastes*

— Paper waste, recycled paper products.

— Mica waste (20,000 t).

— Flyash (5 Mt in 1968-69, today around 22.016 Mt).

— By product of salt industry (50,000 t KCI).

— Wastes from steel plants (slags).

VII. *Construction Material from Wastes*

— Waste from sugar factories.

— Agricultural wastes.

— Coconut, Arecanut and Cashewnut after extracting oil.

— Redmund from aluminium industry.

— Water works silt.

The foregoing list provides an idea about the extent and nature of wastes available in the country. This is only an indicative list. These waste materials can be rearranged keeping in view their end use, namely animal and poultry feed; industrial chemicals; paper, pulp and board; fertiliser, fibre, etc. The urgency of intensifying work in the direction is no doubt obvious but in order to give a thrust, the Department of Science and Technology need to take a view on the recommendations of the NCST (1976) Report. The inportant recommendations have been as follows:

— Declare DST as the nodal agency to coordinate the work with various Ministries/ Departments and agencies in Centre and States.

— To institute an appropriate High Level Advisory Body to look after waste management, advise the Government on short and long term policy on the subject and update the 1976 date regarding locational and seasonal availability. Start R &

D through an All India Coordinated Project on specific subjects, generate new technology, put up Pilot Plants, etc.

— Build up awareness regarding the waste collection, disposal, utilisation and employment generation potential.

THE COSTS OF SOLID WASTES

The Social costs of solid wastes have been not easy to determine. Open dumps may be responsible for a small amount of disease and incinerators add to the air pollution costs of a community. Most disposal methods involve some aesthetic displeasure. In rare circumstances the costs can be much higher.

Litter along roadsides constitutes a solid waste problem with considerable psychological impact, judging from the interest in antilitter and highway beautification programs. Occasionally there are massive efforts to pick up litter along highways, generally resulting in the recovery of several hundred kilograms per kilometre. Since there are approximately 5 million km of roads in this country, highway litter might amount to roughly one million metric tons at present-a small but important part of the solid waste problem, and one which is generally believed to be growing.

The major cost of solid waste disposal at present comes from the collection, transportation, and transfer of the wastes, not from the final stages of disposal.

COLLECTION AND TRANSPORTATION OF SOLID WASTE

Introduction

Of the total, 90 per cent solid waste is produced from agriculture and mining. The solid wastes could be usually disposed of with little difficulty in the rural areas where they are generated. Solid-waste collection, processing and disposal have been far more of a complex management program in built-up urban communities. Urban solid-wastes sources have been of two principal categories: industrial and municipal. Municipal waste is a collective term referring to residential, commercial, institutional, and demolition solid wastes. These subcategories will be described individually.

Residential Wastes

In urban areas, the volume and composition of residential waste may very considerably in different communities and climates. In general, per-capita waste production has increased with per-capita income, although in some localities solid-waste production has actually decreased while presumably income has not. The quantity of residential solid waste in a city also varies with the season of the year. Wastes like garden trimmings, autumn leaves, clean-up campaigns, summer produce, Christmas residues, etc., are all seasonal in their generation.

The type of housing also influences the waste within a city. High-density areas produce less solid waste per capita; apartments produce solid waste with little or no grass and leaves.

Areas having large numbers of garbage-disposal units produce much less food wastes. Typical residential solid waste by weight has been found to range from 35 to 45% paper, 10 to 20% food waste, 5 to 10% glass and ceramics, 5 to 10% metals, 5 to 15% grass and leaves, 2 to 6% taxtiles, and usually less than 4% wood, plastics, rubber, leather, and inert materials.

Commercial Wastes

The aggregate composition of commercial solid waste has been found to be similar to that of residential waste, but individual generation points have tendency to produce a more homogeneous waste stream. That is, restaurants obviously produce more food and related packaging wastes, markets produce some food and mostly paper of carton wastes, whereas hotels produce considerable textile wastes, as well as more normal trash. The amount of solid waste generated per collection unit has been found to be much higher for a commercial solid-waste source than for a residential solid-waste source; because of this, collection vehicles in commercial areas fill up rapidly.

In dense urban areas, commercial solid-waste generation has been more than in less-populous land-use zones. Collection routes in which commercial and residential wastes may be picked up by the same organization may be combined so that vehicles could efficiently make both residential and some commercial stops. In many cities, commercial wastes are collected by private contractors with some municipal regulation. It is desirable that local administrators must be able to control the standards for all collections with in a community to insure uniform, and sanitary service conditions.

Industrial Wastes

These include food-processing residues, boiler ash, wood, plastic and metal scraps, chemical sludges, packaging wastes, debris, dirt, etc. Increased industrial application of air and water-pollution controls will also cause an increase in the solid wastes which have been destined for land disposal. Industrial waste collection and disposal are usually undertaken by contractors, and less often by the producer, Industrial waste-generation quantities in a given area depends upon the types and sizes of the industries present rather than area population.

Demolition Wastes

Solid wastes due to demolition and construction activities have been including mixed lumber, roofing and sheeting scraps, stone, rubble, broken concrete, asphalt, brick, plaster, wallboard, pipe, wire, and other residual building materials.

Bulky Wastes

These are including trees, telephone poles, bushes, abandoned autos and auto parts, tires, stoves, refrigerators or other large appliances, furniture, crates; mattresses, etc. Conventional collection vehicles fails to transport such materials. Special open truck vehicles have to be used for collection of bulky wastes which otherwise may be dumped and abandoned on public or private property. Two trucks with trollies could be used to tow away abandoned vehicles. Vehicles collecting shrub and tree branches have been generally more efficient if equipped with volume-reduction equipment such as a chipper of leaf mulcher to reduce the cellulose wastes.

Hazardous Wastes

Wastes having pathological, explosive, radioactive, or toxic materials are regarded hazardous. Dangerous liquids and solid wastes sometimes get containerized and included in the solid-waste stream. Collection disposal crews must have the training to identify all hazardous wastes. Once hazardous wastes are identified, they must be handled separately, with appropriate safety procedures. Fire-fighting equipment and protective clothing must be present on special and regular collection vehicles for handling hazardous materials and coping with emergencies when accidents requiring emergency action take place.

Storage

Storing solid wastes prior to collection disallows offending esthetic tastes, attraction of vectors, and excessive odors. Storage devices must be convenient for the user and fecilitate, safe, efficient collection, processing, and disposal. Types of storage devices must be described as follows:

Containers

Rigid containers must be made from durable metal or heavy plastic, and covers should fit tightly. Pick-up handles should be strong because if a handle may break it could cause in serious injury to the user or collector. Damaged containers having sharp edges should be immediately replaced to avoid cuts and bruises. Containers should be stored on a flat concrete base or elevated above the ground for reducing metal container corrosion and spilling. Some vinylplastic containers may get oxidized and crack with time or coldexposure; polyethylene and certain other durable plastics may have a superior useful life.

Bags

Plastic and paper storage bags and holders are having certain advantages and disadvantages. The various advantages have been as follows:

(1) one-way disposal without retrieval:

(2) a universal and standard-size, cleaner container;

(3) faster collection because they do not have to be uncovered and returned after being emptied;

(4) reduced spillage and resultant litter;

(5) easier storage and handling as liners or in suspended bag holders;

(6) reduced back strains and similar collector and user injuries because bags are far lighter, will not hold heavy loads, and are easier to handle than rigid cans; and

(7) more sanitary and cause less noise, dust, odor, and microorganism release; and adaptable to mechanized collection by truck equipment.

In summary, plastic bags have been simpler to store, more expandable, less expensive, and are easier to tie and handle than comparable paper bags. However, both paper and plastic bags are more easily punctured by sharp objects, and dogs, raccoons, rats and the like can tear these open. Bags have been a continuing cost and contribute to solid waste in themselves. Thirty-two to 40-gallon bags and rigid containers are generally the standard for solid-waste collection. Plastic storage bags (polyethylene) have been commonly 1.5 mil thick while paper bags have been commonly two-ply and have plastic additives for wet strength. Solid-waste generation points that produce large amounts of solid waste between collections should use large bins (1.5-15 cu yd) rather than numerous smaller bags or containers. The larger bins also may use stationary compactors to reduce storage volumes and collection and disposal costs. Vehicles capable of hoisting and dumping bins should service them as required.

Twice-weekly collection has been desirable since fly propagation may be reduced, over once-per-week collection by 75%. The bins should be seamless, watertight, sturdy, and maintained frequently to prevent vermin infestation and odors. Bin collection has been useful for handling large volumes of industrial, commercial, agricultural and high-density residential wastes.

Drop Boxes

Drop boxes have been larger storage bins and are useful for servicing rural, industrial, and demolition sites. Large bins with or without stationary compactors cost less per unit of solid waste than small bags, cans or bins. Centrally located boxes of 20- to 40-cu-yd capacity can service a large rural area at less cost than a small sanitary landfill. Vehicles capable of hoisting these boxes onto their chassis can be used for collection and transport. An empty box is normally dropped off when a full box is collected to avoid leaving the area without a solid-waste-storage device. This method needs one vehicle pickup trip for each collection, but could be an economical procedure if property planned and managed.

Other Storage Devices

Backyard solid-waste shelters that need emptying by a collection crew using hand tools are inefficient, cause odors, support rats, flies, and other vermin, and create a fire hazard. Underground pit storage of containers is also undesirable because spilled garbage in difficult to reach and clean out, resulting in odors and rat and fly propagation. Hard-to-reach containers in pits may also need collectors to use dangerous lifting positions.

Compactors

Storage containers equipped with heavy-duty compaction devices can be able to reduce space requirements by increasing solidwaste densities severalfold. In communities which have outlawed private on-site solid-waste incinerators,k compactors can get installed in the former incinerator locations. Household compactor units, a fairly new innovations, are currently being used only in more affluent residences. Compactors, shredders, and other volume-reduction devices obviously do not decrease the total weight of generated solid waste.

Source Segregation and Reclamation

Producer-provided segregation of reclaimable materials at the source has been a simple method of solid-waste separation. Segregated materials have minimum contamination, but need special equipment or extra trips for handling and collection.

Resource conservation has been presently the major motivation for waste segregation. For communities incinerating solid waste, the separation of combustibles and non-combustibles may be desirable. The dominant use of sanitary landfills for solid-waste disposal reduces the need for source segregation. Class-I disposal sites having little potential for groundwater contamination require no waste segregation since they have no limitation on acceptable wastes. Class-II disposal sites (located above the groundwater table and protected from surface runoff, with controllable surface drainage) can be able to accept ordinary household and commercial solid waste, including putrescible organics and most common scrap metal.

Source segregation of metal cans, paper and other valuable components may be becoming a public responsibility in the future.

Many communities are having in the past practiced separate collection of garbage, cans glass and paper products, but the trend has been toward combined collection to reduce collection and disposal costs. Unlimited collection of all generated solid waste at the source should causes a cleaner community; reduced service pickup or restricted costly collection systems result in open dumping and litter.

Collection-System Parameters

Many combinations of collection vehicles, routes, crew size, collection frequency, service type, work scheduling, set-out location, etc., have been used. The most appropriate system will be depending on local conditions. Total costs of refuse collection will be a function of population.

In planning and implementing a solid-waste-collection system for any particular area, collection-system design and operation parameters need to be carefully considered. Once implemented, a collection system must be continually monitored for assuring that maximum cost-effectiveness is being realized. Best-known productivity performances of various types of collections systems, have been a useful start for determining the degree to which a designed collection system is reaching its potential.

Crew Size

The number of men in a collection crew is dependent on the system, equipment, and type of service offered. In general, cost savings and high service levels can be realized by employing a minimum-sized crew; there seldom is justification for using more than a two-man crew for curb-alley collection or a three-man crew for backyard collection. Except for the driver, the crew-member time has been mostly non-productive during disposal trips. Some communities try to also utilize crew members for street or alley clean-up or to set out or set back containers during filed-truck disposal trips. Systems using a single crew members as a driver/loader with a sideloading or front-loading vehicle are shown to be efficient for curb and alley collection routes. Mechanical equipment to eliminate manual pickup and utilization of single-operator trucks have been new and significant developments.

Task Assignment

Crew scheduling and collection-vehicle routing has been interdependent. The type of routes used may be fixed routes, weekly routes, single-load routes, or other systems. Routing must be flexible enough to absorb seasonal waste-generation variations and should avoid heavy traffic or left-hand turns. Collection services which need customer set-out allow less flexibility in routing. Crew scheduling basically involves a choice between a fixed-working-hour system and an incentive-work system; the former involves a fixed work day (e.g., 8 hours) with overtime if necessary; the latter generally involves an assigned task rather than a specific number of hours worked. An assigned task system in which early-finishing crews assist other crews until all tasks have been completed has been used successfully in Inglewood, California. Although incentive systems may result in more users serviced per worker-day, there are disadvantages. Workers who are hurrying to finish may be less careful and consequently handle containers roughly and create excessive spillage. Good route-collection management supervision has been always important, but particularly so with incentive systems if high work standards have to be maintained, poor service (such as missed stops) avoided, and customer complaints god minimized.

Type of Service

Frequency of collection. Solid waste must be collected at least once per week. In general, the amount of solid waste collected per stop receiving twice-a-week collection would be greater than the total amount of solid waste per stop receiving once-a-week collection. This phenomenon must be taken into account when determining collection frequency. It may, in fact, get related to more thorough disposal by the waste generators rather than burning on-site or increased littering.

Backyard collection. Collectors remove refuse from the backyards of premises and return any containers. collectors often transfer a household containers, in the backyard, to a larger receptacle which is carried out and emptied into the vehicle. Burlap bags and plastic "swamper" containers are generally used for such transfers. The disadvantages of this service have been higher costs; invasion of privacy by collectors entering private property (trespassing and robbery complaints have been received with this system), accidents to workers, who often fall down in icy or wet weather or are bitten by dogs; and damage to fences, landscaping and so forth as collectors generally try to use shortcuts to the adjacent lot rather than walk longer distances with a partial load. Backyard pickup service provides a maximum degree of routing flexibility, but low productivity and high costs frequently forces a decision to use a different collection system.

Modified versions of the backyard collection system have been a set-out service and set-back service. In set-out service, containers are carried out to the curb by collectors, generally preceding for regular collection vehicle crew. Customers are responsible for returning their own empty containers. Ideally, full cans would therefore on the street only a short time, thus reducing litter. The lids should be replaced on the empty containers by the crew. Set-back-only service needs customers to bring their containers to the curb. Collection crews return the empty containers to the yard. This method is having less routing flexibility than the other two methods.

A danger involved in carring bags or containers has been the tendency for collectors to carry several containers at one time to reduce extra trips, resulting in injuries or accidents. Careful worker training, supervision and good equipment have been necessary to obtain competent collection.

Curb collection. Curb collection has been the most productive type of service. Collectors are not needed to carry waste containers a significant distance. Pickup of the solid waste, however, makes many injuries; solid-waste collection is a work-intensive occupation with high illness risk. Crowded solid-waste truck-loading areas such as have been common with large collection crews should be avoided: they cause collectors to dump while in awkward positions.

Curb collection is occasionally done with pickups by two or more crew members working opposite sides of the street. This technique has been generally less efficient than a single collector alternately working one side of the street, than the other. Collecting both sides of a street at the same time should generally get avoided; an exception has been streets with very little traffic and few parked vehicles. Traffic-related accidents have been a danger on more crowded streets.

Alley collection. Solid-waste collection from alleys has been common in many older cities. It has been similar to curb collection except that both sides of the alley are generally picked up simultaneously. Dead-end alleys should not be entered if the vehicle will have to back up to exit, as backing a large vehicle is dangerous. Alley collection generally gets incorporated with backyard service by requiring containers to be stored in proximity to the alley.

Equipment

Equipment maintenance. Proper collection-equipment maintenance reduces annul system operating costs. Maintenance yards should be provided by the collection system itself since outside vehicle-maintenance facilities are frequently less dependable. Using inefficient past its useful life should be avoided.

Equipment design. Collection vehicles must be enclosed and watertight for all-weather service. Open trucks must be used only for collection of bulky wastes, and must be covered with a tarpaulin during waste transport to prevent littering. A low loading height has been desirable to increase work productivity. Safety devices are essential, including backup lights, noise-making and other warning devices. There must be emergency stop controls on compaction equipment and a safe place for crew members to ride during short trips.

The optimum vehicle capacity has been determined by the routing scheme. Packers involved in slow inefficient collection systems are generally small (less than 20 cu yds); high-density systems with greater productivity generally require larger vehicles (20- to 40-cu-yd packers). The turning radius, local highway load weight (for trees, overpass clearance, etc.) must be taken into consideration before choosing vehicles.

Operation. Compactors have been the preferred waste-collection equipment. Compactors significantly increase solid-waste density from 100 to as much as 600 lbm per cu yd, thus increasing collection route work capacity (Fig. 5.19). Many vehicles cannot be driven during the compaction cycle; a frequent cause of this has been compaction equipment driven by the same engine that propels the vehicles. Dual mechanisms for separate compaction and road operation have been becoming popular, as they increase collection productivity.

The design and location of the loading hopper has been an important vehicular parameter. Small one- or two man crews in which the driver takes part in the loading operation may employ sideloading vehicles. Front-loading vehicles need the dangerous practice of collectors working in front of the vehicle. Rear-loading vehicles make communication between the driver and the collectors difficult also a dangerous situation. All loading hoppers must be large enough to avoid crowding while emptying waste conlainers; and they should be low enough for easy access. Equipping hopper-load vehicles with "dustless" or other hoist equipment to empty bins can be able to reduce overall fleet requirements and eliminate retracting routes.

Collection equipment designed designed for mechanical retrieval of solid waste containers is receiving considerable attention. Under ideal collection conditions, this equipment can be very efficient; however, the presence of hindrances like parked cars and low trees can be able to reduce the effectiveness of mechanical collection. Mechanical collection vehicles must be used in conjunction with uniform containers. (One system uses 80-gallon containers; another uses 32- to 40-gallon polyethylene bags). The community should supply or need these standard containers to achieve uniform operation.

In some communities a relay collection system may be efficient. Smaller vehicles (in some cases, nothing more complex than scooters) having one or more attached bins pick up household waste; when full, they return and empty their waste into a larger 'mother' vehicle. One large 'mother' vehicle can compact and haul several small 'satellite' vehicles. This system is having the cost and environmental advantages inherent in the use of smaller vehicles.

Other Transport systems. Pneumatic-pipeline solid-waste-collection systems are being used successfully for short-distance transport from hospitals and commercial, industrial and large apartment building complexes. These systems generally use 18-inch — minimum piping, automatic inlet hopper valves, heavy-duty vacuum fans, and air-pollution-control devices. Particular care must be taken in the design of pneumatic systems so as the disallow fires. Liquid carriage pipelines for the transportation of solid-waste slurries are being studied extensively for long distances, and the preliminary results are promising. Such a system is having a high initial capital cost which could be compensated in part by lower annual operating costs.

Table 5.19
Variation of Equipment Capacity with City Size and Number of Weekly Collections

Populations class (1000's) From To less than	No. cities	Average capacity equipment once/week collection (cu yd)	No. cities	Average capacity equipment twice/week collection (cu yd)
10 — 20	3	17.6	9	18.2
20 — 30	6	14.7	5	16.5
30 — 40	3	24.5	3	14.0
40 — 50	2	15.4	3	17.3
50 — 60	4	16.5	3	21.7
60 — 70	2	25.5	1	30.0
70 — 80	2	16.0	0	—
80 — 90	1	18.0	1	18.0
90 —100	2	17.4	2	23.4
Combined average	25	18.6	27	18.5
100 —200	5	24.5	2	18.9
200 —300	2	17.0	2	15.8
300 —400	2	17.8	3	17.8
400 —500	0	—	1	20.5
Combined average	9	21.4	8	17.8
500 and over	5	19.4	3	18.77

Transfer Stations

In order to determine the feasibility of using a transfer station, cost-benefit studies of alternative collection/transport systems must be worked out and trade-offs determined. If sufficient savings have been derived from decreased collection and haul costs, than a transfer station must be practical. Whenever possible, adjacent collection areas must be consolidated into centrally-located regional transfer stations, to take advantage of any economics of scale. Transfer stations would be more likely to be economical if initial conditions have been including traffic congestion, large solid-waste quantities, and the disposal site is located so as to require long hauling distance.

Solid- waste Processing

Volume reduction processing frequently can be carried out at a transfer station. Methods like shredding and high-density baling can also increase the haul-vehicle carrying capacity and thus reduce total operating costs. An economic analysis becomes necessary to determine if waste-densification systems have been justifiable.

Baling. Solid waste at present has been baled in several locations. Baled solid could be more easily transported as bales are more easily handled, or fairly uniform size (e.g., 4 ft long, 6 ft in cross-section), and of high density (58-64 lb/cu ft). Higher density can be able to increase the amount of solid waste transported in volume-limited systems such as railroads, and can produce less litter than alternative transport systems. This method is quite cheap. This method allows greater control over the transfer vehicle load.

Conditions at transfer stations have been typically odorous, noisy, and dusty. If stations are located in urban areas, they preferably should be enclosed. Enclosed transfer station also eliminate problems associated with weather conditions. Vehicular traffic control and automatic weighing facilities must be part of basic transfer-station design.

Long-Distance Transport

Rail Haul. Rail haul generally becomes competitive if the following conditions have been met:

(1) round-trip distances must be greater than 100 miles;

(2) at least 1000 tons per day have been transported;

(3) baling takes place at the transfer stations; and

(4) railroad cars are used exclusively for solid waste.

Barging. It is a special case of long-distance transport which is limited to a few suitable geographical areas. Barging costs have been relatively low.

Semitrailer trucks. Semitrailer costs have been found to vary widely for solid-waste transportation because factors like labor rates speed limits, truck size, and truck operation and maintenance requirements differ by area.

Waste loads have to be covered during transport. Mechanical unloading or drag lines placed to loading a fixed-bed truck can be employed to unload, using a bulldozer to pull the solid-waste lines. Care should be taken because snapped line can bring about serious injuries.

Pipeline transport. Pipelines are mentioned briefly already; their potential for low-cost, long-distance solid-waste transport could not be demonstrated. A possible advantage of a pipeline system has been that it could be used be for both solid and liquid wastes transport. Future waste-management systems may combine solid and liquid disposal. Pipeline transport of liquid sludge has been found to be economical.

Street Cleaning

City streets should be kept clear due to the many reasons:

(1) to prevent disease, injury, and annoyance from street dirt;

(2) to prevent vehicle damage from sharp, metallic objects;

(3) to promote safety in reducing fires from dry leaves and litter, and slips and skids from wet leaves;

(4) to increase community appearance;

(5) to disallow the clogging of sewers and storm drains; and

(6) to decrease water pollution from street runoff.

A proper solid-waste collection system will help reduce street dirt and litter, but some street cleaning will still be needed.

Manual

Street cleaning in congested areas of large cities may require manual methods. Bright, fluorescent uniforms must be worn by clean-up personnel for their safety. New equipment, like vacuum sweepers and automatic brooms, are replacing the conventional "white wing" armed with a broom and dust cart. Such mechanical equipment permits one operator to cover a larger area at less cost in the more open locations. The collected sweepings are normally stored in bins: solid-waste bins along the street are unattractive and become litter sources. Using a small relay vehicle for collecting the sweepings from several street cleaners should be considered. Vehicular or house flushing (usually with fire hydrants) has been costly, generates water pollution, and is not recommended except in a few special situations.

Flushing Machines

Large trucks with pressure nozzles are used for street flushing in many communities. The advantages of this system have been as follows:

(1) rapid and thorough pavement cleaning despite parked car or other obstructions;

(2) continued operation with wet pavements;

(3) acceptable operation at night; and

(4) the suppression of dust.

The disadvantages to the system are:

(1) cost has been generally higher than for dry cleaning systems;

(2) it may be increasing drainage and water-pollution problems;

(3) flushing is not a complete method; it moves light dirt to the curb and leaves it there;

(4) splashing of vehicles and pedestrians cannot always be avoided; and

(5) flushing may not be possible during the peak traffic periods.

Well-trained personnel have been necessary for a successful system; sewer or storm-drain requirements get increased by the system. The maintenance of catch-basin residue disposal is a part of the costs of flushing. The trend has been away from simple flushing and toward mechanical brush-sweeping and water-sprinkling machines.

Sweeping Machines

Conventional sweeping machines are having a rotary broom, vacuum pumps or air-circulating pumps, sometimes water-sprinkling tanks, and a storage compartment for the sweepings. As most dirt and litter gets accumulated along the edges of the street, a right hand-drive vehicle with two passes per street can be quite effective. As parked cars prevent cleaning, temporary parking restrictions are frequently imposed. Sings restricting parking must be clear and well-posted to avoid citizen complaints. Noisy vehicles should not be allowed in residential areas during normal sleeping hours, and vehicles should be well-lighted to avoid accidents.

Street sweepings consists mainly of paper, leaves, and inerts, and consequently pose normal disposal requirements. The individual collection vehicles may be able to transport the sweepings directly to the disposal site or transfer vehicles may be used. The practice of street-dumping loads of sweepings for subsequent loading onto trucks must be avoided as sweepings may be dispersed by wind, traffic, small animals, or pedestrians. Street-sweepings collection has been a far more costly unit process than on-site-solid-waste collection.

DISPOSAL OF SOLID WASTES

Nearly 75-80% of all collected residential and commercial solid wastes are sent to open dumps. Less than 10% is buried in sanitary landfills. A small but locally significant quantity is dumped at sea. The remainder is incinerated, partly to obtain energy and recover metals and partly to reduce the volume of wastes and the amount of land required for dumps and sanitary landfills.

A number of different collection and disposal methods are in use and will be discussed in this section.

1. Hog Feeding

At one time, hogs were extensively used for garbage disposal. In colonial times hogs wandering in the streets served as scavengers of the wastes in the gutters. As urban areas developed, hog farms nearby found it desirable to collect urban garbage for feeding to hogs.

It was known that the feeding of raw garbage to hogs led to high incidences of trichinosis in humans, but the practice continued until the mid-1950's. During the years 1953 to 1955 the rapid spread of a virus disease of hogs, vesicular exanthema, largely through raw garbage, led many countries to institute regulations forbidding the feeding of raw garbage to hogs, Although heat treatment (such as exposing the raw garbage to live steam for a half hour or more) cured the problem, the feeding of garbage to hogs decresed in the late 1950's and 1960's and is presently almost significant.

2. Open Dumps

Disposing of solid wastes in open dumps is the most common method. Much of the uncollected refuse is disposed in a similar manner. This method is cheaper than other disposal methods and have the advantage of avoiding the acute pollution problems associated with discharging wastes into waterways or polluting the air from incineration. However, open dumps produce health and air pollution problems and are not an acceptable method of disposal. They can cause public health problems by encouraging the growth of populations of flies (which can transmit typhoid fever, cholera, dysentery, tuberculosis, anthrax, and other diseases, rats (which can transmit plague, murine typhus fever, leptospirorsis, rabies, rickettsialpox, and other diseases), cockroaches, mosquitoes (which can transmit malaria, yellow fever, dangue, mosquito-borne encephalitis, and filariasis), and other pests. Air pollution problems arise when the dumped wastes are burned so as to reduce their volume and conserve space or when spontaneous combustion or arson leads to fires.

The blight of open dumps is aesthetically intolerable and depresses land values.

3. Sanitary Landfills

According to the Sanitary Engineers, sanitary landfilling is "a method of disposing of refuse on land without creating nuisances or hazards to public health or safety, but utilizing the principles of engineering to confine the refuse to the smallest practical area, to reduce it to the smallest practical volume, and to cover it with a layer of earth at the conclusion of each day's operation, or at such more frequent intervals as may be necessary." When properly run a sanitary landfill is odorless smokeless, and ratless.

In the area method of sanitary landfill operation the solid wastes are spread out over an area, compacted, then covered with soil or similar materials on the top and the sides.

In the trench method trench 10-15 feet deep and upto 20 feet wide is dug and the refuse is pushed into it, compacted by a bulldozer, and covered with dirt.

The area and the trench methods need about an acre to handle the trash and garage of 10,000 people for a year.

In the ramp method gulleys, ravines, or quarries are used, the solid wastes are dumped along the slopes, and covered with earth.

In a well-run sanitary landfill operation the soil covering has been deep enough to keep out rodents and flies, and it prevents the escape of noxious odors. A carefully sited sanitary landfill should not be near a community's ground water reservoir, because rain-water may leach pollutants into the water supply. Organic material in the fill decays and the internal temperature rises as high as 150-160^{0}F, high enough to kill heat-sensitive germs in the wastes. Gases such as carbon dioxide and methane are also generated, with peak production occuring within the first 2 years. As methane is highly combustible, a danger exists to buildings built over such landfills. Sanitary landfills can undergo settling for a long time, which also makes it risky to build too quickly over filled areas. Generally 90% of the ultimate settling happens in the first 5 years; the remaining 10% may take several years longer. Landfill areas are better set aside for playing fields, parks, arboretums, and the like, not used for building dwellings.

Inside the filled region, there could be aerobic reactions to the extent that oxygen is present and anaerobic reactions (with production of methanes, H_2S, CO_2, and water), both involving bacterial action. The inside temperature may rise to 70°C in a few days from the heat produced. The amounts of air and water present influence the speed of decomposition with wet landfills sometimes showing rather complete decomposition of organic matter within a few years, while dry landfills show little decomposition after several decades. Gas production can sometimes be dangerous.

The advantages of a sanitary landfill, as opposed to an open, uncovered dump, have been (1) the public health problems are minimized because flies, rats, and other pests are unable to breed in the covered refuse; (2) there is no air pollution from burning and none form dust or odors; and (3) fire hazards are very small; (4) there is danger of ground water or surface water pollution, however, if the landfill site is improperly chosen or if it is dug out too deep.

Landfills, if properly planned, can be used later for construction sites or recreational facilities such as parks and golf courses.

Land disposal of any kind requires large tracts of land. The increasing shortage of land and the ever-growing mountain of waste make the cost prohibitive to many communities. Leaching by surface water or ground water may contaminate streams and other water, especially in open dumps but also in landfills if the wastes are improperly covered.

Site Selection

Site selection for sanitary-landfill requires, in addition to technical engineering competence, attention to the economic, social and political aspects of land-use planning. When several possible landfill sites are being compared, professionals from these various fields can be used to evaluate independently the practicability of a particular location. A rational analytical site-selection methodology can employ a matrix-type analysis based on a weighted scoring system.

Planning and Land Use

Land-use plans and planning goals have been established by the appropriate local and regional agencies; they, obviously,must be consulted as the preliminary step in site selection. A local or regional master plan, for example, may preclude the use of an otherwise technically suitable size. Present and future zoning of adjacent land must be taken into consideration in determining the suitability of a site for receiving waste and its ultimate use. For example, the proximity of present or future airports has been a matter of concern since birds may be attracted to an operating landfill thus and endanger low-flying aircraft.

Climate, Topography and Natural Features

The site-selection process must take into consideration all climatic characteristics such as prevailing winds, precipitation, evapo-transpiration and temperature variations as they relate to the odors, dust, leachate generation, blowing litter, cover soil, erosion, etc. The effect of snow, ice and rain on the ability of vehicles to traverse access roads has been an important operating factor. Steep roads should be avoided as they also decrease vehicle maneuverability. The frost-penetration depth has been also an important factor in excavating cover soil. Frost- protection measures such as the use of insulating material (e.g., straw or leaves) to cover soil will increase operating costs.

The topography of the site is having important implications for landfill capacity, drainage, ultimate land use, surface and ground water pollution control, site access and related operating factors. Whenever possible, floodplain land must be avoided because of the high cost of preventing water pollution and because costs for protective measures such as dikes and levees have been high. Stormdrainage and flood records should always be consulted when choosing a potential landfill site.

Cost Considerations

The square-foot cost of the land has been secondary to the cost of the land per ton of waste disposed. Possible savings could accrue by leasing land from a private owner with a partial consideration being the ultimate improvement to the land. However, since landfills may need hundreds of years of maintenance due to settlement, methane-gas generation and leachate pollution, it may be desirable to provide for a perpetual private or public upkeep. A greater tonnage could be disposed at low costs into steep pits or canyons, rather than flat areas.

Development costs will normally include: clearing the timber and brush, diversion of surface drainge runoff or streams, providing the necessary utilities (remote sites may require electrical generators, chemical toilets, water storage, tank, trucks, etc.), fencing, access roads, buildings and scales. Haul distance is another major economic factor in site selection. Landfill sites should be as close as possible to the collection area.

Soil characteristics and cover-soil availability have been important, practical considerations. Importing cover-soil material may be hard to realize, except at great cost.

Public Acceptance

Public acceptance must be a constant key consideration both during and after the site-selection process. A site having minimal potential for public opposition, all other things being equal, should be chosen over a site generating considerable local controversy.

Site Planning and Design

Many of the basic engineering data requirements for good landfull planning, have been complementary to those for proper site selection. The optimum development plan, however, should include comprehensive site data, upon which would be developed the following facility plans:

(1) Detailed drainage facilities, all-weather access roads, fencing, buildings, sanitation, water, lighting, telephone, scales, equipment-maintenance yard and all other site improvements;

(2) A series of cut-and-fill plans presenting the landfill at various stages of completion;

(3) A detailed description of daily operating procedures;

(4) A phased analysis of equipment and manpower requirements for the life of the site;

(5) An estimate of projected capital and operating costs including final finished landfill landscaping improvements;

(6) An ultimate land-use plan for the site.

The engineers preparing the site-development plan should be knowledgeable of all applicable regulatory requirements. Necessary, statutory permits should be obtained early in the planning stage to avoid costly delays.

Site Improvements

The access roadway leading into the gate-house should have at least two paved lanes of sufficient width to accommodate heavy vehicles; grade of the access road should never exceed 8%. Lime, portland cement, or asphalt and oil binders are desirable road-surface materials that can help control dust in dry weather and prevent heavy machinery from being mired in wet weather.

Permanent or portable chain-link fences have been found to be suitable for protecting the site and catching windblown papers. An attractive entrance area having a lockable gate should be provided along with appropriate signs identifying the site and presenting pertinent information.

Drainage ditches to divert surface waters away from the site should be fully designed channels and appurtenant structures. Surface slopes of the completed landfill should not be less than 2% grade. Cut earthen slopes should not exceed 2:1, and local soil characteristics and rainfall quantities may dictate flatter slopes. Debris basins outside the fill area should be constructed to collect and pond drainage waters, thereby allowing suspended material to settle.

Appropriate sanitary facilities should be provided for landfill and other employees; toilets, sinks, showers, lockers, and lunch-room facilities are desirable amenities. Equipment storage and maintenance facilities should also be provided.

At some Sites, truck-scale facilities may be installed. The scale size and type will depend on local requirements. The scale should be capable of weighing the largest vehicle that may use the landfill; a 50-to- 60-ft platform is usually sufficient. Automatic printouts and computer data control have been at large sophisticated installations. Scale-house installations can be made portable for ready removal to another location.

Progress Plans

A series of plans should be undertaken for the phased landfilling program and include the following elements:

(1) a base map which shows the original topography and initial site improvements;

(2) several interim plans depicting the site topography, excavations and improvements at various construction stages; and

(3) a final plan illustrating the ultimate contours.

Appropriate survey-control monuments must be established to reference the horizontal and vertical locations. A time schedule should accompany the plans.

Equipment and Manpower Requirement

A minimum and two persons is desirable to work at a sanitary landfill for safety reasons.

Equipment requirements will vary with landfill characteristics. Typical solid waste-handling equipment used a landfills includes crawlers or rubber-tired tractors, together with accessories including dozer blades, buckets, multipurpose buckets, and landfill blades. More specialized heavy vehicles often used at larger landfilling include draglines, scrapers, steel wheel compactors, water trucks and other conservation equipment.

Cost

The site-planning and design process must consider capital, operating, and maintenance costs. In many ways, unit costs are affected by the volume of the landfilling operation. Obviously, with a greater population serviced, there have been increased weights of solid waste disposed daily. The economic ability to provide the site improvements listed above is also size-dependent in that small operations typically cannot afford more than minimum essentials.

The sanitary landfill has been generally a relatively economical waste-treatment and disposal process when compared with other acceptable processes and high standard of operation should be maintained.

Dry versus Wet, versus Cold

The planning and design factor mentioned above are on a general level. specific planning will also take into account a variety of local site conditions. Of particular concern are the operating requirements for extremely cold and wet areas. Extended low temperature operations will need the following procedures:

(1) Excavation and stockpiling of cover fill before frost conditions prevail;

(2) Placement on prepared ground of insulating material such as leaves, hay or straw to a suitable depth, as much as 3 ft;

(3) Excavation of trenches or cut/fill areas as needed, including insulated cover material;

(4) Storage of equipment in a heated shed;

(5) Equipping tractors and other landfill equipment with grousers if work on ice is necessary; and

(6) Provision of a heated cab for the operator.

Extremely wet areas should have paved access roads and suitable dumping area.

Environmental Factors

The critical environmental factors which must be considered in sanitary-landfill design have been vehicle traffic, leachate, gas production, odor, noise, aesthetics, air pollution, dust, fires, vectors and birds.

Leachate

Though some leachate has been produced by organic solid waste decomposition and compaction of wet refuse, the main factor causing leachate generation has been inflow of water. Excessive rainfall or other water-caused saturation of solid waste typically results in leachate generation and possible water pollution. Leachate may contain dissolved and suspended materials of all sorts, within the limits given in Table 5.20. Table 5.21. shows leachate constituents derived from organics and ash fill under several types of environmental parameters. Water inflow to a landfill can be reduced by the following measures:

(1) locate the landfill so as the avoid direct ground-water or surface-water contact;

(2) use proper surface-cover grading to increasing rainwater runoff;

Table 5.20
Range of Pollutant Concentrations in Leachate from Typical Domestic Refuse

Ion	*Concentration-mg/litre Range*
Iron	200-1700
Zinc	1-135
Phosphate	5-130
Sulfate	25-500
Chloride	100-2400
Sodium	100-3800
Nitrogen	20-500
Hardness (as $CaCO_3$)	200-5250
Chemical oxygen demand	100-51,000
Total residue	1000-45,000
Nickel	0.01-0.8
Copper	0.10-9.0
pH	4.00-8.5

Table 5.21
Constituents in Leachate from Refuse and Incinerator Ash. (Based on weight of refuse as received)

	Per cent leached					
Material leached	1	1	2	3	3	3
Permanganate value						
30 min	0.039					
4 hr	0.060	0.037				
Chloride	0.105	0.127		0.11	0.087	
Ammoniacal nitrogen	0.055	0.037		0.036		
Biologic oxygen demand	0.515	0.249		1.27		
Organic carbon	0.285	0.163				
Sulfate	0.130	0.084 (as SO_4)		0.011	0.22	0.30
Sulfide		0.011				
Albuminoid nitrogen	0.005					
Alkalinity (as $CACO_3$)				0.39	0.042	
Calcium				0.08	0.021	2.57
Magnesium				0.015	0.014	0.24
Sodium			0.260	0.075	0.078	0.29
Potassium			0.135	0.09	0.049	0.38
Total iron				0.01		
Inorganic phosphate				0.0007		
Nitrate					0.0025	
Organic nitrogen	0.0075	0.0072		0.016		

(3) plant final fill site-cover with vegetation which will avoid erosion while enhancing runoff and evapo-transpiration;

(4) use a none-permeable cover material with maximum lateral-flow characteristics;

(5) delay leachate generation by placing solid waste and soil cover with low moisture contents and high compaction densities; and

(6) install drainage facilities to divert off-site surface and subsurface waters. It is possible to determine the local rainfall and solid-waste parameters in order to calculate the estimated field capacity which will cause leachate generation.

If fill soil and solid waste receive enough water to reach their respective field capacities, leachate may start to be discharged. However, some localized leachate may get discharged due to concentrated channeling effects. If underlying solid or rock strata are impervious, this leachate will "rise" and discharge on the ground surface either at the landfill toe or somewhere down the slopes.

Gas

During the decomposition process, landfills generate several types of gases. During the first several days the decomposition process is aerobic, considerable heat get generated, and the principal gas produced is carbon dioxide. Once all the landfills' free oxygen has been depleted, organic decomposition then becomes facultative and finally anaerobic so that the principal gases generated are methane and carbon dioxide. Some hydrogen-sulfide generation has been noted in landfills which have substantial quantities of sulfate salts.

The impacts of carbon dioxide and methane upon the environment differ. Carbon dioxide is water soluble and can be absorbed into the leachate, increasing its corrosiveness and acidity. Methane diffuses readily and is explosive at atmospheric pressure in concentration from 5 to 15% by volume. Table 8.10 shows the physical properties of methane and carbon dioxide.

Gas control can be both difficult and expensive. Two common techniques used to control gas movement are clay barriers and gas vents. Gas vents have been necessary whenever the ultimate use of the landfill site involves covering with an impervious material.

Odors, Fires and Vectors

The problems of odors, fires, and vectors can all be minimized by proper soil cover. Earth mover-compactors should remove burning solid waste from the fill area and cover it with dirt immediately. Rats and other mammalian vectors are discouraged from feeding on or propagating in solid waste which receives a well-compacted 6 in. layer of low-clay soil cover daily. This type of soil cover also discourages bird foraging; crows and seagulls are especially likely to feed on uncovered solid waste. Flies, another potential vector, have difficulty emerging from well-compacted soil cover. Continuous grading of soil cover to fill in low spots is essential to prevent the development of stagnant pools of water in which mosquitoes may be able to breed.

Noise and Aesthetics

Routing vehicle traffic through industrial, commercial or lowdensity-population areas decreases the adverse noise impacts of landfill-related vehicle traffic. Construction of noise barriers surrounding the landfill, such as earth berms or walls, is good practice, and these with trees and planting can also improve landfill aesthetics.

Dust and Air Pollution

Both main and access roads should be of all-weather type, which should be sprinkled with water or another suitable dust contorl agent. Cover soils should be of a type not susceptible to wind erosion; be planted. Wetting to control dust should be limited to the working zones. To control noise and air pollution, all landfill equipment should be fitted with appropriate abatement devices.

Public Acceptance

Engineers must be prepared for the possibility of negative local public response to a sanitary landfill. A program designed to increase public acceptance for the landfill must be included in the early planning stages. An unnecessarily defensive attitude should be avoided, however, opposition to locating and operating a landfill can come from any of the population groups listed below:

(1) Property owners who believe a nearby landfill will lower their land values and social status.

(2) Citizens opposed to the expense.

(3) Citizens who resis change.

(4) Citizens concerned about conserving (not wasting) resources and natural areas.

(5) Neighbors concerned about traffic, safety and health, blowing paper, insects, rates, noise, unaesthetic landfilling and other environmental effects.

(6) Politically opposed voters.

(7) Citizens who believe a prior agency record of poor dump operations will continue.

Points to be stressed in any public-relations program should be geared toward technically responsible answers to the questioners. Tools that may be used to gain public acceptance are listed below.

(1) *Accurate information.* Sanitary landfills should be well run.

(2) *Comprehensive plans for regulating truck traffic.* The prospect of heavy truck traffic raises genuine and justifiable concern. Plans for regulating speed, weight loading, noise and exhaust odor should be given.

(3) *Increases citizen involvement.* Citizens advisory groups, workshops and civic group involvement will develop positive attitudes among residents.

(4) *Use of media.* Films or solid-waste management (available from federal, state and local agencies, as well as private industry), radio and television spots and news releases can be used to spread information to a large number of concerned citizens. Attractive landfill models. appropriate symbols and slogans can be skillfully used.

Operating Procedures

Method

Landfilling procedures are of two basic types: the area fill and its modifications, and the trench fill and its modifications. The trench fill has been common in flat or rolling areas. Solid waste is placed in excavated trenches, and the excavated earth is used for cover soil. The area method has been common in ravines, canyons, open pits, or similarly concave areas. Soil cover is obtained by cutting down the fill wall areas. The solid waste is confined by equipment compaction, and solid waste or cover soil admixture. In both methods, distance from the fill working face to cover-soil material must be minimized to conserve time and fuel. The new working face created each day must be kept to a minimum (as little as 20 to 30 ft) to conserve cover soil and reduce litter. The pattern of lift construction must be carefully scheduled to route vehicles over previously filled areas for additional compaction of the solid waste and soil (such routing should be avoided in wet weather, however).

Collection vehicles must place their loads of solid waste as close as possible to the fill working face. Solid waste should be spread and compacted on an inclined working face, with a slope of 20^0 to 30^0, in layers no more than 2 ft thick after spreading and compaction by more than 2, and preferably 5, passes of the dozer or compaction equipment. Unloading, spreading, and compacting must be directed toward maintaining thin, even waste and soil layers and realizing uniform cell heights, slopes and widths.

As the fill cell progresses, a minimum of 6 in. of compacted soil cover should be continuously placed so that only a minimum working face remain uncovered at any time. At the end of the working day, the remaining working face should be covered and the soil compacted with 2 or more passes of the equipment.

Indiscriminate salvaging and storage of solid waste (scavenging) should not be allowed at the landfill site, primarily due to health, safety and aesthetic consideration.

Special or Processed Wastes

Special Wastes

Crushable, bulky items, such as automobile bodies, can be deposited at the toe of the working face. Demolition debris is often directed to a separate storage area, used for access sub-base road construction, or placed directly at the toe of the working face.

Reinforcing bars, pipe, rolls of cable, and coils of wire often may become entangled in the landfill equipment and cause mechanical damage, or injure personnel. These entrapping and sharp objects should be covered immediately with dirt and solid waste. Dead animals should be kept in excavated holes and covered immediately; when they are received in large numbers, they may be buried in a specially prepared trench or pit. Rendering plants should be utilized to reclaim animal products whenever possible.

Processed Wastes

Landfills have been tending to utilize more shredded waste, baled waste, and residues from reclaimed processes. These processed wastes can greatly influence landfill practices, Mechanical shredding pretreatment includes cutting, tearing, ripping, and impact-shattering processes which reduce wastes to particles to a dimensional range of 2 to 6 in. Shredding may improve landfill aesthetics. Reclaimed solid-waste residues usually have most of the ferrous and other metal removed. Compost residue has a reduced biodegradable content.

Shredded wastes and other process residues should be handled in the same manner as in a normal sanitary landfill. However, aerobic decomposition can proceed rapidly in shredded wastes giving little methane and rapid settlement. Incinerator and pyrolysis residues must be handled carefully to prevent landfill fires due to hot embers in the residue.

High-density baled solid waste is produced by compressing solid waste into large blocks. Advantages to landfills receiving baled solid waste may include: reduced transfer, haul, and landfilling costs improved aesthetics; greater landfill solid-waste capacity due to higher in-place density; lower cover-soil requirements; reduced litter and vector problems; and considerably less fill settlement.

At a landfill disposing baled solid waste, bales are placed in uniform layers, 2 or 3 bales high per lift using a forklift or front-end loader. Plots of iso-concentration levels may assist in determining gas migration patterns.

Gas collection devices very from a simple "dry well" to an inverted gas capturing container. The most accurate instrument for analysis of landfill gases has been a gas chromatograph. Portable measuring devices, such as explosimeters, are used for rough in situ gas detection.

Other

In settlement monitoring, it is advisable to run a survey-control benchmark-elevation circuit around and within the site to provide necessary references for measuring fill settlement. A good method for measuring differential settlement is to use plates set within the fill connected to a vertical pipe or bar which extends to the surface. First year settlement may be as much as 20% of the initial fill depth and may continue at reduced rates logarithmically for many years.

Cover-soil erosion should be monitored if the ultimate use of the site will require vegetation cover. The burial locations of specific problematic or hazardous wastes within the landfill should be recorded as the landfill is being constructed.

Cost

Monitoring is relatively low in costs compared to the total solid-waste expenditures.

Final Land Use

The best final use of a landfill has been one that creatively takes full advantages of existing site conditions and landfill operating procedures to produce an optimum design which satisfies community needs. Ideally, the final land use should be planned prior to designing the landfill.

Construction of building on completed landfills is usually avoided due to potential problems from gas migration, corrosion, foundation-bearing inadequacies and settlement. Areas of undisturbed ground must be reserved for permanent structures. Extra cover-soil depth becomes necessary in areas in which utility lines, trees or deep-rooted plants are to be located. Existing trees on a site are a valuable asset if the fill can be designed to preserve them. The completed site must be attractively landscaped with native plants when ever feasible. Special assistance is usually needed when preparing a detailed landscaping plan.

4. Pyrolysis

Certain wastes can be pyrolyzed to provide chemicals. Pyrolysis has been as operation using intense heat to cause chemical changes but not combustion. Trash rich in paper (as most trash is) and other organic materials yields combustible gases when pyrolyzed and the gases can be burned for fuel. About 180 million rubber tires have been discarded annually. Firestone Rubber Company has found that the pyrolysis of rubber tires can yield two marketable products, a petroleum-like mixture of gases and liquid (55%) and a carbonized residue (45%).

5. Incineration

There have been several hundred municipal incinerators in use accounting for about half the tonnage burned. True incineration involves the burning of solid wastes at high temperatures, to leave ashes, glass metal, and unburned combustible amounting to perhaps one-fourth the original weight, which must then be disposed of in a landfill or other dump. Air pollution has been often a problem, because of their emissions of pollutants into the atmosphere. Incinerator technology has been developing in recent years, with air pollution control a particular concern.

The important factors to be observed have been the maintenance of a minimum temperature of 850°C in the combustion chamber to burn up the smoke produced, and the presence of grit extractor to remove finely divided dust before it gets discharged into the atmosphere.

Incineration reduces the volume of waste by 60 to 80 per cent and reduces the public health problems associated with the accumulation of refuse. This method is adaptable to a wide range of capacities from small domestic incinerators to large centralized municipal plants. In can handle a mixture of garbage and rubbish, and the physical nature of the residue-called clinker-aids in the disposal of the waste. Much of the heat that is generated can be recovered with modern equipment. However, other forms of pollution are created. Air pollution from the products of fuel combustion must be taken into account, but the principal air pollutant is fly ash. Several methods are used to reduce the emission of fly ash. One such method entails the use of water with devices called scrubbers; but the large amount of water with devices called scrubbers; but the large amount of water needed introduces a water pollution problem. The increasing quantities of plastic products and plastic packaging materials aggravate the air pollution problem due to the formation of gaseous decomposition products that will require research on methods for special treatment of the stack gas. The ultimate requirements have been difficult to predict.

Use of small incinerators

This method of disposing of refuse is widely used but has not been found wholly satisfactory. The reason is that many of the plants used are poorly designed and cause serious local air pollution. Many states have now laid down stringent design criteria for incinerators, and some ban small-scale incinerators absolutely. There is, however, no reason why small-scale incinerators should be inefficient, and there are many one the market now which are completely satisfactory. The important factors to be observed are the maintenance of a minimum temperature of 850°C in the combustion chamber to burn up the smoke produced and the presence of grit extractors to remove finely divided dust before it is discharged into the atmosphere.

The Use of waste disposal units

Waste disposal units in kitchens can normally deal only with some 10 per cent by weight of the refuse generated in a dwelling. There are considerable limitations regarding the types of materials that can be placed into them. Plastics, textiles and even fibrous food remains can clog them up-in fact these are potentially more harmful than glass bottles which are crushed quite satisfactorily. The ground up remain are then carried by the sewage system to the sewage works where they are removed in standard thickener equipment.

6. Controlled Tipping

By this technique, which is still by far the cheapest in many places, refuse has been permanently deposited on land, sealed in cells isolated from the surface with layers of earth. Controlled tipping is generally practised in conjunction with land reclamation schemes, where it serves the useful purpose of reclaiming marshes, exhausted quarries and gravel pits for agricultural, industrial and recreational uses. But, on the whole, the availability of areas suitable for such controlled tipping close to the big cities, has been diminishing. Furthermore, controlled tipping, unless very carefully carried out, can soon create a considerable nuisance and even a danger to health. Some local authorities demand that refuse is pulverized prior to disposal and also specify the minimum thickness of earth cover to be used.

Today controlled tipping is no longer regarded as good solution to the problem of refuse disposal for a number of reasons. There is far less heavy material like cinder and ash in refuse today than there used to be. Instead, refuse has more lightweight materials such as paper and, particularly, polythene film. In additions, plastics in the refuse do not rot and therefore remain unchanged in the soil for ever. This implies that the subsequent usefulness of the soil for agricultural purposes has been less than it used to be when all the components of refuse were able to rot fully.

Local authorities, with a ratepayers looking over their shoulders, have to decide which method of waste disposal they are going to select. From a very short-term point of view tipping usually costs somewhat less than incineration, but with the rapidly increasing cost of land this is unlikely to be so for very much longer. unfortunately, once the decision has been made to look at the problem from the short-term point of view, only the damage has been done and much of the land has been ruined for agriculture.

7. Pulverization of Refuse Combined with Controlled Tipping

Pulverization of refuse prior to controlled tipping obviously adds to the cost of refuse disposal, but has many advantages:

(a) There is far less likelihood that pieces of rubbish below away, as happens with normal tipping practice;

(b) Less earth cover is required than with unpulverized waste;

(c) Rats, mice and insect larvae get destroyed;

(d) The rate of rotting is very much accelerated, so that less time elapses before the material has been converted into humus.

Three types of pulverizing machine are in common use:

Rotating Drum Machines

These pulverize the refuse by attrition. Water is added to the crude refuse and the mix is then passed into the revolving drums which have either a circular, an octagonal or a hexagonal cross-section. The drums operate either on a batch or on a continuous basis, but the number of revolutions and the time it is necessary to have the refuse in the revolving drum to obtain good disintegration vary according to the detailed internal design of the plant.

There has been usually a separating device for materials which cannot be broken down-mainly plastics, rubbers and metals. The final material that comes out of the drums normally has a moisture content of 40-50 per cent and a density of 0.5-0.6 kg/cm³, much higher than the density of the refuse originally treated. Such material can be tipped more easily and more satisfactorily than typical low density town refuse with its large quantities of bulky and loose materials. The disadvantage of wet grinding is the fact that the large quantity of added water increases transportation costs. Drumtype pulverizers usually have capacities of about 10-12 tonnes per hour and require about 35 kW of power to operate. Magnetic separators have been always provided to remove iron and tin cans.

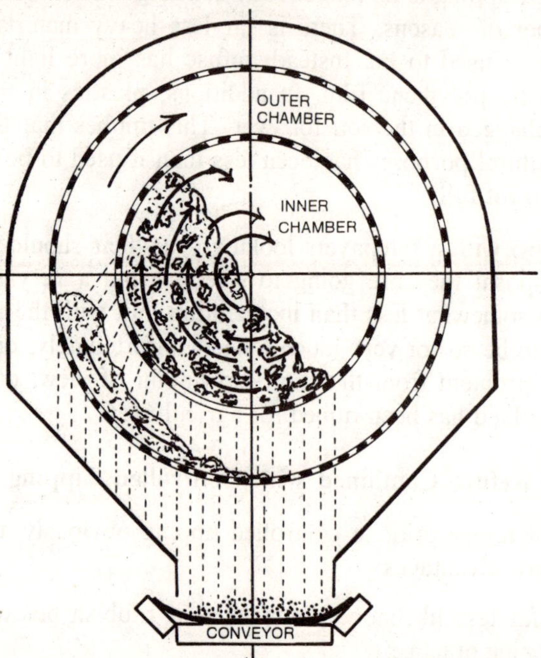

Fig. 5.1. Rotating drum pulverizer.

Hammer Mills

Hammer milling is operated either as a dry process or as a wet one, water being added to produce a denser product which is more suitable for tipping. The mills themselves vary considerably; there are both fixed hammer and swing-hammer types of mills, and either single or double rotor models as available. The biggest hammer mills are able to process up to 30 tonnes of refuse per hour.

Hammer mills consist basically of a horizontal rotor, which carries a number of swinging or fixed hammers. There are impact plates and grates through which the processed refuse material is finally ejected. Fore refuse treatment the grate bars, which determine the final size of the pulverized particles obtained, are rather far apart so that on average the maximum size of the pulverized refuse is about 5.8 cm in diameter. Feed openings with many of the plant used for refuse pulverization are up to 1.2m x 2.3m, so that even such large items as old chairs and old sofas can be conveniently torn up into sizes suitable for controlled tipping. Metals are also fed into the pulverizers and are shredded without difficulty. Iron and steel components are recovered on the discharge conveyor belt by magnetic separators. Hammer mills require very powerful motors, to cover instantaneous peaks when they are fed with large awkward units such as refrigerators or old gas cookers. Motors with a capacity of 150-200 kW are recommended. On average the power consumption is fairly modest, amounting to about 4 kW per tonne processed per hour.

The Dutch Rasp

This consists of a steel drum which has a perforated end, which is fitted with hardened steel pins. The refuse is led past these steel pins and the ground-up components drop through on to a conveyor belt beneath. Hard and bulky objects which cannot be reduced in size in this way finally come out at the side. Plants are made capable of dealing with 10 tonnes of refuse per hour; they have a power requirement of about 60 kW and operate at about 5 revolutions per minute. The disadvantage of the Dutch rasp is that the plant is unable to deal with harder substances and is consequence there is a fairly high proportion of material which does not pass through the size reduction process.

8. Ocean Dumping

Ocean dumping of wastes, from garbage to radioactive wastes and discarded chemical warfare agents, has been, in the past, a favorite disposal method under the prevalent notion that the ocean in an inexhaustible sink. We are now learning that the ocean does not have an infinite capacity for covering, holding, absorbing, or decomposing materials, nor the capacity for anywhere near the load of pollution that would result if the rate of ocean disposal were continued as in the past.

A study made in 1968 identified 126 ocean disposal sites used by twenty coastal cities. There were seven major categories of waste: harbor dredgings, industrial wastes, municipal sewage sludge, refuse and garbage, construction and demolition debris, military explosives and chemicals, and miscellaneous wastes-a total of 38 million tons per year in coastal and estuarine areas. Other resources of waste include agricultural land run-off, pesticides, fertilizers, sewage and garbage from vessels and oil spills. Often overlooked is the fact that a large part of the air pollutants eventually settle out and end up in the ocean. A major concern is that ocean dumping has increased fourfold during a period of twenty years and is expected to greatly increase relative to other means of disposal in the future. Based on a 1970 study, the Council on Environmental Quality recommend a series of drastic actions to control ocean dumping.

Little is known about either the immediate result or the cumulative effects of ocean disposal. By the time adverse effects become evident, it may be too late to discontinue the disposal practices in time to prevent long-term pollution problem. We have a preview of what could happen in the case of DDT, plasticizers such as the polychlorinated such as the polychlorinated biphenyls (PCBs), and more long-lasting and potentially dangerous substances such as lead, mercury, cadmium and selenium. The amount of wastes transported and dumped into the ocean is now relatively small compared to the total volume of pollutants that reach the ocean. Industrial and municipal liquid wastes are the largest source of pollutants.

The oceans cover 70 per cent of the earth's surface. The 140 million square miles of water surface are crucial to the equilibrium of the world's atmospheric environment-the oxygen and carbon dioxide balance and the global climate. The oceans are a major part of the earth's hydrologic system. They are an important part of the biosphere, supporting a rich biota of marine plant and animal life, and are economically valuable to man for the production of food, minerals and other necessities.

Some coastal cities dump solid wastes into the ocean. New York formerly dumped wastes off the New Jersey shore but the State of New Jersey in 1933 won a Supreme Court decision forbidding this practice. New York still dumps wastes over an area several square kilometres in size 20 km out into the Atlantic Ocean; this area has been described by critics as a "dead sea" of muck and black goo, largely because of the sewage sludge disposed there.

The solid and liquid wastes are dumped into the ocean by means of barges and pipes from shore. Raw sewage and other wastes pumped through underwater pipes to offshore locations are included among materials dumped into the ocean, but only in a broad sense of the term.

Dredge spoils make up the largest single category of wastes numped into the ocean, amounting to 6-7 times as much tonnage as all categories combined. Dredge spoils are materials removed from a channel to maintain or improve navigation, and they consist of rocks, detritus, clay, silt, sand, and municipal or industrial sludges.

In 1968, nearly 90% of the 4.5 million tons of sewage sludge deposited in the Atlantic came from the New York area, and nearly all the rest came from Philadelphia. Sewage sludge is a by-product of the treatment of waste water. Most sludge if either incinerated or dumped on land.

Industrial wastes include many heavily polluted liquids such as waste acids, refinery wastes, paper mill wastes, pharmaceutical wastes, pesticides, and many others. Those that are especially toxic are packed into drums that are dumped well offshore into deeper zones. The U.S, Army Chemical Corps once disposed of some poison gases by loading them into an old cargo ship and sinking the vessel into a deep ocean trench. Some radioactive wastes used to be dumped into the ocean, but that no longer is done by the United States. Some industrial wastes are burned in special incinerator ships cruising offshore.

Construction and demolition debris, a minor fraction of all wastes dumped into the ocean, consists of building stone, masonry material, pipe, wood, and cellar dirt for which on-land disposal sites may not be economically accessible.

Solid wastes-garbage, trash and other refuse picked up by trash removal operations in coastal cities-are not disposed of in the ocean to any great extent.

Problems of Ocean Dumping

Very little is confidently known about the effects of ocean dumping on marine life, but much is suspected and feared. Because the wastes include organic matter that is degraded by organisms requiring oxygen, they cause some oxygen depletion in the waters at the dumps sites. The water column at the dump site may uninhabitable by fish or shellfish, but natural mixing action and surface aeration eventually restore the oxygen. Solid wastes that are not biodegradable change the habitats for fish in the dump areas. The most serious threat to fish and shellfish is the presence of a number of toxic metals, pesticides, and other poisons such as the polychlorinated biphenyls (PCBs). Sediments in the New York Bight, for example, contain abnormally high concentrations of copper, chromium, lead, zinc and nickel. Little is known of the long-range effects on marine life of exposure to toxic substances.

No part of the New York Bight where ocean dumping occurs is a biological "desert", a region devoid of marine life. Plankton, bottom-feeding organisms, and commercial fish and shellfish are present, but fin rot disease occurs among several species. Fin rot disease is a progressive erosion of the fin rays and overlying tissue starting at the outer edge and working to the base of the fin. The disease itself is a symptom that something is wrong with the water.

Sewage sludge and polluted dredge spoils usually contain bacteria, viruses and intestinal parasites. These are taken up by shellfish and consequently parts of the shellfish-producing wastes of the American coastline-about one-fifth of the 10 million acres are closed to shellfishing. Residents along Long Island Sound fear that bacterial contamination of waters off their beaches is occurring. Ocean currents circulate dumped material, but they do not carry the wastes in huge globs. Some of the Long Island beaches have been fouled, but the origin of the pollution has not been proved.

In 1972, a multi-government conference agreed to an International Convention on the Prevention of Marine Pollution by Dumping of Wastes.

The EPA assumes that toxic wastes must be mixed and diluted, and the establishment of criteria on levels of toxic substances allowed in the ocean-dumped wastes has been intended to prevent changes in the ocean beyond those that might be expected from natural causes. For synthetic pollutants the criteria are designed to limit the concentration in the waste to 0.0001-0.01 of the lethal concentrations and in the ocean these levels would fall to much lower levels because of mixing.

A total prohibition has been placed on dumping biological, chemical, and radiological warfare agents, high-level radioactive wastes, and substances that are persistent, nonbiodegradable, or naturally floating (e.g., plastics). Solid wastes may not contain more than 0.75 milligram per kilogram of mercury or 0.6 milligram per kilogram of cadmium; and liquid waste are limited to 1.5 and 3.0 milligrams per kilogram of mercury and cadmium, respectively. Oils and greases may not produce a visible seen of the surface at a dilution of 1 : 100. Organohalogens (e.g., certain pesticides and the PCBs) are also sharply limited. Other strictly regulated wastes requiring special handing are those containing arsenic, beryllium, chromium copper, lead, nickel, selenium, Since many of the larger cities and metropolitan areas of the world are on near the coastlines, the ocean offers a ready disposal site for certain types of wastes when inland or costal resources are difficult to obtain. We are therefor concerned with finding a methodology for solid-waste disposal at sea which meets the following criteria:

1. The port facilities must efficiently handle a large number of trucks and also provide docking services for the sizable barges; material storage, reduction by grinding, baling, and encapsulation at these facilities must be donc in an efficient, nuisance-free manner.
2. The transport of the compacted wastes through harbor waters must be in a secure manner: refuse shall not fall off even during period; and
3. The disposal at sea must be in an approved area; the bales must sink to the bottom and stay there, with no breaking off and flotation of particles; adequate electronic fix and photographic records of location and depth, as well as date and time of day, must be kept. Provision must be made for frequent or constant monitoring.

Disposal costs at sea, conforming to the above criteria, have been higher than for sanitary landfill but less than for incineration. The method is therefore attractive.

The situation as it now stands with regard to disposal at sea could be summarized as follows:

1. Specific guidelines, criteria, or standards governing disposal at sea are not available at this time. Interim guidelines announced by several agencies specify a moratorium to develop pilot programs and monitoring systems.
2. Dumping will probably be restricted to designated areas beyond the continental shelf. The water depth will be at least 1000 fathoms (6000 feet) and the site would be at least 25 miles from the nearest land.
3. Disposal of solid wastes at sea would be restricted to inert material which have a bulk density greater than sea water. (At 1000 fathoms this is about 70 lb/cu ft.) All other solid wastes having organic materials shall be packaged or containerized, transported, and deposited at an approved site using methods designed to protect the public health and the marine environment.

Ocean as a Resource

The ocean is a ready dumping ground for waste materials from the large cities located on or near the coastlines. Sewage sludge, so-called acid and industrial wastes, and explosives get dumped for years and are continuing to be disposed of at designated and approved locations (Table 5.22).

However, the concern with the ecology of the sea and of the coastal zone and the recent activities of aroused citizens are bringing these practices under sharp scrutiny. There is a gradual diminution of the life activity as the depth increases; the temperature at 6000 feet is steady at about 4°C, and the pressure is over 2500 Ib/sq in. Since they are the bottom of the food chain, they are of vital concern, particularly in the long-range aspects. There must be no doubt that ocean dumping will have no harmful effects on the ocean ecology.

Conversations with biologists working in the marine environment disclosed some interesting facts of solid-wastes dumping. The submarine "Alvin" got accidentally sunk a few years back in a depth of about 4500 feet. When it was recovered about one year later, it was found that the apple and the bologna sandwich on the mess table had not deteriorated. This points up the theory that bacteria functioning under these temperatures and pressures are not amenable to such land-based foods. We except that the methane producging bacteria already in the solid waste will continue to function, but at a greatly reduced rate.

Table 5.22
Number of Marine Wastes Disposal Areas (By Region and Waste Type).

Type of Waste	Disposal Areas			
	Pacific	Atlantic	Culf	Total
Dredge Spoil	10	11	17	38
Industrial Wastes	7	9	6	22
Sewage Sludge		2		2
Refuse	3			3
Radioactive Wastes	6	11	1	18
Explosive Wastes	14	16	10	40
Total	40	49	34	123

The ocean has been a resource which should be used. Of course, this use cannot imperil the ecology of the marine environment.

Possibility of Dumping Inside the Continental Shelf

Consideration has been given to disposing of baled refuse at locations closer in shore. The possibilities of contamination and damage to sea life are greater at such locations. Nevertheless, properly anchored bales could provide artificial reefs so necessary to the improvement of fish habitats.

The moratorium previously mentioned relates not only to dumping on the continental shelf but also to dumping in the deep ocean (depths of more than 6000 feet). It is interesting to note that the distance from Boston, Massachusetts, to the nearest possible disposal area off the continental shelf is about 230 miles. The distance to the continental shelf from other East Coast metropolitan areas in much shorter-in most cases it is under 100 miles.

COMPOSTING OF REFUSE

Abstract. It is biological process for converting organic solid wastes into a stable, humuslike product whose chief use has been as a soil conditioner. Modern composting has been aerobic and combines mesophilic and thermophilic temperatures. As a biological process it has been subjected to the constraints of all biological activities, i.e., limitations imposed by microbial population and genetic traits, and by environmental factors. Costs have been comparable to those for incineration.

Definition. Simply states, composting has been the biological decomposition of the organic constituents of wastes under controlled conditions. The term "decomposition" is employed instead of "stabilization", because when it has been applied in practical usage, the process rarely is carried on to the point at which the waste gets completely stabilized. In fact, the cut-off point may at times by short of "temporary stabilization". The term "biological" distinguishes composting from other types of decomposition, like chemical or physical (e.g., incineration, pyrolysis, etc.). "Organic" has been applicable because, with few exceptions, only the organic portion of wastes has been subject to biological breakdown. A very important term in the definition has been "controlled". It has been the application of control that distinguishes composting from the natural rotting, putrefaction, or other decompositions, that occurs in an open dump, a sanitary landfill, a manure heap, an open field, etc.

Classification. Compost systems could be classified on three general bases, namely, oxygen usage, temperature, and technological approach. If oxygen usage has been the basis, the division has been into aerobic and anaerobic. When temperature forms the basis, the division becomes mesophilic and thermophilic. Finally, if we use technology as the key, the classification has been into open or windrow and mechanical of "enclosed" composting.

Aerobic Composting takes place the activity of aerobic microbes, and hence the provision of oxygen during the composting process. The opposite prevails in anaerobic composting-i.e., anaerobic bacteria accomplish the decomposition and oxygen (air) has to be excluded from the composting mass. Aerobic composting generally gets characterized by high temperatures, the absence of foul odors, and has been more rapid than anaerobic composting. Anaerobic decompositions gets characterized by low temperatures (unless heat is applied from an external source), the production of odorous intermediate (reduced) products, and generally takes place at a slower rate than does aerobic composting. The advantages of aerobic composting over anaerobic composting arise due to the differences in their characteristics-in other words, aerobic composting has been more rapid, allows high temperatures, and has been not odorous. The main advantage in anaerobic composting has been that the process can be carried on with a minimum of attention, and as such it can be sealed from the environment. Due to its many advantages, most modern compost processes have been basically aerobic.

The mesophilic composting takes place at intermediate temperatures (15°C to 25°C), which is most cases has been the ambient temperature. Thermophilic composting has been conducted at temperatures from 45°C to 65°C. In practice, most processes include the two ranges.

Compost systems falling under the category of "open" or "windrow" have been those in which the entire process is performed in the open. The material is usually stacked in elongated windrows. In mechanical systems, on the other hand, the greater part of the initial composting activity occurs in an enclosed unit, the digester. It is to be remembered that most mechanical process involve windrowing towards the end of the process to permit the composting material to "mature".

General Constraints due to Biological Nature. A point of major importance in designing, planning, and evaluating a compost process as a waste-management device has been that as it is a biological operation, factors and requirements peculiar to the maintenance of biological activities in general also influence the compost process. Therefore, composting has been subject to well-defined biological limitations: (1) a suitable microbial population must be present; (2) the rate and efficiency of the process have been its functions of the rate and efficiency microbial activity; (3) the capacity of a given operation has been limited by the size and nature of the microbial population; (4) the substrate subject to composting generally must be organic; and (5) environmental factors are of key importance.

The origin of the first constraint is quite obvious. As the process is a biological one, living organisms have been the agents for accomplishing it. The limitation to microbial organisms has been indirectly implied by the term "decomposition", in that in nature decomposition in the commonly accepted sense of the term has been carried out by microscopic organisms. Not only must a microbial populations be present, but it must also be one that gets suited to the task.

The limitation on rate implies that no matter how well a piece of compost equipment has been designed mechanically, composting will take place at a pace commensurate with that of the bacterial activity permitted by the particular set of conditions provided by the machine. The microbial limitation on capacity is also having practical ramifications. It implies that once the maximum-sized microbial population gets reached under the conditions provided by a given piece of compost equipment, the loading cannot get increased without giving rise to nuisances. Loading in excess of the maximum permissible will inhibit the process either partly or entirely, and at the least, will cause only partial treatment of the waste material.

The constraint arising from the influence of environment factors generally manifests itself through the mechanism of the limiting factor. If any factor has been present in less than optimum concentration or level, the functioning of the entire process gets inhibited in proportion to the extent of the deficiency of the factor. For example, for a given set-up in which the temperature has been optimum, aeration has been complete, and a suitable microbial population has been present, but nitrogen has been deficient, the only way to increase the efficiency of that set-up would be to add nitrogen in an amount sufficient to make up for the nitrogen deficiency. If we add more bacteria it would only aggravate the problem, because the nitrogen is not sufficient even for the existing population. Building a more elaborate digester would not suffice, since the effective components of the process, namely the bacteria are not having the nitrogen to meet their metabolic requirements.

Microbiology

Utility of Isolates. In general, it can be safely enunciated that there is an overall aura of futility in attempts to isolate and identify all of the organisms to be found in a mass of composting material, at least down to species of even genera. The reasons have been many. An important one has been the difficulty of knowing with certainty that an isolate has been the correct one or has been "artifact" due to the isolation technique. An "artifact" has been an isolated organism or organisms other than the one sought for. Such an even could take place through the use of a medium not suitable to the organism or organisms most responsible for the breakdown. By the same token, simple because a particular group gets isolated in greated numbers than other groups does not imply that it is actually more numerous than the others; certainly it has been no proof that it has been the most active. The nature of the medium has been not the only factor in promoting. the error; the lack of interaction of other types of microorganisms in an isolation culture has been importance. The absence of competition under the isolation conditions combined with the use of a favorable medium may help an organism to flourish. Such activity, however, in a compost pile actually exerts only an insignificant effect on the process. In fact, most of the isolation procedures have been at least in part enrichment techniques or modifications thereof to encourage the growth of one organism at the expense of others.

If one is sure that he has correctly identified the organism in a given compost heap and has unmistakenly assigned a proper role to each of the isolates, one has no assurance that the same overall microbiological set-up would exist in another operation in a different locale. A number of factors could play part to bring about a difference. The type and proportions of the wastes in the second operation might be different. This would imply the existence of a different substrate and hence a different population composition. Physical conditions could differ e.g., the moisture content may be significantly higher, the ambient temperature lower, etc. Another possibility has been that the indigenous populations may differ.

Isolating and identifying organisms would be useful because if one knows the important microorganisms in a compost pile, one would be able to determine the optimum growth conditions for them as well as their growth kinetics by examining pure culture. With this information, one is then able to design the plant and operation accordingly. However, all one learns from pure-culture studies is the nature of the optimum set of conditions for a given organism in an "artificial situation," which may or may not be the same as that which might occur during the compost process in which interaction which other organsims has been an important factor. It means that, the ecology in the actual application is not taken into consideration. Furthermore, the optimum conditions for the key organisms in their "natural" environment, i.e., in the compost pile, could be more readily and more accurately determined by working with the material as it comes with its indigenous microbial population. The optimum conditions have been those at which decomposition takes place most rapidly and in the desired manner. This can be performed without identifying a single organism. The resulting set of conditions would represent the integration of all factors in situ.

Inoculums. Sometimes, a proposal for commercial composting includes the use of some special inoculum, the composition of which would be known only to the discoverer who claims it to be fundamental to the successful operation of the process. Some such Inoculums would be as mixtures of several pure strains of laboratory cultured organisms especially prominent in the decomposition of organic matter and in nitrogen fixation. Others have been purported to contain, as well, such things as "enzyme system", "hormones", "preserved living organisms," "activated factors", "biocatalysts," etc.

The idea of inoculation is applied to the finished compost. The addition of organisms is supposed to be of inestimable value to the soil. While it has been true that certain bacteria promote solid fertility, the mere addition of such organisms is of no use if environmental conditions are not appropriate in the soil. On the other hand when environmental conditions have been suitable, the addition of the organisms would be superfluous as the native population would already be present in more than ample numbers.

The number of bacteria has been rarely a limiting factor in composting, because bacteria have been always present in great abundance on all exposed objects, especially on municipal refuse. They could be eliminated only by a drastic measure like sterilization at high temperature and pressure. When environmental factors become appropriate, native bacteria, because they have been better adapted than forms attenuated under laboratory conditions, multiply rapidly. Environmental conditions would then govern the rate of composting. A large number of enzymes involved in decomposition, as well as the difficulty and expense involved in isolating and synthesizing them, would be making composting with enzymes alone highly impractical even were such a preparation available. The addition of enzymes to raw refuse has been not necessary, however, because bacteria are able to synthesize efficiently and rapidly all of the enzymes required. For a time, "hormones" was popularly used to designate growth factors and vitamins needed by bacteria. However, the organic constituents of refuse have been containing all the growth factors and vitamins needed for normal growth.

Inoculation would be important to the composting process only if the bacterial population in any emerging environment were not able to develop rapidly enough to take full advantage of the capacity of the environment to support it. Such a case would cause a time lag, which could be overcome by supplementing the initial population indigenous to the refuse. Composting has been a dynamic process, which is representing the combined activity of a wide succession of mixed bacterial and fungal populations associated with a wide succession of environments, one overlapping the other and each emerging gradually because of continual change in temperature and substrate. The substrate changes have been due to a progressive breakdown by bacteria of complex foodstuffs into increasingly simple compounds. It is also found temperature increases steadily in proportion to biological activity, so that initial mesophilic conditions soon get superseded by thermophilic ones. As the process has been dynamic and any individual group of organisms can be able to survive in a rather wide environmental range, one population starts to emerge while another is flourishing, and yet another is disappearing. Inasmuch as any group of bacteria has been capable of multiplying at a pace equal to that of its developing environment, the addition of similar organisms as an inoculum would become superfluous. The microbiologist usually thinks of inoculation as minute additions, while the sanitary engineer thinks of what may be termed as "mass inoculation."

Types of Microorganisms

Considerable effort is done in identifying organisms which occur in composting materials. In the various studies, certain groups of microorganisms are found to be apparently associated with different stages of the process.

Composting of refuse occurs in two ways, which are described as aerobic or anaerobic according to the biochemical nature of the bacteriological processes involved.

In *aerobic* decomposition living bacterial organisms utilize oxygen to feed on the organic matter in the refuse, and produce their own cell protoplasm from the carbon, nitrogen and phosphorus present. Much of the carbonaceous matter is converted to CO_2 and an appreciable amount of the rest gets changed to simpler compounds which are extremely useful to plants for growth. Aerobic fermentation occurs very widely in nature and is the main way in which waste products from field and forest are converted into humus.

In *anaerobic* fermentation, the process of decomposition occurs in the absence of air. Any carbon present in the materials which ferment is not converted into CO_2, as happens with aerobic fermentation, but into methanes. This process occurs in nature at the bottom of marshes. Sulphur is not oxidized to sulphuric acid as happens with aerobic fermentation, but to evil-smelling hydrogen sulphides and mercaptans. Anaerobic fermentation produces far less heat than does aerobic fermentation. During the decomposition of carbohydrates anaerobic fermentation only offers about 5 per cent of the heat which aerobic decomposition of the of the same quantity of carbohydrates would provide. This is having certain disadvantages from the public health point of view. The large quantities of heat evolved during aerobic fermentation tend to kill many harmful organisms. This does not occur during anaerobic fermentation. Furthermore, aerobic fermentation is odourless, but anaerobic fermentation is not.

Nature is Material Suitable for Fermentrtion Treatment

The most suitable materials for fermentation treatment are, above all, human and animal excrement, as well as slaughter-house waste; these substances contain very high proportions of fermentable matter. Farm residues also have a very high proportion of fermentable material. The following have been typical compositions:

	Human or animal excrement %	*Slaughter-house waste %*	*Farm waste %*
Moisture content	66-68	75-80	10-60
Composition calculated on a dry basts			
Organic matter	88-97	80-95	25-35
Nitrogen	5-7	8-11	0.4-0.8
Phosphorus (as P_2O_5)	3-5	3-3.5	0.2-0.5
Potassium (as K_2O)	1-2.5	2-2.5	0.8-1.5
Carbon	40-55	14-17	12-17
Calcium (as CaO)	4-5	3-3.5	4-7.5

Sewage sludge also has the sort of composition which makes it eminently suitable for fermentation treatment. The following are the ranges of composition of different types of sewage sludge, calculated on a dry basis:

	General %	*Domestic %*
Organic content	35-85	75
Inorganic content	15-65	18
Nitrogen	1-10	7
Phosphorus (as P_2O_5)	2-4.5	3
Potassium (as K_2O)	0.1-4.5	4

Sewage sludge has been best composted with normal town refuse.

Industrial wastes have been found to vary considerably in their composition, and the decision as to whether any specific group of wastes is suitable for composting has to be made on an individual basis. The best kinds of wastes for such purpose have been again animal residues, but waste products from woodworking process have been also suitable for incorporation into fermentable matter.

Normal town refuse has only about 55 per cent of compostible material on a dry basis, although this varies from place to place and from season to season. In technologically advanced countries the amount of compostible material collected per head per year is about 100-105 kg.

Methods of Composting

Separation of refuse into compostible and non-compostible. The first stage in the preparation of refuse for composting has been segregation of its different constituents. Tin cans and other iron objects are first separated by magnetic devices. Such objects as rags, bottles and non-ferrous metals can then be picked out by hand. Often the segregation of refuse into compostible and non-compostible components can be carried out at source; for example, in schools and hospitals, food remains, which are eminently compostible can be separated from tins and bottles, which are not. Such separation is not practicable with normal domestic refuse collection.

Shredding. Once most of the non-compostible materials have been removed, the rest is shredded. This is necessary to give the material a sufficiently large surface are for bacterial attack to proceed readily. Shredding normally reduces the size of the refuse components to about 20-5 mm diameter. Hammer mills are usually employed for this part of the process. In some cases much finer grinding is practised.

Blending for composting. A very important factor for satisfactory composting is the ratio of carbon to nitrogen- the C/N ratio. If the C/N ratio is much above 35 the material composts only very slowly. The optimum C/N ratio is about 27; ;below this ratio, the speed of fermentation is slow because there is inadequate carbon present to provide the energy for the chemical changes induced in the nitrogen-containing compounds.. Above about 15, the final humus is of poor quality.

For this reason materials with a poor nitrogen content (waste plant products, sawdust, etc.) are blended with substance that have a very high C/N ratio (human and animal excreta, slaughter house remains, fish scrap, etc._. Soil is often added to the composting materials to provide a batter 'body', although this is not done universally. The moisture content is usually adjusted to about 50 per cent, but it has been suggested that this limit depends mainly upon the nature of the material which is being composted. If quantities of straw are present, which stop the composting materials from settling too much so limiting oxygen available, the higher moisture content may be higher than if the composting materials have a tendency to become soggy.

The optimum composting temperature used has been around 60°C. Temperatures above 50°C must be maintained in order to destroy eggs and cysts of harmful organisms such as the pathogenic roundworm Ascaris lumbricoides. Aeration of the composing substance is most important; this is usually carried out mechanically. The main organisms which permit composting to take place are known by the following Latin names.

Streptomyces species,

Micromonospora species,

Penicillium dupontii,

Aspergillus jumigatus.

Composting effectively destroys virtually all harmful pathogens and parasites, as hardly any of them are able to survive ambient temperatures in excess of 50-60°C for any length of time. Aerobic composting effectively destroys the eggs, larvae and pupae of insects, and leaves the processed material in a fit state to be used for fertilizing the soil. The average time needed for composting is about 1-6 months with traditional field methods. With commercial processes using mechanically aerated digesters the period needed for full composting is markedly less-between 3 and 7 days only are normally required. A typical final compost will have a composition in the following range:

Organic matter	25-50%
Carbon	8-40%
Nitrogen	0.5-3.5%
Phosphorus (as P_2O_5)	0.5-3.5%
Potassium (as K_2O)	0.5-2.0%
Calcium (as K_2O)	2-7%
Incombustible matter	20-60%

The Dano Composting Method

The Dano method has been a composting process where the wastes have been kept in continuous motion, and an advantage of this process is that the waste material does not require any preliminary preparation such as pre-crushing since this is achieved within the system.

There have been many types of this plant throughout the world.

Composting process. The composting process has been carried out in units known as Dano Bio-Stabilizers. A Bio-Stabilizer is a horizontal cylinder which rotates at a maximum of one revolution per minute, and the size of the cylinder is determined by the amount of refuse to be composted. Usually the cylinder is large enough to retain refuse for a period of approximately three days, during which time biological oxidation of the organic matter takes place. The discharge of compost and feed of refuse occurs continuously through out the operating period.

The abundant present in the refuse of all types of micro-organisms conducive to the decomposition of organic matter, together with regulated moisture control and addition of air, leads species of these organisms to commence aerobic decomposition of the easily decomposable organic carbon compounds. During this initial decomposition period heat is released and an increase in the population of the organisms is achieved together with further release of heat.

The first phase of fermentation produces temperatures up to approximately 40-45°C and is called the mesophilic phase. Whilst this process is going on the Bio-Stabilizer constantly rotates and refuse is broken down physically quite rapidly by attrition, thus improving the conditions for efficient fermentation. When the temperature of the fermenting mass rises beyond 45°C the activity of the mesophilic organisms ceases and thermophilic micro-organisms develop, thc maximum activity of which is in the range 45-55°C.

Decomposition of the refuse produces mainly CO^2 and water vapour, which are extracted and passed through a washing device before discharge to the atmosphere.

The fermentation process has been controlled by aeration of the mass in various sections of the Bio-Stabilizer through a lowpressure blower and valves which are fitted along the length of the cylinder.

Not all of the organic matter gets decomposed in the Bio-Stabilizer, however, and the more resistant types of carbon compound are decomposed during a secondary composting stage in windrows. The rotation, pulverization and fermentation in the Bio-Stabilizer, however, cause a rapid and considerable volume reduction of the refuse to approximately one-third of its original volume. During the fermentation process in the cylinder a loss of weight takes place of the order 5-10 per cent due to the evaporation of moisture and the reduction of carbon matter to CO^2.

When the process in the Bio-Stabilizer is complete the material is discharged and transferred to screens where the larger inorganic fraction is extracted.

Secondary composting process in windows. It is possible to use the 'fresh' compost, when screened, immediately under certain conditions, providing necessary precautions, are taken to prevent nitrogen depression of plants. However, to produce material which is uniform in quality it is necessary to mature the 'fresh' compost further. The 'fresh' compost is stored in long windrows approximately 1.6 m high on a drained hard storage area, and these windrows are turned approximately every three weeks by a special turning machine which also adds moisture to compensate for loss through evaporation. After approximately three or four turnings, depending on the climatic conditions, the compost reaches maturity; it is recommended that it is screened a second time through a fine mesh to remove glass particles and other inorganic materials, after which the mature compost is ready for use.

Substrate (Nutrients)

Nature. The nature of the substrate in composting obviously has been that of the wastes being processed. The waste, must be organic, which in the chemical sense covers a multitude of materials. In terms of composting, "organic" implies in the popular sense., e.g., paper, wood, manures, food-preparation wastes, crop wastes, etc. In terms of microbial nutrition all of these forms have been highly complex substances, and thus are not available to a large number of groups of microbes. It is only as the materials get décomposed to simpler and yet more simple forms that the spectrum of potential microbial users broadens. The course of breakdown for the protein content has been protein-amino acids-ammonium compounds-bacterial protoplasm and atmospheric nitrogen or ammonia. Of course, this outline of breakdown has been highly abbreviated and does not take into account the many possible intermediate and side reactions. For example, each step actually gets accompanied by the synthesis of bacterial protoplasm, inasmuch as when as organism decomposes a substrate, some of the nitrogen gets transformed into protoplasm. The following scheme for the complex corbonaceous part of the substrate also has been oversimplified: carbohydrate-simple sugar-organic acids-CO_2 and bacterial protoplasm.

A main feature of a substrate as far as composting has been concerned has been the availability to the microorganisms of the nutrients tied up in the substrate. Thus, the carbon in the raw wastes might just as well not be there as far as the greater part of the indigenous population has been concerned, simply since in its existing form, the carbon cannot get assimilated. Therefore, if, the first members of the microbial succession had been missing, nothing much would take place. The important point has been that the simpler the form in which the waste occurs, the wider the array of bacterial species to which it has been subject to attack-and hence the more rapid the pace at which composting proceeds. Consequently, vegetable trimmings from a supermarket compost much more rapidly than do straw or sawdust.

Nutrient balance. In order to keep reproducing and thereby bringing about decomposition, microorganisms, all organisms, must prossess a minimum supply of all the elements of which their cellular matte has been composed. In addition, they require a minimum amount of certain elements that may involve into the metabolic activities of the organisms as an energy source or enzyme constituent and do not necessarily contribute to an increase in the mass of the organisms. We know that the amounts needed vary from element to element, and at a somewhat constant ratio of one to another. In other words, a balance is struck. The balance arises from the fact that growth would get limited by that factor (or factors) which is (or are) present in less than the required concentration. The balance would be important as far as the macronutrients has been concerned.

The more important balance with respect to composting has been the carbon-nitrogen balance or ratio (C/N ratio). As part of the carbon gets lost as CO*2*; and because, depending upon the type of organism, carbon has been present in the cellular material in greater concentration than is nitrogen, the amount of carbon needed considerably surpasses that of nitrogen.

The optimum C/N range with most wastes decreases within 20 or 25 to 1. The more the carbon-nitrogen balance deviates from the optimum, especially in the upper range, the slower would be the process. However, the actual upper limit for an individual appli-cation has been reported to be depending upon the degree of availability of the carbon. If the carbon has been present in a form highly resistant to bacterial attack, it would be of little use to microbes. Hence, if a waste is having a large percentage of carbon in a resistant form, the permissible carbon-nitrogen ratio can be higher than 25/1. Examples include wastes which are having a heavy concentration of paper, fiber, wood, or straw.

The nitrogen concentration of a sample of the refuse to be composted can be found out by the Kjeldahl method. If the apparatus required for the carbon analysis is not available, the carbon content can be roughly found out according to a formula given below:

$$\% \text{ carbon} = \left(\frac{100-\% \text{ ash}}{1.8}\right)$$

In municipal wastes generally nitrogen content has been to low for best, or even practical, composting. The reason has been simple: very few food products reach the "garbage" can. The packaging of food decreasing the amount of food-related wastes generated at the home; and much of that gets ground into the sewers. This nitrogen deficiency can be overcome by adding nitrogeneous wastes like manures and digested sewage sludge.

The main effect of too low a C/N ratio has been the loss of nitrogen through the production of ammonia and its subsequent volatilization. Apparently, the excess nitrogen ends up as ammonia. As far as the Composting process has been concerned per se, such an event has been not detrimental. However, since nitrogen conservation should be one of the objectives of composting, the loss then becomes significant. The loss becomes greatest when high-rate composting has been used because high-rate composting involves a great deal of aeration, generally takes place under thermophilic conditions, and gets characterized by a pH level of 8.0 or above. The high pH is able to forter ammonia formation; the high temperature accelerates the volatilization of ammonia, and the aeration drives off the volatilized ammonia. For example, we could lower the pH by adding some type of acid. Of course, such a procedure would be impractical from the standpoint of processing and expense. Without aeration, the process no longer has been high-rate composting. The temperature could be lower, but this would only reduce the rate of loss, and not necessarily the total amount to loss. consequently, the most appropriate approach has been to avoid the low carbon-nitrogen ratios if possible.

Rate-Controlling Factors

General. As the compost process has been biological, ratecontrolling factors in the process have been those influencing biological activity in general, namely, environmental factors. In fact, organism-related factors have been in reality the ultimate ratelimiting factors, because they determine the rate of growth and degree of activity of the microbial populations, and hence the rapidity and nature of the compost process. The main environmental factors in composting have been moisture, temperature, pH level (hydrogen-ion concentration), nutrient concentration and availability, and oxygen concentration.

Moisture content. The theoretical optimum moisture content in composting has been 100%. The practical moisture content has been a function of aeration capacity of the process equipment and of the structural nature of the material being composted. In accordance to the strictly biological point of view, the ideal arrangement would be to slurry the refuse and then to vigorously bubble air through the slurry at a rate at which the slurry would become completely aerobic, much like the conventional activated-sludge process in sewage treatment. Due to technological and economic reasons, the aerated-slurry approach has been impractical for most waste-composting applications.

The usual practice is to resort "dry" handling in composting, i.e., the material would not be suspended in water. With this type of handling-in other words, with composting in general-the structural strength of the material to be composted decides the upper permissible moisture content. The relation between structure and moisture content. The relation between structure and moisture content comes from the fact that modern composting has been an aerobic process. Oxygen has been made available to the bacteria by way of the air contained in the interstices at the time the windrow was formed. In a mechanical composter, the supply has been by way of the air introduced between the particles as they are tumbled or stirred in the digester. In either case, if the interstics have been filled with water, obviously the air gets displaced and no oxygen is available to the organisms. Because of this the mass of waste becomes anaerobic; the composting process slows and produces foul odors. The size of the interstices has been determined by the particle size of the material. Their stability has been determined by the structural strength of the composting material-i.e., the capacity to resist compaction and consequent obliteration of the interstices.

Inasmuch as a 100% moisture content has been the desideratum, the maximum permissible permissible moisture content ipso facto becomes the optimum moisture for a given material. The maximum permissible moisture contents for various wastes are listed in Table 5.23. From this table it can be seen that the maximum permissible moisture content for composting wastes that have been largely "fibrous" (straw, hay, dry leaves, etc.) or woody (sawdust, small wood chips, wastes having mostly paper or green vegetation (lawn clippings, vegetable

Table 5.23
Maximum Permissible Moisture Contents

Type of waste	*Moisture content %*
Theoretical	100
Straw	75-85
Wood (sawdust, small chips)	75-90
Paper	55-65
"Wet" wastes (vegetable trimmings, lawn clippings, garbage, etc.)	50-55
Municipal refuse	55-65
Manures	55-65

trimming, wet garbage, etc.) All having a maximum permissible moisture content within the range of 50% to 60%. If the maximum permissible moisture content for a given waste has been excessively low, it can get raised by adding an absorbent waste to the material. Therefore, if we add straw to vegetable trimmings, the maximum permissible moisture inereases in proportion to the amount of straw added. The addition of an absorbent also is needed if the moisture content of waste in its "raw" state has been excessively high, as for example, cannery wastes.

The minimum moisture content at which bacterial activity occurs is from 12 to 15%. Thus, the closer the moisture content of a composting mass approaches these low levels, the slower would be the compost process. As a rule-of-thumb, the moisture content would become a limiting factor when it drops below 45 or 50%.

Temperature. Temperature is considered as one of the key environmental factors which are affecting biological activity. In general, each group of organisms is having an optimum temperature and any deviation from the optimum gets manifested by a decline in growth and activity of the organism. For convenience of reference, the total range of temperatures at which life would be possible generally has been divided into three "subranges" into which have been grouped those organisms whose optimum temperature falls within one of the sub-ranges. The three sub-ranges have been cryophilic, mesophilic, and thermophilic. The boundary temperatures of each of the ranges are somewhat assigned as being from about 5°C to about 10°C for the cryophilic range; from 10°C (or as high as 15°) to 40° or 45°C for the mesophilic range; and from 40° or 45°C to 70°C for the mesophilic range. Modern composting processes have been operating within the mesophilic and thermophilic ranges.

Much has been cited in favour of composting at either of the two temperature ranges-mesophilic and thermophilic. The proponents or mesophilic composting (i.e., in the upper range-above 35°C) claim the mesophilic bacteria have been more efficient than thermophilic bacteria, and that composting therefore takes place more rapidly. An important advantage which has been claimed by the proponents of thermophilic composting has been that pathogens and weed seeds get killed at the high temperatures. This latter feature probably has been considered as one of the main reasons why most modern composting processes have been involving thermophilic composting at some stage in the overall process. Some evidence has been cited that the process may be taking place more efficiently in the lower thermophilic range, viz., 50°C to 55°C. All groups agree that the operating temperature must be at least 35°C.

The process becomes less efficient when the temperature becomes greater than 60°C if for no other reason than that sporeforming organisms start to enter the spore or resistant stage at temperatures of 60°C and higher, and activity diminishes to practically zero. Hence the composting process has been correspondingly retarded.

The range of optimum temperatures for the composting process as a whole has been quite broad, i.e., from about 35°C to about 55°C, due to the many groups of organisms taking part in the process. Each group is having its particular optimum temperature which may of may not coincide with that of another group. Therefore, the optimum temperature for the process as a whole has been an integration of, or perhaps better expressed, a compromise among the optimum temperatures of the various forms of microbes involved in the process. Of course, unless a closely controlled digester has been used, a uniform temperature does not prevail throughout the mass of composting material at any one time-except at the start of the process when all material has been at ambient temperature. The existence of a plateau from the mesophilic into the thermophilic range has been due not only to the involvement of many types of organisms, but also to the adaptation of organisms to the temperature range.

At temperatures lower than 35°C, the efficiency and speed of the process gets increased with increase in temperature. The rate of increase starts decreasing as the temperature exceeds 30°C and approaches 35°C. The slope of a curve showing efficiency of speed of the process as a function of temperature must be practically a plateau between 35°C and about 55°C, with perhaps some declination between 50°C and 55°C, upon the temperature exceeds 55°C, efficiency and speed start to drop abruptly and becomes negligible at temperatures in excess of 70°C. At temperatures above 65°C. Sporeformers start to lose their vegetative forms and assume spore forms. In the spore form, very little activity occurs. Non-sporeformers simply die off.

In a practical operation, the desired temperature range has been including thermophilic temperatures. The reasons has been: (1) some of the organisms involved in the process posses their optimum temperature in the thermophilic range; (2) weed seeds and most microbes of pathogenic significance fail to survive exposure to thermophilic temperatures; and (3) unless there are definite countermeasures, a composting mass of any appreciable volume will assume high temperatures.

Hydrogen-ion (pH) Level

It can be stated that fungi tolerate a wider range of pH than do the bacteria. The optimum pH range for most bacteria has been between 6 and 7.5; whereas for fungi it could be between 5.5 and 8.0. In fact, the upper pH limit for many fungi is reported to be a function of precipitation of essential nutrients from the growth medium, rather than of any inhibition due to pH per se.

In a practical operation, it is generally not possible to alter the pH level prevailing in the pile. The reasons have been due to economics and the need for nitrogen conservation. Changing the pH needs the addition of a reagent, which in turn involves a twofold cost-that of the reagent and that of applying the reagent. Nitrogen conservation becomes a factor when trying to raise the pH level of the compositing mass. The least expensive reagent for the purpose has been lime. Unfortunately, the addition of lime has been generally accomptained by an increase in ammonia formation and consequent loss of nitrogen.

Availability of Oxygen (Aeration)

The modern composting has been almost universally aerobic. Therefore, oxygen availability is an important environmental factor. In recent years, a study on the rate of oxygen uptake has been one of the first of the routine steps of a newcomer when undertaking research concerned with composting. Among the earlier reports on oxygen uptake have been those by Schulze and Popel. Schulze estimated that an air supply equivalent to 18,000 to 22,000 cu ft per ton of volatile matter per day was needed in his studies to maintain aerobic conditions during peak oxygen demands. Chromettzka names oxygen requirements ranging from 9 mm³/gram/hour for ripe compost to 284 mm³/gram/hour for "fresh" compost (4 weeks old). "Fresh" refuse (7 days old) required 176 mm³/gram/hour. He also found out that moisture content has been a determinant of the oxygen requirement. For example, the requirement for fresh compost having a 45% moisture content was 263 mm³/gram/hour; whereas with moisture at 60%, it was 306 mm³/ gram/hour.

The difficulty of determining true or actual oxygen requirements in terms of rate stems from the influence exerted by temperature, moisture content, size of bacterial population, and availability of nutrients on oxygen uptake. If we want to translate oxygen uptake into amount of aeration required to assure a sufficient supply of oxygen, the problem is even more complex. Now, in addition to the preceding factors, the aeration equipment and physical nature and arrangement must be taken into account. The upshot of the preceding constraints has been that, while the determination of the absolute oxygen requirements of a composting mass may be an interesting research undertaking, the findings are strictly limited in their application, namely, to the material being tested and to the conditions of the test.

As a control measure in an ongoing operation, a monitoring of oxygen present in the exhaust air has been found to be useful for mechanized composting. When the oxygen content of the exhaust drops below a few percent concentration, the rate of air input could get increased correspondingly. Monitoring could be done with the use of any reliablc oxygen-analyzing equipment. The monitoring of oxygen in a windrow type of operation would be on a far more crude level. In such a situation, the olfactory sense becomes the "intrument," and the basic principle has been the development of putrefactive odors indicates the onset of anaerobiosis, i.e., and insufficient oxygen supply.

Genetic Traits

While not an environmental factor, genetic traits is regarded as the ultimate rate-limiting factor. The significance of this statement has been that regardless of how close to optimum all environmental conditions may be, the rate of decomposition depends upon the capability of micro-organisms to break down the material. The capability of the microorganisms is dependent upon their genetic make-up; and the environment permits the expression of the genetic make-up.

A practical aspect of genetic limitations has been that no amount of sophistication of equipment will hasten the decomposition of resistant materials beyond the limit permitted by the genetics of the organisms involved. It is this limitation that should prompt one to be skeptical of claims for one to three-day compost schemes, or of inoculums that will greatly acceletate decomposition beyond that normally encountered. Another practical aspect has been that maintaining environmental conditions beyond the optimum level has been a waste of effort, inasmuch as the potential as determined by the genetics of the organisms cannot be surpassed. On the other hand, there has been no doubt that the process could be considerably accelerated beyond present-day rates by making a more effective provision for optimum conditions. However, the gain would not necessarily warrant the expenditures.

The Process

Regardless of type, composting has been employing certain sequential steps: sorting, grinding, composting, storage. Some systems demand another grinding after the composting step, and others demand screening as well as a second grinding.

Sorting

As the grinding of the raw material has been one of the essential steps in aerobic composting, some segregation of municipal refuse seems unavoidable. In cities using separate containers, the responsibility for segregating refuse could be imposed upon the individual household by requiring that only specified organic material like garbage, paper, natural rubber, leather, and rags be kept in the container intended for compostable material.

Materials normally needing removal before grinding include tin cans, miscellaneous metals, glass, and ceramic ware. Excess paper could be removed for salvage, or in some cases to decrease the C/N ratio. Rags of natural fiber could be removed for salvage; whereas those of synthetic fiber should be removed because of their adverse effect on the appearance of the compost product.

Tin cans and other ferrous metal objects could be commonly removed by a magnetic separator as the refuse moves along a conveyor belt. Rags and non-ferrous objects could be removed by hand. When the necessary technology has been developed, hand sorting will get replaced by mechanized sorting. Excess paper would be removed by a blower with its suction intake located directly above the conveyor belt. However, bottles and other glass and ceramic objects creates a more difficult problem. Here, hand-picking has been an unprofitable undertaking; however, pulverizing glass mixed well as excessive energy requirements. A project designed to compost municipal refuse must certainly envision the removal of metals, and until better equipment has been available, the removal of glass as well.

Grinding

Grinding or shredding of refuse gives rise to a number of beneficial results which hasten decomposition. The material becomes more susceptible to bacterial invasion, made quite homogeneous, and given a beneficial aeration. The material acquires a structure which facilitates handling and increases its response to moisture control and aeration. None of these characteristics adversely gets affected by ground glass, but such material increases the already phenomenal abrasiveness of refuse, which has been able of destroying the cutting edges of hammers and of blades in a single day's operation. Unless the glass gets pulverized, shards left in the material detract from the quality of the compost product.

The main aim of grinding has been to chop refuse into small pieces. The ideal particle size would be the minimum size at which undue compaction of pulping does not occur. The larger the particle size, the slower will be the rate of decomposition. For municipal refuse, a particle size within the range of one to two inches would have been more suitable.

Composting

Description. It is the step in which the procedures for windrow and mechanized composting differ.

Immediately after the material is ground and is either stacked in windrows or placed in a digester, bacterial activity starts to proceed at an acelerated pace. The first easily recognized manifestation of this activity has been an increase in the temperature of the mass-provided the volume has been not dissipated by blowing, etc. The temperature rises at a rate determined by the environmental factors and the nature of the material.

The temperature rise is also noticed in mechanized composting, because the digesters have been constructed such that heat does not readily escape, and because the composting material has been present in large enough masses to bring about some self-insulation. However, the distribution of temperature level has been more uniform throughout the mass.

The high degree of bacterial activity has been indicted by the fact that the temperature drop accompanying the turning operation in windrow, composting has been very minor, and the temperature gets regained within a very short time after turning gets completed. We know that the heat is due to the excess energy released by the bacteria. Inasmuch as the heat energy does not escape from the composting material at rate faster than it is generated, the temperature of the mass starts rising. The high temperature is maintained until the readily decomposable material is broken down. As that point is approached, bacterial activity starts to decline, and the temperature drops correspondingly.

The physical appearance of the material undergoes a change during the process. Municipal refuse is having a greyish-green color after being ground. This color gradually is changed to brown or brownish-gray, the transistion being generally completed within four or five days after grinding. Corrugated cardboard, unprinted paper, and other organic items lose their identity, while printed newsprint could be seen throughout the process. The composted manures look much like dehydrated fresh manure. Bedding material in the manure retains its identity, but becomes quite brittle and fragile. Cannery waste loses its characteristic appearance and assumes the dark color associated with compost.

A noteworthy change has been that of the odor of the composting material. Ground municipal refuse is having a somewhat pungent odor.

Other characteristic change occur are an increase in oxygen uptake, a shift in the redox potential, a rise in pH level, the development of olfactorily detectable ammonia, and a drop in the C/N ratio. The shift in radox potential is that which characteristically takes place when conditions change from anaerobic to aerobic. Typically, the pH level decreases during the first few days of composting due to the synthesis acid by bacteria. If the composting material is having anaerobic zones, some butyric acid, acetic, and propionic acid may be formed. However, the acids are soon used as substrate by the bacteria, and with the formation of ammonia, the pH level of the pile begins to rise. The lowest pH levels observed in the university studies were on the order of 5.0 to 6.0; and the highest from 8.5 to 9.5. The drop in C/N ratio during the compost step has been due to the release of CO_2 into the ambient air. The CO_2 has been that formed by bacteria when oxidizing the carbonaceous material in the pile. Very little, if any, nitrogen fixation occurs.

Determination of degree of stabilization. The net result of the composting step has been the stabilization of the organic matter in the wastes. The extent of stabilization has been relative, in that the material has been not completely stabilized, i.e., rendered inert. Ultimate stabilization of organic matter would give rise to the end products CO_2, H_2O, and mineral ash. Obviously, this does not take place in conventional composting, nor is it to be desired either in terms of utility of the product or of practicality of attainment. The impracticality arises from the excessive time factor involved. The utility of the compost product, would be depending upon its further breakdown in the soil.

The desired degree of stability has been one at which the material will not give rise to nuisances when stored-even if it should become wet. The problem in a compost operation has been one of determining when this point is reached. It is a important in a practical operation, because unnecessary handling could be avoided through removal of the material from the plant directly after the proper degree of stabilization is attained. Another factor is that an accurate and uniform procedure of stability determination makes it possible to evaluate the various compost systems on the market. Thus, how can one weight the claims of one entrepreneur for a machine which composts refuse in 6 days for another machine? The one may term a material composted when it takes a dark color; while the more modest one relies on final temperature drop as the indicator of stability.

Finding a solution to the problem has challenged several researchers. Change in appearance is not a criterion. The dark color typical of composed material may be attained long before the necessary degree of stability is reached. The same may be said about the development of the earthy odor which is characteristics of actinomycetes. Consequently, the earthy odor would be expected when those organisms have been active. As the actinomycetes need a decomposable substrate, their presence, at least at their first manifestation, would "indicale the existence of "unstable" organic matter.

Parameters of stability, other than appearance and odor, have been final drop in temperature, degree of self-heating capacity, amount of decomposable and resistant organic matter in the material rise in redox potential, oxygen uptake, growth of Chaetomium gracile, and the starch test.

The principle underlying Rolle and Orsonic's method of determining decomposable and resistant organic matter has been the ascertaining of the amount of oxidizing reagent used in the test by treating the sample with a potassium dichromate solution in the presence of sulfuric acid. A certain amount of dichromate added in excess is thereby consumed in the eradication of organic matter. The oxidizing agent left at the end of the reaction is back-titrated with ferrous ammonium sulfate, and the amount of potassium dichromate used up gets established. The amount of decomposable organic matter may be found out as follows:

$$\text{Dom} = \text{ml N}\left(1 - \left(\frac{T}{S}1.34\right)\right)$$

in which DOM has been decomposable organic matter, ml has been ml of $K_2Cr_2O_7$ solution; N has been normality of the $K_2Cr_2O_7$. T is ml of ferrous ammonium sulfate solution for back titration. S is ml of ferrous ammonium sulfate solution for blank test. Quantitatively, resistant organic matter is that amount of loss by combustion that is not degraded in the oxidation reaction.

The oxidation-reduction potential determination of stability has been based on the fact the presence of decomposable organic substances gives rise to an intensification of microbial conversion, which in turn brings about increase in oxygen uptake and accompanying drop in oxidation-reduction potential. Moller shows that the material in a pile of compost may be considered as sufficiently stabilized when the redox potential of the core of the pile is <50 mV below that of the outer layer.

The *Chaetomium* method has been based on the measurement of growth and formation of fruiting bodies of the fungus Chaetonium gracilis cultured on a solid nutrient medium having pulverized refuse or compost. After an incubation period of 12 days, the fruiting bodies have been counted. Growth and number of fruiting has been functions of the amount of decomposable material present bodies in the tested material.

The starch-determination method given by Lossin is depending on the assumption that the more advanced the degree of decomposition, the less would be the amount of decomposable organic matter. He states that three types of carbohydrates occur in composting material, namely, sugars, starch, and ceilulose. In windrow Composting, sugars disappear within a week after the start of the process. Starch passes through its maximum degradation in the fourth to fifth week of composting.

A characteristic of all of the preceding tests, excepting perhaps the two depending upon temperature level or drop, has been that they are relative, in that the determination rests on a change in values, rather than on a standard, consistent number. Hence, they cannot be employed as standard, consistent number. Hence, they cannot be employed as standards for wide-scale application. The measurement of the radox potential has been still not sufficiently accurate and has been subject to a variety of interfering factors. The measuring procedure for self-heating is a slow one and may require several days to complete. The Chaetomium test has been time-consuming, and has been dependent to a large extent on the skill of the tester. The success of the starch always is present in refuse remains to be demonstrated.

Duration of composting stage. As indicated earlier, the required duration of the composting step has been a function of procedure and environmental conditions. Some purveyor of compost equipment name times as short as 3 to 6 days with their systems. However, if one carefully studies their claims, we can find that the complete directions call for a 2-to 6-week "maturation" period after the "active" composting has been completed in the digester. In reality, the material prior to "maturing" is far from being ready for use or for storage.

In a practical operation, speed of the process has been not usually a critical factor in terms of economics, as the market is such that the material must be stored. It becomes important if the type of equipment used has been expensive to operate. In such a situation, the less the exposure time in the composter, the more that can be processed per day, the lower the cost per ton. Another answer has been to invest in a less-expensive composter, even if the required time for composting is thereby lengthened by a few days.

Volume reduction. The extent of volume reduction has been a function of the nature of the waste and the duration of the compost time. The degree of reduction gets increased in proportion to the amount of readily decomposable material in the wastes. The reduction in volume with garden debris had been as much as 60 to 65%. On the other hand, when municipal refuse having a heavy concentration of newsprint was composted, the reduction in volume had been from 30 to 35%.

Final Processing and Storage

After the material has been adjudged sufficiently stable to store, it has been ready for "rough" applications, i.e., large-scale agriculture, land, reclamation, etc. In practice, the usual procedure has been to sort the compost into fractions on the basis of quality, usually by screening the material. The coarsest material has been destined for the "rough" application, while the finest is reserved for the home gardener for "luxury" crops. Generally material remaining after the separation of the coarser fraction is ground and screened. Grinding compost is less a chore than grinding raw refuse due to the fact that composting the material reduces the physical strength. Of course, the major benefit of the second grinding has been increase in "eye" appeal which results.

We have already stated that properly composted material can be stored without danger of subsequent generation of nuisances. Although decomposition will occur during storage unless the moisture content is too low for bacterial activity, the rate of decomposition will be too low for bacterial activity, the rate of decomposition will be very low, and no odorous intermediates will be formed.

Technology

Development. The development of a truly systematic approach to composting began with Lord Howard's work in Indore. India in the early 1930's in which he systemized the traditional procedure in collaboration with Jackson and Wad. His process became known as the Indor process due to the locate of his activities. The Indor process involved piling on open ground to a height of about five feet, or placing in pits, alternate layers of readily putrescible materials such as garbage, night solid, animal manure, or sewage sludge, and relatively stable organic matter such as straw and leaves. The mass usually turned twice during the composting process.

A variety of mechanical devices were designed and patented during the decade 1920 to 1930. Some of the devices were developed hopefully to speed the process.

Among the early mechanized processes, was one developed by G. Giovanni Beccari of Florence, Italy. The Beccari process is a combination of an initial anaerobic fermentation stage and a final aerobic stage. The anaerobic fermentation occurred in an enclosed cell designed to prevent the escape of foul odors usually associated with the initial breakdown of putrescible material under anaerobic conditions. As time passed, vents were opened to admit air, and thereby to permit further decomposition to take place under partially aerobic conditions. The amount of aerobic decomposition under such conditions would get limited to the upper inch or two of the surface layer, and therefore the greater part of the mass continues in an anearobic state, unless it gets stirred or disturbed in some manner. Originally, the Beccari cell consisted of a simple suboidal cell having a loading hatch on the top and unloading door on the front. Valves or other closures got incorporated in the structure as air vents. The process was later modified to provide for recirculation of gases or of drainage liquors. The modified process became known as the Verdier process.

In 1931, Jean Bordas further modified the Beccari process by eliminating the anaerobic stage by introducing forced air into a fermentation silo.

The concept of aeration was carried a step further by the development of a multiple-grate digester to produce compost by a continuous aerobic process. The process was patented in 1939 by Earp-Thomas. In the Earp-Thomas silo, aeration was carried out by a combination of rotating plows and forced air. A feature insisted on by Earp-Thomas as being essential had been the use of special cultures of bacteria furnished by Earp-Thomas.

A later variation of the digester type of enclosed cell has been designed by the Ralph W. Riker Company of Lansing, Michigan. It consisted of a double-walled silo having multiple floors.

A different approach to digester design was represented by the one patented by Eweson in 1949. The Eweson (or Frazer) process had been a fully mechanized and continuous one, in which compost was supposedly produced rapidly and under aerobic conditions. Organic matter got shredded and introduced into a machine where it was kept continuously agitated as it moved from one level to another and was brought into contact with the gases of decomposition a salient feature of the patent.

The Dano process had been a step in the direction of a more sound approach to the design of equipment for composting. Later developments in the process involved the use of a slowly rotating horizontal drum suitably equipped for injecting air, applying moisture, and controlling temperature.

Present Technology

Windrow systems. Types of windrow systems have been ranging from the rather primitive one designed by to the relatively advanced one. The windrow or open composting systems have been characterized by having the composting take place in the open by placing the ground refuse in elongated piles, i.e., windrows. Aeration gets accomplished by periodically turning the piles in a manner such that all particles have been exposed to comparable conditions at some time during the course of the active period of the compost process.

Windrows may be of any convenient length, but the depth of the pile has been somewhat critical. If piled too high, the material will get compressed by its own weight, pore space will be lost, and the mass would become anaerobic. In some instances, the maximum practical height could be governed by the equipment used for stacking the gronnd refuse, or by a tendency of the pile to get excessively hot-above 70°C. A pile that has been too shallow loses heat too rapidly, and is unable to attain optimum temperatures for thermophilic organisms. In addition, loss of moisture has been excessive, especially near the edges of the pile, and the rate of composting has been thereby retarded.

Although experience quickly demonstrates the most suitable height of pile for any particular refuse, a maximum of five or six feet has been recommended for freshly ground municipal refuse. As the material loses volume during decomposition, any desired height of pile could be maintained by reducing or expanding the width of the windrow at the time it is turned.

For convenience in turning, the initial width of a windrow probably should not exceed eight to ten feet at the base. In dry weather, the cross section has been usually made trapezoidal, with the top width governed by the width of the base and the angle of repose of the material, which is something like 30 degrees from the vertical. In rainy climates or in wet weather the cross section of the windrow must be approximately semicircular like a haycock in order to shed water. In that case, the maximum permissible height of pile would be able to govern its maximum width.

Other than maximum and minimum heights of pile, there exists nothing critical about the stacking of ground refuse for composting; hence in each individual case, experience with the materials-handling equipment used will establish the best practice to be followed.

Turning. As aeration in windrow composting has been accomplished by turning, the pile is turned fairly frequently so as to get the rapid, nuisance-free decomposition characteristic of the thermophilic aerobic process. Uniform decomposition essential to rapid composting could be insured by turning the outer edges into the center of the pile at each turn. In this manner, any fly larvae, pathogens or insect eggs which might survive at the cooler surface get exposed to the lethal temperatures of the interior of the pile. This implies that neither a simple overturn nor a vertical expansion of the pile constitutes adequate turning. An inward mixing of the material from both sides of the pile is needed.

Secondary reasons for a particular turning operation might be reduction of the initial moisture content, and relamation of a compost pile which is becoming anaerobic. An aerobic conditions usually result from too great a moisure content for the normal turning schedule practiced. Excess moisture could be inherent in the particular refuse being composted, or it may result from continued soaking by rain, especially during periods of turning, when composting is done in the open. Anaerobic conditions may also exist because of equipment breakdown, a long period of unusually bad weather, or other interruption of a normal turning schedule.

The frequency of turning and the total number of turns required during the composting process have been governed largely by the moisture content. If the initial moisture has been less than 70%, the first turn should be made on the 3rd day. Thereafter turn as follows until the 11th or 12th day:

Moisture 60 to 70% ; turn at 2-day intervals.

Number of required turns-about 5.

Moisture 40 to 60% : turn at 3-day intervals.

Approximate number of turns-4.

Moisture below 40%: Add moisture.

If the moisture has been more than 70%, turn every day until the moisture content gets reduced to less than 70%; than follow the above schedule.

Turning is advocated as a means of dissipating the heat of a compost pile in which the temperature exceeds 75°C. A shallower pile would control the maximum temperature without the labour involved in turning.

The turning program is geared to "high-speed" composting. Where time is not an important constraint, the turning schedule could be modified to allow longer intervals between turning. The same directive applies when nitrogen conservation is an important consideration, because turning accelerates nitrogen loss.

Bottom aerations. Various substitutes have been proposed for turning as an aeration method. Among these have been placing the ground refuse in wire containers, stacking the material on slotted or perforated floors, and forcing air into the pile by way of bottom aeration. Placing the material in wire bins succeeds in aerating only the outer layers of the mass. Practically, the same exposure has been obtained by stacking the material. Resorting to slotted or perforated floors allows a little additional exposure to air, but not enough to warrant the expense and inconvenience.

While the concept of forced aeration may be a sould one in many applications, many difficulties combine to diminish its utility in the windrow composting of municipal refuse Probably the major difficulty has been that of diffusing the air throughout the pile uniformly. In a practical application involving municipal refuse, channels form in the pile and the forced air then short-circuits through these channels. Moreover, the evaporation brought about by the passage of air, the material in the immediate vicinity of the channels becomes excessively dry. Even were forced aeration successful in its purpose, the piles would must be turned occasionally so as to bring about the necessary exposure of all particles to the high interior temperatures and to insure uniform decomposition.

Moisture control. The control of moisture in windrow composting may range from eliminating excess moisture to applying water when a pile becomes too dry.

A number of additives could be used with varying degrees of effectiveness to control excessive moisture. If excess paper is being segregated, some degree of control could be obtained by leaving a part of the paper in the refuse. Because paper is of limited effectiveness, however, there has been danger of using too much and thereby producing an excessively high C/N ratio. While straw could be effective in controling moisture and has the added advantage of improving the structure of the material, the increased cost involved in buying and handling it, as well as the uncertainty of its availability at some seasons of the year, make its use impractical in ordinary Composting. Dry soil or sawdust are other additives effectives in lowering moisture content. Here again, the problem of availability and cost of handling limit their use.

In areas where rainfall is heavy, material being composted in the open may become excessively wet. Therefore, preventive measures must be taken. If economic conditions make impossible the obvious remedy, namely, sheltering the piles with a roof, then the tops of the piles or windrows can be rounded so that the outer surface will shed water much as a thatched roof. The water merely flowed off the piles without penetrating them. If the piles get waterlogged, turning will remedy the situation. Of course, no turning should be done during a rainfall, because all parts of the open stack will get exposed to rain, increasing the moisture content.

In order to increase the moisture content of a pile, water can be added most effectively during the turning operations, at which time the exposed material absorbs water. The outer part of unsheltered piles are having a tendency to dry out during sunny weather, in which case surface sprinkling should be practised to promote uniform decomposition.

Design consideration. An experienced designer could readily develop a plant layout for receiving, segregating, and grinding refuse, and for stacking, turning, regrinding, and stock-piling or storing finished compost. The first there of these operations would presumably be done under cover. The cheapest industrial type of building should suffice.

In determining the necessary land area one may take the length of compost piles as multiples of one day's production, with interruption in continuity, perhaps, to give flexibility to the turning schedule. Spacing of piles could be determined from the proposed turning procedures and the maneuvering space required for the stacking and turning equipment. These considerations together with the period of composting anticipated would determine the land area necessary for any particular composting operation. At least 100% excess land should be available for contingencies such as inability to finish compost on schedule because of breakdown of equipment or excess rain. The land area could be as little as 8 or 10 acres for a population of 100,000, although more perhaps be advisable.

Mechanized Systems. The design of most of the mechanical composters has been based on providing aeration by some type of tumbling or stirring action. Thus, some digesters are equipped with moving plows to stir the material. A horizontal digester is having vanes to carry the composting material to a point at which it drops or tumbles to the bottom of the cylinder and again is carried up. Other types provide the tumbling by having the material drop from one floor to the next in a multistoried digester. A successful system (Metrowaste) combines bottom aeration and aeration by tumbling through passing a mobile endless belt from one end of the compost chamber (bin) to the other. As the belt moves, it picks up compost at its front, lifts, it to a height of about 3 feet and allows the material to drop behind it. Another method (Fairfield-Hardy) has been combining stirring with forced aeration by having large hollows augurs rotating through the composting mass. The augurs are having nozzles out of which air gets forced into the material as the augurs are rotated. In addition to providing aeration, mechanical digesters have been equipped to insure adequate temperature and moisture control.

Land requirements. The land requirements for a mechanical digester operation have been of the order 2.5 to 3 acres for a 100 ton-per-day plant. This number allows for about a 6-month's stockpile to cover periods when the demand for the product has been low.

Special applications. Composting has been owing an attractive potential in the disposal of animal and cannery wastes. These two types of wastes pose major problems in disposal, which will get aggravated as the rate of production of these wastes increases. Their physical characteristics have been such that they readily give rise to such nuisances as objectionable odors and flies. Both types are characterized by a moisture content high enough to make incineration an expensive operation. Buying the material in a sanitary landfill not only is a noisome operation but also increase the danger of pollution of ground and surface waters.

Due to their highly putrescible nature, animal and cannery wastes are especially amenable to composting. The main draw back has been their high moisture content. The problem is more severe with cannery wastes because their moisture content generally far surpasses that of animal wastes. Space does not permit a description and evaluation of the various methods that have been proposed to solve the problem, and hence only two are treated herein.

One of the methods is for the treatment of dairy manures containing no bedding materials. In essence the process entails airdrying the manure until its moisture content drops to 60% or less. In an on-going process, the drying step need be done only once i.e., at the initiation of a given operation. The dried material gets mixed with fresh material in a proportion such that the mixture does not "cake," and is placed in a bin. The bin has been equipped for forced bottom aeration. No problem is encountered in aerating the material in this manner, provided the moisture content is not excessive. By the time the material has been composted, its moisture content is low enough to permit it to serve as a moisture absorbent for fresh manure. A variation has been to spread the composted material in the cows' stalls to serve as a bedding for the animals and to absorb moisture from fresh manure. This system is applicable to any animal waste.

Table 5.24
Thermal Death Points of Some Common Pathogens and Parasites

Organism
Salmonella typhosa : no growth beyound 46°C; death within 30 min at 55° to 60°C.
Salmonella spp : death within 1 hr at 56°C; death within 15 to 20 min at 60°C.
Shigella spp. : death within 1 hr at 55°C.
Escherichia coli: most die within 1 hr at 55°C; and within 15 to 20 min at 60°C.
Endamoeba histolytic : thermal death point is 68°C.
Taenia saginata : death within 5 min at 71°C.
Trichinella spiralis larvae : infectivity reduces as a result of 1 hr exposure at 50°C; thermal death point is 62 to 72°C.
Necator americanus : death within 50 min at 45°C.
Brucella abortus or suis : death within 3 min at 61°C.
Micrococcus pyogenes var. aureus : death within 10 min at 50°C.
Streptococcus pyogenes : death with 10 min at 54°C.
Mycobacterium tuberculosis : var. hominis : death within 15 to 20 min at 66°C; or momentary heating at 67°C.
Mycobacterium diptheriae : death within 45 min at 55°C.

Health Aspects

The early concern with the public-health aspects of composting has been over the survival of pathogens and parasites from the night soil.

The Agricultural research Council in England noted also that fly larvae were sharply curtailed in the compost operations. A similar observation was made by Van Vuren.

It has been reported that while the refuse coming into the process is having a heavy concentration of fly larvae, by the time the material had been passed through the grinder, all larvae get destroyed. No fly activity was noted in the composting material. A table has been complied by various research groups which has been showing the thermal death points of some common pathogens and parasites, and reasoned from the information in the table that disease-causing organisms would not survive in a well-managed compost operation.

Knoll studies the survival of strains of *Salmonella (typhimurium, cairo, infantis and typhi)*. Under the conditions prevailing in the experimental compost set-up, no organisms survived over a span of 14 days.

Principles

The major factors in the destruction of pathogenic organisms in the compost process have been heat and antibiotic reactions. The temperatures in the upper 60s (C) and lower 70s are normally reached during the course of the process. These temperatures exceed the thermal death point of the organisms listed in Table 5.24. Not only are the high temperatures reached, they persist for a matter of days, i.e., in excess of the time required to kill the organisms.

Antibiotic phenomena also play an important part. One would expect the synthesis of antibiotic substances to occur, in view of the wide array of microorganisms, especially of actinomycetes and fungi. Of course, competition for nutrients must also have a part in the destruction. However, the effects of competition for nutrients would probably be more inhibitory than lethal.

The adverse effects on flies in their various stages of development have been both thermal and nutritional in origin. The higher temperatures have been considerably in excess of the thermal death point of the fly in all of its stages of development. Chemical and physical changes in the nature of the materials being composted soon render them unfit as a nutrient source for flies. Consequently, while adult flies may alight on a pile, they rarely remain long enough to deposit eggs.

It must be emphasized that the beneficial effects of composting could be achieved only be careful control of all phases of the process. Care must be taken that all particles are exposed to high temperatures and that all particles receive full exposure to the process. In view of the well nigh impossibility of maintaining perfect control at all times, the composting of night-soil or of raw sewage sludge with refuse must be undertaken with great reluctance. The resulting compost product must be subjected to heat sterilization. If not sterilized, its uses should get restricted to applications that do not involve human contact, directly or indirectly.

Use of the Product: Nature and Value of Finished Compost

Value as a soil conditioner. Finished compost could be designated by the general term "humus". When used in the soil, humus is having many characteristics beneficial both to the soil itself and to growing vegetation. In conjunction with commercial fertilizers, humus is exhibiting certain additional and very desirable characteristics.

The physical effects of humus on the soil have been perhaps more important than the nutrient effects. Soil structure may be as important to fertility as is its complement of nutrients. Soil aggregation or crumb tendency as promoted by humus improves the air-water relationship of soil, thereby increasing the water-retention capacity, and encouraging more extensive development of root systems of plants. Aggregation of soil particles has been brought about by cellulose esters (cellulose acetate, methyl cellulose, and carbomethyl cellulose) resulting from bacterial metabolism. Other beneficial effects of bacterial metabolism associated with humus include an increased ability of the soil to absorb rapid changes in acidity and alkalinity, and the neutralization of certain toxic substances.

Chemical Composition

While qualitatively the chemical composition could be fairly similar from one compost to another inasmuch as organic matter is the substrate, quantitatively the variation will be great. The variability is due to differences in type or combinations of types of material being composted. In one instance the material may be having a high nitrogen (low C/N) content; while in another, nitrogen may verge on being limiting. The gamut of trace elements also varies. Unfortunately, the broad array of metals in the municipal refuse may include a metal or toxic substance of public-health significance. Some idea of the chemical analyses of typical composts may be gained from the information listed in Table 5.25.

Table 5.25
Chemical Composition of Various Composts

Element	1	2	3		4
			Sludge added	No sludge	
Carbon	38.3	23.51	33.07	32.89	—
Nitrogen	1.57	0.97	0.94	0.91	0.57
Potassium	0.04	0.21	0.28	0.33	0.22
Sodium	—	—	0.42	0.41	—
Calcium	—	—	1.41	1.91	1.88
Phosphorus	0.40	0.16	0.28	0.22	0.26
Magnesium	—	—	1.56	1.92	0.12
Iron	—	—	1.07	1.10	—
Aluminum	—	—	1.19	1.15	—
Copper	—	—	<0.05	<0.03	—
Manganese	—	—	<0.03	<0.05	24 ppm
Nickel	—	—	<0.01	<0.01	—
Zinc	—	—	<0.005	<0.005	606 ppm
Boron	—	—	<0.0005-	<0.0005	33ppm
Mercury	—	—	e	e	—
Lead	—	—	e	e	—

In general, compost is having a nitrogen-phosphorus-potassium (NPK) content not high enough to permit it to be designated a fertilizer in the legal sense. Therefore, unless, fortified with one or all of the three elements, it cannot be legally sold as a fertilizer. The usual designation has been "soil conditioner."

Applications

Crops. For crop production, compost compost has been used as much and as a soil conditioner. As a mulch, it is having many of the characteristics of peat moss and bark. The advantage of utilizing compost as mulch in comparison with the use of peat moss or bark, is that the compost when eventually incorporated into the soil is releasing absorbed water much more readily than does peat moss, and imposes a lesser demand on soil nitrogen than does bark. While peat moss may be having a higher water-holding capacity than compost, the tenacity with which it clings to the water makes the water of little use to the crop plants being produced.

The C/N ratio of the compost has been an important determinant of its immediate utility in crop production. If the C/N ratio is too high, i.e., above 20 : 1, the danger of nitrogen "robbing" would be imminent. Nitrogen robbing could be manifested by stunted growth and a chlorotic condition of the higher plants making up the crop being cultivated.

In general, it might be stated that regardless of species of plant, growth is always enhanced by the addition of compost to a greater degree than by mineral fertilizer alone-provided of course, that a proper C/N ratio was maintained in the soil receiving compost. Another advantage is that nitrification in the soil gets reduced almost to zero. This characteristic is of value in preventing the nitrate contamination of ground waters. He also noted that compost brought about a reduction of nematodes in soil.

A problem asiring in the sue of large amounts of compost in crop production has been the accumulation of objectionable elements in the soil.

Land reclamation. The compost has been successfully used for land reclamation in areas where the top soil has been lost due to strip mining. The problem in such areas has been to stop erosion and to supply a substrate on which a plant cover can take hold. Compost would supply both needs very readily. Another type of reclamation is the protection of hillsides denuded of vegetation by fire. Compost has been also useful for retaining soil on highway cuts and other excavations.

A potential use for compost has been the reclamation of lands covered with mining tailings, flyash deposits, and various chemical sludges.

Potential of Composting in Wastes Management

Present status

The present status of composting in solidwaste management has been far from promising in most parts of the world.

The reason for the unsuccessful status of composting in terms of extent of practice have been not the lack of utility of the product nor of technology, but the unfavorable economics when judged by today's short-horizon pragmatic standards. The unfavorable economics comes from the inherent expense of the process and from the absence of a demand for the product. With sound incineration procedures, the volume reduction has been on the order of 90%, whereas with composting, it has been on the order of 30 to 40% at the most. Further, it is possible to build an incinerator on a more compact scale, and land requirements would be less because little or on storage is required in an incineration operation.

In the heyday of composting, the sale of salvage and the compost product has been expected to pay for a substantial portion, if not all, of the cost of the undertaking. In order to absorb the entire compost production the product would have to be used in large scale agriculture. Large-scale agriculture is not interested in distributing bulky material over its cropland, whether it be for purposes of conditioning the soil or of fertilizing, it especially when inorganic (mineral) fertilizers can be applied by "crop-dusting" techniques, or anhydrous ammonia by way of irrigation water. The large-scale framer even considers the manure produced on his holdings as a liability. The application of bulky fertilizer or conditioner needs more labor and machinery expenditure per unit of fertilizer value, than does that of inorganic fertilizer.

Another difficulty in marketing is its seasonality because product can be applied to cropland only at the end of, or prior to, the growing season-not during the season. Consequently, provision must be made for storing the daily production of a plant for long periods to time. The low monetary value of the product is another handicap.

Future

The unfavorable status of composting would alter as people become sufficiently environmentally conscious to demand that waste management include social and environmental benefits. Thus, we would expect composting becoming a way of salvaging the remaining organic resources to maintain the productivity of the land. The expected uses would be land reclamation, sludge "stabilization" or "conditioning", and recreation.

The vast acreage of land despoiled by strip mining has been a waste of land resources. The land could be readily reclaimed through the liberal application of compost. Further large expanses of land become useless as they serve as receptacles for chemical sludges. Where compost has been tried as a method of stabilizing these sludges, results have been favorable.

The consensus at present has been the composting would find its niche as regional management becomes the order of the day. When this time comes, agricultural waste disposal would become a public responsibility to a much greater extent than at present. Most of these wastes are readily composted with municipal refuse. Furthermore, because available sites for wastes burial have been rapidly dimishing and therefore must be must be employed at maximum efficiency, and as air-quality standards become stricter to a point at which airpollution control becomes so expensive as to rule out incineration, the only technology available for solid-wastes processing would be composting. In fact, many cities are already taking the first step needed, namely, grinding-albeit they grind wastes solely to extend the life of a landfill site.

10. Salvage

A little salvaging of solid wastes still takes place in this country but it accounts for only a small percentage of the total weight of all solid wastes produced. Some "recycling" or salvaging of materials before they become solid wastes, also takes place.

11. Chemical Processing

Chemical rocessing of solid wastes is appealing because of the prospect that energy and usable materials can be recovered (see Table 5.26). However, chemical processing is usually costly, even where technically feasible. In addition to the economics of processing and marketing, there are often problems of uniformity, impurities, and the quantity and composition of the residue. However, the salvage of some wastes by chemical methods is now economically feasible. The reclamation of non-ferrous (non-iron) scrap metals has long been an established industry amounting to many millions of dollars annually, involving thousands of business engaged in the collection and processing of scrap metal and metal containing products (see Table 5.27).

Table 5.26
Average Composition of Municipal Refuse (% by Weight)

Rubbishe (64%)	
paper, all kinds	42.0
wood and bark	2.4
grass	4.0
brush	1.5
cutting	1.5
leaves, dry	5.0
leather goods	0.3
rubber	0.6
plastics	0.7
oils, paint	0.8
linoleum	0.1
rags	0.6
street refuse	3.0
dirt, household	1.0
unclassified	0.5
Food Wastes (12%)	
garbage	10.0
fats	2.0
Non-combustible (24%)	
metals	8.0
glass and ceramics	6.0
ashes	10.0
	100.0

Table 5.27
Foreign Metals in a U.S. Automobile

	Total Pounds	*From* Foreign Countries
Iron	3,705	36%
Copper	52	38%
Lead	24	58%
Aluminum,	48	89%
Zinc	123	59%
Other materials	415	—
	4,367	

New Methods

New methods of waste disposal are under trial in various parts of the world. Some of the more promising methods are:

1. The transformation of the organic content into sugars or proteins.
2. The heating of organic refuse under anaerobic conditions (out of contact with air) to convert it into useful gases such as methane; which is usable as fuel, or into liquid products.
3. The compression of refuse into building blocks that can be sheathed with more durable materials. One such plant is in operation in Japan.
4. The compression of refuse into briquets that can be used as fill. A mixutre of shredded refuse, fly ash, dried sewage, incenrator residue, and river and lake dredgings is under study.
5. The use of degradable plastic bags instead of trash cans in the collection and transport of municipal refuse. Collection is speeded up; the bags resists water and keep out flies.
6. The transport of refuse as a liquid slurry in pipelines, a method now under study.
7. The transport of dry refuse in pneumatic tubes for short distances, a method now being developed in Sweden.
8. The use of trains to haul solid urban refuse to land reclamation areas, a method now under study study in the United States.

DISPOSING OF LARGE WASTE PRODUCTS

For very large components such as old furniture, old machinery and particularly old car, special equipment is necessary to get the waste products down to sizes which can be handled by pulverizers. Furniture is usually reduced in size by the use either of impact crushers or of hydraulic shears. Such equipment is very expensive. Hydraulic shears are mainly used for cutting down to convenient sizes awkward objects like motor cars, truck and aircraft tyres.

Processing of Scrap Cars

Three basic processes are commonly employed:

Barbecueing

In this process, the derelict car is clamped into a furnace and rotates in a series of heating chambers capable of raising its temperature gradually from about 400°C to 1000°C. At the lowest temperature all carbonaceous products start burning away, and low-melting-point metals such as aluminium, zinc, etc., melt and can be recovered from the bottom of the furnace. Copper alloys melt just below 1000°C. In this way nearly all the non-ferrous metals present in the car can be recovered, a bare shell of iron being left at the end. This is then dropped into a shear press, and leaves it in the form of a block of reasonably pure iron, which can be reprocessed in a steel works. This process is operated by the Japanese Fezuka Kosan company, and can deal with some 12 cars per hour.

Fragmentization

In this method engines, tyres, seats and interior trim must first be removed from cars; this has to be done by hand. The car is then picked up by a grab and placed into a large hammer mill, which pulverizes it into small scraps. The scrap, which contains a mixture of broken glass, dirt, pieces of plastic as well as iron and non-ferrous metals, is emptied on to a moving belt and the iron is extracted magnetically. Non-ferrous metals are separated out by hand, afterwards. Such a plant not only deals with cars but can also deal with equipment like gas cookers, refrigerators and similar objects. A plant with a power consumption of about 2.2 MW can deal with as many as 120 cars per hour, or an equivalent amount of scrap machinery. A typical plant of this type is the U.S. Proler-Cohen Installation.

The Pusher-ram System

As in the previous technique, the engine, etc., is first removed from the car. The car is then lifted by a crane into a box which has a hydraulic ram fitted. The ram pushes the car into a shredding machine, where it is separated into fragments about 50-80 mm in diameter. These are passed over a vibrating screen where dust is removed, into a rotary furnace, where paint and plastics are burnt off. The material is then passed along a moving belt fitted with a magnetic separator for ferrous scrap, which is compressed into a suitable block for reprocessing. Non-ferrous components are also reclaimed. A plant of this type is the West German Lindermann plant which has a capacity of up to 200 cars per hour.

Incineration

Incineration provides a means of disposing of refuse by high-temperature oxidation. Its primary function is that of reducing the volume of wastes. It can in addition serve the secondary roles of energy production or destructions of putrescibles or hazardous wastes.

Combustion Characteristics of Solid Wastes

Questions that are pertinent to the combustion of solid wastes are: What volume reduction is attainable? How much air must be supplied for efficient combustion? How much air must be supplied for efficient combustion? What are the emissions leaving the furnace enclosure which must be removed by air-pollution-control (APC) devices? How much energy can be recovered from the combustion gases or how much cooling must be provided to reduce the gas temperature to a level at which it an be treated in conventional APC device? A number of these questions can be answered by considering the ultimate analysis of the constituents of solid waste. It must, of course, be recognized that they composition of solid wastes varies enormously for commercial operations from practically all inert to all combustible.

Weight and Volume Reduction on Incineration

Incineration of solid wastes would leave a residue approximately equal in mass to the inert content of the waste. Small decreases in the mass of residue so estimated may take place due to the vaporization or entrainment of some of the inerts and small increases may result as a consequence of either oxidation of some of the metals or incomplete combustion of the organic content of the wastes.

The residue from the combustion of wastes would vary widely depending on the composition of the wastes. For example, waste paper and cardboard will give ash which may amount to as little as 1% of the mass of waste; furthermore, since the ash from paper waste consists of the finely divided fillers and pigments, the incinerators for such specialized wastes may be designed to have all the ash carried over with the combustion products for collection in an APC device. At the other extreme, most of the inert material in municipal wastes consists of relatively bulky glass and metal objects which are left as residue. Conventional incineration provides a means of extending the lives of landfill areas approximately tenfold. Commercial units burning compounds having low proportions of inert constituents can yield much larger volume reductions.

Table 5.28

Proximate Analyses and Heating Values of Solid Wastes (and Their Approxlmate Percentage Contribution to an Average Municipal Refuse).

		Proximate analysis "as received" basis, mass%				*Btu per lb*	
	Percent of total refuse	*Moisture*	*Volatile mater*	*Fixed carbon*	*Noncombustibles*	*Dry basis*	*Dry ash-free basis*
	(1)	*(2)*	*(3)*	*(4)*	*(5)*	*(6)*	*(7)*
Rubbish 64%							
Paper, mixed	42.0	10.24	75.94	8.44	5.38	7572	8055
Wood and bark	2.4	20.00	67.89	11.31	0.80	8613	8700
Grass	4.0	65.00	—	—	2.37	7693	8250
Brush	1.5	40.00	—	—	5.00	7900	8600
Greens	1.5	62.00	26.74	6.32	4.94	7077	8135
Leaves, ripe	5.0	50.00	—	—	4.10	7069	7700

(Contd..)

	(1)	*(2)*	*(3)*	*(4)*	*(5)*	*(6)*	*(7)*
Leather	0.3	10.00	68.46	12.44	9.10	8850	9850
Rubber	0.6	1.20	83.98	4.94	9.88	11,330	12,600
Plastics	0.7	2.00	—	—	10.00	14,368	16,000
Oils, paints	0.8	0.00	—	—	16.30	13,400	16,000
Linoleum	0.1	2.10	64.50	6.60	26.80	8310	11,450
Rags	0.6	10.00	84.34	3.46	2.20	7652	7844
Sweepings, street	3.0	20.00	54.00	6.00	20.00	6000	8000
Dirt, household	1.0	3.20	20.54	6.26	70.00	3790	13,650
Unclassified	0.5	4.00	—	—	60.00	3000	8000
Food Wastes, 12%							
Garbage	10.00	72.00	20.26	3.26	4.48	8484	10,100
Fats	2.0	0.00	—	—	0	16,700	16,700
Non-combustibles,							
24%	8.0	3.00	0.5	0.5	96.0	124	12,000
Metallics	8.0	3.00	0.5	0.5	96.0	124	12,000
Glass andce ramics	6.0	2.00	0.4	0.4	97.2	65	8000
Ashes	10.0	10.00	2.68	24.12	63.2	4172	14,000
	100.0						

Table 5.29
Characteristic Properties of Refuse and Selected Fuels

Fuel	*Moisture (%)*	*Ash %*	*Density (as fired) (lbm/cu ft)*	*Lower heating value (Btu/lbm)*	*Lower heating value (Btu/cu ft)*
Refuse	28.16	20.82	17.0	3935	67,000
Wood	46.9	1.5	60.5	3910	236,000
Peat	64.3	10.0	25.0	2390	59,800
Bituminous coal (high-volatile)	8.6	8.4	50.0 11,	735	585,000
Fuel oil (space 6)	1.5	0.08	61.5	17,280	1,060,000
Methane	0	0	0.042	21,555	1050

Heating Values of Solid Wastes

The wide variation in heating value of solid wastes may be inferred from the values in Table 5.28. The heating value has been reported on a dry basis, and is for case that the moisture in the product gases is condenses (this is referred to as the higher heating value or HHV and corresponds to the net energy relase when the waste and air at 60°F are burned and the combustion products are cooled to 60°F): The heat release for the moist constituents could be obtained from the dry values by multiplying the dry values one minus the moist fraction. the volatile matter refers to the weight loss observed when samples are heated in a covered crucible at 950°C for 7 minutes following an ASTM standard procedure. As such, the volatile matter provides a rough measure of the fraction of wastes that are driven off as pyrolysis products; the actual amount of pyrolysis products evolved has been a function of the heating rate and final temperature and will therefore vary with variations in the design and operation of different units. Examination of the items in Table 5.26 show variations in heating value from the negligible amounts in the non-combustible fraction to values in plastic and rubbers which have been close to those of a bituminous coal.

The characteristic properties of a municipal refuse are compared with those of selected fuels in Table 5.29. Relative to coal, refuse has been a low-sulfur, low-density, low-heating-value, high-ash content, high moisture-content fuel. The heating value of refuse on a volume basis has been much closer to that of peat than that of coal. Classification (i.e., separation of the combustible, light fraction from the inert,-heavy fraction) of refuse may be used to provide a cellulosic fuel with heating values close to those of wood or paper. Two processes for reclaiming the combustible content of solid wastes are the Black-Clawson and the CEA process.

Commercial and institutional refuse often is having high heating values and physical attributes which permit the wastes to be burned in suspension-fired units with heat recovery. More attention will be the recovery of energy from such wastes in a subsequent section.

Air Requirement for Combustion

The principal elemental constituents of wastes have been carbon, hydrogen, and oxygen. If the elemental hydrogen/carbon and oxygen/carbon atios have been designated by x and y, the stoichiometric equation for the complete combustion of refuse would be given of follow:

$$CH_xO_y + \left(1+\frac{x}{4}-\frac{y}{2}\right) O_2 = CO_2 + \frac{x}{2} H_2O$$

In words, $\left(1+\frac{x}{4}-\frac{y}{2}\right)$ moles of oxygen are needed to burn on mole

one mole of CHxOy, or $32\left(1+\frac{x}{4}-\frac{y}{2}\right)$ pounds of oxygen are need to burn (12+x+16y)

pounds of waste, or $137\left(1+\frac{x}{4}-\frac{y}{2}\right) / (12 + x + 16y)$ pounds of air are required to burn a pound of waste. A typical municipal waste is having an elemental composition of $CH_{162}O_{0.62}$ on a day basis or $CH_{2.45}O_{1.03}$ with a typical moisture of 24 mass %. The air requirements for

a pound of ash-free refuse has been therefore $135\left(1+\frac{2.45}{4}-\frac{1,03}{2}\right)$ /(12 + 2.45 + 16 x 1.03) or 4.86 lb of air. The air requirements for a pound (or kg) or typical refuse with 21% inert content has been 3.8 lb (or kg).

A good rule of thumb for calculating the air requirements of organic compounds has been that I cu ft of air at standard conditions (60°F, 1 atmosphere) is required for every 94 Btu of net energy released on combustion, computed assuming water in the combustion products to be in the vapor phase (1060 Btu are subtracted from the HHV for every pound of water in the combustion products). Thus a pound of municipal waste with a lower heating value of 3900 But/lb would need 41.5 standard cu ft of air or 3.2 lb in rough-agreement with the earlier estimate.

Fate of Trace Constituents during Incineration

The preceding section treated the principal constituents of wastes, i.e., their carbon, hydrogen and oxygen contents.

Sulfur. A typical municipal waste is 0.12% by weight sulfur. Part of this sulfur gets oxidized to SO_2 during incineration and part is left in the residue. Kaiser and Carotti in a series of tests on the emissions from incinerators observed SO_2 concentration (corrected to 12% CO_2) ranging from 56 to 132 ppm with an average of 78 ppm. If all the sulfur in an average municipal refuse were converted to SO_2 the expected concentration would be 180 ppm. Based on these observation it is concluded that only about 45% of the sulfur in the municipal waste would be emitted in the flue gases. The amount of sulfur tied up with the ash and residue will, of course, depend on the concentration of alkali or other base elements in the ash and residue.

For coals, in contrast, most of the sulfur gets converted to SO_2. In some situations the use of dual firing of refuse and coal may cause greatly reduced emission of SO_2 due to the dual effects of a replacement of a high-sulfur coal with refuse and the absorption of the SO_2 by the ash derived from the refuse.

Nitrogen. In recent years the emissions of nitric oxide is of increasing concern. It is known that organically bound nitrogen it oxidized to nitric in high yields. The nitrogen in solid waste is contributed by compounds of both natural (meats, leathers, silks, wool, etc.) and synthetic (polyurethane, nitrite-containing polymers, urea-formaldehyde resins, etc.) origin. Typical concentrations of 0.50% by weight nitrogen would yield, if converted stoichiometrically to nitric oxide, concentrations of 2100 ppm of nitric oxide. A very small fraction (<0.038) of the nitrogen in solid wastes is converted to nitric oxide.

Chlorine. Chlorine be present both as inorganic chlorides or in polymers, most notably polyvinyl chloride (PVC). It is estimated that chlorine is present in municipal waste in concentrations of 0.5% by weight, less than half of it in plastics. The major contributors to organically bound chlorine have been PVC resins which account for 0.48% of the solid waste (this value is based on an estimate that plastics contribute 1.5% of municipal waste and that PVC constituted 32% of the plastic). The chlorine in the flue gases has been practically exclusively in the form of hydrogen chloride. Tests conducted by Kaiser and Carotti with the addition of up to 4% PVC yielded concentrations of HCI of up to 3030 ppm (corrected to 12% CO_2), corresponding to emissions of 60 to 85% of the chlorine in the PVC. Similar high-percentage convergence to the halogen acids would be expected with other halogens tied up in polymers (hydrogen fluoride, for example, would he emitted during the incineration of teffons).

Traco Metals. It is expected that metals or metallic oxides with low boiling points would be distilled out of the residue and be carried over preferentially with the fly ash. The high content of aluminum, lead and zinc found in fly ash and furnace deposits support such a postualte but the data are too incomplete to permit the closure of a material on the trace metals.

Design Criteria

The open-burning dump and backyard incinerator have been the crudest forms of incineration and are no longer permitted because of their air-pollution potential. The objective in the design of incinerators has been therefor to (i) provide a furnace to contain the burning refuse, (ii) supply air in the quantities required for combustion, (iii) mix the combustion air and pyrolysis products on a molecular scale in order to insure complete burnout of the combustibles before the combustion gases are quenched, (iv) temper the gases to prevent damage to the refractories and ultimately to reduce their temperature to a level at which they can be effectively cleaned, (v) remove particulate and noxious gases from the flue gases, (vi) feed the refuse and remove the residue from the fuarnce without the escape of combustion products (vii) achieve the above requirements economically in a trouble-free unit, (viii) in residential areas provide units that meet certain minimum aesthetic standards.

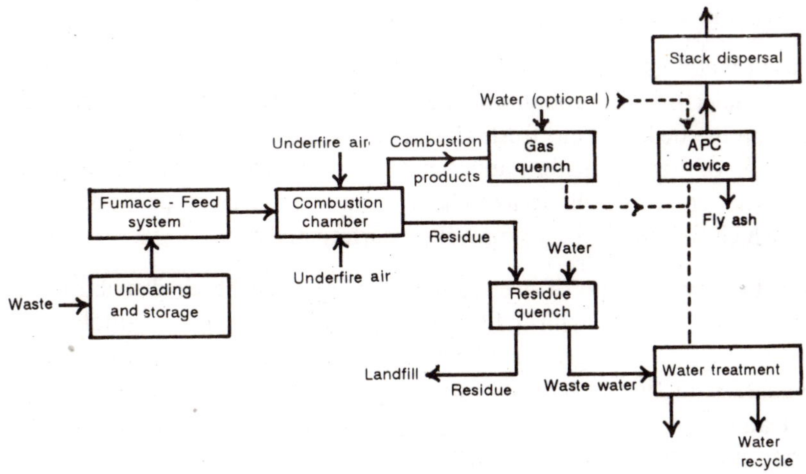

Fig. 5.2. Schematic diagram of an incinerator.

The design details will of course differ widely with scale of operation and type of waste being treated. The schematic flow of refuse in an incinerator unit shown in Fig. 5.7 identifies the different components needed.

Incinerator Components

Receiving and storage areas. Commercial wastes like wood chips, plastic shavings, or other relatively fine homogeneous materials could be stored in bins and conveyed pneumatically. Where such systems can be used economically they offer a preferred method of handling solids. Although pneumatic conveying was initially limited to specialized wastes, such systems are developed for the handling of shredded wastes in large industrial complexes, and as a demonstration unit, for municipal refuse.

The large volume required for storage, the high initial cost and maintenance on cranes, and the high salaries of crane operators make the storage and handling of refuse on of the major costs of incineration. Clearly there has been a room here for the invention of a better system.

Refuse-feed system. The feeding of refuse could be either batch or continuous. Many commercial units have been batch fed. The trend in municipal incineration has, however, been towards continuous units.

Continuous feeding is usually achieved by a hopper and a refractory-lined or water-cooled gravity chute.

Grate and air supply. The functions to be performed on a grate are discussed here. The refuse fed to the furnace has been first dried and preheated by the radiation from the hot combustion gases and refractory. The refuse, as it is heated further, first pyrolyzes and then ignites. Combustion occurs both on the grate to burn out the residue and in the gas space to born out the residue and in the gas space to born out the pyrolysis products. Overfire air jets greatly help the mixing and combustion in the overfire air space.

It is apparent that a grate must undertake a number of functions: provide support for the refuse; admit the underfire air through openings; transport the refuse from feed chute to ash quench; and agitate the bed to bring fresh charge to the surface. Agitation is valuable both in increasing the rate of heat transfer, and hence combustion in the bed, and in covering "blowholes" on the grate (small areas where complete burnout or bulky objects leave spaces through which the underfire air may flow perferentialy).

The ability to vary the ratio of over-to-underfire air and the air distribution under neath the grate provides the operator with a flexibility of operations that is rarely exploited. The recognition that increased under-grate air velocities result in increased particulate entrainment is used to advocate reduction in underfire air rates. But reduced underfire air rates burning rates and may lead to hot spots on the grate (with consequent grate warping or burnout). The underfire air rate must be selected to balance the conflicting demands of high air rates for high burning rates and of low air rates to minimize the particulate loading or the APC system. In many incinerators 60 to 100 percent of the stoichiometric air requirements is supplied under the grates.

The overfire air jets must fulfill the dual function of distributing the air over the bed in order to satisfy the stoichiometric air requirements locally and to promote the turbulent mixing required to complete the combustion process. For a typical municipal refuse it has been found that adjustment of the air rate to provide about twice the stoichiometric amount is near optimal. If this value is exceeded significantly the probability of quenching the gas-phase reactions is increased. If the excess air gets reduced significantly some pockets of fuel-rich gas will escape the combustion chamber.

The amount of excess air required by different fuels has been a function of the ease of mixing the fuel and air. The above values of 100% excess air recommonded for the combustion of solid waste in conventional grate units is somewhat larger than the amounts of 30% excess air used for lump-coal combustion, 15-20% for pulverized-coal combustion, and 1-10% for gas and oil combustion. The burning rates that are conventionally attained in incinerators range from 50 to 70 lb/ht/sq ft normally encountered in coal-burning stokers.

The residence time of the solid waste for complete combustion is, of course, a function of both burning rate and fuel-bed loading. The burning rate in fuel beds is determined primarily by the under-fir-air supply (augmented at times by the penetration of the bed by overfire air) provided that the refuse is not oversized.

Starved-air incineration. In order to avoid the entrtaiment of excessive amounts of particulate from the burning refuse bed, the air rates through the grate can be reduced considerably; this produces a so-called starved-air incinerator. In these units the combustion of the pyrolysis and gasification products leaving the fuel bed is not completed above the burning bed but in a secondary combustion chamber. Sometimes a secondary fuel supply has been provided in the secondary combustion chamber for use in periods when wastes with low heating values are burned or during start-up and close-down. The starved-air incinerator has been successfully used in commercial operations.

Kilns. The tumbling action within a rotating, inclined kiln has been a means of transporting and agitating solid wastes with out the use of grates or as a substitute for a final grate. Kilns are particularly useful for handling substantial amounts of thermoplastic wastes which would melt and clog grates of conventional incinerators or a mixture of solid and liquid wastes. Kilns are successfully used for handling mixed wastes in large industrial complexes, for completing the burnout of residue in conventional incinerators, and are also incorporated in the Monsanto Landgard pyrolysis process which will be discussed later.

Suspension-fired units. Wood chips, plastic wastes, or shredded refuse when reduced to a size at which their burnout times have been lower than residence time of the combustion gases in a boiler (ca. 2 seconds) may be burned in suspension. This criterion for the size of wastes which could be burned in suspension may be relaxed for vortex burners in which the centrifugal forces act to keep suspended particles in the combustion chamber for several gas-residence times. When all the waste to be burned has been finely divided, suspension-firing in a conventional boiler or vortex burner is justified. If a fraction of the waste is too large to burn in suspension, spreader stokers provide a means of combining the merits of grate and suspension fired units. The waste is injected in suspension above a grate stoker onto which the larger particles fall and are burned as in a conventional incinerator.

Open-pit burning. The potential of overfire-air jets for essentially eliminating the combustible content of gases rising from a burning refuse bed has been exploited to reduce emissions from this simple form in incinerator.

Vortex incinerators. In all these, the combustion air is injected tangentially above the buming bed, spirals down through the outside of the bed and up through the inside. The burning rates reported for this design have been only slightly lower, per unit cross-section of the incinerator, than the rates encountered in conventional grate units. Such units can be operated continuously without provision for residue removal from the combustion chamber.

Combustion chamber. The combustion chamber serves to confine the flames and to augment the radiative energy transfer to the fuel bed. Refractory-walled chambers are most commonly used. In new units, water-cooled walls are being increasingly substituted for the refractory walls. Corrosion may, however, be a serious problem with the water-wall units. Tubes have had to be replaced after as little as 1000 hours of service. HCI evolved during the incineration of chlorinated plastics is often blamed for the corrosion. The high content of alkali metals, lead, and zinc in the tube-wall deposits of refuse burning furnaces has also a major influence on the rate of corrosion. Even though refuse is a low-sulfur fuel, there is a tendency for sulfur to accumulate in the wall deposits, increasing the possibility for alkali-sulfate corrosion. The extent of corrosion is dependent, in addition, on tube-metal temperature and on whether the conditions in the furnace gases are reducing or oxidizing. Several measures taken to reduce tube-metal corrosion include (a) the elimination of fuel-rich conditions by proper placement of over-fire jetc, (b) maintaining tube-metal temperatures above the condensation temperatures of H_2SO_4 or HCl/H_2O (300°F should be adequate) but below 500°F to minimize the severe attack that is sometimes encountered at higher temperatures, (c) protecting tubes subject to severe corrosion with silicon-carbide or stainless-steel shields.

Certain precautions should also be taken in the protection of refractory surfaes from slag attack. Silicon-carbide refractories offer the best wall protection at the grate level.

The volume of the combustion chamber is determined by the time required to complete the combustion of the pyrolysis and gasification products leaving the burning refuse bed. The time required to complete the combustion of gases evolved by wastes burned on a grate is limited by both kinetic and mixing constraints.

Incinerator combustion chambers should be well sealed. Many problems in operation can be traced to air in-leakage of pressure maldistribution resulting from openings made in the furnace enclosure.

Residue handling. The residue, as a first approximation, is the composite of the ash of the components of the waste burned. Examination of the analyses of residue and slags show that these are close to the value predicted from the components of a typical municipal waste. It should be noted, however, that the value of steel recovered after incineration is diminished by the diffusion of coatings into the metal.

The residue from municipal units is usually quenched in water and is either dragged out of the water by a traveling chain or is displaced by a reciprocating arm onto a conveyor. By judicious choice of angles and rates of displacement the residue can cool and drain before being finally discharged to the conveyor.

Air-Pollution Control

A critical requirement of modern incinerators is that they meet the most stringent air-pollution codes. The primary concern with emissions from incinerators has and continues to be the particulate matter emitted, which although relatively small on a national basis can account for as much as 20% of the total ambient loading in some cities. The emission of combustible pollutants, such as carbon monoxide, hydrocarbons, and soot, may be controlled in well-designed units by improvement of the combustion process within the furnace itself. The emission rates of other gaseous pollutants, such as nitrogen oxides, hydrochloric acid, and sulfur dioxide, have been small and have not, so far, presented a problem.

Particulates

The particulate emission has been related to the fine ash content of the refuse, to the incinerator design, and, for grate units, to underfire-air rates. For a typical municipal incinerator the particulate emission from the combustion chamber has been about 15 lb/ton of refuse charged.

Particulate is defined as any finely divided liquid or solid, other then uncombined water, collected using the standard EPA sampling train.

Partial control of emissions may be achieved by reducing the undergrate-air velocity and/or bed agitation, thus reducing the emissions from the combustion chamber. The most effective treatment, however, is by removal of the particulate from the combustion products. In order to assess the effectiveness of APC devices it is necessary to first have estimates of the particle-size distribution leaving the furnace enclosure.

Gas-quench Systems

The gases processed by bag filters, mechanical cyclones, or electrostatic precipitators have to be cooled to temperatures in the range of 450 to 700°F depending on the construction of the particulate-collection device. In water-wall units this cooling is attained by radiative and convective heat transfer to the cooled tube surfaces. For refractory-wall combustion chambers, the gases are cooled by the use of excess air to about 1800°F in the combustion chamber (to minimize slag attack of the refractory surfaces); further reduction in the gas temperature to the values imposed by the constraints in the particulate-collection equipment can be achieved by dilution with air or vaporization of water. Air quenching is impractical in modern units since the flue-gas volume imposed by the dilution with air or vaporization of water. Water quenching, however, results in an approximately 40% decrease decrease in volume, since the decrease in temperature on water vaporization more than offsets the increase in mass of the flue gases. Two designs wet-bottom and dry-bottom, are available for the water-quench system. In the wet-bottom system, water in amounts exceeding twice that needed for cooling is pumped through banks of coarse-spray nozzles located in the flue leading to the APC system. The excess water collects in the bottom of the flue, and must be treated before disposal or recycle. The treatment of the water stream, which is highly acidic (pH <2.0) due to the absorption of HCI and organic acids, and which is contaminated with particulates, is a major disadvantage of the wet-bottom system. It is, however, cheaper to install and easier to operate than the dry-bottom system in which in which the water is injected, in the amounts required, as a fine enough spray that complete droplet evaporation is achieved. To provide the residence time required for evaporate, cooling towers 30 to 90-ft high are provided. The dry system needs high water pressures, from 100 to 600 psig, and filtered water to prevent clogging of the fine spray nozzles.

Gaseous Pollutants

No absolute regulations presently exist on the emission of gaseous pollutants from incinerators, but guidelines have been issued. Some measure of the problem may be obtained from considering the emission standards for other stationary sources. The national emission standards for sulfur dioxide from klarge stationary sources has been 0.8lb/million Bru for oil-fired boilers and 1.2 lb/millions Btu for coal. The emission of SO_2 from refuse-fired incinerators corresponds to less than 0.3 lb/million Btu, sufficiently below the standards that refuse can be considered to be a low-sulfur fuel. The standards for emission of nitrogen oxides (expressed as NO_2) vary from 0.2 lb/million Btu for gas-fired boiles and 0.70 1b/million Btu for coal. The average emission from municipal incinerators is about 0.4 lb/million Btu which is intermediate to the recommended standards for coal and gas. The total emissions of nitrogen oxides from incinerators represents only 3% of the national total. The emission of halogen acids could be a problem if unabated. Fortunately, the halogen acids can be readily absorbed by a water wash, and scrubbers for their removal can be readily incorporated into the air-pollution-control system. The variability in furnace-gas composition is large, depending on the feed compositions and incinerator design and operation (Table 5.30).

Table 5.30
Furnace-Gas Compositions

	Concentration Ranges		
	Miami County	*Occanside*	*Norfolk*
SO_2ppm	2-303	0-100	0-10
HCI, ppm	5-115	1-330	14-75
HF, ppm	Traces-0.6	Traces-1.1	Traces-6.0
Nitrogen Oxides, ppm	10-138	107	4-7
Organic Acids, ppm	35-136	Not measured	150-340
Oxygen, vol%	15.0-9.5	17.0-8.0	16.5-15.0
CO_2, vol%	6.0-10.8	3.5-12	3.9-5.6
CO, vol%	Traces-0.2	0.005	0.1-0.2

Heat Recovery and Energy Generation

Energy has been recovered from the combustion of specialized wastes. The use of municipal wastes as a source of energy for steam and electricity production has been of greater interest in countries with more expensive fuels. As a consequence of the projected deficiencies in the supply of domestic oil and gas, and the concomitant increase in the cost of energy, the recovery of energy from solid wastes has been becoming increasingly attractive. The in cineration of solid wastes can be a significant source of energy but cannot, by itself, provide a solution to the energy crisis. It might also be noted that incineration returns carbon to the atmosphere faster than does landfill, thus accelerating the natural process of recycling by means of photosynthesis.

The fraction of the heating value in refuse that can be recovered as energy will depend upon the heating value of the refuse, the excess air used in combustion, the stack-gas temperature, and the heat losses in the processing equipment.

The economics of energy recovery from the disposal of solid wastes depends to a large extent on energy cost, on the local demands for energy, and on the ease of distribution of the recovered energy. Steam can usually be utilized only in the immediate vicinity of the generating plant (unless a stem distribution system is available). A fraction of the steam generated may be used in-plant for heat, turbine-driven blowers, and turbine-generators. Sale of residual steam is facilitated when an incinerator is located next to major users such as municipal facilities, hospitals, residential or office complexes, electric utility power-plants, or military installations.

An alternative to steam generation is the production of transportable energy. Processes are under development for the conversion of the energy in wastes to gaseous or liquid fuels by pyrolysis or to electricity by means of gas turbines driven by the products of the high-pressure combustion of refuse. These will be discussed later.

From a practical viewpoint, incineration has been first and foremost a means of reducing the solid-waste problem and energy recovery should be considered to be a beneficial by-product when local circumstances make the economics attractive.

INCINERATION COSTS

The costs of incineration have changed greatly during the period 1975-1986 for a number of reasons.

Air- Pollution-control requirements have been made much more stringent than they were, resulting in large additional costs.

Interests rates have risen sharply, particularly affecting incineration because of the high capital costs of incineration plant.

Labor rates have also increased sharply.

Inflation has increased the first costs through positive-feedback of some of the above factors and through other mechanisms.

Energy costs have risen sharply, increasing the credit from sale of steam or hot water.

NEW DEVELOPMENT

Space does not permit description of the large number of concepts and invention proposed for the thermal disposal of solid wastes. The discussion here will be restricted to those processes that have been developed at least to a pilot-plant stage.

Pyrolysis Units

The pyrolysis of solid wastes strictly refers to the thermal decomposition of the wastes in an inert atmosphere. Under such conditions a mixture of gaseous products, tars, water-insoluble oils, and an aqueous solution of acetic acid, methanol and other organic compounds is evolved and a solid residue composed of the inert content of the waste and a char is produced. The amounts of the different products that are produced are dependent upon the rate of heating and the final temperature to which the wastes are subjected. In general, the higher the heating rate and the higher the final temperature, the greater the fraction of the initial waste that is converted into the gaseous and liquid products. The products of pyrolysis of refuse confined in a retort and heated externally to different temperatures support the above generalization. The gaseous yield is highly variable but is about 25% by weight of the air-dried, ash-free refuse and has heating value of about 300-350 Btu/cu ft. The solid product or char resulting from refuse pyrolysis is an impure carbon (excluding the ash), very similar in proximate analysis to coal. The yield of char is nearly 17 to 25% by weight of the air-dried, ash-free refuse, decreas in with both heating rate and increasing pyrolysis temperature. The calorific value of the char has been about 11,000-12,000 Btu/pound of ash free char. The ash content of the char if inerts are not removed from the refuse would probably vary between 50 and 60% by weight. The aqueous liquid phase is having about 70 % 80% water and has a heating value of 1000 to 2000 Btu/pound of combustible refuse.

The advantages of the process are that the gas volume to be treated is a small fraction of that needing needing treatment in a conventional incinerator; fuel by products are produced, and the slagging operation provides a 97% volume reduction and a putrescible-free residue. A 200-ton/day pilot plant has been successfully operated at Charlestion, West Virginia.

Another pyrolysis unit at the pilot-plant stage is that developed by the Garrett division of Occidental Petroleum. The process involves the coarse shredding and drying of the solid wastes followed by air classification to separate the combustible from the inert. The advantage of the process is that it produces a readily-transportabe fuel.

Slagging Incinerations

The enormous size reduction achievable in slagging units provides the major motivation for their development. An additional small benefit may gained from the value of the slag as aggregated. The high temperatures required for slagging have been achieved by the use of oxygen, by the use of preheat air (Torrax), and by the use of supplementary fuels (American Thermogen). The gases from the shaft furnace contain gasification and combustion products which are burned in a secondary combustion chamber. The resulting gases are cleaned prior to discharge to the atmosphere.

RESOURCE AND ENERGY RECOVERY FROM SOLID WASTE

Introduction

Solid wastes may be recycled be recycled if one of two cost categories rises to beyond an economic level. One cost category has been that of material cost, which rises due to scarcity, or due to distance from production sites, for instances. (Material cost may also be forced to rise through taxation). The second cost category has been that for disposal of solid wastes. The disposal cost depends on the nature of the wastes, on the location where the wastes are produced, and on the regulations governing permissible disposal practices.

In general, materials of high unit cost (gold, copper, lead for examples) have been recycled simply because the recycling costs have been usually lower than the costs of buying and refining virgin materials. Substances of low unit cost (paper, animal manures) have been recycled only when the disposal costs have been high either because of antipollution regulations restricting the use of low-cost disposal methods or because of the non-availability of disposal sites within reach of local transportation (for example, in metropolitan areas).

There is a range of materials of intermediate value (e.g., ferrous metal, glass) which have been found to be responsive to either or both cost categories.

Table 5.31
Resources Lifetimes

Resource	*Known reserves*	*Static index (years)*	*Projected rate of growth (% years)*	*Exponential index (years)*	*Exponential index calculated 5 × known reserves (years)*
1	2	3	4	5	6
Aluminum	1.17 × 10^9 tons	100	6.4	31	55
Chromium	7.75 × 10^8 tons	420	2.6	95	154
Coal	5 × 10^{12} tons	2300	4.1	111	150
Cobalt	4.8 × 10^9 lb	110	1.5	60	148
Copper	308 × 10^8 tons	36	4.6	21	48
Gold	353 × 10^6 troy cz	11	4.1	9	29
Iron	1 × 10^{11} tons	240	1.8	93	173
Lead	91 × 10^6 tons	26	2.0	21	64
Manganese	8 × 10^8 tons	97	2.9	46	94
Mercury	3.34 × 10^6 flasks	13	2.6	13	41
Molybdenum	10.8 × 10^9 lb	79	4.5	34	65
Natural gas	1.14 × 10^{15} cu ft	38	4.7	22	49
Nickel	147 × 10^9 lb	150	3.4	53	96
Petroleum	455 × 10^9 bbl	31	3.9	20	50
Platinum group	429 × 10^6 troy oz	130	3.8	47	85
Silver	5.5 × 10^9 troy oz	16	2.7	13	42
Tin	4.3 × 10^9 long tons	17	1.1	15	61
Tungsten	2.9 × 10^9 lb	40	2.5	28	72
Zinc	123 × 1^6 tons	23	2.9	18	50

Solid wastes can be considered as composed of non-renewable resources on the one hand and of other materials such as over-burden (mining wastes), manures (animal-husbandry wastes), and wood products and paper (sawmill and domestic wastes, for instance) on the other hand.

Incentives toward reclamation of non-renewable resources will increasingly be due to the increase in price that will accompany the inevitable long-term decrease in availability.

There have been various estimates of the known useful reserves. Table 5.31. shows known reserves estimated by the U.S. Bureau of Mines in 1970 together with the number of years that these reserves would last at present rates of consumption. The rates of growth of consumption have been also shown and the number of years which each resource would last if this rate of growth of consumption continues. Also shown is the number of years which each resource would last under present rates of growth of use if the known reserves were multiplied by five-that is, if there were enormous new finds. The effects of the steady rate of growth can be seen to be dramatic. In the case of coal, the principal fuel resources presently available, there has been enough to last 2300 years at 1970 rates of utilization. However, if the rate of growth continues to be 4.1% year (and it is likely to be more as other fuels are used up) the expected resources life-time drops from 2300 years to 111 years. If there were five times as much coal available, it would last only another 39 years at the projected rate of growth.

The Reclamation Industry

The overall scrap industry has been showing a picture of great complexity, because of the multitude of alternative ways recycled materials can move and be processed. However, individual movements and processes are usually quite simple. Fig. 5.8 shows a number of alternative paths for wastes to pass from consumers (indicated as "sources") to scrap-consuming industries. Sometimes for some materials and in some areas all these alternative paths may be open. For other materials and in other areas only one or two sanctioned. The structure of the market could be governed partly by economics and partly be restrictive agreements. These restrictive agreements, where they exist, can be explained by the desire of certain sections of the industry to protect themselves during a long period of falling or fluctuating scrap prices, rising labor costs, and falling virgin-material prices.

Within the scrap-consuming part of the industry, the movement of scrap has been typified by the situation for ferrous materials, shown in Fig. 5.8. Ferrous solid waste from consumers can pass through the scrap industry to the steel industry for production of new iron and steel. During the production process some scrap is produced during, for instance, the cutting and trimming of the steel products, and this is termed "home" scrap. It has been of high quality and can be immediately returned to the furnaces. The products of a steel mill-billets, forgings, sheets and so forth-are used by fabricating industries producing refrigerators, automobiles and so on, and from these operations so-called "prompt industrial" scrap is produced. This scrap is normally of high quality, especially when produced during mass production, because it is then worthwhile to keep different grades of material segregated. A high proportion of prompt industrial scrap is, and always has been recycled.

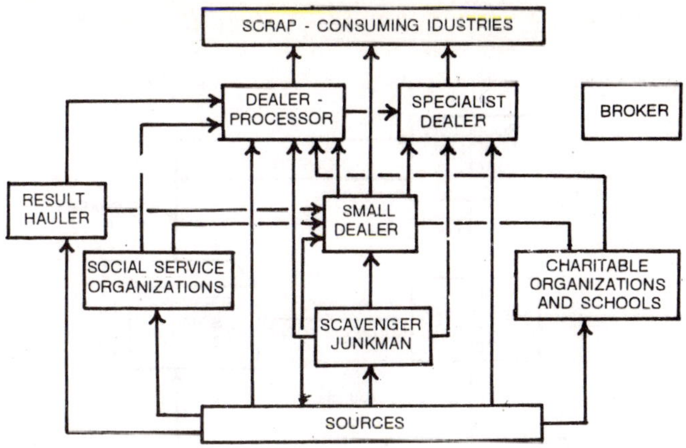

Fig. 5.3. Structure of the salvage industry and flow of commodities.

When the steel product (still using steel as an example typical for many other industries) reaches the consumer and, eventually, the end of its useful life, it is discarded along with an almost infinite variety of other products and materials. It is the mixture of materials which causes great difficulties in the return of consumer wastes termed "obsolescent" scarp-to the scrap industries. Small quantities of impurities in most categories of scrap can render them valueless. In steel, a proportion of 0.02% copper can reduce the value of scrap steel of zero.

The existence of these "internal" routes for scrap (i.e., "home" and "prompt industrial" scrap) give rise to confusion in representing the actual proportions of solid wastes which are being recycled. Someone wishing create the impression that there has been a large proportion of wastes being recycled will include home and prompt industrial scrap, whereas if it is desired to show how small a proportion of scrap is being salvaged, only post-consumer or obsolescent scrap will be considered. Definitions change with changing practices. Material that is normally not considered as wastes should not be classified as being in the recycled waste stream. However, there are many examples of industries where material which was once regarded as valuable and even essential and therefore was not "recycled," but was utilized as a normal part of doing business, is now a severe solid-waste problem because of changing economics and technology. Animal manures constitute an example.

Specifications or Recycled Materials

Quality standards for categories of recycled materials are usually governmental concerns.

Separation of Mixed waste for Reclamation

If we are able to collect secondary materials in a reasonably homogeneous form (a description which needs careful definition they can normally be reclaimed and returned either to original used (such as ferrous and non-ferrous metals) or to modified uses (such as many grades of paper; wood wastes; most plastics; and so forth). When we collect wastes in the

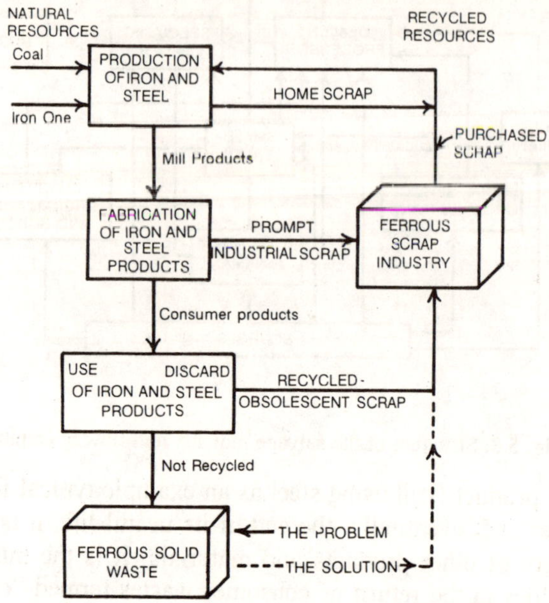

Fig. 1.4. The scrap cycle and ferrous solid waste.

mixed state, some form of separator is required. There are a few processes which can take unmodified mixed wastes and produce useful values in a single operation. The mixed wastes can, for example, be incinerated to produce thermal energy; pyrolyzed (thermally decomposed in the absence of oxygen) to produce oils, tars, chars and fuel gases: fermented to produce alcohol; and so on. Each of these processes has been in itself a form of separator. These processes would, in general, work more efficiently if mechanical separation were effected first. Preliminary mechanical separation of mixed wastes provides maximum freedom to choose among alternative uses for the homogeneous materials produced.

Hand Sorting

The traditional method of separation is by hand. Over the last few decades, hand separation has generally been accomplished by extracting useful materials from mixed refuse on a moving belt passing "picking" stations. A magnet has usually been employed to remove ferrous materials.

Mechanical Sorting

For reclamation from mixed municipal wastes to be generally feasible, either the value of secondary materials must be increased or the proportion of labor costs must be reduced by some form of mechanization or automation. There are many alternative possible approaches, not all of which can be listed here. However, the broad categories of possible approaches to mechanical separation have been as follows:

1. The refuse may be sorted in the as-received condition or it may be comminuted-shredded, hammer-milled or otherwise reduced in size.

2. The whole bulk of the refuse may be subjected to a selective force field-magnetic, electrostatic, electrodynamic, or aerodynamic-or individual items may be sorted one by one.

3. Sorting can be done by a series of binary sorts until a sufficient degree of purity has been achieved; or a stream of incoming items may be switched among a possibly large number of categories in one operation.

Central-station Separation of Mixed Municipal Refuse

Introduction. Any municipal solid-waste management system which includes the collection of municipal refuse and transportation to a central station for processing prior to final disposal could be designed to include some degree of separation for reclamation and refuse. This selection reviews available and proposed techniques for carrying out such central-station separation, including methods which range from manual sorting from a conveyor to fully automated mechanical separation plants.

Central-station separation began as an adjunct to the operation of large incinerators and shredding stations. At the very least, it has been necessary to remove oversized or unshreddable objects as well as items likely to explode during incineration. Automatic removal of ferrous scrap with magnetic separators is also widely used, either before or after incineration. In the past five years, there has been a rapid increase in the development of separation techniques specifically designed to remove a major fraction of municipal refuse for reclamation and refuse. Some of these methods, such as composting and pyrolysis involve the conversion of solid waste to other uses, and are treated elsewhere in this volume. This section focuses on techniques of direct separation of items from municipal refuse, with the identity of the items (i.e., as glass, fiber, metal, etc.), being unchanged in the separation process.

Coding and switching. A sorting process can be thought of as requiring two steps: an initial step in which an item in as input stream gets coded according to its proper category or classification, and a second step in which the item is physically switched into the output stream corresponding to that category. Some sorting devices combine the coding and switching functions into a single operation, while others have been involving separate coding and switching operation.

Manual coding and switching. The simplest kind of separation process, and the one most widely used, has been manual separation. Refuse from collection vehicles is loaded on to a conveyor and carried past workers who hand-pick items which either must be removed from the stream (such as explosive aerosol containers) of which have salvage value, such as scrap metal and clean paper. The worker is in this case performing both the coding and switching functions, although the speed and sophistication of the coding decisions (i.e., recognizing individual items) far exceeds the speed and dexterity of the switching function (i.e., manual removal of each item from the conveyor).

Manual coding and machine switching. A variation of the manual sorting method involves taking advantage of the human's remarkable skill at recognizing and coding individual items, but providing mechanical methods for performing the switching functions.

Machine coding and switching. An even more automated version of the sorting process relies on machines to do both the coding and switching functions. Which the separation machines to be described below could be expected to handle a much greater volume of material tham humans, they can perform. Due to this fact, it requires many different kinds of machines to synthesize a system capable of performing automatically separation tasks which a human could do in one step, albeit at a slower rate.

Automatic bulk sorting. The bulk sorting of refuse, unlike the item-by-item sorting carried out by manual methods, foregoes any attempt to classify every item on a refuse stream. Instead, average variations of inertial, electrical or other properties of the items belonging to different categories are used as a basis of classification. Bulk methods depend on prior size reduction by shredding and size separation by screening.

Separation by pulping. The foremost example of automated bulk sorting of municipal refuse is the Black Clawson plant in Franklin, Ohio. The technology of this process is based on methods for the manufacturing of paper. The central purpose of the process is to recover a large fraction of the fiber in municipal refuse in a form which is well suited for recycling (as paperboard or felt), with the remaining fiber being prepared for efficient incineration in a fluid-bed reactor. Ancillary to this treatment of the fiber component of refuse, which accounts for about half the mass of municipal refuse, is the recovery of other reusable fractions such as glass, ferrous scrap and non-ferrous scrap. A schematic diagram of the plant is shown in fig. 5.5; a step-by-step description of the process follows.

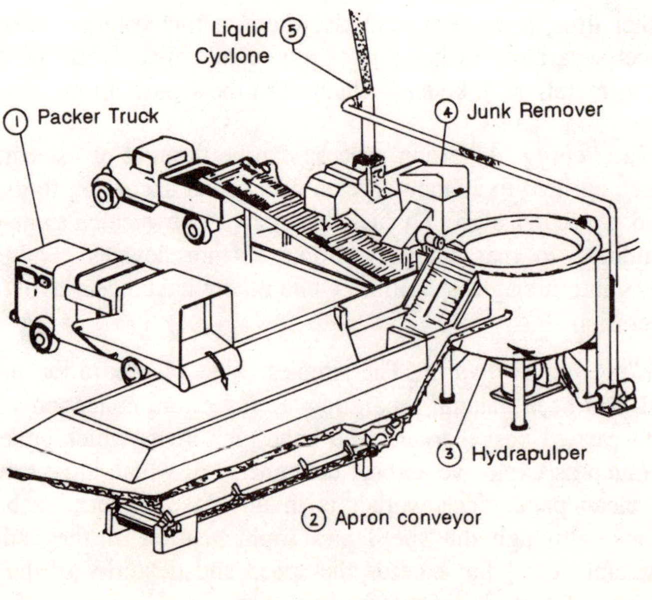

Fig. 5.5. Pulping reclamation plant.

Refuse from the packer truck gets discharged onto an apron conveyor from which it is fed to a pulper. All pulpable, grindable, and friable materials like food waste, paper, plastic and rubber, rags, glass, wood, and leaves, are disintegrated and removed through a screen in a water slurry. Heavy metallic items get ejected from the pulper and removed by the junk remover. The slurry is pumped through a liquid cyclone, which removes broken glass, sand, and other small dense items such as metals. The slurry is then sent through the Fibreclaim R process, developed by Black Clawson specifically for the reclamation of reusable fiber from such a slurry. The slurry first gets screened, then further cleaned in centrifugal cleaner, and then sent to fiber selectors which remove long and intermediate length fibers which are dewatered and baled. The residue is then dewatered and sent to a fluid-bed reactor for incineration, the ash being taken to a landfill. The rejects from the liquid cyclone are having about 75% glass mixed with metals, plastics, sand and dirt. Stages are being added which will sort glass, aluminum, and ferrous metals from this residue. The glass is then sorted by color using the Sortex process. This system, at the time of writing, was recovering about 15% by weight of the incoming refuse, as pulp for use in felt manufacture and as steel cans.

Separation of Incineration Residues

An automatic continuous process has been developed for the recovery of resources from the residues or municipal incinerators. Like the Black-Clawson process, the Bureau of Mines process involves many binary sorting operations. The basis of the technology, however, is the ore-handling and ore-separation technology of the mining industry.

Incinerator residues, which are a soaking-wet complex of metals, glass, slag, charred and unburned paper and mineral-containing ash, are first screened in two stages, the larger fractions being shredded in a hammermill before further screening. All screening operations involve continuous spraying with water. The plus-4 mesh component from the shredder is subjected to a 30-in-diameter by 14-in-wide permanent-magnet drum-type separator, the non-magnetic fraction being further shredded in a second hammermill. The minus-4-plus-20-mesh fractions from all steps are combined for an additional stage of magnetic separation in a 3000-gauss magnetic separator is able to remove the weakly magnetic colored glass, leaving a product that is primarily flint glass contaminated with small amounts of light amber glass and carbon. Optical glass separation of glass by color using the Sortex process may replace this method of glass separation. The fines from all screening operations are combined, the waste solids being separated from a waste slime which is thickened to a pulp.

Fluid Classifiers

The movement of shredded and size-classified items in an air or water stream has provided the basis for several kinds of separation methods. The Stanford Research Institute has investigated a zig-zag classifier (Fig. 5.6) in which shredded refuse is fed into an air stream in zig-zag inner compartment. The heavier fraction, which cannot be supported by the air stream, falls out the bottom, while the lighter fraction is carried by the airstream past the zig-zags and out the top. This device has been strictly a binary sorter, and many such devices would be needed to accomplish separation of mixed municipal refuse. The rising-current separator achieves a similar result in a different configuration using water.

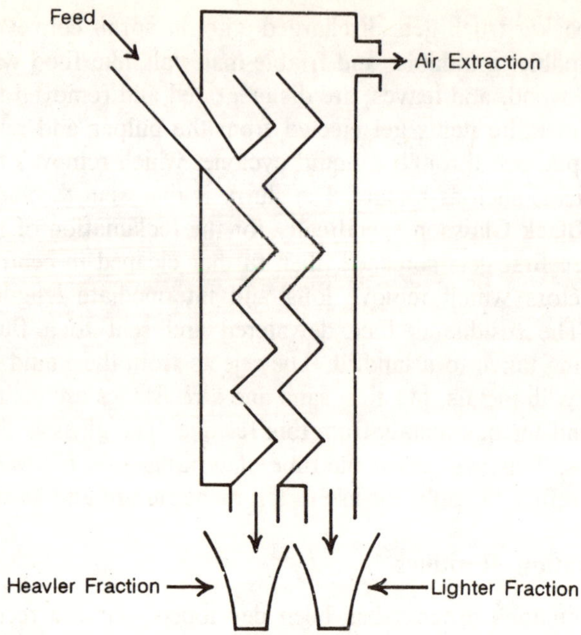

Fig. 5.6. Zig-zag Classifier.

Inertial Classifiers

Related to the air classifiers described above are sorters which attempt to use inertial properties of sized particles for sorting. Three kinds of devices have been developed principally for the purpose of removing contaminants from compost. In the ballistic separator, the lighter, organic particles experience more of a drag force and fall into the bin nearest to the rotor. In the secator, particles are bounced onto a counter-rotating pulley. Those that stick, the lighter and inelastic particles, are dragged by the pulley into a bin while the heavier and resilient particles bounce off.

Eddy-current Classification

The principle of electromagnetic induction can be used to separate particles on the basis of conductivity and density. In a process being developed at Vanderbilt University, the magnetic force on a conducting particle in which currents have been induced by a time varying magnetic field is used as a basis for the sorting of non-ferrous scrap. Tests of several different methods showed initial success in separating aluminum, which is highly conducting and of relatively low density, from other non-ferrous scrap components.

Automatic Item-by-item Sorting

There have been several time-by-item sorting devices which may play an important role in the overall design of a central reclamation plant. The most prominent of these has been the Sortex process for separating glass by color. Particles are driven on a high-speed belt and allowed to fall through an optical chamber where the reflectivity of the particle on all sides is compared with standard backgrounds by a set of photocells. If the particle does not match the back ground, an air jet blows the particle into a reject chute. Successful operation of the machine depends on rather close control of sizing.

An item-by-item sorter that attempts to simulate the sophistication of the manual sorter is developed at M.I.T. In this process, the larger items are screened from incoming refuse for sorting, large in this case being items are screened from incoming refuse for sorting, large in this case being items larger than a few inches in diameter. The items have been automatically loaded individually into conveyor carts and sent part a set of sensors each of which obtains a signal, or "signature," from each item. These signatures are fed to a computer which collects all the signatures for a given item, and makes a classification decision based on standard signatures stored in the computer memory. It then directs the switching of the item to the appropriate output category. Items which defy classification are rejected.

Two kinds of sensors are developed. The first is an infrared reflectance spectrometer, which measures the relative intensity of light reflected at a set of infrared wavelengths. The second is an impact sensor, which measures the declaration of an impact tools as it strikes the item under known impact conditions.

REFUSE INCINERATION

Small and Medium Sized Installations

On the face of it, incineration of refuse should be the most satisfactory way of getting rid of the mounds of unwanted by products created by the living processes of our advanced civilization. A large proportion of the waste we produce is combustible. It includes newsprint, paper, card board and other package materials. PVC, etc.

Each of these materials is having an appreciable calorific value.

The nature of refuse varies considerably according to the type of dwelling. People who occupy houses in areas without smoke control seem to burn a lot of their waste newsprint and packaging materials. It has been found that paper and cardboard only comprise some 35 per cent of the refuse in such areas. People who live in small houses which are equipped with central heating produce refuse which is noticeably richer in packaging materials-41 per cent of their refuse consists of paper and cardboard. Refuse from flat dwellers is having on even higher proportion of paper and cardboard-over 65 per cent. The probable reason has been that many occupants of small houses burn paper in garden bonfires even if they are unable burm it inside the house. But for all classes of dwellings there is a definite rise in the proportion of packing materials in refuse. For this reason, increasing attention has been focused on the possibilities for incinearting waste matter-preferably with recovery of the heat produced.

Problems of Refuse Incineration

In most of the highly developed industrial societies the calorific value of refuse fluctuates between 6 and 9 MJ/kg. This compares with the calorific values of various fuels as follows:

Coal	32-36 MJ/kg
Wood	18-20 MJ/kg
Lignite	20-25 MJ/kg
Coke	30 MJ/kg
Oil	42-45 MJ/kg

It can be seen that the calorific value of domestic refuse is not particularly high, and because of its very high moisture concentrations, combustion of refuse is often difficult. The usual proportion of water contained in domestic refuse has been 40 per cent or more.

The principal factors which make combustion difficult have been as follows:

(a) The material has been very variable in size, moisture content and calorific value;

(b) The density of the material has been only about 0.15-0.25 kg/dm^3, and, in addition, varies widely;

(c) The material has been a mixture of very soft and very hard constituents, containing anything from dust to large furniture sections, steel, springs, wires, etc., which have been often difficult to separate magnetically;

(d) Modern refuse contains much polyvinylchloride waste, which produces highly corrosive hydrogen chloride gas on burning.

This attacks metal parts of the incinerators and causes high maintenance costs.

It can be seen, therefore, that the installation of refuse incinerators has been by no means a simple matter in all cases. The refuse incineration cannot always be defended economically because it has been found that costs are invariably greater than those of controlled tipping. It is only when maintenance of the environment and its amenities are taken into account that it becomes obvious that incineration scores heavily. Incineration is able to reduce the volume of refuse to about 10 per cent of its original volume, and produces a compact ash which does not smell and is not putrescent. The ash produced in small and medium sized installations has been not usually hard enough for use as hard-core in building and road construction. It can, however, be used for earth filling operations, like dam and embankment construction. To get clinker suitable for building purposes it is necessary to attain much higher temperatures-not usually possible with small medium sized installations.

Small and medium sized installations generally operate without using heat produced, because it is rarely an economic proposition to reclaim the small quantities of heat generated. The reasons for this have been twofold. The actual heat produced varies according to the calorific value of the refuse being put through at the time in question. In addition it has been difficult to abstract heat from a flue gas containing high proportions of corrosive gases Sulphur trioxide, oxides of nitrogen chlorine and hydrogen chloride. Heat exchangers must be made from special corrosion resistant alloys, which adds considerably to the capital costs involved.

In plants having a capacity of less than two tons of refuse per hour, it simply does not pay to attempt to utilize the heat. Exceptions have been where the incinerator is being fed with refuse of more uniform composition, such as industrial waste. There have been many examples of small incinerators being fed with sawdust, wood shaving, waste packaging and the like, which can be burnt economically with proper utilization of the heat produced for works purposes.

Advantages and Disadvantages of Smaller Incinerators

From the cost point of view it has been found that the capital and operating costs per unit weight of refuse processed fall sharply as the size of the plant increases to about 1000 to 1500 kg/hour. Very small incinerator have been not found to be economical except in special cases where it is preferable to burn fixed quantities of refuse instead of employing a refuse collection service.

If the plant is to be larger, say, having a capacity of 10 or even 20 tonnes per hour, refuse incineration for an entire large city will be possible.

Nor are large refuse incineration plants necessarily cheaper per unit of output. With smaller plants a certain amount of rationalization in design and production of components is possible; but large incinerators are stricty 'one-off' plants which are intrinsically much more expensive. On the other hand, the area of land required for the erection of a refuse incinerator complex and its auxiliaries has been less per given quantity of refuse the larger the incinerator. Waste heat utilization-uneconomic with small incinerator plants becomes a viable feature. In fact, most of the large refuse incinrator plants are operating in such a way as to produce steam which drives back-pressure turbines, and the waste heat is afterwards used for district heating purposes.

Temperatures inside large incinerators have been high enough for partial sintering of the ash to occur; this can produce a valuable clinker suitable for building foundation work and road construction.

Classification of Small and Medium Sized Incinerator Systems

Refuse incinerators have been sub-divided into three basic types:

A. Fixed grate systems.

B. Moving grate systems.

C. Grate-less systems.

All incinerator systems must be capable of heating the feed to at least 750°C, the lower critical limit below which exhaust fumes are no longer odourless. In practice this combustion temperature varies according to the nature of the feed employed, and usually fluctuates between 800°C and 1000°C. The hot flue gases are then cooled, using dilution with external cold air to obtain an exhaust gas temperature of around 300°C.

A. Fixed Grate Systems

Fixed grate systems has been only suitable for small refuse incinerators. They are used preferably for the combustion of industrial wastes, which are more homogenous than municipal refuse. The advantage of fixed grate systems has been cheapness, but they suffer from the fact that in the absence of proper agitation, combustion of high moisture-content and high ash-content materials with a large volume and often a low density is not easy. This difficulty could be overcome by provision of high combustion temperatures using generously dimensioned gas or oil burners. Some fixed grate systems do use agitators, which are able to turn over the refuse; most, however, seem to rely on the tumbling action inherent in a step-grate.

Examples of Incinerators with Fixed Grates

(a) *Heinicke*. These incinerators burn between 100 kg and 250 kg of refuse per hour, and the units are mainly used in hospitals, schools and homes for old-age pensioners. The basic structure of the system has been a sheet-metal shell of 3 mm thickness, rectangular in cross-section. This is lined internally with fireclay bricks. The angle of the grey cast-iron grate can be adjusted manually to vary the rate of throughout of refuse according to its nature. The refuse has been loaded manually from the top, and combustion is initiated by means of a gas burner. The grate bars have been cooled by incoming primary air. The combustion gases from the primary chamber are them allowed to pass into a secondary chamber also fitted with a gas burner. This raises the temperature of the exhaust gases to between 800 and 1000°C, at which they become odourless. The supply of secondary air to this afterburner chamber has been controlled by flaps. The flue gases are then cooled by admission of air from the outside to around 350°C, and pass into a centrifugal fly-ash separator, from which they emerge to the outside. The ashes could be removed by hand from smaller units and mechanically from larger units. Larger units also employ mechanical feed methods.

There are several systems on the market, which differ slightly in design and lay-out from Heinicke systems but operate basically on the same lines.

(b) *The TPT Ltd. waste Paper Incinerator*. Dust from paper mills presents a very difficult disposal problem. TPT Ltd. has developed a plant able to burn some 140 kg/h of paper dust and contaminated paper cut-offs as well as general waste. The plant is completely automated and the volume is reduced to 2 per cent of its original figure. This system incorporates a waste heat boiler, able to utilize the heat produced to raise about 900 kg of works steam per hour. The plant consists of a supplementary fire-box around the gas burner quarrel, within the main combustion chamber. 20m^3/h of natural gas supplies rapid heat to this, and paper dust is metered into the flame path from a 2.5m^3 storage hopper by means of four 150 mm diameter worn screws. A 1.5 kW fan maintains a constant primary air feed pressure. The paper dust system has been able to handle 140 kg per hour, and there is also a separate bucket-type elevator designed to handle up to 90 kg/h of paper off cuts, plastic waste, oily rags, etc. The incinerator has been at first-floor level and the area beneath houses the collection hopper for general refuse at the foot of the elevator, as well as the large enclosed scoop into which ash and small clinkers fall via a sealed chute.

The incinerator is having three chambers. The first has been the combustion chamber, which has a holding capacity of 2 m³ and a rating of 240 kg/h. The total grate area has been 3.5 m² including a hydraulically-operated reciprocating hearth which progressively moves burning materials on to the fixed bars and thence down the ash chute. The second chamber gives a flue connection to the waste heat boiler and also permits the installation of an afterburner if so required. The third chamber is having two water troughs, the first of which includes a water sprayed chain screen to eliminate fly-ash as the waste gases pass to the exitflue. If the furnace temperature should ever rise to exceed 1200°C, the feed lines and the gas supply are stopped automatically. The chamber temperature then falls rapidly under the influence of the primary air flow.

(c) *Cremater for burning small quantities of animal remains.* This unit is able to burn up to 13 kg/h of animal remains. It is of three pass design to give maximum travel within a comparatively small enclosure for the high temperature flue gases. The feed door is mounted on parallel hinges so that it shifts sideways on opening, without exposing the operator to the hot refractory lining. This lining is made of special light weight refractories so that the door can be opened with one hand for rapid loading of carcasses. The burner has been gas or oil fired, automatically controlled to provide operating economy, and is positioned underneath the fire bars.

These are particularly popular for the combustion of reasonably high calorific value refuse such as wood wastes, certain chemical wastes, etc. Recovery of waste heat is usually worthwhile under these circumstances even if the plant has been of small capacity.

B. Moving grate systems

A large number of moving grate systems are available on the market, and these are far better for the combustion of variable composition and variable size classification refuse than the fixed grate burners described above which do require a reasonable uniformity of material. Moving grate systems have been subdivided again into the following:

(a) Chain grate systems.

(b) Roll grate systems.

(c) Forward feed grate.

(d) Backward feed grate.

(e) Agitated grate.

(f) Rotary grate.

(g) Rotating plate grate.

(h) Combustion cone.

(a) *Chain grate systems.* Chain grate systems for refuse incineration have been different somewhat from traditional chain grates. Special precautions must be taken at the sprocket side that protruding pieces in the refuse do not clog the

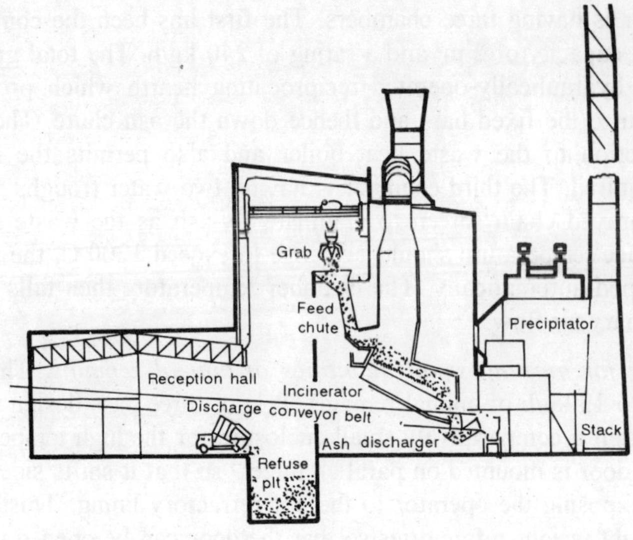

Fig. 5.7. The refuse incineration plant using an inclined chain grate

teeth. Chain grates, where the top surface has been continuous, are used instead of open grating which is used when the fuel is coke or coal. Often number of such chain grates are kept one above the other thereby serving to turn over the refuse to enable previously unburnt portions to burn out.

(b) *Roll grate systems*. Such a system is having typically of series of rolls, usually positioned at different vertical levels. These carry a continuous grate-bar chain complex. As the refuse passes along the grate it gets turned over several times enabling thorough combustion to take place. Such a system can be used for both small and large incinerators.

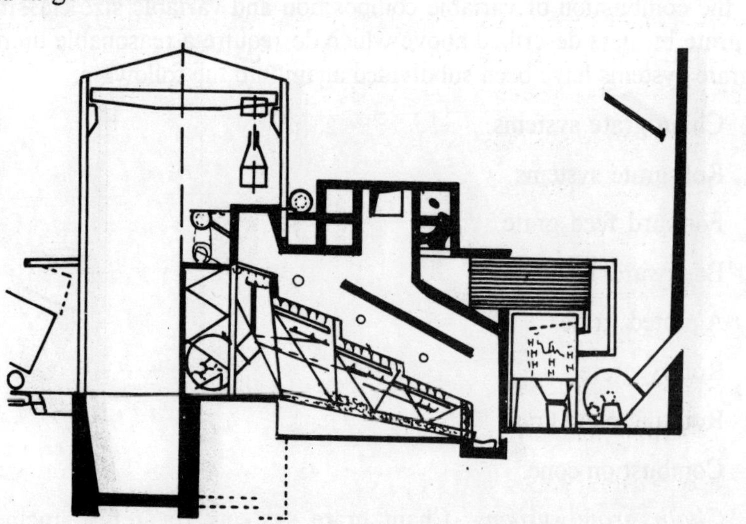

Fig. 5.8. Forward feed grate incinerator

(c) *Forward feed grate*. In this the refuse has been pushed by a continuous movement from the feed zone into the combustion zone of he incinerator. The movement at the same time serve to agitate and turn the burning refuse.

(d) *The backward feed grate*. This type of grate is having inclined set of movable grate bars. The feed takes place at the top and the fresh refuse slides down slowly. At the bottom partly burntout refuse has been grabbed by the motion of the grate-bars and pushed upwards again underneath fresh refuse thereby serving to dry it out. The ashes fall down between the grate bars to a bunker beneath.

(e) *Agitated grates*. These usually take the form of slightly inclined step grates, where each grate member independently moves the refuse along and turns it over. This is effected by means of rods actuated by cams on a drive shaft. During this process the refuse slowly drops down and is burnt out, the ash being extracted at the bottom of the furnace.

(f) *Rotary grate*. Rotary incinerators are having a cylindrical combustion chamber with a fixed cast-iron grate at the bottom, through which the ash is able to fall. At the periphery of this circular grate, the grate openings have been bigger, to enable the discharge of larger incombustible particles (stones, bricks, tins, etc.) to take place. The refuse is fed in at the centre being mixed and turned over by means of a system of arms which rotate about a central axis. The burnt-out refuse gradually shifts towards the periphery of the circular grate.

(g) *The rotating plate grate*. Rotating plate grates are having horizontal perforated or segmented plates, which rotate about their own axis. The refuse can be fed in at the centre, wherein is turned and agitated continuously by the rotation of the grate assisted by a number of static bars. The ash finally drop between the grate bars or disappears at the circumference. In an alternative design, the feed is at the edge of the rotating grate, the ash being extracted at the centre. The rotating plate grate could either be flat, concave or convex, depending on the precise way it is designed to operate.

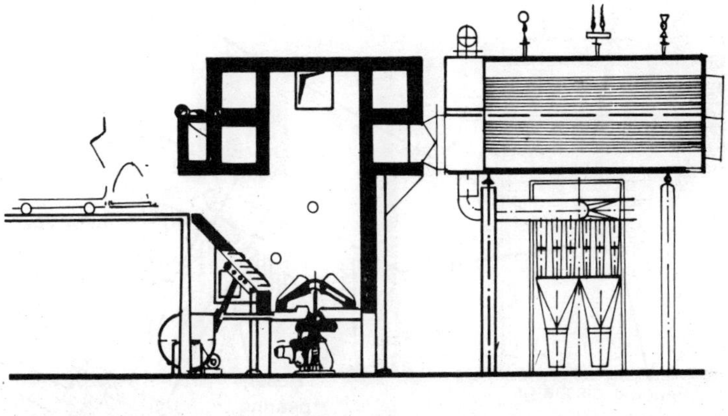

Fig. 1.9. Rotary grate incinerator.

(h) *The combustion cone system.* This system is having an inclined plate with a perforated bottom, which rotates slowly so that the refuse gets turned over constantly. This plant has been less affected by solid components in the refuse such as pieces on brick and stones, than are most other incinerators. Ash falls through the apertures in the bottom of the plate, but large unburnt components are removed by tilting the plate over periodically, the feed having been discontinued for some minutes beforehand to ensure a proper burn-out of the waste.

Examples of Moving Grate Incinerators

There are many different types of moving grate incinerator on the market, all operating on one or other of the principle described above. The following are a few examples, chosen mainly because information is readily to hand.

(a) *The paland chain grate burner.* This incinerator is made with capacities of 1 to 10 tonnes per hour and is contained inside a light-weight 3 mm sheet-steel casing. The inside of this casing is lined with fireclay brickwork. Refuse is fed from the top on to a special moving grate, which travels upwards for most of the time, but downwards for three minutes in every hour. Secondary air is brought into the underside of the grate through special distributors, while primary air is brought in through flaps at the top. The incinerator is fitted with three burners; a primary burner to initiate combustion and maintain combustion conditions at the moving grate, a secondary burner fixed in at the fixed lower sloping grate to burn the unburnt refuse after it has left the moving grate, and an afterburner which is fitted into the flue gas outlet. This is used to keep the temperature of the outlet gases at about 800°C to ensure that they are odourless. Inside the combustion chamber the temperature is about 1000°C. The system is not generally used for waste heat recovery, but the heat produced can be used to dry fresh refuse before it enters the incinerator.

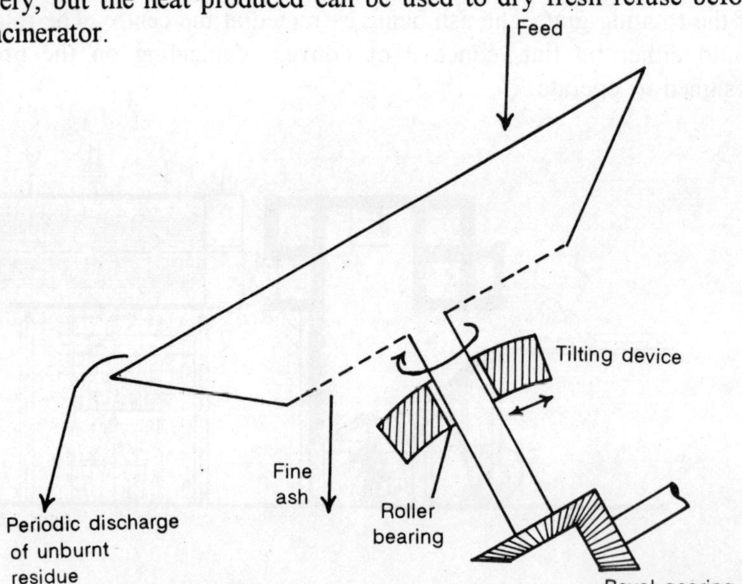

Fig. 5.10. The combustion cone system.

Ash is segregated into three size ranges, and the unit is suitable for all kinds of refuse including waste tyres which are normally difficult to burn smokelessly.

(b) *The Simon-Boulger incinerator*. This refuse incinerator is able to treat a maximum of 5 tonnes of refuse per hour. The combustion chamber is cylindrical and refuse is fed in from the secondary hopper of the plant. The only major moving part inside the plant is the stoker. They are driven by a medium sized electric motor via bevel gears. The rotating movement of the arms turns the refuse over the mixes it. It moves outwards during its burning circle towards the periphery of the chamber, where there are eight pairs of dumping gates controlled by a hydraulic mechanism, through which the ash is drawn off. Both forced and induced draught fans are used for combustion air supply, and the air supply system is automatically regulated by probe-controlled dampers. These maintain the charge hopper, ignition and combustion chambers as well as the ash hopper under a slight negative pressure to prevent outward leakage of fumes and dust.

The products of combustion have been drawn off into a separate cylindrical combustion chamber for completion of the burning process. The flue gas cleaning equipment used in conjunction with the plant consists of a water-spray gas cooling chamber and two groups of multi-cyclone dust collectors. The gases which emerge from the chimney contain less than 460 mg/m^2 of solid particles. Ash and tins from the ignition chamber hopper are discharged into a quenching conveyor which rises at an angle to deliver to a rotary screen above an ash bunker. The screen removes most of the ash and water from the tins, which are then collected by a magnetic separator and sent for baling.

It is estimated that the plant is far more expensive than controlled tipping in the area. Even if the operating costs of the incinerator are offset slightly by obtaining revenue for disposing of trade wastes and for the sale of baled waste metals, the disparity in costs between incineration and tipping is still very marked.

(c) John Thompson moving grate incinerator. This organization provides incinerators of varying capacities. The moving grate incinerators use three banks of grate bars for capacities of up to 5 tonnes per hour and four for capacities beyond this. Each section of the grate gets inclined at about 15^0 to the horizontal, and the sections are arranged with a definite fall of level between them. The refuse gets fed on to the top grate section. Each of these consists of alternating moving and fixed grate bars, the moving ones being actuated from below. Fine ash falls between the grate bars to hoppers underneath. The partially burned refuse then topples over on to the next section during which operation it is turned over, so that previously unburned sections have been able to burn as well. The larger unburnt remains are then taken off at the bottom and are eventually conveyed to the ash hopper for discharge into lorries. The flue gases pass on to a secondary combustion chamber and then to the centrifugal dust separation plant, before being released to the chimney. A variation of this plant, which is suited for incinerators having an hourly capacity of 6 tonnes, is a combination of three moving grates with a rotary kiln, in which the burn-out of the refuse is completed. The final rotary kiln is having a drum with its axis inclined at about 5^0 to the horizontal.

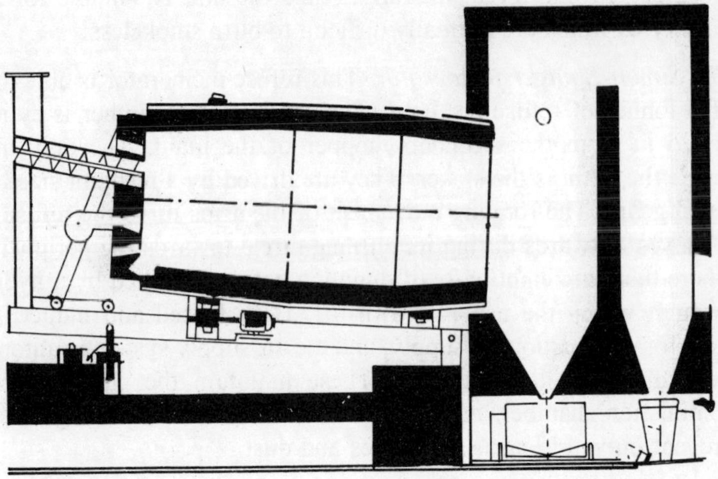

Fig. 5.11. Rotating drum incinerator.

C. Grateless incinerators

(a) *The shaft oven*. The simplest form of grateless incinerator has been the shaft oven in which refuse is burned in a vertical chamber, various methods of agitating the refuse being employed. Ashes have been withdrawn at the bottom by means of a conveyor belt.

(b) *The horizontal drum system*. In this a rotating drum is used with its axis inclined at a few degrees to the horizontal. The drum has been made of steel and is lined with refractory bricks. The rotation of the drum produces agitation and turning over of the feed, which burns out during its passage towards the exit of the drum at the back. Sometimes several drums have been used in series. The system can also be combined with other methods of feed agitation.

(c) *The fluidized cyclone system*. For fluidized operation of refuse incinerators it has been necessary to pre-treat the material first by grinding or other methods of size reduction. The feed is injected tangentially into the top of a conical combustion chamber and meets a stream of secondary air from the bottom. The material burns in this fluidized state, a very high efficiency of combustion being obtained. And is having a greater density than unburnt refuse and eventually sinks to the bottom, where it is then discharged to the ash extractor system.

Although fluidized combustion systems could be used for properly pre-treated town waste, they are generally considered to be more suitable for the combustion of industrial waste.

Large Scale Refuse Incinerators

With the rapid increase of combustible refuse collected in big cities, large scale refuse incineration becomes an atractive proposition. Plant can be positioned outside the most densely populated area, and can be equipped with a high chimney and efficient dust scrubbing equipment. It is able to collect and process the refuse of an entire town. Although the cost of installing and operating such a plant is high, some of this outlay can be recouped by using the waste heat for steam generation, the steam in turn being used for power production, for district heating or for both. It is best to combine power generation and district heating by the use of back-pressure turbines; ITOC or pass-out turbines are seldom employed because it is difficult to obtain high enough inlet steam temperatures from refuse incinerators.

Apart from money earned in these ways, cash can also be recovered by the sale of incombustible materilas. It is usual to equip the incinerator ash extractor with a magnetic belt for the removal of ferrous metal scrap, which can then be baled in an automatic baling press.

The construction of a modern refuse incineration plant with its auxiliary equipment has been in expensive matter and a suitable plant able to deal with between 10 and 25 tonnes of refuse per hour will cost several million pounds.

A few of the most important makes of large scale refuse destructors have been discussed as follows:

The vereinigte Kesselwerke System

This system, is also known as the VKW system or the Dusseldorf system. It is probably the most widely used large scale refuse incineration system in the world.

The heart of the system has been a series of roller grates-six or seven in number. These are arranged in a step formation and rotate slowly. Incoming refuse gets dumped from the vehicle into a reception bunker from which it has been fed by grab crane into the incinerator hopper and thence into the incinerator at a controlled rate by using a push-plate conveyor. The refuse gets deposited on top of the first roller, which passes it on to the next roller as it rotates, and so on, down to the last one in the line. Throughout its journey down the sequence of rollers, hot gases flow continuouslly over the refuse, first drying it and then igniting it. Movement from one roller to the next agitates the refuse and helps ensure complete combustion. Scraper plates have been fitted between each pair of rollers to keep them clean. At the end of its passage through the incinerator, the refuse is transformed into sterile clinker and ash, which is discharged through a quenching tank on to a plate convey or to a storage bunker for further treatment and disposal. The quenching water can be re-cycled to reduce costs. Control of the burning refuse within the incinerator has been effected by varying the speed of rotation of the rollers individually, and also by controlling the volume of combustion air entering the incinerator. An auxiliary oil burner has been fitted for starting up the plant and can`be brought into use to aid combustion at any time.

The construction of the roller grate. This consists of a hollow roller on which interlocking heat-resistant fire bar segments are kept. The fire bars have a very long life, and it has been shown that even after 22,000 hours of rubning there is little deterioration or wear of the material. The complete roller is 1.50 m in diameter and its length varies in accordance with the capacity of the plant, as follows:

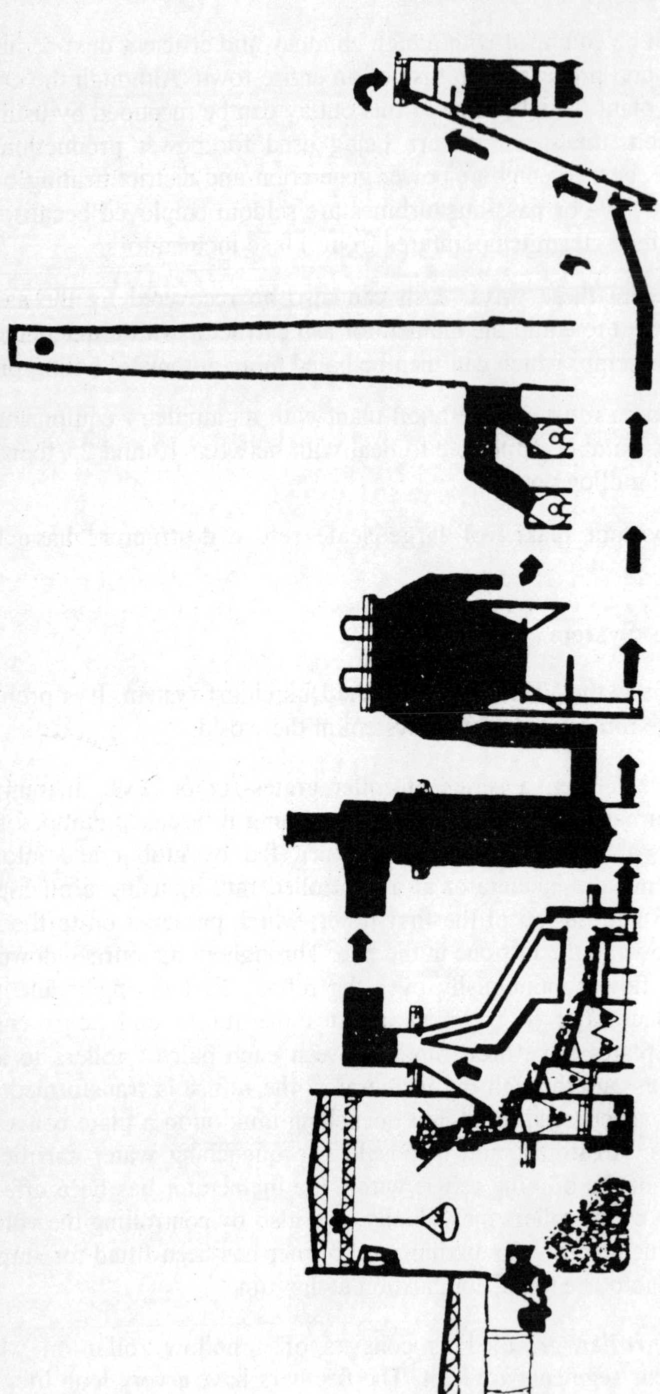

Fig. 5.12. The Dudley refuse incineration plant uses multiple chain grates.

Refuse throughout in tonnes/h	*No. of rollers*	*Length of rollers (m)*
4	6	1.50
6	6	2.10
8	6	2.60
10	6	3.00
12	7	3.00
16	7	3.70
20	7	4.10
25	7	5.20

The driving mechanism of each roller and its bearings have been outside the fire area. Each roller is having its own drive motor and gearbox, allowing speeds to be adjusted individually within an operating range of a half to five revolutions per hour. Air for the combustion chamber enters through the rollers and the quantity of combustion air which is allowed to enter can also be controlled individually.

The Edmonton incineration plant. This plant has been designed to handle a normal continuous throughput of 1333 tonnes of crude refuse per 24 hours with a maximum of 1667 tonnes. The plant has been also designed to utilize the heat from refuse incineration for power production. The normal output is between 25 and 35 MW with a peak output of 45 MW.

The capacity of each incineration unit is 14 tonnes of refuse per hour. The gases from the combustion chambers then raise steam at 43 bars and 450°C, which is used in four 12.5 MW turbogenerator sets and two 2.5 MW house sets. The power produced has been used for supplying the works and any excess is then sold to the Central Electricity Generating Board network. No pre-sorting is necessary with the plant, and the feed opening is large enough to admit the largest object considered likely to be fed in-a piano.

Details of design. The grates used with the Edmonton incineration plants have been of the Dusseldorf type, using seven rollers at an angle of 30° to the horizontal. Each roller is 1.50 m in diameter and 3.50 m long. Primary air has been blown through the rollers into the combustion chamber, having previously has been pre-heated from the hot exhaust of the steam turbine. The side walls of the grate are protected by air-cooled heat-resisting cast iron side plates. Air for these purposes has been drawn from the vicinity of the grate, passed through a heat-exchange labyrinth within the castings and then discharged under pressure at the furnace throat as secondary air. About 20 per cent of the total air needed for combustion is secondary air. The temperature of gas leaving the combustion chamber has to be high enough to reduce the putrescible and paper char matter to less than 0.3 per cent, but not high enough to cause slagging problems. The temperature at the outlet has been commonly controlled to between 980°C and 1040°C. About 60 per cent excess air is normally found to be the optimum quantity needed, but far capacity is available to step this up to 80 per cent if required.

Design of steam raising units. The corrosion problems involved in waste-heat boilers used in conjunction with refuse incinerators have been worse than those experienced with conventionally fired plant. Apart from SO_2 and SO_3, which are also present in the flue gases of oil-fired and coal-fired furnaces, there have been chlorine and hydroged chloride derived from the breakdown of PVC plastics. As can be seen from the table, the percentage of plastics in refuse will probably treble within the next ten years, thereby further exacerbating this problem. Corrosion primarily occurs at comparatively low temperatures and could be minimized by keeping the metal surfaces above the temperature at which any of these acid compounds condense. This in turn implies that chimney exhaust temperature is kept above 220°C-significantly higher than with conventional combustion equipment. Naturally, some sensible heat gets lost but this cannot be prevented. High temperature corrosion of flue gases chlorine or hydrochloric acid again becomes appreciable whenever the temperature of metal in contact with such gases exceeds about 480°C. All normal steels readily get attacked under these conditions. As a result, the practical limit of steam temperature produced by a refuse incineration plant has been about 450°C. It can therefore be seen that the twin phenomena of low temperature and high temperature corrosion effectively fix the temperature limits.

In the Edmonton plant there have been five steam raising units each of which has been fully operotional for 11 months in the year with the twelfth month being allocated for overhaul. A vertical radiation shaft above the combustion chamber reduces the temperature of the flue gases to about 820°C. These gases then impinge upon superheater tubes, which have been arranged in two steam/gas passes with an interstage spray type de-superheater to enable steam temperature to be controlled within about 6°C. Long stroke blowers have been arranged to remove soot and other deposits from the tubes and permit to fall into hoppers beneath. It has been found to be quite effective in reducing corrosion risks-indeed, only the last five loops of the second superheater had to be constructed from more expensive 1 per cent Cr ½ per cent Mo steel. The main boiler has been of simple construction with 55 mm o/d pipes arranged with a 125 mm pitch across the gas flow. The economizers used have beenfinned steel tubes, and the feed to these is at about 125°C. A combination of shot cleaning and soot blowing is used to keep them clean.

Before exhausting, the flue gases pass through economizers to reduce their solids content to below 0.115 g/m^3 of dry gas at NTP, to comply with the requirements of the Clean Air Acts.

Ashes and incombustible residues. The residues from the combustion chamber get quenched in a water bath, then discharged on to one to two transverse belt conveyors leading to a special building. The residue is then fed on to a vibratory screen, where oversize materials have been taken off. An overhead magnetic crane attachment removes any iron and drops this into a baling machine. The rest has been allowed to pass to a rotary screen/crusher. Once more iron and steel waste are extracted magnetically using both a magnetic overhead separator and a magnetic pulley. The baled metals have been finally supplied in neat cubes with 300 mm side length and a density of about 1.8 kg/dm^3, and are transferred from a storage hopper to road vehicles by a special overhead magnetic crane. The broken-up clinker is having a density of 0.72 kg/dm^3 and is also stored in an overhead hopper from which it can be discharged to lorries.

Unfortunately the Edmonton plant is not operating on a total energy principle, and because of the limitations on steam temperature imposed by corrosion consideration, power production using low inlet-pressure steam turbines is not particularly efficient. Continental refuse incineration stations invariable couple power generation with district heating using back-pressure turbines, and this considerably increases profitability. At Edmonton revenue from the sale of metals, hardcore and electricity contributes only 39 per cent of the total cost of running the refuse incineration plant.

Rotating Drums

Rotating steel drums, lined on the inside with refractories, have been popular for the construction of large scale refuse incinerators. Volund plants are found all over the world.

Design of a Volund type refuse incinerator. From silos the refuse is lifted by using a crane grap and dropped into the furnace through a funnel-shaped chute. The first part of the furnace is having a drying grate on which the fuel is exposed to radiation fromthe incandescent refractory walls of the ignition chamber grate. This drying process generates steam and some other gases too, which have been led away to by-pass the main rotary kiln. The speed at which the refuse moves along has been controlled by the grate and adjusted in such a way that the refuse has been well dried befor passing into the next part of the plant.

The well-dired refuse has been then dropped on to the ignition grate, the thickness of the layer being controlled by the relative velocities of the two grates. Combustion air is allwed to blow in by a 30-35 kW fan, for a plant with an hourly capacity of about 10 tonnes. The air is pre-heated, as it has been used as a cooling medium for the cast iron sides of the lower part of the furnace room and is blown in at about 1 kN/m^2 pressure. Once the refuse has been well alight, it is led into the rotary kiln, whose diameter for a 10-t/h plant decreases from 2.5 m to 1.5 m in a length of 7.0 m. The speed fo rotation has been normally about four revolutions per hour. The axis of the kiln gets inclined, the angle of inclination being variable, depending upon the nature of the refuse being burned. During the slow passage of refuse through the kiln, new portions of refuse have been exposed constantly, and burned. The charge of refuse passes through the kiln in about 90 minutes. The rotary kiln of a Volund furnace has been lined with fireclay refractories and insulated with calcined diatomite/clay bricks. The refactory linings have a life of up to 10 years. The temperature inside the rotary combustion chamber must not fall below about 850^{0}C because not only is there then a danger that smells will not be eliminated,but also that the ash obtained will be worthless instead of useful hard clinker. The upper limit of temperature-1000^{0}C in summer and 1050^{0}C in winter-has been conditioned again by the nature of the clinker required. If internal temperatures are raised beyond these limits, becomes too hard and brittle.

In order to adjust the temperature of combustion two techniques have been available. The first has been to add coal as a 'sandwich' fire, the coal being made to lie below the refuse. The second method has been to use an oil burner just below the by-pass, its flame directed towards the lower part of the drying grate. In summer very little additional fuel is required as there fuse is fairly dry. In winter as much as 200-300 kg of oil may be needed for every 10 tonnes of refuse.

Analysis of Refuse by Weight

	Actual percentage now	*Probale percentage in 1980*
1	*2*	*3*
Dust and cinders	12-28	9-12
Vegetable and putresent matter	17-35	19-28
Paper and cardboard	30-32	37-47
Metals	6-8	0-9
Rags and textiles	2	1
Glass	8-10	7-8
Unclassified debris (wood, straw, stones, etc.)	2.5-6.5	2
Plastics	0.5	3
	now	*in 1980*
Density as collected	150-178 kg/m³	120-130 kg/m³
Density after storage in hopers	200-240 kg/m³	160-180 kg/m³

Chemical Analysis of Refuse

	Actual percentage now	*Probable percentage in 1980*
Water	20-31	20-26
Incombustible matter	27-33	26
Carbon	21-27	25-28
Hydrogen	2.8-3.0	3.3-3.7
Oxygen	16-17	18-21
Nitrogen	0.3-0.5	0.3-0.4
Sulphur	0.2-0.25	0.2
	now	*in 1980*
Gross calorific value	8.5-9.8 MJ/kg	9.7-10.5 MJ/kg

As in other refuse incineration plants, the temperature of steam produced in a Volund furnace has been not high, as the furnace temperature itself is comparatively low. Thus the overall thermal efficiency of the refuse incineration plant as a heat generator is poor compared with other types of furnace/boiler constructions.

The temperature of the flue gases is varying between 175° and 200°C and the percentage of CO_2 has been between 8 and 12.

The heat losses can be summarized as follows:

	Per cent
Chimney losses	12-15
Sensible heat loss in slag	10-18
Radiation heat loss	5
Other heat losses	5

In the Volund type refuse incineration plants little attempt has been made to obtain particularly high steam pressures from the waste-heat boilers.

In Denmark sale of waste products usually raise the following sums (1978):

Iron —about 1.5p per ton of refuse;

Clinker —about 35p per ton for large clinker; and 25p per ton of small diameter clinker;

Steam —about f 1 per ton sold.

In general, it has been found that the sale of waste heat and waste products covers 50-70 per cent of the added capital and running costs of refuse incinerators, making incineration by far the cheapest method for disposing of refuse.

The Lyons plant. The French city of Lyons had installed a Volund type refuse incinerations plant in 1963 to burn a maximum of 10 tonnes of refuse per hour in each of two units. The handling chamber has been 120 m long, 30 m high and 10 m wide and contains the travelling cranes and feed hoppers for the refuse. The refuse in Lyons has an unusually low calorific value-only 3.5 MJ/kg or less. Due to this the refuse must remain inside the furnace for no less than six hours. The total surface of all the mechanical grates is 22.5 m².

The heat has been recovered using a vertical tube boiler in each unit, with a total heating surface of 950 m² for the boiler and 75 m² for the superheater. The maximum steam output of the plant is some 20 tonnes per hour at 20 bars. The steam is piped to a neighbouring district heating centrale where it first drives a back-pressure turbine, with the condenser water from this feeding the district heating network.

Multiple Hearth Refuse Incineration

This method of refuse incineration has been the most widely used method and has been the basis of the Combustion Engineering multiple grate system.

The refuse is lifted by a crane grab to fall through a large funnel into the incinerator. This receiver shaft has been water cooled. The first section of the incinerator has been an inclined chain grate which serves to dry out the refuse. The speed at which this chain grate operates is dependent upon the moisture content of the refuse. There are three further grates thereafter. The first grate is having the function of igniting the refuse. Once the refuse gets passed along it, it falls down to the second grate where main combustion takes place. The refuse then falls to the last to the grates where burning out of the residue continues until the putrescent matter content falls below 0.3 per cent, the desirable maximum figure. Each of the chain grates has been driven separately and speeds can be adjusted so that the material thickness or the surfaces has been optimal. Similarly, the amount of secondary air which is blown through can be varied independently. To start the plant and also to provide extra calorific value when the throughput of highly combustible waste materials falls-there has been a number of oil-fired burners. These has been run largely on waste sump oil from garage and machine shops. The temperature inside the combustion chamber has to be maintained at about 1000°C. About one-third of the air necessary for cumbustion comes from below the grates, while the rest has been blown in through the side wills as secondary air. In order to attain combustion chamber temperatures 1000°C the air of has to pre-heated; if this is not done it is difficult to get chamber temperatures in excess of 750-800°-too low to ensure that the flue gas will be odourless.

Multiple hearths for incinerations of sewage sludges. Multiple hearth incinerators are widely used for sludge incineration and the combustion of industrial wastes. A typical plant of this type is the one constructed by Bartlett-Snow-Pacific Inc. It is a cylindrial refractory-lined steel which contains a series of horizonal refractory hearths located one above the other. These hearths, which have alternate in-feed and out-feed directions cause the sludge to move completely across each hearth as it drops from one level to the next. Mechanical stoking has been provided by a motor-driven revolving central shaft to which are attached radial arms. Teeth, somewhat similar to the ploughs of a circular clarifier, get attached, which are able to move the material across the hearth, and the whole assembly is cooled by air blown across it. The furnace could be divided into three zones, with the drying zone at the top, followed by the combustion zone where the temperature rises to between 750°C and 900°C, and finally the cooling zone, where the ash gets cooled down prior to discharge into hoppers.

The Mannheim refuse incineration plant (Wast Germany). This plant is known to serve three distinct functions-to burn the refuse produced by the town of Mannheim and its surrounding industrial areas, to supply steam throughout the year to the industrial area on the Friesen-heimer island of the Rhine, and to supply both electric power and hot water to the town of Mannheim itself.

The plant is not only a refuse incinerator as combustion appliance but also four conventional oil-fired boilers. It has been equipped both with back-pressure turbines and ITOC turbines having a total power generating capacity of 64 MWh. The refuse incineration plant uses multiple hearths, and has an annual capacity of up to 220,000 tonnes of refuse. Industrial refuse in the area is having a calorific value of about 8.5 MJ/kg, while the domestic refuse varies between 5MJ/kg insummer and about 4MJ/kg in winter. A refuse bunker having a capacity of 3000 m³ is available and there are facilities to accommodate 40 industrial refuse trucks and 30 municipal refuse trucks each hour. Very large refuse components can be ground up in a mill, which is part of the plant. The capacity of the refuse incineration grate is some 18 tonnes per hour.

The waste heat boilers which have been heated by combustion of refuse have been also capable of being heated by oil and are designed to produce variable quantities of steam at 525°C. By operating the refuse-fired boilers in conjuction with normal oil-fired boilers an operating flexibilty has been achieved which is not usully possible with plants where refuse has been the only fuel.

The Takuma Incinerators

The Japanese company has built a number of very large refuse incinerators in Japan. Their total capacity has been nearing 10,000 tonnes of refuse per day.

Two types of grates are used-the H-type being a multiple hearth grate consisting of six inclined chain grates, and the S-type consisting of twin step grates together with a long horizontal chain grate final burner. Air for combustion has been drawn from the refuse pit by a forced draught fan, pre-heated by an air heater, and then distributed to the various combustion stokers. Steam raising and hot water facilities have been built into the waste gas system. The refuse pit has been of air-tight construction and air for combustion has been drawn from the area thus producing some negative pressure to eliminate offensive odours from the plant.

Conclusion

There have been naturally, several other important systems for refuse incineration on a large scale, but-by and large the general design criteria described apply to these as well.

Large scale incineration has been an efficient and desirable way of getting rid of the waste produced by large communities. The capital investment needed is certainly great, but provided refuse incineration can be combined with power and hot water or steam delivery on the total energy principal, costs per ton of refuse treated have been probably lower than with any alternative system available.

SOLID WASTE MANAGEMENT

In recent years new and better solid waste management methods have been suggested and/ or developed which are discussed as follows:

1. Utilization

In some cases solid wastes might be put to direct use. Fly and bottom ash from industrial installations now total about 30 million metric tons annually, of which only some 5 to 10% is used commercially, largely as a cement substitute in concrete for dams, highways, and other major construction. New uses are being developed for fly ash to make brick, to dewater industrial waste sludge, as a land cover, etc.

Fly ash from coal fired power stations has been a major industrial waste product. India is surplus in coal and there are several coal-based power stations in operations and are in the process of being installed. About 22.016 Mt of fly ash at 80% collection would be available during 1985-86.

While exact statistics for India are still a measure of conjecture, for Japan it is known that power generation through coal is going to be nearly 53 MKW in 1990 consuming 82 Mt of coal and generating nearly 9 Mt of fly ash. About 55 t of coal are needed to generate one kW of power and fly ash constitutes about 15-20% of the total coal consumption.

The different uses to which fly ash is put are: Cement and concrete including (light weight cement, foam light weight cement, etc.), land and land foundation (land development, fragile ground improvement materials, fill type dam materials, asphalt filters, blocks for civil engineering, agriculture and fisheries (soil improvment, potash fertiliser, gathering ground for fish, compost for night soil treatment), building materials (rock wool, glass wool, adiabatic molds), and sewer treatments (deodrants, sludge treatment materials, sewer treatment materials) There is an increasing shift to coal as a source of thermal energy. The use of fly ash is already been heat-treated during the burning of coal (Anonymous, 1984; Grant, 1984). Obviously, less heat is needed to fire (heat-harden) bricks which may be partly made from the fly ash. On account of the inconsistent standards of the finished products, bricks and ceramics made from fly ash have not been acceptable. The reason is that ash produced from different coals as also from the same coal with different generating conditions differes in particle size (2-18 micron), shape, and chemical composition. These impart varying physical properties. Obviously, when it is mixed with different clays, it produces very variable products. One way to standardise the end products is to process ash to adjust properties to the required specifications before using. The ash has to be ultrafine and homogenised.

The advantages of bricks prepared fromfly ash have been as follows:

— These are fired at lower temperature because fly ash has already been heat tested. This gives one-third of saving in cost of bricks.

— Bricks take less time to dry and fire and in simpler kilns.

— Fly ash is cheaper than clay.

— Clay content being less, the landscape takes longer to get spoiled.

— Coal being a realistic power source in many developing countries, its waste material can utilised which otherwise could be an environmental hazard.

— Fly ash can also be used beneficially in pottery industry.

— Likely to lead to brith of new industry at the local place of generation of fly ash.

The work in Great Britain as also in West Germany has progressed considerably and it would be not long, before bricks of fly ash-clay mixture would be marketed. In Britain nearly 12 Mt of the fly ash is available per year.

Another use for ordinary municipal solid wastes would be the production of heat or electricity during incineration. Recently, there has developed a fluidized-bed incinerator which burns solid wastes at high pressure producing hot gases which power a turbine and generate electrity. Municipal solid wastes have a heating value (heat per unit mass) about one-third that of coal, and a typical community might be able to obtian 10% of its electric power requirements by burning its own refuse.

2. Recovery

In some circumstances, valuble materials can be recovered from solid wastes, lessening disposal problems and finanical costs. Over the years, salvaging of materials from solid wastes has been practised at times, abandoned when it became uneconomical, and sometimes revived.

3. Re-use

With affluence, people in the developed country want to discard and throw away garments which are out of fashion and may even be less than one year old but are still useful to others and are being sold as such or for providing raw materials for hosiery industry. One of the best examples of reuse of waste materials is seen in Chandigarh at the Rock Garden. It indicates, that given mind and an eye, what use can be made of materials which have been otherwise discarded by the society.

Relevance to developing countries. While in developed countries, resource-recovery has been a high technology area, in many cities in developing countires garbage is sorted out by hand for reuse as tin, glass, paper, plastice, rags and bones which generate new materials and in the process also generate employment. In fact, for a long time in our country waste paper, rags and empty bottles have a regular market. The recycling psychology has been, therefore, already inbuilt in us. It is now being followed by several other countries like East Germany where there are long lines at collections centres.

The reasons for waste utilisation have been economic environmental, resource conservation, employment generation and provision of basic necessities of life. Developing countries with constraints of resources cannot afford to call any material as waste. Paradoxically it is developed countries like Japan, FRG, UK, USA, that are in the vanguard of waste utilisation. in India due to out environmental ethos several materials get recycled and reused, but by and large waste utilisation has still not taken roots.

Utilisation of wastes would constitute direct and indirect contribution to economic development. For example, only 10% of biomass in paddy has been grain and 90% is straw. In the grain, 22—75% is husk and 5—7% is barn. If the annual rice production is 40 Mt. bran would constitute 2 Mt which would yield 0.4 Mt of bran oil which amounts to Rs. 1,500 M to national resources (Lahiri; 1980). Many such examples could be cited where wastes can make significant contribution.

Wastes is allowed unused can act potential environmental hazards in spreading diseases and leaching of unwanted chemicals into life support systems. Resource conservation has been yet another advantage that accrues from the waste utilisaiton. Further more, with decentralised management of wastes, employment would get generated, by way of collection, handling and processing which may not require high technical skills. Many useful products would generate from waste utilisation which are of value (as food, fuel, fertiliser, fibre, paper, pulp, industrial chemicals, metals, etc.), from agricultrual and agro-industrial, wood and forestry residues, animal resiudes, mineral wastes, textile and textile processing wastes, and metallurgical, metal utilites, paper and plastic wastes.

Some obvious constraints have been the dispersed nature of wastes and the consequent collection would be costly. This can be solved with the public participation and by cooperative movement. The plus point is that instead of making it totally mechanical, waste collection could be partly labour intensive with its utilisationalmost at the source end. This would save on transportation as well.

So as to encourage industry to take to waste utilisation, there need to be fiscal incentives for using non-conventional raw materials and penalties for not recycling materials.

A crash programme for technology generation in the case of higher priority items requires to be undertaken. In fact, there is a need for a coordinating body at the Centre to give a push and a focus on waste utilisation. This aspect should take care of all facets of the problem right from technology generation, engineering designs, equipment, demonstration plants and in fact, whole package on almost a turn-key basis.

A system requires to be build to collect, validate and disseminate relevant information on the subject for policy and decision makers to enterpreneurs, researchers and customers.

Utilisation of primary resources in conjuction with waste / residue utilisaiton could thus materially help in the socio-economic develoopment of the country. This will not come about unless and until the Government is partnership with people attempt to make a determined effort in this matter. Waste utilisation is something that we cannot avoid intime to come, so why not take it up where we are, and with what we have and make efforts to perfect the system. There are blocks, often mental ones, say in utilisation of sewage and night-soil which is a very important resources. Chinse regard every humn and animal stomach a fertiliser factory. Thus we shall require, crash programme on education onthis aspect. In the ultimate context success will depend on the political will to take up programme on waste management because every material has some value.

World has been witnessing a major crisis in resource availability. The developed countries want to conserve as much as they can and import resources from the developing countries, in return for some hard currency for their development plans. In this game the developing countries have been the major sufferers. This is occurring in Latin America, Africa, and parts of Asia.

Time has come for the developing countries particularly their leaders as also their policy and decision makers to think and ponder as to what they are getting in return of their irreplaceable resources. The efficient utilisation of indigenously available resources is very important. Using resources only once and that too in an efficient manner, is something that is to be increasingly avoided, and we are in the midst of an era where recycling and re-recycling will be the order of the day. They compelling reasons have been realisation about finiteness and dwindling stocks and minimising environmental pollution. However, due to the dispersed nature of resources, their collection and transportation is a problem which may call for an in situ utilisation in a decentralised manner for local use.

For instance, in recent years due to increasing cultivation of paddy in Punjab, Haryana and Western UP, huge quantities of paddy biomass are available. The grain constitutes only about 10% of the total biomass. The remaining biomass, in the form of straw, can be used to a limited extent as cattle feed, manure, fuel, and bedding, packing and thatching material. Today in Punjab where rice has followed wheat, the farmers want to getried of this material in shortest possible time and, more often, they burn it. It is a common site to see huge fires and thick clouds of black smoke in fields throughout Punjab.

India being an agricultural country, much of the wasters are in the form of biomass and hence easily biodegrable. Here then is a major resource which needs utilisation economically. Hence one of the important constraints in lack of R & D on recycling. This is not a fashionable research area and there is a general lack of technological availability and reliable feasibility reports. This stems from a general lack of awareness about wastes as an environmental hazard, as also socio-economic implications and cost benefit about the residues. Equally important is the lack of data about availibility of wastes/residues both location and season-wise. The major problem is the lack of a nodal agency overseeing this interdisciplinary and inter-agency area.

Research and Development

On an average an Indian is able to generate 0.5 to 0.7 kg of waste and nearly 204 million tonnes per year of residues. Besides, 1000-1050 million tonnes of wet dung is generated annually by 237 million cattle (ABE, 1985). Dung along equals to 22,425 Mm^3 of gas. This is biomass based waste besides there is waste from non-biomass sources. The present day waste disposal by dumping and/or burning has been not environmentally sound. Most of the developed countries convert waste into energy for heating water and generating steam and then electricity or process industrially.

The crux of the matter has been the collection of waste as also its classification. This forms the basis of recycling. The waste could be divided at the starting point with the help of women, old people or others and sent to recovery traders. This method would enable recoverng resources of high degree of purity. There is need for a regular recovery campaign.

At present in India, three major uses of waste are as follows:

— Composting for agricultral use is being practiced at several places in India like Baroda, Jaipur etc.

— Energy Genration.

— Biogas from sewage at Okhla, Kashavpur and Padrona.

— Heat Genration. A pilot plant is coming up at Delhi. This is a common use all over the world.

— Incineration is being practiced at many places. Actually, it is waste of resource.

— Senitary land fill. In the strict sense it is not recycling. This is again the most common use throughout the country. However, the way it is actually done is in fact insanitary land fill.

A national survey must to be undertaken regarding quantity, availability, seasonality and other characteristics of wastes and residues in the country.

Various routes which can be followed for bioconversion into biofuels through fermentation have been digestion through aerobic/anaerobic and enzymatic method. In all such cases an additional bonus is sludge which can be useful as nitrogenous fertiliser.

The vast amount of lignocellulose available requires to be utilised. It needs concerted effort through application of genetic engineering. The relevant genes which can degrade cellulose, hemicellulose and lignin in wood rotting fungi need to be cloned in the genome of the yeast for using ligno-cellulose as a substrate. Once successful, this would be a major breakthrough and solve a number of problems in one step (see also Khoshoo, 1982). This can help in utilising the lingocelluloses into food (single cell protein, microbial biomass protein) fuel and fodder. Such processes have been inter-diciplinary and involve at least 12 disciplines namely, microbial physiology, genetics and ecology, chemistry, fermentation and enzyme technologies, agricultural and chemical engineering, fodder evaluation, food sciences, economics and sociology.

Such problems have been far more relevant to a country like India than to USA or other developed country because of the immense benefits that such technology promises.

4. Recycling of Waste/Residues

During the recent years, it is very clear that the resources of the world have been indeed finite and in order to maintain a stable economic growth and a standard of living in the future, it has been very necessary that we use resources very carefully and evolve technologies for recycling wastes and residues thereby save both resources and energy. Another reason to recycle has been that limited resources of the world are indeed a sacred trust of the present generation which it owes to the genrations that will follow.

Waste is a part of human activity from time immemorial. For instance, the activities of humankind like the baking of bricks for the Great Wall of China saw the end of the forests all round and thus damaged the environment and impoverished that country. Felling forests in historical past for use in iron ware, in mining, jhumming, grazing, etc., have accelerated the ecodegradation. Similarly, natural events such as the volcanic eruptions send huge quantities of gases, (SO_2CO, Nitrogen and Hydrogen and volcanic ash in atmosphere) that influence the temperature regime.

In the earlier times, the population was small, requires were few and resources were abundant, the generation of waste was such that it got naturally recycled being mostly biodegrdable. However, after the industrial revolution, particulary during the recent decades, the resources have been used recklessly and there has been generation of very diverse type of wastes which are often both non-biodegradable and hazardous. An ecosystem cannot absorb them in the natural course. The piling of such wastes poses an environmental problem often on account of chemicals which the ecosystem has never been used to. Thesc affect its propor functioning and in turn at the globallevel are likely to affect even the stability of the biosphere.

In fact, nothing has been really waste and much of the waste is "vegetable matter in wrong place." Today's waste is considered as tomorrow's raw material and waste is wealth and "trash is cash" are the things one hears these days. A wide variety of organic waste, of diverese origins, like agri-horticultural, agro-industrial, forestry, cattle, municipal and other urban wastes like rags, newspapers, etc., have been available and all of these in the ultimate analysis could be traced to a plant origin.

World evolved towards "a Throw-away Society" when population was less, energy was inexpensive and raw materials were abundant and recycling of the used materials was not feasible (Brown and Shaw, 1982). The shape of things to come is different and it is now certain that a nation that will not be able to recycle used materials, will not be able to sustain itself and there is global realisation of the fact that single use of some of the important materials like metals, glass and paper would lead to scaricity of such materials as their feedstocks would get exhausted.

Discarding materials after a single use means also wastage of energy. For instance, energy saving in recycling aluminium in comparison to bauxite, the original ore, is 96%, these figures for copper are 90%, for steel from scarp 47%, newsprint 23% and glass 8%. An added advantage of recycling newsprint is reductions of pressure on forests, a ton of recycled newsprint saves a ton of wood i.e., one dozen trees. Glass when reced saves 8% energy but re-use of glass containers saves far more (Brown and Shaw, 1982).

Waste in some situations can be a big health hazard and must be disposed off so as to prevent such a situation as also be put to proper use. At present, heaps of waste has to be disposed for no use other than sanitary (or should we say insanitary) land fill. Every year new sites for such land fill must be found. In USA, nearly 14,000 land fill sites on about 476,000 acres are in use. Even so, several hundred new sites have to be formed (Brown and Shaw, 1982).

Recycling in Japan. Japan alone has a definite culture of garbage collection and disposal dating back to 17th and 18th century AD. In fact, it was in 1900 that country promulgated Garbage Clearnce Act. The collection and transport system for garbage has since been modernised, by using pipeline system. The garbage is carried on an air stream at a speed of 20 to 30 m/second into garbage separator. It is then compressed by discharger before carried by the container mover to the outlet. The air separated from the garbage by separator goes through the dust screen and is then deodorised and discharged into atmosphere as clean air.

The example of Japan is indeed noteworthly. The amount of waste materials recycled in Japan was 16% in 1974 and went up to 48% in 1978. In 1979 Japan consumed a total of 1931 Mt of resources, out of which 330 Mt ((17%) were consumed as fuel, 80 Mt (40%) as export goods, 1200 Mt (62%) as consumed or stocked goods and the remaining 320 Mt (17%) was waste (CJC, 1984).In 1980 Japan produced 292 Mt of industrial waste out of which 124 Mt (43%) were recycled, 100 Mt (34%) were reduced through dehydration and incineration and 68 Mt (23%) were used as land fill. During 1981, out of 43 Mt of garbage (at 0.828 kg/person/day), 65% was incinerated, 32% was used as landfill and 3% was used for composting and feeding animals. The number of treatment plants where garbage is incinerated for generating electricity is nearly 50. In order to run the whole process more efficiently, planned group collection is used in increasing manner.

Japan's work on recycling has had most marked effect on automobile industry. The country purchases scrap from USA for recycling and, in fact, a saying goes round that "Last years Buick (a US car) is the years Toyota". Such a policy has, in fact, displaced USA from its position of being the world's biggest automobile producer. On the other hand, Britain which like Japan imports most of the resources, has not made use of wastes and residues and it still dumps 90% of waste in landfills. The economics of the two nations, offer contrasting pictures. Even in East Germany the emphasis is on recycling rather than imports of virgin materials.

Even a country like Sweden which has been endowed with large forests and has long forestry traditions, is now recycling newspaper.

In homes, offices and other establishments, separation of waste needs to be done into water paper, kitchen refuse, leaves and other types of garbage. Many of the organic materials can be composted giving rich organic manure. If this is not possible, separation may be done in resource-recovery centres. The major problem is the collection of waste and sometimes more energy may be spent in its collection than it actually generates. Collection and sorting of material should be tackled at the source. To be economical, such centres should have fleet of trucks which, of course, will create moreodours and add to traffic congestion.

Clean Japan Centre. Increased industrial development leads to increased economic activity and associated with this is discharge of waste materials of a very diverse variety. Accumulation of waste is a hazard and the Clean Japan Centre aims at recycling the waste with dual objective. Firstly, to conserve the resources which are finite, secondly to keep the country environmentally clean. Here then is an activity where environment and economics meet. Industrial culture of Japan is import of resources and export of processed products.

The Centre was established as an autonomous body with the support of private and Government organisations for purpose of searching methods of making good and effective use of waste materials and thereby help to keep the environment pleasant. The Centre is sponsoring a number of projects in various governmental, public and private sector organisations. From this points of view, Japan is indeed a head of many developed countries. This Centre has done commendable work in developing technologies for recycling a number of organic and inorganic materials including discarded tyres, plastics, chemicals, home appliances, etc. The recycling techniques followed in Japan are:

— Crushing of materials like discarded furniture, electric appliances, etc;

— Selection for magnetic, non-magnetic and specific gravity based materials;

— Thermal decomposition technique for recovering oranic substances in the forms of gas and oil;

— Food sources from organic wastes to meet the need for livestock feed;

— Composting;

— Methane fermentation plus residual sludge (60%);

— Conversion into solid fuel: Wood chips and saw dust produced into solid fuel;

— Melting blast furnace slag for making artificial jewellery and ammoniac alum for synthetic sapphires;

— Alcohol fermentation;

— Use of incineration heat: The average calorific value of garbage from rural areas is 1000-1200 K cal/kg, for medium cities it is 1200-1400 K cal/kg and for large cities it is 1500-2000 K cal/kg. This energy is being used for generating power and supply of hot water. Out of the 100 boilers in operation in Japan, 50 are equippped for power generation;

— Land fill: The waste after being treated, for prevention of pollutionis used either for landfill or even reclamation of sea. Normally, land fill is done in a very unenvironmental manner and the whole place not only stinks but is infested with flies and rates and emits very foul smell, thus effects the environment;

— Secondary pollution has to be prevented. Among other things this includes: smell and noise, visual pollution, dust and litter, pathogentic organisms, insects and rodents small fires (due to generation of methane), birds, surface water pollution, ground water pollution, etc.

British example. Britain has recently got into the area of recycling and there is a major campaign spearheaded by Warmer Bull. Warmer stands for "Warmth and Energy from Refuse." The campaign intends making public and public representatives conscious and avail of the potential economic and environmental benefits what is popularly called as DUSTBIN ENERGY with the assistance of the combined Heat and Power Association. The major use Britain has planned is combustion for community heating schemes which are in operation elsewhere in Europe. The watch word is "Rubbish is Fuel of the Future" and refuse-derived fuel (RDF) is becoming common.

The other aspect of the campaign is to "Keep Britain Tidy". At Edmonoton, the incineration plant despite its problems in early years has, during the last 10 years, disposed off between 3,00,00 to 4,00,000 tonnes of garbage per year. The technology for mass incineration of crude refuse seem to have been proven all over the world. With regard to reliability of energy recovery, the plants show reasonably high merit order. "On a comparable basis, in 1982/83, all nuclear power stations had an average rating of 74.2% Edmonton refuse incinerator achieved 76% and a plant in USA had rating of 85%" (Warmer Bull, 1984).

The recovery of energy and warmth by mass incineration process of rubbish is going to be economic where it is not possible to find place to dump rubbish near the cities/towns. For instance, New York generates 27,000 tonnes of garbage/day enough to fill 15 city blocks, 20 storeys high. In the London plant average cost of disposal per tonne ahs constantly decreased on account of increasing cost of energy and cost of disposing waste at distant sites has risen with inflation. This plant alone could save about 5 Mt of coal equivalent per year and this may contribute to the economy when North Sea oil and gas start to run out. The waste disposal by energy recovery has shown profit (Warmer Bull, 1984).

There are several cases of utilisation of municipal solid wastes, either alone or in combination with coal, as a power source in number of American cities like Chicago, Milwaukee, St. Louis, Connecticut, Iowa and New York. It is now a matter of common knowledge that municipal, agricultural and light industrial wastes are used for conversion into energy by direct burning in refuse-fired energy systems at 450 places in the world.

At Saugus (Masschusetts), one plant burns 1200 tonnes of residential and commercial refuse per day to produce steam for nearby General Electric manufacturing plant. If this plant were to produce electricity, the steam would provide enough energy for half a million homes (Khoshoo, 1982).

Other examples of usefulness of diverse agricultural and horticultural wastes, are the utilisation of straw in Denmark, where thee are 6000-7000 straw operated furnaces and in West Germany 150,000 farm houses are heated by burning straw (Khoshoo, 1982).

Recycling conserves natural resources but in too many cases it is not economical (accoding to our customary standards of what is economical) to recycle materials compared to the exploitation of virgin resoures. There is no reason why government policies that have encouraged the exploitation of natural resources might not be changed to encourage recycling instead.

5. Avoidance of Solid Wastes

The real reason for the phenomenal growth of solid waste tonnages has been the trend to the "throw-away" saciety : throw-away bottles, throw-away cans, disposable diapers, disposable medical supplies, increased packaging, planned obsolescence, etc. Until people realize that apparent conveniences involve less apparent inconveniences, present and future, and espouse a program of environmental stewardship, the solid waste problem will continue to increase. India badly needs efforts and talents directed at reversing the trend toward needless consumption.

PACKAGING

One solid waste that is worth looking at in closer detail is packing. There are several reasons for the phenomenal growth of the packaging industry. One is the continuing rise in self-service merchandising, which requires packages that will help to sell the product by themselves. Another is the use of paper and plastic packages for small items (batteries, razor blades, picture-hanging hooks, and countless others-look over any grocery or dedartment store) that might otherwise be easily shoplifted. Another is the trend toward the use of non-returnable bottles and cans for beverages.

It has been estimated that the average returnable glass container is used 19 times over an average lifetime of almost exactuly 1 year.

The number of non-returnable bottles produced has beenrising sharply in recent years, bringing higher profits to leading glass manufacturers. The major growth has evidently been to nonreturnable beverage bottles. Milk containers show a similar trend, with returnable glass containers being insignificant in numbers today, having been displaced by non-returnable paperboard and plastic containers.

The types and nature of packaging will greatly affect the collection and disposal of municipal solid wastes. Most packaging materials are not very degradable. Paper, the most degradable of the major categories, has been known to persist unchanged for over 60 years in landfills. The tin in "tin cans" is a thin coating that protects the steel from corrosion and provides a surface that is attractive and easily soldered and decorated; the tin accounts for only 0.5% of the weight of the can but its use makes the steel unsalvageable. Some wastepaper is presently salvaged for refuse in paper and paperboard production.

[**Note**. The contents of this chapter have been adopted from Presidential address of T.N. Khoshoo, Indian Science Congress, 1985.]

Pollution by Pesticides

Farmers and gardeners have been engaged in a never-ending struggle to avoid the ravages of insects and other pests that damage vegetation. Public health officials have been constantly trying to control numerous pests that have been vectors of human disease. The number and quantities of chemical compounds used in these campaigns have increased sharply in the past 2 decades. There is great controversy about whether or not the benefits produced by various uses of pesticides outweigh the disadvantages and possible long-term risks.

In medieval times, diseases that had reached epidemic proportions were termed as pestilences, and places where diseased people lived or were kept were called pestholes and pesthouses. They were frightiening words that reflected the broad meaning of the word pest. The modern custom has been to categorize the crawling, walking, or flying creatures separately from the pestiferous microorganisms that cause diseases. We call the latter, collectively, microbes or germs, and the chemicals that are used to control them are medicinals and related toxicants in the categories of disinfectants, antiseptics, germicides, and internal medicines such as the antibiotics, for killing bacteria and other internal parasitic organisms. Most of the larger enemies of man—the so-called pests—have been generally usually combatted in similar ways.

With the increased production in agriculture all over the world and especially in India, it is essential that maximum efforts should be geared to protect the health of workers engaged in this vital field of activity. Herbicides are used more or less in the same scale as pesticides and the effects of the former are comparatively less harmful than the latter.

But indiscriminate use of herbicides has serious repercussions as they often attact the nontarget plants, even killing the soil binders and accelerating soil erosion. They may cause siltation of the fresh-water system and pollute freshwater lakes. Several herbicides are also health hazards. Some of them can cause genetic damage affecting the chromosomes of nontarget plants (Sikka K and Sharma AK, Proc. Ind. natl. Sci. Acad., 42B, 299, 1976) influencing ultimately the productivity of crops. The herbicide 2, 4, T is teratogenic and may initiate birth defects. The problem of weed control needs to be tackled with great caution as allelopathic or weed interaction forms a part of the ecosystem and is beneficial to crop production.

In tropical countries like India pest control is required on an extensive scale in the view of occurrence of a large number of pests with regular periodicity. Insecticides or pesticides, including the rodenticides and fungicides, pose serious problems of pollution in agriculture.

They interfere with the reproduction of birds and disrupt the ecological balance. Some insects may become resistant through mutation to such chemicals, thereby jeopardizing agricultural production. In view of the growth of these mutant strains, extremely toxic compounds such as methyl parathion instead of DDT are often employed to kill the insects. They are disseminated in large quantities in the environment and pose hazards to the health of agricultural workers. The poisioning capacity of the residue is normally measured by its acute toxicity. The response varies in different populations, which is evidently under genetic control. A relationship between DDT and liver tumcur, cirrhosis and hypertension has been recorded. Similar studics have been carried out with its metabolite DDE (dichlorodiphenyl ethance-Rudd RL in Environment ed. Murdoch WM Sinauer Associates USA, 1975). Several plant compounds such as rutinones, pyrethroids, nicotine, chlorinated hydrocarbons, mercurials affect the kidney, respiratory system and central nervous system. Even the endoctrinal and gonadal systems have been affected.

History of Pesticide Usage

Prior to the 19th century, lye, lime, soap, turpentine, tobacco (for its nicotine), pyrethrum powder [made from pyrethrum (chrysanthemum) flowers], mineral oil, and arsenicals were used as pesticides, but their use was never great.

The first real use of a chemical compound as a commercial pesticide had been that of Paris gren (also known as copper acetoarsentie or copper acetate metaarsenate), whose approximate composition has been Cu $(CH_3COO)_2$. $3Cu(AsO_2)_2$, for control of the Colorado potato beetle in the western United States about 1868. Within some years it was also being used against cankerworms and codling moths on apple tres. Several other arsenic compounds were soon in use; in the early 20th century, lead and calcium arsenates $[Pb_3(AsO_4)_2$ and $Ca_3(AsO_4)_2]$ were the preferred choices.

During the late 19th and early 20th centuries several other insecticides and fungicides such as bordeaux mixture (a mixture of copper sulfate and lime), sulfur, oil, nicotine, and rotenone (from derris root (were used).

A major advance was achieved in 1939 with the discovery of the insecticidal properties of DDT (for dichloro-diphenyl trichloroethane) by Paul Muller, a research scientist with the Swiss chemical firm of J.R. Geigy. DDT itself had been known since 1874. After demonstration during World War II of the value of DDT in the control of typhus (a rickettsial infection transmitted by lice and fleas) and malaria (a aprotozoan infection of the misquito that can be transmitted to humans), Muller won the Nobel Prize for Physiology and Medicine in 1948.

DDT was the first of a number of chlorinated hydrocarbons to be developed as pesticides. Another important group of pesticides called the organic phosphates, got developed as by-products of German research on nerve gases about the time of World War II. The herbicides 2, 4-D and 2, 4, 5-T were also discovered (in the U.S.) during World War II.

The past 30 years have seen the development of various types of chemical pesticides discussed in the following sections. There has also been progress in the area of nonchemical methods of pest control, some of them quite old compared to the chemical pesticides; these have been discussed in a separate section.

Pests in the Ecosphere

Parasites and predators, undesirable though they may be to the victims, have been a normal part of the ecosphere. Few organisms escape being victimized by others higher in the food chain. Humans are among the worlds worst pests from the viewpoint of the victims. In turn, we have ben pestered by other living organisms since the dawn of history, and before that judging from what takes place in the animal and plant world around us, our ancestral relatives were almost certainly plagued with numerous kinds of parasites, predators, competitors, and destroyers of their food, health, and comfort.

Dysentery had been a common afflication in ancient Greece and Rome and it was perceived by early writers that it was spread partly by flies falling into food. Later, in 1498, Bishop Kund of Denmark associated high populations of flies with an approaching plague. Thomas Sydenham, a seventeenth-century English physician, reported that an abundance of flies in the summer was an unhealthy sign.

Pests and Human Welfare

The variety of injurious effects caused by pests-from the human veiwpoint-has been somewhat almost beyond description. Among the major pests have been those that consume crops and stored food products; others destroy fibers, forests, and structural materials ; and, among the worst offenders have been those that cause or carry diseases.

1. *Carriers of disease*. More than one hundred human diseases get transmitted to man by arthropod pests (arthropods are invertebrate animals such as insects, having "jointed" legs and other appendages). Malaria is still a minor problem in most parts of the world, but it was not always so. Even today, millions of people suffer from malaria, and the disease is on the increase again due to ineffective control measures against the malaria mosquito carriers (several species of the genus Anopheles). Malaria mosquitos become infected with the malaria blood-parasite, Plasmodium, when they such blood from a malarious person ; they then transmit the disease to a healthy person when they bite a new victim.

The tsetse fly is one of the most dangerous animals in Africa. It ranges across 4 1/2 million square miles of tropical Africa, an area that is nearly 1 1/2 times that of the United States. Together with the organism it carries, the trypanosome blood parasite that causes sleeping sickness in man and nagana in animals, the teste fly dominates much of this vast domain by destroying people and their animals, and choking the economy. An epidemic of sleeping sickness that swept the shores of Lake Victoria killed over 200,000 people, two-thirds of the population, while the nagana disease of domestic cattle has been the direct cause of much of the malnutrition in Africa today.

Bubonic plague is always present in the Orient and other parts of the world, and is under close watch by health authorities guarding against its spread. It was the disease of the great epidemics of the Black Plague that occurred during the Middle Ages when nothing was known about its cause. We now know that the black rat teams up with an accomplice, the oriental rat flea, which carries the plague germ from rats to humans. Rats are called the reservoirs of the disease, and fleas the vectors.

2. *Pests and Food.* Pests consume much of the food intended for humans because in some parts of the world, pests harvest as much as 25 to 50 per cent of the crops before the crops ever reach maturity, and afterward, insects and rodents destroy much of the remaining food while in storage. In a world where one-fourth of the people have been suffering from hunger, much depends on who gets to the dinner table first. Hundreads of pests-insects, mites, fungi, worms, weeds, and rodents and other mammals-cut down the harvest of crops by reducing the yield, cause losses in storages, and degrade the quality of cereals, fruits and vegetables. All this increases the cost of food in the markets.

The food consumed by some species of pests is prodigious. A single tobacco worm larva (a cousin of the tomato hornworm consumes in 28 days of growth, food weighing about 50,000 times its birth weight, during which time the larva increases 12,000 times in size. A single cabbage aphid weighs about 1 miligram (1/28,000 of an ounce). If it could find enough cabbage, this tiny aphid could produce, in one season, enough aphids to form a mass weighing 822 million tons, five times the weight of the total human world population. This does not happen. But, because of the vagaries of the natural checks and balances, the cabbage aphids and other pests often get out of hand. The nourishment need not be human food. A single rainbarrel has been observed to produce more than 100,000 mosquitos in one season.

A disease of potato plants called "late blight," caused by the fungus *Phytophthora infestans*, was sresponsible for one of the greatest social upheavals in history when it wiped out the potato crops in Ireland and caused the frightful four-year famine during 1845-1848, now known as the Irish Potato Famine. Many thousands of those who survived moved to the United States in one of the largest migrations in history. The fungus is still present in potato fields through-out the northern hemisphere, and waits for suitable weather conditions when it turns the plants black almost overnight.

Agriculture got developed about ten thousand years ago. With it came concentrations of single crops, irrigation, disturbance of the soil, and other practices that intensify old pest problems and create new ones. Highly intensified agriculture brings about drastic changes in the environment by which minor pests may become major pests or become so destructive that the effort must be abandoned unless practical means can be found to combat them.

Until recent times pest control had been simple : either mechanical means were employed, or the pests were allowed to share in the harvest. As the pests had first choice, there was hunger and periodic famine. Even today, in many areas where primitive methods of agriculture are being followed, there is chronic malnutrition and starvation, partly due to pests.

Insects. The insects have been occupying a special place in the category of pests due to their abundance and their impact on human health and nutrition. Three-fourths of all species of animals have been insects. All the land animals on the face of the earth-including man-do not weigh as much as the earths insects. Some of these insects have been carriers of disease. Others destroy food, fiber, stored products, and man made structures. Insects steal or destroy one-third of everything that people grow, build, or store. Except for man, they have been more destructive to the natural animal and plant environment than any other group of animals

There are about 10,000 species of insects that are considered to be pests in the world. Most of them are minor pests much of the time, but they go on sporadic rampages when conditions become favorable for a surge in numbers. About 100 species are major pests year in and year out, and these types cause 80 to 90 per cent of the insect-pest damage. About 80 to 90 per cent of the total insecticide use has been for their control.

Pests and Man

The role of pests without the intervention of man could be seen in those remote areas of the world that have been uninhabited by humans. Populations of plants, animals, parasites, and predators have been usually well balanced before man's arrival. But when people appear, even in moderate numbers, they change the course of rivers and streams, big new channels, make artificial barriers, impound water, and irrigate fields where only rain sufficed before. They cut down the trees, burn the shrubbery, rip open the soil that supports a great diversity of wild plants and animals, and put in their place expansive fields of one kind of crop. The disruption of the natural condition upsets the balance and encourages the buildup of what then comes to be known as pests. Instead of people living in harmoney with nature, there ensures a battle, and forever after there has been unceasing strife betwen man and pest.

An example had been the opening of the prairies to settlement and agricultural development. The "corn belt" was originally a vast grassland with what is called an Andropogon Climax vegetation. The grass had been the natural food of a small sucking insect called the chinch bug, which lived of the juices sof the plants and hibernated duringh the winter in the clumps of dead grass. People saw the potentialities of this rich grassland and quickly found that here they could grow luxuriant cornm and wheat . They plowed out the Andropogon, planted crops of one or two species, cultivated and fertilized them, and created a succulent growth that was highly favorable to the chinch bug. Wheat in the spring and corn in the summer were made to order for the habits of the insect, which has two generations, a year, and enabled it to build up to enormous populations. Despite the settlers annoyance and their vigorous battle against the bugs, it was inevitable that the insects would be able to destroy the crops.

Coping with Pests

As soon as people could think of better ways to deal with pests than by scratching or clubbing them to death, they took recourse to whatever means were available. Flailing with bundles of sticks, a method of attack still in use against grasshopper invasions, was supplemented by the use of scoops, ditches, and traps of various kinds.

People of primitive cultures kept great store in ceremonial practices to control pests. When Europeans came to the east coast of America, they found natives with an advanced agriculture who were engaged in a continuous fight against insects, birds, and large animals. The Indian remedy for cutworms was for the squaw, who was responsible for planting and tending the garden, to step around the gardedn naked under a full moon, dragging her garment in a magic cricle that would keep out the pests. A similar custom got practiced in ancient Rome, but there, the quarter of the moon made no difference. However, there was an advantage if the woman was menstruting. The ritual was used, according to Pliny, to keep out canker worms, caterpillars, beetles, and "all such worms and hurtful vermin." The invention, said Pliny, came from the Cappadocians who, however, were more mannerly inasmuch as they had their women go only barefoot with their hair hanging loose. But they had to be careful not to do it at sunrise or the crop would wither away to nothing.

The early American settlers were inclined to experiment with more sophisticated remedies than those used by their neighbors to protect their crops and fruit trees. Vinegar, cow dung, and urine were applied in various ways.

There is no evidence what the extravagant measure recommended by Josselyn caught on with the thrifty, and thirsty, Yankees.

Pest Control by Chemicals

The use of chemicals for pest control was an early innovation. An Egyptian scroll dating about 1500 B.C., of provides formulas for preparing pesticides. Homer spoke in the Iliad, about 1000 B.C., "pest averting sulfur", perhaps in reference to the early practice of burning sulfur to rid the premises of pests and disease germs. Sulfur dioxide, which is evolved from the burning of sulfur, is highly effective and is still used for fumigation of stored fruit. The magically curative effect of sulfur when pounded into a fine powder was inevitably discovered in the early days of the need for a palliative against crawling creatures. Elemental sulfur in finely divided form, as well as sulfur in inorganic chemical combinations, is still used in large quantities as a pesticide.

Democritus, a Greek philosopher of about 400 B.C., described a remedy for "blight" of plants. Cato (184 B.C.) tole of rubbing crude oil, or the waxy portion of it, to cure scabies or sarcoptic manage, a skin ailment caused by a small mite. He described a pesticidal formulation consisting of a mixture of chemical substances for burning to produce ana aerosol for killing pests of the vineyard. Dioscorides, a Greek physician of the first century A.D., said that scorpions could be paralyzed by putting larkspur on them. Pliny, a prominent Roman of the first century and the most popular science writer of ancient times, told how to control "mildew" of wheat seeds with a pesticide and described several concoctions for protecting plants and killing lice, nits, rats, and mice. Among Plinys favorite remedies were sulfur and bitumen, as were materials of botanical origin such as olive extracts, powered larkspur seeds and flowers, wild mint, and hellebore. The latter continued in use until modern times.

Many of the control measures for pest infestations in use during the past three hundred years have included ancient herbal remedies as well as more recently discovered botanicals. The more modern botanicals include the ground-up roots of certain tropical plants having rotenone, the ground flowers or extracts of pyrethrum (a daisy originating in the Near East), and the powdered leaves or water infusions of tobacco. Urine had been a popular cure-all throughout the ages but was too rustic for most modern pest fighters. The antibiotic property of urea was rediscovered in 1947 by American plant pathologists, George Zentmeyer and James Horsfall.

Upon discovery of the pesticidal effect of the more poisonous minerals, farmers and gardeners often turned to substances other than those of plant the animal origin. The Chinese were using arsenic to kill garden pests more than one thousand years ago. Arsenic, mercuric chloride, copper sulfate, soap, and turpentine came into use during the seventeenth and eighteenth centuries. By the time lead arsenate was invented in 1892 for use against a devastating invasion of the Gypsy moth that was defoliating forest and ornamental trees in Massachustts, the beleagured populace had little compunction against using all manner of knwon poisonous materials to protect vegetables and fruit trees against insects and fungi.

About fifty years after the introduction of lead arsenate, a fresh attack was made on the Gypsy moth with DDT, a material much safer to humans. The widespread revulsion against its indiscriminate use from airplanes in residential areas contributed to a popular distrust of pesticides from most parts of the United States and Western Europe.

Modern Pesticides

There are about 600 different basic chemical compounds used for pest control, and as of June 1974, there were 34,029 formulations of these registered for use in the United States by the Environmental Protection Agency. They have been mainly insecticides, herbicides, fungicides, algicides, and rodenticides, But they also include chemicals that are used against other pests, for example, nematocides (microscopic worms), molluscicides (snails and slugs), acaricides (mites and spiders), avicides (birds), piscacides (fish), Lampreycides (parasites of fish), and echinocides (sea urchins and sea stars). Materials related to pesticides include repellents, attractants, and synergists. The latter have been substance that enhance the effectiveness of a toxicant.

Pesticides include a variety of chemical types. The most commonly used:

1. *Botanicals*. Botanicals-materials derived from plants although among the oldest of pesticides, still find useful applications because of their unique biological effects. The active ingredients called pyrethrins, extracted from the flowers of a species of chrysanthemum, are used widely in household insect sprays because they are practically nontoxic to humans, though extremely small amounts will kill fish in aqaria. Nicotine, an alkaloid extracted from the leaves and stem of tobacco, was once a favorite insecticide for use on both agricultural and ornamental plants because it kills a wide variety of pests quickly. Though nicotine is having the disadvantage of being extermely poisonous to humans and other higher animals, it decomposes or volatilizes within a few days to a few weeks and is no threat to the environment. Rotenone, which in pure form is a crystalline material, is extracted from the roots of several species of plants found growing in parts of Southeast Asia, Indonesia, and Central and South America. It was discovered by the aborigines who employed the material to catch fish by beating the sides of their canoes with shredded roots, releasing the poison into the water. The paralyzed fish could be easily picked up by hand. Other botanicals that find limited use have been ryanodine, contained in the roots and stems of ryania, a plant growing in Trinidad, and the alkaloids in the seeds of sabadilla, a plant that grows in the highlands of northern Mexico.

2. *Important types of pesticides*. Pesticides may be classified in a number of different ways: (1) by their target [insecticides, herbicides, fungicides, rodenticides, algaecides, molluscicides, acaricides (miticides), nematocides etc.); (2) by their chemical nature (natural organic compounds, inorganic compounds, chlorinated hydrocarbons, organophosphates, carbamates, etc.); (3) by their physical state (dusts, dissolved solutions, suspended solutions volatile solids, etc.); or (4) by their mode of action (stomach poisons, contact poisons, fumigants, etc). In this section a classification based on the target has been adopted, with several subclassifications by other means.

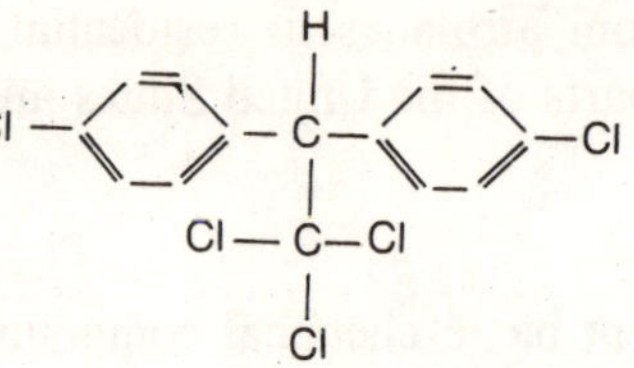

"DDT" or 2, 2-bis (p chloropheny1)-
1, 1, 1-trichloroethane

A. *Insecticides*. Insecticides have been generally classified as stomach poisons (including systemic poisons), contact poisons, and fumigants. Stomach poisons have been those that are taken internally through the mouth of the insect (and are thus effective against those with biting mouth parts, such as cater-pillars) and then absorbed into the body through the digestive tract. Systemic poisons have been those that are absorbed into the plant upon application and then ingested by the insect when it eats the plant or its sap. Most of the newer synthetic insecticides have been contact poisons, which affect the insect upon contact, entering through the body wall or the respiratory systems and affecting the nerve or respiratory system or the blood stream. Fumigants have been gaseous chemicals that kill insects and other pests by entering the respiratory system; these are usually employed in closed spaces.

1. *Stomach poisons*. The most important stomach poison insecticides have been the arsenic and fluorine compounds, although several other types have been in use. Arsenicals are generally arsenites (AsO_2-) or arsenates (AsO^3_4). Arsenites have been readily easily solube in water and are very toxic to animals and plants; arsenates have been less phytotoxic (toxic to plants) and their poisonous effects generally follow reduction to arsenite. A wide variety of arsenicals have been in use today : acid lead arsenate ($PbHAsO_4$), basic lead arsenate [$Pb_4(Pb)H(AsO_4)_3$. H2O], calcium arsenate [$Ca_3(AsO_4)_2$], Paris green arsenious oxide (As_2O_3), and various other compounds.

Inorganic fluorine compounds are used as stomach poison insecticides in certain situations. Major members of this group have been cryolite (sodium fluoaluminate, Na_3AlF_6), sodium floride (NaF), and sodium and barium fluosilicates (Na_3Sif_6 and $BaSif_6$). Sodium fluoride has been not so suitable as the other compounds because it has been toxic to plants; it is also very soluble in water and therefore does not release fluoride as slowly as the other compounds, which are effective over longer periods of time. The fluorides appear to exert their toxic action on pests by complexing with and inhibiting many enzymes that contain metals such as iron, calcium, and magenesium.

Other stomach poisons used as insecticides have been mercury compounds (such as $HgCl_2$ and HgCl), boron compounds (such as borax, $Na_2B_4O_7$, and boric acid, H_3BO_3), antimony compounds, thallium compounds (such as Tl_2SO_4 and CH_3 COOTI), yellow phosphorus, and formaldehyde (HCHO).

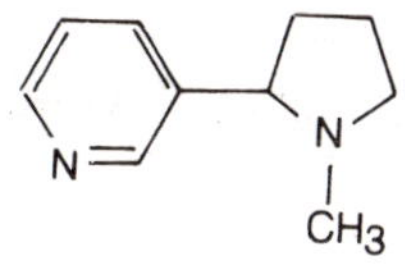

Nicotine

2. *Contact poisons*. Most of the synthetic organic insecticides fall into the category of contact poisons which includes the chlorinated hydrocarbons, the organophosphates, the carbamates, and several other important groups, including the natural organic compounds obtained from plants.

Certain botanical compounds are having insecticidal properties and are extracted from plants for use as contact poisons by man. Pyrethrum Insecticides, are extracted from the dried powdered flowers of certain chrysanthemums; they are safe to plants and only sligthly toxic to animals but skin contact or inhalation may give rise to allergic reactions in certain sensitive persons; they cause rapid paralysis in insects, apparently affecting nerves or muscle, but their mode of action and selectivity to insects are a mystery. Nicotine is a very toxic heterocyclic amine which is obtained from the tabacco plant and its relatives. It produces tremors, convulsions, and eventually paralysis in insects; the mean oral lethal does in adult humans is about 60 mg (1 mg/kg), but a few milligrams may produce severe illness and occasionally death. Rotenone which can also act as a stomach poison, gets extracted from the roots of derris and other leguminous plants; it interferes with the biochemistry of oxygen utilization and is very toxic to insects, fish, and pigs but is of low toxicity to most mammals.

Another family of contact poison insecticides, the chlorinated hydrocarbons includes DDT, its metabolites DDD (also known as TDE) and DDE; BHC (for "benzene hexachloride" although it is really hexachloroclohexane); heptachlor; dieldrin; endrin; aldrin; methoxychlor; toxaphene (manufacured by chlorination of camphene compounds); mirx; chlordane; tetradifon; and many others. DDT has been characterized by a very low vapor pressure (not true of all the chlorinated hydrocarbons) and is very insoluble in water, with a limit of solubility of about 1.2 ppb. DDT has been appreciably soluble in oils, from which it can be absorbed readily through the skin to accumulate in the fatty tissues of the body; if gets released slowly when the fat is regarded as a source of energy. DDT and other chlorinated hydrocarbons all appear to have their Primary effects upon the central nervous system in vertebrates and invertebrates but there is no good understanding of the mechanism of action. In the environment,

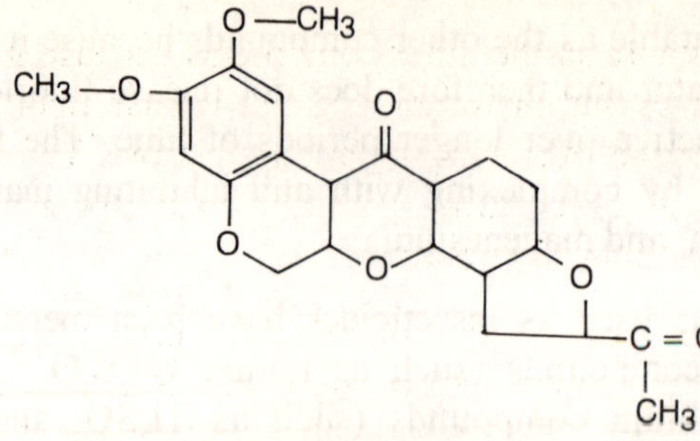

Rotenone

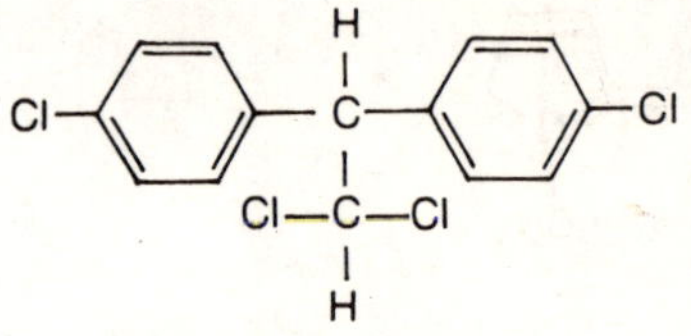

DDD (TDE)

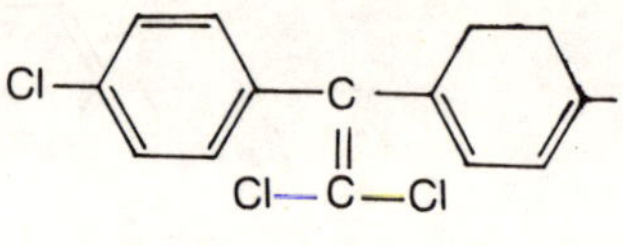

DDE

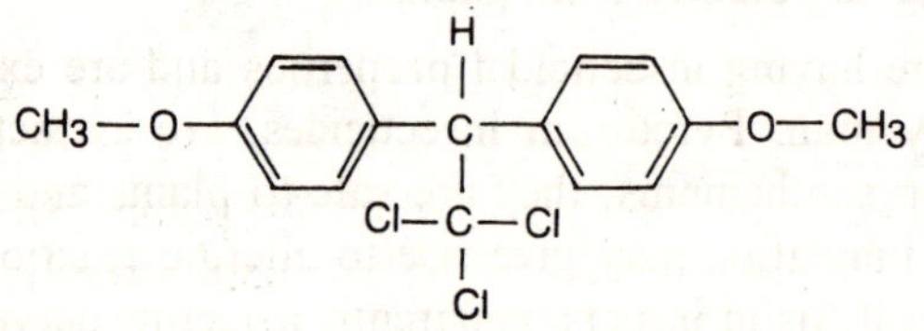

Methoxychlor

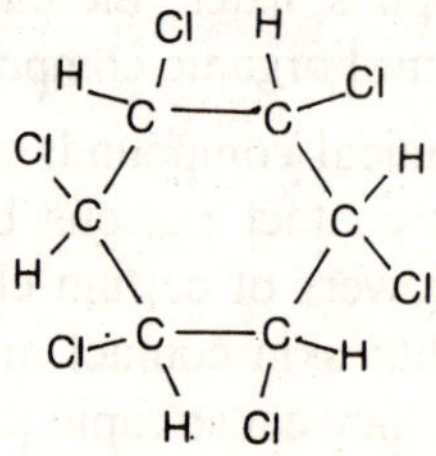

BHC

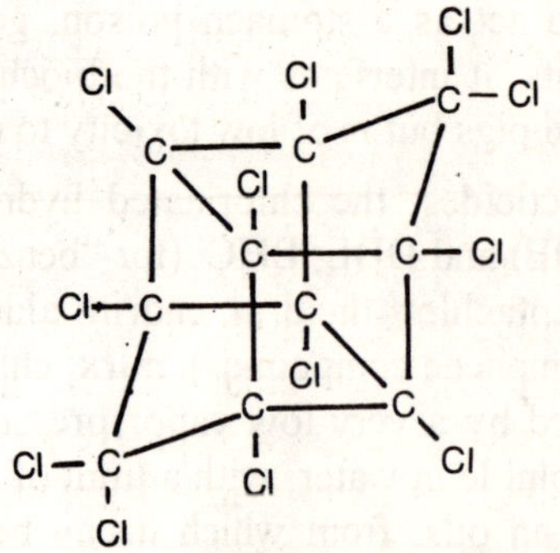

Mirex

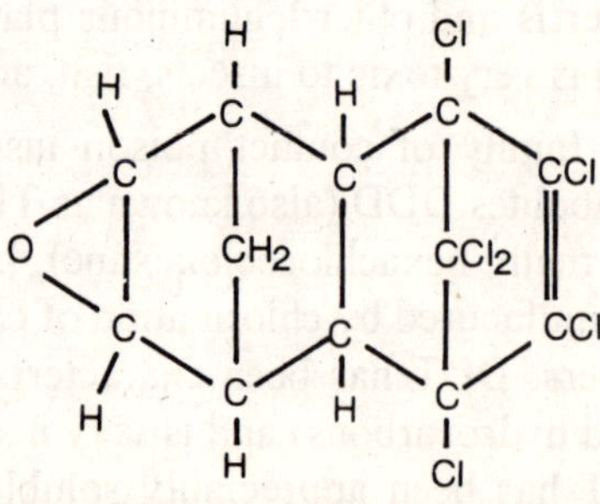

Dieldrin and endrin

Representative chlorinated hydrocarbons. Mirex has been a nonplanar molecule resembling a distored box. Dieldrin and endrin are stereoisomers; i.e., their three-dimensional configurations are different.

DDT gets degraded or metabolized into DDD, DDE, DDA[($ClC_6H4)_2$ CHCOOH], Kelthane [$ClC_6H_4)_2$ C(OH)CCl_3], and similar products. Aldrin is known to be converted into dieldrin, an unusually persistent insecticide. The chlorinated hydrocarbons have been for the most part persistent broad-spectrum insecticides; i.e., their residues persist in the environment for long periods of time (a few months or years) and they affect a large number of different species. These are the qualities that made them so desirable, but that are also responsible for their undesirable side effects, discussed later in this chapter.

Many of the chlorinted hydrocarbons, are highly stable chemical compounds. Many of them are broad spectrum pesticides, so called because they are effective against a large number of species. This property gives a pesticide an economic advantage in that it can be broadly useful for controlling a wide variety of pest problmes; but is has the disadvantage that the lack of a high degree of specificity may make it seriously injurious to beneficial organisms such as parasties and predators.

Characteristically, the chloribnated hydrocarbons are persistent in the environment, have an affinity for lipids (fatty tissues), are stored in the body fat, are metabolized (decomposed) within the body slowly, and therefore eliminated slowly, and are prone to be taken up by organisms and passed along in the food chain. Some of them, such as DDT, can be found in living organisms at concerntrations many times that in the environment.

The continued use of chlorinated pesticides, which have a lasting effect on pest management in agriculture, has been responsible for the presence of the residues in grains, milk, eggs, vegetable and human tissues. Significant amounts of residues of chlorinated hydrocarbons have been recorded in human placentae, cord blood and breast milk in the tests conducted at the Industrial Toxicology Research Centre, Lucknow. The contaimination of wheat flour with gammaxene has led to the development of epilepsy and other complications. Several peculiar paralytic attacks have been reported following the consumption of crabs caught from rice fields which had received massive doses of pesticides (CR Krishnamurti, ITRC workshop NCIRTC, 1979).

Organophosphorus compounds have been the most toxic insecticides, dangerous to insects and mammals alike. Many of them (parathion, thimet, TEPP, tetram, Phosdrin, paraoon, pyrazoxon, HETP, di-syston, and others) have been in the "super toxic" category of human poisons for which human fatal doses are less than 5 mg/kg, along with arsenic, cyanides, heroin, strychnine, digitoxin, LSD and others. As small a total does as 2 mg of parathion has been known to kill children. Several of these compounds are shown below. It is widely accepted that the mode of action of organophosphates in vertebrates and invertebrates is the inhibition of the enzyme cholinesterase with consequent disruption of nervous activity caused by accumulationof acetyl-choline at the nerve endings. Symptoms are nausea, vomiting, diarrhea, cramps, sweating, salivation, blurred vision, and muscular tremors. In general, organophosphates are more easily biodegraded than the chlorinated hydrocarbons and are thus not as persistent in the environment, however, some of them (such as paraoxon) are quite stable and can persist for years.

Amongst the peach and strawberry pickers phosphate poisoning after parathion application has been reported through the conversion of parathion into paraxone, a very toxic and stable

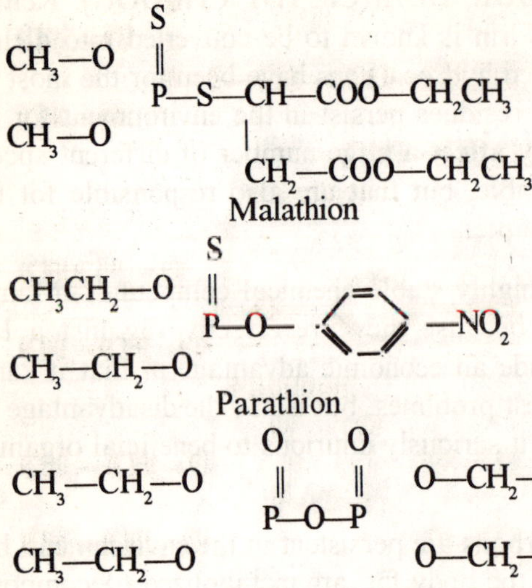

Some common organophosphate insecticides.

compound. The carbamates are strong cholinesterase inhibitors and may even lead to mental aberrations. Nearly all the organophosphorus, or OP, compounds are used exclusively for their effectiveness against insects and mites. They differ greatly in their insecticidal and miticidal action. Most of them are broad spectrum with only limited specificity. Parathion, for example, tends to be disruptive of the ecosystem by reason of its effectiveness against beneficial species. However, the chemical decomposes relatively quickly and does not constitute a significant long-term hazard to the environment.

The OP compounds also differ in their safety to users. Parathion is extremely poisonous to humans and other vertebrates as well as to invertebrate pests. Users must be instructed in appropriate precautions and safety measures. malathion, on the other hand, is about as toxic as aspirin and is therefore an effective insecticide that can be used safely around the garden and on household pets. Most of the other OP insecticides are intermediate in their mammalian toxicity.

The OP insecticides indirectly affect the nervous system and for that reason have been called "nerve gases" or "nerve poisons." The primary action of the OP toxicants is to inactivate the enzyme acetylcholinesterase, the enzyme that is responsible for breaking down acetylcholine (A Ch), one of the nerve transmitter substances, after it has performed its function of bringing about transmission of a nerve impulse across the junction between two nerve cells or between a nerve cell and an affector organ such as a muscle fiber. Thus it is accurate to speak of the OP pesticides as "enzyme poisons," although they are customarily called enzyme inhibitors or, more specifically, *choline terase inhibitors.*

Other classes of pesticides include a miscellany of chemical types, having a variety of effects on living organisms. It would be possible to classify them according to their biological effects, but the classification would be complex and overlapping.

A fourth group of contact poisons, the carbamates, have been derivatives of carbamic acid (NH_2COOH), like carbaryl ("Sevin") shown. They are broad-spectrum insecticides

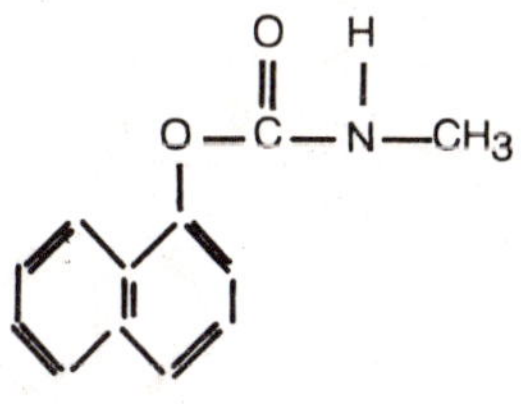

Carbaryl ("Sevin")

with relatively low toxicity to mammals and even to some insects. Like the organophosphates, they are inhibitors of cholinesterase and produce similar effects on the nervous system.

3. *Funigants*. These are gases (or the vapors from heated liquids or solids) that are used in sealed spaces to kill pests of stored products or nursery stock; sometimes they are used in the open under tents. Popular fumigants have been hydrogen cyanide (HCN) released from calcium cyanide (Cyanogas) by atmospheric moisture, methyl bromide (CH_3Br), carbon disulfide (CS_2), carbon tetrachloride (CCl_4), nicotine, naphthalene, and a number of other compounds.

B. *Fungicides*. These have been used to prevent or eradicate fungal diseases of plants. Protective fungicides used to prevent fungi are sulfur, organic mercury compounds (such as methyl) mercury used on seeds), and copper compounds. Eradicant fungicides include lime sulfur, mercury, formaldehyde, and several dinitro compounds.

C. *Herbicides*. These are chemicals that kill plants. Non-selective herbicides such as sodium arsenate, sodium chlorate, sulfuric acid, and some oils kill all plants, while selective herbicides kill only certain type of plants, such as ferrous sulfate for dandelions. The most commonly used herbicides include 2,4-D (for 2,4-dichlorophenoxyacetic acid) and 2,4,5-T (for 2,4,5-trichlorophenoxyacetic acid) and their esters, growth-regulating substances that kill many broad-leaved plants while not injuring many narrow-leaved plants; thus they are valuable weed killers. The human median lethal dose is estimated at 300 to 700 mg/kg of 2,4-D, and several deaths have resulted from use of the herbicide. Figure shows the chemical formulas for these herbicides.

D. *Rodenticides*. These are used against rodents like rats and mice. Some have a direct, immediate action, such as strychnine, sodium fluoroacetate ($CH_2FCOONa$, "Compound 1080), phosphorus, thallium, and ANTU (alpha-napthylthiourea). Others are anticoagulants, such as the commonly used warfarin. Of special interest has been norbormide, a very selective toxicant for rats that is much less toxic to other animals. The LD_{50} (median lethal does) of norbormide for rats has been 5 to 15 mg/kg taken orally, while for mice it is 100 to 1000 mg/kg and for most lab animals it is well above 1000 mg/kg. This poison seems to constrict the small peripheral blood vessels in rats to an extent that does not take place in other species.

Biology of Pesticides

1. *Selectivity*. The primary consideration in the use of a pesticide, as with any

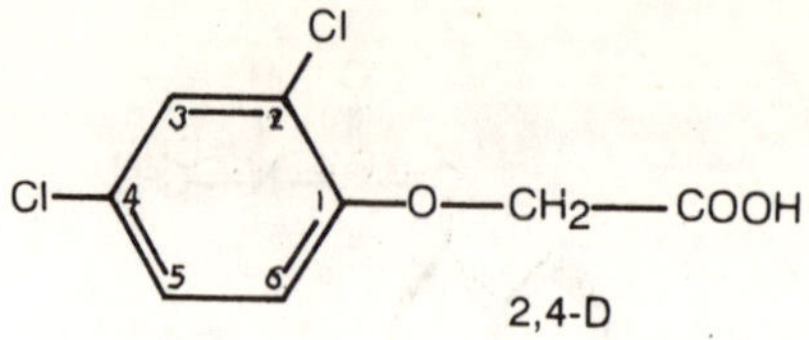

2,4-D

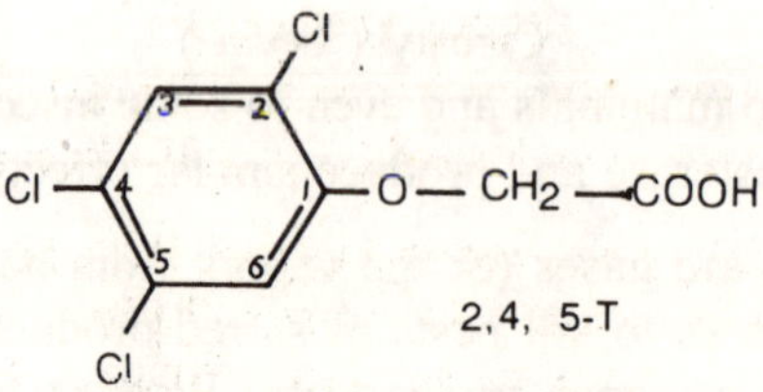

2,4, 5-T

"2,4-D" and `.2 4. 5-T."

biocide, has been to find a substance that will have maximum effect against the offending organisms and minimum effect against the organism to be protected : plants, animals, the beneficial organisms, or humans. This property of being effective but safe is called selectivity. It is an ideal that is rarely achieved to perfection. In medicine the undesirable toxic effects are called "side effects." In pest control the injurious effects are referred to as phytotoxicity (effect on plants), mammalian toxicity (effect on mammals), human toxicity, or whatever specific undesirable effect may be of importance. The goal of future research on pesticides is to find materials that will be more highly specific in their action against the particular pest under attack, while being noninjurious to beneficial species, safe to humans, and nonpersistent in the environment.

2. *Resistance*. Many pests are having the ability to develop resistance to the effects of certain chemicals. Populations of organisms commonly become resistant to chemicals following repeated treatments that do not completely wipe out the population but leave some survivors to reproduce. This biological trait has been a direct threat to health when disease-causing bacteria, become resistant to medicinal drugs. As with medicines and bacteria, pest populations often become resistant to chemical treatment, and when this happens, it creates difficult problems that may make it necessary to discontinue using the material and to find an effective substitute.

More than 230 species of insects have become resistant in different degrees to various pesticides. It is a major problem in malarial dominated areas where mosquitoes have developed resistance to commonly used materials. However, the ability to adapt to such artificially imposed conditions as the intrusion of chemicals in the environment is not resticated to insects. Rats became resistant to the anti-coagulant poison Warfarin. Resistance to pesticides has been developed by insects, bacteria, fungi, mice and the rodents, any by six species of weeds. Malaria took 2, 500, 000 lives annually in the mid-1950s. It seemed that control of the number one killer was in sight, but today thirty-eight species of anopheline mosquitoes, carries of the protozoan parasite that causes the disease, are resistant to insecticides.

Resistance is also characteristic of bacteria and fungi. During the summer of 1970, a new race of the fungus that causes southern corn leaf blight reached epidemic proportions, and before the season was out, destroyed an estimated 710 million bushels of corn worth one billion dollars. Reliance had been placed on a strain of corn (with T-type cytoplasm) widely used in hybrid seed production. The incident was a dramatic demonstration of a shift in the genetics of the host-parasite balance.

A familiar example is the danger of penicillin and other antibiotics resulting in the appearance of resistant strains of bacteria for which no control is available. Resistant strians of staphylococcus bacteria can develop within a few hours in the laboratory by artificial selection. Only 5 to 15 per cent of staphylococcus strains found generally outside hospitals are resistant to penicillin, but astonishingly, 90 to 95 per cent of those found in hospitals are resistant. It was found that 45 per cent of a large number of children studied under the age of five years had penicillin-resistant staphylococci acquired from hospital nurseries, Penicillin resistance in Staphylococcus acquried is common enough to cause the abandonment of the drug in all serious systematic "staph" infections.

Many strains of microorganisms formerly sensitive to streptomycin also have become resistant in some degree to the drug. Infections from resistant strains are more dangerous because higher doses of less desirable drugs may have to be substituted. With pesticides, second choice materials may require very heavy applications, involving greater dangers to users and the environment.

The use of antibiotics in treatment of several diseases in animal husbandry including mastitis in cows is responsible for the problem of residues in milk. Their application in stimulating the growth of poultry and pigs is also practiced. It is claimed that the development of bacterial resistance to drugs in man is related to the presence of antibiotic on food chain in our ecosystem.

Mechanisms of Resistance. The development of resistance on the part of a species or strain is genetic in nature. In some cases a single pair of genes is responsible. A population of house files, for example, consists of individuals that differ in their susceptibility by reason of inherited characteristics. Even though most of the members may be highly susceptible to a particular insecticide, there will be a few individuals that have the ability to survive. The survivors will pass that ability on to some of their offspring, while the least resistant individuals of the original population will not have the opportunity to do so because they do not survive. The process is one of Darwinian selection, though artificially so by the introduction of human made substance into the files' environment.

This selection process taking place over several generations of insects has been known to develop resistance to a pesticide many fold over the normal resistance of the original wild population. For example, DDT was initially highly effective against houseflies. Now, strians of houseflies that are completely unaffected by 100 per cent DDT are used in research to find new insecticides that can be used against resistant insects. The result of selective pressure often involves simultaneous resistance to chemical toxicants other than the one that brought about the selection process. This is called cross resistance. But following removal of the toxicant from the insects' environment, the original susceptibility returns to the populations slowly.

The problem of resistance is not easily solved, with medicines or pesticides. Resistance has developed to all pesticides that have systemic action. Even those materials that exert their effect by physical action, such as petroleum oil, may in time manifest genetic pressure to bring about strains that are more difficult to kill. However, in general, physically acting pesticides, cultural methods, and other ecological approaches to controlling pests are apt to have more long-range dependability than chemicals that exert highly specific modes of physiological action involving enzymes or other biochemical reactants.

Biomagnification. The chlorinated hydrocarbons are notoriously persistent under most conditions in the environment, and even living organisms generally find them resistant to breakdown and hard to get rid of. Any chemical that fails to degrabe to harmless end products is apt to give trouble. A persistent chemical may be taken up by plants or animals and thus enter the food chain of both man and wildlife. When this happens, there is often a progressive increase in the concentration of the chemical as it passes from one organism to another in the food chain, resulting in what has been called biomagnification. Sometimes extremely high concentrations are found in the tissues of animals high in the food chain. For example, analysis for chlorinated hydrocarbons in several parts of the ecosystem of Lake Michigan showed that these substances were several hundred times as concentrated in herring gulls as in the fishes on which they fed, and several thousand times as concentrated as in small invertebrates that served as food for the fishes. Even the small invertebrates contained within their bodies almost 50 times as much as that in the bottom sediments of the lake.

Effects of Pesticides

The effects of pesticides on pests and other wildlife and on man have been studied extensively during the past 2 or 3 decades. Some have been quite desirable but others have been very undesirable and even alarming. In this section we shall review some of the most important questions that have been raised about pesticides.

Effects of Pests. Nearly 1% of the insect species can be regarded harmful to man because of their adverse effects on crops, livestock, and poultry. The number of economically important insect pests might be 10,000 and the number of plant species affected 1000, but the 200 major pests cause almost all the damage. About 80% of the insecticides used annually have been used against the 100 major insect pests. In parts of the world in which pesticides are not common, agricultural production can be reduced by large amounts; for example, it has been estimated that the fraction of total production lost to insects in Asia is 41% for rice, 18% for cotton, 14% for maize, 13% for vegetables, and 6.5% for wheat, oats, barley, and rye. The use of pesticides has reduced this damage by more than the cost of the pesticides, although it is not always certain that the use is limited to quantities that are economically justified.

A number of pests have been vectors of human diseases and pesticides have been of great use in controlling some of these diseases. In recent years the majority of the DDT is used especially into the campaign against malaria. The Wold Health Organization estimates that the antimalaria campaign has saved over 5 million lives and prevented over 100 million illnesses in less than a decade, the annual malaria death rate in India was reduced from 750,000 to 1500, In Ceylon the incidence of malaria in the early 1950's was over 2 million cases a year, but DDT spraying reduced it to almost zero.

The main problem that has been encountered with pesticides is the development of resistant strains of pests. An insecticide that kills a large fraction of the population of some insect will after several generations be less effective. As early as 1947 it was realized that DDT was losing its effectiveness in controlling houseflies. The DDT resistance of houseflies has been widely studied and is apparently due to the presence in resistant strains of highe concentrations of an enzyme (called DDT-dehydrochlorinase) that dehydrochlorinase the DDT. Exposure to one pesticide sometimes leads to cross resistance-resistance not only to the pesticide used but also to other pesticides, including even chemically dissimilar compounds.

Resistance to rodenticides can also occur, even though the life-times of rodents are much longer than those of insects and the selection process is therefore much slower.

Another problem with pesticidesis the harm which sometimes caused to beneficial species, such as bees. In addition, pesticides can cause great changes in an ecosystem, sometimes actually increasing the amount of harm caused by a pest. The mathematics of a two species system, one species being a predator (or parasitie) of the other, was worked out by the noted mathematician volterra, who showed that in the absence of disturbances the numbers of the two groups would fluctuate periodically. Volterra also revealed that if for some reason the same proportion of the two species were suddenly destroyed, different fluctuations would result with the average (over time) number of the predators being decreased and the average number of the prey being increased. It is observed in practice, a pesticide sometimes increasing the average numbers of a pest because some natural enemy has also been affected. Likewise, the reduction of a natural enemy sometimes gets increased the numbers of an insect of no economic importance of the point at which it is of economic importance.

Effects on Humans. Every year some persons die due to pesticides, although the number is much less than from other accidental causes or even from other types of poisoning, such as carbon monoxide poisoning. Almost all the pesticides in use can be fatal in sufficiently large amounts but the organophophates are the most toxic. While the organophosphate parathion could be lethal at an oral dose of 10 to 20 mg and the mean lethal adult does is probably 300 mg (roughly 4 mg/kg of body weight), it takes 10 mg/kg of DDT or dieldrin to produce untoward symptoms and the man lethal oral doses are perhaps 20 to 70 mg/kg for dieldrin and 250 mg/kg for DDT.

No exact lethal doses can be fixed, of course, as there are many variables that can affect human respose to pesticides—age, sex, diet, state of health, and many others.

Food products and water are often preserved in plastic containers due to their ready availability and low cost. The possible leaching of thalates, lead salts and other plasticisers has injurious effects on health against which proper precautions are desirable (CR Krishnamurti, ITRC Report 1979).

Upon occasion, large numbers of persons have died from eating food contamined accidentally with pesticides. These cases have included 330 deaths in Turkey from eating seed grain treated with hexachlorobenzene, 80 deaths in Colombia from eating flour contaminated with parathion, and 17 deaths in Mexico from eating sugar containing parathion.

Chlorinated hydrocarbons are present in the body fat of human beings, the main source being small amounts present in food supplies. Studies have revealed that the daily content of DDT and DDE in meals might be about 0.04 to 0.5 mg, while concentrations of DDT and derived materials in body fat might be 5 to 12 ppm. There are no detectable effects from these concentrations, and no correlations noted with any disease rates or other adverse health effects. No detectable health effects could be reported in volunteeras who took 35 mg of DDT daily for 21 months, even though their body fat concentrations increased greatly.

Since chlorinated hydrocarbons were first used commercially only a quarter of a century ago, there could very well be important effects that have not yet showed up, although there is no basis at present for expecting such effects. Some disturbing effects on wild life have been noted, which are discussed below.

Long-term effects that might occur could be related to carcinogenic, mutagenic, and teratogenic properties-the abilities to produce cancers, mutations, and congenital in utero malformations. Sufficiently large quantities of DDT, dieldrin, hepatachlor, and mirex are able to induce tumors in mice, including malignant tumors. Several pesticides have been discovered to be capable of producing mutations or chromosomal aberrations in plants but much more study is need along these lines.

Pesticides may be capable of causing other effects. One interesting medical report dealt with importance (for up to a year) that occurred in four or five agricultural workers (the fifth was single and was never questioned) who got exposed to a variety of toxic chemicals, including malathion, dieldrin, and 2, 4-D in April and August 1977 when their troubles began; each worker was unaware of his colleagues' impotence, and it was accidental that one doctor became aware of the cases.

Several rodenticides, including cyanide, strychnine, sodium, fluoro-acetate are highly toxic and cause convulsion and respiratory failure. Pesticides remain for long periods in the environment. For DDT 39% residual aftereffect has been recorded even after 17 years. Non-persistent pesticides are comparatively less toxic. In order to minimize the effects it is often recommended that spraying or dusting should be carried out following the direction of prevalent wind and a protective clothing against skin contact is required for the operator.

Effects on wildlife. The effects of pesticides on wildlife are better known and often alarming. Pesticides are often concentrated in food chains, leading to levels high enough in birds and fish to damage their health or reproductive abilities. One example is the killing of robins by DDT used against Dutch elm disease, the pesticide being concentrated in the chain from leaf to litter to earthworms (to which apparently no harm is done) to robins. The dangers to birds and mammals of acute pesticide poisoning below) are not known but there is evidence that is could be important. In an experiment with Japanese quail, it was revealed that when diets of from 2 to 250 ppm dieldrin were fed, deaths occurred over a period of time but tended to occur earlier with the higher-dosage diets; brain residues of dieldrin correlated well with death, typically being 4 or 5 ppm and higher.

Chlorinated hydrocarbons get implicated in the decline of several bird populations through interference with their reproduction. The famous peregrine falcon, which survived the warfare of farmers, hunters, and egg collectors over the centuries, began a sudden, drastic decline in propulsion in both Europe and North America and similar trends occurred with grebes, woodcock, bald eagles, ospreys (fish hawks), kestrels (sparrow hawks), and other species at about the same time. The cause has been related to a drop in the reproduction of the birds, rather than to their killing. Breeding was delayed or no eggs were even laid; the eggs that were laid were thin and easily broken; and fledgling mortality was high. Reproductive failure was greatest in areas in which persistent pesticides, especially the chlorinated hydrocarbons, were used most widely.

The decrease in the thickness and weight of the eggshells of repeaters occurred immediately after the introduction of chlorinated hydrocarbons into general usage.

The chlorinated hydrocarbon pesticides are not the only chemicals that might be having adverse effects on wildlife. The polychlorinated biphenyls ("PCB") used as plasticizers and in the manufacture of paints, resins, electrical insulators, and other products, are now being widely distributed over the world and are being concentrated in the food chains of marine ecosystems..In one experiment it was found that mallard ducklings fed PCB in the diet for 10 days at levels from 25 to 100 ppm suffered no apparent clinical intoxication but when challenged with duck hepatitis virus they suffered significantly higher mortality than unexposed birds.

Fish also exhibit acute responses to pesticides. One notable example, still a matter of some controversy, had been the series of major fish kills that began in the late 1950's along the Mississippi and Atchafalaya Rivers in Louisiana. Catfish and other fish exhibited hemorrhaging and death in great number s despite the fact that the water did not give unusual measurements of parameters such as pH, dissolved oxygen, or temperature, and there had been no evidence of pathogens. Investigations that attempted to reproduce the symptoms with viruses, toxic, metals, etc., were fruitless until it was discovered that chloroform extracts of bottom sediments would reproduce them exactly. Eventually traces of endrin were identified as the source of the intoxication. It was know that catfish in the vicinity exhibited acute effects when endrin got sprayed over the Louisiana cane sugar fields; although this spraying seemed to be a logical source for the endrin, levels were keeping the cane field water out of the river systems in question and the fish kills in the rivers were not occurring at or near spraying time. Federal officials identified the source as the endrin plant of a chemical company at Memphis, tenn., several hundred kilometers upstream, although the company denied being responsible; when the company stopped the discharges, however, the fish kills ended.

The question of sublethal effects in fish from pesticide contamination could not be investigated very much but behavioural changes are believed to occur. One experiment on brook trout that DDT affected the central nervous system and interfered with responses to temperature.

The possibility that chlorinated hydrocarbon pesticides also may affect photosynthesis and growth in marine photosynthesis has been raised by experiments. It has been found that DDT residues present in marine phytoplankton collected in Monterey Bay, Calif, increased by three times in the period from 1955 to 1969.

Defoliation

Few pesticides have caused more divisive controversy than the so-called defoliants (actually weedkillers) used as chemical warfare agents during the Vietnam war. During that episode of jungle warfare, the United States flew 6,539 missions between 1965 and 1971, including 4,561 defoliation missions and 858 crop destruction missions, for a total of 17.6 million gallons of weedkiller spray. Several formulas were used in which the most important active ingredient was a weedkiller compound based on 2,4,5-T.

There were many complaints from people in and outside the United States. At first, objections to these of defoliant sprays were based on the fundamental opposition to the use of chemical warfare agents in any form or for any purpose. But the opposition became even stronger and more vocal when it was discovered that 2, 4, 5-T could cause deformed offspring of animals in laboratory experiments. Subsequent studies showed that nearly all of the fatal effects were due to an impurity called dioxin, or TCDD that was formed during the manufacture of the weedkiller. TCDD is one of the most toxic substances known, and the fact that a potent chemical capable of causing fatal abnormalities was being sprayed over the Vietnamese countryside in wholesale quantities was cause for concern.

Several surveys were made in an effort to determine whether reported birth defects in children and animals born in the sprayed jungle areas were related to the use of the defoliants. The most comprehensive study was conducted by an international committee of seventeen scientists under the auspices of the National Academy of Sciences, which was instructed to proceed by Congress in 1970. The panel found "no conclusive evidence" linking defoliants to birth defects, but did not rule out a relationship between defoliants and symptoms of human illness, reports of which, in the words of the panel, "are so striking it is difficult to dismiss them as simply the effects of propaganda, high normal death rates, or faulty understanding of cause and effect."

The study group found that the ecological damage to the warravaged country of Vietnam had been easier to assess, and that if was extensive. They concluded that the mangrove forests would not spontaneously recover for nearly a century if at all, but that a massive reforestation program might cut the recovery time to two or three decades.

About 35% of South Vietnam's 57,000 km² of tropical hardwood forests have been sprayed one or more times, killing merchantable timber equal to the country's current demand for 31 years. Damaged forest lands are now being invaded by tenacious bamboo and grasses that may take over the area for many decades (Another result of the war is that shrapnel in the timber has become a serious problem for the Vietnamese lumber industry, causing sawmills to lose 1 to 3 h daily due to damage of saw blades.

The ecological effects of the military difoliation program have been extensive and generally harmful, and may Vietnamese-especially the Montagnard tribesmen of the Central

Highlands are deprived of food. Many ecologists and other scientists have argued that this program is much more inhumane than it would appear to be on the surface since the civilian population will be adversely affected for decades to come.

In Sweden, the agricultural use of seed-dressing by mercury and industrial effluents led to the accumulation of high levels of mercury in fish and birds. As a result, the use of mercury was banned. Similarly, thinning of egg shells in Peregren falcons was noted as mainly due to DDE, dieldrin, PCB and mercury. A complete ban on the use of such pesticides, where toxic effects may spread through leaching and is not confined mainly to the site of application, is necessary. The use of pesticides having persistent effects for a long period is also dangerous. Chemicals with non-persistent residual effects (one to two days), such as "Tapp", are comparatively less harmful.

The aftermath of accidents in the use of pesticide was shown in a disater in 1976 in Sevesco, near Milan, In a pesticide factory, the safety valve blew off due to high pressure, emitting a highly toxic substance. Following several deaths due to the chemical, the town was evacuated.

In the last Asian Congress of Rural Medicine, held in Japan (Usuda), several cases of poisoning amongst greenhouse cultivators have been reported. They were mainly attributed to the compounds which affected through inhalation or skin absorption. Dullness, headache, excessive perspiration as well as contact dermatitis are rather common ailments. The use of copper and arsenic and other toxic compounds has led to several accidents.

The heavy demand and comparatively lower production of food in terms of population has prompted adulteration of grains, pulses, and milk, which is continually on the increase. Several samples of lagumes, spices and vegetables contain dyes which have an injurious effect on health. Similarly the seeds of Argemone mexicana, growing together with rape seed and mustard and often used as adulterants, are well known for their carcinogenic and neurotocic properties. The use of processed food and beverages requires a strict guideline for their production as they often contain toxic antioxidants and cyclamates.

A problem often affecting agricultural and veterinary workers is the increasing speed associated with the agricultural operations including the use of implements. The heavy load on the engine generates heats influences the microclimate of the cabin, and emits carbon monoxide and hydrocarbons which pervade the atmosphere of the operator. The stresses of working at high speed lead to loss of hearing. The stresses of working at high speed lead to loss of hearing, cardiovascular and nervous diseases (recorded by Kiev Occupational Health Research Institute, USSR). In Japan, a survey has revealed that amongst the agricultural workers, hypertension, rheumatism, neuralgia, shoulder stiffness, lumbar, pain, nocturnal polyurea, breathlesseness-all manifestations of chronic, fatigue attributed to cumulative effects of strees and strain-are rather common. Chronic bronchitis is one of the prevalent diseases amongst the crop threshers.

Alternative Methods of Pest Control

A number of alternative pest control methods have been developed or are being developed that may involve less environmental pollution than the chemical pesticides presently in use.

The alternative methods, some of them rather old, generally involve biological controls of various types or selective chemicals.

Possibly the best way through which the agricultural production can be augmented without the use of pesticides is biological control. This practice is based on the principle of introducing natural enemies to the environment without harming the crop. The progress technology in relation to biological control has been rapid and several methods are now in vogue. These are the culturing, breeding and release of already existing natural enemies in the environment, introduction of new enemies from the country of origin of the pest; the use of microbes to kill the pests and lastly sterile mating for their ultimate elimination. In several cases, such enemies may require other plants for the completion of their life cycle. The necessary environment needs however to be provided. For example, the leaf hopper in vines, in the United States, has been controlled by growing blackberry plants near the fields.

Autocide. Insect sterility was developed by raising large numbers of flies in the laboratory, sterilizing them, and then releasing them (during the season in which the screwworm fly population was at its lowest) to mate with the natural population-without offspring. The noted geneticist H.J. Muller postulated the use of X-rays to sterilize the laboratory flies. The success of the method was proved in 1954, when it was used to eliminate the screwworm fly from other land and therefor insulated from invasion by migrating flies. Since then the screwworm fly gets suppressed in the Southwest, and sterile flies are released from time to time at a buffer zone along the Maxican border and wherever the flies reappear.

The same technique has been used along the California-Mexico border to control the Mexican fruit fly and is being tested in California and Arizona against the pink bollworm. It is also used against other fruit flies and may be developed for use against the cabbage lopper, the boll weevil, the corn earworm, and several other insect pests.

While ordinary chemical pesticides have been most useful when the pest population has been high and have their effectiveness reduced when the pest population has been decreased (especially if some resistance is occurring), autocide is most useful when the pest population has been very low and is therefore of tenused after chemical pesticides or natural causes have reduced the population. The method also requires careful mass rearing and sterilization in the laboratory.

Chemosterilization. A similar method involves the sterilization of natural pest populations by applyies a chemosterilant, leading to a reduced reproduction rate. It has sometimes been difficult to get the pests to take the chemosterilant or the effect on reproduction might be only temporary; these consitute disadvantages of this method. One study has shown that the feral pigeon common in urban areas could be controlled by the use of synthetic grit containing the synthetic steroid mestranol, which is capable of inhibiting reproduction in birds.

Sex attractants. Chemicals called pheromones get secreted into the environment by animals and elicit specific reactions in receiving individuals of the same species. Sex attractants provide some promise as pest control techniques as they may be used to with traps baited with insecticides or may be used to permeate the atmosphere to prevent males from locating females. The first sex attractant was isolated from the female gypsy moth in 1960 and several hundred have since been indentified, some very complex molecules (cis-7, 8-epoxy-2-methylocatdecane for the gypsy moth and some rather simple (phenol for the grass group beetle). A synthetic lure, methyl eugenol, had been used to eradicate the oriental fruit fly on the island of Rota in the Marianas; the attractant was combined with an insecticide on pieces of fiberboard distributed by airplane. Sex attractants are being tested for a number of insects.

Juvenile Hormones. Juvenile hormones and ecdysones have been internal secretions of insects that are able to to regulate their growth and metamorphosis from larva to pupa to adult. These and related compounds provide another promising approach to pest control. Although they are vitally necessary for proper development at certain stages in the insect's life cycle, they must be absent at other periods or else abnormal development will occur-eggs fail to hatch or immature insects die without reproducing. Exposure to these hormones (or to synthetic compounds of similar effect) at the appropriate time can thus help in controlling the insect. These compounds have no effect on other types of life and therefore may not be having dangerous effects on humans, fish, and wildlife, although this could not be proved. They have the important advantage that insects cannot develop resistance to them as they are essential to the insect's life cycle.

It is believed that the juvenile hormone of each insect has been a specific chemical compound, and isolation and identification of these compounds has been accomplished. Fig. (a) shows the chemical structure of the juvenile hormone of the Cecropia moth (the adult stage of the silkworm). Synthetic compounds which are having identical activity, and therefore just as useful as the true juvenile hormone, have also been discovered. Although these affect only insects, as is far as known, they generally influence beneficial or harmless insects as well and hence have the disadvantage inherent in any broadspectrum pesticide.

The possibility remains that chemical compounds may be found that affect only particular insect species. One such compound was discovered at Harvard University by Karel Slama and Carroll Mr. Williams. Slama had brought some European linden bugs (pyrrhocoris apterus) from Czechoslovakia but it was impossible to raise them to the normal, sexually mature adult stage; the larvae continued to grow as larvae or molted into adult-like forms with larval characteristics.

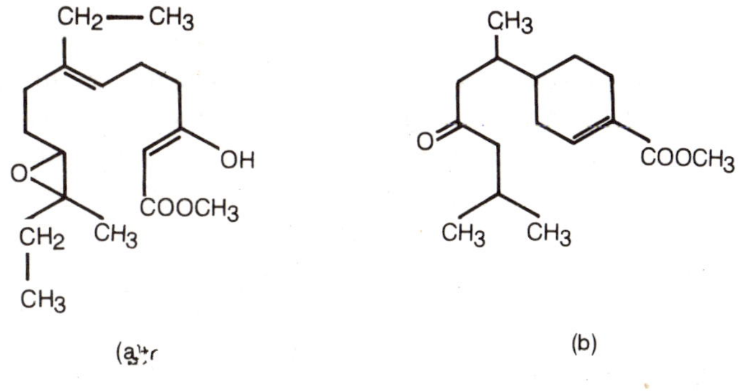

Fig. *(a)* The juvenile hormone of the Cecropia mothe.
(b) The "paper factor" found in balsam fir, a potent juvenile hormone for the insect family pyrrhonoridae.

Evergreen trees like the balsam fir synthesize and emit many different terpenoid materials, for no apparent reason, but the "paper factor" and other compounds may have been evolved as a defense against natural enemies of the trees and may include useful pesticides.

Natural Predators and parasites. As early as 1888 the ladybug bettle *Rodolia cardinalis* was used successfully to control the cotton cushion scale of citrus plants. In Mexico the citrus black fly has been controlled by the release and establishment of hymenopterous parasites. The establishment of several parasites of the spotted alfalfa aphid, together with the development of resistant species of alfalfa, may constitute a permanent solution to the alfalfa aphid problem.

Bacteria and viruses that affect insects offer another possibility for control. The insect Bacillus thuringiensis has been against a variety of insect pests, including the cabbage looper, the diamondback moth, the alfalfa caterpillar, the corn borer, and the tobacco hornworm. Its toxin has been isolated and bacterial toxins may be even more useful as infecticides than the pathogens, at least if commercially economical synthesis becomes possible. Many viruses have also been developed for use against such pests as the corn earworm (also called the cotton bollworm).

Resistant crop varieties. These are yet another alternative to chemical pesticides. The outstanding example has been wheat resistant to the Hessian fly. The first such variety was introduced in 1942, and some 22 now exist. Although occasionally one variety will loss its effectiveness against the Hessian fly in some area, the problem is not so serious as the development of insect resistance to chemical pesticides, and another resistant variety can be substituted. A variety of alfalfa that resists the spotted alfalfa aphid has also proved successful, as mentioned above in the discussion of parasites. A number of other agricultural crops and trees with insect resistance are developed and released. The juvenile hormone produced by the balsam fir mentioned above gives protection against an insect that is now extinct or has learned to avoid the balsam fir entirely.

The development of a resistant variety of crop may take 10 or 20 years but the benefits can be enormous because the use of pesticides is then avoided year after year for long periods of time. Even partial resistance, reducing (but not eliminating the need for pesticides, can be valuable.

Cultural practices. These were used by farmers before more sophisticated pesticide methods, cultural practices can be of great help in reducing the damage caused by pests. Included have been proper sanitation, destruction of crop residues, deep plowing, rotation of crops and animals on the land, planting of crops in strips, and proper timing of planning and escape major pest damage. For example, the corn rootworm is easily controlled by crop rotation-avoiding planting corn on the same acreage 2 years in a row. These have been difficult to use as effectively today as in the past since U.S. agriculture is based on the monoculture system, with each crop and livestock produced in the region of the country best suited as far as soil and climate are concerned.

Quarantine. Another practice that has been helpful in preventing invasions of exotic pests and diseases is quarantine.

Integrated control programs. Programs for important pests have been developed by entomologists. They involve a variety of pest control methods in proper sequence at the proper times to make the best use of natural controls and create the least hazard to humans, wildlife, and non-target species. The cost of developing and using such a program has been generally regarded as rather high but perhaps only because the benefits to the environment are being neglected.

For exotic pests, biological control is not very difficult. But for endemic once the problem is much more complicated. In the practice of biological control extreme caution is, however, desirable as alterations in environmental chain following pest control may lead to injurious effects on wild plants and animals. The recent report of rat plagues in the central provines of Henan and Hubi on the Yangtze and yellow rives at China is an example of harmful effects of disturbance in ecological balance. Due to the use of rodenticides, cats and birds, feeding on poisoned rats, have been exterminated. It has led to a tremendous increase in rat populations, which has resulted in severe damages. This is an example of the adverse effects of upsetting the ecological balance and natural mechanism of biological control.

Other New Methods

Although the radiation sterilization method is perhaps the most spectacular to date of the unorthodox methods of pest control, other systems that are still largely in the experimental stage have potential for practical application. Some of those having various degrees of promise are described below:

(i) *Chemical sterilization* and release of males of the southern house mosquito eliminated the indigenous population of the pests from an island off the coast of Florida. Chemicals that can be used for this purpose are called chemosterilants.

(ii) *Hormones* are substances that are secreted by the glands internal secretion, called endocrin glands. Insect hormones that control growth and development have been called "third generation insecticides" because of the prospect that such hormones, or synthetic materials that have similar action, can be used to interfere with the normal development of insects. In insects there are two interacting hormones, ecdysone, which stimulates normal development, and the juvenile hormone, which causes the insect to retain its immature form. Giant moth larvae can be produced by applying an excess of juvenile hormone, but the adult stage is never achieved, or if it is, the adult is abnormal and incapable or reproduction.

(iii) *Pheromones* are substances that are excreted externally by organisms and have the function of affecting the behavior of other individuals. Sex attractants, trial markers, chemicals warnings, replants, or similar chemical signals are pheromones. The term is most commonly used to refer to those substances secreted by an animal to influence the behavior of other individuals of the same species. Many species of animals secrete sex attractants or stiumlants. They are the aphrodisiacs of the animals world. In insects, females of some species secrete male attractants that are attractive in extremely small quantities. A trap containing a virgin female of the introduced pine sawfly attracted more than 7,00 males during a 5-hour

period, and after that, about 1,000 males per day for 5 days. Sex attractants have been used successfully in insect surveys. Knowledge of the chemical structure of attrctants provides information that is useful in the development of synthetic pheromones.

(iv) *Genetic* control by the release of individuals having lethal genes is a possibility for bringing about the self-destruction of a population. Genes might be lethal only under certain conditions, causing the organisms to succumb, for example, to heat or cold.

(v) *Sonic* may have possibilities for mimicking the sounds of other organisms that are either attractive or repellent. The high-frequency shriek of insect-feeding bats has been suggested as a possibility for frightening insects away.

(vi) *Irradiation* of wheat by a high-intensity gamma radiation source such as cobalt-60 has been demonstrated to be effective in killing insect eggs.

(vii) *Photoperiod* manipulations may be used to disrupt the life cycle of organisms that the sensitive to day length. Prolonging the length of the day with artifical light interrupts the development of the European corn borer and codling moth by preventing the larvae from going into diapause (inactive period).

(viii) *Lures* (attractants) have possibilities that have been incompletely explored, especially for use in combination with toxicants and chemosterilants.

Alternate Pest Control Methods

Some of the more successful methods of controlling harmful organisms do not involve the use of either chemicals or parasites and predators. Such control measures have been practiced for many years. Some of them are only partially effective; some are effective only under favorable conditions; and some of them can be relied upon for complete protection under suitable circumstances. Timetested methods include;

(i) Cultural practices such as the rotation of crops to provide a period during which the pest has no susceptible host plant to attack.

(ii) The selection of varieties of plants that are resistant to attack or that are more tolerant of attack.

(iii) The use of resistant root-stock on which the desired variety is grafted.

(vi) Timing the planting date to come before or after the period of high susceptibility to pests; for example, avoiding those periods when the soil is so damp that fungus growth is encouraged, or when insect pests are normally emerging.

(v) Plowing or otherwise disturbing the overwintering environment needed for completion of the life cycle of the organism.

(vi) Flooding the area temporarily until pests such as soil insects and fungi are killed.

(vii) The drainage of water from the breeding areas of aquatic organisms, such as mosquitos, which require water for the development of the immature stages.

(viii) The removal of diseased plants from a field to prevent or impede further spread of the causative organism.

(ix) The construction of barries which the invading organisms can cross only with difficulty, such as ditches and fences of various sizes and kinds to impede insect migrations, and protective devices around the trunks of trees to keep out rodents.

(x) Weeding to eliminate alternative host plants on which the pests may be breeding either in the field or nearby.

(xi) Traps, such as attractive light-traps, bait-traps, and electorcutting traps.

(xii) Heat, such as that used for the direct sterilization of clothing and bedding.

(xiii) The removal of breeding material such as manure, lawn clippings, or other organic material in which pests, such as flies, multiply.

(xiv) Covering, breeding material, such as that in garbage cans, with lids or other devices to prevent access to organisms.

CONCLUSION

The major pesticides in use today have been the synthetic chemical pesticides developed in the past quarter century. Although they are of great help in reducing the damage caused by pests that affect agricultural crops and livestock and are vectors of human disease, they have often had undesirable effects on beneficial species and they have often lost their effectiveness as pest resistance has evolved.

The question of the value of different levels of pesticide use has been an important one but which has been difficult to resolve at present. Studies have revealed that the use of pesticides has been economically justified in many situations but in some cases farmers may be overusing pesticides; i.e., their total benefits may be greater than their total costs but their marginal benefits may be less than their marginal costs. Farmers generally get their information about pesticides from the chemical dealer (rather than from agricultural extension agents) and are sometimes clearly oversold on the use of particular chemicals.

From the point of view of society as a whole, it also becomes necessary to ask which uses have been most important and which are frivolous. Disease control would certainly be a high priority use of pesticides; perhaps we should continue DDT usage in malaria control programs, although we should actively search out safer substitutes. Agricultural employs probably rank somewhat lower in importance—at least in the development countries with perennial agricultural surpluses. The importance keeping roadsides free of weeds has been much harder to appreciate, however, especially for someone who prefers natural roadsides to unnecessarily formal and well-womed highway rights-of-way.

Many alternative pest control methods are available. Some of these methods have (or could have) some of the same disadvantages are chemical pesticides but other might not.

In the meantime, if chemical pesticides have been used, it is desirable to replace broad-spectrum, persistent pesticides by those that are more selective and less persistent in the environment whenever possible. Some of the latter pesticides (such as the organophosphates) have been more toxic and it is clearly necessary to exercise great care in their application and storage.

It would also be wise to ask whether or not we are being a little too fastidious in some of our eating habits and thereby making use of pesticides to an excessive extent. No one would be happy to learn about rodent hairs or other remains in his or her jam because of the eating habits of rodents and their ability to transmit diseases to human beings. But the presence of bits of larvae that fed on the fruit cannot reasonably be objected to except out of fastidiousness. Insects that feed on fruits and other crops relished by man can also serve as food. Moses, permitted some insects to be eaten by the Israelites, although others were forbidden. John the Baptist lived upon locusts and honey in the desert. The ancient Greeks and modern American Indians relished cicadae. Roman epicures are beetle larvae. Today, some people eat snails, grass-hoppers ant, wasp group, and such creatures that other people regard as disgusting.

In India, where the principal occupation is agriculture there is an immense possibility for the development of Ergonomics—a synthetic science combining biomedicine and engineering. The design of tools and methods of work, commensurate with the ability of the individual, lead to reduction of unnecessary effort and fatigue. In this subcontinent, where developments in agriculture are progressing at a rapid pace, it is essential to utilise the knowledge of ergonomics for agricultural workers, to provide them with a congenial environment.

6

Vapour Extraction for Remediation of Soil Contaminated by Gasoline

Introduction

The release of gasoline, fuel oil, and other types of contaminants from underground storage tanks has become as major environmental issue across the country. Released contaminants migrate through the soil with surface water infiltration, often to the groundwater below, where the potential to impact drinking water resources is great. Remediation techniques for handling this problem have typically centered on removing gasoline that is floating on the water table surface through water table depression, pump, and recovery methods. This represents only a partial solution to the recovery problem, in that the majority of gasoline adsorbs to soil particles in the vadose zone where it will continue to contaminate groundwater for years to come. The remediation of these discharges can be accomplished using a variety of proven techniques, all with limiting characteristics associated with site-specific conditions, regulatory involvement, and particularly, cost-effectiveness. One technology developed in the early 1980s following studies by the American Petroleum Institute is induced soil venting, or extraction, of volatile contaminant vapours, an in situ technique that has been used for decades inventing explosive methane gas from landfills. Although a well-documented technology for methane removal, vapour extraction has only recently been applied to remediating soil and water contaminated by volatile chemicals, particularly gasoline.

Vapour extraction involves installation of vapour extraction wells and application of a vacuum to volatilize contaminants from soil and groundwater. Extracted vapours are treated or exhausted into the atmosphere. Vapor extraction has proven in some cases to be more cost effective than typical solutions of soil excavation and disposal, and a faster remedial method than groundwater recovery and treatment. Capital costs and operation and operation and maintenance costs are generally much less for vapour extraction systems than other remedial technologies. Treatment duration is significantly less than groundwater treatment systems; documented cases have shown up to 99% removal of contaminants in periods as short as three to six months, where groundwater treatment systems typically take years to clean up the contaminant source. Vapour extraction can also be used as an emergency response technique for venting contaminant vapors from structures. Other benefits of this technique include its suitability to inaccessible locations, remediation beneath structures, and the ability to treat large volumes of soil with little or no visible above-ground disturbance and without transferring the problem elsewhere. The main benefits, however, are low initial overhead and operating costs, minimal equipment requirements, and small impact on facility operations. Many gasoline stations at which this remedial technique has been applied have been able to maintain operations during remediation of the site.

Background

As public awareness grows and as local, state, and federal authorities are forced to act, the problem of contamination from leaking underground storage tanks has become a major environmental issue. Concurrently, costs of remediating such situations grow daily, as disposal locations reach capacity, and as state and federal guidelines for cleanup become more stringent. Although there are many alternatives to disposal, most are cost-prohibitive or unproven technologies. Obtaining approval from regulatory agencies is usually difficult for these new methods, and in most cases, the time delay associated with approval is prohibitive. With the growing concern and public familiarity with the hazardous waste problem, on-site treatment has become an attractive method of remediation.

Leaking underground storage tanks have been the focus of the Environmental Protection Agency (EPA) in ongoing legislation. An estimated number of 1.4 to 3 million underground storage tanks exist in the United States, of which ten to thirty per cent are thought to have leaked. In response to this problem, Congress devoted a section (Subtitle I) of 1984 Hazardous and Solid Waste Amendments to the Resource Conservation and Recovery Act (RCRA) to regulating all underground storage tanks storing petroleum products and hazardous substances. Out of these developments have been numerous studies, publications and conferences sponsored by the American Petroleum Institute and other organizations addressing leak detection and remediation of leaking underground storage tanks.

Soil and groundwater contamination by gasoline from a discharge underground may be a result to tank corrosion and failure, pipe ruptures or simply spills during filling that have accumulated over the years. In the past, gasoline station records were usually inaccurate or non-existent; the potential for continuous leakage to go unnoticed was great. In addition, the general public was not concerned or aware of the potential impact of these discharges. This attitude has changed significantly since the 1984 RCRA amendments. Under these amendments, the EPA is required to develop regulations addressing leak detection, corrective action, closure, recordkeeping, reporting, and new tank performance standards. The "Interim Prohibition" requires all underground storage tank owners to investigate the potential for leakage (by conducting tank-tightness tests), to install corrosion-prevention and leak detection devices, and to install all new tanks with secondary containment features. These regulations are designed to eliminate future releases of hazardous substances to the subsurface. These regulations will, for all intents and purposes, eliminate the need for remediation of soil contaminated by future gasoline discharges. However, they do not change the current problem.

The hazardous substances typically stored in underground storage tanks range from gasoline, fuel oil, and to waste solvents and raw material used in manufacturing. In instances of gasoline spills, the monoaromatic hydrocarbons benzene, toluene and the three isomers mp-and oxylene have relatively high pollution potentials due to high aqueous solubilities and estimated toxicity. Chlorinated volatile organic compounds such as trichloroethylene and tetrachloroethylene likewise have high solubilities. These types of contaminants have been responsible for severe contamination of drinking water supplies. Due to their chemical and physical characteristics they are also highly suitable compounds for vapour extraction, as they volatilize well.

Developmental History

Vapour extraction, or induced soil venting, was developed primarily from methane venting technologies used at sanitary landfills. Volatilization products from a gasoline spill will behave in the same manner as methane gas generated in a landfill: the same migration routes are followed. Methane was originally vented to reduce the possibility of gas explosions within landfills, and to control off-site migration of methane gas from the refuse pile, through surrounding soils and into nearby structures such as buried utilities or basements. The basic design principles of soil venting have been borrowed from the solid waste disposal industry, where control, recovery, treatment, and even sale of landfill gas generated as a by-product of bacterial decomposition has long been practiced. This evolution was slow to develop; methane has been vented from landfills for at least fifteen years, and only recently has the technology been applied to removal of volatile organic contaminants from the soil.

The initial application of soil venting to remediation was to remove noxious fumes from buildings contaminated by migrating plumes of hydrocarbon vapors, usually originating from underground gasoline storage tanks. The success of ventilation for removing vapors from landfills and buildings led to the current applications fro treating soil on-site rather than excavation and disposal off-site. Basic venting technologies have also been applied to removal of radon gas, control of vapors during pertroleum marketing operations, and in remediating pipeline leaks.

Methane Gas Generation, Migration, and Control

Methane gas is generated by anaerobic microbial decomposition of organic solid waste. In an ideal situation, complex organics are consumed by acid-forming bacteria. The by-product of this consumption is a simple organic acid, which is in turn attacked by methane-forming bacteria, with methane gas and carbon dioxide the end product. Small amounts of nitrogen, hydrogen, ammonia, carbon monoxide, and hydrogen sulfide can also be produced. This reaction will occur only under anaeribic conditions, at a low pH, and with minimal temperature fluctuations. These conditions typically occur deep within a sanitary landfill.

Vapors from a gasoline storage tank discharge will naturally migrate through soil following the same pathways as methane gas. Gases generated in a landfill will migrate upwards through the refuse and soil cover and diffuse into the atmosphere. However, when upward movement is impaired, with frost or well-compacted cover, the gases may move laterally ;along the path of least resistance. Asphalt and concrete structures will impede gas migration and cause it to seek a more permeable alternate route, such as backfield utility trenches or foundation walls, as well as poorly compacted refuse cells. The theories of gas migration are discussed further.

Many documented cases for control of methane gas at landfills exist that have described theory and design parameters in detail. These theories have been used in the design of vapour extraction systems for gasoline spills. Systems for forced venting of methane gas consist of perforated vertical or horizontal PVC pipes, packed with gravel, connected to a main headerline, and finally to the suction side of a blower or positive suction pump. The blower creates a vacuum stream throughout its radius of influence, which entrains methane gas for exhaustion to the atmosphere. Some of the design parameters of a forced methane ventilation system consist of the diameter of the pipes, depth of gravel casing, diameter and depth of ventilation well, and spacing of pipes or ventilation wells. System design is based on gas production rates, concentration, and migration tendencies, which are largely dependent on refuse permeabilities. Incorporated into the design is maximum well spacing to maintain a radius of influence that will control migration at a designed pumping rate. Gas flow measurement within the collection header and at the blower is usually accomplished with the use of a pilot tube or other flow meter. The methane collection system typically surrounds the outer limit of the landfill. Off-gas is usually collected for energy recovery. In many systems, the discharge is incinerated, flared, or exhausted to the atmosphere. Depending on its BTU value, collected methane gas can be retained, concentrated, and treated for use as recovered fuel. All of these design parameters have been used in implementing vapor extraction remediation.

Theory

The basis for design, the theoretical models for the physical and chemical behavior of contaminants, is directly related to the study of methane gas movement in forced extraction systems at solid waste landfills. From these studies, physical parameters were recognized and mathematical constants were established. In turn, countless bench and pilot scale tests have been performed and theoretical models developed for vapor migration under natural and forced conditions. Fig. 6.1 shows the relationship between the vacuum developed in a methane recovery system compared to a vapor extraction.

The conceptual theories of gas migration under natural and induced flow conditions follow the substantiated theorems of water flow through porous media. That is, chemical-specific characteristics, coupled with media characteristics, govern the rate and direction of gas flow.

In addition, modeling of test results has shown that under most circumstances this flow can be considered laminar, and therefore is considered Darcian in nature.

Soil venting uses the principle that organic hydrocarbons contained in gasoline products have high vapor pressures and will volatilize at temperatures typically found in soil, until a state of equilibrium is reached in the soil pore space. These vapors will migrate along the same paths as methane gas: upward in the avenue of least resistance. Hydrocarbons that remain in the liquid phase will adsorb to the surrounding soil particles until leached out by percolation or replaced with more air to allow volatilization. The soil venting technique accomplishes both increased volatilization by continued influx of air, and removal by entraining of vapors into the air stream created by the vacuum. By creating a vacuum, the pressure in the pore spaces is decreased and a state of non-equilibrium is reached. The lowering of the static pressure in the soil will induce air flow through the soil and enhance continual volatilization. Continued application of the vacuum will eventually volatilize all adsorbed hydrocarbons.

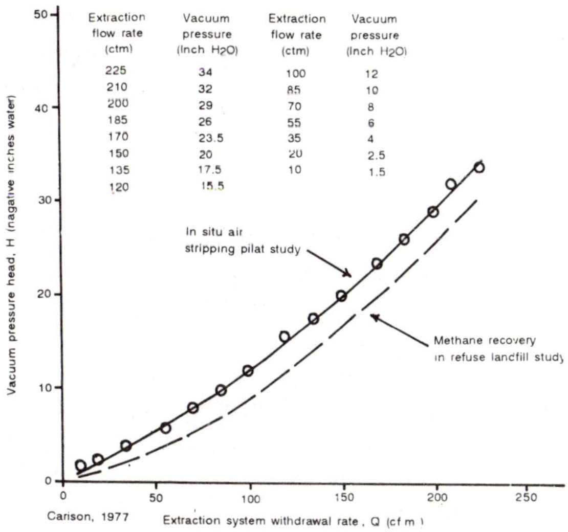

Fig. 1. Typical Vacuum-flow rate trends[13].

Based on this assumption hydrocarbons existing in fluid or dissolved phases can also volatilize. Vapor extraction has been demonstrated to work in remediating volatile organic-based product (insoluble fraction) on the groundwater table as well as within the soluble dissolved phase.

Chemistry of Gasoline Constituents

As stated previously, the remediation of pertroleum hydrocarbon discharges is the focus of this report. This due in most part to the fact that these compounds present the greatest threat to groundwater quality due to their physical nature and toxicity, and therefore have been most widely studies. Petroleum-based fuels are essentially hydrocarbon mixtures of natural gas and petroleum compounds. Gasoline constituents can be divided into five major groups: alkanes, cycloalkenes, aromatics, and additives. Because each of these groups, except for the additives, possesses similar behavior (and toxicological properties) and because of their ability to solubilize and migrate in groundwater, one aromatic chemical, benzene, will be used as an example in modeling fate and behavior in the subsurface. Aromatics constitute approximately 25% of a gasoline mixture, whereas larger, less mobile alkanes comprise the bulk. In a typical gasoline approximately 2 percent is ethylbenzene, 12 percent toluene and 8 percent xylene, by weight. The behavior of additives such as methyl tertiary butyl ether (MTBE) are not as well-documented, although MTBE is highly soluble in water and very mobile, and has been found to be cosoluble with benzene, toluene, and xylene(BTX) at high concentrations.

Fate of Volatile Organic Chemicals

A chemical released into the subsurface will fill soil interstices in a liquid state and continue migration until disequilibrium occurs and volatilization ensues or all liquid is adsorbed onto soil particles. Volatilization occurs when a contaminant has a high vapour pressure, usually greater than 1 torr. With infiltrating surface water, the likelihood of continued migration, resulting in groundwater contamination by dissolved compounds increases. These chemicals will migrate due to the physical and chemical properties of the contaminant and physical characteristics of the subsurface, such as depth to groundwater, geology, etc.

Fluid Migration

When a discharge occurs, gasoline will migrate downward under the force of gravity and capillary action, with some lateral dispersion. During this process, some gasoline will be adsorbed onto the surface of the soil particles in the vadose zone. The adsorbed hydrocarbons will volatilize into the air space between soil particles. The amount of liquid hydrocarbon adsorbed will remain constant once equilibrium is reached between the liquid and vapour. This liquid will slowly be reduced by leaching and natural diffusion. Leached hydrocarbons will continue to migrate downward until an impermeable barrier is reached, such as a low permeability clay layer, or until it reaches the water table. Because pertoleum-based fuels are largely immiscible, some of the fluid will spread in a lens on top of the water table while the remainder solubilizes in groundwater.

On the water table surface, gasoline will migrate in the groundwater flow direction. The mechanisms of migration both within the soil and on the groundwater surface include partitioning between solvent and solubilized states, and between solvent and gaseous states.

The rate at which a hydrocarbon a product reaches the water table can be estimated using the following epuation:

$$D = \frac{RvV}{A}$$

Where D = maximum depth of penetration (meters)

V = volume of infiltrating hydrocarbon (cubic meters)

A = area of spill (square meters)

RV = a constant reflecting retention capacity of the soil and viscosity of the product.

Lateral dispersion of migrating hydrocarbons occurs due to capillary action. The weight of the product will exceed that of the water contained within the soil interstices in the capillary zone, and temporarily displace the water and possibly the water table. As buoyant forces act to restore the water, the product will be pushed into horizontal directions. In addition, seasonal water table fluctuations or those due to changes in pumpage can aid in lateral migration.

Vapor Migration

Several factors control subsurface transport of vapors. These factors, described by Weston, fall into three categories: chemistry, soil and environment. As discussed previously, gas transport mechanisms are modeled after fluid flow equation. Chemical properties, such as water solubility, partition coefficient, vapor pressure, and absolute viscosity will govern the compounds ability to volatilize and become entrained in the extraction flow.

Chemistry

When a chemical is introduced to the subsurface, it will partition between soil grains, organic matter, water, and air. The concentrations are different in each phase and are multiples or "partition coefficients" of each other. The partitioning of a chemical between air and water is a Henry's Law Constant. Calculation of the partitioning coefficients of chemicals involves concentration, chemical properties, volume of contamination and factors affecting transport. Partitioning of air has been described by Kuhlheimer, who gives a rate equation for describing the mass transfer from solvent to vapor phase. This rate is a function of solute concentration, liquid or gas phase resistance, and specific surface area.

The Henry's Law Constant is typically used as an indicator of stripability. It is dependent on vapor pressure and water solubility of the chemical. Compounds with higher solubility and high vapor pressure volatilize most easily. Fig. 6.2 represents the relationship between Henry's law, vapor pressure, and aqueous solubility of some common gasoline components. In general compounds at the lower right hand side tend to stay dissolved in groundwater, those on the lower left are more likely to remain sorbed to soils, and those in the upper left are most likely to volatilize with induced venting.

Because gasoline is a mixture of many components, the equilibrium concentration of the vapor constantly changes as vapors are moved. Hoag et al. have summarized equations to calculate vapor phase equilibrium concentrations at time of extraction.

Soil

Soil factors such as moisture content, permeability, clay content, and adsorption site density will also affect mobility. A higher soil moisture content will decrease the diffusion rate of vapors because vapors compete with liquids for pore space. Low soil permeability due to high clay content will decrease volatilization. Other physical properties of the soil, such as grain size, stratification, disturbances from excavations,and natural cracks or fissures also play an important role. Finally, an increase in sorption sites or organic content will result in a decrease in volatilization.

It has been demonstrated that the permeability of the soil and existence of confining layers will have an effect on the feasibility of vapor extraction as a remedial method. The development of a vacuum in clayey soils is much more difficult to establish than in sandy soils. Fig.3 compares the effectiveness of volatilization with the coefficient of permeability of soil, as well as Henry's Law Constant. As shown, extraction is much more difficult in soils with permeabilities of 10-7 cm/sec and lower. The presence of a confining layer above the extraction media will also have an effect on the efficiency of the system. This barrier may be a clay layer, pavement, or a structure. Depending on the permeability of the barrier layer and the extraction media, the presence of the less permeable layer overlying the extraction soil will enlarge the radius of influence of the blower and the vacuum at the extraction well will increase.

Sorption of hydrocarbons can occur in solvent, dissolved, or gaseous states. In the absence of an air source, infiltrated hydrocarbons will adsorb to the surrounding soils, depending on several soil conditions. The sorption capacity of a soil depends on several factors: the

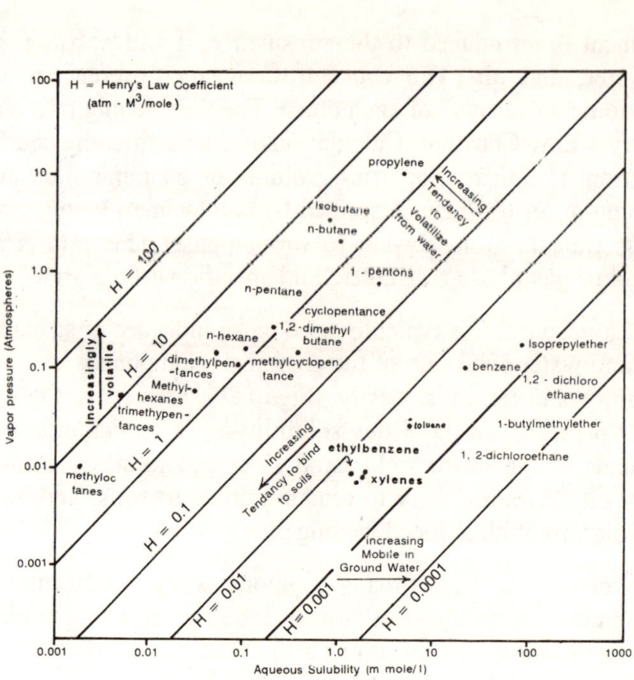

Fig. 6.2. Important physicochemical characteristics of representative gasoline components.

number of sites available, the organic content of the soil, surface tension between the hydrocarbons, liquid and soil, moisture content, grain size, porosity, permeability, and temperature. Because of the vast number of controlling parameters, quantifying the amount of adsorbed hydrocarbons is difficult. The adsorption coefficient, Kd, of a compound can be calculated using the following equation:

$$\text{Kd} = \frac{\text{Conc. Soil}}{\text{Conc. Water}}$$

Compounds with Kd greater than 1,000 have a greater tendency to stay sorbed to the soil.

The organic content of the soil will also affect a chemical's tendency to volatilize when sorbed. The organic carbon (or octanol-water) partition coefficient, Kow, describes this tendency. Chemicals with high Kow (>4) are more likely to adsorb onto the ;surrounding soils and less likely to volatilize. Ford, et al. have developed a model for predicting the mobility of a chemical based on the organic carbon partition coefficient:

$$\text{MI} = \frac{\log(\text{WS} \times \text{VP})}{\text{Koc}}$$

Where MI = mobility index
WS = water solubility (mg/1)
VP = vapor pressure (mm Hg)

Based on the model, chemicals with a mobility index of >5.00 are described as extremely mobile and those with a mobility index of <-10.00 as very immobile. Numerous other models exist for vapor phase transport and volatilization rates.

Darcy's Law

Darcy (1856) proved that water flows through porous media, provide it is laminar, following the equation:

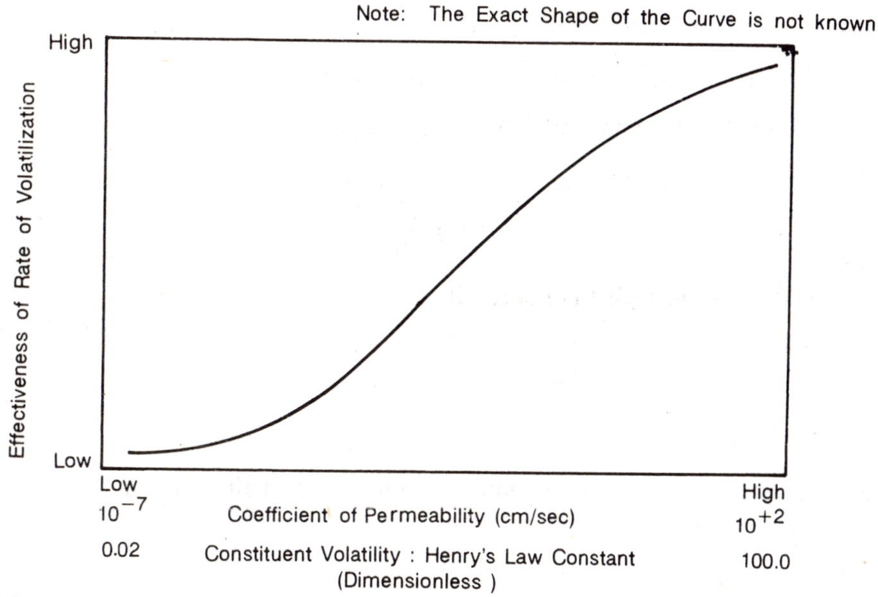

Fig. 6.3. Major factors influencing in-situ volatilization.

$$Q = KA \frac{dh}{dl}$$

Where Q = flow rate (square feet per second)

K = hydraulic conductivity (feet per second)

A = cross-sectional area in direction of flow (square feet)

dh/dl = hydraulic gradient, or change in pressure head (feet).

The determination of laminar flow of gases is made using the Reynolds number. As with water flow, if the Reynolds number of a gas is less than one, flow is laminar. For gases, laminar flow is sure to exist when the mean grain size of the porous media is less than 0.2 cm. Turbulent flow will exist in media with cracks or fissures, or in very fine or very coarse grained soils, where flow ill follow Knudsen diffusion.

Darcy's law for radial flow of gases toward a recovery well is expressed as follows:

$$Vr = K \frac{dh}{dr}$$

Where Vr = apparent gas velocity of r in a direction toward the well (feet per second) r = radial distance from the recovery well (feet)

K = coefficient of permeability (feet per second)

h = piezometric head (feet).

h can be expressed as:

$$h = \frac{p}{\gamma} + z$$

Where p = total pressure at distance r(feet)

γ = Specific weight of the gas (pounds per cubic foot)

z = elevation above some arbitrary datum (feet).

Expressed in terms of velocity, Darcy's Law can be written another way:

$$V = K\frac{dh}{dl}$$

Where V = velocity of gas (feet per second)

K = hydraulic conductivity (feet per second)

l = distance to end of influence (feet)

h = piezometric head (feet).

The determination of the above variables is important in final design parameters and calculations of feasibility and extraction duration.

Environment

Environmental factors such as temperature, wind velocity, evaporation rate, and precipitation affect the volatilization rate of hydrocarbons. Volatility increases with increasing temperature and increased wind speed. Increased barometric pressure will decrease volatilization. A higher evaporation rate at the soil surface will increase the tendency for a capillary rise of vapors within the interstices. Finally, increased precipitation and infilitration will leach hydrocarbons, thus increasing the chance for volatilization.

Vapor phase transport is ultimately controlled by convection and diffusion. Convection occurs when a total pressure gradient exists. Vapors under a higher total pressure will flow toward areas of lower pressure. Convective flow is affected largely by barometric pressure changes, soil porosity, wind fluctuation, and percolation. Convective flow is considered to be laminar flow and is therefore described by the Darcy equation. Diffusive flow occurs when the concentration of a gas is not uniform. Similarly, however, diffusive flow is from areas of high concentration toward areas of low concentration. Without application of a vacuum, hydrocarbon vapors will diffuse based on the concentration gradient, causing the contaminated vapor to distribute itself radially round the discharge source. This transport process is negligible with creation of a pressure gradient (or within the influence of the vacuum source), and therefore convective flow will persist in areas under vacuum.

Vapor Extraction Techniques

Numerous innovative types of extraction systems have been successful in remediating soil and groundwater contaminated by petroleum hydrocarbon spills. Different techniques include induced in situ venting through installation of one or multiple extraction wells, enhancement of remediation through biological additives combined with venting, and passive venting using vertical wells or horizontal extraction trenches. Other techniques involve excavation of contaminated soil, and venting through controlled windrowing or venting of a contained cell. Another method widely used is draft fan induction, which is a pressure system.

Vacuum Extraction

The basic vacuum extraction system consists of installing one or multiple extraction wells and applying a vacuum. These wells are vertical and connected to one blower via header line, or single or multiple horizontal trench systems. Construction of these systems is similar to that of methane recovery systems discussed, and detailed design of this type of system is discussed further. The soil to be treated remains in situ. fig. 6.4 is a schematic showing the typical layout of a vapor extraction system.

Enhancement of Biological Degradation

The circulation of atmospheric air through a contaminated soil during the venting process will enhance microbial degradation of hydrocarbons, a process that would occur naturally without remedial efforts. Some remedial techniques combine vapor extraction with the aid of bioadditives to hasten the process, particularly where semi-and non-volatile contaminants are present in addition to gasoline such as in fuel oils. Bacteria that feed on gasoline compounds are present in all soils, and an adequate supply of nutrients and oxygen complete the process. The oxygen introduced by forced venting increases microbial activity significantly. It has been demonstrated that hydrogen peroxide is the most successful oxygen to inject into venting systems.

Several techniques combine venting with the enhancement of biodegredation by addition of nutrients such as phosphates, nitrogen, and potassium, along with the oxygen introduced through forced venting. Strains of bacteria have been found that have adapted to persistent petroleum contaminations that have been genetically modified to, when injected, become very efficient "petroleum eaters" Nutrients are typically added through infiltration galleries or by injection.

Passive Venting

The simplest form of venting consists of turning over the contaminated soil periodically to expose it to the atmosphere, and is limited to shallow contamination. The passive venting technique, borrowed from the municipal sludge treatment process of windrowing sludge, enhances volatilization, oxidation and photolysis, and microbial degradation. Although emissions are not controllable, steps can be take to effectively isolate contaminated soil from infiltration by use of liners and containment structures.

Passive venting is most effective in removal of vapor buildup in basements and other structures. Robitaille et al. describe a passive venting system to alleviate gasoline vapors

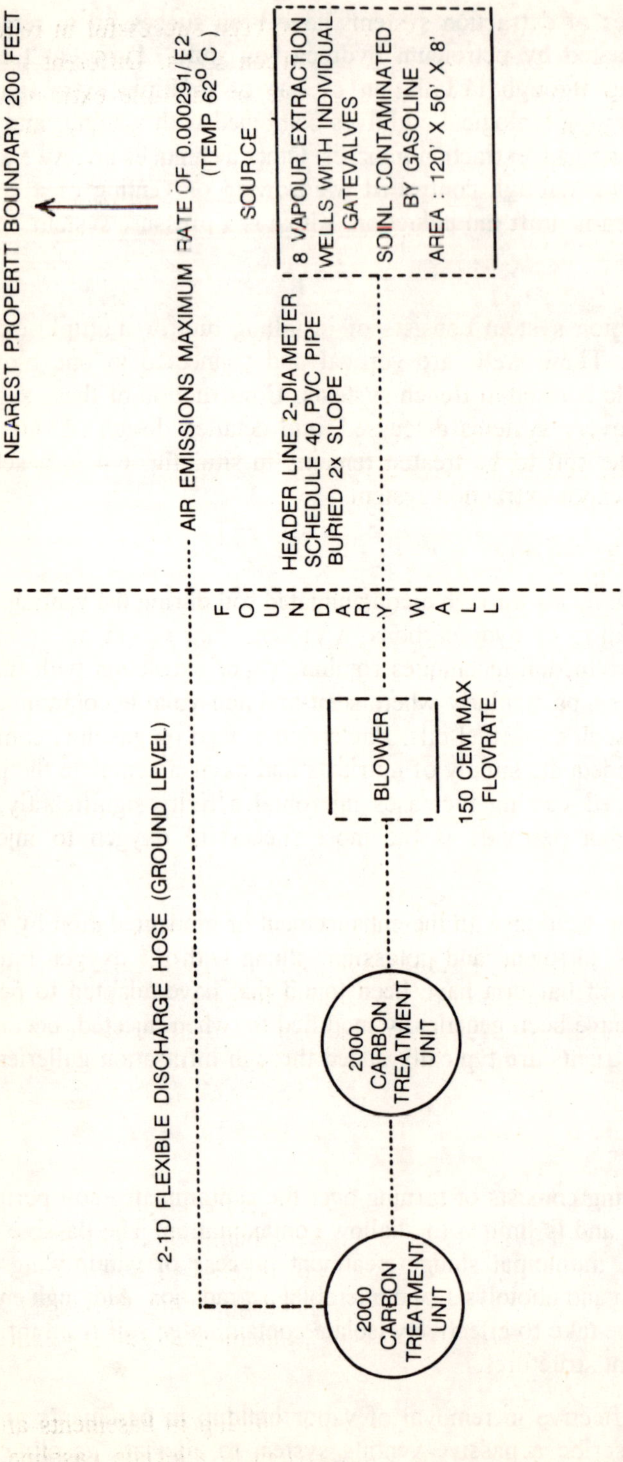

Fig. 6.4. Line drawing for vapour extraction system.

from beneath a building before they reach explosive levels. This system employed construction of a header within a gravel-filled trench, which surrounded the perimeter of the building. Vapors collected in the trench and were vented to the atmosphere through the roof of the building. Similar systems have utilized gravel subbase beneath concrete slabs and existing utility trench backfill to aid in passive venting systems. In addition, Ames has documented significant reduction of volatile organic levels in the forced venting of a basement within an eight-hour period.

The use of passive venting to prevent vapor plume migration is described in detail by the EPA. These control systems usually entail construction of trench surrounding the spill area, to the depth of groundwater. The trench is backfilled with coarse gravel, and vertical vent pipes are installed at intervals to aid in the venting process.

Containment Cell

Foglio et al. have reported the successful implementation of extraction of volatile organic vapors from a contained cell, at an estimated 75% cost reduction over excavation and disposal. Contaminated soil was excavated and stockpiled, and a double-lined treatment cell installed in the excavation. This treatment design was implemented to satisfy regulatory requirements and enable construction at the site to proceed.

The treatment cell was constructed of two synthetic liners with a leachate collection system and vapor extraction header line surrounded by crushed stone. The contaminated soil was placed into the cell, covered with geotextile fabric, and a forced air header line placed on top. The cell was sealed from the atmosphere with a synthetic liner and pavement. During extraction, contaminated vapors were pulled down through the filter fabric to the header line. Fresh air was pulled down through the air intake line, so the contaminated soil was constantly being flushed. The cell was constructed to conform with RCRA standards for permanent on-site disposal of solid hazardous waste. Forlio reported several advantages to this type of treatment, one of note. By excavating the soils and mixing them, and natural stratifications were eliminated and the replaced soils achieved uniformity, thereby promoting equal volatilization and flow throughout the cell.

Air Injection

Many successful systems have combined contaminated vapor extraction with atmospheric air injection. Continual forced injection of air effectively strips adsorbed volatile organics which otherwise may not volatilize if passive or vacuum extraction is applied. These systems have been proven effective by Coia, who described a successful system comprised of heated air injected into a series of vertical wells, and vapors extracted from a different system of vertical wells.

Design

Design of a successful vapor extraction system will depend on several parameters. Initially, a feasibility study should be performed that considers such items as contaminant type, soil characteristics, quantity of contaminated soil, potential threat to groundwater supplies, and cost. This study may include performance of a pilot test if insufficient geologic data is available for the site. Once feasibility has demonstrated, a pilot test will depend on budget constraints; many extraction systems used in pilot testing can be modified to function as the final system. During the pilot test, data such as vapor concentration of contaminants, radius of influence of the blower, and air permeability are determined. Other considerations are agency involvement and the associated permitting process. The applicability of vapor extraction as a remedial method will depend on the agency's familiarity with the technique, which is considered a new technology, and the permits that may be associated with approval to operate, which most likely will require treatment of extracted vapors.

After the necessary design data is collected and feasibility established, calculations for final design are made, and the system is implemented.

Feasibility Study

The feasibility study should consist of fact-gathering tasks. All available information on site-and region-specific studies is obtained. This includes available supply ell or monitor well logs, soil and groundwater sample analytical results, and geological/hydrogeological information, especially groundwater elevation data. If this information is not available, an initial site investigation should be performed. This investigation will include, at a minimum, drilling of test borings to characterize soil properties, and collection of soil samples for ;laboratory analyses to characterize the contaminants. It is advisable to delineate the vertical and horizontal extent of contamination. Delineation can be done by screening all soil samples with a volatile organic analyzer such as a photoionization detector (PID) calibrated to benzene, an organic vapor analyzer (OVA), or a flame ionization detector (FID), combined with laboratory analytical methods. As discussed previously, only contaminants with high vapor pressures and low water solubilities will desorb and volatilize under induced venting. Therefore, proper identification of the contaminants is a key task in performing the feasibility study. Definition of the areal extent and concentration of contamination will enable calculation of the quantity of contaminated soil, contaminant vapor concentration and eventually the duration of extraction necessary for remediation.

During the initial subsurface investigation, soil sample should be retained for laboratory permeability testing. Calculation of permeability will aid in determination of the radius of influence of the blower and overall feasibility of the system.

The determination of the depth to groundwater or presence of non-uniform geologic units can also be made during this phase, and can significantly impact the feasibility of vapor extraction. If groundwater is very shallow, vapor extraction may not be the most cost-effective means of remediation. Clayey soils with low permeability and gravels with very high per-meabilities are not very suitable for extraction. Instead, sandy soils are ideal for this technique. Conversely, if an impermeable clay cap overlies a contaminated more permeable soil unit, the feasibility of using vapor extraction increases. All these site-specific conditions must be accurately identified prior to choosing vapor extraction for remediation.

In addition to adequate knowledge of geological conditions, the locations and depths of subsurface structures, such as underground utilities, building footings, or underground storage tanks should be identified. These structures will become obstructions and will decrease extraction efficiency and may affect design parameters.

Finally, and most critical, is calculation of a reasonable cost estimate, which should include comparisons to alternative remedial methods. This factor alone may determine the feasibility of extraction. Typically, the in situ method of vapor extraction is inherently less expensive than alternate methods. This determination is based largely on quantity of contaminated soil and presence of structures that may inhibit or prolong excavation. If the amount of soil to be remediated is small and considered non-hazardous it may be more appropriate to excavate and dispose of contaminated soil. Similarly, if groundwater is contaminated and poses an immediate threat to public health, vapor extraction may have to be combined with groundwater remediation to speed up the remedial process.

Pilot Test and Design Variations

Once feasibility is established, a pilot test is performed. Pertinent design data is obtained during the pilot test, including determination of radius of influence, air permeability through soil media, and vapor concentrations. Conducting the pilot test requires installation of an extraction well and vacuum monitoring probes. Choice in location and spacing of the well and probes is critical for a well-designed system. A vacuum is applied to the extraction well using an injection or regenerative type explosion-proof blower, and the extracted vapors are passed through treatment apparatus, if necessary. Flow rate of the system is monitored and pressure drop across the blower and treatment system recorded.

Many different designs for pilot tests have been applied. Several of these systems have pending patents and many components vary considerably between designs. There are also many differing ideas on optimal spacing of extraction wells, well design, etc. However, the basic premises and design parameters are the same.

Extraction Well and Vacuum Probes

For the purposes of conducting a pilot test, and extraction well is installed in the area of highest contamination. The depth of the extraction well is determined by two factors. Obviously, the well should be screened to a depth below contamination. Screening the well to the water table in cases where groundwater contamination or floating product exist reduces condensation buildup throughout the system. Several factions exist that believe the extraction well should be screened into the water table to maximize product removal. This increases the condensation rate, present the possibility of actually drawing water into the system, and has not been proven to be anymore effective at volatilizing product than a well screen at the water table depth.

The extraction well is typically constructed in a fashion similar to a groundwater monitoring well. Schedule 40 to 80 PVC screen and riser pipe are installed into a six-to eight-inch diameter borehole. Two-to six-inch diameter pipe is the most effective diameter for the extraction well. The well screen should be sized so as to maximize flow through the porous media without entrainment of fine soils. The screen should be the length of the contamination depth and an equal vacuum will be achieved across the length of the extraction well. A gravel filter pack is installed around the screen in a fashion similar to monitor well installation. Wrapping the screen with filter fabric will reduce entrainment of fines.

Following well and filter pack installation, the annular space is grouted to seal the well screen from the atmosphere. Atmospheric influence must be eliminated at the extraction wellhead to enable exact vacuum readings. Construction of the pilot test extraction well in this manner will enable its use in the final extraction system.

Vacuum monitoring probes are installed using the same construction method. Smaller diameter screen and riser, anywhere from one-to two-inch diameter is used, and screen length can be as little as two feet. Tygon or plastic flexible tubing is inserted through an airtight wellhead cap using a Swagelock™ or ferrule-type fitting. The tubing is inserted to the middle of the vacuum probe screen and is used to monitor vacuum during extraction.

The positioning of the vapor probes should be such that at least one is located outside the influence of the blower, and in linear or grid pattern. Although it is not possible to foresee the exact influence of the blower on the air within the soil, based on soil characteristics (i.e., coarse grained versus fine grained) a spacing of vapor probes can be estimated. A linear or grid pattern spacing is desirable to more easily construct "drawdown" or vacuum curves with distance from the extraction well. This will also aid in calculation of influence if all vacuum probes fall within the blower's influence during the pilot test.

Spacing of vacuum probes should also correspond to the soil permeability. In tight, impermeable soils, the radius of influence or vacuum achieved will be less than in coarser soils. Other factors such as areas of backfill, underground structures, or presence of an atmospheric barrier will affect vacuum extent. In some cases, probe clusters have been installed at varying depths within one borehole. The purpose of clustering probes is twofold; variances in concentration and vacuum with depth can be determined.

Equipment

The equipment set up for the pilot test includes a blower, treatment system, and assorted valves and gauges. An airtight connection is made between the extraction well and blower. Of critical importance is measuring the flow rate at the blower; from flow rate measurements influence is calculated and blower sizing requirements established. Blower size can range from 1.5 hp to 10 hp, at flow rates ranging from 50 cfm to 1,500 cfm, depending on air permeability. Knowledge of flow rate and pressure drop across the treatment unit may be required for permitting discharge. A large loss in vacuum occurs across the treatment unit; therefore, the blower chosen should be explosion-proof and placed in series before the treatment unit.

Depending on permitting requirement and in calculation of vapor concentrations and duration of extraction, air samples are taken prior to treatment (influent) and after treatment (effluent). Sampling tees can be constructed at the well head, blower, and/or at the treatment unit discharge. Another requirement the pilot test is to obtain vacuum readings at the vacuum probes. This can be done using digital or u-tube type manometers.

Pilot Test

Depending on soil type air permeability, and the subsurface conditions at the site, the duration of the pilot test may last anywhere from one to several days. During this time, critical measurements of flow rate, vacuum at probes and extraction well, and vapor concentrations are determined.

Uniform soils with high permeability and a remediation site with an atmospheric barrier such as asphalt achieve a constant vacuum in a relatively short time period. Conversely, it has been observed that fine soils with low permeability may take days to stabilize under vacuum. Once stabilization is reached, vacuum will not change at any point in the system unless "short-circuiting" occurs. The system can short-circuit if extracted air is replaced with air from the atmosphere. This can occur through leakage under conditions where there is communication through the soil between the extraction well and atmosphere. Short-circuiting may also occur through the backfill of utility trenches. The vacuum will stabilize in the highly permeable unit, and will not extend into the less permeable formation until all air in the trench has been exhausted. Therefore, identification of all underground structures must be made prior to conducting the pilot test. In addition, paving or placement of a synthetic liner over the treatment area will increase vacuum extent and prevent short-circuiting of the system.

It has been found that in coarse grained sandy soils, vacuum will stabilize in a matter of minutes at a constant flow rate. In horizontal extraction trench systems or finer grained soils, it may take several hours or days to stabilize.

Once stabilization has been achieved, vacuum readings are taken at the vacuum probes and extraction well, and flow rate at the blower recorded, air samples of influent and effluent vapors can be obtained. If a u-tube or similar type manometer is used to monitor vacuum, readings must be corrected for barometric pressure changes. If a discharge permit or treatment system is not required, determination of vapor concentrations can be made using headspace analysis and a portable gas chromatogram. This can be done at any time before or after the extraction test and calculation will determine rate and duration of extraction. If a discharge permit require treatment of vapors, effluent vapor concentrations as well as effluent vapor temperature must be determined. Air samples can be obtained from sampling tees using syringe, bag, or charcoal tube collection methods.

Design

Design calculations are based on data derived during the pilot test and are based on those equations presented priviously. The applicability of these equations will vary due to the sitespecific nature of the system and the success of the pilot test. Many models have been derived for both gas flow at landfills, for which data is readily available, and for vapor flow in extraction systems. The latter are not available in print due to the competitive nature of the technology. Therefore, calculations are performed with the knowledge that conclusions are subjective, and upon implementation, adjustments to the system may be necessary. Based on the uncertainty of the calculations, overdesign steps are recommended to allow for flexibility in the field. The easiest way to overdesign the system is to choose a larger blower than calculated.

Calculations

The data obtained from the pilot test include flow rate, radius of influence, vapor concentration, and vacuum differential. In some cases, the radius of influence is not determined in the pilot test due to an insufficient number of probes located outside the influence of the blower. The following calculations enable sizing of design blower capacity, calculation of radius of influence, treatment system design, and duration of extraction. Carbon adsorption is used for treatment because it has been well-documented and has widely used application.

If the radius of influence (R) is known, the intrinsic permeability of air in soil can be calculated using the following equations:

$$ks = \frac{Qu\ln\frac{R}{rw}}{2tth\Delta p}$$

Where ks = intrinsic permeability (square feet)

Q = flow rate (cubic feet per minute)

R = radius of influence (feet)

rw = radius of extraction well (feet)

h = depth to water or depth of extraction well (feet)

ΔP = vacuum difference between extraction well and distance R (pounds per square foot).

From calculation of intrinsic permeability, the permeability of vapor in soil can be approximated using the following equation:

$$K = \frac{\gamma}{u} ks$$

Where K = permeability or hydraulic conductivity (feet per second)

γ = specific weight of vapor (pounds per cubic foot)

u = absolute viscosity of vapor (pounds-seconds per cubic foot)

The specific weight and absolute viscosity of the vapor should be based on characteristics of the gasoline mixture at test-specific temperature and pressure. The properties of benzene can be used for estimation purposes. Changes in temperature, pressure and moisture content will occur frequently during extraction and will affect permeability of the vapors. Many models describe variations due to these conditions and these variations are thought to be negligible. The equations presented here are generalized to represent ideal (no change) conditions.

If radius of influence R is not known, but data is available on the changes in vacuum over a known distance, R can be estimated by plotting the natural logarithm of known radius versus vacuum. Determination of intercept and slope from this graph will enable estimation of R based on the following equation:

$$P = B\ln r + A$$

Where P = change in vacuum (inches of water)

B = slope

r = distance from extraction ell (feet)

A = intercept

Once this calculation has been performed, A and B can be applied to the equation:

$$R = e - A/B$$

to approximate radius of influence of the blower used in the pilot test.

Using the calculated radius and a determination of the linear extend of contaminated vapors to be extracted, a blower can be chosen. Based on manufacturer's performance curves and using the above equations, an adequate capacity blower is chosen. Again, due to the theoretical nature of the above equations and the similar costs between smaller and larger capacity blowers, a larger blower should be chose for the final system.

Based on air sample results of samples taken during the pilot test and headspace analysis in the field, carbon capacity and duration of extraction are calculated. Carbon capacity can be determined using the following assumptions:

1. Carbon will adsorb 10% by weight of contaminated vapor
2. An average, or worst case, vapor concentration occurs across the entire treatment site
3. All contaminants will be volatilized and entrained in the flow.

A worst case concentration should be used in this determination. Fluctuations in groundwater level and related change in treatment area should be recognized. Duration of extraction is calculated using flow rate, area and vapor quantity estimations.

Design

Based on these calculations, extraction well number and spacing, blower size, and duration of extraction are determined.

If extraction through the soil is difficult, or the designed radius of influence large, more than one extraction well is used. Individual blowers can be installed at each well, or wells can be connected via header line systems, which can be connected above or below ground. Header lines are typically three inches in diameter or greater, depending on flow and vacuum requirements and are made of PVC, polyethylene, or fiberglass, depending on cost and chemical compatibility. Connections between header line, vent wells, and blower must be made airtight to assure effective vapor removal.

Considerations in design should take into account provisions for gate valves at extraction wells, flow meters, vacuum gauges, air or noise filters, and condentsate traps. The air filter, which prevents fine particles from entering the blower is typically placed at the blower influent. One quick-connect fitting can be used for the vacuum gauge and flow meter. A condensate trap or cyclone unit is typically placed at the beginning of the header line. If no provisions exist for the trap, sloping of the header line should be toward the extraction well or well system.

The entire blower and treatment system should be housed in a structure close the extraction well to reduce pressure losses through the system and minimize system exposure to climatic changes. Depending on the season, a heat source is applied to prevent condensate from freezing and to optimize vapor temperature, which enhance volatilization and extraction efficiency. Exhaust is controlled by venting through the structure.

Design Options

Several design options have been employed at different sites, and documents but not compared. These variations include installation of an air inlet well into the system, installation of many interconnected extraction wells, and application of individual blowers to each well.

Air inlet wells have been demonstrated as effective at reducing hydrocarbon concentrations at distance from extraction points by allowing continued influx of clean air into the recovery system. During one venting study air flow rates were compared between the inlet and recovery wells. Data showed the inlet air comprised only a fraction of the total volume of subsurface air displacing hydrocarbon vapors, indicating the balance of air was drawn from surrounding soils. The presence of an air inlet well, however creates a barrier to extraction; in this same study no vacuum was observed downgradient of the inlet well. This suggests that although soil "washing" appears to be enhanced by the addition of an inlet well to the system, there must be certainty that his well is located sufficiently outside the remedial area, to enable complete remediation.

Off-Gas Treatment Alternatives

As concern mounts over discharge of potentially toxic compounds into the atmosphere, and as municipalities and cities fail to meet the requirements of the Clean air Act (1977), regulattory agencies have become stricter in the control of emissions from hazardous waste treatment, Where extracted vapors and methane vented from landfills could be released to the atmosphere to degrade by photolysis, without treatment, this type of discharge is being approved less frequently.

Three alternatives for off-gas treatment are suitable in conjunction with vapor extraction. Vapor phase carbon adsorption, although cost-prohibitive under certain conditions where a large amount of soil needs treatment, is the most easily permitted and easiest to utilize. Other methods of treatment are flaring and incineration.

Carbon Adsorption

Carbon adsorption is efficient in the removal of a variety of organic compounds associated with gasoline to up to 99.99%. These compounds will adsorb in vapor phase on the activated (or charged) sites available on the carbon. To allow this process to occur, flow rate through the carbon must be controlled, but such a wide range is allowable that the typical range of flow rates used in vacuum extraction is suitable to ensure adequate contact time. Concentration of vapors entering the granular activated carbon (GAC) unit must also be known to predict the life of the carbon, and to size the appropriate bed to maximize removal. Removal rates will also depend largely on the presence of humics, which concentrations obtained from air samples enable calculation. Small (150 pounds) carbon cannisters can be included in the pilot test at a low expense and can provide valuable design and permit information. Samples of effluent air will be identified when all carbon sites are filled an breakthrough has occurred.

In general, granular activated carbon will adsorb an average of 10% by weight of contaminant. This assumption is useful in calculation of carbon capacity and to predict breakthrough. Breakthrough occurs when the adsorptive capacity for a particular compound is exhausted, and that compound begins to appear in the effluent. However, most manufacturers will use models to predict breakthrough and carbon capacity, based on concentration and flow rate, and can design suitable systems to fit site-specific conditions. Many commercially available units range in size from 150 pounds to several thousand pounds. In small to medium size gasoline spills, these treatment units can usually be designed such that no replacement of carbon is necessary during treatment. One way to eliminate replacement is to place a number of units in series, whereby a second unit can adsorb contaminants immediately upon breakthrough of the first unit. Carbon amount should be overdesigned to ensure there is no uncontrolled discharge. Overcompensation in carbon capacity also allows for changes in flow rate or concentration that may occur under site conditions. Sampling ports can be installed between units to obtain samples to predict breakthrough.

When treatment is complete, carbon can be regenerated or disposed. Because it is typically considered a hazardous waste, transport and disposal are governed by RCRA and adequate documentation must be obtained. Regeneration of carbon can be performed at a facility that specializes in large-scale regeneration. Documentation regarding the nature of the contaminants in the carbon is required by these facilities. Regeneration involves desorbing contaminants by thermal regeneration in multiple hearth furance, steam treatment, solvent extraction, or biological processes. Regeneration on-site has been successful, as demonstrated by Connor, where gasoline vapors were extracted from the soil, treated by GAC and the carbon regenerated on-site using an automatic steam conversion unit housed in a trailer.

As will be discussed, costs increase dramatically and become infeasible if a large prohibitive of soil has to be treated. However, in the most typical sites, where underground storage tanks have leaked at gasoline service stations, carbon treatment becomes more viable. In addition, this type of treatment, because of its "passivity" and propensity for control and proven success, is the preferred method by regulatory agencies.

Incineration

Controlled high-temperature incineration of off-gas in the presence of oxygen has been widely studied and has been approved in several states for treatment of emitted vapors. The process involves complete destruction of hydrocarbon vapors by combustion into carbon dioxide, water, and trace by-products. Incineration is only cost effective when concentrations of vapors are sufficient to maintain combustion temperatures with little or no fuel addition. If an addition fuel source is needed,. natural gas or propane is usually used. Under lower concentration conditions, catalysis of oxidation can be enhanced with the use of catalytic converter. The use of catalysts can significantly lower combustion temperatures thus eliminating supplemental fuel addition.

The process of catalytic oxidation has been used in the solid waste and wastewater treatment industry for some time. The basic process involves catalysis of the oxidation of the vapor in a controlled reactor, and results in physical destruction of the hydrocarbons. Catalytic oxidation is limited to a certain range of vapor concentrations. At excess concentrations, the oxidizer can destruct, and the presence of chlorinated organics has been proven to cause catalytic fouling. The major limiting factor is regulatory approval and permitting. Approval to operate an incineration facility, regardless of the size, is a very lengthy, time-consuming, and expensive process, and is not practical at highly visible sites such as gas stations.

Flaring

Flaring technology has also been widely used in industry for the destruction of emissions from stacks and methane gas from landfills. Lower heat value and lower concentration vapors can be successfully flared without supplemental fuels. However, this scenario is atypical of the very dilute vent streams produced by in situ volatilization. Flaring falls under the same regulatory constraints as incineration.

Implementation

Once feasibility has been established and design specification completed, permit application procedures can begin. These procedures usually involve submission of vapor sample concentrations, which can be determined using such sampling methods as charcoal tubes and syringes or PID and FID readings. Vapor concentration monitoring of the treatment system may be required by the permit. Once the permit is obtained and the system operational, it should be monitored to allow for flow rate or other adjustments during operation. At the end of extraction, soil sample should be obtained to demonstrate remediation.

In many instances, untreated effluent concentrations may be less than federal or state discharge limits, and off-gas treatment and the associated air discharge permits may not be required. On the other hand, insufficient treatment in some states will preclude approval. Adequate knowledge of federal and state discharge guidelines, as well as local attitude, will ensure fewer obstacles in gaining discharge approval. Permit application may involve groundwater quality bureaus as well as air pollution control. In New Jersey, for instance, in situ treatment is considered a continuous discharge to groundwater, and appropriate Pollution Discharge Elimination System applications must be made for the treatment duration, as well as appropriate air discharge permit applications.

Permit approval is typically a lengthy process, and usually results in some kind of mandatory sampling program to monitor the system's efficiency. This program may entail installation and sampling of monitor wells to monitor impact on groundwater during and after treatment, collection of influent and effluent air samples to monitor extraction and treatment removal of volatile organics. Continuation of the permit may also be made conditional upon achieving designated cleanup levels.

Vapor concentrations have been shown to decrease with increased extraction time. This is because removal rates decrease as immiscible phase composition equilibrates toward the mixture with fewer volatile organics. Vapor sample concentrations can be determined using a number of different sampling procedures and analytical methodologies. Changes in vapor concentrations throughout treatment can be monitored a the treatment system as well as the vacuum extraction well or vacuum probes. Samples can be collected at a sampling tee, with an air sampling pump, using standardized charcoal tube samplers. The air pump is calibrated to an established flow rate and the sample collected through the charcoal type for an established duration. The types are then laboratory analyzed for volatile organics by gas chromatography (GC), and total weight of adsorbed compound reported. From this analysis, the total amount of contaminant extracted per day can be calculated for the operational flow rate. Samples can also be collected at sampling tees using gas-light syringes and analyzed on-site using a portable gas chromatograph equipped with a flame ionization detector, which allowed semiqualitative analysis of hydrocarbon vapors. Portable laboratory units are also available which offer laboratory-quality results. Applegate et al. established on-site Gc analysis using the headspace method as 100 times more sensitive than the methane-extraction method used by laboratories.

Monitoring of vapor concentrations at the vacuum probes can also be accomplished using a photoionization detector (PID) or flame ionization detector (FID). The readings obtained from the PID are more qualitative than quantitative. Connor reports PID concentrations at one ;fifth to one tenth of what actual concentrations were, based on gasoline concentrations adsorbed onto carbon. FID readings are regarded as more reliable, with some measure of variability. Therefore, it is prudent to establish relative concentrations if either instrument is to be used during extraction by analyzing samples using laboratory methods as well. It should be noted that readings obtained using the PID are in weight per weight per ;weight units and must be converted to weight per volume units for laboratory comparison.

A soil gas survey can also be performed to monitor progress of treatment or in post-treatment sampling. These surveys, which entail installation and sampling of shallow gas probes or adsorptive tubes, enable qualitative or quantitative analysis.

Monitoring of the treatment system through vapor concentration measurements at the extraction well and vapor probes, and flow rate measurements, enable determination of efficiency and serve as indicators for needed adjustments. Modifications may be necessary due to change'sin concentration, temperature, or flow rate. These modifications may include installation of additional extraction points, increasing the capacity of the blower of addition of a heating fan to increase temperature of volatilization and prevent pipes from freezing. Sample ;results usually show a decrease in vapor concentration as extraction proceeds, indicating treatment or flow rate reductions can be made.

Once sampling result have indicated that hydrocarbons have been removed from the spill area, a subsurface investigation would be performed to prove the success of the treatment system. This is done through sampling soil at discrete intervals for volatile organics analysis, or sampling groundwater from monitoring wells. Extraction must continue until hydrocarbons are removed to the level indicated in the permit or until agency approval is obtained. Vapor extraction systems have remained on-line anywhere from six weeks to three years to achieve cleanup goals.

Case History

Literature review reveals numerous types of vapor extraction systems in use. Highlights of several cases, where diverse systems were successful, have been discussed. Many more designs were documented in which only pilot tests had been performed than those in which successful remediation was achieved. A pilot test for extraction of gasoline vapors has been performed at a site in New Jersey, the results of which are previously uncocumented. The following discussion describes the pilot test design and presents actual filed data used in calculations for final design. Based on the final design, application was made to the New Jersey Department of Environmental Protection (NJDEP) for permission to discharge treated vapors to the atmosphere. Permit approval was granted and implementation of the final system is ongoing.

Background

The site a former iron foundry where all underground storage tanks have been removed. At one former gasoline storage tank location, contaminated soil was observed during tank removal. Post-excavation sampling of the soil beneath the tank revealed contamination by petroleum hydrocarbons in concentrations ranging up to 11,000 ppm. Contamination extended to the groundwater table, at a depth of approximately eight feet below ground surface. During excavation and sampling activities, gasoline a product was observed entering a brook approximately one hundred feet away. An exploratory boring program was performed to delineate the extent of contamination. This program consisted of 19 borings that extended to the groundwater table. Extent of contamination was determined from visual observations a and PID readings.

Soil encountered throughout the discharge area is very loose, highly permeable foundry sand consisting of ashes and cinders. This fill is uniform and extends to the groundwater table. Approximately 90% of the spill area is paved with asphalt. Due to the estimated quantity of contaminated soil (approximately 2,000 cubic yards), and the high permeability of the soil, a pilot vapor extraction test was performed to determine the applicability of extracting volatile organic vapors from the contaminated soil in lieu of more costly excavation and disposal. The pilot test consisted of installation of vapor probes, which can also function as extraction wells, and application of a vacuum to determine the influence of the pilot test vacuum pump under in situ conditions and enable design calculations and sampling of extracted air for both design and permit applications. Extracted air was filtered through an activated carbon unit to determine the carbon's effectiveness in adsorbing the contaminants in the vapor.

Eight vapour probes were installed at locations shown on Fig. 6.5. Probe spacing was chosen in a linear pattern, radiating outwardly in two directions from the center of the spill area. This spacing was chosen to more easily predict uniformity of soil permeability, identify underground structures and establish a radius of influence without complete dependence on calculations. All probes were constructed in the same manner. All probes were installed to a depth of 9.5 feet (1.5 feet into the water table) and were constructed using 2-inch diameter Schedule 40 PVC pipe with 10-slot screen.

The depth of the borehole was screened from 2.5 to 9.5 feet below ground surface. The annular space adjacent to the probe was backfilled with a sand filter pack to a depth of 1.33 feet below ground surface. A granular bentonite seal was then placed to ground surface and wetted. Each probe, except PP-1 (the extraction well), was fitted with an airtight wellhead assembly. An 1/8-inch diameter tygon tube was placed to a depth of seven feet through each wellhead assembly for measurement of vacuum differential.

Pilot test equipment consisted of a single Phase, 21/2 horsepower, 220 volt vacuum blower. The inlet of the blower was connected via 2-inch diameter flexible hose to the outlet

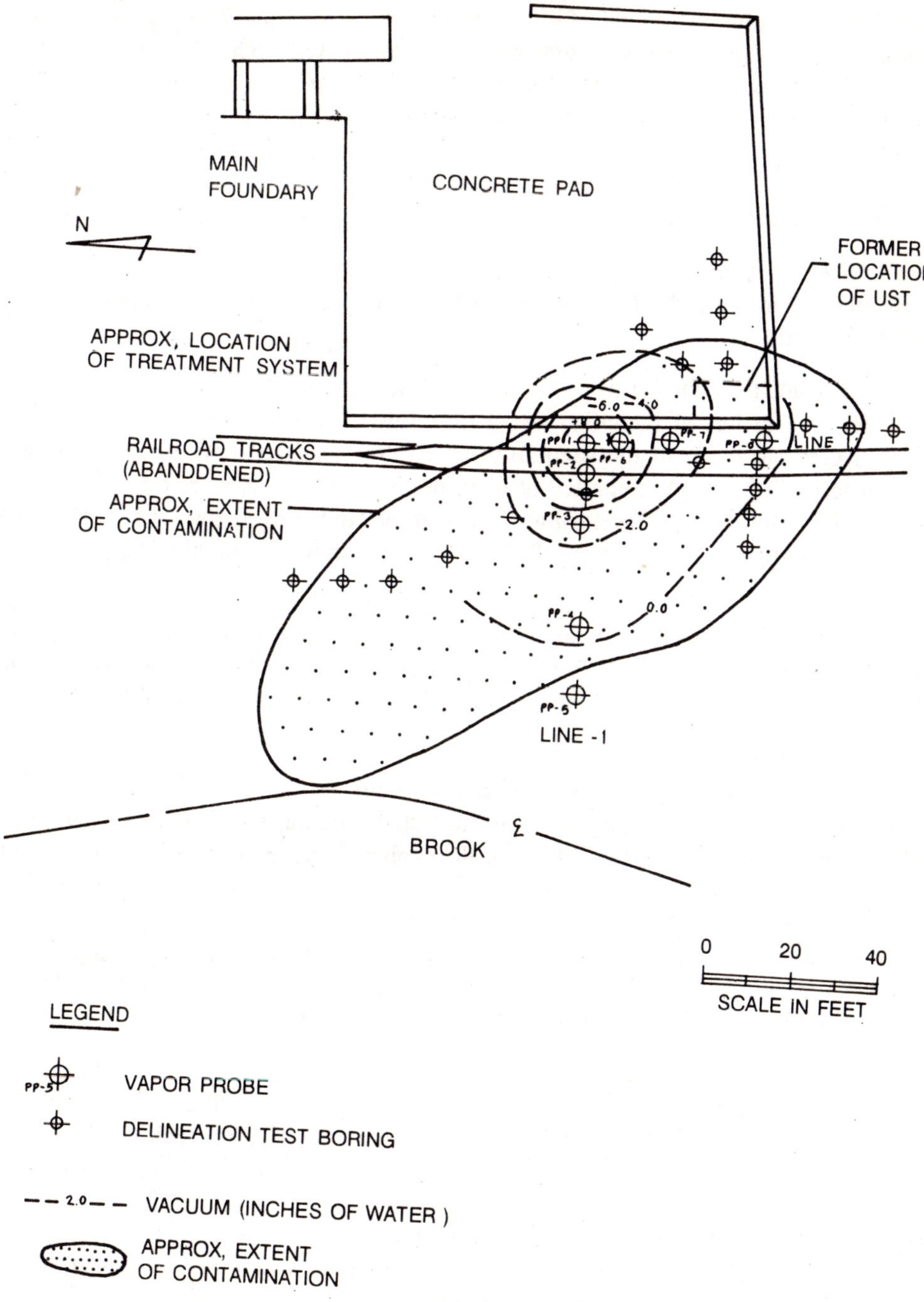

Fig. 6.5. Vapor extraction pilot test at UST.

of a vapour phase, granular activated carbon cannister. The cannister had a 55 gallon volume, containing approximately 150 pound of activated carbon. A flexible discharge hose from the blower exhausted extracted vapours outside. The blower was placed in series after the carbon cannister because it was not explosion-proof. The entire unit was mounted inside a van. Pressure losses between blower and cannister were evident during the test. An airtight wellhead assembly was placed at PP-1. This assembly was equipped with a quick connect for a vacuum gauge. A 2-inch diameter flexible hose connected the extraction well with carbon unit inlet. Vacuum gauges were also placed at the carbon inlet and blower inlet.

Prior to startup, static pressure readings were obtained at all vacuum probes using a digital manometer. Throughout the test the manometer was calibrated to account for changes in barometric pressure.

During the pilot test, the blower was operated at negative pressure at an intake of 42 inches of water (3.2 inches of mercury) and a flow rate of approximately 90 cubic feet per minute. Vapor concentrations at the vacuum probes were measured both prior to and during the extraction test using a PID calibrated to benzene. Measurements were taken by attaching the PID tip to the tygon tube. Concentrations prior to startup ranged form 13 to 194 ppm-v. Effluent was constantly measured during the pilot test to monitor for breakthrough.

Upon starting the blower, several rounds of vacuum readings were obtained at the probes during the first few hours of operation. Based on these readings, vacuum stabilized at all probes in a very short time, approximately fifteen minutes. This is due to the highly permeable fill from which the vapors were extracted. Contours of the steady state vacuum readings in inches of water are presented in Fig. 6.5.

During the test, which was conducted over a 24-hour period, vacuum at the carbon unit and blower were constant, stabilizing early in the test. Changing the flow rate of the blower resulted in an almost immediate stabilization in the probes. From manufacturer's performance curves flow rate was calculated for the blower to be constant at 90 cfm. Based on vacuum readings, pressure losses did occur between the extraction well, carbon unti, and blower. However, in considering the system as a whole and assuming final design will not vary widely from pilot test design and for the purposes of sizing a final blower, these losses were considered negligible. Also by overdesigning the final blower capacity these losses can be ignored.

As part of the pilot test, vapor concentrations were measured at the carbon effluent and the extraction well(influent). This was done to confirm the PID readings, calculate quantity of contaminated vapors, provide identification and concentration of the individual hydrocarbons present, provide information necessary for calculation of duration of extraction and carbon sizing, and for permit applications. These samples were obtained by collecting vapors using an air sampling pump, at a flow rate of one liter per minute, and charcoal collection tubes. Sample duration was 50 minutes, following state guidelines for volatile organic sampling using charcoal tubes. Each sample was analyzed for total volatile organics. results of these analyses are presented in Table 6.1. The blower was shut down during sampling, as it would have overcome the capacity of the air sampling pump.

As part of the test, the blower flow rate was decreased and the effect of vacuum in the probes recorded. This was done to substantiate theories of air flow and related calculations, and to enable comparisons to be drawn between theory and actual field data. A graphic representation of vacuum recorded at the probes versus distance from the extraction well at the different flow rates is shown in Fig. 6.6.

Data Summary

The pilot test was a success in that data obtained supported theoretical calculations, and that it proved extraction of gasoline vapors in this site-specific situation was the best remedial alternative.

Vacuum readings obtained at the 90 cfm flow rate were used in calculations of radius of influence and final blower sizing. Vacuum was observed to decrease substantially with increased distance from PP-1, as expected, and as shown in Fig. 6.5 and 6.6. During the entire

Table 6.1
Summary of Volatile Organics Analysis (μg), Air Samples, UST, Pilot Vapor Extraction System

Sample ID :	S-1 Air Infl.	S-1 Eff. Front	S-1 Eff. Back
Sample Duration:	50 min	50 min	20 min
Parameter:			
Chloromethane	ND	ND	ND
Bromomethane	ND	ND	ND
Vinyl Chloride	ND	ND	ND
Chloroethane	ND	ND	ND
Methylene Chloride	ND	ND	ND
Acetone	ND	ND	ND
Carbon Disulfide	ND	ND	ND
1, 1-Dichloroethene	ND	ND	ND
1, 1-Dichloroethane	ND	ND	ND
Trans-1, 2-Dichloroethene	ND	ND	ND
Chloroform	ND	ND	ND
1, 2-Dichloroethane	ND	ND	ND
2-Butanone	ND	ND	ND
1,1,1-Trichloroethane	ND	ND	ND
Carbon Tetrachloride	ND	ND	ND
Vinyl Acetate	ND	ND	ND
Bromodichloromethane	ND	ND	ND
1,1, 2,2-Tetrachloroethane	ND	ND	ND
1,2-Dichloropropane	ND	ND	ND

Contd.....

Sample ID :	S-1 Air Infl.	S-1 Eff. Front	S-1 Eff. Back
Sample Duration:	50 min	50 min	20 min
Trans-1,3-Dichloropropene	ND	ND	ND
Trichloroethene	ND	ND	ND
Dibromochloromethane	ND	ND	ND
1,1,2-Trichloroethane	ND	ND	ND
Benzene	4.0	ND	ND
cis-1-3 Dichloropropene	ND	ND	ND
2-Chloroethylvinylether	ND	ND	ND
Bromoform	ND	ND	ND
2-Hexanone	ND	ND	ND
4-Methyl-2Pentanone	ND	ND	ND
Tetrachloroethene	ND	ND	ND
Toluene	ND	ND	ND
Chlorobenzene	ND	ND	ND
Ethylbenzene	21.0	0.20	ND
Styrene	ND	0.24	ND
Total Xylenes	5.0	ND	ND
Total Dichlorobenzene	ND	ND	ND

Notes:

ND-Not detected.

Infl.-influent to carbon treatment (raw)

Eff.-Effuent from carbon.

Front-first carbon sample in series.

Back- second carbon sample in series.

test positive pressure and not vacuum was observed in one probe, PP-5, shown of Fig. 6.5. This presented a visual extent of radius of influence of the blower used in pilot test. From the data shown on Fig. 6.6, several conclusions can be drawn. A higher vacuum and smaller radius of influence occur at a higher flow rate. As flow rate is decreases, radius of influence spreads out and vacuum in probes closest to the extraction well decreases. By spacing a probe outside the influence of the blower, one can visualize the extent of influence and can compare observed with calculated extent. Finally, the presence of underground structures was demonstrated by the different slope of Line I vacuum readings from Line II. The slight variance in radius of influence in the horizontal direction was due to the presence of a concrete wall footing along the length of Line II. Based on the vacuum contours shown on Fig. 6.5, vacuum did not vary in the vertical direction between Lines I and II.

Calculations

Based on vacuum readings obtained during the 90-cfm test, the zone of influence of the blower is that area experiencing a vacuum of 0.3 inches of water or greater. The radius of influence of the blower was also calculated using the equations presented in the Theory

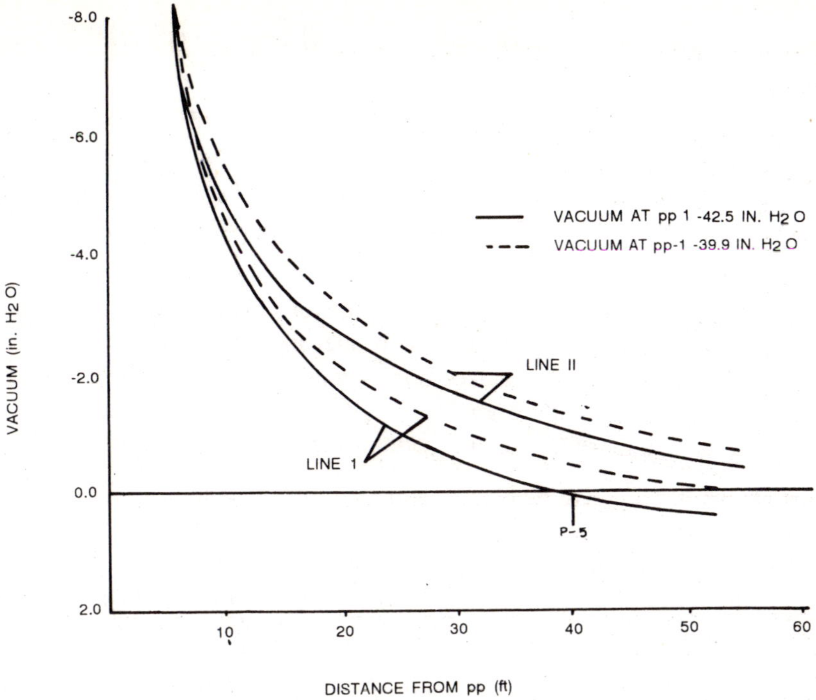

Fig. 6.6. Case study: vacuum/flow rate trends.

section. As shown on Fig. 6.5, a radius of influence of 39 feet was observed in the field . A radius of influence was calculated to be 50 feet. The intrinsic permeability of the soil was calculated from the observed radius of influence, and was used in calculations for final blower sizing. Observed and calculated radii would have been closer if laboratory permeability values had been used in calculation. The delineated spill area is shown on Fig. 6.5. Based on this area, a design radius of at least 100 feet was used to size the final blower.

During the pilot test, PID readings of the effluent from the extraction wells (influent to the carbon adsorption system) indicated total hydrocarbon readings of up to 194 ppm-v. These readings were used in sizing of carbon treatment because they were much higher than laboratory results of charcoal tube analyses. The reason for the difference in concentration was not determined. Assuming a molecular weight of 78 grams per mole for benzene, this translates to a mass flux of 0.23 pounds per hour of total volatile hydrocarbons, based on PID readings. The laboratory analytical results indicated that 4 micrograms of benzene were collected in the charcoal sampling tubes over the course of 0.83 hours sample collection time. This converts to a sample concentration of 4.8 milligrams per cubic meter (mg/m3), or 4.5 X 10^{-5} pounds per hour of benzene. Because PID concentrations were higher, they were used in final design calculations.

Final Design

Based on the calculations derived from the pilot test, the proposed extraction system involves use of a larger capacity blower and interconnection of existing probes to serve as a series of extraction wells. an activated carbon unit will treat vapors prior to exhaustion into the atmosphere.

From calculations of intrinsic permeability, observed radius of influence, and from determination of desired radius to extract all contaminant vapors from the spill area, a larger blower was sized for the full-scale system. The design blower is capable of achieving a radius of influence of up to 234 feet at a flow rate of 150 cubic feet per minute and vacuum of 70 inches of water. The actual operating flow rate will be fine-tuned in the field.

The installed probes will function as extraction wells. They will be interconnected via an airtight two-inches diameter, Schedule 40 PVC header line. Airtight tees will be placed at the head of each extraction well to enable access for water levels or monitoring. Gate valves will be placed at each extraction well to enable regulation of the extraction system, and quick connects placed to enable measurement of vacuum using a vacuum gauge.

The airtight header line will be placed below ground surface, on a 2% slope. Sloping the header line will allow for water collected in the pipe as condensate to drain back into the extraction wells. A condensate collection reservoir will also be placed above ground at the blower. Tees will enable drainage of condensate at this point. A small gravel bed will be constructed below this relief valve for drainage of water back into the ground.

The vapors extracted from this system will be channeled through two 2,000-pound activated carbon units for treatment. These units will be installed in series. Based on vapor concentrations measured in the pilot test, approximately 233 pounds of total volatile vapors were calculated to be present. This quantity of carbon was calculated to be sufficient to adsorb all vapors and was overdesigned to ensure that breakthrough does not occur. The blower and treatment system will be housed inside the main foundry building.

Final design calls for obtaining air samples of effluent to monitor breakthrough, which does not appear likely due to excess design carbon. Monitoring of the system will be by permit, and includes weekly smapling of effluent using the PID.

Based on permeability of air calculations, the volume of contaminated vapors, and the proposed flow rate to achieve cleanup, it is estimated that complete cleanup of contaminated soil will occur before the end of a three-month period. Extraction will continue for one week after influent samples indicate no contamination is present. At this time, soil samples will be obtained from across the treatment area to substantiate clean up.

Cost Comparisons

Inherent in determining the feasibility of using vapor extraction in remediation of soil contaminated by pertroleum hydrocarbons is performing a cost comparison between various remedial alternatives and vapor extraction. In performing this analysis, it appears that type of contaminant, site-specifics including soil type, and site characteristics and quantity of contaminated soil will govern the use of vapor extraction over other methods. Other alternatives to soil venting include excavation and disposal in a landfill, bioremediation, excavation and incineration, or in situ physical or chemical fixation.

Alternative Methods

The most widely used remedial alternative is excavation and disposal. This method is easily approved by regulating agencies and the most widely accepted practice. However, the long-term liability associated with moving contamination elsewhere instead of destroying it remains a potential problem. Excavation and disposal is best limited to soils contaminated with petroleum hydrocarbons in low concentrations, which will eventually biodegrade on their own. Present mean cost for excavation, hauling, and disposal of non-hazardous soils is approximately $ 250 per cubic yard. Hazardous soils are remediated for approximately $400 per cubic yard.

In situ bioremediation is becoming a more widely accepted remedial technique. This method is limited to contaminants that can be biodegraded, which includes most of the components of gasoline and other petroleum products. Biodegradation is controlled greatly by soil conditions, temperature, and pH. An advantage to mocrobial degradation is that treatment is in situ and that hydrocarbons are destroyed without long-term liability associated with disposal. This technique has been demonstrated to be 99% efficient. Microbial degradation is one of the least expensive options, with unit costs ranging from $66 to $123 per cubic yard treated. Combination of microbial degradation and forced aeration will decrease unit costs. The distinct disadvantage of using bioremedial techniques over venting is the great dependency of microbial populations on ;conditions not easily controlled by mn, such as temperature and moisture content. In addition, enough documented data exists to conclude that this method is not well understood because of the many uncontrolled variables. Also, microbial degradation of hydrocarbosn takes much longer than extraction of contaminated vapors.

Incineration of contaminated soils is one of the more difficult techniques to gain approval for, is much more expensive, and requires excavation ;and most likely transport. However, incineration destroys contaminants up to 99% in complete oxidation. Estimates for incineration of contaminated soil range from $200 to $640 per cubic yard, and increase dramatically with less than 20,000 cubic yards to be treated. These estimates are based on on-site incineration and do not include excavation ;and transport.

Physical and chemical treatment methods include flushing soil with water, and chemical fixation. Treatment using these methods is often less complete than other methods and ideal treatment conditions are difficult to achieve. Typical costs for ;soil washing (which is the least expensive physical/chemical treatment method), range from $150 to $200 per cubic yard. Other techniques, such as in-situ steam injection, and radio frequency are very dependent on site-specifics and range in costs of at least $200 to $300 per cubic yard.

Vapour Extraction Systems

The costs associated with design, construction, and operation and maintenance for a vapor extraction system are summarized in Table 6.2. This estimate represents the cost estimate per-formed for the system described in the case history. Included in this estimate is a quantity for addition carbon in case total vapor concentrations exceed estimates. The costs presented represent a conservative estimate for an extraction duration of 3 months. This estimate does not include costs associated with the pilot test; the installation of the vapour extraction wells is not included.

Table 6.2
Cost Estimate for Vapor Extraction

Activity	Item	Quantity/Units	Unit Cost (S)	Item Subtotal	Total
Capital Costs:					
Materials/ Equipment	Piping, fittings, tees, valves, gauges	—	—	7,000	
	Vacuum Blower	—	—	1,400	
	Carbon Units	4,000 lbs	7	28,000	
	Air Sampling Equipment Rental	—	—	850	
Installation	Entire System	—	—	4,500	
Operation & Maintenance	Entire System, 3 Months	—	—	7,800	$ 49,550
Sampling and monitoring:	Monitoring and sample Collection	10days	1,350	13,500	
	Post-Extraction Drilling	2 days	3,250	6,500	
	Laboratory Analysis-Air	20 days	350	7,000	
	Laboratory analysis-Soil	6 tests	360	2,160	
					$ 29,160
					$ 78,710
	Total capital and sampling costs				
	Engineering (20%)			15,700	
	Legal consultation (5%)			4,000	
	Administration (5%)			4,000	
	Contingency (10%)			7,900	
	Total Cleanup Cost				$110,310

Notes: The above are based on a cleanup duration of three months, and are inclusive.

Generally, there is little cost variation between the several types of extraction systems. However, there is marked increase in remedial cost with increased extraction time due to the difficulty of extraction (due to soil permeability). Off-gas treatment will significantly raise the overall cost of remediation using vapor extraction.

In general, soil venting ranges in cost from $245 to $320 per cubic yard of soil treated, without off-gas treatment. This can also be translated to a cost of $70.60 per gallon for every 1,000 gallons of fuel spilled. If treatment of extracted vapors is required, the the above range can be significantly increased depending on the cleanup levels to be achieved, allowable discharge concentrations, flow rate, and vapor concentration. As stated previously, the major disadvantage of using carbon adsorption for off-gas treatment is prohibitively high costs if large quantities at high concentrations ar to be treated. The major advantage is ease in permit approval and handling. The major advantage of incineration and flaring is complete destruction of contaminants, ability to treat large quantities, and reduced long-term liability. The major disadvantages are cost and regulatory approval.

Costs for carbon are approximately $7 per pound, which includes changing, handling, transport, and disposal at a regeneration facility. Fir activated carbon a rule-of-thumb cost is approximately $20 per pound of hydrocarbon or approximately $130 per gallon of fuel spilled.

This fig. does not include changing, transport, or regeneration. Carbon changes involve the bulk of the maintenance costs associated with this treatment. EPA has presented an equation, which is based on flow rate, to use in estimating operation and maintenance costs for carbon treatment:

$$OM + 12.2\ Q$$

where OM = annual operating and maintenance costs, \$1,000 per year

Q = vapor flow rate, 1,000 cfm.

Costs associated with incineration, catalytic oxidizers, and flaring depend largely on concentration of vapors and the needs for supplemental fuels, and can range up to \$10,000 in capital costs. Operational costs for these systems are high due to the need for constant monitoring and fuel addition, and can run up to \$300 per day for operation.

Conclusion

There is little doubt that forced extraction of contaminant vapors from petroleum or volatile organic discharges is an effective way to remediated soil, up to 99 + %, in a relatively short time. Several supporting facts, as well as questionable observations of available documentation, are worth noting.

Vapor extraction has been a proven technology for removal of landfill gases for many years, and has only recently been applied to the remediation of hazardous substance discharges into the ground. Petroleum hydrocarbons related to gasoline and fuel oil, as well as chlorinated solvents, have proven most suitable to forced extraction due to chemical properties of high vapor pressure and low water solubility. Common examples of these chemicals are benzene, toluene xylene, trichloroethylene, and tetrachloroethylene. Vapor extraction has been proven to be effective in volatilizing floating and dissolved product from the groundwater table. Concentrations of vapors typically decrease dramatically within the first few days of operation, indicating that most contamination is removed early in treatment.

The feasibility of vapor extraction not only depends on contaminant type, concentration, and physical properties, but also largely on site-specific soil, conditions. Although demonstrated to be effective in less permeable soil, the increased cost of extraction due to more equipment and power requirements and longer extraction times may make vapor extraction infeasible. Vapor extraction is most cost-effective when greater than 500 cubic yards of soil are contaminated, and off-gas treatment is not required. Treatment of vapors will raise remedial costs significantly and treatment methodologies depend upon vapor concentration and regulatory involvement. Incineration or flaring methods are much harder to obtain regulatory approval for than carbon, but in cases where vapor concentrations are very high can be more economical than carbon.

Several observations were made during literature review. There are tow schools of thought as to installation of the extraction well into or above the water table. It does not appear that it makes a difference. The argument for depth within the water table is that greater floating product could be volatilized, but installation to a depth close to the water table achieved the same results. Similarly, it was not clear what the advantage was to installing clustered vacuum probes at different depths other than to sample vapor with depth or to monitor vacuum differences if non-uniform soils were present. As long as vacuum was measured in the deeper probes, the vacuum differential at shallower depths is insignificant to the remediation of the contamination. The need for the air inlet will used in several cases is still unclear. A larger radius of influence and higher vacuum will be created, provided there is no short circuiting with the atmosphere. The presence of air inlet wells or forced draft air injection severely limits the area capable of treatment. It would appear that creation of an impermeable atmospheric barrier would be more effective.

Theories of air flow through porous media have proved reliable through calculations from vapor extraction testing and can be depended upon to provide adequate design data. That is, the system as a whole is well understood. However, theories of volatilization and dependency on physical and chemical characteristics of vapor and the interaction with individual soil particles are not as well understood, and generalized equations are used. Consequently, overdesign of the system should be performed.

Few documented cases used any treatment other than carbon adsorption for off-gases. Therefore, it appears more research (or more regulatory approval) of incineration systems is warranted. Without destruction, contaminants adsorbed onto carbon are moved to a different location after carbon is exhausted and the potential for long-term liability is created. In addition, the potential for uncontrolled emissions exists upon breakthrough.

Vapour extraction is definitely more effective in more permeable soil, where cleanup levels can be achieved more quickly than in low permeability soil, with fewer extraction well installations. However, the technique was demonstrated to be ultimately effective in any type soil with the proper design. Less permeable soil may make the treatment cost-prohibitive.

7

In Situ Technologies for Vadose Zone Remediation

Introduction

Contaminated soils in the unsaturated (vadose) zone can act as sources for release of contaminants into the groundwater and surrounding soils. Unless these sources are eliminated or controlled, remedial measures can be ineffective or costly. Thus, while groundwater extraction and treatment can prevent or redirect the migration of contaminant plumes, cleanup may be slow due to continued leaching of contaminants into the groundwater. Effective "source control" is necessary to expedite cleanup. He traditional approaches to source control at slurry walls, liners, and caps and/or removal and disposal of contaminated soil and waste at off-site locations. Despite considerable improvements in the design, construction, operation, and monitoring of waste isolation and containment systems, uncertainty remains regarding the ability of these systems to provide long-term environmental protection. Waste relocation and on-site waste isolation are also considered by many as shortsighted strategies that merely transfer the problem to future generations or to new locations.

Because of the above considerations, there is a substantial interest in treatment and destruction technologies that reduce toxicity, mobility, or volume of the contamination rather than merely provide temporary protection via isolation and cotainment. This interest is reflected in the Superfund Amendments and Reauthorization Act (SARA) of 1983 that encourages selection of alternative remedies and authorizes funds for applied research, development, and demonstration support programs. EPA is now engaged in the Superfund Innovative Technology Evaluation (SITE) program to demonstrate, evaluate, and report on technologies that are applicable to on-site or off-site cleanup of hazardous substances spills and releases at Superfund sites. The effort is cooperative between EPA and the technology developers/vendors and involves 10 to 15 demonstration projects per year.

Treatment of contaminated soils and wastes can be accomplished via off-site or on -site processing of the excavated materials or by in situ in-place treatment (i.e., without excavation). In situ treatment offers number of advantages, the most important of which is the elimination of safety, environmental, and the public health risks associated with excavation, transportation, interim storage, and handling of hazardous wastes.

In situ technologies (ISTs) are generally based on processes and systems that have been developed and are used in conventional water and waste treatment and mining, oil and gas, and chemical process industries. Biological, chemical, physical, or thermal methods are used to degrade, detoxify, extract, or immobilize contaminants in-place. Treatment agents such as heat, air, steam, or solutions containing surfactants, chelating agents, bioactive materials, and oxidizing/reducing chemicals are delivered to the subsurface via gravity or forced systems (or, in the case of in situ vitrification, by inserting electrodes in the soil for in situ heating). Recovery of the by products and unused chemicals (if any or if implemented) is again via gravity or forced or vacuum systems. Some ISTs require aboveground operation support. Examples of such support are carbon adsorption for treatment of waste streams resulting from in situ treatment and equipment for addition of treatment agents.

ISTs are in various stages of development. Each technology has the potential to serve specific site clean up needs and would be most suitable to specific categories of contaminants, target cleanup levels, and cleanup schedule. The following technologies are discussed:

- Soil Vapor extraction
- Steam stripping
- Bioremediation
- Soil washing
- Solidification/stabilization
- Vitrification.

Soil Vapor Extraction

The soil contamination at many sites is the result of accidental spills of petroleum-based fuels or solvents or leakage from underground storage tanks containing such products. Because of their relatively high volatility, these contaminants can be removed from the soil by vacuum extraction or by steam stripping. Soil vapor extraction (SVE), also referred to as vacuum extraction, soil venting, soil vacuuming, and enhanced volatilization, is a fully developed commercial technology and has been used widely for cleanup of soils contaminated with petroleum hydrocarbons.

Principle of Soil Vapour Extraction

Volatile organic compounds (VOCs) in the soil are removed by application of vacuum to extraction wells to induce a flow of clean air from the atmosphere into the subsurface. The vacuum not only draws vapors from the unsaturated zone, but also decreases the pressure in the soil voids thereby releasing additional VOCs. The highest rate of VOC recovery is experienced initally, but that rate drops as the extraction proceeds, with the total cumulative recovery typically reaching a plateau in less than 100 to 200 days (see Figure 7.1).

Depending on the VOC content of the gas and the applicable emissions regulations, the extracted gas can be handled in one of the following ways:

- Directly discharged to the atmosphere.
- Processed for VOC removal (e.g., via activated carbon adsorption) and then discharged to the atmosphere.

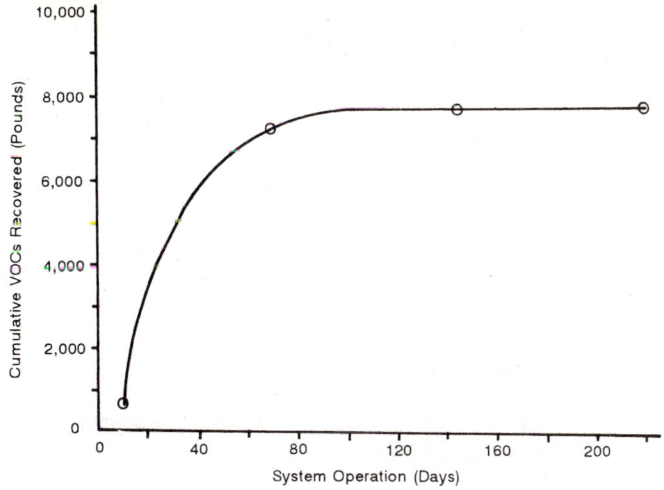

Fig. 7.1. Total VOC recovery with time.

- Processed for VOC removal and then injected into (recycled to) the subsurface.
- Disposed of via combustion in existing boilers that are operating on a continumous basis. (This may only be possible at sites located in or near industrial installations).

The performance of a SVE system may be enhanced by placing appropriately designed air inlets at strategic locations near the extraction wells to promote entry of fresh air to the subsurface. Some SVE systems feature air injection wells to injected air or to recycle the extracted gas into the subsurface after treatment. Returning the treated gas to the subsurface offers a potential for reducing air emissions and is attractive for use in locations with strict air pollution control regulations. Supplementary air injection/recycling increases the air flow through the contaminated zone and is believed to bring about the following benefits:

- Minimize possible "dead" zones (i.e., areas where adequate soil-air contact is not achieved)
- Improve air flow at greater distances form the extraction wells.
- Reduce possible short-circuiting of the extraction air through the ground surface.

Vapor extraction and air injcction wells differ form the conventional groundwater monitoring wells in that they are installed in the unsaturated zone with the screened (perforated) sections extending above the water table surface. Both extraction and injection wells are perforated and grouted along predetermined lengths to reduce short circuiting and loss of the injected air to the ground surface. In cases of high groundwater table and /or when the contamination is confined to shallow depths, perforated pipes buried in horizontal trenches would be used in lieu of wells. The extraction wells are connected to the suction side of a positive displacement blower through a surface collection manifold. A schematic diagram of a SVE system with provision for returning the extracted gas to the subsurface is shown in Figure 7.2.

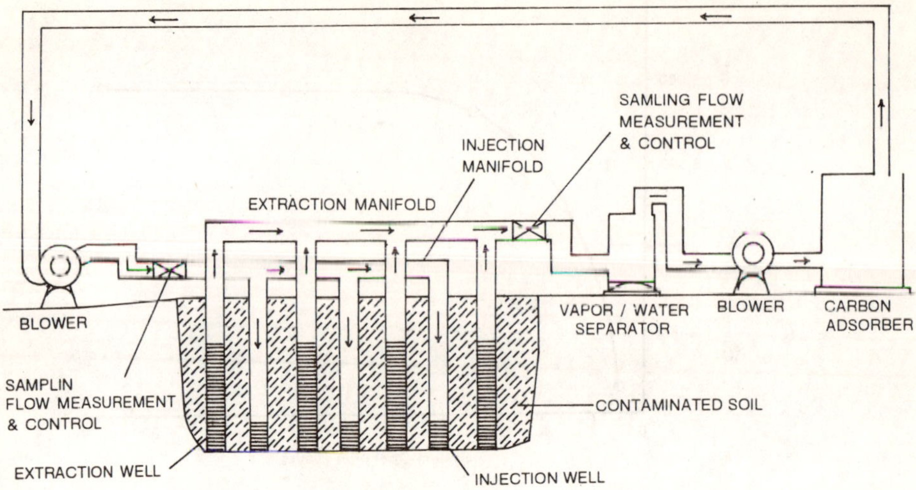

Fig. 7.2. Soil vapor extraction system with provision for returning the extracted gas to the subsurface after treatment.

Applicability and Limitations

SVE systems are best suited to sites where VOCs are the primary target contaminants (e.g., in cases of solvent spills or leaks from underground gasoline storage tanks or pipes); the contamination is in the unsaturated zone in relatively permeable formations; and site conditions and regulatory requirements accommodate a cleanup schedule entailing months or years. As a first order of approximation, compounds that exhibit vapor pressures of about 0.5 to 1.0 mm Hg or more at 20°c may be considered good candidates for removal by vacuum extraction. Cleaning solvents such as percholroethylene (PCE),trichloroethylene (TCE), and 1,1,1-trichloroethane and most constituents in gasoline can be effectively removed by vapor extraction.

SVE systems would be particularly attractive for sites that are located in highly developed areas and/or in cases where the contamination has spread to adjacent properties or underneath buildings. In these situations a remedial option involving excavation and removal of contaminated soil may be impractical or very costly. Once installed and in operation, SVE systems generally require little maintenance and operating attention; system deployment thus may not interfere with existing site uses or planned site development (e.g., construction and use of buildings, parking lots, etc.).

Compared to most other in situ technologies that are still under development and await a successful full-scale field demonstration, SVE has been successfully employed in a number of full-scale applications. As with all in situ technologies, the successful use of SVE requires a good knowledge of subsurface conditions which may necessitate some fairly detailed sampling and analysis for site characterization. Even then, and with the best of designs, uniform and complete treatment may not always be attainable at every site due to subsurface nonhomogeneity that can result in channeling. Thus, unless verified through actual soil sampling, a significant reduction in and the leveling off of the VOC from the soil. Finally, because it can only remove VOCs, SVE by itself would not bring about complete remediation for many sites where a spectrum of contaminants including inorganic compounds and non-volatile or low-volatility organic substances are encountered. In these cases VOC removal via vacuum extraction can only serve as an element of the total remediation effort.

The applicability of a SVE system and its use as an element of a total site remediation plan is best illustrated by the "Verona Well Field" Superfund project near Battle Creek, Michigna .

Verona well Field supplies potable water to most of the city of Battle Creek. In 1981, the well field was found to be contaminated with VOCs, principally chlorinated solvents, with a contaminant plume encompassing an area of approximately a mile by a half-mile. Two facilities operated by a local solvent wholesaling company were identified as major sources of contamination. VOC levels as high as, 1,000 ppm were found in both groundwater and soil at these properties. It was estimated that contaminated soils and groundwater on these properties contained approximately 1,700 pounds and 3,900 pounds of VOCs, respectively. After evaluating a range of treatment alternatives, SVE was selected by EPA for use in conjunction with groundwater treatment for cleanup of this site.

Since groundwater contamination was exclusively caused by VOCs, use of a stripping tower followed by processing of the stripper off gas in a granular activated carbon adsorption unit was determined to be the most appropriate method for groundwater treatment. Several source control alternatives were considered for eliminating further release of contaminants into the groundwater. Construction of a clay cap to reduce infiltration and leaching was considered but rejected since it did not constitute a permanent cleanup. Excavation and proper disposal of the 9,300 cubic yards of contaminated soils were eliminated because of the lack of sufficient space for on-site disposal and because it, too, did not constitute a permanent solution. In situ soil washing, in which clean water would be percolated through the soil to flush the contaminants into the groundwater for treatment in the groundwater extraction and treatment system was also considered and rejected because it was projected that eight years of soil washing would be required to reduce contamination to acceptable levels.

Given the above limitations of the alternatives considered, SVE was selected as the preferred soil cleanup option for the Verona Well Field application.

Design Considerations

The SVE system design is highly case-specific. Design features such as number, location, and depth of extraction and injection wells; size of the blower(s) and the piping system; equipment and monitoring instrumentation for management of the extracted gas; and other considerations must all be tailored to the conditions and requirements at a particular site. A discussion of the technical factors that must be considered and cost saving possibilities that should be explored in project planning and design is contained in a paper by Bennedsen et al. These factors include the following:

- Characteristics of the contaminants, primarily from a volatility viewpoint

- Lateral and vertical concentration distribution of VOCs in the soil
- Subsurface characteristics including type, uniformity, and variations in permeability
- Emissions control requirements
- Schedule for cleanup
- Site factors such as location, distance to water table, and physical constraints of the site.

Opportunities for cost saving and incorporation of adequate monitoring systems in the design to permit performance evaluation and appropriate future modifications to the design and operating conditions must be explored during the project planning stage. These opportunities can include use of existing or planned monitoring wells as components of the SVE system; completing soil boring used in site characterization as air inlets, injection/extraction wells, or monitoring stations equipped with probes and sensors for depth-specific gas sampling and/or measurement of in situ vacuum/pressure levels; and disposal/ treatment of the extracted gas in existing boilers or gas processing units.

In a novel application at a southern california site the operation of the SVE system is integrated with that of a moderate vacuum steam striping Aqua Detox process for VOC removal from contaminated groundwater. In this arrangement, shown schematically in figure 7.3, the activated carbon adsorption units which treat the extracted soil gas before it is recycled to the subsurface also handle the vent stream from the secondary condenser of the Aqua Detox process. This vent stream contains the non-condensables and an equilibrium quantity of VOCs. The carbon adsorbers are regenerated on-site using steam. The VOC-laden condensate from steam regeneration is sent to the Aqea Detox stripping tower for coprocessing with the tower feedwater.

As noted previously, an assessment of the feasibility and success of a full-scale SVE system requires a fairly detailed knowledge of the subsurface characteristics. A detailed site characterization, however, can be very costly and often impractical. Because of this, a prudent approach to feasibility assessment and design of full-scale systems (at least for large sites) is a phased approach whereby a partial or a pilot scale system is first installed and operated for a short time to develop a more reliable data base and design criteria for a full-scale system and an estimate of the time frame required to achieve site cleanup. A pilot-scale testing would have the following objectives:

- Determine of vacuum levels required for significant volatilization of the contaminants in the subsurface.
- Develop estimates of the rate of extraction of the primary constituents of the VOC and of total VOC, the decrease in the removal rates with time, and time required for site cleanup.
- Estimate the radius of influence of the extraction wells.
- Demonstrate the effectiveness and safety of the VOC recovery system.

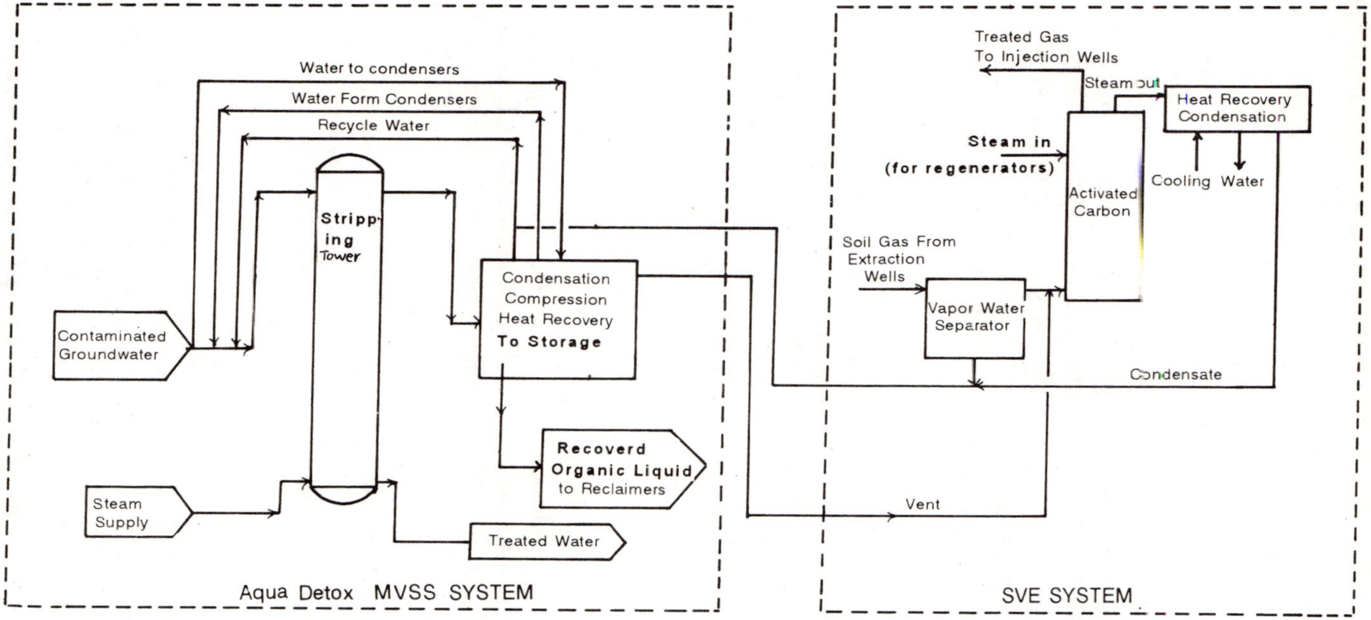

Fig. 7.3. Simplified block flow diagram for Aqua Detox moderate vacuum steam stripping for VOC removal from groundwater and SVE for vadose zone cleanup.

A pilot facility consisting of about four wells and operated for a period of about three weeks should generally be adequate to provide the data base for assessing project feasibility, designing a full-scale system, and projecting the time frame for overall cleanup. Each well should be capable of independent operation with provisions for monitoring of subsurface vacuum and contaminant concentrations at the well head under a range of design and operating conditions.

The design of the full-scale system should also allow maximum operating flexibility. Large sites may thus be divided into smaller areas, each with its own independent extraciton system and provisions for gas sampling, and for flow and pressure measurement and adjustment. The discharge side of the blowers for the individual systems can be connected to a common header pipe that would discharge to the emissions control system for the site. Blowers typically used in these applications have ratings of 100 to 1,000 cfm, at vacuum ratings of up to 10 in. Hg gage; ratings of the electric drive motors are 10 Hp or less.

SVE System Example

As noted previously, SVE systems are currently in use at a number of sites for the cleanup of VOC contaminated soils. Two such systems are in operation at two separate areas (Site "D" and Site "G") of the Twin Cities Army Ammunition Plant (TCAAP) in New Brighton, Minnesota. These two systems, for which considerable design documentation is available, are briefly described here to illustrate a systematic approach to and the considerations used in design.

Site Characteristics

Site conditions data used in the design are as follows. Site D is located on a sand deposit extending to a depth of approximately 165 ft with the water table apparently near the bottom of the sand deposit. Soil samples indicated VOCs, predominantly TCE, at concentrations of up to 8,000 ppm. Most observations were in the upper 10ft of soil. However, a deep boring showed 1,000 and 400 ppm of TCE at depths of 120 and 140 ft, respectively. Till and sandy deposits underlie site G to depth of approximately 135 ft, with the water table found at a depth of 130 ft in one boring. Samples taken from the upper 20 ft of soil showed VOC levels of up to 1,000 ppm, with TCE and trans-1,2-dichlorethylene being the predominant compounds. An 18-in. compacted clay cover had been placed on both sites D and G.

Pilot Plant Feasibility Studies

From November 1984-February 1985, a 14-week pilot scale field demonstration was conducted to evaluate the feasibility of SVE for remediation of VOC-contaminated soils at TCAAP. Based on site characterization results, two areas within the contaminated site, representing low and high VOC levels, were selected for the feasibility study. (Borings in the test plots had generally encountered sandy materials with traces of silt and gravel.)

Two distinctly different pilot systems were installed to test several designs and performance variables. The study focused on the removal of TCE as an indicator of VOC contamination. To prevent freeze-up in subzero conditions, the injection air for both systems was preheated to 45-55°F. The injection and extraction wells consisted of 20-ft, 3-in. PVC pipes, assembled in 5-ft sections. The injection and extraction pipes were slotted (slot size of 0.06 in.) in the bottom 15 ft and 5 ft sections, respectively. The injection vent pipes were left open at the base, whereas the extraction pipes were capped at the base. The annular space outside the PVC casing was packed with gravel along the slotted interval, and the solid wall intervals were grouted with cement/bentonite. The extracted air was passed through canisters of activated carbon before discharge to atmosphere. Other features of the two systems and the results obtained are summarized in Table 7.1.

Table 7.1
Design and Operating Features and Results for the SVE Pilot Systems

Parameter	System 1	System 2
Test plot size	50×50 ft	100×100 ft
Vent pipe spacing	20 ft	50 ft
Extraction flow rate	40 to 55 cfm	200 to 220 cfm
Extraction vacuum	0.5 to 1.3 in w.g.	17 to 20 in w.g.
Injection flow rate	55 to 60 cfm	46 to 223 cfm
Injection pressure	0.9 to 1.4 in w.g.	10 to 11.9 in. w.g.
Pre-test soil TCE	0.05 to 2.3 mg/kg	0.06 to > 4,000 mg/kg
Post-test soil TCE	<0.05 to 4.9 mg/kg	0.06 to 5,000 mg/kg
Pre-test soil moisture	3 to 13%	3 to 13%
Post-test soil moisture	1 to 36%	2.2 to 12.5%
TCE removal in 14 weeks	1 kg/8,000 ft^3 soil	730 kg/50,000 ft^3 soil
Initial off gas TCE	30-70 ppm	250-300 ppm
Final off gas TCE	<5,000 ppb in 10 wks	250-300 ppm
TCE removal, calculated	75% of initial	25% of initial

The TCAAP study demonstrated the effectiveness of the systems tested in removing TCE and other solvents form porous soils. The carbon adsorption system also proved effective in removing TCE, other chlorinated solvents, and PCB from the extracted gas.

Data from the TCAAP study suggested that under the experimental conditions no benefits may have been derived from supplementary air injection. During a 10-day period when the injection air for one system was shut down, no change in extraction air flow rate or TCE concentration was observed. Although it is possible that an effect could have been observed at higher flow rates and/or different vent spacings and designs, the impacts of such changes on the VOC removal levels were not evaluated. Because of the lack of conclusive evidence that supplementary air injection indeed brought about an increase in VOC removal, the two full-scale systems that were subsequently installed and are currently operating at TCAAP do not include air injection and are operated as SVE systems.

Design and operation of Full-scale System

The full-scale systems were designed to cover a total area of about 80,000 square feet. Vent spacing is a key design parameter for SVE systems. A design vent spacing of 25 ft was selected based on results from the pilot plant tests at Site D, which provided a discharge rating curve relating the vent flow rate to required vacuum pressure head at the vent and a family of curves depicting the vacuum drop as a function of the radial distance from the vent for various vent flow rates. Based on the pilot plant data, a vacuum pressure of 0.5 inches of water had been suggested as the minimum vacuum pressure at the radius of influence of the vent for producing reasonable assurance of positive flow direction. The design curves indicated a flow rate of 110 cfm per vent and a required vacuum pressure head at the vents of 13 inches of water for a vent spacing of 25 ft.

Other criteria considered in the design of the full-scale systems were the need for operational flexibility, vent penetration depths, and emissions control requirements. Flexibility was built into each system by grouping vents along lateral collectors feeding a central header, and by including valves and sampling ports at strategic locations. Soil data suggested a vent depth of 30 ft should generally be sufficient to encompass the region of contamination. The emissions control and monitoring requirements were established based on anticipated VOC levels in the extracted gas and discussions with the cognizant state air quality regulatory agency. Some key features of system configuration and design are listed in Table 7.2.

Table 7.2
Features of the Full-Scale SVE Systems at TCAAP

Feature/Parameter	Site D	Site G
No. of Vents	39	89
Manifolding		
Lateral collectors	2 100-ft 3 75-ft	9 75-ft to 225 -ft
Central header	1 150-ft	1 225-ft
Blowers	4 20-hp, 2,200 cfm	4 40-hp, 5,700 cfm
Vent system	3-in. PVC, slotted gravel packed from a depth of 7 ft to at least 30 ft, cement grouted above	same as Site D
Vent Capacity	110 cfm each	110 cfm each
Treatment of extracted gas	Carbon adsorption	direct discharge

Operating Results

The two SVE systems at TCAAP have been in operation (on an on-and-off basis) since early 1986. During the first day of operation, the amount of total VOCs removed with systems D and G was approximately 1,200 lb and 4,600 lb, respectively. Site D blowers were run for 37 days and removed a total of approximately 20,000 lb VOCs. System G was run for 14 days and removed 24,000 lb VOCs. Initially, no air emissions control had been chosen for either system, with the extracted gas being discharged directly into the air without treatment. Because of the high VOC removal rates experienced, however, both systems were shut down while evaluating air pollution control needs and requirements. The evaluation indicated that the operation of system D can be resumed under a controlled type of air emission rate discharge. Site G operation was resumed with carbon adsorption units added to the air discharge. Presently the spent carbon is removed and trucked to an offsite facility for re-activation.

Ambient air monitoring has revealed no air quality deterioration resulting from the operation of the VES systems. VOC removal rates obtained with the systems at Sites D and G are shown in Figures 7.4 and 7.5, respectively. Plans are being developed for a soil sampling program to verify treatment effectiveness and uniformity.

Other Reported Applications and Results

Several full-scale applications of SVE systems have been reported in the literature. Terra Vac Inc., San Juan, Puerto Rico, reports on the successful use of SVE for monitoring leaks from underground storage tanks and for removal and recovery of a range of volatile compounds, including solvents, form contaminated sites. Extraciton wells as deep as 300 ft have been installed. Based on these applications, the radius of influence of the negative pressure applied to the extraction wells has been estimated to generally extend outward from the wells for a distance of 10 to 40 ft, depending on the soil conditions. The extracted gas typically passes through separator tanks, condensors, or after burners to either recover or destroy VCOs.

The U.S. Environmental Protection Agency recently completed field demonstration testing of the Terra Vac in situ vacuum extraction process at a site in Groveland, Massachusetts. This site has been contaminated mainly by TCE, which was used as a degreasing solvent by a machine shop that still operates at the site. Results of the demonstration test indicated that site remediation can be accomplished with this technology. Shallow soil gas concentrations were reduced by 95%. Whellhead VOC concentrations showed a decline with time, which was controllable. Soil concentration of VOCs were reduced substantially

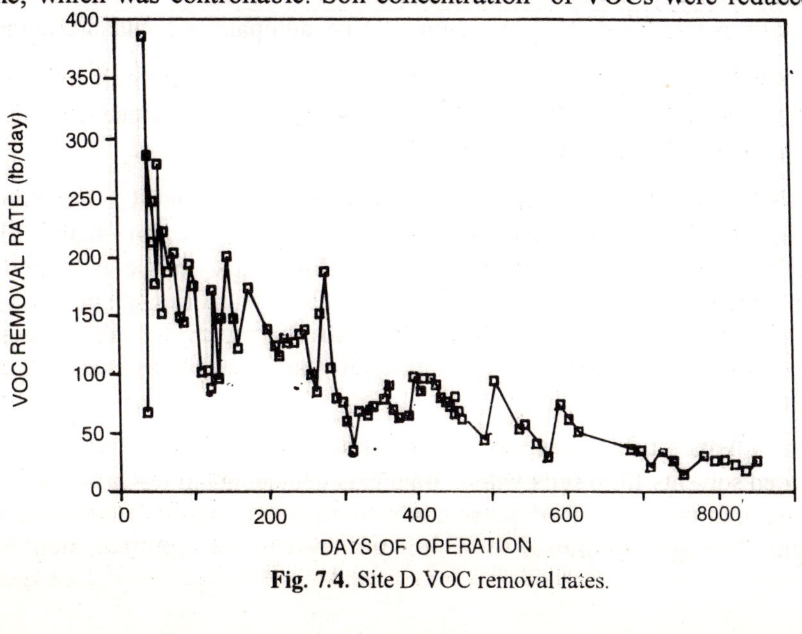

Fig. 7.4. Site D VOC removal rates.

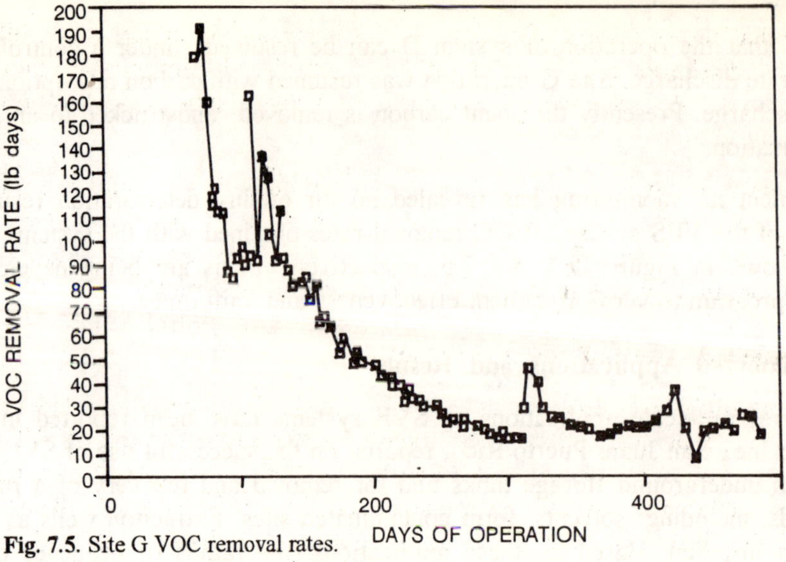

Fig. 7.5. Site G VOC removal rates.

and evidence pointed to the fact that the achievement of the target remediation levels is only a matter of time.

Midwest Water Resources, Inc. (now part of Envirosafe Services, Inc., Valley Forge, Pennsylvania) markes its patented Vaportech process for VOC removal from soil, primarily at gas stations. The Vaportech system reportedly differs from other soil vacuuming systems by using injection wells (typically located at the clean perimeter of the site) and some form of nearly impervious surface cover such as a layer of compacted clay. The surface cover is believed to promote horizontal flow pathways and provides for a confined soil atmosphere, there by increasing the effective radius of influence of an extraciton well. The Vaportech system has been reportedly used in remedical projects involving several hundred thousand cubic yards of contaminated soil. Three case studies involving soil contamination with perchloroethylene, trichloroethylene, and paint solvents are described.

Hoag has reported on the use of SVE for cleanup of soils in California and Connecticut that were contaminated with gasoline, petane, n-naphthalene, methylene chloride, methyl ethyl ketone, tetrachloroethylene, and 1,1,1-trichloroethane.

In one SVE application involving soils contaminated with spent 1,1,1-trichloroethane, more than three years of operation has resulted in the removal of more than 12,000 lb of VOCs. During this time, the concentration of trichloroethane in the extracted gas has decreased from more than 2,000 ppm to about 50 ppm. The soils at this site are predominantly alluvial clayey silts and sands that are considered relatively impervious. With the particular well design used, a vacuum of about 3 in. Hg generally sufficient to induce 100 cfm of air through the soil.

Morrow and Vanderpoel described a case study using a high efficiency blower to remove chlorinated solvents from soils with a high clay content and a low permeability. Initial pilot tests showed that the zone of influence can be as high as a ten-foot radius at 16 inches of water vacuum. Average withdrawal for a four-well system was approximately 5.2 lb/day of TCE. Based on the pilots tests, a 156-well system was designed to be implemented at the site.

Regalbuto et al. have reported on the use of the Vaportech system at a site where a catastrophic fire had destroyed a large paint warehouse storing an estimated 1.5 million galons of paints, paint thinner, and associated products. The site is located within one of the municipal well fields serving a large city in Ohio. VOCs were detected in city wells a few days after the fire began. The 6-acre contaminated area was divided into four cells based on the air flow rate which could be provided by a Vaportech pumping unit. Over the entire area, the vacuum withdrawal extended to the depth of the groundwater (about 40 to 45 ft), for a total volume of about 425,000 cubic yards. After 56 weeks of operation, over 8,000 lb of VOCs had been recovered from the site, with most of the recovery occurring during the first three months of operation.

Costs

The cost of SVE systems varies with the particular design, which in turn reflects site characteristics, cleanup objectives, and emissions requirements. The cost of soil cleanup with SVE is estimated at $20 to $35 per cubic meter of soil treated. This cost can increase somewhat depending on the VOC air emission control required.

In general a complete blower assembly for a SVE system can be purchased for less than $5,000. However, this may represents only a small fraction of the total project cost which typically includes costs associated with the initial engineering and subsurface site characterization studies; permitting; installation of extraction wells and system appurtenances such as pipes, valves, and monitors; emissions control equipment; and performance monitoring and data acquisition systems. The cost for sampling and analysis of the extracted gas to monitor system performance is the major element of the operating cost. Except for routine data recording, inspection, and servicing, SVE systems will normally operate continuously and unattended.

The following cost information has been reported for the SVE systems at TCAAP. The capital costs of the installation at Site D and at Site G (which includes carbon adsorption) were $167,000 and $470,000, respectively. Operating costs for site D have varied form $2,000 to $3,000 per month. Operating costs for Site G with carbon regeneration are about $10,000 per month. During the first 300 days of operation the removal at both sites has been in excess of 160,000 lb of VOCs. For site D, without carbon, the total costs are estimated at $243,000. With the removal of 87,000 lb of VOCs this is $2.80 per lb of VOC removed. For site G, the capital costs of $470,000, and an additional $194,000 operating costs for the time of operation (300 days) would bring the cost per lb of VOC removed to approximately $8. Figures 7.6 and 7.7 show the removal costs versus amounts of VOCs removed.

Based on the pilot tests, Morrow and Van Derpoel estimated the cost of a proposed 156- well VES system at approximately $100,000 which is significantly less than the $8 to $12 million estimated for excavation, transportation, and incineration of that volume of soil. As described previously, this system uses high efficiency blowers to remove chlorinated solvents from soils at a site with a high clay content and low permeability.

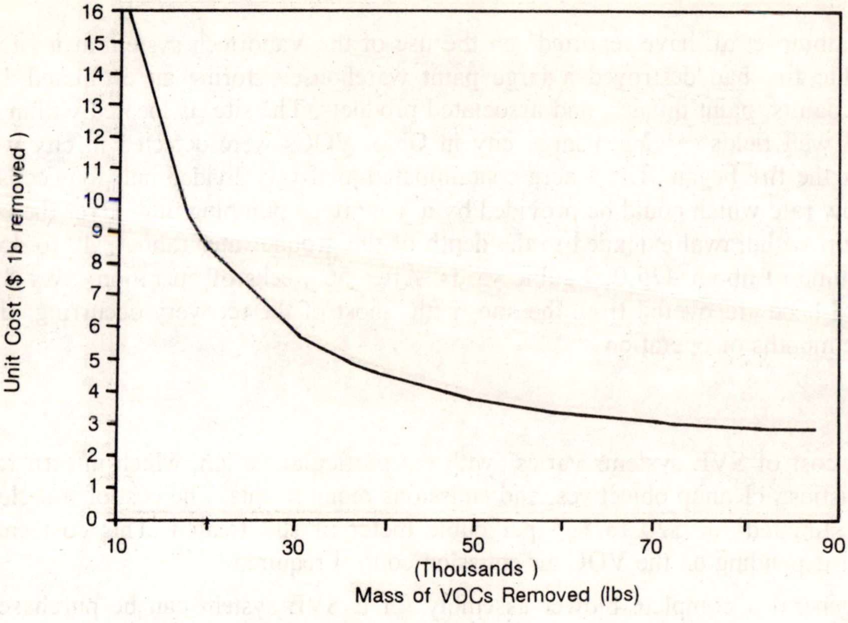

Fig. 7.6. Site D VOC removal costs.

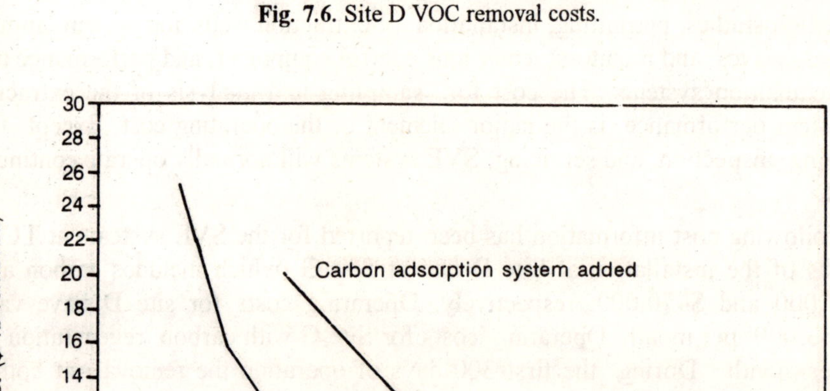

Fig. 7.7. Site G VOC removal costs.

Regalbuto et al. reported a final cost of less than $10 per cubic yard of soil for site cleanup with the Vaportech VES system at a 6-acre site in Ohio containing 425,000 cubic yards of VOC-contaminated soil.

Hoag estimates soil decontamination costs at a minimum of $30,000 per acre, and prices a typical service station spill cleanup at $40,000 to $80,000.

Outlook and Information Needs

SVE has been proven to be an effective and economic alternative to other methods for the cleanup of VOC-contaminated soils. Despite a relatively wide-use of the technology, little information is available on the design and performance of many of the operating systems. A number of studies are currently in progress or planned which should improve the data base currently available to aid in proper design of new systems. Under the auspices of the Site program EPA recently completed demonstration tests of the Terra Vac, Inc. VES system at the Groveland Wells Superfund site in Groveland, Massachusetts. The first complete report on the operation of the system is to be issued in 1989. The U.S. Air Force, working with the Department of Energy's Oak Ridge National laboratory in Oak Ridge, Tennessee, is working on the application of in situ soil venting for cleaning up jet fuel, gasoline, TCE and other hazardous, volatile liquids spilled in porous soil. The project is expected to lead to the development of a "how-to" manual for cleaning up a site and estimating the cleanup costs . The American Petroleum Institute is sponsoring a demonstration of the soil decontamination via vacuum extraction method.

The TCAAP pilot -scale field demonstration project described above was not designed as a parametric study to determine the benefits of air injection. The observation that the VOC removal remained the same whether or not air was mechanically injected into the subsurface was an incidental one. On the other hand, supplementary air injection is a feature of the Vaportech VES system that has been successfully used in a number of installations. To properly assess the role of air injection and the effectiveness of SVE, the current data base must be supplemented with data from fundamental studies of air movement and VOC striping in porous media and from additional pilot-scale studies conducted under a range of field conditions. Fundamental studies are essential for developing a good understanding of the air flow and contaminant partitioning and movement in the subsurface. One such fundamental study is currently underway at Colorado State University (Fort Collins, Colorado). Through a contract with the Argonne National Laboratory, the Colorado State University is carrying out an air-flow-through -soil-media modeling program for Usathama in support of the operation of the SVE systems at TCAAP.

Steam Stripping

Technology Principle and Technical Considerations

Steam stripping uses heat in the form of steam to enhance volatilization and stripping of VOCs. The effectiveness of steam for removing VOCs from contaminated soils has been demonstrated in both laboratory and pilot scale field tests. Steam can be delivered to the subsurface through injection wells or by use of special drilling and subsurface injection equipment. The excess steam (i.e., the steam that is not condensed in the soil) can then be collected at the surface or in a network of extraction wells for treatment and disposal/ recycling. Heidemij Uitvoering BV of the Netherlands has reportedly employed a well-point system for injection and recovery of steam at several sites. Steam is injected at 150-200°c. Volatile organic contaminants are extracted under vacuum. It is reported that it takes about a few months for the ground to be heated to initiate steam stripping of contaminants; the actual cleanup takes a few additional months.

Although no detailed technical and engineering data are available on the Heidemij method of steam stripping, purely technical considerations would indicate difficulties with the use of a conventional well-point system for steam injection and recovery in a field situation. These difficulties can steam from several sources, including the following:

- Nonhomogeneity of the soil, which can result in channeling and nonuniform or incomplete treatment
- Slow treatment rate, especially in low permeability soils
- Difficulty of containing the steam and volatilized contaminants within the target zone and potential for spread of contamination, including through fugitive emissions to air
- Lack of a practical means for continuous monitoring at the treatment zone so that necessary adjustments can be implemented as the treatment progresses.

A very innonvative steam stripping system, which attempts to address some of these difficulties, is the "In Situ Detoxifier," a patented system for in situ injection of steam, air, and other remediation agents in liquid or powder form. The system is marketed in the U.S. by Toxic Treatments (USA) Inc. (acronym: TTUSA) of San Francisco, California. A prototype of the In Situ Detoxifier was recently field demonstrated and a production model is being field tested in a full scale site remediation project.

Description of the In Situ Detoxifier

The Detoxifier, shown in Figure 7.8, is a mobile unit and includes an aboveground system for processing and recycling of the off gas in a closed loop operation. The heart of the system is the process tower which is essentially a drilling and treatment agent dispensing system, capable of penetrating the soil/waste deposit to depths of 25 ft or more. The process tower consists of two cutter/mixer bits connected to separate, hollow kelly bars. The bits overlap and rotate in opposite directions. The rotating action provides for simultaneous cutting and mixing of the soil/waste material. Steam or other treatment agents are conveyed through the kelly bars and ejected through feed jets to the mixing area. (Orifices are used for dispensing powders or slurries.) A rectangular shroud covers the mixing area and captures the off gas which is then treated in the off gas process train and recycled to the treatment zone.

The off gas treatment train is trailer-mounted and can typically include induced draft fans for gas conveyance and recycling, scrubbers for removal of the particulates, a cooling and refrigeration system for condensing bulk of the volatiles, heating coils, activated carbon

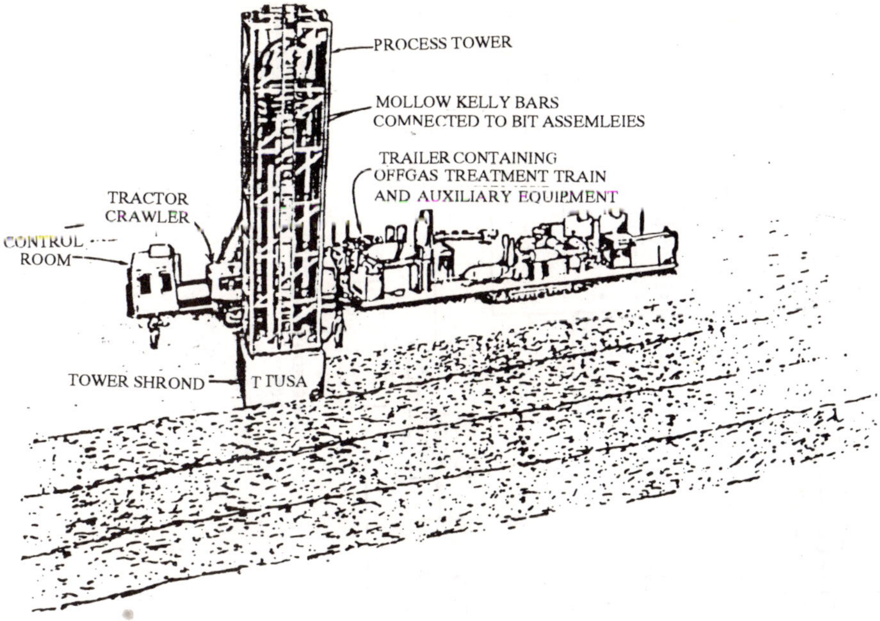

Fig. 7.8. The In Situ Detoxifier. Courtesy of Toxic Treatments USA, Inc., San Francisco, CA.

adsorption units, and gas monitoring systems. A schematic diagram of an off gas treatment system for VOC removal is shown in figure 7.9. The off gas from the shroud is monitored continuously for total hydrocarbons using a flame inonization detector. The output is used to adjust the treatment conditions, including the length of treatment, to achieve desired treatment objectives.

The treatment of an area is on a block-by-block basis with the process tower moved to an adjacent block after the treatment of a block is completed. The effective treatment area per block is about 24 ft^2.

Field Demonstration Result

A full scale field remediation application of the in situ Detoxifier was recently undertaken at a 10,000-square-foot site in southern California, using prototype equipment. The soil at this site was contaminated with petroleum hydrocarbons from leaking underground fuel storage tanks. The total petroleum hydrocarbon level in the top 25 ft of soil was generally in the 100- to 1,000-ppm range; one segment of the site had values in excess of 15,000 ppm.

A mixture of hot air and steam was used for stripping. For some blocks, an oxidizing solution was also included in the mixture. The off gas from the shroud, which contained the exit air, steam, and volatilized hydrocarbons, was cooled and passed through a scrubbing system (to remove particulates), a cooling system (to condense and remove excess moisture), an activated carbon adsorption system (to remove hydrocarbons). Following carbon adsorption,

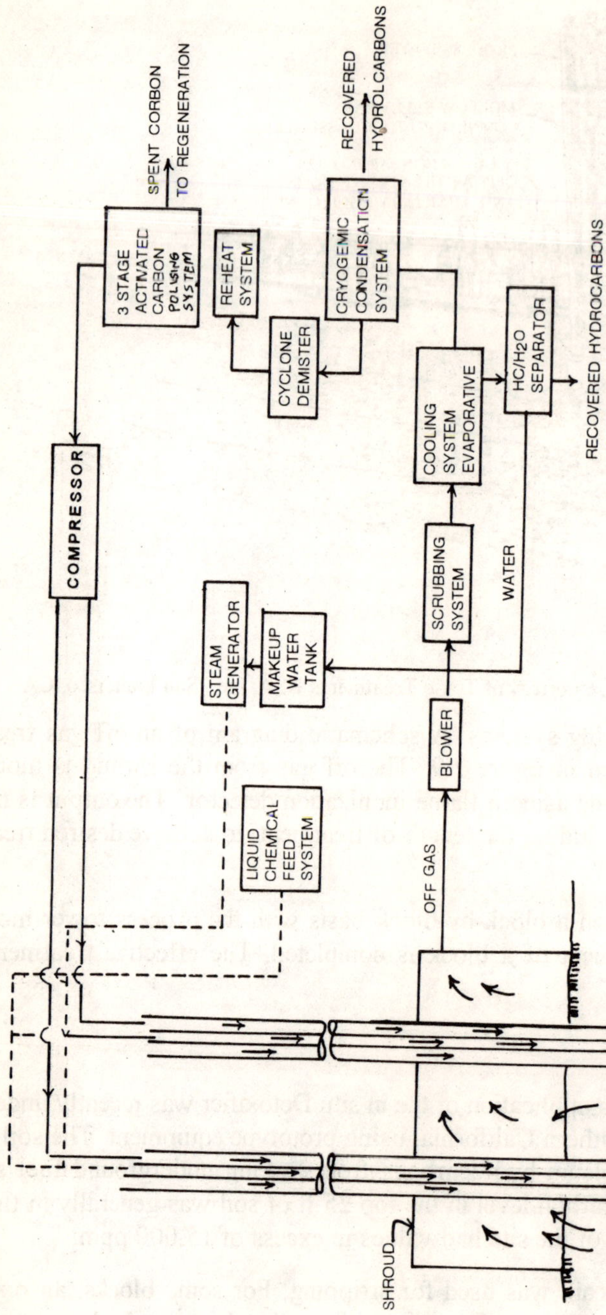

Fig. 7.9. Process flow diagram for the in Situ jDetoxifier steam stripping process.

the gas was compressed, reheated, and recycled to the treatment zone through the two kelly bars in the process tower.

The project had the following objectives:

- Demonstrate the capability of the equipment for dispensing, mixing, and recovering remediation agents in an in situ steam stripping application.
- Evaluate the adequacy of the various components of the off gas treatment train under a range of loading conditions.

The results indicated that by adjusting the treatment conditions (i.e., the amounts of air and steam used, and the rate and duration of treatment in terms of speed and number of up and down movements of the drill bit), the total petroleum hydrocarbon level in the soil could be reduced to less than 100 ppm via hot air/ steam stripping. The results also indicated that off gas monitoring via on line total hydrocarbon analysis can provide a reasonably accurate basis for assessing the completeness of in situ treatment and hence decision by the operator to move the equipment to a new block. One key problem area was the quick overloading of the carbon adsorption system when treating soils with very high levels of hydrocarbons. This necessitated frequent carbon replacement.

Production Unit Testing

Based on the demonstration project results, a production unit of the Detoxifier which is more compact than the prototype model and incorporates several design improvements has been fabricated. the first full-scale site remediation tests using this production unit are currently under way at a former bulk chemical transfer and storage facility site in San Pedro, California. The soil in certain areas at this site is contaminated with chlorinated VOCs, with total VOC levels ranging from less than 100 to more than several thousands mg/kg. The remediation at this site will consist of treating some 2,000 blocks to a target depth of about 6 ft. Baseline calibration involving remediation of some 85 test blocks is currently in progress.

The Detoxifier in situ steam/air stripping technology is being demonstrated under the EPA site program.

Bioremediation

When the contamination in the subsurface is limited to the soil in the unsaturated zone, the need for oxygen (for aerobic bacteria) and water restricts the use of in situ bioremediation to near surface contamination. In this application, the contaminated soil at or near the surface is manipulated by addition of supplementary nutrients and special bacterial cultures (if necessary) and by periodic mixing to enhance the biodegradation rate. Many case studies of reclamation of lands contaminated with petroleum wastes and products using bioremediation have been reported in the literature. Use of the surface soil as a medium for treatment of wastes (a technique known as landfarming or land treatment) is an established technology which has been widely used in the petroleum industry.

When both soil and groundwater are contaminated or when the contaminants in the soil can be flushed into the groundwater for subsequent treatment, bioremediation of soil and

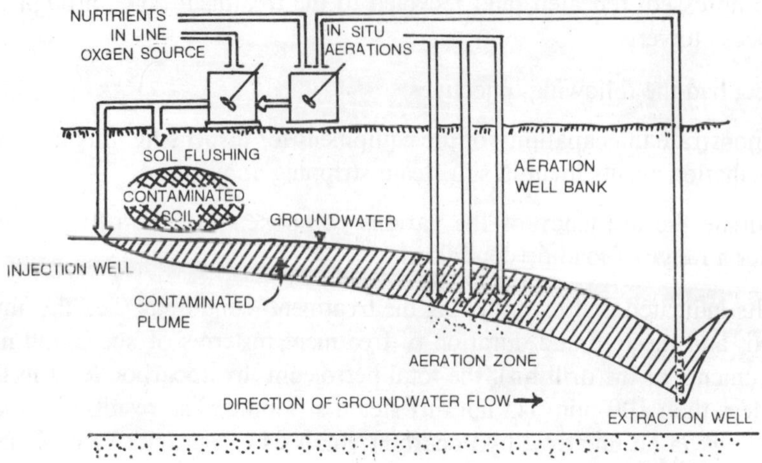

Fig. 7.10. Bioremediation of contaminated soil and groundwater.

groundwater can be carried out simultaneously by recirculating the groundwater using a system of injection/percolation and extraction wells (see Figure 7.10). This is a recent application of the in situ bioremediation and there have been a few small-scale field applications of the technology, including some recently completed and ongoing demonstration projects. The discussion of the in situ bioremediation presented here is for a combined system for the treatment of both soil and groundwater.

Operating Principle

In situ bioremediation relies on the concept of stimulating the indigenous subsurface microbial population, or special cultures of microorganisms that would be introduced into medium, to degrade the organic contaminants to inert or less harmful end products . Although biodegradation can take place in both aerobic and anaerobic environments, the aerobic systems are preferred because they are faster, more controllable, and can bring about a more complete degradation of contaminants. Provided that the contaminants are biodegradable and not present at toxic levels, and that proper conditions exists or can be developed in the subsurface to promote biological activity, biological treatment can be a potentially effective, inexpensive, and safe method for soil and groundwater decontamination. Because the in situ method treats contaminants both in the soil and in the groundwater, aquifer restoration can be much faster with in situ biological treatment than with surface processing methods.

Enhancing the biological activity in the subsurface may require pH adjustment and/or addition of supplemental nutrients such as nitrogen, phosphorous, and trace metals, and sometimes organic carbon. The optimum treatment conditions and nutrient requirements can be established through bench-scale studies prior to implementing a field program. In general pH values between 6.5 and 8.5, temperatures between 85°F and 95°F and a ratio of organic carbon to available nitrogen and phosphorus (i.e., C:N:P) of 300:15:1 are considered suitable for biological activity. Considerable work has been done and is currently in progress to develop bacterial strains and enzymes most suitable for degrading specific contaminants.

Detox Industries (Houston, Texas) has reportedly selected and bred a bank of 200 naturally occurring, nonpathogenic soil microorganisms for particular waste destroying capabilities, including degradation of PCBs, pentachlorophenol, and creosote.

Oxygen Sources and Delivery Methods

In most subsurface contamination cases involving simple hydrocarbons, oxygen transport is almost always the rate-limiting step in the biodegradation. Thus ability to deliver the required amount of oxygen to the subsurface is essential for effective biological treatment. Oxygen can be supplied to the subsurface via air sparging in aeration well banks (see Figure 7.10), aeration in surface facilities, or addition of hydrogen peroxide. In certain situations, in situ treatment can be used in conjunction with and to supplement groundwater treatment for VOC removal via air stripping in an aeration tower. In such an application, the aeration tower serves the dual purpose of removing the volatile organic compounds from the water and aerating the water to achieve oxygen saturation before the water is returned to the subsurface.

The amount of oxygen that can be introduced by air sparging or aeration in an aeration tower is limited to the solubility of air at the temperature of the ground or the recirculation system, typically corresponding to 8 to 10ppm of oxygen. This amount is frequently inadequate to sustain rapid and continued biodegradation. The oxygen supply limitation can be eliminated when hydrogen peroxide is used as the oxygen source. Since breakdown of 1 ppm of hydrogen peroxide releases about 0.5 ppm of oxygen, and the hydrogen peroxide is added to the water in concentrations of up to 1,000 ppm, a significantly greater amount of oxygen can potentially be delivered to the subsurface via hydrogen peroxide addition than the use of air spargers or other mechanical means that can only deliver 8 to 10 ppm of oxygen. In a demonstration program of in situ bioremediation of a jet fuel contaminated aquifer at Eglin Air Force Base (AFB) in Florida considerable peroxide instability was noted which limited the amount of oxygen which could actually be delivered to the water via hydrogen peroxide addition. At a 500-ppm level, the observed half-life for hydrogen peroxide was in the 30 to 90 minutes range. Catalase enzymes released by aerobic bacteria near the points of injection, humic materials, and inorganics such as iron are suspected as causes of rapid peroxide decomposition and oxygen off gassing. In other studies, hydrogen peroxide addition has been shown to increase bacteria activity by several orders of magnitude over air sparging.

When hydrogen peroxide is used as the oxygen source, control of hydrogen peroxide addition is very important as too much hydrogen peroxide can exert toxic effects on the microorganisms and cause precipitation and “blinding” in the iron-rich soils. Blinding, which can result in channeling and inadequate and nonuniform treatment, can also result from chemical precipitation involving nutrient components (e.g., precipitation of phosphate as calcium phosphate) and this precipitation can also reduce the availability of the nutrients to the microorganisms. Several technology vendors market proprietary nutrient formulations designed to reduce precipitate formation and retain nutrient availability.

The possibility of using nitrate as an oxygen source (electron acceptor) in bioremediation of gasoline-contaminated groundwaters via denitrification has been investigated in laboratory and field experiments. No direct evidence, however, was generated to indicate that denitrification was indeed the primary biotransformation route responsible for the observed contaminant removal.

Reported Applications and Results

Spill Cleanup Case Studies

Much of the data reported in the literature on use of in situ bioremediation are from private projects and are provided by the technology vendors. The technology has not been used in major site remediation applications and an adequate data base on the technology does not currently exist. Vendor-reported information indicates success in several applications for cleanup of spill-impacted soil and groundwater involving single contaminants (e.g., spills of specific industrial solvents) or hydrocarbon mixtures such as gasoline. These applications have also generally involved small areas or zones of subsurface contamination. These cases include contamination with methylene chloride, ethylene glycol, isopropanol, n-butyl alcohol, dimethyl aniline, phenolic compounds, acetone, tetrahydrofuran, and petroleum-based hydrocarbons. The following two example case studies illustrate the reported applications and the results obtained.

At a 4.5 acre industrial plant site, approximately 1,200 cu.yd. of soil in a 1.75 acre area were contaminated with methylene chloride, n-butyl alcohol, dimethylaniline, and acetone. Three years of operation of an in situ biodegradation system resulted in a 90% reduction in the size of the contaminated plume. It was estimated that a total of five years of operation will achieve complete remediation compared to 15-20 years for groundwater pumping and off-site disposal. Capital and operation and maintenance costs are documented at $926,000 and $226/day, respectively. About one half of the overall capital cost was for preliminary research and pilot plant treatabily studies.

At a site where a buried methylene chloride line had ruptured, an undetermined amount of the solvent had been released into the soil and groundwater. Following removal of 160cu.yd. of highly contaminated soil, elevated temperature air stripping had been used for the treatment of the contaminated groundwater. Air stripping at temperatures between 80°F and 120°F had resulted in lowering of the methylene chloride concentration from 9,300 mg/l to 300mg/l. Air stripping, however, was ineffective in achieving further reductions in the level of residual methylene chloride in the effluent. After conducting treatability studies to establish optimum conditions for biological treatment, a system was designed and placed into service for in situ biological treatment. A rapid reduction in methylene chloride was noted in the groundwater following the start of the biological treatment. Within eight days a 50% reduction was observed, and after 240 days, the methylene chloride in the monitoring wells ranged form less than 1.0 to 3.0 mg/l, a reduction of 5 orders of magnitude in groundwater concentrations.

Field Demonstrations

Although these and other reported case studies indicate the effectiveness of in situ biological treatment for small site cleanup applications involving soil and groundwater contamination with a single contaminant or category of contaminants, the viability of the technique for large-scale cleanup of hazardous waste sites has not yet been satisfactorily demonstrated and is questionable.

The first field demonstration of technology at a waste disposal site involving a complex mixture of both organic and inorganic wastes was recently completed at Kelly Air Force Base (AFB), Texas. The site originally was used as a disposal pit for chromium, sludges and other electroplating wastes. Prior to closure in 1966, it had been used as a chemical evaporation pit for chlorinated solvents, cresols, cholorobenzenes, and waste oils. The site was selected for the project because it offered a number of advantages with respect to demonstrating in situ biodegradation. These included the presence of a highly adaptive and substantial microbial population, location in an area with near optimum soil and groundwater temperatures, presence of a perched aquifer not used for drinking water, and the biodegradable nature of the organic contaminants present at the site. A small portion of the site (about 60 ft in diameter) was used for the demonstration project.

The design of the treatment system and selection of the operating conditions at the Kelly AFB site were base on results of geological investigation and site characterization, determination of the contamination profile, microbial investigation, and biodegradability studies. The system used to circulate groundwater within the treatment area consisted of nine 4-in. diameter pumping wells and four 6-in. diameter injection wells. One upgradient and two downgradient monitoring wells were also used to determine any effects of the system outside the perimeter of the treatment area . The saturated zone within the treatment area was at a depth of 15 to 20 ft below the land surface. The nutrient solutions used consisted of chloride, nitrogen, and phosphorous. Hydrogen peroxide addition was used as the method for providing oxygen to the subsurface.

The system at Kelly AFB was operated and monitored form June 1985 to February 1986. The results from the demonstration project indicated a need for additional studies and demonstration works to develop the technology as a viable method for application to large sites with a range of hydrogeological and contaminant characteristics. In the demonstration project, the total concentration of PCE and TCE in site groundwater decreased from 4.0 mg/l to 0.93 mg/l, while the concentration of trans-1,2, -dichloroethylene (a decomposition product of PCE and TCE) increased from 0.03 ppm to 1.4 ppm. Mobilization of heavy metals (antimony, lead, arsenic, cadmium, silver, and thallium) from soil was indicated but was not detected in the groundwater. However, heavy metals were found in sediments deposited in several surface pipe. The introduction of nutrients and hydrogen peroxide to the system resulted in an almost immediate and significant decrease in permeability and heavy precipitation of calcium phosphate.

Henchee and Downey have provided interim results on a demonstration of in situ bioremediation of a jet fuel contaminated aquifer at Eglin AFB in Florida. The relative effectiveness of three systems (injection wells, infiltration galleries, and spray irrigation) for oxygen and nutrient injection were evaluated. The infiltration galleries and injection wells have shown a tendency to clog, with a resultant reduction in the flow-through. Regular treatment with acid and vigorous scrubbing have been used to clean both recovery and injection well screens. In the case of infiltration galleries the problem has been remedied by increasing their number. The spray irrigation system has generally performed well. Occasional rototilling of the soil improved permeability and reduced overflow.

Process Cost

The cost for in situ Bioremediation is expected to be highly dependent on site-specific conditions which establish the basis for design and system operation. Site characterization and treatability studies, which precede system design, and subsequent monitoring of the operation can constitute significant cost elements.

The cost for treating the Kelly AFB site were $36 per ton of contaminated soil, with analytical cost accounting for 25% of this cost. Based on results form the demonstration project, it was concluded that in situ treatment can be performed for $50 to $100 per ton of contaminated soil for a typical site, and this would be less than the typical cost for removal and redisposal of soils.

Results form the Eglin AFB demonstration project indicate that unless in situ peroxide stability is greatly improved, hydrogen peroxide addition may not be cost competitive with other methods of supplying dissolved oxygen (e.g., use of industrial grade liquid oxygen) for in situ bioremediation.

Soil Washing

Operating Principle

Techniques such as those used in solution mining and mineral extraction have been proposed for removal of contaminants form in-place soil. Extractant fluids such as water or solutions containing surfactants or chelating agents can be delivered to the subsurface via gravity, using systems such as flooding, surface seepage, and ponding, or via forced systems using injection wells or specialized equipment. (One example of such specialized equipment is the In Situ Detoxifier described earlier in connection with steam stripping.) The spent solution containing the flushed contaminants as well as any excess treatment agents can then be collected for treatment, recycling, or disposal. The collection of the spent solution can be via gravity (e.g., using porous drains, trenches, and open ditches) or forced systems such as well points. Where the contaminants in the soil have already seeped into the underlying groundwater necessitating groundwater restoration, and under favourable site geohydrology. Soil washing can be used to acceperate contaminant migration through the vadose zone and into the groundwater, which can then be treated in situ or in aboveground treatment units.

Status of Development and Potential Applications

To date, in situ soil washing has been primarily used in laboratory and small-scale field pilot tests. Results indicate that while the general concept of in situ soil washing remains a valid one, a large number of problems can be anticipated in a field application. The most critical problems stem from the nonhomogeneity of the soil which prevents uniform distribution of the Extractant in the subsurface, penetration of clumps, and adequate contact with fine particles which typically contain most of the target contaminants. The results are incomplete and non-uniform treatment, and slow processing rate. Other problems with in situ soil washing relate to the difficulty of monitoring and controlling the in situ treatment, potential for the spread of contamination to clean areas through redepositon of extracted contaminants, potential for groundwater contamination, loss of the Extractant is side reactions, less than quantitative recovery of the excess chemicals, and generation of a large volume of a dilute waste requiring treatment for material recovery prior to recycling or disposal.

Soil washing would probably find its greatest site remediation application in situations where the soil is contaminated with a single type or class of contaminant(s) that is amenable to extrication with a "manageable" volume of a single extractant fluid which would then be processed for material recovery prior to reuse/disposal. Example applications would involve soils that are contaminated with heavy metals (e.g., lead-contaminated soils at sites formerly used to recover/recycle lead form batteries) or with petroleum hydrocarbons (e.g., at fuel spill sites).

Soil washing has been considered by EPA for possible use for cleanup of a number of former battery dumping/processing sites. Laboratory and field soil washing/flushing tests have been conducted on the contaminated soils from certain of these sites to determine technology applicability . At a battery dumping pit in Leads, Alabama, The leaching of lead-contaminated soil (47,000 ppm lead) with a 2% aqueous solution of EDTA resulted in a residual lead level of 80 ppm in the soil. Despite this high extraction efficiency, the soil washing was considered unacceptable as the treated soil did not meet the target cleanup goal of 5 ppm residual lead in the soil. Soil washing for lead removal at this site was conducted in a closed loop operation; the extracted lead was removed by sulfide precipitation and the regenerated EDTA solution was recycled. Soil washing was also considered for cleanup of lead contaminated soils at Lee's Farm, a former battery manufacturing site in wisconsin. Sulfide precipitation and electrolysis were considered feasible for removal of lead form the EDTA extraction solution.

Based on laboratory and field experiments, soil washing in general would probably be of limited applicability in site cleanup service where a range of contaminants and soil characteristics are encountered. Test results indicate that while certain contaminants such as simple phenols can be readily extracted from both organic and inorganic soils, other contaminants such as PCBs or arsenic cling tenaciously to the soils and are released less readily into the washing fluids.

Reported Field Application

In situ soil washing technique has reportedly been successfully used to clean up cadmium contaminated soil at a site in the Netherlands. The treatment site contains 30,000 m^3 of soil within an area of 6,000 m^2 with an estimated total cadmium content of 725 kg. Cadmium is disrobed form the soil by leaching with a 10^{-3} molar solution of hydrochloric acid (pH=3.5). Wells and drains were installed to establish a system for infiltration and withdrawal. A cross section of the system is shown in Figure 7.11. The cadmium-containing spent extractant is treated on-site via ion exchange to remove the cadmium and to regenerate the acidic extractant for recycling.

Operationally, the site was divided into five compartments. The cleanup of the first compartment was completed within a 5-month period. Soil samples collected after three months of operation showed cadmium concentrations of 1 mg/kg or less (dry weight basis); the cadmium concentration in the extractant was 10 ug/l. Acidification of the extractant was stopped and neutralization of the soil with NaOH (pH 8.5) was initiated. When the cadmium concentration in the extractant from every drain was less than the detection limit (10ug/l), neutralization was stopped and the operation moved to the next compartment.

Related Experiencc from Aboveground Doil Washing

Because of the many potential problems with large-scale field application of the in situ soil washing method (including the very slow processing rate), research, development, demonstration, and commercial use of the soil washing technology have emphasized treatment of the excavated soil in aboveground facilities and development of systems and equipment for soil handling and extractant processing. The aboveground systems offer the advantages of higher throughputs and more effective process control. Most of the aboveground systems rely on size classification to achieve volume reduction before soil cleanup. Thus, physical separation and removal of fines, which usually contain bulk of the contaminants but account for a small fraction of the total volume, reduces the load on the scrubbing/washing systems and produces a small volume of "concentrated" stream which can be treated /disposed of by means other than soil washing (e.g., via incineration in cases involving soil contamination with organics).

Several fixed, mobile, and transportable aboveground soil washing installations ranging in rated throughput from 11 to 30 tons/hr are reportedly operating in the Netherlands and the Federal Republic of Germany. Particle sizing is typically the first step in these operations with the goal of removing the fines (<15-100 um), which are typically disposed of as a sludge by-product, while the coarser fractions are cleaned via washing/scrubbing with detergents, water, or other extractants and discharged as clean soil. The spent extractant fluid is treated before recycling or disposal, Equipment used for size fractionation, scrubbing, and solids-liquid separation typically includes dry and wet sieving devices, mixers, hydrocyclones,froth flotation systems, sedimentation tanks, sand and carbon filters, and filter presses (for sludge dewatering). The Harbauer soil washer, which is considered among the best soil washers in Germany, uses a vibratory system that can achieve soil cleanup for particles down to 15 microns in size.

Process Cost

The total project cost for the cleanup of the 30,000 m^3 of cadmium-contaminated soil at a site in the Netherlands cited above has been estimated at \$2.5 million or \$83/m^3.

The operating costs reported for the aboveground soil washing installations in Europe range form \$53 to \$155 per ton, excluding the cost of sludge disposal which can be as high as \$182 per ton of sludge.

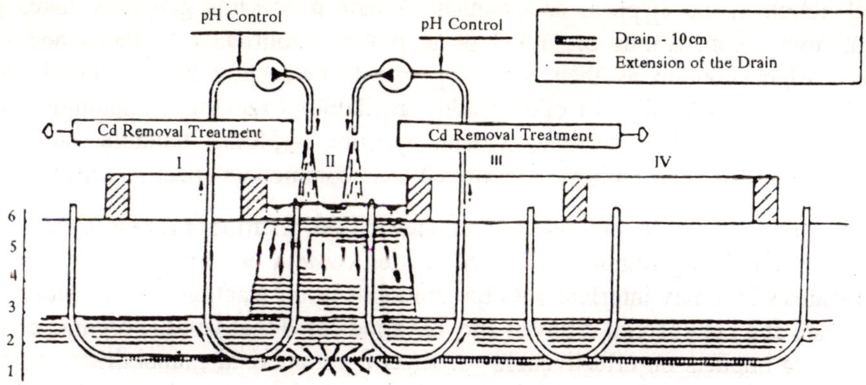

Fig. 7.11. Cross section of an in situ soil washing system for cadmium removal.

Solidifications/Stabilization

Steam stripping, vacuum extraction, and soil washing discussed above involved removal of the contaminants form the target zone. Solidification/stabilization, on the other hand, does not remove the contaminants but rather renders them less mobile (i.e. less leachable) or converts them to inert or less harmful products. This is done by adding to and mixing with the soil certain physical or chemical agents that "fix/stabilize" the contaminants in place. Vitrification, which is a form of in-place stabilization and involves application of thermal energy to convert the contaminated soil into a vitrified product, is discussed in the next section.

Basic Technology and General Applications

The term stabilization, fixation, solidification, and immobilization have been used interchangeably in the literature to describe processes whereby physical or chemical agents are added to and mixed with a waste (typically, sludges in pits, ponds, and lagoons) with the primary objective of improving the handling or leaching characteristics of the waste destined for land disposal. Treatment agents used have included the following:

- Precipitating agents such as lime or limestone
- Sorptive agents such as clay or agricultural products or by-products (e.g., straw, saw dust, bark)
- Solidification / encapsulation agents such as Portland cement.
- Strongly adsorptive complexing agents such as ethyenediaminetetraacetic acid (EDTA).

The application of these systems has been largely to heavy metal sludges with the objective of rendering the metals less leachable and mobile in the disposal environment. Most commercial immobilization systems do not provide complete control or a very high degree of long-term effectiveness due to composition changes that may be brought about by microbial or chemical action in the disposal environment. Waste processing generally takes place in aboveground units (e.g., drums or tanks) or in ponds. Proprietary methods and chemical formulations, which contain as their basic ingredients cement or silicate-based materials (e.g., fly ash, blast furnace slag, or other readily available pozzolanic compounds), are used. Silicate-based processes also use lime, cement, gypsum, and other suitable setting agents . The procedure and formulation are often altered to accommodate specific wastes.

Commercially available cement and silicate-based fixation processes are generally ineffective in immobilizing organics or when wastes contain certain nontarget inorganic or organic substances that may interfere with the immobilization reactions. Tests with a range of hazardous waste at a commercial disposal site have indicated that volatile organic compounds can have adverse impacts on effectiveness of several commercial immobilization processes. Several companies now market technologies which are claimed to be capable of treating wastes and soils containing high levels of organics, Hazcon, Inc., of Brookshire, Texas, markets a solidification process that uses certain proprietary compounds that, when added to cement, interact with organic and inorganic components of the waste to render them compatible with the cement. The Hazcon solidification process for high organic wastes was recently demonstrated on exeavated soils at he Douglassvile, Pennsylvania, Superfund site. The tests showed successful immobilization of metals and significant reduction in product permeability with feedstocks containing up to 25 wt% oil and grease. International waste technologies (IWT), Wichita, Kansas, markets certain proprietary chemicals (designated as "HWT" compounds) that the company claims are suitable for fixation of both organic and inorganic contaminants. According to IWT, a variety of organic and inorganic wastes have been effectively stabilized by treatment with HWT-20 formulation. For example, the results shown in Table 7.3 were obtained when treating an API separator bottoms (24.9% oil and grease content) with 15% to 25% by weight of the fixation agent.

Table 7.3
Reported Results for Waste Fixation with HWT-20 Formulation

Compound	Concentration in Waste, ppm	Leach Value
Chromium	630	0.03-0.04
Lead	250-332	0.05
Ethylbenzene	10	ND
m-Xylene	40	ND
O- and p- Xylene	43	ND
Antbxacene	19	ND
Chrysene	29	ND
Methylnaphthanlene	170-470	ND
Naphthalene	13-93	ND
Phenanthrene	110-206	ND

Note: Leach value determined using "toxicity characteristic leaching procedure" (TCLP); ND indicates not detected at a detection level of 110 ppb.

In Situ Applications

To date, the in situ application of the stabilization method has been limited to cases involving closure of liquid or slurry ponds. It has been used to impart load-bearing capacity to industrial sludges in ponds, pits, and lagoons so that the sludges can be buried in-place or removed and transported to offsite disposal sites. Large volumes of low reactivity, solid chemicals are mixed with the pond content and the product is allowed to remain in place. The present state of technology limits application of in situ mixing to the treatment of low solids content slurries or sludges.

In situ application of solidification/stabilization techniques to site remediation is a fairly new concept. Under the sponsorship of the EPA, considerable research and development has been carried out to develop and evaluate techniques for bottom sealing of hazardous waste sites. Ideally, bottom sealing can be achieved by injecting or insering and inert impermeable and continuous horizontal barrier in soil below the source of contamination. Bottom sealing could be used in conjunction with other containment techniques such as slurry walls, caping, and counterpumping to insure that the contaminants do not move from the site into surrounding soil or groundwater. The EPA-sponsored bottom sealing effort has centered on the evaluation of various grouting formulations from the standpoint of non-reactivity and resistance to hazardous wastes and the determination of the effectiveness of direct injection or jet grouting techniques in forming a grout barrier. Grout formulations tested have included acrylate, 30% silicate, urethane, and portland cement. Tests in various soil media indicated that the present designs do not permit close enough control to assure a bottom seal to be formed in all media.

The Oak Ridge National Laboratory is demonstrating an "in situ grouting" technique where a grout consisting of a water slurry of portland cement, fly ash, and clay is injected into shallow trenches of buried low level radioactive waste.

Demonstration in the Site Program

The effectiveness of in situ stabilization in bulk processing of contaminated soil or buried wastes would depend to a large extent on availability of equipment and procedures capable of delivering, distributing, and mixing the treatment agents with the soil in the subsurface, all at a reasonably fast rate. Several systems has been developed or proposed for this purpose. One such system is the In Situ Detoxifier described earlier in connection with in situ steam stripping. Another system is the Geo-Con/Deep Soil Mixing system which was recently demonstrated under the EPA Site program at a PCB-contaminated site in Hialeah, Florida. The equipment and procedures used in the demonstration prograin have been described by EPA as follows:

The Geo-Con/Deep Soil Mixing system was used for in situ slurry injection and mixing. The system consisted of a drilling auger containing one set of cutting blades and two sets of mixing blades attached to a vertical drive shaft. The blades were rotated at approximately 15 rpm. Two conduits were constructed within the drive rod. Injection ports were provided at the bottom of the shaft so that the additive slurry (containing IWT fixation chemicals) and supplemental water could be injected into the zone being agitated by the rotating blades. To create a vertical column of treated soil, the auger was advanced to the desired maximum depth of the treatment (14 and 18 ft). The additive was injected in slurry form on this downstroke and mixed into the background soil and limestone. Additional mixing was provided during auger withdrawal. Columns were positioned to overlap. The diameter of the treated soil column was 36 inches and a total of 72 columns were constructed in the demonstration program. Soil sampling was performed before and after test remediation with the objective of evaluating mobility of contaminants and uniformity of solidified soil mass. The data obtained are under evaluation at EPA.

Process cost

Because in situ stabilization has not been used in full-scale site remediation applications, a data base does not currently exist to develop realistic cost estimates. Cost data reported for stabilization of industrial wastes indicate significant dependence on the nature of waste, which determines the type and dosage of stabilization chemicals required, and the mixing method. Based on results from the Site demonstration program, the cost of stabilization of excavated soil using the Hazcon solidification process is estimated to range from $90 to $200 per ton of soil.

Vitrification

In situ vitrification (ISV) is a thermal treatment and destruction process that achieves stabilization by converting contaminated soil and inorganic wastes into chemically inert, stable glass and crystalline products, resembling obsidian. The process was originally developed by Battelle, Northwest Laboratories (Richland, Washington), for the U.S. Department of Energy (DOE) for management of radioactive wastes. Battelle was granted a license by DOE to apply the technology to the management of nonradioactive hazardous wastes. Battelle in turn has established Geosafe Corperation (Kirkland, Washington) for worldwide commercial application of the technology. A mobile system is available for testing and demonstration of site remediation applicability. The first commercial unit capable of processing 1,000 tons of contaminated soil per setup has been constructed.

Process Description

The ISV process is illustrated in Figure 7.12. The process principle and sequence are described by Battelle as follows. A square array of four electrodes is inserted into the ground to the desired treatment depth. Because the soil is not electrically conductive once the moisture has been driven off, a conductive mixture of flaked graphite and glass frit is placed among the electrodes to act as the starter path. An electrical potential is applied to the electrodes to establish an electric current in the starter path. The resultant power heats the starter path and surrounding soil up to 3,600 °F, well above the initial melting temperature or fusion temperature of soils (normally between 2,000 and 2,500°F). The graphite starter path is eventually consumed by oxidation, and the current is transferred to the molten soil, which is now electrically conductive. As the vitrified zone grows, it incorporates nonvolatile elements and destroys organic components by pyrolysis. The pyrolysis products migrate to the surface of the vitrified zone, where they combust in the presence of oxygen. A hood

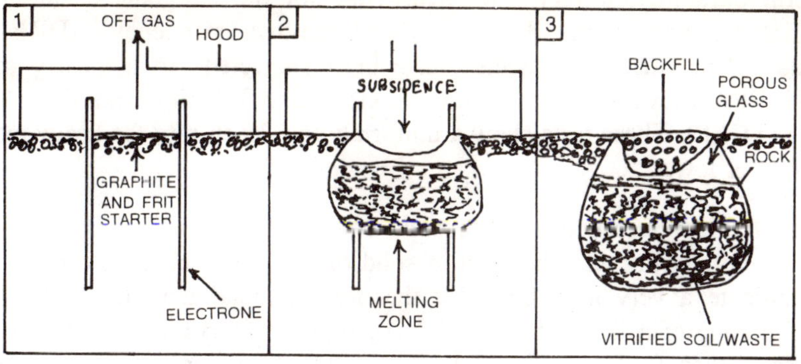

Fig. 7.12. In situ vitrification process sequence

placed over the processing area provides for collection of the combustion gases which are drawn into the off gas treatment system.

Process Equipment, Field Setup, and Operation

ISV process equipment and field setup and operation are described by Battelle as follows . Except for the off gas hood and the connecting lines, which would be transported to the site on flatbed trailers and assembled at the job site, all components are contained in three transportable trailers. These are the off gas treatment, process support, and process control trailers. The process off gas is cooled, scrubbed, and filtered in the off gas treatment trailer before discharge to the atmosphere. The process support trailer contains the glycol cooling system that interfaces with the scrub solution in the off gas treatment trailer and extracts and rejects to the atmosphere the heat removed from the process off gas. The monitoring and process control and done from the process control trailer.

Site cleanup is on a block-by-block basis. The three processing trailers are typically coupled together and moved form one block to another as a unit. The hood is moved form one position to the next with a crane. Moving from one position to another takes about 16 hours, which can also allow time for routine equipment maintenace. A large-scale ISV is capable of processing an average of 3 to 5 tons of soil per hour. On the average, processing at one processing position would be completed in about 150 to 200 hours, depending on the treatment depth and the electrode spacing. For processing of depths of 50 ft, single processing operations can run in the range of 300 to 400 hr and result in a vitrified mass of greater than 1,000 tons.

Evaluation Results

Much of the currently available information on ISV process capability has been obtained in tests with actual or simulated radioactive wastes. Battelle reports that it has conducted over 40 bench-, engineering-, pilot-, and large-scale tests in support of its ISV technology development effort. The large scale tests have used an electrode spacing of 11.5 to 18 ft and produced a vitrified mass of 400 to 800 tons per setting. According to Battelle, these tests indicate outstanding product characteristics from the standpoint of leaching and long-term durability, a very low release rate for the inorganic contaminants during processing, and a very high destruction and removal efficiency (DRE) for organic contaminants . The vitrified products have been subjected to a variety of leach tests, including the EPA's Extraction Procedure Toxicity (EP Toxicity) and Toxic Characteristic Leach Tests (TCLP). These tests show a uniformly low leach rate for heavy metals of about 110^{-4} lb/ft²/day or lower.

Battelle has conducted tests with soils contaminated with PCBs and electroplating sludges, or containing buried containers of solid or liquid combustible organics. These tests reportedly indicate a very high DRE for the organics, and that the DRE can be further enhanced by adding of a layer of clean soil. Tests with PCB-contaminated soils showed a DRE of greater than 99.5% in the melt. The small release in the process off gas (about 0.05%) was effectively retained in the off gas treatment system (a dual activated carbon filter), yielding an overall system DRE of >99.9999 %. Limited amounts of PCBs (0 to 0.7 ppm) were detected in the surrounding soil and none were found in the vitrified block , indicating that migration away from the vitrification zone would not be a significant concern.

Engineering analyses by Battelle indicate that, except for the impact on cost, the ISV process can be implemented at sites with high groundwater table, as long as the rate of recharge does not exceed the vitrification rate. The melt typically proceeds at a rate of about 3 to 6 in./hr. Thus, soils with permeabilities in the range of 10^{-5} to 10^{-9} cm/sec can be vitrified even in the presence of groundwater or in the water table, without lowering the water table or installing underground barriers.

A demonstration of ISV technology for the stabilization of radioactively contaminated soil sites at Oak Ridge National Laboratory (ORNL) was successfully completed in July 1987. This demonstration was the first application of the process not performed at the Hansford Site, where the technology was originally developed. The pilot-scale demonstration was performed on a 3/8 scale trench (2 m deep × 1 m wide × 10 m long) that was constructed to simulate a typical seepage trench used for low-level radioactive waste disposal at ORNL from 1951 to 1966. In the ORNL demonstration, a 25,000-kg mass of melted rock approximately 1.2 m thick × 2.1 m wide × 4.9 m long was formed during 110 hours of operation which consumed approximately 29 MWh of power. Data being obtained on the operational performance of the tests and product durability will be used to assess the feasibility of applying technology to an actual waste trench.

Process Costs

Detailed cost analyses have been carried out by Battelle for the ISV process. The amount of moisture in the soil and the cost of electricity have been identified as the two factors that most significantly affect the total cost. The cost of ISV of large-scale site cleanup involving contaminated soils is estimated at $100 to $250 per ton.

Costs have been calculated for treating saturated and dry contaminated soils, and a wet industrial sludge, as a function of the local electric rate. The data indicate the maximum cost for all three cases occurring at an electric rate of 8.25 cent/kWhr, beyond the use of a portable diesel generator power is recommended.

Outlook and Information Needs

Based on the existing database developed by Battelle, the ISV technology appears to be a suitable remediation method for sites where the soil is contaminated with heavy metals and inorganics. Considerably less test data are available to assess applicability to and readiness for use at typical superfund sites containing a variety of contaminants including a range of organic compounds.

In general, much of the available process, emissions, and cost data are based on results from tests involving relatively simple systems and under controlled conditions which may not be representative of those likely to be encountered in actual site cleanup applications. Additional information is needed to assess treatment depth limitations, potential for fugitive emissions, and the degree of treatment uniformity and product homogeneity. Also, the developer's data and process effectiveness claims have not yet been independently verified.

The performance evaluation of the recent pilot-scale field demonstration of the application of ISV to low level radioactive waste disposal trenches is still in progress. ORNL is planning a demonstration of the technology on leaking underground storage tanks containing mixed wastes. It is hoped that ISV can create a single, stable waste form from the tank, its piping and contents, and the surrounding soil without excavating the material. The ISV technology has also been selected for independent evaluation under the U.S. EPA superfund Innovative Technology Evaluation (SITE) solicitation program.

SUMMARY

Situ site cleanup technologies have been review here these technologies are listed in Table 7.4 along with a summary of their operating principles, desirable features, and major limitations.

Except for the soil vapor extraction, the ISTs have not yet been subjected to the rigorous testing and independent engineering and environmental performance and cost analysis that are required for a satisfactory demonstration of readiness and suitability for use in large scale site cleanup applications. Much of the performance and cost data reported for ISTs have been supplied by the technology developers /vendors and have not been independently verified. This and the fact ISTs vary in their applicability preclude any meaningful comparison among ISTs based on the reported data and claims.

A successful IST application requires a good knowledge of the subsurface, including vertical and horizontal distribution of contaminants. Some ISTs by themselves do not provide complete treatment and can best be used as an element of an overall remediation program.

Table 7.4
In situ remediation technologies

Principle /Description	Desirable Features	Limitations
Soil Vapor Extraction Application of vacuum to the subsurface to remove volatile containants. The vacuum may be applied through vertical extraction wells (low water tables) or horizontal extraction systems (high water table). The extracted air may require treatment before discharge to the atmosphere. Treated air can be reinjected into the soil to provide for "closed-loop"operation Supplementary air injection is featured in some systems to increase air flow through the contaminated zone.	Simple, possibly low-cost and safe method (with proper controls). Repuires little maintenance and operating attention. Has been used in full-scale applications. Can treat significant soil depths in the unsaturated zone. Can be an effective "source control" measure for containing the spread of contaminan in the soil and preventing release or further of contaminants to grpimdwater Attractive for use at sites in highly developed areas and/or where contamination has spread to adjacent properties or underneath buildings.	Applicability limited to cases involving volatile compounds; low groundwater table, and loose sandy formations. Proper location and design of wells require good knowledge of subsurface characteristics. Channeling and uneven treatment may result due to nonhomogeneity of the subsurface. Not recommended for low hydraulic conductivity soils (requires conductivities perhaps exceeding 0.001 to 0.01 cm/sec. Difficult to monitor or control Treatment progress and completeness. Limited field experience and operating data available. Treatment rate can be very slow.
Steam Stripping Injection of steam and/or hot air into the subsurface to remove volatile contaminants. Off gas collected at the surface or through extraction wells for aboveground processing. One system, known as the Detoxifier, uses a novel adaptation of the drilling technology for in situ steam injection and mixing.	General: Enhanced volatilization and removal of a broader range of compounds due to application of heat. For the Detoxifier: Equipment design permits in situ mixing and hence uniform treatment. On-line monitoring allows for process control and adjustment of treatment conditions to achieve desired level of cleanup. Closed loop and mobile nature of operation. Prototype unit demonstrated in an application involving decontamination of soil contaminated with hydrocarbons. First production unit currently being employed at a site contaminated with chlorinated VOCs.	General: Better understanding needed of the behavior of the steam and of contaminant distribution among various phases. Potential for groundwater pollution exists. For the Detoxifier: Treatment depth possibly limited. Limited field experience available.
Bioremediation Degradation of organic contaminants to inert or less harmful products via action of suitable microorganisms that are developed in or added to contaminated soil or groundwater. Oxygen and nutrients are added to the system to support biological growth.	Simple, low-cost and safe method. Effectiveness demonstrated at many solvent and hydrocarborn spill sites. Simultaneous treatment of contaminants in the groundwater and soil. Generation of little or no treatment residues requiring disposal or air emissions.	Generally limited to applications involving localized groundwater contamination (spills of single solvent or petroleum based hydrocarbon fuels). Not applicable where the contaminants are refractory or are present at toxic levels. pH-adjusting chemicals, oxygen, and nutrients may have to be provided.

(Contd...)

Principle /Description	Desirable Features	Limitations
		Extensive treatability studies and subsurface characterization are generally necessary to establish optimum treatment conditions and the basis for design. Effective, proven method for uniform distribution of oxygen and nutrients in the subsurface (especially to great depths) does not exist. Channeling and uneven treatment may result due to nonhomogeneity of soil and/or formation of precipitates and biological deposits. For contaminated soils, the application seems to be limited to permeable soils (requires hydraluc conductivities perhaps exceeding 0.001 to 0.01 cm/sec). Regulations may require treatment of water withdrawn in surface facilities before reinjection. Introduction of nutrients can adversely affect water quality (groundwater or nearby surface waters). Difficult to monitor or control.
Soil Washing Flushing or rinsing of contaminants from soil using suitable soulutions (e.g., surfactants or chelating agents). Treatment solutions can be delivered and collected via gravity or forced systems similar to techniques used in solution mining and mineral extraction.	Can potentially provide for recovery of chemicals in cases chemicals in cases involving spills of individual chemicals or chemical types.	Systems for site cleanup primarily conceptual and little field experience or operating data available. Not feasible when complex wastes containing a range of contaminants with different solubility characteristics are involved. Difficult to limit reactions to target contaminants. Channeling and uneven treatment may result due to nonhomogeneity of the subsurface. Treatment rate can be very slow. Difficult to monitor and control treatment progress and completeness. Difficult to meet current requirements for residual levels in the treated soil. Recovered solutions can be very dilute and large in volume and hence costly to treat and dispose of.
Solidificaiton/Stabilization Rendering contaminants in soil/waste less mobile by addition of physical or chemical agents that react with or encapsulate contaminants.	Has been successfully used in applications involving closure of liquid or slurry ponds. Both organic and inorganic wastes reportedly can be handled with new proprietary formulations.	Proposed systems for in situ delivery and mixing of chemicals not demonstrated at contaminated sites. Difficult to assess long-term effectivness based on short-term leachin tests.

(Contd...)

Principle /Description	Desirable Features	Limitations
Vitrification A soil-melting technology whereby electric current is passed between electrodes placed in the ground. The soil and inorganic are converted to a stable glass. The organic contaminants are pyrolized and the pyrolysis products are combusted. Process off gas is collected under a hood and treated before discharge.	The product glass is very stable and leach resistant. May result in significant volume reduction. Considerable data from bench-, engineering-, and large-scale contaminated with metals and inorganics.	Limited data available for cases involving organic contamination. Many of the tests conducted to date have been in connection with radioactive-waste management. Available process and emissions control information must be verified in actual application to contaminated sites.

ENCYCLOPAEDIA OF
ENVIRONMENTAL SOIL AND MARINE POLLUTION

ENCYCLOPAEDIA OF ENVIRONMENTAL SOIL AND MARINE POLLUTION

[Vol. 2]

Edited by

G. R. CHHATWAL

ANMOL PUBLICATIONS PVT LTD
NEW DELHI- 110 002 (INDIA)

ANMOL PUBLICATIONS PVT LTD
4374/4B, Ansari Road, Daryaganj
New Delhi- 110 002

Encyclopaedia of Environmental Soil and Marine Pollution

First Edition 1997

ISBN 81-7488-434-3

PRINTED IN INDIA

Published by J. L. Kumar for Anmol Publications Pvt Ltd, New Delhi and Printed at Mehra Offset Press, Delhi.

Contents

Vol. II

8

A Microbiologist's View on Long-Term Fertility

Introduction

Up until three decades ago, most developing countries utilized various kinds of organic wastes and residues to maintain or improve tilth, fertility, and productivity of their agricultural soils. When inexpensive energy sources and increased mechanization became available, organic recycling practices and crop rotation were largely replaced by chemical fertilizers for plant nutrition, pesticides for crop protection, and energy-intensive monoculture systems for cash grain production. Unfortunately, the vital importance of organic matter in crop protection, and its proper use in soil management and conservation practices were neglected. As a result, there has been excessive soil erosion, nutrient run-off, loss of soil organic matter, and impairment of environmental quality due to sedimentation and pollution of surface and ground waters by agricultural chemicals. The excessive soil erosion and the concomitant loss of crop productivity are now most serious agricultural problem in regard to both short and long term fertility. Consequently, there is a renewed interest in the use of organic wastes and residues to protect agricultural soil from erosion and to maintain and restore soil productivity. In agricultural soil the organic matter plays an essential role, often merely described as an improvement of physical and, indirectly, chemical fertility. Actually, soil organic matter not only improves but is a precondition for soil fertility. A decrease in organic matter starts an involution of soil fertility. Modern practices of agriculture may only partially contrast this process through chemical and mechanical aids, often involving high energy costs.

Organic fertilizers are the basis for the biological stimulation of plant production and soil activities sometimes depressed by an excess of chemicals. Most chemical fertilizers are relatively easily soluble. The concentration of ions therefore may often exceed the tolerance area for many of the useful organisms, so that they stop functioning. One example of this is the nitrogen-fixing microorganisms that become inactive when quick-solving nitrogen is added. There also seem to be other uncertain effects on plants, concerning their ability to utilize readily soluble nutrients such as phosphorus and potassium.

Providing new organic matter to agricultural soils is one of the major tasks that modern agriculture should face to re-establish soil fertility, especially seriously endangered long-term fertility.

There is no generally accepted concept covered by the term "soil fertility." Soil fertility is a frequently used term for indicating the "inherent capacity of soil to supply nutrients to plants in adequate amounts and in suitable proportions".

This broad definition, summarizing the belief of some scientists, claims that soil fertility depends on chemical, physical, mechanical and biological factors which affect soil nutrient status and nutrient availability for plants. One century ago, Dokuchaer stated that soil is a living entity deriving from parent rock and slowly developing under the influence of climate, organisms, topography, and time. (See Figure 8.1).

Whether or not one entirely agrees with this it seems essential that soil should be studied in relation to its origin and environment. The natural fertility is a biophysical rather than a physicochemical consequence. "It concerns the transformation of energy into living organisms: essentially the transformation in plants of heat and light into chemical energy which is returned to the soil to provide energy for the edaphon (the soil flora and fauna) to live and work to construct a habitat which concerns the transfer of energy".

In the natural environment, the most determining factor in the formation of a fertile soil is the immense work performed by soil organisms. Among them, soil microflora play a particular role. The living mass of microbes (bacteria, fungi, actinomycetes, algae, protozoa) processes considerable amounts of different mineral and organic substances, decomposing plant and animal residues and participating in the transformation of the decomposition products. Therefore a microbiological outlook is necessary for a better understanding of the dynamic processes that condition soil fertility. (See Figure 8.2).

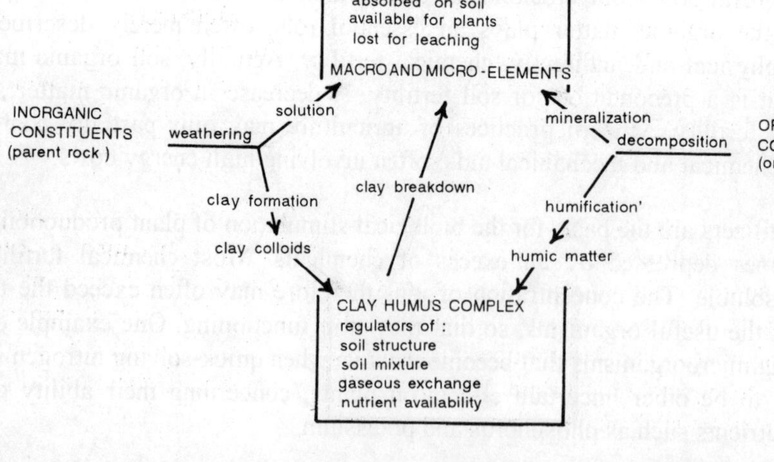

Fig. 8.1. The formation of the soil from organic and inorganic materials

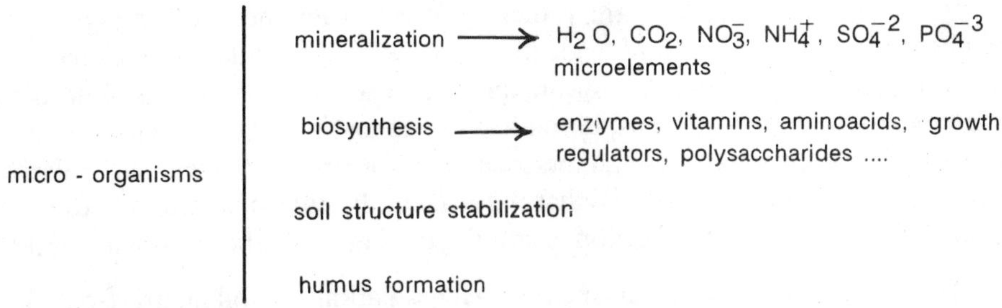

Fig. 8.2. Microbial activities in the soil.

Sludge in Agriculture

In recent years, sewage sludge has been supplied more and more commonly to croplands. Sewage sludge from urban plants may represent an important source of nutrients once the non-toxicity to public health and the environment has been ascertained. The possible harm of such materials is linked to the presence of pathogens and polluting substances, especially heavy metals. When sludge is supplied to croplands, a major question should be posed: "Should sludge be considered a fertilizer or a waste, and is it reasonable to separate beneficial and harmful effects of sludge application?"

Sludge disposal on the land involves risks of phytotoxic effects on plants, of toxic effects on man and animals, of pollution of ground water, and of damaging the environment if it is carried out in an uncontrolled manner. There is a need for a more scientific basis for criteria to provide adequate protection against harmful effects without incurring undue cost. It is necessary to consider both the direct and more immediate effects on agriculture, and the indirect effects on animals and on man. However, it appears that for most countries, sludge is considered a waste with fertilizing value, especially regarding its nitrogen, phosphorus, and organic matter content. Many questions arise concerning the fate of these after their shredding on land. In particular, looking at each component separately, conclusions can be drawn as follows:

Nitrogen

Although nitrogen is the most important factor in fertility, its short term effect is only now beginning to be understood. The nitrogen contained in sludge is not readily usable by crops, and research and technical skills are required for its proper application.

- Not enough is yet known about the after-effects and long-term effects
- Very little is known about loss through vaporization, in particular denitrification, or as a result of leaching

An understanding of these phenomena is vital if proper advice is to be given and legislative action taken.

Phosphorus

The problems of phosphorus are rather different from those of nitrogen, in that an excess of phosphorus in the soil (although the term 'excess' is rather vague) has no adverse effects on crop production. Generally speaking, if the use of sludge is based on the nitrogen supplied, an excessive supply of phosphorus results; almost all of this accumulates with little or no migration. There is nonetheless an increased risk for water-courses, particularly through runoff, but here again it must be stressed that the scale of the eutrophication varies considerably from one country to another. Clarification is still required on a number of points, in particular:

- The effects of a large phosphorus accumulation in soil apart from the risk of leaching
- The heavy metals-phosphorus relationships.

Organic Matter

It must be admitted that there is a general lack of practical and usable information especially about:

- The effects of sludge on the physical properties of soils
- The effect of sludge on soil biology, in particular on soil fauna and microbiology, especially plant pathogens

When sludge is supplied to soils, two kinds of effects can be distinguished:

- The short-term effect, mainly related to directly available plant nutrients and to rapidly mineralizable materials
- The long-term effect, mainly related to the transformation of the remainder and to its effect on soil properties

Some important questions are connected to these topics, especially to the second one:

1. Sludge is often supplied to cropland on the basis of total nitrogen content. Only about 50% is mineralized in the first year. What is the fate of the remainder in the long run, and when heavy amounts are repeatedly supplied?
2. Sludge is more organic than mineral. What is the fate of organic matter and how is humus produced?
3. Sludge improves the physical properties of soil. Is that only a physicochemical process or are microorganisms also involved in increasing soil aggregates, e.g., via polysacharides?
4. Sludge supply could modify biological equilibrium in soil. Can biological cycles give information about the health status of soil and its fertility after sludge treatment?
5. Soil microorganisms not only mineralize organic matter, but are responsible for several processes of biosynthesis. As a consequence, different kinds of metabolic products are excreted into the soil. What is the role of microbial metabolites in crop improvement?

Soil Microorganisms and Higher Plants

Living roots provide a localized and continuous source of nutrients for the soil microflora in the form of live tissues, exudates, and sloughed-off dead cells. They also provide a continuously expanding surface for colonization and by their metabolic activities change the soil environment in their immediate vicinity. It is not surprising, therefore, that great changes in microbial activity occur in the root region, which may have both beneficial and harmful effects on plants. It is well known to plant and crop physiologists, microbiologists, and agronomists that soil fertility is strictly linked to microorganisms which, particularly in the rhizosphere, play a fundamental role in the plant nutrient uptake.

The special composition of the microflora, like the soil, has a definite and considerable influence on the growth and development of plants and, consequently, on crop yield. The microflora of the root zone is of great importance in plant nutrition. Growing near or on the roots, microorganisms, together with the plants, create a special zone: the rhizosphere. Soil of this zone differs in its physical, chemical, and biological properties from soil outside the root zone. It possesses different characteristics in respect to the absorption and excretion of substances by the root, prepares various organic and inorganic nutrients for the plants, and by its regulation of ionic absorption, of ten determines the induction of several enzyme complexes in the plants. The new equilibria between soil microorganisms and plants, modifying plant capacity in nutrient absorption, determine the agronomic value of sludge with regard both to short and long-term fertility.

The knowledge of life in soil and the investigation of biological and biochemical processes taking place in it are inseparably connected with the knowledge of soil fertility. The importance of soil microorganisms on growth and development of higher plants and the strict relationships between them must be taken into consideration to guide agriculture in the desired direction and to eliminate harmful occurrences, thus, in other words, to restore the health of soils. This is particularly true for the free-living microflora which inhabit the root environment.

An understanding of the effects of microorganisms on the growth of plants is essential if we are to exploit the possibilities of altering crop productivity and maintaining soil fertility by the introduction of exogenous materials into soils, and by harnessing the activity of the indigenous microflora. Microorganisms can bring about changes in plant growth which may be stimulating or inhibitory. These may be due to:

1. The use of microbial metabolites as major nutrients
2. The effect of growth regulators produced by microorganisms
3. The liberation of otherwise unavailable nutrients from soil organic matter and minerals
4. The suppression of plant pathogens
5. The production of phytotoxic substances by saprophytes and parasites
6. The production of enzymes
7. The competition of microorganisms with plants for essential nutrients.

Not surprisingly, these direct interactions are frequently attributed to organisms in the root region of plants. However, changes taking place in root-free soil are equally important since they may influence subsequent root growth into that soil. Indirect promotion and inhibition of plant growth can be caused by changes in soil structure partly brought about by soil microorganisms, e.g., aggregation of soil particles and crumb formation. All plant-microbe interactions are difficult to study because the changes in plant growth are similar to, and may be confused with, changes in growth due to solely physical and chemical events. These events may, in turn, affect microbial growth, which can influence plant growth and the environment further. Other difficulties arise because most microbe-plant associations, apart from symbiotic and parasitic ones, are rather unspecific. Microbial activities and the complex relationships in the plant-soil system can be influenced and, sometimes, altered, when materials not related to the soil ecosystem, e.g., sewage sludge, are supplied. The effects could be strongly enhanced when heavy amounts are supplied on cropland.

Due to its biological origin, sludge is a material very rich in microorganisms, enzymes, and precursors of many microbial metabolites, and can stimulate the biosynthesis of biologically active substances by soil microorganisms. Apart from an amendment in soil properties, sludge has been proven to improve crop yield and quality of many different cultures.

Sludge supply influences the physical, chemical, and biological properties of soil, affecting its permeability, porosity, and structure, and modifying the redistribution, movement, and transport of nutrients. Consequently, several degrading and biosynthetic microbial activities are favored, resulting in an improvement of soil fertility, especially in relation to the supply of nutrients to the plants.

Nutrient uptake takes place in the rhizosphere and it is greatly stimulated by microbial metabolites, some of which able to induce, or increase the induction of, many enzymes, thus enhancing mineral nutrition, protein synthesis, and carbon assimilation and so promoting whole plant productivity.

Furthermore, more exudates are released into the soil by the roots. The new energy sources, in turn, will re-affect microbial growth and processes, and therefore also plant metabolism. (See Figure 8.3)

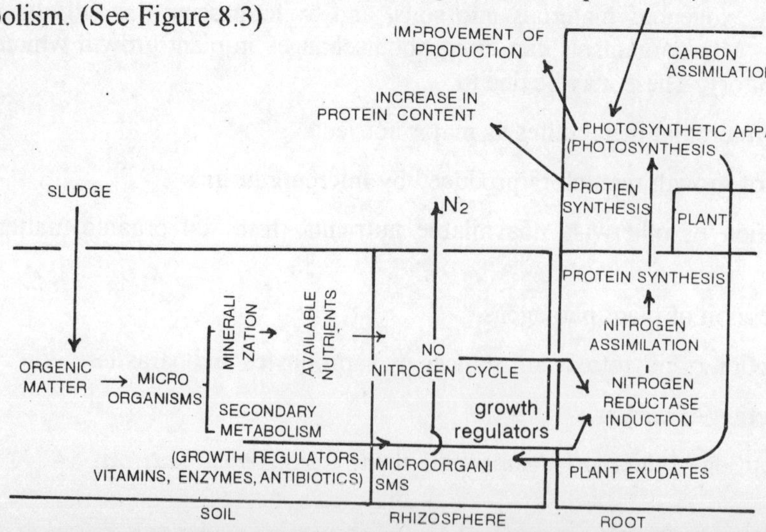

Fig. 8.3. The nitrogen cycle in the plant-soil ecosystem as influenced by organic waste.

Fertilization Value of Sludge

Many large-scale field experiments concerning the effects of sludge on the fertility soil started to study the positive effects, mainly due to nitrogen, phosphorus, and organic matter content, and possible negative effects, due to content of polluting substances, especially heavy metals.

Research agrees in assuming that the most interesting effect of sewage sludge, like that of organic fertilizers in general, may be a long-term one, which is an effect on yield that cannot be achieved with mineral fertilizers alone applied in the optimum amount. This so called rest-effect may be an"organic matter effect" due to slow release of nutrients, or an effect on soil physical characteristics, including a better availability of mineral nutrients supplied by the mineral fertilizers. The rest-effect is generally positive, but may be negative when an excess of micro-elements, particularly heavy metals, is present. The long-term effect on soil fertility and crop production is a consequence of the slow mineralization that sludge underwent over an extended period of time. Its availability is ruled by a whole series of known or unknown parameters, among which the degree of sludge stabilization and C:N ratio, as well as soil and climatic factors, are the most important.

The general conclusions that can be drawn by the consistant body of experiments are as follows:

1. The main effect of sewage sludge is dependent on its nitrogen content, whose fate cannot be predicted exactly. Nitrogen contained in sludge is mainly in an organically-bound form: soil, climate, and application time are the most important factors conditioning its mineralization.
2. There is a positive after-effect of sludge application made evident by the increased crop. This could be due to an improved nitrogen status of the soil, but an involvement of microbial metabolites could better explain the effect both on crop yield and quality so often recorded.
3. Application of sewage sludge results in an enrichment of the soil in micro-elements. This is positive in the case of copper, zinc, boron, molybdenum, and the other elements involved in plant metabolism, but strongly negative in the case of mercury, cadmium, lead, and the other elements which are toxic for plants.
4. Repeated and regular applications of sludge increase the pool of mineralizable nitrogen and of nutrients in the soil. Sludge applications at different times act differently on plant growth. Generally relatively small effects attributable to a loss of nitrogen are recorded when sludge is supplied in the rainy season.
5. Long-term application of sewage sludge can increase humus content, nitrogen content, cation exchange capacity, and heterotrophic soil microorganisms and their activities to a higher degree in a light soil than in a heavy soil. On the other hand, high applications of sewage sludge can decrease autotrophic microorganisms (algae) more emphatically in a heavy soil not only on the surface, but also in the deeper layers. The effects are strongly dose-dependent.

6. The aerobic sludge is the most effective in increasing soil organic carbon. The composition of the humified fraction of soil organic matter, derived from the determination of humic and fulvic acids and gel-filtration, was practically the same.
7. As for crop production and quality, an improvement has often been recorded that cannot only be explained by physical and chemical factors of fertility. Biological factors, mainly microbial metabolites, could be a key for a better understanding of crop productivity over short and long times; in other words, for better understanding soil fertility.

The fertilizing value of sewage sludge has been the object of a consistent body of research, mainly pointed at the influence of treatment on chemical and physical factors of fertility, both in the short and long runs. By contrast, little information is available about the changes in biochemical and microbiological properties of soil, and even less is known about sludge effects on plant-soil systems.

Although limited data concerning the relationships between microorganisms and soil fertility are available, we try to discuss soil fertility on crop production from the point of view of a microbiologist. We underline that many effects recorded after sludge is supplied cannot be explained only by the presence of nutrients or by an improvement of soil structure. They can be better understood when the action of microorganisms and microbial metabolites is taken into closer consideration.

The agronomical importance of sludge is mainly related to its nitrogen content. Therefore, we will deal principally with the activities connected to the nitrogen cycle in both soil and plants, to justify the increased soil fertility.

Sludge Effects on Soil Microflora and Microbial Activity

The most common way to assay the fertilizing value of a sludge is the so-called "crop test," in which crop yield is the reference term to evaluate soil fertility. A more sophisticated method able to clarify some factors that could influence the growth and yield of cultures consists of detecting as many parameters as possible, in both soil and plants, to determine which of them can provide useful information. As sewage sludge can alter, either temporarily or permanently, soil and rhizosphere microbial activities, thus influencing plant metabolism, it seems reasonable to choose activities in the plant-soil system as an index of sludge ability in affecting plant productivity. The agronomic value of sludge can be distinguished as a long-term fertilizing value, dependent on storage of nutrients released from mineralization, and an actual fertilizing value, dependent on nutrients immediately available for the plant. Soil and rhizosphere activity can be considered as indexes of these two kinds of soil fertility: they may supply, respectively, information about future soil fertility, because of the role in metabolizing the complexes and matcrials supplied, and about global metabolism in the plant-soil system.

We will discuss the long-term effect of sludge on soil fertility mainly in terms of soil and rhizosphere activities and plant metabolism, considering the strong relationship between them.

We will take into greater consideration four points in the long-run experiments:

1. Sludge effect on the plant-soul system at the first stage of the developing culture
2. Sludge effect on the plant-soil system during the whole vegetative cycle of the culture
3. Changes in soil and rhizosphere parameters after repeated sludge treatments
4. Sludge effect on soil structure and stability

Both aerobic and anaerobic urban sludge in varying doses (4-20 ton/Ha as organic matter) were applied to soils, with resulting effects found to be dependent on both doses and sludge origin. Wheat (Triticum aestivum) and maize (Zea mais), as respectively, winter-spring and spring-summer cultures, were considered.

Sludge Effect on the Plant-Soil System at the First Stage of the Developing Culture

The beginning of the rhizosphere effect is an important factor governing the future plant development and productivity. Many plant enzymatic activities can be enhanced by a stimulated biosynthesis of active microbial metabolites. The rhizosphere effect starts and begins to increase immediately after seed germination, reaches a peak around the time of flowering and fruiting, and then declines sharply as root senescence begins. The rhizosphere effect is greatly variable according to soil properties and climate. Within any one soil, environmental changes seem to have less effect on the rhizosphere microflora than on the populations in soil far from roots. Addition of fertilizers can alter soil and rhizosphere populations and their activities.

The beginning of the rhizosphere effect is not easy to investigate in the open field, since many environmental factors, mainly climatic, disguise and conceal it. IT is suitable to study the rhizosphere effect contemporarily through pot trials, which may complete the information derived from the field experiments.

Therefore, to better evaluate the soil and rhizosphere activities at the first stage of plant growth as influenced by sludge treatment, pot experiments are here illustrated (Table 8.1). Soil used was collected from the field where aerobic sludge has been shredded yearly. Plant microorganism interactions start immediately at seedling time. As roots develop, increasing their biomass and their contact surface with the soil, a selection in microflora probably occurs, and starts the rhizosphere effect. The increase of microflora growth, strongly stimulated by root exudates, provokes an enhancement in biological activities in regard to mineralization and biosynthesis of microbial metabolites.

Not surprisingly, microbial activity is more intense in the rhizosphere than in the soil far from roots. Looking at each parameter separately:

Microflora growth, oxygen consumption, and heteroauxin production. Microflora growth, especially bacteria, depends largely on the amounts of sludge supplied; the growth is greatly enhanced in the rhizosphere. The rapid development of microflora confirms that the micro-organism-plant interaction starts immediately after the sowing. As the root system develops and increases in contact with the soil, microorganisms in the immediate vicinity are benefically influenced (Figure 8.4). Their growth is probably selective and anticipates the initiation of the rhizosphere effect enhancing it.

Table 8.1

Pot Experiments

Conditions	T°C: 25°/18° day/night Relative humidity: 60% Photoperiod: 14th Irradiation intensity-fluorescent and incandescent lamps: 20.000 lux
Plant	Corn CV Saturno TV 34
Pot Mixture	Alluvional-sandy-loam soil from the fields where sludge has been yearly shredded in 20, 10, 6 tons/Ha doses
Samples	Collected after 7, 12, 21, 27 days from sowing

The behavior of oxygen consumption is very similar to that of the microflora, especially bacteria, and may be considered as an index of soil metabolism efficiency. The greater oxygen consumption in the rhizosphere may also be related to both an enhanced microbial population and activities. (See Figure 8.5a)

Heteroauxin production could be considered as an index of the secondary microbial metabolism. It is greatly enhanced in the rhizosphere, whereas no differences were detected in soil far from roots. This fact is of particular importance because many metabolic activities in the plants are influenced by microbial growth regulators (Figure 8.5d).

Nitrification and denitrification. The large oxygen consumption depending on biodegradation of organic compounds may considerably alter the ratio aerobic/anaerobic microorganisms, favoring the aerobic ones. The behavior of denitrification and nitrification seems to confirm this alteration. Nitrification is more depressed in the rhizosphere than in the soil. No evident differences were detected either in soil or in the rhizosphere (Figure 8.5b, c).

Protease activity. Protease activity is largely dependent on the supplied amounts of sludge. In the soil it quickly decreases, whereas in the rhizosphere it remains strongly active (Figure 8.6a).

The high protease activity recorded in untreated rhizosphere soil is a sign that the protease activity is also dependent on root exogenous proteolytic enzymes. This is usually beneficial for the root environment because it permits a more continuous mineralization of N-containing substances.

Acid phosphatase activity. Acid phosphatase activity is very similar to that of bacterial growth and seems to confirm the hypothesis of a correlation between the two.

As with all the other activities, acid phosphatase activity is largely dependent on sludge amounts. Its more enhanced activity may be explained by the high amount of phosphate required by microflora and plant development (Figure 8.6b).

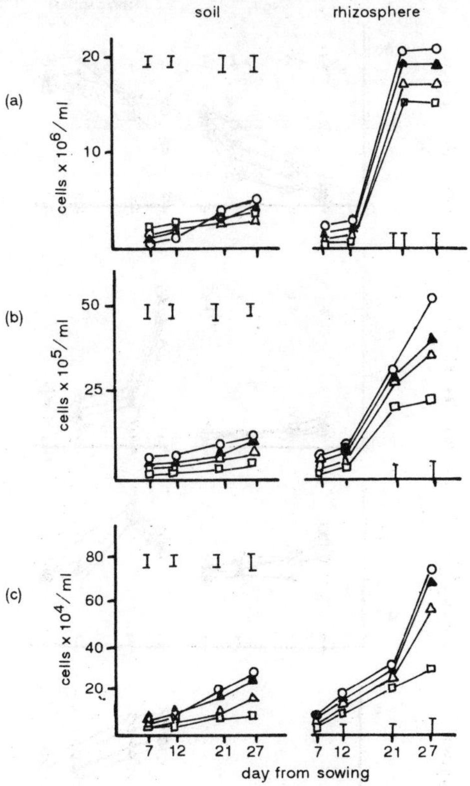

Fig. 8.4. Sludge effect on microflora in soil and maize rhizosphere; pot experiments. a) bacteria; b) actinomycetes; c) fungi. (□) = mineral; (Δ) = sludge (6 ton organic matter/ha); (Δ) = sludge (10 ton organic matter/ha); (O) = sludge (20 ton organic matter/ha); vertical bars = LSD (P = 0.05).

Nitrate content. Nitrate content in soil and in the rhizosphere supply further information about nitrogen mineralization and nitrate utilization. Nitrate concentration in the soil is higher than in the corresponding rhizosphere. The enhanced nitrate content in the soil is an index of good sludge mineralization. The lower nitrate concentration recorded in the rhizosphere indicates its utilization both by microorganisms and plants.

In agronomic terms, soil activities and nitrate content in soil could be considered as an index of the potential soil fertility, whereas higher rhizosphere activities and nitrate consumption could be an index of actual fertility (Figures 8.7a, b).

Phosphate content. Phosphate content increased in soil and in the rhizosphere. Diagrams show a larger phosphate consumption in the rhizosphere (Figures 8.8a, b).

Plant metabolism. Not surprisingly, sludge supply alters plant metabolism, as the enhancement of the rhizosphere activities shows. In particular, an increase in phosphate content in roots and in leaves (probably dependent on sludge-phosphate available for plants) is recorded (Figure 8.8c, d).

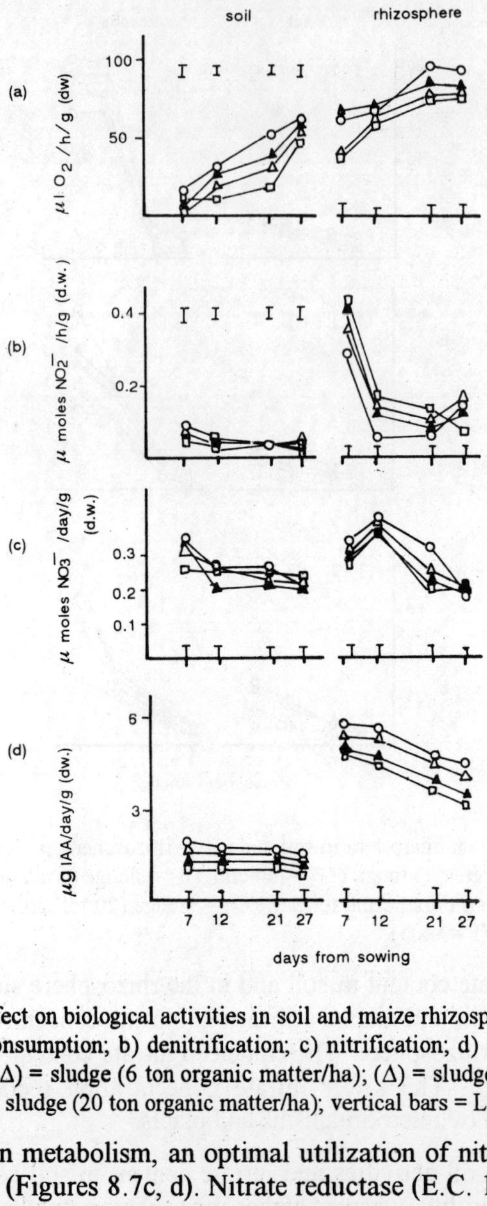

Fig. 8.5. Sludge effect on biological activities in soil and maize rhizosphere; pot experiments. a) oxygen consumption; b) denitrification; c) nitrification; d) auxin production. (□) = mineral; (Δ) = sludge (6 ton organic matter/ha); (Δ) = sludge (10 ton organic matter/ ha); (O) = sludge (20 ton organic matter/ha); vertical bars = LSD (P = 0.05).

Concerning nitrogen metabolism, an optimal utilization of nitrate is confirmed by plant nitrate reductase activity (Figures 8.7c, d). Nitrate reductase (E.C. 1.6.6.1) may be considered the first step of nitrogen assimilation. Its activity is controlled by nitrate content and hormone-like substances, and inhibited by many heavy metals. In recent years, nitrate reductase activity has been proposed as a predictive test for crop productivity.

The sometimes recorded loss in nitrate content in roots and leaves could be explained as a result of competition between microorganisms and plants for nitrate.

In addition to a general beneficial effect on plants and soil, it can also be hypothesized that sludge supply is not harmful at the first stage of plant development: no changes were in fact recorded in the typical behavior of the plant-soil system.

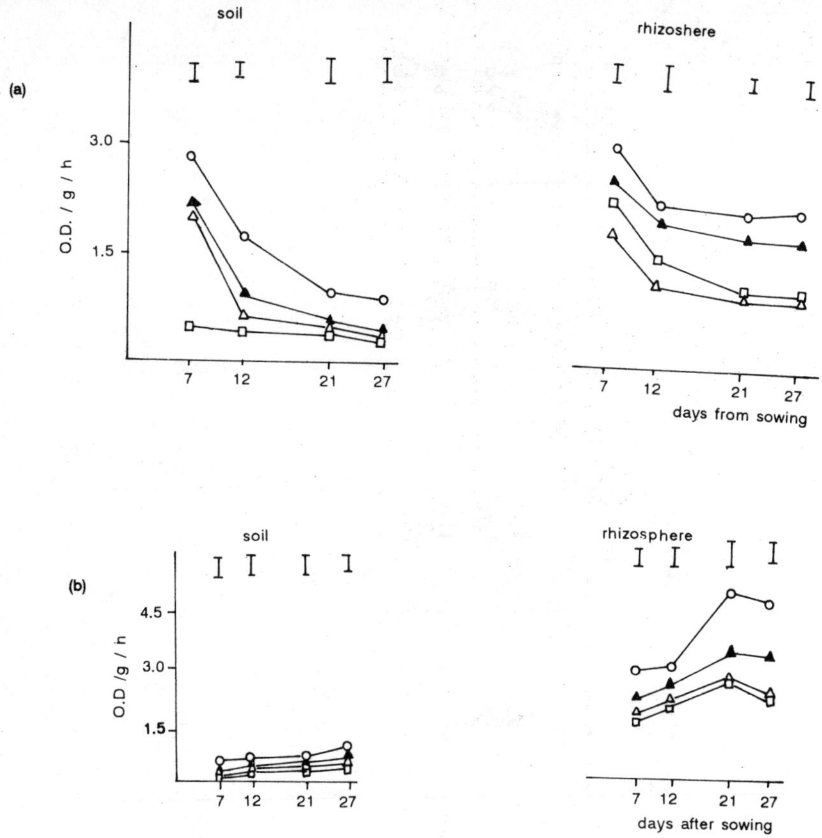

Fig. 8.6. Sludge effect on biological activities in soil and mazie rhizosphere; pot experiments. a) protease activity; b) acid phosphatase activity. (□) = mineral; (Δ) = sludge (6 ton organic matter/ha); (□) = sludge (10 ton organic matter/ha); (O) = sludge (20 ton organic matter/ha); vertical bars = LSD (P = 0.05).

Aerobic sludge is more effective than anaerobic sludge, especially when high doses are supplied.

General remarks can be made as follows: The enhanced oxygen consumption, also related to the organic compounds breakdown, brings a change in the ratio of aerobic to anaerobic microorganisms, benefitting the aerobic ones. This seems to be confirmed by the depression of denitrifying activity, and by the stimulation of the nitrifying one. As a consequence of the nitrifying activity and mineralization, soil becomes enriched in $N\text{-}NO^{-}_{3}$ and this way becomes available for cultures. The marked decrease in $N\text{-}NO^{-}_{3}$ in the rhizosphere proves an active N-uptake by plants.

An interesting effect of sludge application is the increased biosynthesis of phytohormonal substances recorded both in pot and field experiments, especially where minerals were not supplied.

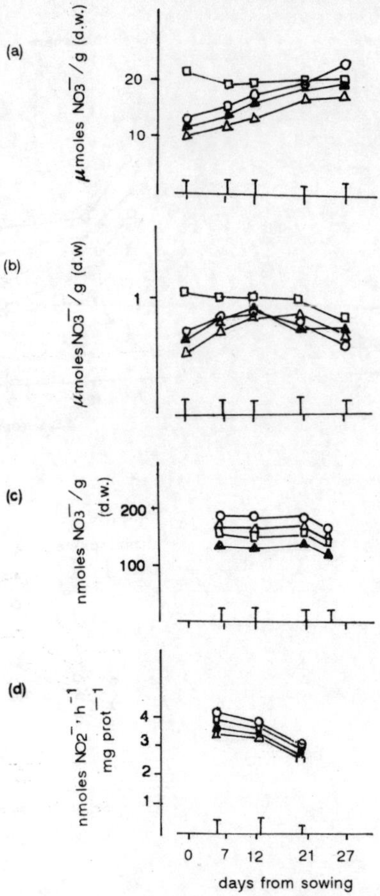

Fig. 8.7. Nitrate content in a) soil, b) maize rhizosphere, c) leaves, and d) nitrate reductase activity in leaves after sludge treatment; pot experiments. (□) = mineral; (Δ) = sludge (6 ton organic matter/ha); (Δ) = sludge (10 ton organic matter/ha); (O) = sludge (20 ton organic matter/ha); vertical bars = LSD (P = 0.05).

Microbial growth regulators are important factors acting on plant metabolism and are critical in determining the fertilizing value of sludge. Because of its origin, sludge is very rich in nucleic acids, aminoacids, purines, and other organic compounds which are precursors in microbial growth-regulator synthesis. Phytohormonal action on the uptake and utilization of nutrients by plants is well known. The effect of nutrient uptake and utilization on the stimulation of and/or the induction of several enzyme complexes of the plant is also familiar.

Sludge Effect on the Plant-Soil System During the Whole Vegetative Cycle of the Culture

The action of aerobic or anaerobic urban sludge on soil and rhizosphere activities becomes clear in field experiments that follow cultures during their entire vegetative cycle. Experiments were conducted in an open field to which two doses of sludge (6 to 20 tons of organic matter) were supplied every year.

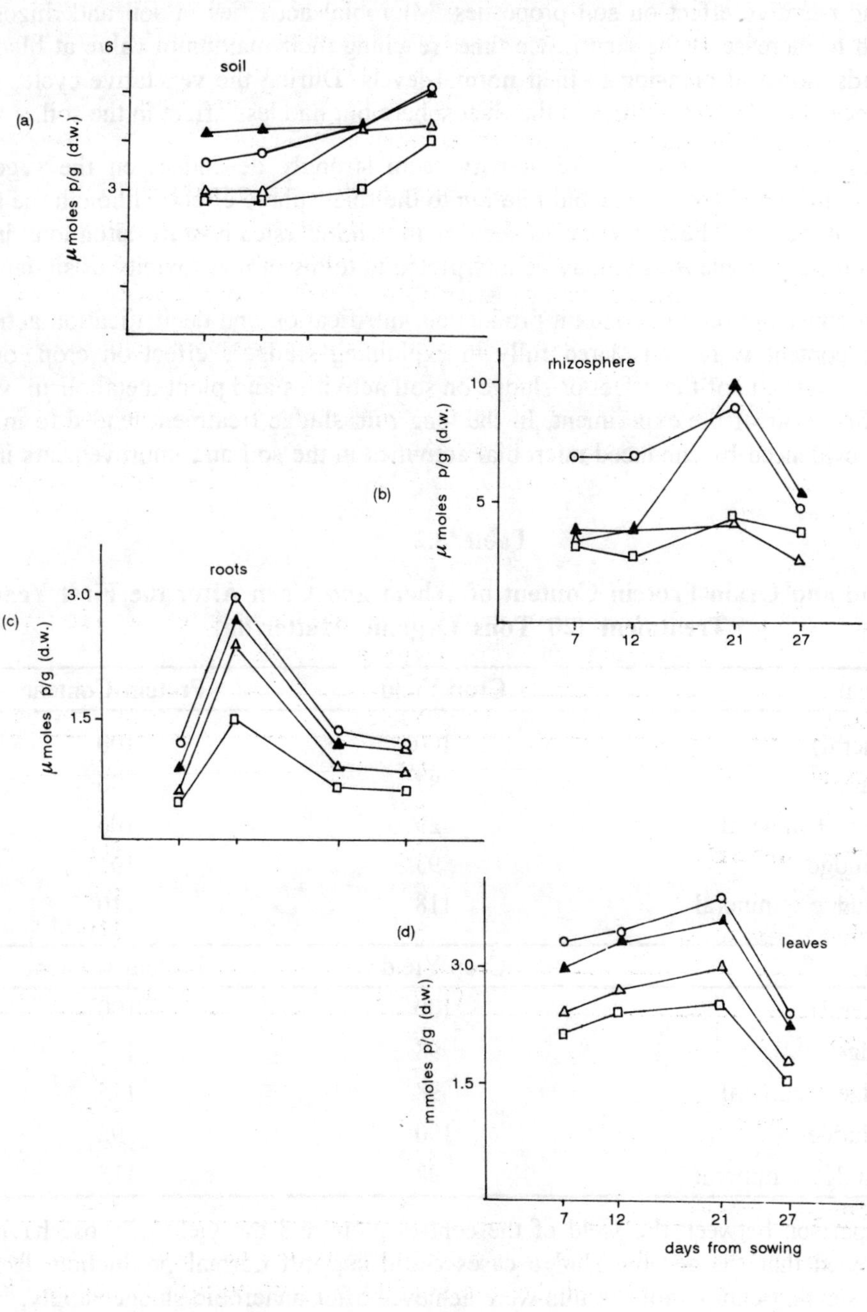

Fig. 8.8. Phosphate content in a) soil, b) maize rhizosphere, c) roots, and d) leaves after sludge treatment; pot experiments. (□) = mineral; (Δ) = sludge (6 ton organic matter/ha); (Δ) = sludge (10 ton organic matter/ha); (O) = sludge (20 ton organic matter/ha); vertical bars = LSD (P = 0.05).

Sludge did not affect the shape of soil and rhizosphere activities, thus confirming that sludge has no negative effect on soil properties. Microbial activities in soil and rhizosphere usually begin to increase at the emergence time, reaching their maximum value at blooming, and afterwards slowly decreasing to their normal levels. During the vegetative cycle, sludge greatly enhanced microbial activities in the rhizosphere but had less effect in the soil.

Changes in soil and rhizosphere activity seem strongly dependent on the vegetative status of the culture and suggest partial relation to the rhizosphere effect, although the sludge effect is incontestable. The tendency to restore that initial steady-state situation, in part dependent on the soil homeostasis, may be interpreted in terms of non-toxicity of sludge.

Oxygen consumption, heteroauxin production, nitrification and denitrification activities, and nitrogen content were considered fully in explaining sludge's effect on crop, or soil, fertility. As an example of the effect of sludge on soil activities and plant metabolism, we can discuss the first year of the experiment. In the long run, sludge treatment tended to improve soil fertility, evidenced by enhanced microbial activities in the soil and improvements in crop production.

Table 8.2

Crop Yield and Grain Protein Content of Wheat and Corn After the First Year of Treatment (20 Tons Organic Matter/ha)

Wheat	Crop Yield	Protein Content
Control (mineral)	100	100
Aerobic sludge	87	96
Aerobic sludge + mineral	129	106
Anaerobic sludge	93	108
Anaerobic sludge + mineral	118	110
Corn	**Crop Yield**	**Protein Content**
Control (mineral)	100	100
Aerobic sludge	98	110
Aerobic sludge + mineral	98	115
Anaerobic sludge	100	92
Anaerobic sludge + mineral	98	115

A comparison between the yield of the control plots and the yield of those having no minerals showed that the aerobic sludge cases could support normal production. Refer to Table 8.2 to see that comparable results were achieved after anaerobic sludge supply.

No differences were recorded regarding protein content, even though a weak increase appeared after anaerobic sludge treatment. When minerals were supplied together with the sludge, a similar behavior in rhizosphere activities was made evident, but an improvement in both crop yield and quality was assayed.

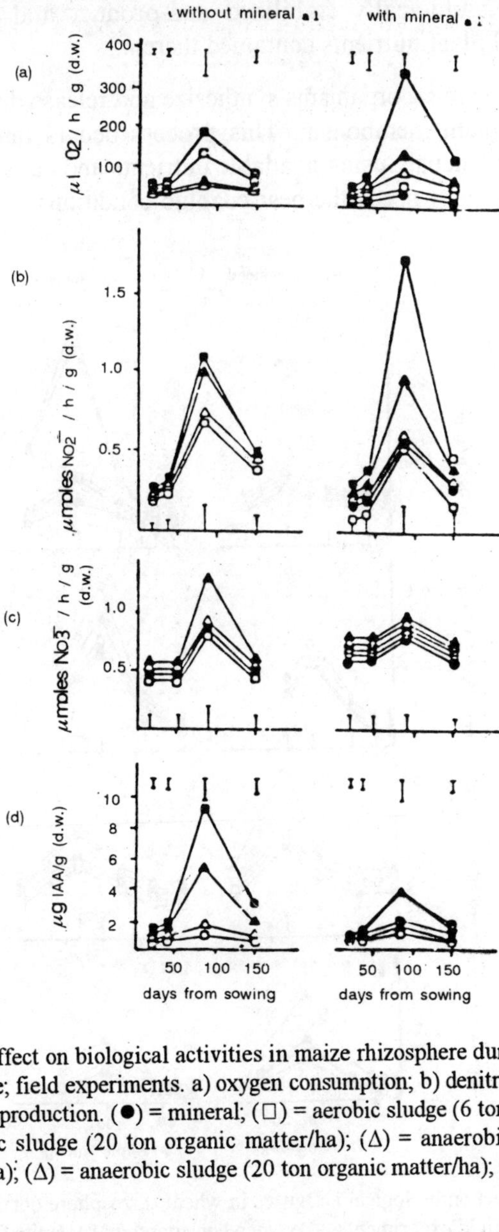

Fig. 8.9. Sludge effect on biological activities in maize rhizosphere during the vegetative cycle of culture; field experiments. a) oxygen consumption; b) denitrification; c) nitrification; d) auxin production. (●) = mineral; (□) = aerobic sludge (6 ton organic matter/ha); (■) = aerobic sludge (20 ton organic matter/ha); (Δ) = anaerobic sludge (6 ton organic matter/ha); (Δ) = anaerobic sludge (20 ton organic matter/ha); vertical bars = LSD (P = 0.05).

In both cases, results could be explained on the basis of biological factors of fertility (mainly microflora and microbial growth regulators), whose production was strongly increased by sludge treatment. In fact, although no effect was recorded on O_2 consumption, when supplied to soil sludge strongly increased heteroauxin production (Figures 8.9a, d and 8.10a, d). Microbial growth regulators, which are able to stimulate nutrient uptake, particularly nitrogen, could in part justify both the consistent yield when plants grew under conditions of nitrogen shortage, and the crop improvement after the addition of minerals. When supplied to soil, sludge is slowly mineralized by the living mass of microorganisms, transforming the organic matter into a biologically stabilized end-product and allowing more complete conservation of potential plant nutrients contained therein.

At the same time, the microorganisms synthesize and release different metabolic products capable of stimulating plant metabolism. This process occurs during the whole vegetative cycle, with the plant continuously using available nutrients and active metabolites, particularly at the emergence time when it needs the best edaphic conditions.

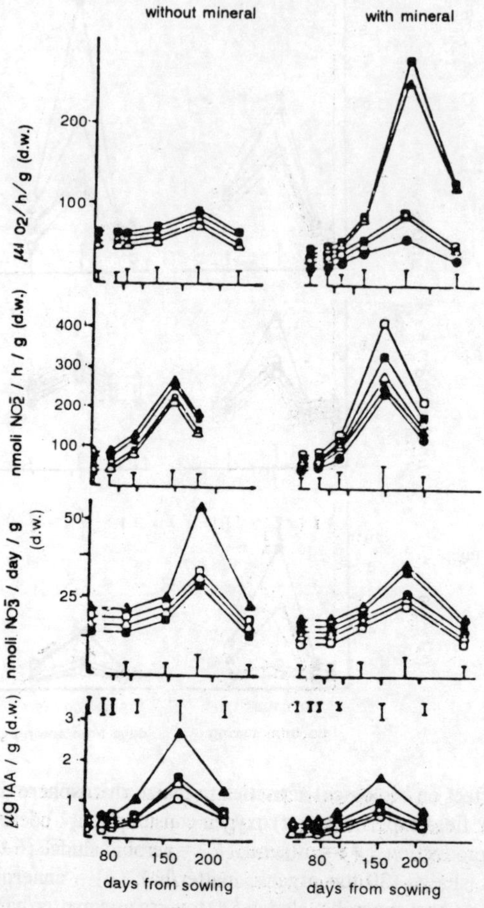

Fig. 8.10. Sludge effect on biological activities in wheat rhizosphere during the vegetative cycle of culture; field experiments. a) oxygen consumption; b) denitrification; c) nitrification; d) auxin production. (●) = mineral; (□) = aerobic sludge (6 ton organic matter/ha); (■) = aerobic sludge (20 ton organic matter/ha); (Δ) = anaerobic sludge (6 ton organic matter/ha); (Δ) = anaerobic sludge (20 ton organic matter/ha); vertical bars = LSD (P = 0.05).

Therefore, the fertility value of sludge depends mainly on its decomposition in the soil. This fact plays an important role in long-term fertility, mainly related to the transformation rate of the remainder substances.

Nitrate concentration, monitored during the whole vegetative cycle, is strongly increased in the soil away from the roots, whereas a significant decrease in nitrogen in the rhizosphere was recorded (Figures 8.11, 8.12).

In terms of soil fertility, the increase and storage of nitrogen in its different forms in soil can be thought of as a source of nutrients available on a long-term basis.

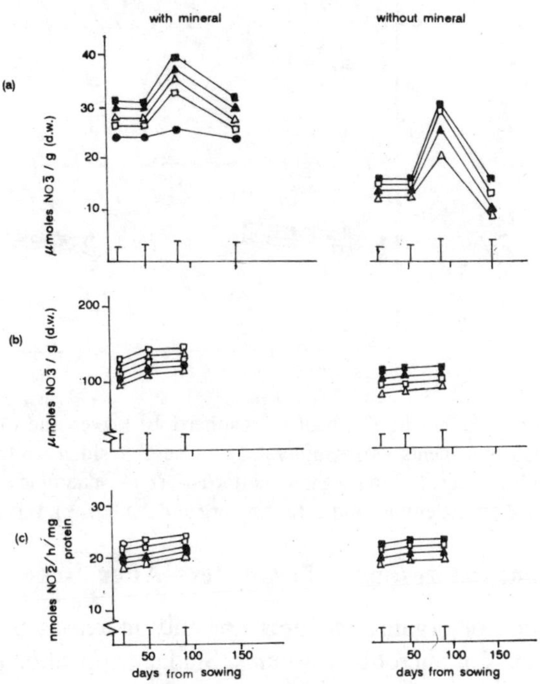

Fig. 8.11. Nitrate content in a) maize rhizosphere, b) leaves and c) nitrate reductase activity after sludge treatment; field experiments. (●) = mineral; (□) = aerobic sludge (6 ton organic matter/ha); (■) = aerobic sludge (20 ton organic matter/ha); (Δ) = anaerobic sludge (6 ton organic matter/ha); (Δ) = anaerobic sludge (20 ton organic matter/ha); vertical bars = LSD (P = 0.05).

The decrease of nitrogen, particularly in its available forms in the rhizosphere, indicates the actual fertility or the soil's ability to support the plant culture.

General remarks can be made as follows:

- During the vegetative cycle, microbial activities, especially heteroauxin biosynthesis, are enhanced more in the rhizosphere than in the soil.
- A marked nitrification occurs, whereas no differences are recorded in the denitrification (Figures 8.9b, c, 8.10b, c).
- The increase of total and available nitrogen in the soil and its strong decrease in the rhizosphere than can be regarded as indexes, respectively, of the long-term and immediate fertility of the soil.

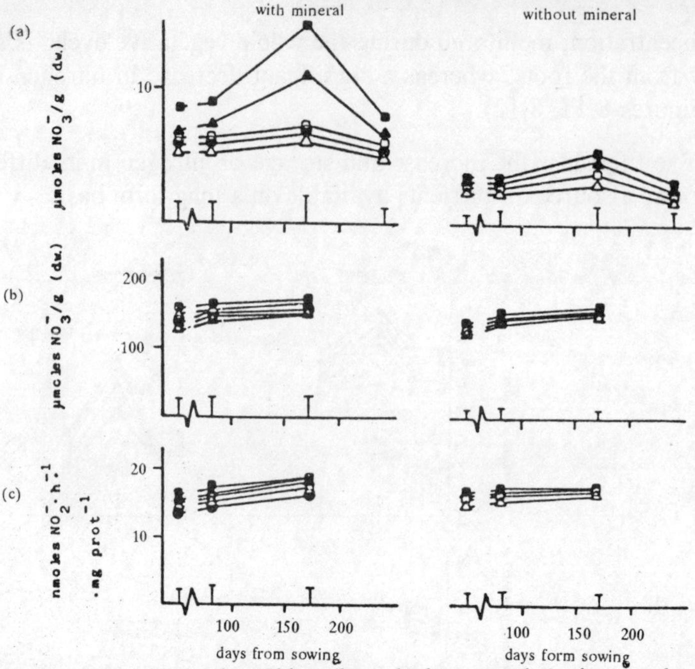

Fig. 8.12. Nitrate content in a) wheat rhizosphere, b) leaves and c) nitrate reductase activity; field experiments. (●) = mineral; (□) = aerobic sludge (6 ton organic matter/ha); (■) = aerobic sludge (20 ton organic matter/ha); (Δ) = anaerobic sludge (6 ton organic matter/ha); (Δ) = anaerobic sludge (20 ton organic matter/ha); vertical bars = LSD (P = 0.05).

Changes in Soil and Rhizosphere Parameters After Repeated Sludge Treatments

Land application of organic manures not only increases plant production, but also the humus content and the soil microbial activities. Sludge application usually has a great beneficial effect, made more and more evident as mineralization and humification occur. The objectives of the investigations presented in this section are mainly aimed at studying the effect of long-term heavy applications on microflora and related activities (oxygen consumption and auxin production) and on soil nutrient content, particularly nitrate and total nitrogen.

Contemporarily, some of the plant's metabolic parameters were tracked for a general view of the conditions created in the plant-soil system by sludge treatment. Data collected every year before the treatments, and at the end of the emergence time, when microbiological parameters are most enhanced, are discussed here.

The temporal sequence of the parameters tested every year before the treatment or at the end of the emergence time clearly shows the long-term effect as dependent on the doses and the origin of the supplied sludge, with the aerobic one being the most effective.

Among the components of the soil microflora, only bacterial population is enhanced; no significant differences were determined for fungi and actinomycetes. This becomes more evident over time when the ratio of bacterial to fungi increases. The improvement in such a ratio is a positive statement about soil fertility (Figures 8.13a-c).

Auxin biosynthesis and oxygen consumption also increased, showing an improvement in the biological conditions of the soil. This is confirmed by nitrogen content, especially in its available forms (Figure 8.13b).

Plant analyses seemed to support the better edaphic and environmental conditions created by supplying sludge. Nitrate content in the plant increases, showing a stimulation in nitrate uptake. As nitrate reductase activity value confirms, the plant is able to control $N\text{-}NO_3^-$ uptake and its organicity (Figures 8.14, 15).

As for crop yield, there are no remarkable differences between yield (as ton/ha and Kg of protein/ha) obtained after sludge and after mineral treatment, although a slight improvement was recorded since the first year after aerobic sludge treatment. As for anaerobic sludge treatment, the improvement became evident after the third year, when crop production reached the control values.

Seven year-long experiments performed by Statelmann and Furrer confirm that longterm application of sewage sludge improves soil and microbiological activities (Table 8.3). An increase in soil humus content, nitrogen content, cation exchange capacity, heterotrophic soil microorganisms, and their activities were recorded in heavy and light soil although to different degrees. These effects, more evident in light soils when heavy amounts of sludge were supplied every year, were recorded not only in the surface soil layer but also in the deeper ones to a depth of at least 1 meter.

An increase in soil organic carbon content and in humification was also recorded after 6 years of experimentation by Levi-Minzi et al. who studied the composition of the transformed fractions or soil organic matter through the determination of humic and fulvic acids. They found that the composition of the humified fractions were practically the same as most of the fulvic acid fractions.

The incorporation of sludge into the soil generated humified substances, aerobic sludge being more effective for this than the anaerobic.

Only slight differences among the characteristics of the humified substances, which were comparable to those of farmyard manure, were recorded. The evaluation of type and content of humic substances in sludge was studied more deeply by De Nobili et al.

The increased humus content helps to explain the enhancement in microbial activities and the effects on crops. In fact, humus stimulates nutrient uptake and assimilation, has colloid and buffering properties, a positive influence on protein synthesis, and displays hormone-like activity. All these properties are dependent on the level of humification which is able to justify the multiplicity of the effects on plant metabolism so often recorded in short and long time spans.

An enrichement in soil nutrients, humus, and organic matter after heavy repeated sludge applications has been often recorded. Koskela concluded, after 10 years of experiments, that dewatered sewage sludge is valuable to soil and could have a very positive effect on yield and on the nutrient storage in the soil. Obviously, the effect varies depending on soil, crop, and year.

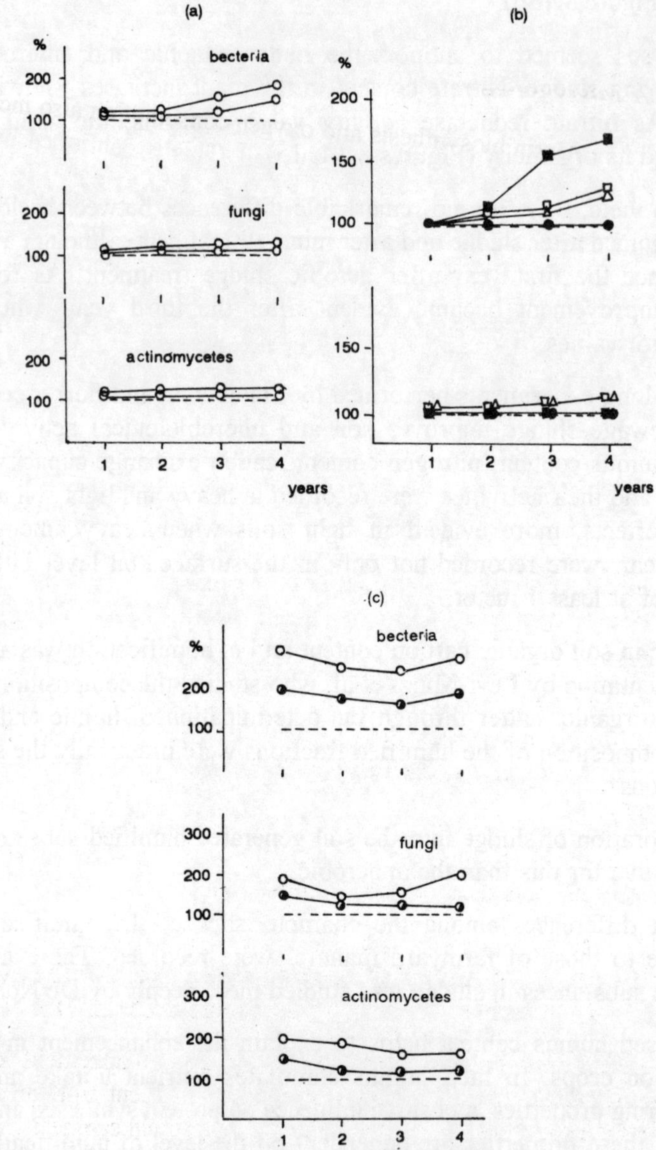

Fig. 8.13. a) Microbial population into soil before sludge treatments;) O = aerobic sludge; O = anaerobic sludge. b) Biological activities into soil before sludge treatments; ■ = total nitrogen; ● = nitrate content; Δ = auxin production; □ = oxygen consumption. I° = aerobic sludge; II° = anaerobic sludge. c) Microbial population in maize rhizosphere after the treatment (at emergence time). O = aerobic sludge, O = anaerobic sludge.

The mineralization of organic matter in sludge stabilized by different processes has been the object of investigation by Dumontet et al. who monitored carbon and nitrogen mineralization, and also followed the variations in the microbial groups responsible for mineralization processes. They found a general improvement in nitrogen mineralization, an increase of soil microbial biomass and hydrolitic activities depending on the origin, the kind of sludge, and on the composting process which the sludge underwent.

The importance of sludge origin and climatic factors on the availability of nitrogen has been studied in relation to its efficiency. Experiments performed by Barideau and Impens showed that nitrogen availability seems to be strongly dependent on sludge origin more than climatic factors, which condition the rate of mineralization and humification.

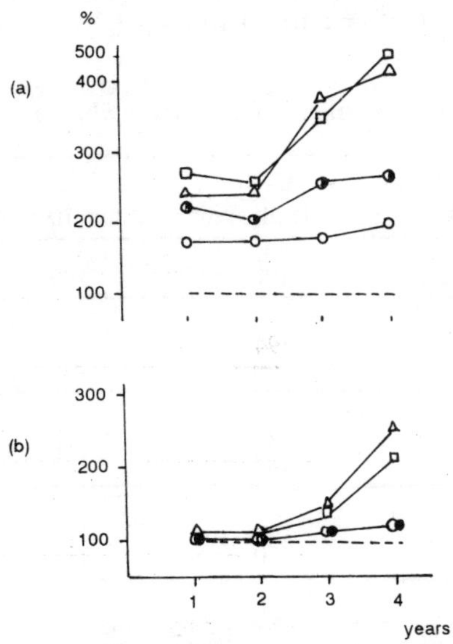

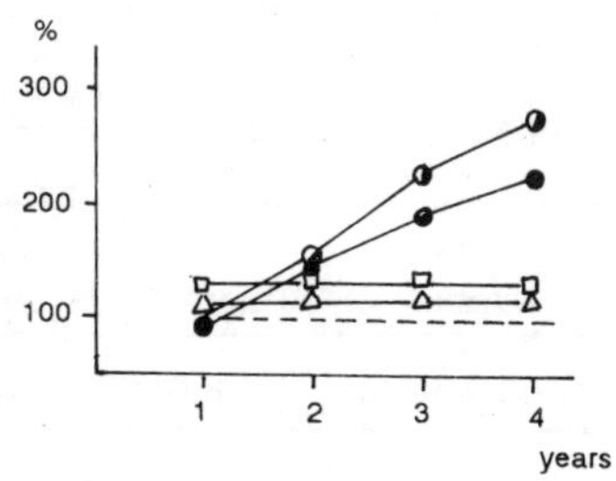

Fig. 8.15. Biological parameters in the plant after sludge treatment (at emergence time). O = nitrate content-aerobic sludge; O = nitrate content-anaerobic sludge; □ = nitrate reductase activity-aerobic sludge; Δ = nitrate reductase activity-anaerobic sludge.

Fig. 8.14. Biological activities in maize rhizosphere after the treatment (at emergence time). a) aerobic sludge; b) anaerobic sludge. O = total nitrogen; O = nitrate content; Δ = auxin production; □ = oxygen consumption.

The possibility of reducing the amounts of additional N-fertilizer required when sludge is applied either on a regular basis to farmland or in a large one-time application has been suggested by Hall. Experiments performed on a light soil confirm that in the long run sludge supply increases the amounts both of available and total nitrogen. A positive after-effect of sludge application through an improved nitrogen and phosphorus content was recorded too. Application of sewage sludge resulted in an enrichment of the soil microelements, most of them heavy metals, whose level remained well below safe limits for soil.

General remarks can be made as follows:

- The main component from the point of view of sludge utilization is the organic matter and the nitrogen contained in it. Sewage sludge is also a useful source of phosphorus for crops over time. Availability of nitrogen obviously depends on the status and composition of the organic matter.
- The long-term effect of sludge is still an object of discussion, since variable residual effects are often recorded. Sludge mineralization is in fact strongly dependent on climatic factors, soul conditions, and application time.
- Residual effects are much more evident in those situations where sludge is applied regularly every year over a period of 3 or 4 years. Then the cumulative residual effects are reflected in improved crop production (Table 8.4). To obtain any large residual nitrogen benefits from a single application of sludge, it must be applied in such a large amount that it could prove harmful to the environment.

Table 8.3
Microbial Soil Properties (Surface Layer 0-20 cm) After 7 Years Application of Sewage Sludge

Fertilizer Treatments	Aerobic Bacteria	Actinomycetes	Catalase Activity	Phosphatase Activity	C-Mineralization	N-Mineralization
Mineral	100	100	100	100	100	100
4 ton/ha org. matter	125	—	110	109	94	102
5 tons/ha org. matter	157	149	127	145	183	121
15 tons/ha org, matter	138	—	121	106	170	114

Table 8.4
Influence of 4 Year Application of Aerobic or Aerobic or Anaerobic Sludge (20 Tons Organic Matter/ha/year) on Maize Crop Yield

Treatment	1 Year		2 Year		3 Year		4 Year	
	Crop Yield	Protein Content	Crop Yield	Protein Content	Crop Yield	Protein Content	Crop Yield	Protein Content
Mineral	100	100	100	100	100	100	100	100
Aerobic sludge	83	113	113	104	113	119	102	111
Anaerobic sludge	92	86	104	86	114	119	105	106

- In the case of phosphorus, sludge has been shown to be capable of maintaining soil phosphate status and appears to be able to improve the "available" soil P values.
- Sludge supply, in the long run, increases microflora and stimulates microbial activities, mineralization, and production of growth regulators. This could be a

direct effect through increased organic matter in the soil, or an indirect effect related to improved soil structure and organic matter humification.

- The increased humus content in the soil stimulates nutrient uptake and assimilation and plays a beneficial role on plant metabolism through a hormone-like effect.

Sludge Effect on Soil Structural Stability

The adhesion of mineral and organic particles, to form soil aggregates, is the most important physical characteristic of soils. The shape and physical packing of aggregates influence, among other things, aeration, infiltration of water, water holding capacity, surface area (and thus availability of channels through which roots may grow), the availability of sites for nutrient uptake by plants, microbial and fungal activity, movement of microfauna, the proportion of aerobic compared with anaerobic sites and volumes, and, through these things the fertility of the soil.

A good soil structure is essential in promoting fertility and plant growth. Soil structure, especially in fertile soils, is a consequence of biotic processes that build up a stable granular structure, rather than of abiotic processes. Soil structure formation is a gradual process requiring the intervention of living organisms and the availability of organic residues as starting material (Figure 8.16).

The ability of soil microorganisms, particularly fungi, to bind soils into stable masses was first observed by Waksman and Martin and after confirmed by McColla, Swaby, Harris et al. Aspiras et al., Gasperi-Mago and Troeh, and Foster.

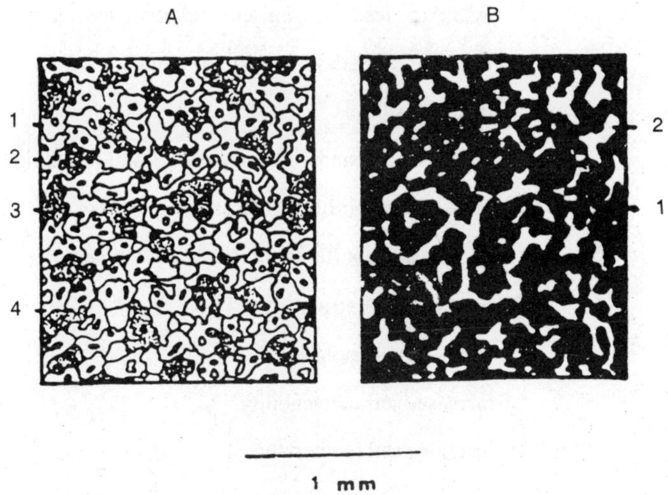

Fig. 8.16. a) Porosity of a well-structured arable soil (scheme) : 1) thin capillary pores; 2) medium sized pores; 3) capillary pores; 4) large pores. b) Reproduction from a microsection of the visible porosity of soil aggregate: 1) micro aggregates; 2) visible pores. Source: Krasil'nikov.

The role of bacterial in soil aggregation seems likely to be providing a cementing action between particles through their excreta, particularly polysaccharides.

Many bacterial extracellular polymers and humic compounds aggregate all soil particles into crumbs, which are then shaped, cracked, and pored by the members of macroedaphon. Many of these creatures, swallowing soil particles to facilitate their passage, contribute to soil formation through amalgamation of organic substances and mineral material. A well-structured soil can be imagined as a porous system where microorganisms live and process organic residues, thus providing organic matter to soil.

Deterioration of a favorable structure often occurs as a result of soil management operations. Permeability, aeration, porosity, volume-weight, and degree of aggregation can undergo marked changes; the reduced extent of aggregation, the reduced volume of pore space, and the increased volume-weight usually provoke a decrease of humus, starting the soil fertility involution.

A close relationship exists between the level of organic matter in soil and its physicochemical properties. However, little is available about the effect of sludge on the complex of properties responsible for physical fertility.

In general, the application of sewage sludge, and of other organic wasters to soil improves its structure. An increase in the number and size of water-stable aggregates has been made evident by Epstein, and Morel and Jacquin in pot studies, and by Kladivko and Nelson, and Pagliai et al. in field experiments. These authors demonstrated that the increase of the structural stability of the soil is strongly dependent on both sludge and soil properties, in which organic matter plays an essential role (Figure 8.17).

	- delays the leaching of macro and micro nutrients
	- favors nutrients availability for plants
	- modifies soil pH
chemical	- reduces the toxicity of heavy metals and polluting organic compounds
	- favors the solubility of inorganic elements
	- increases the buffering properties of the soil
	- improves soil structure and porosity
	- increases soil moisture
physical	- increases soil permeability
	- increases soil temperature
	- favors microbial activities
biological	- increases enzymatic activities
	- stimulates root development

Fig. 8.17. The role of organic matter in the soil.

The dominant role of microorganisms in the improvement of aggregation, when organic matter is added to soil, has been pointed out by Lynch and Bragg. The enhancement in soil structural stability is dependent on the quality and properties of added materials. Easily-decomposed components are more effective over a short time, whereas the slowly decomposed ones are more effective over a longer time, although with less effect.

Relationship between the microbial biomass and the stability of soil aggregates has been shown by Lynch, who studied the relative contribution of different microbial groups to the formation and stabilization of soil aggregates, and the microbial mechanisms involved. Fungi seem to be more effective in stabilizing aggregates than bacteria and actinomycetes. Mechanisms producing soil aggregates are believed dependent on both a mechanical action through fungal hyphae, and secretion of cementing substances of which microbial polysaccharides are the most important factor. (Figure 8.18)

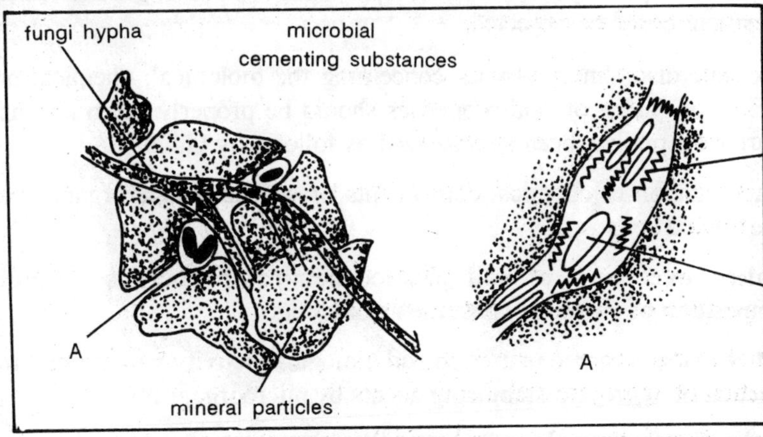

Fig. 8.18. Representation of a soil aggregate

The dynamics of aggregation in a waste-amended soil seem dependent on two phases the first is mainly an additive phase in which the labile part of the substrate undergoes microbial degradation; the second one, a stabilization phase, is characterized by a process which "stabilizes" the aggregates, i.e., links them to clay substances.

In a recent paper Metzger *et al.* discussed the effect of sewage sludge on soil structural stability from a microbiological point of view. They concluded that, although there is controversy about the way in which organic amendments induce the formation and stabilization of water-stable aggregates, microorganisms are a fundamental component deeply involved in it. In particular, fungi and actinomycetes act by holding primary particles of soil mechanically. Bacterial polysaccharides may serve as cementing agents and are responsible for the resistance of aggregates. Changes in soil structure and in microbial population in a long-term field experiment were observed by Guidi et al. Seasonal variation of the stability of soil aggregates, total porosity, pore size distribution, mycorhizal infection, and aerobic cellulolitic microorganisms have been followed. They concluded that seasonal variation in the number of microorganisms may explain the stability of soil structure, which increased coincidentally with the increased activity of celluloytic microorganisms and vescicular-arbuscolar mycorhizae. The microbiological origin of most processes that improve physical properties in sludge-amended soil, and the role of microorganisms in improving aggregation when organic materials were supplied to soil has been stressed by Guidi and has been the subject of an exhaustive review by Lynch and Bragg.

General conclusions can be drawn as follows:

- There is evidence that sewage sludge improves structure, hydraulic characteristics, and cation exchange capacity although the literature dealing with the effect of sludge organic matter on physicochemical properties of soil is not exhaustive.
- Improvement in physical soil characteristics has been recorded in soil dressed with high amounts of sludge. Therefore the use of sludge as soil conditioner should be encouraged, especially when the content of organic matter in soil is low and no harmful effect (mainly linked to heave metals, pollution of ground water, and over-fertilization) could be expected.

In any case, interdisciplinary studies concerning the biological, chemical and physical aspects of the effect of sludge on soil properties should be properly taken into account. The need for the future research has been summarized as follows by Guidi.

- Characterization of chemical constituents present in sludge organic matter and in sludge treated soils
- Variation over time of soil physicochemical properties according to the decomposition of chemical constituents of sludge
- Effect of sludge organic matter on soil biological activity with special regard to the production of aggregate-stabilizing agents by microorganisms
- Pore-size distribution; shape and spatial arrangement of soil pores
- Pedogenetic processes affecting porosity.

Conclusion

Two kinds of fertility-promoting processes proceed simultaneously in all soils: a mineralogic fertility, derived from the composition of the parent rock and from its release rates of plant nutrients, and a biological fertility derived from the rate of decomposition of organic and inorganic components and the consequent release of nutrients for recirculation in the plant-soil system.

The causal agents of biological fertility and, in part, of mineralogical fertility, are soil organisms. Among them, microorganisms have a particular importance, improving the efficiency of energy transport through the plant-soil ecosystem, thus enhancing soil fertility. Applying sludge to soil has several beneficial effects. Since sludge contains considerable quantities of both organic matter and plant nutrients, it acts as a soil conditioner and fertilizer. To study the properties that determine the crop capacity of soil as influenced by sludge treatment it is first necessary to know soil properties and choose, among them, those which determine soil productivity. The principal properties of soil fertility are determined by biological factors, mainly microorganisms. "The notion of soil is inseparable from the notion of the development of organisms living in it... when this life dies or ceases, the former soil would become an object of geology."

The knowledge of life in soil and the investigation of the biochemical processes taking place in it should be a way, probably the most important way, to understand the effect of sludge and determine its fertilizing value.

Results dealing with microflora and its activities in sludge-treated soils show an undoubtedly stimulatory effect both on soil and on rhizosphere activities. The interaction between soil microorganisms and higher plants is very complex. Plants excrete, via their roots, various organic and mineral substances that attract microorganisms that in turn influence plant physiology by their metabolic products.

The beneficial effect of sludge on crop productivity-thus on soil fertility-is demonstrated by much literature, evidencing the increase in crop yield and quality, especially when heavy and repeated does were supplied.

The increase of soil microbial activities and organic substances may explain this effect that in other words may be defined as the long-term fertility of soils. The improvement recorded in microbial rhizosphere activity is an important index showing that, after sludge treatment, the actual fertility of soil tends to increase. The microflora of the root zone is of great importance in plant nutrition. Growing near or on the roots, microbes are able to influence nutrient absorption and plant metabolism.

It is important to note that, after the cultural cycle of the plants, soil and rhizosphere properties tend to return to the original levels existing before sludge treatment. This could be considered in terms of non-toxicity of sludge or in terms of homeostatic power of the soil. As Alexander says:

"... when the environment changes, self-regulatory mechanisms or homeostatic reactions come into play to restore the relationship previously existing among the residents and to maintain the relative constancy in the make-up of the community. Biological alterations occur as a result of environmental stress, but the processes contributing to homeostasis tend to restore the initial steady-state situation."

Anyway, it is evident that further investigation on soil microflora as influenced by sludge treatment over both short and long terms in necessary; the main aspects to look into are the effect of soil and climate characteristics, and the origin and doses of sludge on the composition of soil microflora.

It is important to define microbial ecology, to investigate the specificity of microbial interaction with the plant, and its reciprocal interaction.

All of this can help to explain the positive action of sludge and organic fertilizers on crops, which cannot be justified only by the content of the mineral nutrients therein, but is also related to the presence of microbes and their metabolites. Therefore, it is apparent that mediators exist between the nutrient sources of soil and the root system. Such mediators are the microbes and their secondary metabolites, e.g., phytohormones, which can be considered a very interesting and worthy field of study, not only for microbiologists but also for agronomists.

9

Sorption of Mercury in Soils

Introduction

Mercury is one of the compounds considered most hazardous for the environment. As an element it is persistent, and it may form fat-soluble, readily bioaccumulating forms. Most of the mercury in human diet comes from aquatic food chains, mainly fish. As mercury may be mobilized from soils and transported into aquatic systems it is important to know the behaviour of this metal in soils.

Mercury has been widely used as fungicide and for many industrial purposes as well. Consequently, mercury may be found in agricultural land, municipal sludge, and municipal and industrial wastes. This anthropogenic mercury may be transported via air and water and bioaccumulated in the environment.

Mercury in the Soil

Occurrence

The background level of mercury in air, soil, and water is normally very low. The mean concentration in the earth's crust is 0.05-0.08 mg/kg. Igneous rocks and minerals often contain less than 0.01 mg/kg. Higher concentrations are found from sedimentary rocks and organic-rich soils. The most important mercury mineral is cinnabar (mercuric sulfide: HgS). Most of the mercury ores are situated, geologically, in relatively young areas with volcanic activity.

The mercury content of soil does not generally exceed 0.15 mg/kg except in contaminated soils and soils rich in humic matter. Reviews of mercury concentrations in different soils are given in the literature. Barber and Steel calculated a mean of 0.036 mg/kg from 120 soil samples from the central USA. In Saskatchewan, both cultivated and uncultivated soils usually contained 0.01-0.04 mg/kg of mercury.

Normally, higher contents are recorded from organic soils than from mineral soils, and higher concentrations in topsoil than in deeper layers. In Finnish soils, the mercury concentrations were significantly higher at the soil surface (mean 0.1 mg/kg) than at a depth of 10 cm (mean 0.02 mg/kg). The mercury contents of Greek agricultural soils ranged 0.025-0.1 mg/kg and no significant differences between different horizons (0-180 cm) or soil types were detected. In about 700 humus samples from South Norwegian forest soils, the mercury concentrations ranged 0.02-0.55 mg/kg with an arithmetic mean of 0.19 mg/kg. Di Giulio and Ryan found 0.04-0.19 mg/kg of mercury in North Carolina peat cores without noting any depth-related trends.

Dudas and Cannon noted clear seasonal fluctuations in the mercury contents of L, F and H horizons (Figure 1).

Inorganic Forms

Mercury has three oxidation stages: 0, +1, and +2. Metallic mercury (Hg^0) is volatilc, and is the most important form of the element in the atmosphere. Three fixation sites are possible for soil minerals (in decreasing order of importance): Fe-, Al-, and Mn- oxides/ hydroxides; clay particles; and silica. Mercury has also a great affinity to sulfur (HgS) and chlorine ($HgCl_2$, $HgCl^-_3$, $HgCl^-_4$).

Organic Forms

Most of the mercury in soils is adsorbed on sites with high binding energy, which means that mercury is very strongly bonded to soil humic matter. Both inorganic and organic forms of mercury have a great affinity to SH-groups. This affinity to a biologically active group is the most important explanation for the great toxicity of this metal. Mercury forms monomethylmercury (CH_sHg^+), the covalent binding of which is very strong. Monomethylmercury is often bound to proteins. It penetrates biological membranes and may thus damage the brain and fetus. Monomethylmercury is readily bioaccumulated in aquatic food chains and this compound is mainly responsible for the toxic effects of mercury. Dimethylmercury $(CH_3)_2Hg$ is a volatile compound which is formed mainly under alkaline conditions.

According to Hogg et al., the sorption of three different forms of mercury to soil increases in the order: methylmercury chloride < phenylmercury acetate < mercury chloride.

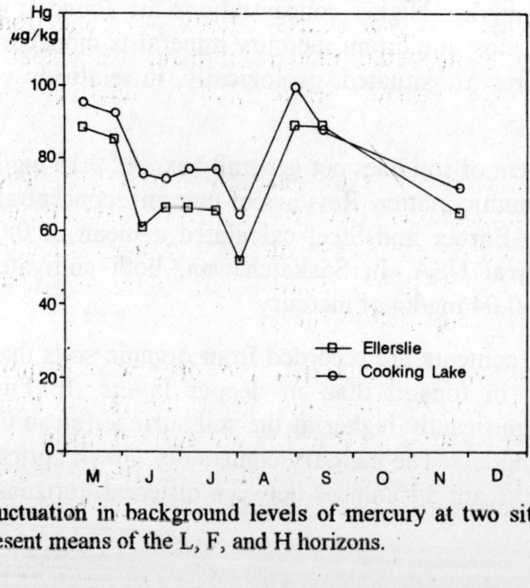

Fig. 9.1. Seasonal fluctuation in background levels of mercury at two sites in Canada. The values represent means of the L, F, and H horizons.

SOIL QUALITY FACTORS AFFECTING THE SORPTION OF MERCURY

Physical Properties of Soil

The physical properties of soil affect the sorption and transport of metals. Coarse gravel has essentially less capacity to bind metals than finer soil materials. Slit contains more mercury than sand, and clay significantly more than silt. Mercury sorbed to quartz sand may be linked to the porous matrix either by clay minerals or Fe, Mn, or Al hydroxides. Significant quantities of mercury are removable from these hydroxides by addition of NaClO.

Farrah and Pickering studied the uptake of mercury by three clay minerals and found the following order for uptake: illite > montomorillonite > kaolin. The amount of mercury adsorbed by three clay minerals decreased in the order montmorillonite > vermiculitechlorite > kaolinite, while the strength of binding followed the order vermiculite-chlorite > montmorillonite > kaolinite. Under acidic conditions ion exchange is the most important form of sorption to clay minerals, while mercury adsorption is assumed to occur mainly in the form of $Hg(OH)_2$ at higher pH-values.

Organic Matter

The content of organic material affects significantly the metal absorption capacity of the soil. In most cases the humic binding is far more important than the mineral retention. This binding is strong but reversible. At low mercury concentration, soil organic matter is responsible for most of the mercury sorption, while mineral components participate in the sorption at higher mercury concentrations.

In studies of the sorption of mercury in soils with different contents of humus, a significant sorption was recorded even in pure quartz soil. However, this mercury-contrary to that bound to humus-containing soils-was easily removed by addition of artificial acid rain. The results obtained by Semu et al. suggest that soil organic matter plays an important role in the sorption of inorganic mercury, while the mineral fraction is more involved in the sorption of organic mercury.

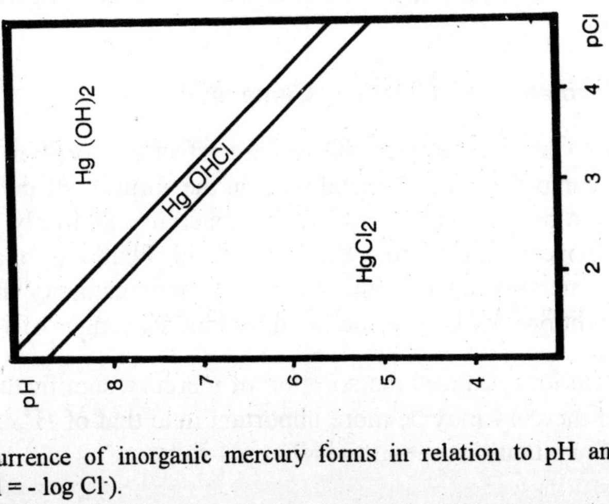

Fig. 9.2. Occurrence of inorganic mercury forms in relation to pH and chloride content ($pCl = -\log Cl^-$).

Mercury is bound to humic matter by exchangeable cationic and by non-cationic sites simultaneously. Using selective extractions diGiulio and Ryan found that most of the mercury was associated with organic matter: humic acids, fulvic acids, and organic sulfidebound fractions. They did not record any mercury from the water-soluble or ion-exchangeable fractions.

In a Iysimeter study, two different mercury peaks were found from a peat soil, indicating the possibility of a transformation of the added mercury acetate and sorption as two different compounds. This transformation was inhibited at low temperature (+5°C) and by addition of acid rain.

Mercury is also strongly bound to thiohumic acid, and Yoshida et al. concluded that the coordinate bond between COS^- and mercury contributed markedly to the sorption of mercury on thiohumic acid. This sorption was not affected by addition of $HClO_4$, Ca^{2+} or Fe^{3+} and only scarcely by addition of HCl or NaCl. Abiological methylation of mercury by humic acid has been confirmed in vitro, but it is known whether this phenomenon is of importance under natural conditions.

pH

Data concerning effects of pH on the sorption of mercury in soils are partly contradictory. The maximum sorption of mercury for several soil types is in the pH range 4.75-6.50. At low pH values the pH-related changes in sorption are insignificant. Leaching of mercury from peat soil decreases with decreasing pH. On the contrary, mercury adsorbed to an inorganic matrix (pure quartz sand) was easily removed by acid rain.

According to Farrah and Pickering, the uptake of mercury by clay minerals decreases with increasing pH. However, Semu et al. found an increasing uptake of mercury from a methoxy-ethylmercury chloride (Aretan), there were only insignificant changes in sorption with pH. Wimmer found a higher washout of mercury from acid soils ($pH < 5$) than from alkaline ones ($pH>7$). Obukhovskaya found an increasing sorption to clay minerals with increasing pH reaching a maximum at pH 6. At higher pH values, mercury complexes (e.g., $Hg(OH)_3^+$) may occur.

Ion Exchange and Formation of Complex Compounds

Other metals (Na^+, Ca^{2+}, or Fe^{3+}) or NO_3^- do not affect the sorption of mercury. Stronal and Huljev studied the competition of 13 metal ions on the sorption of mercury on humic acid in aquatic medium. None of these ions could replace mercury and only the addition of new mercury ions could remove mercury from the humic acid. Haidouti et al. found a positive correlation between the mercury content and cation exchange capacity in Greek agricultural soils. Under certain conditions MnO_2 may be an important scavenger of Hg(II).

Addition of chloride ions reduces the sorption of mercury significantly and influence of Cl^- on the adsorption of mercury may be more important than that of H^+. Addition of chloride also shifts the region of maximum sorption to higher pH values.

Chloride complexes with Hg(II) are possible in aqueous solutions already at concentrations above 10^{-9} M. $HgCl_2$ is formed above 3 x 10^{-8} M. $HgCl_3^-$ and $HgCl^{2-}_4$ are formed above 10^{-2} M. On activated carbon, Hg(II) may be strongly adsorbed as halide complex HgX^{2-}_4 and the addition of halide ions even increases the adsorbability in the order I > Br > Cl > F. EDTA may complex mercury and thus enhance the leachability, but high concentrations of the complexing agent are needed.

Temperature

Little is known about the effect of temperature on the sorption of mercury in soils. For clay minerals the sorption was greater in + 3°C than in +18° C, while the sorption to iron hydroxide increased with increased temperature. Yoshida et al. observed a greater sorption of Hg (II) with increasing temperature (0 - 40° C) in an acid medium (0.1 N Hcl), while a slight decrease was recorded in an alkaline medium (0.01 N NaOH).

In laboratory experiments, differences between sorption in different temperatures have been detected. Significantly stronger leaching was found in a sandy soil at + 5°C than at room temperature.

Under moist conditions (60-100% of field capacity) the sorption of mercury to a grey forest soil significantly decreased with increasing temperature. Under dry conditions (air dry soil and 30% of field capacity) there was no significant temperature effect (Figure 3)

Mobilization of Mercury

Leaching of Mercury from the Soil

Fang studied the sorption of mercury vapor (Hg^0) in dry and moist soil columns and found an increased sorption with increasing soil moisture until a maximum was reached near the maximal water-holding capacity. This might have been due to microbial activity.

Under dry conditions there are no significant effects of temperature on the sorption of mercury in soil. In a Iysimeter study greater amounts of artificial precipitation resulted in increased amounts of mercury being leached out. A complete drying of peat soil enhanced the leaching greatly, probably due to the changes in the physical properties of the peat.

Humic substances increase the solubility of mercury by formation of water-soluble complexes. The binding of soil mercury seems to occur especially to relatively easily water-soluble humic compounds. In a peat soil, significantly more mercury was bound to the high molecular (humic) fraction than to the small-molecular (fulvic) fraction (Figure 4). At pH 7.0 there is an increasing binding to the humic fraction in an aquatic system with increasing chloride content (0-10 mag/1). At lower pH values (5.5 and 4.0) this binding is always strong. In aquatic systems the maximum mercury sorption to humic acid occurs at pH 4-5.

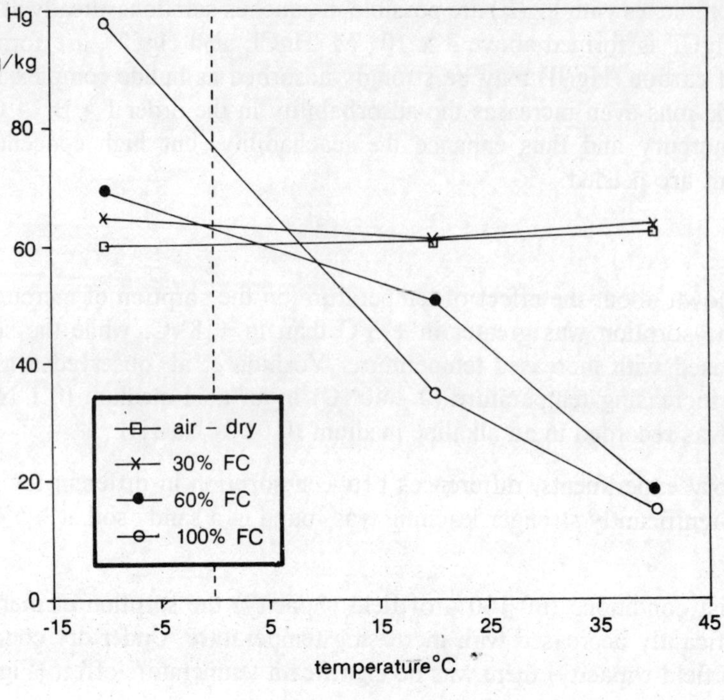

temperature °C

Fig. 9.3. Influence of temperature of mercury to soil after four HCl-extractions. The moistures of the soils were: air dry, 30%, 60%, and 100% of field capacity respectively

Evaporation of Mercury

The retention mercury is much stronger and the volatilization smaller in peat than in the sandy soils. In a soil column experiment, Wimmer found a significant evaporation of mercury from soils with poor sorption capacity (small humus content). Mercury may be evaporated

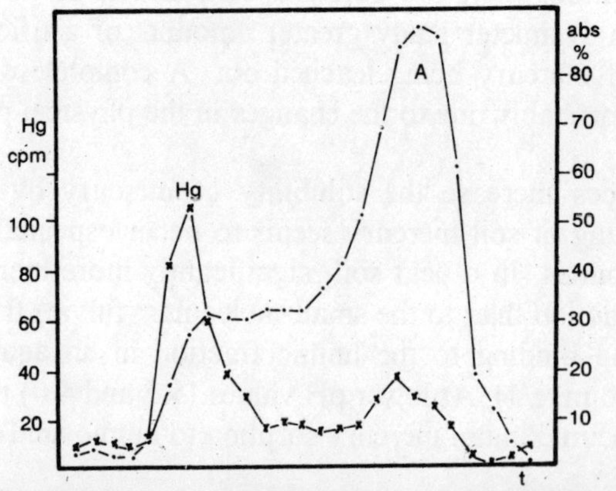

Fig. 9.4. Distribution of ^{203}Hg (as cpm) after get filtration (Sephadex G-100) leached in six days through a peat column in relation to the absorbance (270 nm) of the leachate. High molecular compounds are leached out first.

from soils as both ionic and non-ionic forms (Figure 5). Bacterial and other microorganisms are able to reduce Hg^{2+} to Hg^0, which is readily volatile. Little is known about the effects of liming on the formation of dimethylmercury and other volatile compounds.

Dudas and Pawluk found mercury to be translocated in soil horizons and in many cases lost from surface soil, probably through evaporation to the atmosphere. Soils amended with $Hg(NO_3)_2$ lost 5-30% of the added metal by evaporation to the atmosphere. Soils amended with $Hg(NO_3)_2$ lost 5-30% of the added metal by evaporation. As autoclaving reduced the loss rate and addition of glucose enhanced it, microbial it, microbial activity seems to be an important factor in the evaporation of mercury. The mercury loss increases with increasing temperature.

Mercury emitted from industrial processes or by burning of wastes or coal may spread hundreds of kilometers through the atmosphere. Even if considerable amounts of mercury may be present in landfills, no mercury was detected in the air above a medium-size Finnish landfill.

BIOLOGICAL FACTORS AFFECTING THE SORPTION OF MERCURY

Microbiology and Methylation

Many mercury compounds are strongly toxic to fungi and other microorganisms and have therefore been used as fungicides and bactericides. Mercury compounds may inhibit CO_2 evolution, dehydrogenase activity, and nitrification. Phenylmercury acetate is more toxic to soil microorganisms than $HgCl_2$ and the effects are more pronounced in sandy soils than in

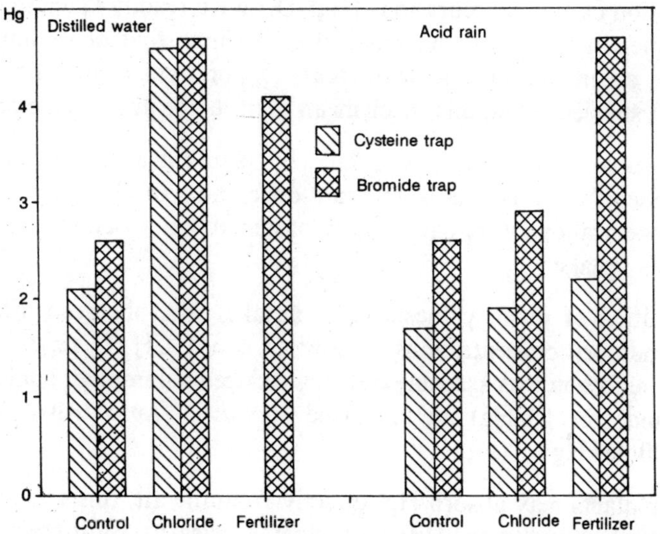

Fig. 9.5. Evaporation of mercury from peat columns. Mercury was trapped by two different traps intended for either ionic mercury (e.g., Hg^{2+} and Ch_3Hg^+; cysteine trap) or non-ionic mercury (e.g., Hg^o and $(CH_3)_2Hg$; bromide trap). The columns were treated with distilled water or artificial acid rain (pH 3.9).

clay soils. Inorganic mercury compounds Hg^0, HgO, $HgCl_2$, Hg_2Cl_2) have a strong inhibitory effect on Azotobacter and nitrifying microorganisms. Ag(I) and Hg(II) are the most effective in inhibition of nitrogen mineralization. However, mercury compounds may even stimulate the growth of certain microorganisms like Actinomycetes.

An addition of fertilizer caused decreased leaching and increased evaporation of mercury from a peat soil, which could have been a result of stimulated microbial activity. However, in a subsequent experiment no effect of fertilizer or sterilant addition to a similar peat soil could be observed.

In an aquarium experiment, significantly higher concentrations of mercury were found from aquatic biota after addition of inorganic mercury at low oxygen conditions, compared to high oxygen conditions. This could be a result of greater availability of inorganic mercury for methylation. Miller and Akagi found also an increasing methylation at decreasing pH. On the contrary, Steffan et al. found that acidification strongly decrease the methylation of mercury in lake sediments. Mercury-resistant strains have been reported for soil bacteria and fungi, which in most cases obviously are able to reduce mercury to elementatl (volatile) mercury.

Vascular Plants

The uptake of mercury in vascular plants is usually poor and the background concentrations in leaves are low. The mercury concentrations are usually lower in fruits, grain, and leaves than in other parts of plants. The background concentrations of mercury in plant foodstuffs usually vary within 0.003-0.1 mag/kg. The uptake is most efficient in young plants, and mercury forms that may be absorbed are Hg^0, Hg^{2+}, CH_3Hg^+, and $C_2H_5Hg^+$.

In a solution culture mercury ions ($HgNO_3$) were readily taken up by plants. The uptake of inorganic mercury seems to be more effective than that of organic mercurials Catalase seems to play an important role in mercury vapor uptake and this is true especially for gramineous C_3 species, while this mechanism is of lesser importance for C_4 species.

When comparing the mercury uptake by plants near mining .areas and geothermally active areas, Siegel et al. noticed significant differences in plant concentrations even in cases where soil concentrations were equal. They assumed that specific local environmental factors must affect this uptake.

An inhibition of photosynthesis (CO_2 uptake) was observed after exposure of spruce seedlings to mercury concentrations as low as 1 nM CH_3Hg or 10 nM $HgCl_2$ in the soil solution. The inhibition was associated with a reduced chlorophyll content and transpiration in the needles and (for Ch_3Hg) an increased shoot/root ratio. Also the root elongations is inhibited at 100 nM Hg.

Vascular plants may absorb Hg^0 directly from the air, this fact is useful for monitoring of airborne mercury pollution. Different plant species have different capabilities to absorb mercury from the air. Most of this mercury is absorbed by the leaves but a small part is translocated to the roots. Vascular plants may also release mercury, but this ability is highly dependent on the plant species.

Fungi

In most macrofungi the background mercury concentrations are low (<o.5 mg/kg d.w.). However, in some lawn decomposer species (especially Agaricus spp.) the mercury content may be as high as approximately 100 mg/kg (dry wt) even in unpolluted environments, while the concentrations in mycorrhizal fungi normally are very low. Part of this mercury is in the form of methyl mercury. The mechanism of uptake is not fully known, but a positive correlation between the uptake of mercury and the catalase activity has been demonstrated.

Animals

Earthworms may readily accumulate mercury from the soil. The estimates of earthwormsoil ratio vary widely: from approximately 0.4 (dry: wet) to approximately 20 (wet: wet). Mercury is present in earthworms partly in the form of methyl mercury and earthworms obviously are an important link in terrestrial accumulation of mercury. Mercury concentrations of 0.5-5 mg/kg in the soil increase both the mortality and the reproduction of earthworms. The concentrations decrease with increasing size (age), and small specimens show faster uptake and excretion than large ones.

Although insects seem to have an effective excretion system for mercury, at least for its inorganic forms, this metal may be toxic to insects. Concentrations of 1-10 mg/kg significantly reduce the embryonic and post-embryonic development of grasshoppers.

Small mammals belonging to terrestrial ecosystems normally contain small amounts of mercury, while similar species may accumulate considerable concentrations from aquatic food-chains.

HUMAN IMPACT ON MERCURY FLUXES

Mercury is released to the atmosphere from industrial sources (chlor-alkali plants, base metallurgy, etc.) and burning of coal and municipal wastes. This mercury may spread over vast areas and elevate the background mercury values even in remote areas.

Acidification of precipitation affects both terrestrial and aquatic ecosystems. It is known that fish in acid lakes usually contain more mercury than fish from more neutral lakes. This is obviously not a result of increased leaching of mercury from the soil, but instead a result of some other chemical and biological processes. Elevated mercury concentrations have also been recorded from fish in man-made lakes, and lakes where the water level is regulated for hydroelectric and forests may contain elevated concentrations. In these cases mercury has obviously been mobilized from the inundated soil, attached to humic compounds, methylated, and bioaccumulated in aquatic food-chains.

Mercurial fungicides have been widely used in the agriculture. Most of the mercury is obviously retained at the soil surface, but in some cases, significant amounts of mercury may leach out. Considerable amounts of municipal and industrial wastes containing mercury have been put into landfills. Niebla et al. found mercury to be much more mobile in the presence of (simulated) landfill leachate than in other solutions. This was thought to be due to the presence of organic compounds. Elevated concentrations of mercury have been found from vegetation in the immediate vicinity of landfills.

10

Agricultural Pollution

Very broadly, agricultural pollution is caused by refuse of any form from agricultural operations of any kind. In common usage the term agriculture refuse generally includes the following types of waste.

(i) Manure and other wastes from farms and the operation of feedlots or poultry houses.

(ii) Slaughterhouse wastes.

(iii) Fertilizers runoff from cropland.

(iv) Harvest wastes.

(v) Pesticides that escape in the atmosphere or into the water supply.

(vi) Salt and silt drained from irrigated land or eroded land.

Air and water pollution has affected agriculture adversely and agricultural methods have themselves contributed to environmental pollution.

Pollutants from Feedlots and Poultry Houses

Scientists believe that liquid wastes from the feedlot industry may seriously contaminate the water table. Ammonia volatilizes from liquid feedlot wastes, but because of its ready solubility in water, it returns to nearby lakes and streams. Feedlot wastes have high biological oxygen demands; their nutrients upset natural ecological communities and nourish plant growth; water purification plant in communities downstream are taxed with increased burdens of sludges and have more difficulties in eliminating objectionable tastes and odors from the water. Feedlot runoffs may also add disease causing organisms to the water.

Farm animal wastes cannot be avoided, however, and their presence poses serious problems of odor and water pollution in addition sometimes to public health problems arising from their role in transmitting disease. The magnitude of the potential animal waste problem dwarfs that of the human waste problem.

The amount of nature of the wastes produced will depend on the type of animal and the diet (including feed additives) on which it is fed. Most of the wastes consist of undigested foods, largely cellulose and fibers. Data on wastes are difficult to collect because of the variations in their nature, the differing moisture contents (the elapsed time until measurements are made can be important), the amount of bedding present, etc. The usual parameters for the description of municipal sewage-BOD, COD, suspended solids, etc.-were developed to charcterize highly liquid wastes consisting mostly of water and are not as applicable to animal wastes, which are quite solid. The total weight values for BOD and suspended solids, however, can be compared with those from the human population to give population equivalent values.

Animal wastes have not been as great a problem in the past as these numbers might suggest because they have traditionally been returned to the soil so their plant nutrients and organic matter would maintain and improve the quality of the soil. The organic matter in livestock manure improves the tilth of the soil and its capacity to hold water, aids in the aeration of the soil, lessens the amount of erosion produced by wind and water, and has a beneficial effect on soil micro organisms. Returning manure to the land - as the Amish in this country are still careful to do-is a form of recycling.

Animal wastes are a major problem once they are permitted to enter water supplies, when confinement areas are cleaned or when runoff carries the wastes into waterways.

Current interest in animal waste control centers around the use of aerobic and anaerobic biologically active systems and the spreading of wastes on the land, although such methods as incineration, dehydration, and compositing are also studied. In Europe, biological methods have found some use but the chief emphasis has been on returning the manure to the land, and research and development work has centered around better methods of hauling and handling the wastes. The availability of inexpensive commercial fertilizers, which are easier to handle because the nutrients are more concentrated, has led farmers to seek ways of disposing of animal waste rather than reusing them, and European farmers appear to be moving in the same direction.

Anaerobic digestion systems have proved successful in the laboratory and in the field under some conditions. Anaerobic lagoons are built with a small surface area, as deep as construction and ground water conditions will permit. Cold weather can interfere with proper operation and solids-with a high nitrogen content (such as chicken manure) may lead to odors from ammonia production. Methane gas is produced by anaerobic digestion, although its production becomes minimal when the temperature of the system is below about 13 or 14°C. The Mother Earth News (No. 10) carried a story about an English farmer who runs his automobile on methane gas produced from chicken manure in a special digestion tank; since this is completely equivalent to using natural gas, which is relatively nonpolluting, his automobile produces very little air pollution.

The purpose of anaerobic lagoons has been to remove and destroy much of the organic matter and they can be combined with aerobic systems to give an effluent that is of adequate quality for discharging into rivers. The anaerobic system accepts slugs from a feedlot operation, for example, and partially degrades and gasifies them, while the aerobic unit degrades the remaining soluble and particulate matter.

Aerobic units are generally shallow oxidation ponds or lagoons that actually have anaerobic conditions near the bottom. Although they work well, they require very large surface areas—perhaps 8 hectares (20 acres) for 1000 head of cattle and 2 hectares (5 acres) for the same number of hogs. Sometimes the amount of land needed can be reduced by the use of mechanical of diffused aeration systems.

Manure disposal on the land is not fundamentally complicated but confinement feeding leads to difficulties in using this method. A poultry unit housing 100,000 hens may produce 11 metric tons of manure daily, and its application to the soil at a rate of 18 metric tons/hectare (8 short tons/acre) annually means that some 220 hectares (540 acres) will have to be found for disposal. The European philosophy seems to be that the manure will have to be disposed of anyway so one might as well return it to the soil and make use of its nutrients.

Insofar as nutrients are concerned, it is most efficient to haul manure into the fields while fresh and work it into the soil immediately. Nutrients can be lost by volatilization (especially of nitrogen and organic matter); leaching of soluble components by rain; or by loss of the liquid urine, which contains about half the nutrients. In general, it is preferable to apply fresh manure to the fields daily rather than to let it rot first; however, this is often impractical. In winter the soil is frozen and unworkable much of the time in many parts of the country, and if manure is spread over snow-covered ground, it is likely to pollute waterways as soon as a thaw occurs. During the growing season, crops are on the soil and manure cannot be worked into it.

Several processes are being investigated which make use of very fluid (diluted) manure which can be spread thinly on the land, injected into the subsoil, or sprayed on the land. Spray irrigation of grasslands can be used for manure as well as for food processing wastes (discussed in Chapter 10). Many feedlots are building detention ponds (which are sometimes required by law) to catch runoff from rainstorms, and the water can be pumped out and sprayed over fields. One hog-feeding facility in Oregon which has about 15,000 hogs at any given time periodically washes the pens with water which is flushed into lagoons, from which it is pumped out after dilution into a sprinkler system onto 200 hectares of grainfields, hayfields, and pastures. On another small farm in Nebraska, wastes from 200 to 300 head of cattle are drained into a lagoon, from which they may be pumped out (even when up to 25% solids) to irrigate and fertilize cornfields.

Fertilizer Runoff

Farmers distribute fertilizers extensively on croplands, pastures, and orchards. These fertilizers are important to high-yield agriculture. It is assumed that one third of the country's annual production of food and natural fiber is attributable to ferilizers. their extensive use, however, creates some problems. Some (but much less than half) of the phosphates entering water come from fertilizer runoff.

Nitrates are especially susceptible to runoff with each rainfall, and they enter groundwater. Where they are extensively used, they appear in well water.

Harvest Wastes

These include the straw left over from harvesting wheat and other cereal grains, corn stalks, the remains of annual plants raised as vegetables, and similar materials. Straw is a commodity used in agriculture as bedding material, as mulch, and as an absorbant for liquid wastes in barns. (It has little food value for livestock). Straw has been used as an absorbing agent in cleaning oil spills from surfaces and shore areas. Much of the wastes from harvests are shredded and worked into the topsoil, where the material eventually decays and provides some enrichment for the soil. Unless this procedure is carefully managed, these wastes may harbor pests. Where the law allows open burning, some harvest wastes are burned. Due to the oil embargo of 1973, there has been increased interest in using fermentation to convert feedlot and harvest wastes into methane or methanol.

Pesticides in Runoff

Pesticides in soil are carried into lakes and streams in the dissolved state, as well as soil particles in the runoff. It is believed that the pollution of ground water by pesticides has been found to be more widespread than commonly has been realised but the concentrations (less than 1 microgram per litre or less than 1 ppb) are very low. Serious poisonings of well water by pesticides usually occur by accident during the handling of the chemicals.

Soil Erosion

Sediments—(soil and mineral particles) get washed into streams by storms. The amount of soil erosion occurring in the world today may be two or three times what it was before man's intervention in nature. Some 3½ billion metric tons of sediments are washed into tributary streams each year, about one-fourth of them eventually being transported into the sea. At least half of this sediment is coming from agricultural lands, with water erosion being a dominant problem. Eroded materials can also be contributed by urban, industrial, and highway construction sites and by road banks, and urban areas can be major sediment sources.

Sediments have been both a blessing and a curse to agriculture. They have added to the fertility of the soil where deposited, as in the Nile River Valley in Egypt, and they have settled in and destroyed irrigation systems such as those of the ancient Sumerians. Sediment can settle on trout spawning beds and suffocate the eggs. It can clog gills of adult fish. It interferes with domestic and industrial uses of water and produce extra expenses for clarification of the water. It can lead to reduced dissolved oxygen levels and thereby adversely affect aquatic life. It affects the recreational values of lakes, reservoirs, and streams and shortens the life of farm ponds by producing an increased rate of filling through siltation.

In recent years it has been realized that sediments carry many inorganic and organic materials with them, often adsorbed to the surfaces of silt particles. In particular, phosphorus adsorbed on silt and carried to the bottom of lakes and rivers may play an important role in eutrophication, and many water-insoluble pesticides (such as the chlorinated hydrocarbons) are carried into waterways by sediment. Control of soil erosion consequently looms as one of the most important present-day concerns in the control of environmental pollution.

Control methods exist that can cut erosion rates down to 10 to 50% of current rates without affecting agricultural productivity. These methods act to reduce the kinetic energy of raindrops and flowing water and keep the soil particles in place. The presence of vegetation, either living or dead, reduces erosion. If the soil is covered with grasses or trees, the vegetation will intercept raindrops and prevent them from imparting their energy to soil particles. Much of the rainfall will be absorbed in the ground instead of flowing over the surface of the land and this, too, leads to reduced erosion. It is well-known that in periods of heavy rainfall, flooding will not occur in forest lands but will occur on nearby land which has been developed commercially.

Control methods are especially desirable on sloping lands since the amount of erosion increases rapidly with increasing slope. In addition to planting suitable vegetation, the land may be shaped to reduce the erosion. Even in ancient times good agricultural practice included the building of terraces that followed the contour of the land and forced downward-flowing water to encounter a series of flat pieces of land. Terraces can take a long time to build and can make it difficult to apply mechanized agricultural techniques, straighter and thus easier to farm. Erosion can also be reduced by growing different crops in alternate strips ("strip cropping"), using broad, gently sloping waterways covered with a thick carpet of grasses, and by preventing overgrazing of grassland by livestock, which destroys the vegetative cover.

Wind erosion is a related problem. Strong winds can carry soil particles into the air and thus produce the air pollution problems associated with airborne particulate matter. Many fields filled up with sand dunes and many homesteads were covered up. These problems are alleviated by a good vegetative cover that can be maintained on the land. Most serious wind erosion is likely to accompany periods of drought where maintenance of vegetative cover is difficult. Residue from the previous crop left on the surface of the soil will reduce soil loss from wind erosion. The growing of hedges and other windbreaks (junipers, ponderosa pines, etc.), has also helped in the deserts.

Plant Residues

Plant residues from crops and orchards can constitute environmental pollution when they harbor plant diseases and pests or when they are burned and emit smoke and hydrocarbons, since both situations correspond to an unfavourable alteration of man's environment. Agricultural burning is a fairly important air pollution problem. It is estimated to be responsible for:

8.3 per cent of monoxide emissions.

8.5 per cent of particulate matter emissions.

5.3 per cent of hydrocarbon emissions.

1.5 per cent of nitrogen oxide emissions.

Agricultural refuse are burned. Most of this burning is for sanitation purposes since plant debris can carry diseases on pests to succeeding crops. Burning of grass residue is the only practical method of controlling blind seed disease by the Oregon grass seed industry. Agricultural burning and the resulting air pollution problems are likely to continue until better methods of controlling the plant diseases are available.

Plant debris can be used as a mulch; e.g., it can be spread over the ground under orchard trees. Mulching on a large scale can be effected on croplands by leaving residue from the previous crop on the soil surface. Mulching is known to reduce wind and water erosion and it sometimes (but not always) leads to improved crop yields. Plant residues are also used to some extent for other purposes, such as bedding for poultry and livestock or in the manufacture of corrugated cartons.

The logging debris are left in forests annually. This debris can act as a reservoir for tree disease and insects, just as elms killed by Dutchelm disease and oaks killed by oak wilt can transmit disease to healthy trees. Controlled burning is the only feasible disposal method at present. It causes air pollution problems, and has interferred with recreational activities and airport operations. Unburned debris from logging operations is a very serious fire hazard, which is known to cause larger-than-average forest fires.

Agricultural Chemicals

In recent years there has been an increased use of agricultural chemicals, notably pesticides and fertilizers. These chemicals have been effective in controlling diseases and pests that affect crops, live-stock, and even humans, and they have enabled farmers to secure high crop yields at relatively low cost. They contribute to air and water pollution, however, and have adverse effects on fish, wildlife, and upon occasion, human beings.

The availability and low price of commercial fertilizers has led to rapid increases in their usage. Problems cab arise from excessive application rates since the fertilizers can be transported into ground water by leaching or into waterways by natural drainage and storm runoff. Nitrates are of special concern but they generally find their way into waterways from feedlots, not from excessive or improperly timed fertilizer use. Nitrates in drinking water can cause methemoglobinemia in babies ("blue babies") because in an infant's stomach the nitrate (NO_3-) is converted into nitrite((NO_2-), which acts on the blood hemoglobin to form methemoglobin. The same process can occur in the stomach of ruminants so that livestock can also be affected by nitrate poisoning.

Plant nutrients also contribute to entrophication. Phosphorous is often implicated in algal blooms in waterways. Phosphates have a great affinity for soil and are generally carried into waterways adsorbed to sediment, in a state unavailable for plant growth.

There are some alternatives to the use of inorganic fertilizers on agricultural crops. One is the return to the methods of years gone by, including use of organic fertilizers (such as animal wastes) and crop rotation (such as planting legume crops occasionally to increase the nitrogen in the soil). Another possibility is that the government might find it necessary to institute controls on agricultural chemical usage.

Miscellaneous Agricultural Pollution

There are several other types of pollution that occur in agricultural areas.

Animal Disease Agents. Infectious agents transmitted by air, water, and soil can affect livestock, poultry, and humans.

Dead Animals. The disposal of dead animals is another problem. Poultry meat decomposes readily and produces objectionable odors, while the feathers are rather inert and stable. Incineration can be difficult and can produce air pollution. Disposing of carcasses by delivering them to rendering plants is probably the best solution. Sometimes, though, dead animals are thrown in old wells, where they may lead to groundwater pollution.

Rural Domestic Wastes. The domestic wastes in rural areas must also be considered. Generally the waste waters are disposed of in the ground through the use of pit privies, cesspools, or septic tanks. Sometimes the absorptive capacity of the soil is inadequate for proper disposal. The main concern is to prevent contamination of groundwater supplies or at least to prevent contamination of wells from which drinking water as obtained.

Food Processing Wastes. Industry in agricultural areas can also cause serious pollution. Of particular note would be industries closely associated with agriculture, such as the food processing industry. These industries can produce both air pollution (such as by cotton gins and alfalfa mills) and water pollution.

Salinity of Irrigation Water. Inorganic salts present in the soils and geologic materials of arid regions can move into waterways and be deposited on farms that use irrigated water, and the salinity of the soil can build up unless effective drainage is provided for.

Conclusion

In conclusion, it should be noted that the greatest agricultural pollution problem at present is probably soil erosion by water and wind, especially since many other agricultural pollutants (such as pesticides and phosphates) are transported by sediment.

The animal waste problem is already important in some areas such as near large feedlots, and could be a problem of overwhelming importance if no animal wastes were returned to the soil.

It should be recognized that our agricultural problems are aggravated by our eating habits. Agricultrual problems would be lessened if we were vegetarians since the use of grains and grasses and other plants for animal food, which is then used for human food, is a rather inefficient process. Putting a given acreas into soybeans for human food would yield over 10 times as much protein as putting it into corn to be fed to livestock and eventually converted into meat protein. Many countries obtain little food from animal sources; meat consumption in China and India is very close to zero.

PREVENTION AND CONTROL OF LAND POLLUTION

The industries should install collectors to remove the particulate waste from the chimneys. Appropriate methods have to be found out to dispose off the use the other type of pollutants. The garbage should not burn in the open, but can be used to produce energy as well as filler for cement, bricks, ashpalt, pavings. some of these wastes, if they could be shifted and separated, can even be recycled as raw material for industries.

The sanitary landfill has been regarded as a better remedy for large objects because it causes inexperative biodegradation of such trash without causing much pollution, disease, etc.

It is also desirable to fill low-lying watery areas and ditches by this stuff and land thus reclaimed may be used for making gardens, parks, playgrounds or even apartment complexes. Realistic management of domestic wastes by municipalities has been a must for proper disposal and economic recycling.

Animals refuse and agricultural waste can find use as manure. They can be used for the production of biogas that in turn can be used to generate electricity. Sludge from biogas plants after proper conditioning may find use as soil builder and fertilizer.

India is operating a pilot project to extract combustible gas and thereby electricity from a garbage dump when the Delhi Energy Development Agency under the auspices of the Department of Non-Conventional Energy Sources of the government of India though of tapping the Sanitary Landfill at Timarpur in Delhi in 1986. The area may be able to generate 8 MW of electricity from landfill which is spread over an area of about 80 acres.

The cost of generating at the project in some what 10 paise per unit compared to 62 paise per unit through conventional sources of electricity generation. It has been estimated that one well will be sufficient to extract gas from one acre of landfill and well has a capacity to produce 100 KW and from 10 acres, it is possible to produce one MW. At present there are ten wells which have been dug up at a distance of 200 feet from each other.

It is possible to extract the gas by drilling perforated pipes into the landfill, piping them together and pumping the gas to a common plant through collecting pipes, This gas can be used to produce power through IC engines or gas turbines or cooking fuels after cleaning.

The department of Non-conventional Energy Sources has commissioned another pilot project where advanced technology has been tried to dispose of municipal solid waste by incinerator. It uses 300 tonnes of garbage per day generating 3.75 MW of power. The engineering work here is entrusted to Bharat Heavy Electricals with Volound Miljoteknik, a Danish firm.

The specialised equipment which is not produced in India has been imported from Denmark. If the project becomes successful the Department of Energy may install bigger incinerators at different places in the city so that the cost of carrying garbage and power can be maintained low.

Making biogas from the garbage will also decrease land pollution to an appreciable extent. A break-through has been achieved experimently in a joint venture which has been launched by the Burn Standard Company Ltd., Central Mechanical Engineering Research Institute (CMERI) of the CSIR and the Calcutta municipal corporation. This works on the fact that any organic matter on biodegradation yields methane, carbon dioxide and trace of carbon monoxide, hydrogen sulphide and some oxides of sulphur and nitrogen accompanied by obnoxious smell. It is a very well known natural phenomenon which is brought about by a group of microbes called methanogenic bacteria, but these bacteria act in an efficient manner only in anerobic condition, that is, in the absence of oxygen and at 37 degree centigrade temperature.

The US scientists are able to convert organic waste such as rubbish, wood, shaving, rags and even cow manure—into clean high energy oil or gas. In this process, there occurs the break down of organic wastes into oil, gas and a combustible hard waste by heating it to temperature in the absence of oxygen. This process is called pyrolysis. The advantage of pyrolysis over the burning of wastes is that it is able to produce a transportable fuel. The U.S. Bureau of mines has developed another efficient method by which it is possible to convert wastes into oil by treatment with carbon monoxide and water under pressure. By following another method they are able to convert manure into a pipeline gas by treatment with hydrogen at high temperature. Thus, we can realise that there has been no waste in the nature which cannot be recycled and put to use for our daily benefit.

11

Environmental Injury to Plants

Introduction

Environmental injury to plants is a study of the various facets of the environment such as drought, heat, chill, freezing, high salinity, and oversaturation of the soil with water that initiate a high-enough level of stress within the plant to cause an observable or hidden injury. Such damage can either be irreversible or countered by an alteration of the physiological and/ or metabolic reactions that normally take place within the plant cell. Such studies are of importance in that ;they determine the environmental limit of valuable agronomic or horticultural plant population boundaties. In addition ;such studies can also help to engineer various strategies that enhance some of the coping mechanisms of the plant to stave off injury.

In the past decade there has been a gradual change in the earth's climatic patterns. These changes are thought to be attributable in part to the large-scale interference and influence of mankind on the global environment. Massive production of carbon dioxide from industrial factories and greater use of the internal combustion engine are causing a gradual increase of the mean global temperature. The temperature increase is thought to occur when radiant solar energy from the sun cannot be easily reflected back into the outer atmosphere due to a gradual increase of a carbon dioxide blanket in the upper atmosphere. This trapped high energy is then converted into a lower energy form represented by hear. Such a process is known as the "greenhouse effect." The latter is also enhanced by the excessive release of fluorinated hydrocarbons into the atmosphere. These compounds react with atmospheric ozone to destroy portions of the ozone layer that protects the earth from excessive high-energy radiation (such as ultraviolet light). Thus more heat results from the extra radiation that is allowed to pass through the partially fragmented ozone cover from the sun.

Normally, much of the accumulated carbon dioxide can either be absorbed into the ocean or photosynthetically converted into oxygen via the vast areas of rain forest foliage scattered throughout the world. However, with the increased population of mankind and the obvious need to clear land for settlement and grazing much of the primeval rain forest areas are being systematically destroyed. Therefore, the carbon dioxide blanket can exist as a stable entity or imperceptibly increase over a finite amount of time.

With such a significant and stable rise in the mean global temperature, the stage is set for gradual but dramatic shifts and changes in climatic patterns that can enhance several kinds of environmental stress responsible for a large-scale injury and death of important horticultural and agronomic plants that supply food and fiber for humanity. In this article, the following types of stress that reflect these global changes of climatic patterns will be emphasized: chill and freezing injury due to the abnormal onset of winter in regions not normally affected by stable climatic conditions; drought and heat stress above that normally encountered in regions now becoming more arid in nature: osmotic stress that occurs after an extensive accumulation of salts on irrigated arable land located in the arid regions; and finally, anaerobic stress produced by soaking and flooding of fertile areas of soil subjected to greater-than-normal rains and ice thaws during the spring season.

II. Chill and Freezing Injury

Plants that originate from warm climes are susceptible to low chilling temperatures (from 0 to 15 C) in that they show signs of injury, such as wilt and bleaching, as well as changes in physiology that effect seed germination, growth, and photosynthesis. In some cases even reproductive development is affected. Examples of valuable crop plants that are cultivated in warm weather are cotton, soybeans, tomato, rice, corn, sorghum, and cucumbers. Besides the obvious visual appearance of ;chilling damage, a more subtle level of damage is evidenced by a multiple array of biochemical and cellular changes. Included are changes in rates of respiration, protein synthesis, protoplasmic streaming, altered membrane permeability, and inefficient photosynthesis. In addition, chilling exposure in the light is more deleterious than under darkened conditions.

In contrast to chilling, freezing damage is much more severe. In plant tissues, crystallization of cellular water as the temperature drops below 0 c can occur anywhere inside the cell and cause lethal damage of a physical nature to the cell. If many cells in the plant are affected in this manner, death of the plant can occur on thawing. In addition to ice formation within the cell (intracellular), the formation of ice can also take place in the intercellular spaces at subzero temperatures. When such an even takes place, liquid water will move from inside of the cell (due to a vapor pressure difference between the water and the ice crystals) and contribute to the growing ice crystals on the outside of the cell. As this process continues, the cell or cells will become dehydrated and undergo a condition similar to that of osmotic shock induced by a high salt concentration outside the cell.

A component of the cell that is initially acted upon by both chilling and freezing events is viewed by most investigators to be the plasmalemma membrane (the membrane on the exterior of the protoplast, in turn surrounded by the plant cell wall); this known as the "membrane hypothesis". This is where the common ground of the two events (chilling and freezing) ends, however. The actual mechanism of lesion formation of membrane hypothesis proposed thermotropic lipid phase transitions of the membrane alteration differs in both cases. For chilling, the original concept of the membrane hypothesis proposed themotropic lipid phase transitions of the membrane that is induced by low temperatures (0 to 15 C) in chill-sensitive plants for all classes of lipids. Such a temperature-induced liquid-gel phase was thought to initiate conformational changes in membrane-biochemical processes. Notable among these processes were a cessation of protoplasmic streaming, respiration rate changes, alterations in photosynthesis, and modification in protein synthesis. The initiating process was shown to be reversible as opposed to the time-dependent and nonreversible secondary or cascading events.

Recently, the membrane hypothesis became refined to a certain extend when the proposed role of the lipids as a whole in chilling sensitivity became more focused on one molecular species of the polar lipid fraction-namely, phosphatidylgltcerol. The influence of this small subpopulation of the membrane lipids on the phase separation event (and thus the relative sensitivity to chilling temperatures) depends largely on the phosphatidylglycerol molecular species composition (i.e., on the specific ordering of saturated and unsaturated fatty acid carbon chains on the phosphatidylglycerol molecule).

It has recently been proposed that the first important and perhaps controlling factor of the "cascading effect" is initiated by the altered permeability of the chill biological membranes. Under normal conditions, both chill-tolerant and chill-sensitive plant cells maintain a homeostatic level (about 10^{-7} M of calcium (Ca^{2+}) in the cytosol. Recent studies show that chilling temperatures may increase Ca^{2+} by reducing the rate at which the chill-sensitive plant cells exclude Ca^{2+} from their cytosol; or alternatively, chill temperatures may activate voltage-dependent cation channels that transport Ca^{2+} into the chill-sensitive cell. A higher-than-normal level of Ca^{2+} in the cytosol is generally toxic and can lead to the death of the chill-sensitive cell due to an imbalance of the calcium/calmodulin activating system for inactive proteins or enzymes.

As with chilling injury, destabilization of the plasma membrane is also a primary site of freezing injury; it too plays a prominent role in cellular-biochemical behavior during a freeze-thaw cycle. Different opinions, however, are presented as to the cause and nature of the investigations measure the expression of injury after the freeze-thaw event. Recent investigations of a direct and ;less inferential nature, in which the plasma membrane of the plant protoplasts was observed during a freeze-thaw cycle, were able to characterize the conditions that lead to the destabilization of the protoplast plasma membrane. Supercooling at a specific temperature range in a given plant-cell cytosol is a function of the rate of heat transfer, which is then relative to the mass transfer of water. Water flux is determined by the water permeability of the plasma membrane. Generally, the greater the cooking rate, the greater is the incidence of intracellular ice formation, which is usually disruptive and lethal to the cell. Thus a plasma membrane with a relatively lower water permeability would tend to negate or slow down the formation of ice crystals in the cytosol.

Under conditions of nonintracellular ice formation, injury or plasma membrane destabilization during a freeze-thaw cycle seemed to be caused by two main factors: (a) osmotic shifts from the inside to the outside of a cell and vice versa, and (b) disorganization of the plasma membrane components induced by serve dehydration during freezing. The extreme changes in osmotic pressure during a typical freeze-thaw cycle usually cause lysis or cell bursting, which is one of the predominant forms of cell injury. Osmotic contraction results in the formation of large unilamellar vesicles or exocytotic extrusions from the plasma membranes of both nonacclimated and acclimated protoplasts. However, during osmotic expansion, the extruded material that is drawn back into the plasma membrane behave differently between both types of protoplasts. The cold-acclimated protoplasts produced extrusions that were continuous with the parent bilayer during the contraction phase so that when expansion took place, the extruded material was smoothly drawn back into the bilayer. The extrusion withdrawal of nonacclimated protoplasts, on the other hand, does not proceed as smoothly as the acclimated protoplasts, and lysis generally results. The differential behaviour of the plasma membrane observed between both types of protoplasts is attributed to changes in lipid composition. Recently, it was demonstrated that alteration of the proportion of a single lipid species in the plasma membrane was responsible for the difference in stability between both types of protoplasts is attributed to changes in lipid composition. Recently, it was demonstrated that alteration of the proportion of a single lipid species in the plasma membrane was responsible for the difference is stability between both types of protoplasts.

If the freezing conditions are such that no ice crystals are formed in the cytosol of protoplasts frozen to temperatures below -5ºC, more than 90% of the osmotically active water is removed from these cells. Under such conditions, lamellar to hexagonal rearaangements are induced in the ultrastructure of the plasma membrane, which results in a loss of semipermeability. The protoplasts are then in an osmotically unresponsive state, which leads to biochemical and physiological imbalances preparatory to injury or death. The exact mechanism that causes dehydration to induce the particular type of membrane rearrangements mentioned above has not yet been resolved. Several attractive theories have been advanced; however, most of them have to do with forces of hydration repulsion between the two bilayer components of the membrane and the pressures required to overcome the hydration forces so as to induce several different types of change in bilayer structure. Although these pressures ar large, they do occur from the osmotic potential increase during freezing of the aqueous solutions that surround and are included within the protoplasts at temperatures ranging from -10ºC to -20ºC.

Cold acclimation for many species of plants in the temperate and alpine regions involves an initial exposure over a given period of time to low nonfreezing temperatures (0-10 C) so that the plants will be able to alter the freezing tolerance of their cells. Although intensive study has been given to the cold acclimation process, not much is really known about the mechanisms that lead to increases in freezing tolerance, especially with regard to distinguishing between the specific-biochemical changes that assume the main role in freezing-tolerance mechanisms from those that take place in adjustments to temperature in the nonfreezing or chilling ranges. In spite of such a lack of definitivde knowledge in this area, there is increasing evidence that implicates the involvement of a genetic component connected with the plants ability to acclimate to chilling temperatures followed by the induction of freezing tolerance. Plants grown and evolved in tropical regions do not have this ability, as seen by experiments in which a plant hormone, abscisic acid, is unable to turn on genes that cause cell culture of freeze-sensitive species to become more freeze tolerant. On the other hand, cell cultures of cold-hardy species can respond to this plant hormone and develop tolerance to chilling and freezing temperatures.

Although widely accepted that cold acclimation follows a well-defined inheritance pattern, there appears to be disagreement as to whether cold acclimation is a multigenic process or whether a few genes control cold acclimation as seen with the heat shock and anaerobiosis models. Since cold acclimation is an inducible trait that depends on changes in cellular physiology, cold acclimation genes can be studied by means of standard molecular and DNA recombinant methods. Most of these studies have dealt with nucleic acid and protein changes in cold-hardy plants. At the onset of acclimation DNA remains constant while an accumulation of RNA occurs. Most of the RNA increase was attributed to ribosomal RNA, which made up the increased level of ribosomes and ploysomes. The latter remain intact and are three times higher in content than that of non-acclimated tissues. This increase can be attributed to the need to synthesize more protein, since enzymatic processes are slowed down significantly at the lowered temperature. It would be of interest to know which mRNSs are being translated by the polysomes with regard to freezing tolerance. Recent attempts to clone low-temperature-induced genes revealed that three low-temperature specific cDNAs were isolated from the poly(A) RNAs (active messenger RNAs) induced by low temperatures.

The soluble proteins accumulate at the start of acclimation and closely follow the changes in tissue freezing tolerance. The appearance of a glycoprotein component in the soluble protein function was thought to be a factor for resistance to ice formation inside the cell. Recent protein labeling studies in plants exposed to low temperatures provide additional evidence for induction of changes in protein content by low temperatures. The pattern of protein synthesis different significantly from the of plants exposed to warm temperatures. Part to this change demonstrated the appearance of new protein polypeptide bands not present in non-freeze-tolerant plant species. It is not clear whether these protein bands are similar to those that appear as heat and osmotic shock proteins and whether they appear in many species of cold-acclimated plants with enough consistency to be termed "coldshock proteins"

In certain cold-acclimated plants there exists a group of soluble proteins of thigh molecular weight (10,000-20,000) capable of protecting membranes against freezing damage. Enzymes from cold-acclimated plants (such as esterases, phosphatases, amylase, and lactate dehydrogenase) also differ in isozyme patterns from those shown by nonacclimated plants.

Thus it appears that the mass of information gained in the study of cold tolerance is still in the descriptive stage and more work needs to be done to demonstrate whether the cold acclimation process is initiated by a few genes as seen in heat shock and anaerobioses.

III. Drought Stress, Injury and Adaptation

Lack of water is a stress factor that has been the source of much investigation, since this factor often limits agricultural productivity in arid and semiarid regions of the earth. Most of the important agricultural crop plants are either classified as hydrophytes (growth where water is superabundant such as rice) or mesophytes (where water is intermediate, such as cotton). Thus with changing global air movement and increasing, the plant population boundary with regard to drought is continually changing in certain parts of the earth. Drought injury manifests itself in plants that do not have built-in mechanisms to ward off the effects of severe lack of water (as opposed to Xerophytes, where water is searce) such as bound water, small transpiring surfaces, leaf abscission, deep and wide-spreading root systems, etc. Irreversible drought-induced injury in a typical plant cell appears to be caused by mechanical forces when the vacuole shrinks during drying with a resultant inward pull and a semultaneous outward pull on the cytoplasm as exerted by the cell wall to which the cytoplasm is attached.

The cell is contracted with a pulling ion of the wall and a visible torn protoplasm (cytoplasm plus plasmalemma membrane).

With a less severe drought where relatively smaller amounts of water are withheld, there is a slowing down or lag in cellular enlargement and subsequently in root and shoot growth in the plant. This lag period is followed by an adaptation period that consists of a sequence of events that cause the plant to resume normal operations on a less efficient but somewhat effective plane. During the lag period there is an alteration of biochemical and physiological processes ranging from protein synthesis to photosynthesis and solute accumulation. Throughout the stress period as a whole, the water and osmotic potential are reduced by equal amounts so that turgor pressure remains the same in the growing regions of the plants, even thought growth first stops and finally continues after the lag period is over.

$$\psi_p = \psi - \psi_\pi$$

where ψ_p is turgor pressure, ψ is water potential, and ψ_π is osmotic potential. In addition, decreases in water potential seem to coincide with the occurrence of growth retardation caused by drought stress. The factors involved in osmotic potential adjustment primarily involve molecules of a molecular weight of about 200, such as the hexoses glucose and sucrose. Both of these moieties take part in rapid osmotic adjustment which proline, an amino acid, is not involved until a longer period of continuous stress has taken place. Stresscaused growth reductions seem to be highly correlated with a reduction of polyribosomes to single ribosomes in the growing areas of the plant. Thus growth seems to be affected at the translational level of protein synthesis.

The anatomy of a plant, particularly in the expanded ad growing portions of the leaf, are critically related to the stress response. Mature and open vessel elements that transport water are present throughout the growing and expanded lead blade regions. Transpirational water will move throughout the organizational pattern of the mature and open zylem vessels. As noted above, a sudden osmotic stress causes an almost immediate growth cessation. Such suggests there must be a lateral transmission of the stress signal from the vessel elements to all of the intervening mesophyll cells. The signal consists of a reduction in the hydrostatic pressure of the xylem cells. The pressure change is probably transmitted through the pits in the vessel walls onto the continuous mass of water in the mesophyll cell walls of the growing region.The only difference between the growing versus the non-growing expanded regions (which are insensitive to the hydrostatic pressure change) is that there are more mesophyll cells between the intervening transpiration vessels in the expanded lead blade than in the area of leaf growth. As a result much of the transpirational water in the expanded lead tends to bypass most of the mesophyll cells on its wy to the stomatal regions. Such is the path of least resistance, since the apoplastic water (water between and in the cell walls of the contiguous mesophyll cells) does not pass easily across the plasma membrane and into the cells. Eventually, the mesophyll cells in the expanded portions of the lead blade will lose turgor as opposed to the growing areas. The blades turgor reduction occurs because the water loss through the stomate exceeds its ability to replace lost water and because much of the limited water is removed by the cells in the growing regions. It is proposed that the apoplastic water potential controls the direction of water flux. Since the growing region maintains positive turgor, normally and under mild stress, it is possible that the net water influx might control the release of factors necessary for cell wall extension. If the stress is great, on the other hand, adequate solute accumulation to reduce the cellular loss of water must occur.

One of the most critical factors that the plant uses to respond to low soil-water availability is the role of abscisic acid (ARA) influence on stomatal closure to control the transpirational loss of water. Rather than increased amounts of ABA in the vicinity of the guard cells to induce closure of the latter as formerly hypothesized, it is now believed that increased sensitivity of the guard cells to ABA is responsible for this phenomena. The initial ;event is a release of ABA from storage areas, such as the mesophyll cells and roots, to the extracellular space (apoplast). Once in this apoplastic region, the ABA can be carried by the transpiration stream toward the guard cells. It is known that guard cells do not produce their won ABA can be carried by the transpiration stream toward the guard cells. It is known that guard cells do not produce their own ABA and in addition do not have any plasmodesmata connections to other cells. Some reports have appeared that give evidence for binding sites specific for ABA on the guard cell membranes. Thus, the sensitivity of the guard cells toward ABA is related to the number and binding ability of these sites to the modest concentration of ABA in the surrounding apoplastic water.

The biochemistry of ABA synthesis in higher plants is much more difficult than that observed in fungi and involves a more indirect pathway in addition to the usual derivation from mevalonic acid through the isoprenoid pathway. Water stress, however, stimulates both biosynthesis and degradation of ABA, but the rate of biosynthesis usually exceeds that of catabolism during the drought period. This excess of ABa is generally high enough to maintain levels necessary to keep the stomata closed. Other ;uses for the high concentration of ABA that are concerned with drought (but which have no relation to stomatal closure) are the senescence and defoliation of older leaves on the plant. This action indirectly aids the plant in conserving much-needed water.

Mathematical models of the relationship between transpirational water loss and carbon dioxide (CO_2) uptake for photosynthetic reduction to starch precursors show that although both processes take place through the same stomatal openings in the epidermis of the leaf, their pathways vary within the leaf and are subject to different constraints. As a result, stomatal closure during drought stress will not affect these two processes in the same manner. It can be shown that stomatal closure diminishes water evaporation more than CO_2 diffusion into the leaf. However, under more severe drought conditions, this relationship does not hold because of the direct damaging effect on chloroplast metabolism. Since factors controlling the ABA release from the mesophyll are determined ;partially by photosynthetic reactions (water stress inhibits electron transport and disrupts the proton transfer process in the chloroplasts, causing a change in pH that in turn facilitates the release of ABA from the mesophyll cells), it is suggested that ABA could function as a "messenger" from the mesophyll to the stomates to also provide a fine-tuning effect on gas exchange in the leaves under normal drought conditions.

Recent investigations, using molecular biological techniques, show that applications of ABA caused the appearance of a new set of messenger RNAs that almost matched the set induced by wilting. Such results indicate that nearly all drought-initiated events are responses due to ABA.

IV. Heat Shock Responses in Plants

Drought is often accompanied or caused by high-temperature conditions, especially in the arid regions throughout our planet. Plants must alter their cellular biochemistry and physiology to survive such an environmental change that often rapidly occurs. Death of a plant cell and the resultant disruption of the protoplasmic structure is most likely to be induced by a rapid rise in temperature. On the other hand a gradual temperature increase causes a protein degradation that releases ammonia, which is toxic to the cell. Plants resistant to high temperatures (Xerophytes) possess high amounts of bound water an high protoplasmic viscosity. Unlike drought or freezing temperatures, plant acclimation (aside from the Xerophytes) to high temperatures is not as extensive or efficient.

In the past decade probably the most investigated of all stress responses is that induced by heat shock factors. the most easily noted response to heat stress, in plants as well as other organisms, is the initiation of heat shock proteins (hsp). It is thought that the stimulus perception for the induction of the heat shock response may involve several changes in the physiology or biochemistry of the heat-shocked cells. A change in the oxidative-reductive environment of the cell toward a predominantly oxidative state as induced by the rise in temperature somehow activates the heat-shock transcription factor so that it binds to the promoter region of the genes. Another possible alteration would be a resultant denaturation of proteins in the cell, which are then combined with a specific protease (known as ubiquitin) for degradation. Such a preferential binding of ubiquitin so that less is available to bind and inactivate the heat-shock transcription factor. A third possibility is the perturbation of the membrane ion pumps by heat shock so that the exchange of protons and potassium ions or the sequestration of calcium ions are in an imbalanced state, especially with regard to calcium, which is now available for the calcium/calmodulin activation of kinases (enzymes that add phosphate groups to proteins). Some of these kinases could probably be instrumental in the activation of the heat-shock transcription factor (through the addition of phosphate) which ;can now bind efficiently to the heat-shock protein (hsp) gene promoter. Even though the heat-shock proteins (in large, intermediate, and low molecular weight classes) are coordinately expressed under heat stress, the induction of these proteins is transient and lasts only for several hours in spite of a continuous omnipresence of the elevated temperatures. During this brief transient period of hsp induction, transcription of normal cellular proteins is momentarily suspended. In addition, the messenger RNAs (mRNAs) for the hsp are preferentially translated. The normal cellular mRNAs, however, remain stable and become reactivated during the recovery process.

In addition to a change in direction of physiological processes, heat shock also changes some aspects of cell ultrastructure that probably plays a role or is linked to some of the overall physiological changes. In certain plant systems, heat shock causes a definite disassociation of the endoplasmic reticulum (phospholipid membranes associated with ribosomes, polysome formation, and protein synthesis). As to the cause of this dissociation, it is proposed by some investigators that heat shock lowers the activity or enzymes responsible for the synthesis of endoplasmic reticulum (ER) or possibly enhances the activity of phospholipases a and d which break down certain phospholipids within the ER membrane. Membranes, as discussed in the sections on chilling, are very sensitive to temperature changes; and membrane fluidity may be a liability at higher temperatures, especially with regard to protein synthesis. It is probable that disassociation of the ER is an attempt to decrease the ER fluidity and thus maintain a reasonable balance between membrane integrity and protein synthesis at higher temperatures. Heat shock also produces a change in the ultrastructural appearance of the nucleoli. The latter exhibit less of a granular appearance and a lower heterochromatin content than under normal temperature conditions. This observation is also consistent with the heat-shock inhibitory action on RNA processing and resultant ribosome assembly.

The function of the heat-shock proteins is not wet clear, but several hsp have had certain roles ascribed to them in cell protection. One of these roles appear to be involved with the stabilization of important cellular constituents. Studies by means of specific poly and monoclonal antibodies in conjuction with cell fractionation procedures have been helpful in localizing the position of specific hsp in the cell. There is a strong association of these proteins with nuclear pellets. Other tests with this pellet fraction suggest that these proteins function in a structural capacity either inside or outside the nucleus as nucleoskeletal elements. Other hsp are localized in the cytoplasm, where they are believed to function in the protection (by sequestering) of normal cellular protein mRNA from turnover; they also ;stabilize important heat-labile proteins during heat shock. No hsp synthesis has been observed in plastids or mitochondria, but there are observations that hsp produced in the nucleus are transported to these particulates as well as to the chloroplasts. There transported proteins might be involved in the important physiological processes of the chloroplast, such as gene activation or photosynthesis. In addition to a stabilization role as discussed above, some of these proteins are involved in cell cleanup or recovery where protein aggregates are dismantled along with a continuation of proper RNA splicing.

Thus the heat-shock response to high temperatures aids the cells to survive and recover from an environmental stress that could do irreversible damage. Additional information on the role and significance of as yet unidentified hsp is an active on-going phase of research.

V. Osmotic (Salt) Stress and Trace Metal Tolerance

Along with drought, salt (or osmotic stress) is another widespread factor in the restriction of growth on large tracts of essential agricultural land on our planet. As salt accumulates in the soil from irrigation water, the plant is confronted with two problems-namely, obtaining water from a soil of moderate to high negative osmotic potential and coping with higher than normal concentration of potentially toxic chloride and carbonate ions. Although salt tolerance or hardiness can be increased in some measure by exposure to saline conditions, it is the accumulation of these ions that prevents salt hardiness to be on a par with drought or cold acclimation. [See Soil Salinity].

Essentially, there are tow basic classifications of plants with regard to the study of salt hardiness. Halophytes (such as *Atriples and salicornia*) grow only where there are high levels of slat in the soil, such as brackish waters or on near sea coasts. There are some halophytes that actually accumulate salt within their tissues. Glycophytes on the other hand perform better at lower levels of salt and include many of the important crop plants, such as alfalfa, cotton, barley, and tomatoes. These plants are usually termed salt-tolerant species.

A similar factor in common between plants that are exposed to either a dry or saline environment is that of growth inhibition. For many decades scientists attributed this growth reduction to a loss in turgor pressure from an inbalance in plant-water relations. Normally, a plant cell is surrounded by water (with a minimal amount of mineral components) as a water potential ( ) close to zero. The cell's water potential on the other hand is determined by the sum of its osmotic potential ( )plus its pressure potential (P), which is related to the cell turgor pressure. Given that the membrane permeability is constant, the rate of water movement into the cell can be enhanced by the lowering of either the osmotic potential (osmotic adjustment) or the pressure potential. The becomes more negative by slat uptake or by production of organic solutes into cytoplasm by either synthesis of certain organic components (such as proline) or by the breakdown of macromolecules (such as hydrolysis of starch to glucose). A decrease in P by cell-wall loosening through auxin-stimulated growth occurs via cell expansion. In this instance, the cell wall is stretched in an irreversible manner that release pressure and enhances water uptake. It much be noted that decreases in either or P during cell growth are more or less negated by the resulting water uptake, so that well-watered plants essentially retain their original and P values. The relationship between ;turgor itself and growth has been approximated by plant physiologists with the following expression:

$$G = \phi(P-Y)$$

where G is growth , ϕ is the extensibility coefficient, P is turgor pressure, and Y is the threshold turgor that must be exceeded preparatory to any growth or cell expansion. This expression shows that constant growth can still proceed even at lowered turgor pressure provide that is increased an Y reduced.

During the past decade, however, studies on long-term effects of osmotic stress on growth have definitely shown that even though osmotic adjustment had led to a complete restoration of turgor pressure in the presence of continued. It was then thought that this was the plant's method of conserving water uptake from its nearly moistureless environment; increased growth meant more water would be needed to keep the turgor pressure constant. With the advent of osmotic stress studies wherein cell suspension cultures (small clumps ;of undifferentiated plant cells shaken in a liquid media) were utilized, the cultured cells that had adapted to the added NaCl by means of osmotic adjustment still showed a reduced growth response even when the turgor pressure was above that in prestress conditions. This finding indicated that stressed cells being surrounded by adequate amounts of water was not a factor in restoring normal rates of growth. A reassessment involves a concept of restricted cell expansion as a necessary adjunct to the high salt environment; this concept postulates that a reduced rate of cell expansion might be the mechanism by which a manageable ion flux rate through the cytoplasm to the vacuole or an adequate organic solute synthesis rate in the cytoplasm is achieved. A rapid rate of cell expansion of adapted cells on the other hand would necessitate a rapid influx through a small cytoplasmic pool (to prevent toxicity) into a large vacuolar pool to provide the necessary osmotic adjustment to the high salt environment outside the cell. Since it is best to obtain as large a size difference between both pools as possible, a greater amount of osmotic adjustment is required. In turn, a greater influx must be available to adequately drive the vacuolar osmotic adjustment. Such an ion flux rate through the cytoplasm would most likely be unattainable, hence the need for a more efficient and adaptable system such as a reduced rate of cell expansion.

Studies on a molecular basis of adaptation to salinity stress with regard to mechanisms of altered or retarded cell expansion have recently focused on the involvement of cell wall biochemistry in cell growth. The cell wall itself is a well-organized structure of very large protein and carbohydrate polymers. The synthesis of these polymers occurs at different stages of the cell's development. Certain polymers originate only during cell expansion. After cell expansion stops, other polymers are made that are responsible for a permanent locked-in configuration of the cell wall pending further differentiation of the cell itself. On comparison of unadapted and salt-adapted cells, the latter fail to gain fresh weight as rapidly as the unadapted cells. Although decreased cell division could account for a lower rate of fresh weight growth in adapted cells, it can be demonstrated that reduced cell division does not account for all the observed reduction in fresh weight gain. In addition, the slowly expanding salt-adapted cells have much less cell wall extensibility (O) than the nonadapted cells.

The nature of reduced cell wall extensibility or cell wall stiffness is not as yet clearly understood. There are, however, some observations on physical and biochemical alterations in the cell walls that somewhat pave the way to a more definitive understanding of this phenomena. Initial investigations showed that osmotic stress- adapted cells partitioned more carbon toward osmotic adjustment (via organic solute synthesis) than toward cell wall polymer synthesis. Despite this preferential diversion of the available carbon, decreased extensibility still took place. recently this anomaly was resolved in part by studies on the tensile strength of plant cell walls. The loss of tensile strength of salt-adapted cell walls as compared to the nonadapted cells seemed to correlate with a diminished mass of the cell wall's cellulose-extension frame-work as a result of preferential use of carbon for osmotic adjustment. As a result the adapted cells remained in a locked configuration wherein non microfibril separation resulting in cell wall expansion could take place. Certain proteins (extensions) that are ionically bound to the cell wall control the extension of the wall. Much of this protein is released into the high-salt media of the salt-adapted cells as compared to the unadapted cells in the same media. Loss of this protein may then result in an inhibited extension process of the salt-adapted cell well. It is of interest that plants regenerated from the salt-adapted cell cultures have leaf cells that exhibit lower rates of expansion. The leaves of these plants are smaller in size than those of normal plants.

As with heat, drought, and water stress, the plant hormone abscisic acid (ABA) plays a prominent role in plant tissue subjected to water stress by a high-salt osmoticum. Endogenous ABA levels in crease in plant tissue subjected to this type of stress and are involved in the induction of proteins termed as osmotins in the presence of salt. These proteins, specifically synthesized and a accumulated in cells that undergo gradual osmotic adjustment to salt stress, are localized mostly in the vacuoles. Although the exact role of these proteins is not fully defined, most of them are believed to play a role for strategies of defense in the plant against the type of environmental injuries mentioned above. The main activity of these proteins seems to be linked to the reduced growth discussed previously and to stress adaptation.

Osmotin synthesis along with all of the other metabolic adjustments mentioned previously consumes additional energy that is reflected by increases in respiration.

Recent investigations on the molecular genetics of salt stress have dealt mainly with the following areas: (a) a limited number of metabolic pathways that are involved in the stress response of the affected plants and that are involved with the plasmalemma. These include the control of salt uptake in the roots, salt transport in the xylem, regulatory control of pathways for organic osmolyte synthesis, maintenance of photosynthesis and growth (cell cycle control and biosynthesis of necessary cell wall components). (b) Mechanisms used by plants to transfer signals from the perception of osmotic changes in their environments to initiate changes in gene expression. Changes in turgor pressure or the osmotic state outside of the cell initiate physical and chemical ;changes in the plasmalemma, which in turn brings about the rapid activation of specific genes. (c) Studies of molecular mechanisms that control gene expression, which in turn finally result in the necessary metabolic adjustments to saline stress. Thus it is postulatd that salt stress acts through a chain of events: perception, signal transfer, and gene expression.

Along with salt stress is often the presence of toxic tracemetal ions present in the soil from a number of sources. These ions are generally introduced into the soil from emissions of fuel-burning power plants and automobiles, from addition of industrial and municipal sewage sludge to the soil as a fertilizer, and from metallurgical mining and processing. These toxic trace-metal ions can be released from the soil in free ionic form through changes in soil acidity. Such a pH change has in some instances led to the contamination of valuable groundwater. Airborne distribution from emissions has also insured a widespread deposit of these contaminants within the soil. Recently information has been gathered as to the biochemical and molecular aspects involved in plant tolerance of these toxic metal ions.

Metal ions have been categorized as to their affinity for different binding sites found within organic molecules synthesized by higher plants. Class A metals (such as aluinum) form stable complexes with organic molecules containing oxygen. Class B metals (such as cadmium and copper) form complexes with compounds that contain a nitrogen or sulfur center.

Aluminum, although not a trace metal, is a common element in certain parts of the earth's crust. Changes in soil pH cause small amounts of the free ion to be taken up by plants. Toxicity occur as a result of several interactions that have an adverse effect on some aspects of the plant's physiology. Exposure to aluminum and phosphorus uptake to levels below that needed for tissue survival. Aluminum tends to be concentrated in the nucleus, where it interferes with DNA synthesis and thus inhibits the cell cycle. Several strategies are used by the aluminum-tolerant plant to overcome the toxic effect of this metal. Changes in the ammonium versus the nitrate ion tend to raise the pH in the cell and thereby render the ;aluminum ion insoluble and ineffective as a toxic agent. Aluminum-tolerant plants also produce different chelating agents, which form a metal complex that again is not longer toxic to the plant. Most of these chelating agents are an overproduction of organic acids, such as citric or malic acid which are normally in the plant at lower concentrations.

Cadmium normally enters the soil from emissions of fossil fuel-burning power plants and mining activity. Cadmium concentrations and soil pH are major factors that influence the uptake of this metal into plants.Cadmium is known to replace copper and zone in critical enzymes that use these metals as cofactors. This ion then interferes with several metabolic pathways that are critical for ;cellular as well as whole plant physiology. Such pathways include formation of chlorophyll, critical steps in the Calving cycle, and respiratory reactions. cadmium tolerant plants are ;able to synthesize polypeptide that have metal binding properties and are able to form cadmium-polypeptide complexes. Although some cadmium-sensitive plants are able to also produce metal-binding polypeptides, the latter are unable to form complexes with cadmium. Once taken into the cell as a complex, cadmium is then accumulated and stored as an insoluble form (as the insoluble cadmium phosphate) and is not released into the environment.

Under normal condition copper, manganese, and zinc are necessary micronutrients for plants. In quantities above micronutrient levels, however, these metals are toxic. Plants tolerant to high levels of these metals are able to synthesize sulfur-containing polypeptides and in some cases also accumulate high levels of organic acids. Thus, either complexing or chelation reactions of these metals take place depending on which metal is being considered.

As with aluminum, selenium is present as a natural component of certain soils. Selenifrous soils exist in a broad section of the central part of the United States and contain plants that accumulate high levels of this element initially in its anionic form. Most of these plants belong to the genus Austragalas (,ilk vetch). The main reason that selenium is toxic to plants is its similarity in structure to sulfur. Even though selenium replaces sulfur in many biological compounds, it cannot carry out the same functions provided by sulfur. For example, selenium accumulators (which are tolerant to selenium) are capable of excluding the seleon amino acids and thus do not incorporate them into the critical proteins (enzymes, etc.) during synthesis. Otherwise, defective key enzymes would adversely affect several important metabolic pathways that control the plant's growth and development.

More knowledge of the mechanisms of metal tolerance in higher plants is important, because without such knowledge it would be difficult to predict the transport of these toxins through the biological chain. Thus, the storage of toxic ions in plant tissues would have a different impact on human health than if the plants could altogether exclude metal ions from their system by a different mechanism.

VI. Anaerobic Stress and Adaptation

During an excessive rainy season or a rapid spring thawing of winter snow build-up, much flooding of ordinary arable land where plants are under cultivation takes place. Many of these economically important plants are nonaquatic in nature and in some cases suffer from oxygen deprivation to their root cells. Such an extreme condition is termed anoxia ;and results in either death or cessation of root cell growth where complete fermentative metabolism is taking place. Some of these plants however are able to partially counter the extreme effects of anoxia by providing limited amounts of oxygen (hypoxia) from intercellular air spaces (aerenchyma) present in the non-submerged tissues.

There are basically two types of reactions from plants under anaerobic stress. The first is a production of ethylene by the roots, which causes structural changes within this organ to allow more oxygen diffusion through the root and also causes the root tips to grow upward through the soil toward a more aerobic environment (as opposed to their natural inclination to grow downward toward a gravitational pull). Secondly, within half a day after flooding, a set of nearly 20 soluble proteins are produced that eventually replace all the proteins produced by the root cells under normal aerobic conditions. The replacement or more accurately the repression of the preexisting aerobic protein synthesis is accompanied by a disassociation of nearly all the polysomes that were involved in aerobic protein translation. This event is somewhat reversible, since 70-90% of the mRNAs originally involved with aerobic protein synthesis are reassembled back into translatable polysomes without synthesis of new mRNAs after restoration of aerobic conditions. Anaerobic protein synthesis takes place in two stages: (1) There is an immediate production of transition proteins. (2) After a lag period, anaerobic proteins appear. Several of the anaerobic proteins are linked to fermentation reactions (fermentative glycolysis). One of these enzymes, alcohol dehydrogenase (ADH) is necessary for the survival of plants under anaerobic conditions. There are two possible end products of glycolytic fermentation- ethanol or lactic acid. To prevent acidosis, which would be toxic for root cells during anaerobiosis, there is a mechanism of pH control. During a short increase of lactate production, the cellular pH drops below 7.0 and activates enzymes that preferentially lead toward formation of ethanol and simultaneously inactivate enzymes that enhance lactic acid formation. Thus if the pH remains stable and slightly below pH 7.0, the formation of ethanol prevails. The importance of ADH was demonstrated when mutant plants lacking the ADH gene did not survive flooding more than 3-5 days.

Under hypoxic conditions, ADH activity increases in root tissue; but as the oxygen supply diminishes the glycolytic flux or rate of reaction of glucose metabolism must increase to keep the ATP (adenosine triphosphate - the "Energy currency" of the cell) at least at a 50-70% level of that present under normal aerobic conditions when the Kreb's cycle of glucose oxidation is operative. ATP concentrations substantially below this level will not allow the adaptive biosynthetic responses to take place in the root cells. Finally, both of these coping mechanisms appear to be linked to one another in that when the ATP levels are maintained at minimum efficiency, the root ethylene levels decrease and vice versa.

It is of interest that the induction of heat-shock proteins is immediate and similar to that of the transition proteins involved in anaerobic stress. Both heat-shock and transition proteins then show a decelerated rate of synthesis. This is where the similarity ends, however, for there are only one or two protein polypeptides that overlap with regard to similar molecular weight. In addition, it should be mentioned in closing that most of the stress topics discussed in this article (osmotic, chill, drought, heat, and anaerobiosis) can be shown to have a few of the induced stress proteins in common with one another. Further research in the area of environmental stress will probably clarify a set of physiological or biomolecular reactions that are directly related to the shock proteins in common with each of these stress categories.

[Source: F.R. Katterman, University of Arizona]

12

Anaerobic Digestion of Rice Crop Residues

Introduction

Agriculture is both a producer and a consumer of energy. Crops convert solar energy, through photosynthesis, into carbohydrates which are a form of energy suitable for consumption by animals and man. As a consumer of energy, agriculture depends on energy-intensive inputs such as irrigation, fertilizer, and machinery. The energy demand for agricultural sector has risen greatly. In a developing country like India, the commercial energy consumption in agriculture has increased from 1.8 million tonnes coal replacement (MTCR) in 1953-54 to 31.32 MTCR in 1978-79. The agricultural sector registered the sharpest growth rate and its energy demand increased from a mere 3% in 1953-54 to about 10.6%, in 1978-79, of the total commercial energy demand in the country. At the present rate of increase it is estimated that by the year 2000 A.D. the agricultural sector would be consuming 300 MTCR of commercial energy in India. With the depleting stocks of fossil-energy and a phenomenal increase in the cost of commercial energy, it is important to meet the demands of the agricultural sector through non-conventional energy sources. The potential of crop residues for energy is significant because the yield of useable grain or other food staples is only about half the yield of total biomass under a wide range of conditions, because substantial quantities of unused stalks, straw, bark, and other crop residues are produced. With the increase in biological harvest, not only has the grain yield incresed, but the availability of residues has also increased. In India, a remarkable increase in yields and change of cropping pattern in some of the irrigated areas in recent years have created a surplus of crop residues such as paddy straw. The increased availability of crop residues in recent years has even posed a disposal problem. In one study conducted in Ludhiana (India) it was reported that most of the rice straw is disposed of by burning it in the field. The burning it in the field. The estimated production of agricultural residues in India is 200 million tonnes per year, and 80 million tonnes of this is attributed to rice crop residues in the form of rice straw and husks. The estimated world production of paddy and wheat straw is 323 x 10^6 and 360 x 10^6 metric tonnes, respectively. These crop residues can be utilized to generate biogas and prove to be a good source of renewable energy.

Crop residues can also be combined with animal manure for generation of biogas. There are reports that availability of crop residues is much greater than that of animal manure. It is estimated that in the United States about 200 x 10^9 kg (dry weight basis) of collectable corn stalks and wheat straw could be available for fermentation as opposed to about 30 x 10^9 kg of collectable manure. Another advantage of mixing crop residues with manure is that, nutritionally, the highly nitrogenous manure complements the highly carbonaceous but nitrogen-deficient crop residue. Crop residues, when subjected to anaerobic fermentation, produce biogas containing 60-80% methane. The material, after fermentation, is a good organic fertilizer.

The technology for producing biogas from animal wastes has made significant progress and is being widely used in countries like India, China, and Taiwan. In India, more than eight lakh family-sized biogas plants (2-3 M^3 capacity) are operating. There are two main designs, namely, the floating-drum type and the fixed-dome type.

The main components of floating drum gas plant are gas holder, digester, inlet, and outlet assembly (Figure 1). The gas holder assembly consists of a central guide frame, a drum made of mild steel sheets and a gas outlet. The digester tank is made of masonry. Through the inlet pipe, a mixture of cow dung and water in the form of slurry is added every day, and based on the principle of hydrostatics some spent slurry called effluent comes out through the outlet pipe. Biogas is stored in the drum, which rises as gas is added and sinks when gas is used.

The fixed-dome type plant is basically a Chinese design and has been adopted in India with some modifications. It is entirely a masonry structure. The plant consists of an underground digester made of bricks and cement having a dome-shaped roof that remains below the ground. There are two rectangular openings facing each other and extending a little above ground level, which act as the inlet and outlet of the plant (Figure 2). Gas is accumulated and exerts pressure on the slurry, displacing it from the digester to the inlet and outlet tanks. Difference of slurry levels in inlet and outlet tanks and the digester depicts the pressure in terms of the slurry column which is acting on the gas. When gas is used, the slurry level in the digester starts going up, while it falls in the inlet and outlet tanks.

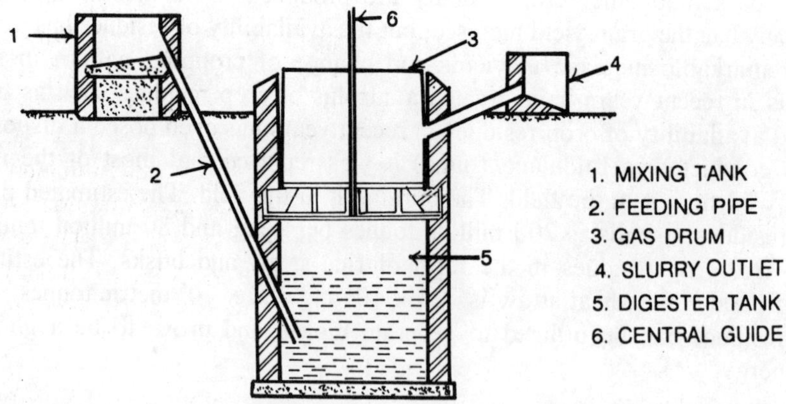

Fig. 12.1. Floating-drum biogas plant.

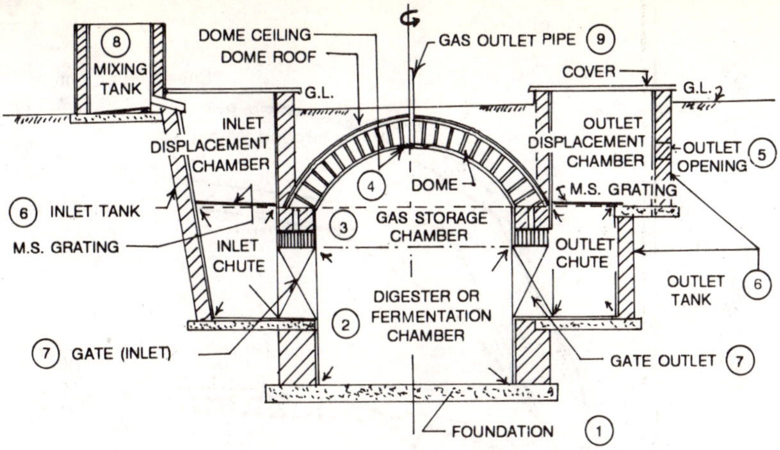

Fig. 12.2. Fixed-done biogas plant.

These biogas plants are of a basically semicontinuous type, i.e., fresh material is charged every day. These plants are not suited for digestion of crop residues due to the peculiar problems associated with crop residues.

The major problems associated with the anaerobic digestion of crop residues are:

1. Crop residues have a tendency to float on top of water when fed to the digester, due to their low density.
2. They need to be chopped into small pieces before being fed to the digester.
3. They stick together and form a lump, and stirring becomes necessary.
4. They do not form a consistent slurry after digestion, thus creating a problem of removal from the digester.
5. It is necessary to add an inoculum to start the digestion process as these residues do not contain microorganisms needed for digestion. Slurry from a running digester or animal manure can be used as starter.

Due to these problems the crop residues are not suited for feeding into daily-fed digesters. The batch system of digestion or a combination of batch and continuous system is more suited for residues.

Digestion of Crop Residues

Work on anaerobic digestion of various crops and agricultural wastes was carried out to determine their biogas potential. Laboratory scale anaerobic batch digestion was carried out at 37°C with various crops and organic solids at loading concentrations of 3-10% total solids for a period of 17-36 days. The effect of the concentration of volatile solids, loading concentration, and pretreatment of the material was studied. The biogas yield from different feedstocks over complete digestion cycles is shown in Figure 3. In general, higher biogas yield was obtained from crops than from wastes. IT was further reported that the yield of biogas decreased as the loading concentrations were increased and, above 5% total solids,

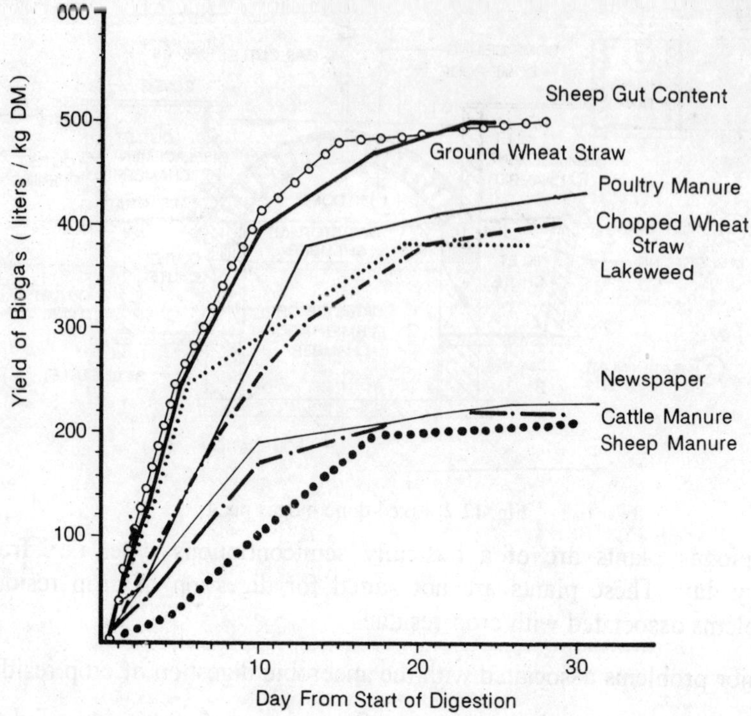

Fig. 12.3. Biogas yield from different feedstocks.

digestion of crops with high contents of soluble carbohydrates required the addition of an alkali to maintain a pH of 7 ± 0.5. The total yield of biogas obtained per kg of total solids from different feedstocks is given in Table 1. It is observed that pretreatment by grinding increased the rate of production of biogas and the total yield. It was reported that during digestion of plant matter, the fresh material floated and only sank to the bottom as it was digested. The most active area of digestion in a digester using plant matter is the top-third, while in digestors using manures and sewage it is bottom-third.

The gas productivity of animal wastes and crop residues like rice straw, husks, and sunflower leaves and stalks is reported in Table 2. It could be observed that biogas yield and methane content of the gas from crop residues were higher as compared to animal wastes. However, the details of the digestion process and treatments given to agricultural wastes were not reported.

Crop Residues—Animal Manure Mixtures as Feedstock

The effect of the carbon-nitrogen ratio on methane production was studied in six 4-liter laboratory digesters. Screened dairy cow manure, which had a carbon to nitrogen ratio of 8.0 was combined with glucose and, later in the experiment, cellulose, to obtain carbon to nitrogen ratios varying between 8.0 and 51.7. The loading rate was also varied: 1.0, 1.5, and 2.0 kg. Vs/m³/day. Results are shown in Figure 4 and it is clear that highest methane production per unit loading rate occurred when the carbon to nitrogen ratio of the feed was 25. This study indicates the possibility of mixing high carbon to nitrogen ratio crop residues with cattle manure (low carbon-nitrogen ratio) to obtain a good digester feed.

Table 12.1

Volatile Solids Content, Volatile Solids Destruction, and Yield of Biogas from Anaerobic Digestion of Various Crops and Organic Wastes at 5% Total Solids Loading Concentration at 37°C

Material	Wt. of V.S. (kg)	Digestion time (days)	V.S. Destroyed (kg)	%	Yield of Biogas 1/kg T.S. Added	V.S. Added	V.S. Destroyed
Grass hay	0.93	30	0.61	73	462	497	587
Grass hay*	0.93	18	0.59	70	507	546	678
Kale	0.89	36	0.60	73	440	493	733
Sugar beet**	0.77	21	0.55	79	380	450	691
Maize	0.95	33	0.69	81	500	526	725
Oats	0.86	18	0.50	66	470	546	940
Wheat straw	0.94	31	0.68	79	412	440	620
Wheat Straw*	0.94	18	0.78	91	530	564	780
Lakeweed	0.79	17	0.44	62	380	481	864
Cow manure	0.83	36	0.29	39	216	260	745
Sheep gut contents	0.86	24	0.44	56	415	482	943
Newsprint	0.99	25	—	—	235	237	—
Lucerne hay*	0.91	17	0.61	69	346	380	567

* Sample ground.

** Leaves only.

V.S. -Volatile Solids.

T.S. - Total Solids (In each expt. T.S. = 1kg).

Table 12.2

Agricultural Waste: Gas Productivity and Methane Content

Material	Gas Produced/ Tonne of Dried Material (m^3)	Percentage Content of Methane
General stable manure from livestock	260-280	50-60
Pg manure	561	—
Horse manure	200-300	—
Rice husks	615	—
Fresh grass	630	70
Hemp	359	59
Straw	342	59
Leaves from trees	210-294	58
Sunflower leaves and stalks	300	58
Sludge	640	50
Water hyancinths	300-500	60-65

It was reported that dairy cow manure could be mixed with field crop residues to obtain a favorable C:N ratio, resulting in improved gas production per unit volume of digester. Six 4.4-liter laboratory digesters with operating volumes of 3.4 liters were tested for 18 months with the dairy cow manure combined with either barley straw, rice hulls, or rice straw in three separate six-month phases. The digestion was carried out in the mesophilic temperature range at 35°C. By adjusting the proportion of field crop residues added to the dairy manure, various digester feeds with different nonlignin carbon to nitrogen ratios were obtained. The weekly mean gas production rates were fairly constant for each digester during the four month steady-state period of each investigation phase. However, daily fluctuations up to 25% were observed. A plot of these mean values for the best performing digester of each phase is presented in Figure 12.5. A summary of digester performance during the steady-state periods is given in Table 3. The third column of this table lists the volume of methane gas produced per unit of COD (Chemical Oxygen Demand) destroyed. This parameter is important because the values suggest the bacterial efficiencies of converting COD into methane gas. In this study values between 0.28 and 0.38 m^3 methane per kg COD destroyed were obtained. The greater values generally occurred in the digester where the feed consisted of material whose combined non-lignin carbon to nitrogen ratio was between 25 and 32.Not only a greater reduction in COD occurred for these feeds;the resulting gas production was substantially greater as well. It is important to note that methane concentration decreased with an increased carbon to nitrogen ratio.

The reduction in volatile solids and COD at different carbon to nitrogen ratio is given in Table 3. The greatest reduction in these parameters occurred for feeds whose carbon to nitrogen ratios were 25 and 30. For mixtures of dairy manure and barley straw or dairy manure and rice straw, increasing quantities of straw produced greater reduction up to a point (C/N > 35), and then a decline. The decline may have been due to the fact that nitrogen became a limiting factor.

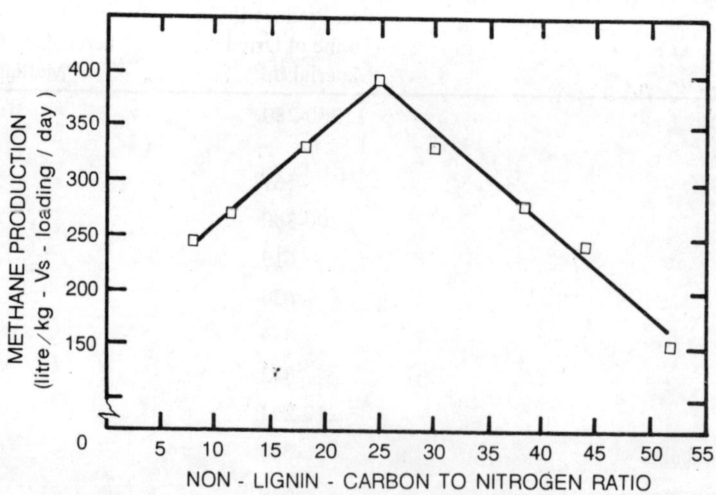

Fig. 12.4. Methane production as affected by non-lignin C:N ratio for dairy manure and cellulose.

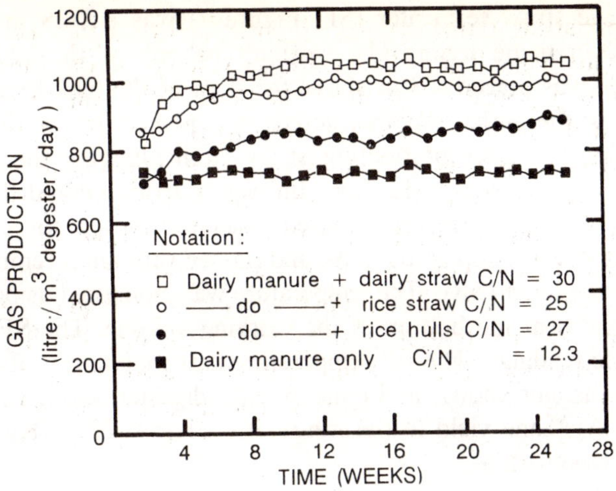

Fig. 12.5. Total gas production at STP for the best performing digesters.

It can be concluded that combination of either of the two carbonaceous field residues (chopped rice or barley straw) with dairy manure *up to C:N ratio of 30) as digester feed resulted in increased methane production. However, chopped rice hulls did not perform well due to their high lignin content (16% as compared to 4.9% of rice straw and 6.3% of barley straw).

Hashimoto studied methane productions from wheat straw-manure mixtures (90% 75%, 50% and 0% straw) using laboratory scale fermenters. Table 4 shows the substrate and fermenter operating conditions used in the study. Both the batch and daily-fed fermenters were 4-dm^3 aspirator bottles 3 with 3-dm^3 working volumes. Table 5 shows the composition of the straw-manure mixtures fed to the fermenters. Methane yield and ultimate methane yield of straw, straw-manure mixtures and manure are given in Table 6. Results showed that the ultimate methane yield (Bo) of straw-manure mixtures was directly proportional to the relative proportion and Bo of the individual components. This relationship has important practical implications since it provides sound justification for estimating Bo of mixtures by adding the products of weight fractions and Bo of individual components. This reduces the need to experimentally determine each mixture Bo when evaluating the fermentability of various mixture combinations. It was further reported that hammer or ball milling did not increase the Bo of Straw.

Hashimoto and Robinson reported the pilot scale operation and economic assessment of a two-stage, straw-manure fermentation system. Figure 6 shows a schematic diagram of two-stage fermentation system. The first stage high rate fermenter (HRF) had a total volume of 5.7 m^3 with a working volume of 5 m^3. It was insulated with 3 cm of polyurethane foam. Four baffles were equally spaced around the inside of the tank. Mixing within the HRF was achieved with a 1.5 kw variable-speed motor two three-bladed stainless steel marine propellers on a stainless steel shaft rotating at 220 rpm. An external, double-tube heat exchanger was used to maintain a 55^0 C fermentation temperature using water heated to 65^{0}C. The HRF was operated at 5 days hydraulic retention time (HRT) and influent VS concentration of about 80 kg VS/m^3.

The second stage straw fermenter (SF, Figure 7) was a 16.8 m^3 steel tank insulated with 7.6 cm of polyurethane foam. The working volume of the tank was 14.2 m^3. An expanded steel screen was placed 0.3 m above the floor of the tank to support the straw and to prevent plugging of the outlet. Another screen was placed 3.5 m above the lower screen (4.0 m above the floor) to prevent floating straw from plugging the gas outlet pipe. The volume packed with inlet straw was 13.0 m^3. Influent slurries entered through a 7.6 cm pipe at the top of the tank, while effluent drained from the bottom of the tank. The biogas produced was collected at the top of the tank and passed through a temperature-compensated gas meter and a pressure relief valve. The temperature was monitored using three thermocouples at the top, middle, and near the bottom of the working volume. The SF temperature was an average of these three readings. The SF temperature was maintained at about 35°C by using the external heat exchanger shown in Figure 6. The digester was fed on wheat-Straw and manure from steers. Methane yield in the range of 0.21 to 0.27 m^3/Kg VS was obtained in different trials from straw.

Table 12.3
Summary of Digester Performance (Mean Values) during Steady-State Periods

C:N†	VS Reduction (%)	COD Reduction (%)	Methane Production*	Hemicellulose Reduction (%)	Cellulose Reduction (%)	Lignin Reduction (%)
			Dairy Manure + Barley Straw			
12.3	24.3	27.8	0.31	27.7	34.1	2.0
20	27.0	27.0	0.33	26.9	36.5	-13.2
25	32.2	32.9	0.38	27.8	41.2	-14.8
30	34.4	34.3	0.37	27.2	41.3	-5.0
35	32.8	31.6	0.35	30.7	36.8	0.8
40	32.1	35.2	0.31	31.4	32.4	-2.8
			Dairy Manure+Rice Straw			
12.3	27.6	28.9	0.32	30.5	28.2	3.0
20	30.9	31.8	0.35	29.8	30.8	-6.2
25	31.6	33.6	0.36	27.7	34.1	11.1
30	32.9	33.2	0.34	24.1	36.3	2.1
35	32.2	31.7	0.31	28.7	33.8	3.5
40	29.1	31.6	0.28	31.2	29.5	0.0
			Dairy Manure+Rice Hulls			
12.3	27.0	28.1	0.31	30.0	31.1	0.0
17	27.1	27.8	0.34	28.3	31.2	7.2
22	26.1	30.7	0.31	32.7	28.4	-11.8
27	30.6	31.2	0.32	32.6	28.7	3.2
32	27.7	39.8	0.36	31.4	18.9	-1.3
35	12.8	16.1	0.31	28.1	7.6	6.1

* Methane production = m^3 methane/kg COD destroyed (STP)

† C:N = non-lignin carbon to Kjeldahi-nitrogen ratio

Table 12.4
Substrate and Fermenter Operating Conditions

Substrate	Type	HRT (days)	Temperature (°C)	Replicates
Manure (M)	batch	96	55	3
Straw	batch	96	55	3
Ball-milled straw	batch	96	55	3
Hammer-milled straw (HMS)	batch	96	55	3
75% HMS-25% M	batch	96	55	3
50% HMS-75% M	batch	96	55	3
25% HMS-75% M	batch	96	55	3
90% HMS-10% M	daily fed	15	35	2
90% HMS-01% M	daily fed	10	55	2
75% HMS-25% M	daily fed	15,10,8	35	2
75% HMS-25% M	daily fed	8,6	55	2
50% HMS-50% M	daily fed	15,10,8	35	2
50% HMS-50% M	daily fed	10,8,6	55	2
Manure	daily fed	15,10,8	35	2
Manure	daily fed	10,8,6	55	2

Table 12.5
Composition of Straw-Manure Mixtures Fed to Fermenters*

Parameter	Percent Straw in Straw-Manure Mixtures			
	0%	50%	75%	90%
Total solids (kg/m³)	73.3±1.8	78.0±5.2	75.5±5.8	78.7±5.1
Volatile solids (kg/m³)	63.2±3.0	66.0±4.8	6.5±4.4	65.1±3.6
Chemical oxygen demand (kg/m³)	73.4±4.2	102.8±26.7	96.3±34.8	78.5±19.9
Kjeldahi-N (kg/m³)	2.82±0.08	1.80±0.11	1.14±0.14	0.88±0.06
Ammonia-N (kg/m³)	0.85±0.03	0.42±0.03	0.23±0.03	0.10±0.01
Total volatile acids (kg/m³)	6.17±0.66	3.14±0.45	1.59±0.23	0.80±0.08
Alkalinity (kg/m³)	3.17±0.52	2.26±0.32	1.73±0.18	0.94±0.18
pH units	5.07±0.30	5.72±0.30	6.41±0.18	6.40±0.21
Cell contents (kg/m³)	50.1	37.3	26.0	24.5
Cellulose (kg/m³)	11.9	16.0	21.7	25.7
Hemicellulose (kg/m³)	6.7	18.2	17.0	16.2
Lignin (kg/m³)	2.6	3.7	5.1	4.0
Carbon nitrogen ratio	11	19	22	25

* Data presented as mean + standard error of mean, except for cell contents, cellulose, hemicellulose, lignin, and carbon nitrogen ratio.

Table 12.6
Methane Yield (B) and Ultimate Methane Yield (Bo) of Straw-Manure Mixtures and Manure

Substrate	Replicate	B* (m^3CH_4/kg VS fed) Replicates	Mean	SEM**	Bo† (m^3CH_4/kg VS)
Straw	1	0.247			
	2	0.275	0.243	0.20	0.26
	3	0.207			
Ball-milled straw	1	0.242			
	2	0.245	0.242	0.002	0.26
	3	0.238			
Hammer-milled straw (HMS)	1	0.224			
	2	0.198	0.232	0.022	0.26
	3	0.273			
75% HMS-25% M	1	0.256			
	2	0.282	0.280	0.014	0.29
	3	0.303			
50% HMS-50% M	1	0.340			
	2	0.324	0.333	0.005	0.34
	3	0.335			
25% HMS-75% M	1	0.330			
	2	0.325	0.322	0.006	0.34
	3	0.310			
Manure (M)	1	0.427			
	2	0.382	0.374	0.033	0.38
	3	0.314			

* B is after 90 days of fermentation at 55° C.

† Estimated by plotting B vs reciprocal of fermentation time and extrapolating to infinite time.

** Standard error of mean.

Pretreatment of Straw

Pretreatment of crop residues to improve their digestibility is important to increase the methane yield and the suitability of this feed stock for bioconversion. Hashimoto studied the effect of pretreating wheat straw with gamma ray irradiation, ammonium hydroxide, and sodium hydroxide on methane yield. Pretreatments used were: gamma ray irradiation, NH_4OH, low-temperature NaOH, and high-temperature NaOH. Experiments were conducted in 4-dm^3 aspirator bottles. The fermenters were maintained at 55± 1°C and the biogas produced was collected in gas-impermeable Tedlar bags. Experiments were conducted to study the effect of various pretreatments on the ultimate methane yield (Bo, m^3 CH_4/kg VS fed). The straw was passed through a hammer mill equipped with a 1 cm opening mesh screen, and ball milled for 24th in a 5-dm^3 porcelain jar charged with 2 cm diameter by 2 cm long porcelain cylinders and rotated at 52 rpm. The straw samples were irradiated at levels 1, 5, 10, 50, and 100 Mrad, but the results indicated that gamma ray irradiations had no significant effect on methane yield. Ball-milled straw (BMS) was added to 4 dm^3 of tap water and

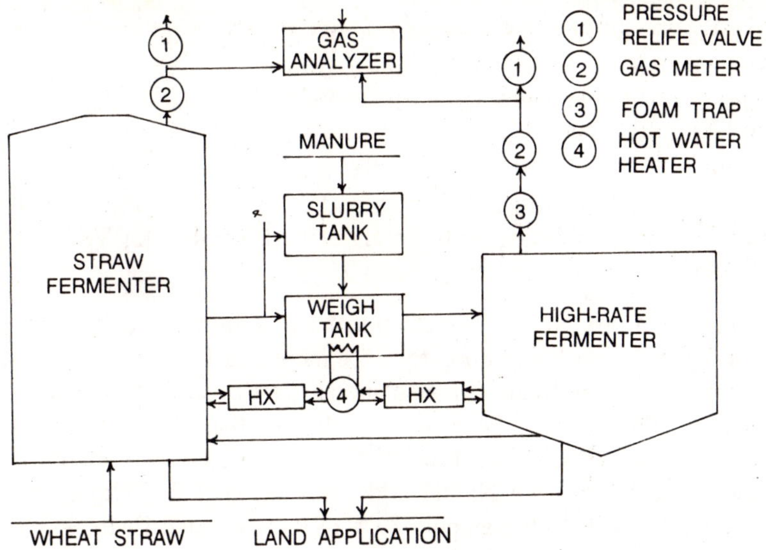

Fig. 12.6. Schematic diagram of two-stage fermentation system.

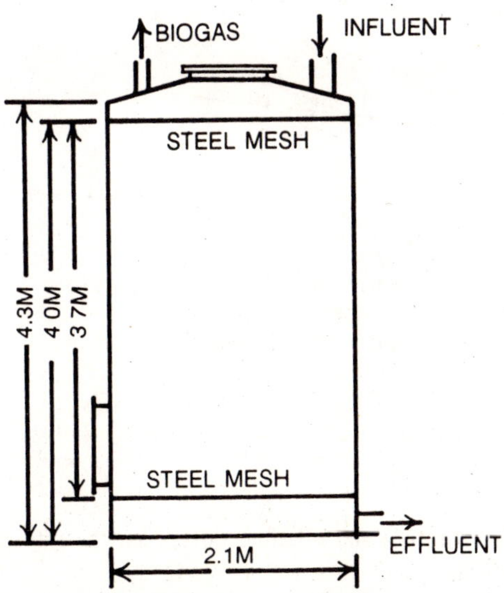

Fig. 12.7. Second stage fermenter.

heated to 80°C and NH_4OH treated straw, and inoculum used in the experiments. In another experiment BMS was added to 4 dm³ of tap water and heated to 80°C and NH_4OH was added to yield dosages of 17.0, 25.5, or 34.0 g OH-/kg VS. Table 7 shows the composition of the bms, nh_4oh treated straw, and inoculum used in the experiments. In another experiment BMS was added to 4 dm³ of tap water heated to 90°C and NaOH dosages of 17.0 or 25.5 g OH - /kg VS and 34.0 g OH - /kg VS were obtained. Table 8 shows the composition of the BMS, NaOH treated straw, and inoculum. To study the effect of high temperature NaOH pretreatmcnt BMS was heated to 200°C with NaOH at 0, 20.5, 41.0, or 61.5 g OH- /kg VS dosages, and BMS was combined with NaOH at 41.0 g OH- /kg VS dosage and heated to 150, 175, or 200°C.

Results indicated that NH_4OH Pretreatment had significant effects on Bo. Pretreatment of 25.5 and 34.0 g OH - /kg VS at 80°C for 24 h increased Bo by 11% and 14%, respectively. Low temperature NaOH pretreatment also increased Bo as NaOH dosage increased. At a dosage of 34.0 g OH- /kg Vs, Bo was 28% higher than from the untreated BMS. The ultimate methane yield of BMS heated to 200°C for 1 h was only 56% of the untreated BMS. For the straw heated to 200° C, NaOH addition increased Bo. However, Bo of the straw treated with 61.5 g OH - /kg VS was only 83% of the untreated BMS. For straw treated with 41.0 g OH - /kg VS, Bo decreased as the temperature increased from 150 to 225° C. These results clearly show the detrimental impact of high temperature treatment on Bo. Results from this study indicate that Bo increased as pretreatment alkali concentration increased. Figure 8 shows that the ratio of Bo of treated to untreated straw increased as alkali dosage increased. For pretreatment temperatures below 100°C, NaOH was apparently more effective than Nh_4OH.

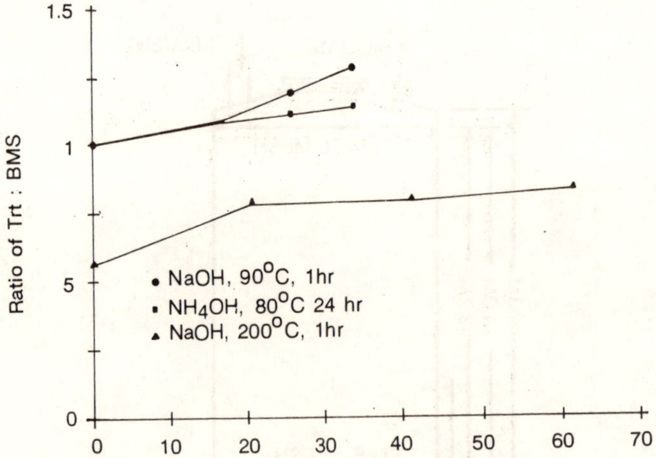

Fig. 12.8. Effect of temperature and alkaline dosage on the ratio of Bo from pretreated and untreated BMS.

Anaerobic Digestion of Rice Crop Residues: Studies Conducted at Iari, New Delhi

Keeping in view the problems of digestion of crop residues as they were listed earlier in the chapter, a batch system of anaerobic digestion for rice crop residues was adopted. Batch digesters were designed and fabricated in 205 liter empty oil drums (Figure 9) with the following key factors:

1. An inverted trough was provided as gas holder. A water Jacket was fabricated all along the periphery of the digestion tank (205 liter empty oil drum) to accommodate water to prevent leakage of gas.

Table 12.7
Composition* of Ball-Milled (BMS) and NH_4OH Treated Straw, and Inoculum

			NH_4OH Pretrement+ (g OH-/kg VS)			
Constituent	BMS	34.0**	17.0	25.5	34.0	Inoculum
Total solids	8.87	8.83	10.71	10.55	9.82	2.74
Volatile solids	86.13	86.18	83.01	83.03	83.81	71.90
Total nitrogen	1.07	3.33	1.76	2.11	3.00	9.49
Chemical oxygen demand	123.00	135.45	117.46	116.40	128.00	113.14
Total organic acids	0.56	0.89	1.57	1.63	1.85	2.08
Cellulose	37.10	29.10	39.30	39.00	38.00	9.60
Hemicellulose	22.80	29.10	17.50	14.80	23.50	5.00
Lignin	8.20	7.40	9.60	10.10	7.70	9.30
pH	7.61	9.71	8.11	8.52	8.93	8.04

* Values are expressed as the per cent of total solids except total solids and pH; total solids are expressed as per cent of wet weight.

† Pretreatment was at 80°C for 24 h.

** NH_4OH was added after 24 h at 80°C just prior to fermentation.

Table 12.8
Composition* of Ball-Milled (BMS) and NaOH-Treated Straw and Inoculum

				NaOH Treatment† (g OH-/kg VS)			
Constituent	BMS	17.0	25.5	34.0	34.0 (liquid)	34.0 (solid)	Inoculum
Total solids	10.29	10.50	10.59	10.66	2.36	6.71	3.69
Volatile solids	88.71	83.55	80.95	81.62	60.42	83.07	69.76
Total nitrogen	0.94	0.91	0.89	0.83	1.42	0.53	8.76
Chemical oxygen demand	105.52	107.60	109.78	111.71	80.55	100.68	90.27
Total organic acids	0.57	1.74	2.01	2.24	5.43	1.54	1.86
Cellulose	32.90	33.70	34.50	34.90	2.10	43.40	5.80
Hemicellulose	26.10	15.20	10.50	6.70	1.30	12.20	8.70
Lignin	9.00	9.70	9.30	8.80	10.00	10.20	8.00
PH	6.59	9.25	10.05	10.48	10.01	10.01	7.80

* Values expressed as the per cent of total solids except total solids and pH; total solids are expressed as per cent of wet weight.

† Pretreatment was at 90°C for 1 h.

2. To keep the material submerged in water for proper fermentation, a screen was provided.

3. The stirring arrangement for proper mixing of straw, water, and starter is very important for digestion of crop residues. A stirring arrangement was designed so that rotary as well as up-and-down movement of the material could be achieved.

Digesters were kept aboveground for experimentation and no external heating system was employed. The bulk density of rice crop residues was found to be 38.9 kg/m³ and a solidwater ratio of 1:20 was selected for these investigations.

While rice husks did not need size reduction, straw was chopped manually into pieces ranging in size from 25 to 40 mm. The feedstock materials were analyzed for carbon, kjeldahl nitrogen, cellulose, hemicellulose, and lignin, and the results are given in Table 9. No pretreatments were given to feed. Daily maximum, minimum, and mean temperatures were recorded.

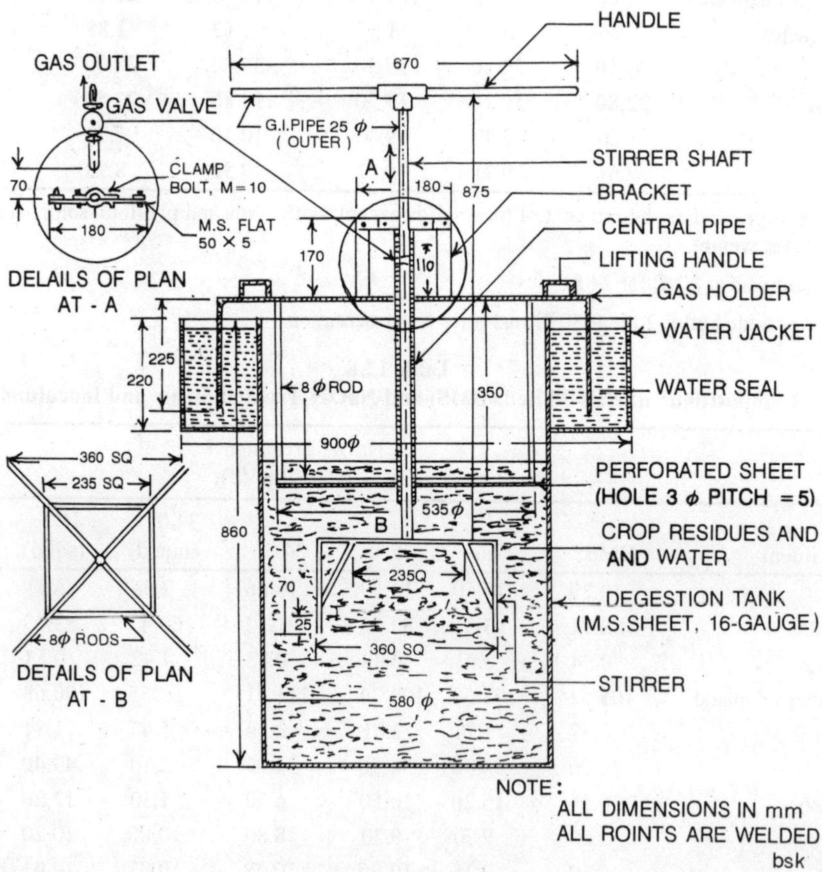

Fig. 12.9. Batch digester for crop residues.

Table 12.9
Proximate Analysis of Rice Straw and Husk

	Rice Straw	Rice Husks
Cellulose	26.7	29.4
Hemicellulose	19.4	17.7
Lignin	4.1	13.0
Kjeldhal nitrogen	0.984	0.355
Non-lignin carbon*	30.8	28.6
C/N ratio†	21.34	80.56

* Calculated from cellulose plus hemicellulose.

† C/N, Non-lignin carbon to Kjeldahl nitrogen ratio.

Values are percentages of total solids.

Five kilograms of dry feedstock were mixed in a ratio of 1:20 with water. Two percent pig droppings, by weight, were used as seed and mixed thoroughly to initate the digestion process. After loading the material into the digester tank, the gas holder was lowered into position. The digester contents were stirred by hand twice a day for 5 minutes and rotary as well as up-and-down movement was imparted. During the initial bioconvention process, carbon dioxide was primarily generated. The gas measurements were started only after the production of combustible gas. Gas was measured by a recording-type gas flow meter.

Husk-Straw Mixture as Feedstock

The availability of rice crop residues varies with the farmers and the milling industry. The farmers have all the rice straw while the rice mills have large quantities of rice husks. Keeping this in mind, three different mixtures of rice straw and husk were used as feedstock for fermentation and their biogas potential was evaluated. The ratios were 100% rice straw, 100% rice husks, and a mixture of 50% each of straw and husks. The experiment was conducted from February to May and the range of mean daily temperature during the period was 13.8 to 32.6°C. The relative gas production, in liters per day per kilogram of dry feedstock, with respect to time and daily mean temperature, is shown in Figure 10. The cumulative gas production during the entire digestion cycle and the peak valve of relative gas production are shown in Table 10.

In comparison, the straw alone produced 456% more gas than the husks alone and 167% more than the mixture. The husks had a very small gas potential, which could be due to very low nitrogen content. The nitrogen content of rice husks was found to be 0.355% as compared to 0.984% in rice straw. The low gas yield could also be due to high lignin content (13.0%) in rice husks. IT is reported that lignin is a complex compound whose structure is still not fully determined, and it is, to all intents and purposes, unattackable by anaerobic bacteria. The lignin, which is intimately connected with the cellulosic structures, acts as a barrier to bacterial attack on the cellulose.

The peak value of relative gas production varied almost in the same proportion as cumulative gas production for the three feedstocks (Table 10). Approximately the same ratio

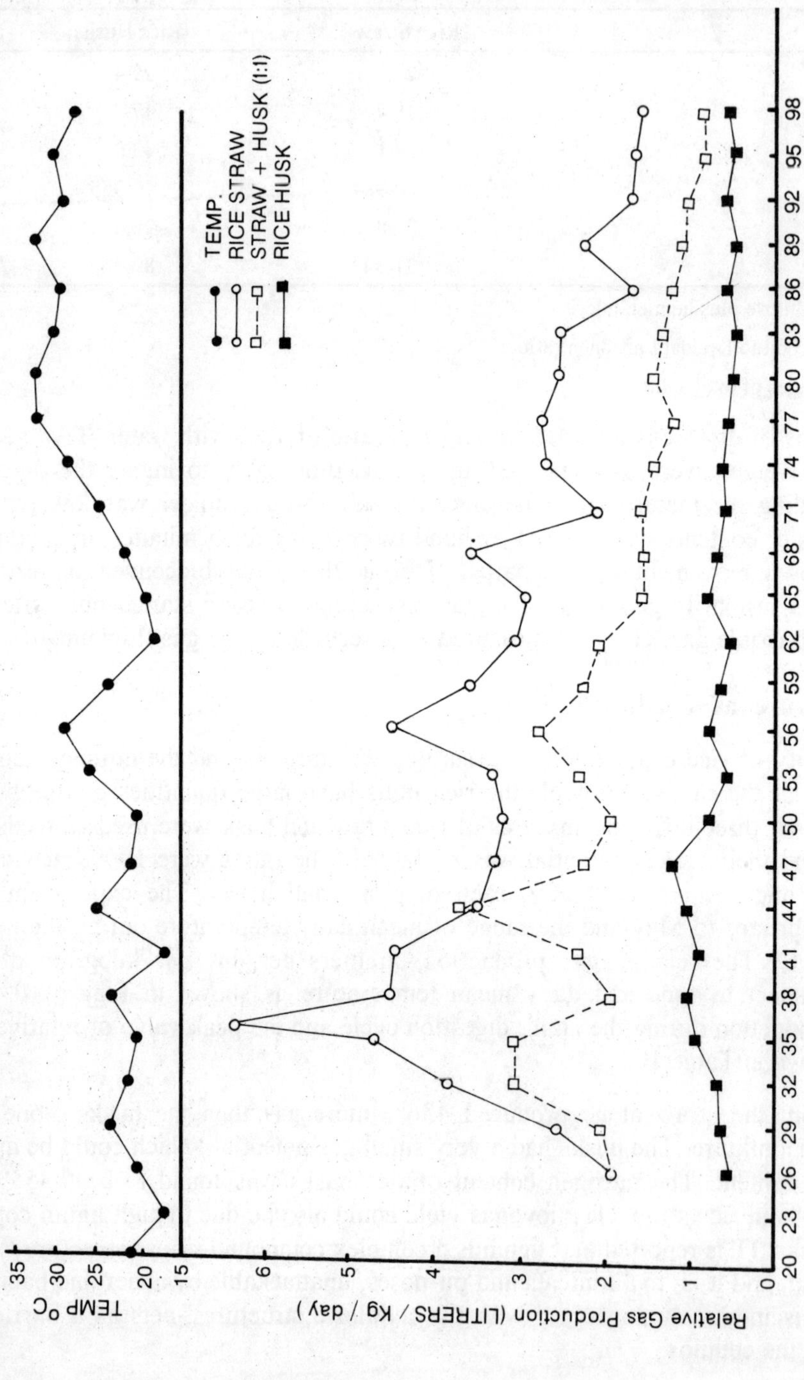

Fig. 12.10. Relative gas production for different husk-straw mixtures

of relative gas production by the three feedstocks was maintained at various stages of the cycle as per the peak value (figure 10). The variations in daily gas yield were due to the changes in temperature and the stage of digestion.

Table 12.10
Data on Gas Production from Different Straw-Husk Mixtures

Feedstock	Cumulative Gas Production (1/kg)*	Peak Value of Relative Gas Production (1/kg/day)	Digestion Cycle (days)	Starting Time (days)**
Straw	218.2	6.3	99	25
Husk	47.84	1.2	99	25
1:1 mixture	130.72	3.6	99	25

* Per kilogram of dry feedstock.

** Days after which combustible gas was produced.

Gas Yield from Mixture of Rice Straw and Cattle Dung

Rice straw and cattle dung were mixed in a ratio of 1:1 on a dry weight basis and fermented during June and July. The dung was from cattle fed on green fodder, mixed bhusa, and mixed concentrates. The daily mean temperatures during this period were 25.3 to 36.1°C. The daily gas yield and daily mean temperature were plotted (Figure 11). The gas yield was greater during the entire digestion cycle from the mixture of rice straw and cow dung than from rice straw alone. The mixture of rice straw and cattle dung yielded 9% more gas than the rice straw alone. The digestion cycle in this experiment was also reduced to only 53 days due to higher ambient temperature during this period. The improvement in biogas generation in the case of the mixture may be due to improvement in C:N ratio of feedstock as high C:N ratio paddy straw was mixed with low C:N ratio cattle dung. Hills and Roberts also reported improvement in gas generation when dairy manure and rice straw were mixed to obtain a C:N ratio of 25.

Further experiments were conducted on rice straw and the best result was obtained in experiments from May to July when the range of daily mean temperature was 24.55 to 37.25°C. Daily biogas yield and mean daily temperature over the entire digestion cycle have been plotted (Figure 12). The combustible gas was generated on the 13th day and the peak daily gas yield of 14.8 liter/kg was obtained on the 22nd day. The digestion cycle lasted for 46 days. In the batch system of digestion, the daily gas yield followed a set pattern. After the generation of combustible gas, the gas yield was almost constant for 10-12 days and then the gas yield decreased towards the end. The experiments were stopped when gas yield fell below 20% of peak value. The cumulative gas yield over the entire digestion cycle has also been plotted (Figure 13). Total biogas yield of 251.9 liters was obtained form one kilogram of dry paddy straw. It is interesting to note that after the generation of combustible gas if we divide the period in two halves, the first half (17 days) has yielded about 75% of the total gas generation. This is important because for continuous supply of biogas a farmer can employ 2 batch digesters and these can be charged in a staggered manner and stopped after the expiry

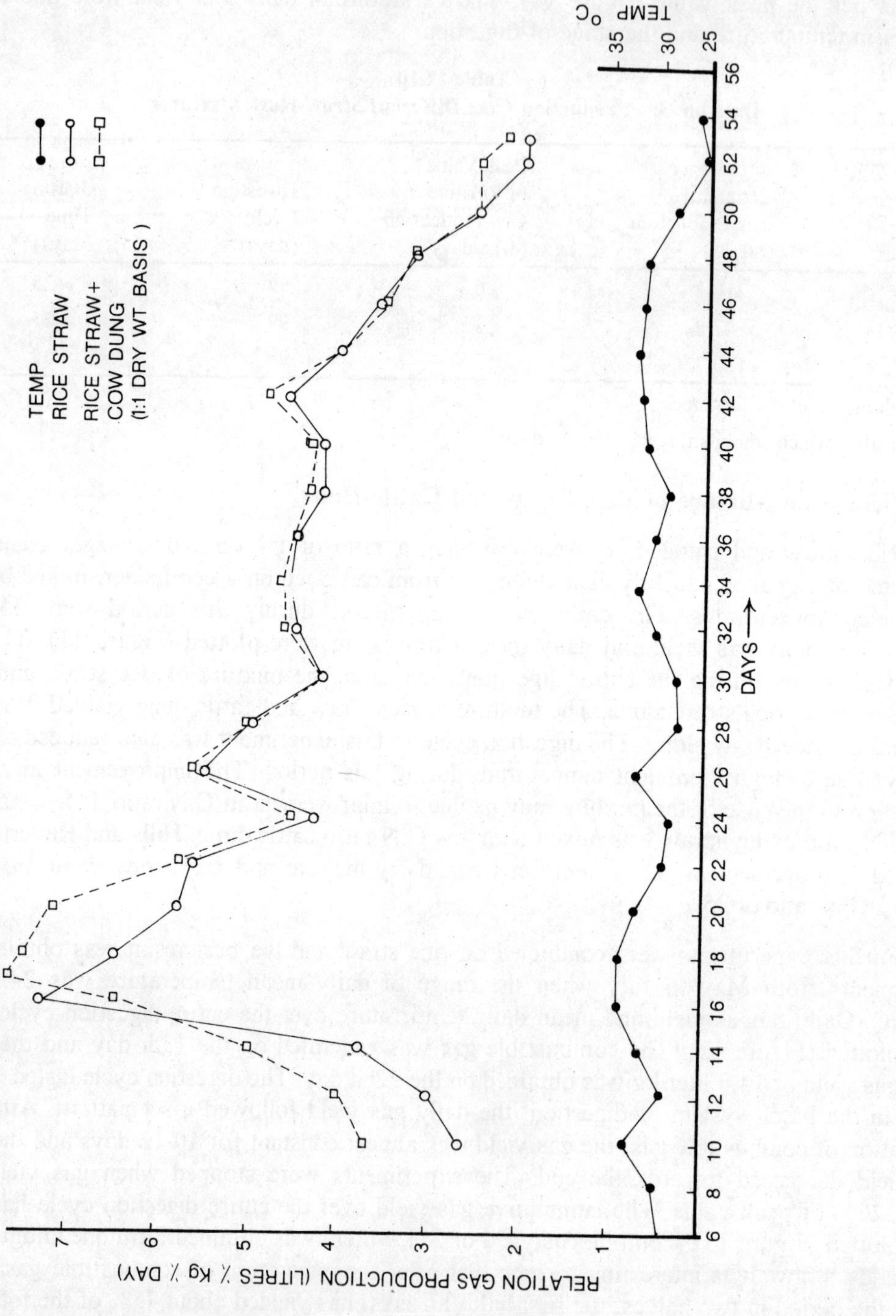

Fig. 12.11. Relative gas production as influenced by mixing cowdung with rice straw.

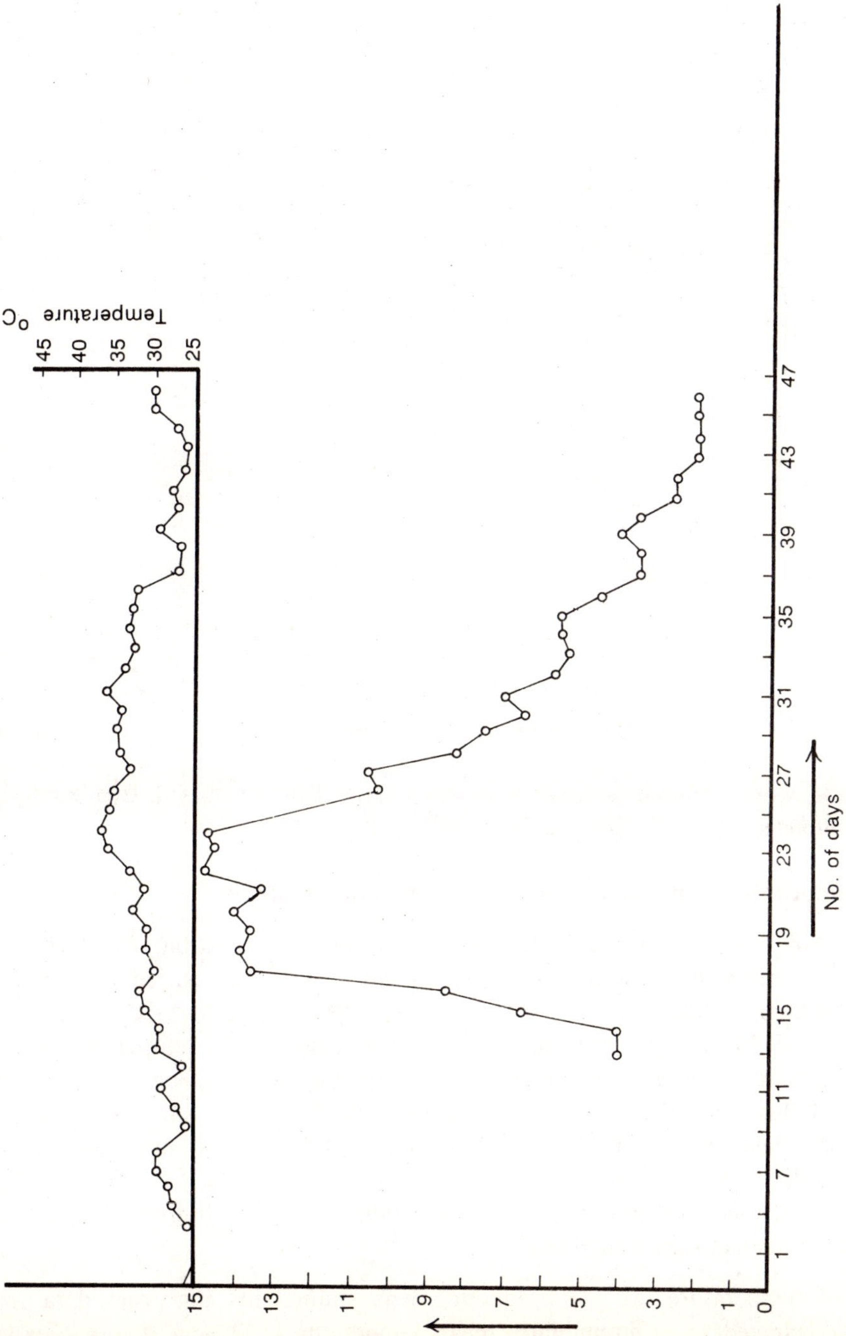

Fig. 12.12. Daily gas yield from rice straw.

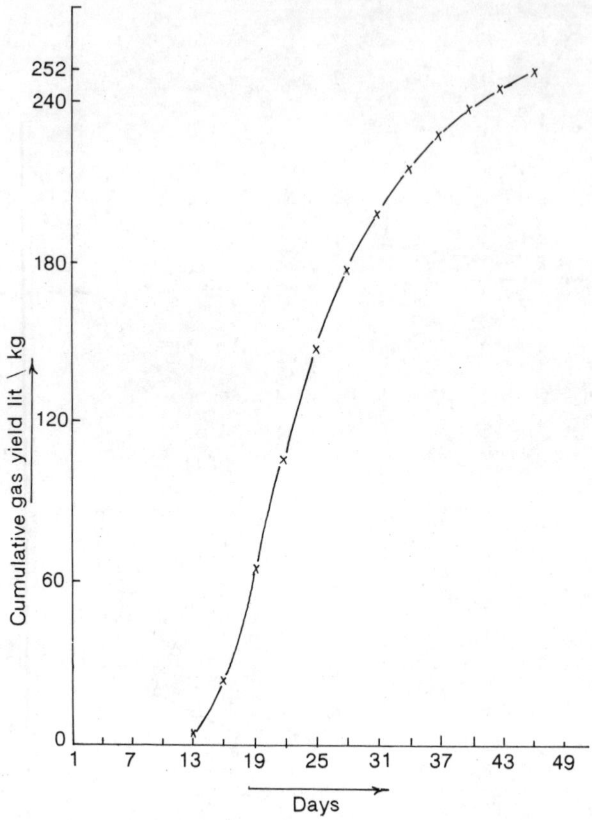

Fig. 12.13. Cumulative gas yield from rice straw.

of the first half, when 75% yield of gas has been obtained. So even with two batch digesters a fairly continuous supply of biogas can be maintained.

Effect of Temperature on Anaerobic Digestion of Rice Straw

Temperature has a significant influence on anaerobic digestion. Cellulytic bacteria, responsible for fermentation of agricultural residues, can be classified into two groups: mesophilic and thermophilic. The former have a temperature range of 5-50°C, while the latter have a range of 55-70°C. In general, the bacterial grow faster as the temperature is increased from the lower limit to a temperature of maximum growth rate, and then rapidly decreases in growth rate as this temperature is exceeded. Most of the investigations on the effect of temperature on fermentation employ external heating systems to maintain a constant digester temperature while, in India, biomass digesters operate under ambient conditions. The effect of temperature on digestion of rice straw was studied when the digesters operated under different ambient temperature conditions.

Effect of temperature on gas production was studied by analyzing data from two experiments conducted at different daily mean temperatures. During the first experiment, mean daily temperature ranged between 13.8 and 32.6°C, and the mean of the daily mean temperatures during the period was 23.09°C. The second experiment was conducted when

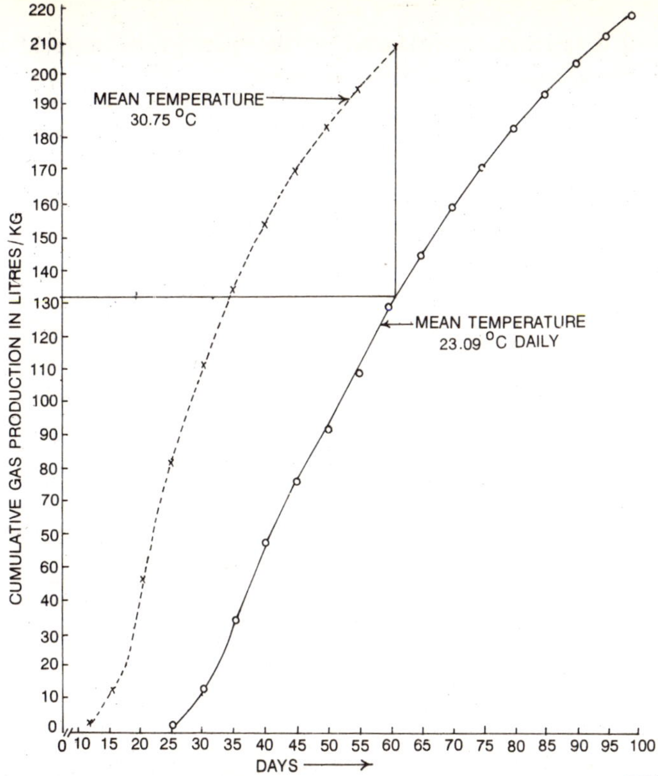

Fig. 12.14. Effect of temperature on cumulative gas production

the daily mean temperature was in the range of 25.0 to 36.1°C, and the mean of the daily mean temperatures during the period was 23.09°C. The accumulated gas production from two experiments versus retention time is plotted in Figure 12.14. The total digestion cycle was 61 days at a mean temperature of 30.75°C, and 99 days at a mean temperature of 23.09°C. The total gas production at higher temperature was 208.77 1/kg and the corresponding value of gas generation during 61 days at lower temperature was only 132.38 1/kg as shown in Figure 14. A temperature rise of 7.5°C in the mean temperature reduced the digestion cycle from 99 to 61 days and 57.70% more gas was generated during the same time period of 61 days. The increase in gas production at higher temperature is attributed to faster growth rate of bacteria. Hawakes reported that mesophilic bacteria had an optimum temperature range of 25-35°C. In a second experiment, which was conducted during mean daily temperature range of 25.0 to 36.1°C, the higher gas yield was obtained due to a faster growth rate of bacterial in the optimum temperature range.

The mean of the daily mean temperatures for the period, after which combustible gas was produced, were computed for both the experiments. The increase in temperature had a significant effect on the initial lag period after which combustible gas was produced. In the first experiment, the lag period was 25 days and the mean of ambient temperature during this period was 17.5°C. The lag period was reduced to half, that is, only 12 days, in the second experiment when the mean temperature rose to about 30°C.

The effect of temperature on fermentation of rice straw is also clear from the graph of daily gas yield versus time (Figure 12.15). It is apparent that methane-generating bacterial are very sensitive to temperature variations as the increase in ambient temperature, at all stages of digestion, caused increase in daily gas yield.

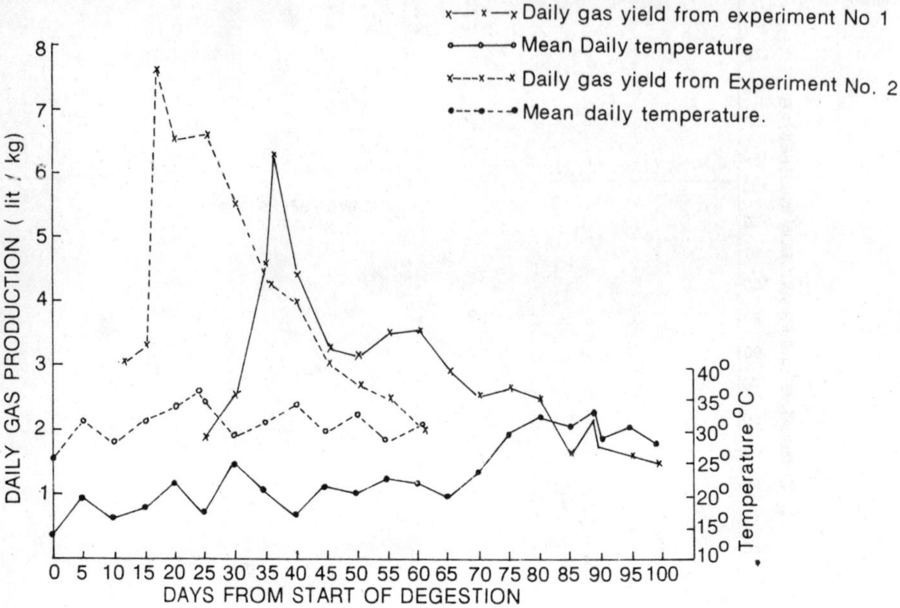

Fig. 12.15. Effect of temperature on daily gas production.

Influence of Seeders on Gas Production

There are two methods of starting the digestion process in animal waste digesters-with and without an inoculum or "seed" - as these wastes already contain the necessary bacteria, although not in correct number or proportion. Crop residues do not contain bacterial required for anaerobic digestion and the use of a seeder or inoculum is essential to operate the digesters. To study the influence of seeders, three different seeders, namely, pig droppings, cow dung, and used slurry from rice straw digesters were used for anaerobic digestion of rice straw. Experiments were conducted in 25 liter (total volume) bottles; and 0.5 kilogram of dry, chopped rice straw was fed to the digester, and a dilution rate of 1:20 was adopted. These bottles were painted black on the outside to prevent exposure of the bacterial in the slurry to direct sunlight, which is harmful to bacteria. Digesters were seeded as follows:

1. Cattle dung, 2% (dry weight basis)
2. Pig droppings, 2% (dry weight basis)
3. Used slurry from paddy straw digester, 5% (volume basis)

The experiment was conducted from April to June when the range of daily mean temperature was 25.3 to 34.5°C.

The different seeders caused variations in total gas yield and the digestion cycle. The total gas yield was greatest when pig droppings were used as the starter and lowest when the digester was seeded with cow dung. The total gas production from one kilogram of dry rice straw was 203.5, 180.96, and 191.93 liters, respectively, when pig manure, cow dung and used slurry were used as seeders. The cumulative gas production, with respect to time, for different seeder is shown in Figure 16. It was observed that initially the gas production was greater in the case of the used slurry seeded digester than in the others. After 37 days, which was two-thirds of the digestion cycle, the pig manure seeded digester produced more gas and maintained this lead until the end of the digestion cycle. However, the digester seeded with cow dung produced minimum gas throughout the digestion cycle. The digesters seeded with pig manure and used slurry produced 12.45 and 6.06% more gas, respectively,as compared to the gas yield of cow dung seeded digester.

The digestion cycle was also affected by the different seeders. The relative gas production with respect to time for digesters seeded with different materials is shown in Figure 17. Different seeders caused different starting periods (period before availability of combustible gas) for the simultaneously charged digesters. The used slurry seeder was quickest in production of combustible gas and the cow dung seeder proved to be the slowest. The combustible gas was produced on the 11th day when the digester was seeded with used slurry. The digesters seeded with pig manure and cow dung produced combustible gas on the 15th and 18th day, respectively.

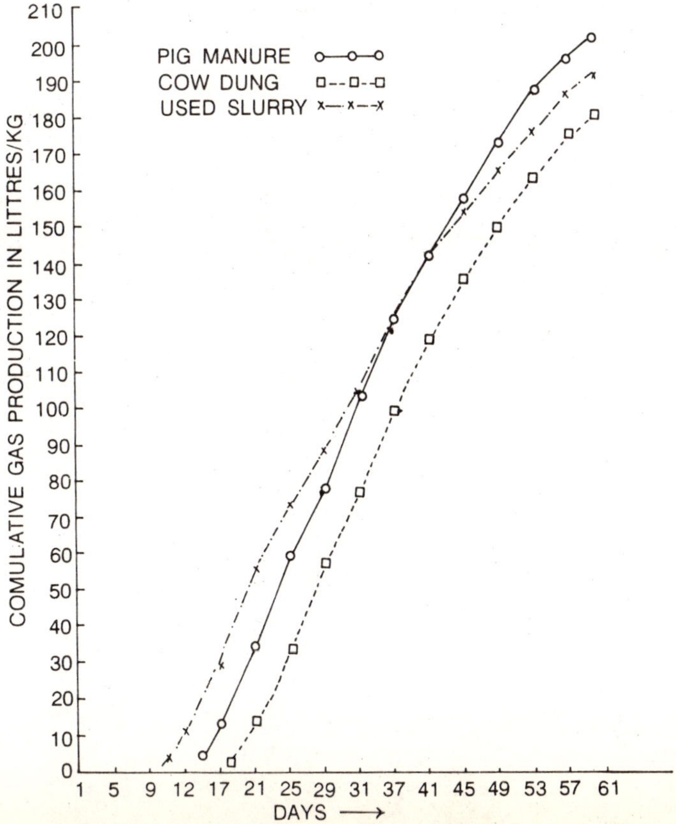

Fig. 12.16. Cumulative gas production patterns due to different starters.

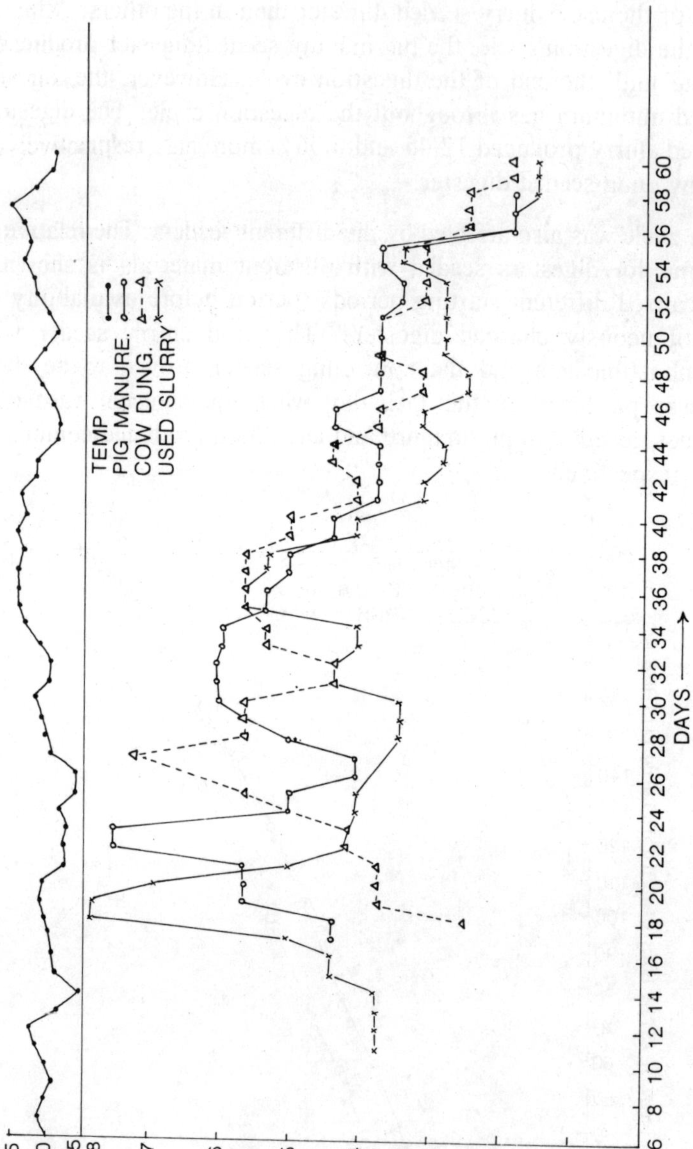

Fig. 12.17. Influence of starters on relative gas production.

The peak value of relative gas production, and the position of the peak in relation to the digestion cycle also varied in case of the three seeders. The digester seeded with used slurry produced a peak in the digestion cycle on the 16th day and the gas yield on that day was 7.85 liters/kg of dry material. The peaks in the digestion cycle in the cases of pig manure and cow dung seeded digesters occurred on the 22nd and 27th day, respectively. The peak values of relative gas production from pig manure and cow dung seeded digesters were 7.54 and 7.23 liters/kg of dry material.

The total gas produced in different segments of the digestion cycle also varied with the three seeders. Table 11 gives the breakup of total gas produced in three one-third digestion cycle segments. Maximum gas was produced by the used slurry seeded digester during the first one-third cycle. In the second one-third segment of the digestion cycle the pig manure seeded digester produced maximum gas, followed closely by the cow dung seeded digester. During the last one-third cycle, almost equal amounts of gas were produced by cow dung and pig manure seeded digesters.

Table 12.11
Total Gas produced, with Three Seeders, in Different Cycle Segments

	Total Gas Produced (1/kg dry matter)		
Period	Used slurry	Pig manure	Cow dung
1-20 days	51.52	28.90	10.09
21-40 days	87.63	109.90	104.93
41-60 days	52.78	64.70	65.94

The varying action of different seeders can be attributed to the presence of varying bacterial populations in the seeders. Used slurry was relatively rich in bacteria required for anaerobic digestion, which explained its quick action is generating combustible gas. In the case of cow dung, it is reported that although bacteria required for acid formation and methane fermentation are present, their number may not be very large. It was suggested that even in the case of cow dung digesters it would be beneficial if methane fermentation could be in creased by artificial seeding. This could be one of the reasons of the long initial lag phase in cow dung seeded digesters and their low total gas yields as compared to digesters seeded with pig manure and used slurry.

However, in anaerobic digestion processes there are many kinds of methanogenic bacteria, but their number and type are not fully known. It is not certain whether a similar variety of bacterial exists in every digester and whether it varies with the feedstock material or digester operating conditions.

Qualitative Analysis of Biogas and the Effluent

Several samples of gas were analyzed to determine methane and carbon dioxide content. The methane content in the biogas was found to be in the range of 65-72% by volume. The residue from the anaerobic digestion of rice straw was analysed for NPK content. The range of the NPK content, on dry weight basis, was as follows:

- N: 1-1.2%
- P: 0.11-0.13%
- K: 1.2-1.4%

It can be concluded that rice straw is a good feedstock for anaerobic digestion either as a sole substrate or in combination with animal manure.

13

Assessing Movement of Pesticides in Agricultural Environment

Introduction

The 1979 discovery of pesticide residues in drinking water wells in California and Long Island, New York sparked many research programs into the movement and degradation of pesticides and other agricultural chemicals. (Although the term "pesticides" will be used in this article, the discussion is equally applicable to compounds used as plant growth regulators. Nutrients are not included, but many of the same transport processes are applicable.) Paralleling laboratory and field research programs was the development of techniques for predicting the potential for movement of pesticides into groundwater.

These techniques for predicting the potential the potential for pesticide residues to inter groundwater can be divided into two categories. The first category consists of simple screening tools. Two of the best known examples are the criteria proposed by Cohen et al. and Allen et al. (DRASTIC). These types of models consist of comparing values of pesticide and/or environmental parameters with proposed criteria or the development of a simple rating based on simple arithmetic operations(usually addition) of weighted values of various important variables. In these types of models, no attempt is made to model the actual physical processes. The other category consists of mechanistic models, in which actual or simplified partial differential equations describing solute movement in the unsaturated or saturated zones are solved analytically or numerically. Examples of mechanistic models include PRZM and LEACHMP and a variety of saturated zone solute transport models.

Advantages associated with most screening models are that they are relatively easy to use and they highlight important variables. Disadvantages of these models include their inability to predict pesticide concentrations, and that vulnerability assessments are valid only for conditions within the original data base. Screening models where predictions are based on simple weighted sums or averages may not be sensitive enough to controlling variables. For example, in an area with essentially no recharge of water below the root zone, no pesticide movement will occur regardless of soil or pesticide properties or the proximity of groundwater or potable wells; yet a screening model based on a simple weighted sum may predict significant groundwater vulnerability under some variable combinations. (One approach to partially overcome this shortcoming, illustrated by Back *et al.*, is to use weighted sums or averages to arrive at rating for several different variable categories which are then multiplied together so that a low rating in one category cannot be completely offset by a high rating in another category.)

Advantages of mechanistic models are that actual pesticide concentrations (or at least the amount of material moving past a certain depth or point) can be estimated. Since the effects of individual variables can be evaluated, these models can be used to develop appropriate management strategies as needed. Also, since the models are based on equations describing the physical processes, mechanistic models can be used to make predictions for conditions outside the available data base as long as the equations or simplifying assumptions remain valid. The major disadvantages of mechanistic models are theat they are often quite complex and usually require large quantities of information describing weather conditions, soil properties, and pesticide parameters. Some mechanistic models require large amounts of computer memory and time to perform the necessary calculations.

The role of data is quite important in both types of models. In screening models, data are needed to define the initial weighting of the various factors and to verify and improve the accuracy of the predictions. Mechanistic models require data for model validation and as a source of input parameters such as degradation rates. It is important to understand the underlying reasons why comparisons between predicted and observed results may not agree. For screening models, these disagreements may indicate an important factor not previously considered or improper weighting of model factors. For mechanistic models, disagreements between predicted and observed results often stem from improper values of input parameters, such as degradation rates and soil properties.

Although both screening and mechanistic models are useful in making qualitative predictions about the probability of residues in groundwater, it is important to remember that no current model can quantitatively predict concentrations for a specific well. These models are most accurate when used to address specific sites. For larger geographical areas, changes in such factors as soil properties and water table depth make predictions less certain, due to the varying levels of vulnerability within the area. To a certain extent this limitation can be overcome by dividing the larger areas into smaller areas and performing assessments for each smaller area. Model predictions are also chemical-specific. Because of differences in degradation an sorption mechanisms, areas which are highly vulnerable for one compound may not be highly vulnerable for other compounds. Even the areas of highest vulnerability may not coincide for two different compounds. In spite of these limitations, models are quite useful for focusing resources on areas of expected vulnerability. As more data are obtained from these areas, the original estimates ca n be refined as necessary and appropriate changes made to enhance the accuracy of future predictions.

Screening Models

A variety of criteria and indexes have been used to identify pesticides or land areas where potential exists for pesticide residues to enter groundwater. Three examples of these screening criteria are given in approximate order of development.

These criteria, developed by Cohen *et al.* and presented in Table 1, consist of ranges for eight pesticide characteristics:

1. Water solubility
2. Soil/water partition coefficient
3. Organic carbon/water partition coefficient
4. Henry's law constant
5. Speciation
6. Hydrolysis half-life
7. Photolysis half-life
8. Soil half-life.

These criteria, or minor modifications of these criteria, have been widely used to identify pesticides with potential to enter groundwater. However, the field conditions portion of these criteria are rarely cited.

Most, if not all, of the pesticides shown to have the potential for entering groundwater as a result of leaching through soils have characteristics that meet at least some of the criteria of Cohen et al., but most of these compounds do not meet all of the criteria. Therefore, a pesticide not meeting any of the criteria probably has low potential for moving into groundwater. However, there are also compounds with low potential for groundwater residues that met some of the Cohen criteria.

Drastic

This is a screening methodology for evaluating potential for groundwater pollution of any hydrogeological setting. There are actually two versions, one for general use and the other for agricultural use of pesticides. The methodology(Table 2) consists of assigning ratings between 1 and 10 to each of seven factors:

1. Depth to water
2. Net Recharge
3. Aquifer media
4. Soil media
5. Topography
6. Impact of the vadose zone media
7. Hydraulic Conductivity.

Each value is then multiplied by its corresponding weight and the products are summed to get an overall score. The ratings for each of the seven factors are given by Aller et al.

When using DRASTIC for evaluating sensitive areas, two limitations should be remembered in addition to those generally applicable to screening models. These arise because DRASTIC essentially assumes that pesticide degradation is unimportant in determining site vulnerability. One of these limitations is that the ranking does not consider temperature which usually significantly affects the pesticide degradation rate. Therefore, potential for groundwater residues at a site in warmer areas is usually less than for a similar location in a cooler area. The other limitation is that DRASTIC does not consider any pesticide-specific effects of soil parameters on the degradation rate. For examples, the pH of surface and subsoils can greatly affect the degradation rate or mobility of specific pesticides.

Table 13.1
Criteria Proposed by Cohen et al. for Indicating Potential for Groundwater Contamination[1]

Pesticide Characteristics	
Water solubility:	Greater than about 30 ug/L
Soil/water partition coefficient(K_d):	Less than 5, and usually les than 1 or 2
Organic carbon/water partition coefficient (K_{oc}):	Les than 300
Henry's law constant:	Less than 0.01 atm m3/mol
Speciation:	Negatively charged(either fully or partially) at ambient pH
Hydrolysis half-life:	Greater than about 20 weeks
Photolysis half-life:	Greater than about one week(this criterion may not be as influential as the others)
Soil half-life:	Greater than about 2-3 weeks
Field Conditions	
Recharge:	Total precipitation and irrigation recharge greater than about 25 cm/year. An important factor in this criterion is the soil's drainage ability; i.e., soils with low moisture holding capacity are conducive to high recharge
Nitrates:	High levels in the groundwater are indicative of pesticide groundwater contamination potential
Aquifer:	Unconfined, porous soil above unconfined aquifer
Soil pH:	pH values providing high stability to the pesticide residues

Table 13.2
Factor and Assigned Weights for Agricultural DRASTIC

Factor	Weight
Depth to Water	5
Net Recharge	4
Aquifer Media	3
Soil Media	5
Topography	3
Impact of the Vadose Zone Media	4
Hydraulic Conductivity	2

Note: Each factor is assigned a value between 1 and 10. The DRASTIC rating is the sum of the products of each value and its corresponding weight.

Half-Life-Organic Carbon/Water Partition Coefficient Correlations

Two screening procedures for determining the potential of a pesticide to reach groundwater based only on the pesticide half-life and its organic carbon/water partition coefficient (K_{oc}) have recently been proposed by Jury et al. and by Gustafson. These correlations, shown in Figures 1 and 2, are based on observations that compounds which have been found in groundwater tend to be located in one region of K_{oc} versus half-life. The Jury criteria are based on the results of simplified model simulations, while the Gustafson criteria are based on a statistical analysis of the properties of 22 pesticides which were classified as "leachers" or "non-leachers" by the California Department of Food and Agriculture. Both the jury and Gustafson procedures appear to classify the 22 pesticides appropriately, with the main difference in the two procedures being the location of the dividing lines at high values of K_{oc} and half-life (this difference is discussed by Gustafson).

The Gustafson screening methodology defines a "GUS," or Groundwater Ubiquity Score, for a pesticide by the equation:

$$GUS = \log_{10}(t_{1/2}) \times (4 - \log_{10}(k_{oc}))$$

where $t_{1/2}$ = pesticide half life in soil (days)

K_{oc} = organic carbon/water partition coefficient (cm^3/g)

The two GUS value of 1.8 and 2.8 appear to bracket the transition region between "leachers" and "non-leachers." Those pesticides with GUS values of greater than 2.8 are usually more likely to be found in groundwater than compounds with lower GUS values however, Gustafson recommends that these compounds with GUS values greater than 2.8 should be subjected to further investigation by applying more sophisticated modeling technologies before concluding that significant movement will occur.

Gustafson also analyzed whether information on water solubility, octanol/water coefficient, and volatility in addition to information on the half-life and K_{oc} would provide additional discrimination between "leachers" and "non-leachers." One important conclusion was that information on any of these three parameters did not provide any additional insight into the leaching potential. Although water solubility has been widely used as an indication of leaching potential, K_{oc} is a better indication of pesticide mobility.

Mechanistic Models (Much of the information presented in this section has been adapted from Jones and Hanks)

A variety of unsaturated and saturated zone models have been used to assess the potential for groundwater residues resulting from pesticide applications. Most of this discussion will focus on unsaturated zone models since unsaturated zone processes determine whether pesticide residues will reach groundwater and, if so, when and what amounts. Saturated zone models can be used to simulate the persistence and movement of any pesticide residues in groundwater to evaluate the impact these residues may have on nearby wells used to supply drinking water.

Examples of Unsaturated Zone Models

This chapter reviews seven models (listed in Table 3) developed in the United States.

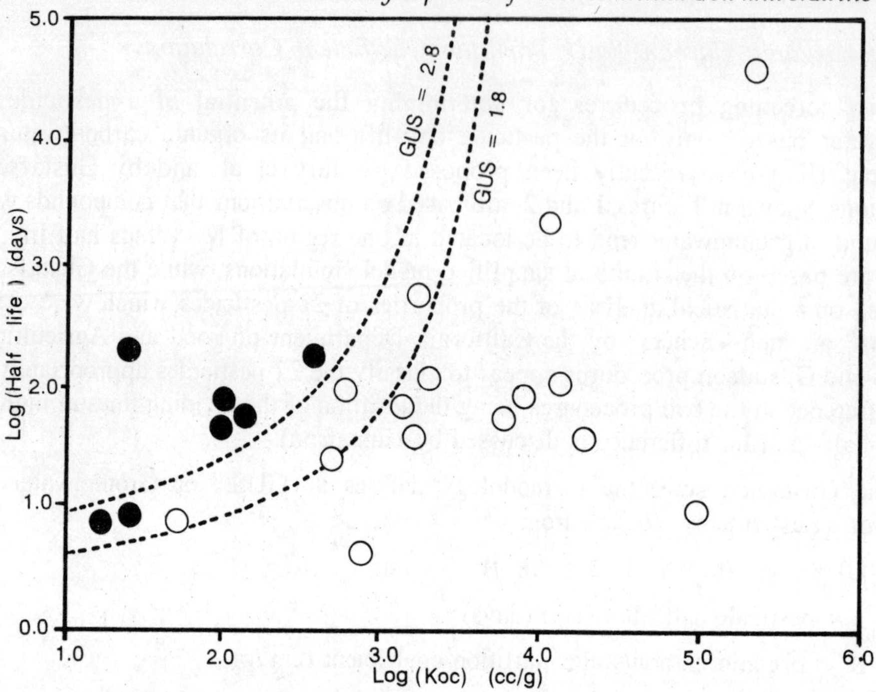

Fig. 13.1. The Jury criteria for determining pesticide leaching potential as a function of soil half-life and k_{oc}. Closed circles represent pesticide found in groundwater via leaching and open circles are for pesticides which have not been found.

These models meet three criteria: First, these models are applicable to non-point agricultural sources, that is, for applications to fields rather than sources resulting from spills or landfills. Second, these models directly simulate the movement of pesticide residues in the unsaturated zone based on physical processes (the models are mechanistic rather than empirical, although each model has some elements) and predict actual movement for a specific set of conditions. Third, these models are available for general use, not just research models developed for a specific application.

Two other models are also of interest: VULPEST (developed by the IRNSEau at the University of Quebec), a stochastic model which can be used to predict the range of movement given the normal variations in climatic conditions at a specified site; and SESOIL(developed by A. D. Little under contract to the U. S. EPA), a statistical model for estimating seasonal solute distribution in the soil.

Information on unsaturated zone models being used in Europe is provided by Vachaud et al. and Willigan et al.

PESTAN

Developed by the United States Environmental Protection Agency, this model became available in 1980 and is the earliest of the seven models reviewed here. It uses a steady-state analytical solution to the convection-dispersive solute transport equation obtained by assuming one-dimensional constant flow in a homogeneous soil profile with constant hydraulic properties, sorption, and degradation parameters. The lack of a user's manual poses little problem due to

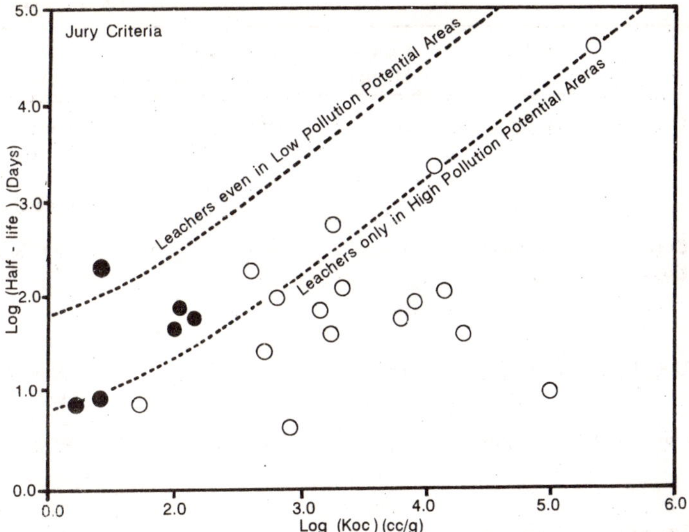

Fig. 13.2. The GUS correlation for determining pesticide leaching potential as a function of soil half-life and K_{oc}. Closed circles represent pesticides found in groundwater via leaching and open circles are for pesticides which have not been found.

the simplicity of the model. Very little input data is required, although one parameter, net recharge velocity, may be difficult to estimate (for model comparison simulations, net recharge velocity is often obtained from the water balance portion of one of the more complex models such as PRZM). Although the model assumes constant soil and pesticide properties, changes in adsorption or degradation rates with depth may be approximated by superimposing simulations made under these changing conditions. One mistake often made by users of this model concerns the appropriate degradation rate. Unlike the other models, PESTAN uses a solid phase degradation rate, which is usually a factor of 3 to 5 slower than the overall degradation rate measured in most field experiments. Therefore, when the overall degradation rate is used in PESTAN, the predicted degradation is a factor of 3 to 5 too fast. This model was widely used EPA and industry scientists before the introduction of more comprehensive models such as PRZM, but today its original one-dimensional form is rarely used.

MOUSE

Developed at Cornell University, this model solves the transient convective-dispersive equation for solute transport using a simplified water balance. Because the principal goal of the model was for user education rather than actual predictions, the solutions not as rigorous as in the other transient models. Also, a simplified saturated zone component is included. The one-dimensional soil model assumes two layers, root zone and subsoil, with fixed values of soil and sorption parameters within a layer. However, the degradation rate can very within a lyer. Daily weather data can be generated by an internal routine. The model was written for an IBM PC and has very good graphical routines for displaying water balances and pesticide movement.

Table 13.3
List of Mechanistic Unsaturated Zone Models Developed in the United States

	Developer	Model Type	User's Manual	Relative Computer Time Requirements	Frequency of Use	Comments
PESTAN	U.S. EPA	Steady-State	No	Minimal	Wiedespread use until about 1985. Currently is rarely used.	
MOUSE	Cornell Univ.	Transient	Yes	Low	Low to moderate	Developed primarily as an educational tool and graphics.
PRZM	U.S. EPA	Transient	Yes	Moderate	Most widely used model.	
CMLS	Univ. of FL	Transient	Yes	Moderate	Moderate	Model focuses on position of the solute peak.
GLEAMS	USDA	Transient	Yes	Moderate	Moderate	Extension of CREAMS surface runoff model. Currently applicable only to the root zone.
LEACHMP	Cornell Univ.	Transient	Yes	High	Moderate	
RZWQM	USDA	Transient	—	High	—	Model currently being developed.

PRZM

Developed by the United States Environmental Protection Agency, this one-dimensional model numerically solves the transient, convective-dispersive transport equation using a simplified water balance. The user's manual for this model does a good job of dodcumenting this code and explaining how required model parameters can be obtained or estimated. PRZM also includes optional estimation routines for some model parameters. The model is flexible enough to allow simulation parameters such as soil properties and pesticide degradation rates to very with depth. However, a program to prepare input files is quite machine-specific and is not used by some people. Although the model requires daily weather data, a weather tape is available providing this information for at least 25 years at locations throughout the United States. Procedures for estimating soil parameters from information in the SIRS data base are also available. The program can be easily modified to allow such changes as varying parameters with time or the use of kinetics other then first order. The available of input parameters, the support provided by EPA to model users, and the use of PRZM by EPA in several regulatory assessments has contributed to considerable use of this model, especially by industry scientists. A new version of this model (RUSTIC) will be released in 1989 which will allow simultaneous prediction of both original compounds and metabolites, use the method of characteristics to improve estimation of dispersion in the soil profile, and include routines for estimating soil volatilization losses. RUSTIC also includes linkages with a vadose zone model based on Richards' equation and a saturated zone model.

CMLS

This one-dimensional model is a refined version of an earlier model, PISTON, which was developed at the University of Florida by Rao et al. The model, written for an IBM PC, calculates the movement of a solute front using a simplified water balance. Although the model does not consider dispersion, retardation of solute relative to the movement of water is considered. Soil properties and degradation rates may be varied with depth and the model requires daily rainfall data.

GLEAMS

This model is an extension of the CREAMS runoff and erosion model. Both GLEAMS and CREAMS were developed by the United States Department of Agriculture. Currently this one-dimensional model is recommended for application only to the root zone. It solves the transient convective-dispersive equation for solute transport using a simplified water balance. Daily weather data is required but may be generated by an internal routine. Soil properties and degradation rates may vary with depth. The model can also calculate the transformation and movement of nutrients.

LECHAMP

Developed at Cornell University, this model is the only generally available one-dimensional unsaturated zone model that solves the water balance using Richards' equation. This additional complexity introduces an interative calculation procedure which must converge at each time step. Under some circumstances, problems with convergence have been reported due to the non-linearity of the surface boundary condition. Computer time requirements are substantially greater for LEACHMP than for models using a simplified water balance. If the appropriate rate constants are supplied, LEACHMP can simultaneously predict concentrations of both original compounds and metabolites.

RZWQM

This model is currently being developed by the United States Department of Agriculture. During the development process, a large number of regulatory, academic, and industry scientists were invited to make suggestions about the various algorithms to be used and how the model should be structured. Since all components of the model will be designed to incorporate the most rigorous estimation techniques, computer time requirements are likely to be quite high. A data base of pesticide and nutrient properties will also be included. The model is scheduled for release in 1989.

Discussion of Unsaturated Zone Models

Comparison of Models

As pointed out in the previous discussion, there are many differences in the seven models. Two of the most fundamental differences are the methods used for calculating the water balance and for estimating dispersion.

The method of calculating the water balance is the primary difference between the seven models. Since PESTAN is a steady state model assuming a constant recharge, it cannot accurately model areas with large variations in seasonal rainfall. However, by appropriate scaling of time and rainfall, this steady state model can be used to approximate a transient situation, especially if the degradation rate does not change significantly with depth. Depending on rainfall patterns and the assumptions used in calculating net water recharge velocity, PESTAN predictions of solute movement may be either greater or smaller than predicted by the other models. Little difference in predicted solute movement would be expected between MOUSE, PRZM, CMLS, and GLEAMS since essentially the same simplified water balance (no movement until field capacity is exceeded, then instantaneous movement) is used in each of these models. Also, no upward movement of water in the soil profile (and therefore no upward solute movement) in response to evaporation losses at the soil surface is considered in the transient models with the simplified water balance. These assumptions tend to result in these models slightly overpredicting solute movement compared to field measurements. Because LEACHMP calculates the rate of water movement using the Richards' equation and accounts for upward movement of water, water movement, and hence solute movement, is slower than predicted by the models using the simplified water balance, especially in heavier soils. However, recent comparisons by Hornsby (personal communication, 1987) have indicated that in sand soils where the rate of water movement is relatively fast, the differences between the predictions of LEACHMP and the other transient models are not significant.

The method for calculating solute dispersion also differs in the transient models. The movement of the solute peak is calculated in CMLS and MOUSE, while in the others numerical techniques are used to calculate the movement in each soil layer assuming some value of dispersivity. When solute dispersion is calculated using finite difference or finite element procedures, numerical dispersion may affect the calculated dispersion of a solute (finite difference techniques are more prone to numerical dispersion tahn finite element techniques). Thus the predicted solute dispersion, and therefore the concentration profiles, may be dependent on the dimension of the depth increment. In MOUSE, the dispersion around the solute peak is calculated using the appropriately scaled error function (similar to the procedure used in PESTAN). In CMLS, no dispersion around the solute peak is considered and therefore soil concentrations are not calculated by the model. However, this is not a serious deficiency compared to other models since calculated concentration profiles from any of these models usually do not adequately match measured profiles. Instead, the preferred way of reporting results is in mass or percent of applied material moving past of specified soils depth because such values are much less sensitive to the amount of dispersion and the shape of the concentration profile (for multi-year or Monte-Carlo simulations, the uncertainty of these values for mass flow can be expressed using a cumulative probability distribution). Also, it may be misleading to report concentration profiles since even in a relatively homogeneous filed, concentrations of pesticides at a specific depth and time are likely to vary by at least two orders of magnitude.

Model Assumptions

Potential sources of error in mechanistic model predictions include the theory and assumptions that comprise a model, the numerical solution of model equations, and selection of input parameters. The user controls these errors by the choice of an appropriate model for the specified application and by correct specification of input parameters. Therefore, it is important for a model user to be aware of the assumptions inherent in a model. Then, when considering the use of a model for a certain application, the appropriateness of these assumptions should be examined. The validity of the values of each of the input parameters should also be considered. The appropriateness of extracting parameters from the work conducted at other locations or other compounds should be reviewed.

Several assumptions are common to most of the models discussed in this chapter. All assume first order kinetics for the degradation rate either for an overall rate of disappearance or for each reaction step. Degradation rates at a specified depth usually cannot be changed with time (for example, due to changes in soil temperature). Sorption is usually considered to be described by a linear isotherm with no difference between sorption and desorption disruption. The models assume that concentrations in the liquid and solid phases are always in equilibrium. The appropriateness of these assumptions should be considered for each application. In some cases, the model can be modified by relatively minor changes in the code to eliminate the need for a particular assumption. For example, changes in the code to eliminate the need for a particular assumption. For example, changes in kinetics or sorption isotherms could be made quite easily if the data were available.

One of the most important factors determining movement of pesticides is the amount of water moving below the root zone. In all of the models, water movement below the root zone, often referred to as the amount of recharge water, is calculated as the amount of rainfall and irrigation minus the sum of evaportranspiration, runoff, and changes in water storage. Usually the last term is insignificant for simulations being conducted for periods longer than several months. In many cases, the sum of rainfall and irrigation and the sum of transpiration and runoff are relatively large compared to the amount of water moving below the root zone. Therefore the accuracy of the water recharge term is quite sensitive to the accuracy of the other terms.

Values for runoff and evapotranspiration losses are usually not measured directly and must be estimated. Some of the models do not consider runoff losses of water or pesticides, while models calculating runoff (PRZM and GLEAMS) use the Soil Conservation Service curve number approach. In sand or loamy sand soils, where potential for downward movement is the highest, precise estimation of runoff is usually not needed since runoff and pesticide losses in runoff water are typically quite small. However, in clay soils, a significant fraction of a rainfall event may move laterally off the field rather than moving downward through the soil profile.

One difficult step in most simulations is to obtain accurate potential evapotranspiration date. Often potential evapotranspiration is estimated from average daily temperature data or from pan evaporation data using an appropriate correction factor (pan factor and/or crop coefficient). Inaccuracies can result using either of the two procedures so the resulting evapotranspiration losses calculated during a model simulation should always be checked for reasonableness. The predicted solute movement has been substantially altered in several PRZM simulations depending on whether the potential evaporation was estimated using average temperature data or pan evaporation data. Often the estimation routine based on average temperature data has appeared to underestimate evapotranspiration and thus overestimate pesticide movement compared to measured field results. However, the crop coefficient approach may also produce inaccurate estimates, especially at times when the crop cover is small and the amount of bare soil is large.

The models described in this article are not designed to simulate movement of pesticides in irrigated fields, but he models can be used to provide estimates under certain conditions. Usually this is accomplished by adjusting the amount of rainfall on days when irrigation is applied. When pesticides are present on plant foliage in flood or furrow irrigated fields, this approach would not be valid due to overestimation of residue washoff. None of the models can be used in situations where irrigation is applied by raising the water table underneath a field and water is supplied by upward movement to the root zone. Also none of the models can be used to simulate movement in or below areas that are temporarily saturated (perched water tables) where significant lateral movement of water may occur. In fields where irrigation water is not applied evenly (such as with furrow or flood irrigation), the location of the pesticide application relative to the location of irrigation furrows influences chemical movement. Therefore, model simulations may either overpredict or underpredict movement of pesticides in fields where irrigation is applied uneventy. Even when irrigation water is applied uniformly using some form of sprinkler system, model predictions may be in error. For example, for those models which consider runoff, excessive values of runoff may be predicted (thus underpredicting downward movement of water). One approach to reducing runoff losses from irrigation is to spread the amount of irrigation over more than one day in the simulation.

Another assumption made by all of the models is that macropore flow is not significant. Since macropores exist in finer-textured soils, while concern with groundwater residues is usually limited to coarser-textured soils, this assumption is not usually of concern. Ignoring macropore flow may result in either underestimating or overestimating movement of pesticides depending on their location. If pesticide residues are located on foliage or the soil surface, water flowing through macropores located at the soil surface may contain residues and the models will tend to underestimate movement. However, if residues are located in the soil (for example as a result of application methods which include soil incorporation), then water moving through the macropores do not come into contact with the residues and the models will tend to overestimate movement of pesticide residues. In general, macropore flow has the potential to cause relatively low levels of pesticides to occur deeper (or faster) than the models predict, although data with aldicarb indicate that in most cases macropores may have relatively small impact on the movement of relatively mobile pesticides in the unsaturated zone.

Model Selection

The appropriate model to use depends very much on the specific application. For example, if an individual needs better understanding of the general effects of various variables, then MOUSE is probably the model of choice because it is easy to use and has good graphical displays. If the application is to estimate whether a pesticide is likely to move below the root zone in a specific location, then the use of PRZM, GLEAMS, CMLS, or MOUSE would be appropriate. If a more rigorous estimate is desired, then LEACHMP, PRZM, or

GLEAMS probably should be used. PRZM, CMLS, MOUSE, or LEACHMP can be used for predictions of movement and degradation in subsoils below the root zone. Although more data and effort is required to produce simulations with LEACHMP, the use of this model is quite appropriate for comparing predicted versus observed movement in details field research studies.

All of the transient models give generally the same results in light soils. However, in heavier soils the predicted movement of LEACHMP will be significantly less than that predicted by PRZM, CLEAMS, MOUSE, or CMLS. In general, movement of pesticides in heavier soils is of little consequence for groundwater residue concerns, since most instances of pesticide movement through agricultural fields occur in lighter soils. LEACHMP will probably provide the most accurate predictions of pesticide movement in heavy soils. The other transient models can be used to estimate an upper bound on pesticide movement as long as the modeler realizes that actual movement may be considerably less than predicted. The increased data and computational requirements of LEACHMP probably preclude its use for assessment techniques involving many different individual simulations. However, if such assessments indicate a potential concern in areas with heavy soils, selected simulations should be repeated using LEACHMP.

Input Data for Unsaturated Zone Models

The mechanistic unsaturated zone models (especially the transient models) require much more input data than the screening models, including daily weather data, soil properties, pesticide properties, crop parameters, and pesticide application information. Table 4 provides a list of sources for obtaining information on parameters needed by PRZM (these parameters are also needed in many of the other unsaturated zone models). As mentioned earlier, the PRZM user's manual describes techniques for estimating parameters used in this and other transient models.

Perhaps the most difficult unsaturated zone model parameter to obtain is the pesticide degradation rate in the unsaturated zone (or the degradation rate in the saturated zone for groundwater models). Because the degradation rate of a pesticide often varies with temperature and soil properties such as soil-watercontent, pH, or microbial activity, estimation of the degradation rate for a specific location is not straightforward, especially in the absence of information from a number of locations. Because of the difficulty of extrapolating laboratory data to field conditions, field measurements of the degradation rate are preferred. Usually degradation rates in surface soils are measured by the pesticide registrant by performing field dissipation experiments at several locations. Degradation rates below the root zone may be slower than degradation rates in surface soils, especially if the major degradation pathways are microbial rather than chemical; however, degradation rates in subsoils are available for only a few compounds. Therefore, a needed (and expensive) step in performing assessments of pesticide mobility may be the conduct of field research studies at a number of locations. Procedures for conducting such studies are discussed by Jones. Such studies may also provide information on degradation rates in the saturated zone.

Although weather records are readily available from the National Oceanic and Atmospheric Administration, the daily weather data requirements of the models are often the greatest deterrent to performing model simulations, especially the 20 year simulations needed to assess climatic variability (weather data is available in monthly pamphlets or can be ordered on magnetic tape, but the latter option can be quite expensive). As discussed earlier, the EPA has made available a weather data required by PRZM for locations throughout the United States. With minor adaptations, this information can be used to provide weather data for other models. Due to the limited number of locations within a state, this tape is useful primarily for general assessments. Weather data for comparing field research results with model predictions will usually have to be entered manually from NOAA or site measurement logs.

Table 13.4
Sources of Information for PRZM Input Parameters (from Carsel and Jones, manuscript in review)

Parameter	Source
Meteorologic data	
daily rainfall	NOAA reports, EPA data base
daily evaporation*	NOAA reports, EPA data base
average daily temperature†	NOAA reports, EPA data base
Hydrology data	
minimum evaporation depth	[19]
pan factor	[19]
snow factor	[36]
runoff and infiltration	[37]
erosion**	[19, 37. 38]
Soil parameters (per horizon depth)††	
bulk density	[19, 20, 39]
field capacity and wilting point	[40, 19]
organic matter	[20, 41] soil surveys
Crop information (per crop simulated)‡	
emergence, maturity, and harvest dates	[42]
maximum root zone depth	[19]
foliage interception of rainfall	[19]
Pesticide parameters§	
application description (foliar or soil)	pesticide label
data and rate or application	pesticide label
pesticide decay rates foliage	[19, 24]
soil (as a function of depth)	[43, 24, 44]
partition coefficient (as a function of depth)	[19, 43. 44]
plant uptake efficiency factor	[19]
foliar washoff	[19]

* Can be estimated from daily average temperature.

† Not necessary if pan evaporation data is available and snow accumulation is not simulated.

** This process can be ignored in certain cases.

†† There are a variety of additional sources including Soil Conservation Service reports and published literature.

‡ Additional information can be obtained from extension agents, crop production reports, and agricultural research facilities.

§ University extension services, pesticide manufacturers, and crop protection manuals are sources of additional information.

Two of the transient models, CLEAMS and MOUSE, try to reduce weather data input requirements by including an option for internally generating weather data. Although long weather records are needed to appropriately estimate the parameters required by the wheather simulation routine, this greatly reduces the effort required in data input. Of course, for comparing predicted versus observed results, actual weather data would still have to be entered. The use of weather generators is appropriate when models are used for training purposes or for drawing general conclusions; however, the use of actual weather records is preferred when making more comprehensive assessments. The weather data generator is an approximation of the actual records, so the use of actual records helps eliminate any biases that might result from the generation procedure.

The required soil parameters for the transient models that use the simplified water balance can be estimated from the information collected by the Soil Conservation Service and published in county soil surveys. The information is also available in the SIRS data base. However, the organization of the data base makes finding the distribution of specific soils within a state quite cumbersome. To address this problem, the EPA has recently completed preparing a subset of SIRS including only data related to the location and use of agricultural soils and their properties pertinent to unsaturated zone movement. This new data base, DBAPE, will allow modelers to identify the amounts and locations of soils fitting specified criteria and to obtain model parameters for these soils.

Saturated Zone Models.

If pesticide residues reach groundwater, then saturated zone models can be used to predict the magnitude and movement of these residues. Note that if an assessment performed using unsaturated zone models indicates no movement of pesticide residues to the water table, then the assessment is complete without saturated zone modeling. Many of the existing saturated zone, solute transport models are suitable for conducting such evaluations if the appropriate pesticide loadings, as a function of time, are provided. An example of using a saturated zone model to predict pesticide transport is described by Rothschild et al.

One approach to modeling pesticide residues in groundwater is to couple an unsaturated zone model directly to a saturated zone model. Examples of such linkages are given in the literature. The RUSTIC model includes a saturated zone transport model which can be linked to either of two unsaturated zone models. As with unsaturated zone models, the choice of models depends on the specific application. Most of these models were developed to simulate site-specific cases and require detailed site-specific hydrogeological information. For more general situations, for example, when a constant water velocity is assumed, the use of simple models (often analytical solutions) is more appropriate. As illustrated by Carsel *et al.*, a simple saturated zone model coupled with an unsaturated zone model can be a powerful tool in the development of management practices, such as restrictions on applications of pesticides near shallow drinking water wells.

Assessment Procedures Using Mechanistic Models (This information has been adapted from Jones and Rao).

The unsaturated and saturated zone models described in this article have been developed primarily to draw generalized conclusions based on existing site-specific data for a broader range of soil and climatic conditions. However, most of the modeling discussion in the literature focuses on model validation by comparisons of predicted versus observed.

Assessments performed using unsaturated and saturated zone models are generally of two major types. One type of assessment is to determine the potential for a pesticide to enter groundwater as a result of specified uses. Another type is to develop management practices for minimizing concentrations of a pesticide in groundwater and preventing its appearance in drinking water wells. For the latter application, linking unsaturated zone models with saturated zone models is often necessary.

When performing assessments, the modeler must design an appropriate set of simulations that cover the likely range in values of model parameters. The selection of input parameters for these simulations usually will have more effect on the assessment results than the accuracy of the individual model simulations. Therefore, the procedures and assumptions used in developing the set of cases to be evaluated should be carefully documented when assessment results are reported.

Because of the variability of soil properties and weather data, results of assessment simulations usually should be represented in stochastic terms, for example, as plots of cumulative probability distributions of loadings to groundwater. The exceptions are when "worst case" situations show no significant movement of a pesticide into groundwater and when "best case" situations still show movement of sufficient amounts of a pesticide into groundwater to be of concern. Defining what constitutes a worst case or a best case, however, may not be straightforward. Assessments based on a simulation of pesticide movement in a "typical" soil profile using a single year of weather data (sometimes without the knowledge of whether the rainfall in this year is typical or abnormal) can often be misleading. Defining a typical soil profile or rainfall pattern is also not straightforward.

One approach to assessing movement of a pesticide in a specific soil has been illustrated by Jones et al. This approach involves conducting unsaturated zone simulations with 15-25 years of weather data and then expressing the results as a cumulative probability distribution; an example is shown in Figure 13.3. The variability in both the amount and the distribution of rainfall are accounted for using this approach. The effect of management practices, such as changes in application timing, can also be evaluated using this procedure. The example presented in Figure 13.4 shows that aldicarb applications to citrus made in June will generally result in greater concentrations of aldicarb in groundwater than applications made in February. However, in one of the 17 years greater concentrations would have resulted from February applications (indicated in Figure 4 by the crossing of the two curves between 5 and 10%). This example illustrates the need for multi-year simulations when performing assessments.

Assessments of the potential for a pesticide to enter groundwater usually are not restricted to a specific soil, but rather encompass a region such as a single country or an entire state. One approach was to perform multi-year simulations on each soil suitable for growing a specified crop. In the example presented in Table 5, the SIRS data base was used to identify soils suitable for growing potatoes in upstate New York. This procedure can also be used to evaluate management practices. As illustrated in Table 5, residues of aldicarb are

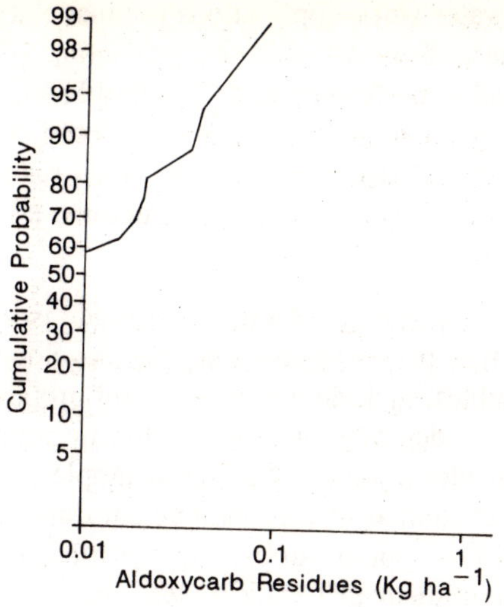

Fig. 13.3. Cumulative probability distribution showing the effect of weather on the predicted amount of aldoxycarb residues moving below 1.5 m from applications to tobacco at Harrellsville, NC.

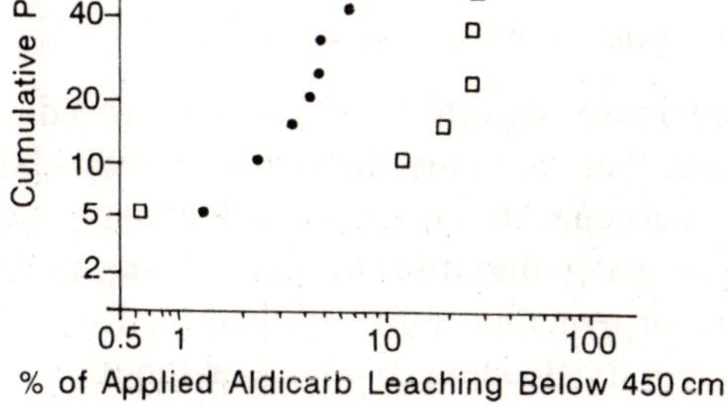

Fig. 13.4. Cumulative probability distribution showing the effect of application timing on the predicted amount of aldicarb residues moving below 4.5 m for applications to citrus in central Florida.

more likely to occur in groundwater when applications are made at the time of planting rather than when the potato plants emerge from the soil. One advantage of this assessment procedure is that an evaluation of each soil is performed and, therefore, soils with the greatest potential for a pesticide to move into groundwater are readily identified. A disadvantage of this approach is that the percentage of applications that may result in a pesticide entering groundwater is not defined without performing work, since the acreage devoted to a specific crop is not the same for each soil.

To obtain estimates on the percentage of time a pesticide is likely to enter groundwater, a Monte Carlo procedure has been proposed involving the development of probability density functions for a number of variables, including soil type, soil properties as a function of depth for each soil type, weather data, degradation rates in the unsaturated and saturated zones, water table depth, and groundwater velocity. For the example presented in Figure 5, 2,000 sets of input parameters were evaluated using a simple saturated zone model linked with an unsaturated zone model (PRZM) to obtain an estimate of aldicarb residue concentrations in groundwater beneath and downgradient of treated fields, and to evaluate the effectiveness of a management practice preventing aldicarb applications near shallow wells.

Table 13.5
Example Assessment Procedure showing Movement of Aldicarb Residues as a Function of Application Timing and Rate for Applications to Potatoes in Upstate New York

Application	No. of Soils with Movement Exceeding 0.1 ka/ha	
	Below 1.5 m	Below 3 m
3.36 kg/ha at planting		
In at least 1 of 19 years	20	6
In over half of 19 years	10	1
2.26 kg/ha at plant emergence		
In at least 1 of 19 years	3	0
In over half of 19 years	0	0

Notes: 103 soils were identified by SIRS as being used for potatoes in upstate New York.
Application rates refer to the amount of active ingredient.

Even though the previously described assessment procedures account for variability in soil properties and weather, there are other limitations which should be recognized. One such limitation is the inability to account for other factors that may contribute to pesticide movement into groundwater, such as water discarded from washing application equipment, a leak or break in an irrigation line, or irrigation water applied for freeze protection not being included in irrigation records. Temporary flooding of a field or runoff of water into low areas of a field may also result in more downward water movement, and thus more pesticide movement, than predicted using an unsaturated zone model.

Conclusions

Techniques for assessing potential for pesticide residues to enter groundwater consist of

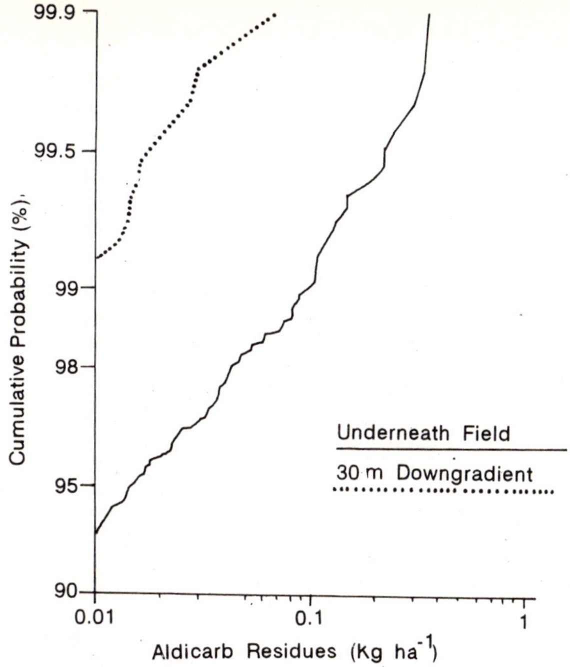

Fig. 13.5. Cumulative probability distribution showing the predicted amount of aldicarb residues in shallow groundwater underneath and 30 m downgradient of treated fields resulting from applications to North Carolina peanuts.

both screening methodology and mechanistic models. Perhaps the best screening methodology for indicating which pesticides have the potential to reach groundwater under some conditions consists of correlations of pesticide half-life and organic carbon/water partition coefficient. Because of the complex interactions between pesticides, climatic conditions, and soil properties, no screening methodology has been generally accepted by the scientific community as adequate, although DRASTIC has been widely used by state and federal regulatory agencies in the United States.

Mechanistic models can be used to perform more precise assessments of pesticide movement when information on the organic carbon/water partition coefficient and degradation rates in the unsaturated zone (as a function of depth) are available. Extensive data on soil and climatic properties are also needed but may be available from existing data bases. When performing assessments using mechanistic models, simulations should be conducted over the appropriate range of input parameters, especially soil types and climatic conditions. Therefore, assessment results usually need to be represented stochastically, for example, as cumulative probability plots of loadings to groundwater.

14

Wood Waste Pyrolysis

Introduction

Pyrolysis

Pyrolysis is the thermal degradation of an organic substance under an oxygen-deficient atmosphere. when applied to wood, the products include gaseous and liquid fractions, together with charcoal. Carbonization refers to a pyrolysis process aimed at the production of charcoal, while the alternative designation of wood distillation is used when the liquid products are sought. The relative yields of the gaseous, liquid, and solid products depend primarily on the pyrolysis conditions, namely temperature and heating rate, as well as on the nature of the material undergoing pyrolysis.

Pyrolysis processes may be divided into two main groups, according to the heating rate: slow pyrolysis and flash pyrolysis. Flash pyrolysis is operated at extremely high heating rates (e.g. 1,000 k/s) in order to maximize the production of gas and/or liquids. The amount of charcoal obtained under such conditions is minimal. At much lower heating rates, the process is called slow pyrolysis and the production of charcoal is maximized at the expense of liquid and gas. In both cases, the yields of gas increase with the pyrolysis temperature, while the yields of the other fractions decrease. The product spectrum obtained from wood pyrolysis is discussed in detail in a later section.

Historical Background

The pyrolysis of wood is a very old process. It was practised by ancient civilizations to produce charcoal, used as a fuel and reducing agent for metallurgy, and tars, for embalming purposes and shipbuilding.

During the 19th century, wood distillation became a major chemical process, supplying charcoal (mainly for iron ore smelting), methanol, acetic acid, and acetone, among other products. Softwoods additionally produced turpentine and various oils. By the turn of the century, wood was the principal raw-material for the organic chemical industry. Figure 14.1 shows a typical diagram for wood distillation.

In the 20th century, wood-derived chemicals were largely replaced by synthetic chemicals. Charcoal production remained important for heating and industrial purposes, mostly in developing countries, and for the production of activated carbon. After the oil crisis, as a result of the

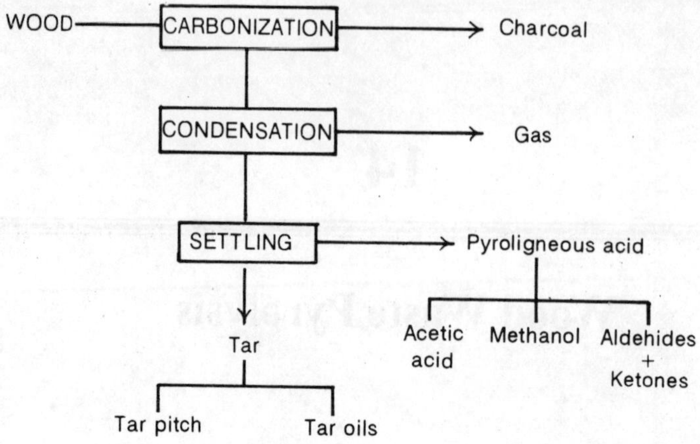

Fig. 14.1. Simplified diagram for wood distillation.

increasing awareness of the limited supplies of fossil fuels, attention focused again on the utilization of renewable raw materials for the production of energy and chemicals, and the thermal conversion of wood revived. An expanding market meanwhile appeared for charcoal briquets intended for barbecues and outdoor cooking.

Much of the current interest in pyrolysis is, on the other hand, directed towards its applications in solid waste disposal. In this context, pyrolysis is particularly attractive, as it can be operated on a small scale, consistent with the disperse nature of many of the wastes and residues available as feedstocks.

Wood Wastes and Conversion Methods

Vast amount of wood wastes are generated each year in harvesting, forestry, and processing operations, which need to be disposed of, but also may be regarded as an important source of energy and chemicals. The magnitude of the problem may be recognized by considering that about 1/3 of the wood harvested in the U.S. is wasted.

Wood wastes may be broadly classified into industrial and logging wastes. Industrial wastes from pulp mills, saw mills, plywood, and furniture manufacture include bark, trimmings, shavings, and sawdust. Logging wastes include branches, bark, stumps, and roots, as well as damaged trees. Additional lignocellulosic wastes are produced in agricultural operations, including shells, fruit stones, husks, and straws.

It is generally estimated that a sawmill will generate about 12 % bark, 10% sawdust and fines, and 8% shavings (weight %) of the dry wood processed. Most of these wastes may be suitable for reutilization (for instance, for pulping), but this is not always feasible, often due to the transportation costs involved. The disposal problem thus created may be solved by several upgrading techniques (Figure 2).

Direct combustion is usually the simplest and most common method of conversion when there is a user for the total amount of energy generated and no market for charcoal. Problems associated with wood waste conversion include large heat losses through the stack gases and particulate emissions, which increase with the moisture content of the wood.

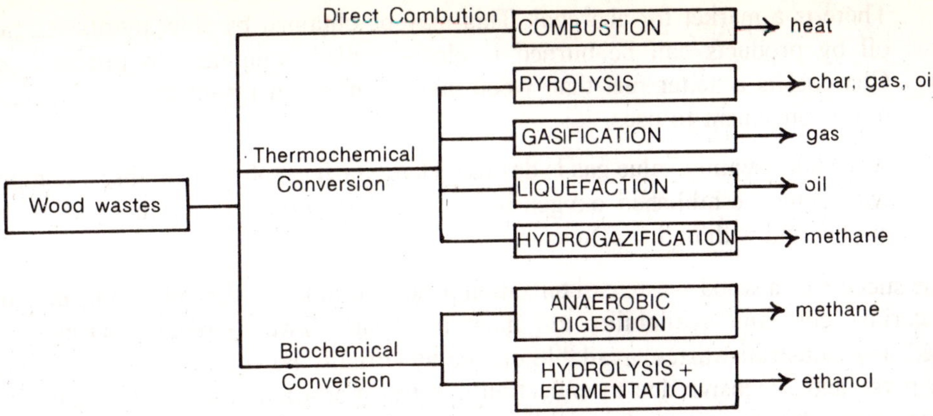

Fig. 14.2. Alternative processes for the production of energy and simple chemicals [illegible] wood wastes.

The thermochemical conversion methods produce not heat but fuels, which can be stored, transported, and used in existing combustion equipment with fewer operational problems. Alternatively, they can be designed to produce chemicals. In principle, wood wastes can be thermochemically processed in scales ranging from 0.25 to 40 ton/h.

Gasification involves reaction with an oxidizing agent (air, oxygen, steam) and consists of a sequence of steps occurring at successively higher temperatures: pyrolysis, originating volatile matter and char, gas phase reactions of the volatiles, and char gasification. Complete gasification of the wastes can be accomplished in this way, producing a mixture of CO, H_2, CH_4, CO_2 and N_2 (if air is used). This can be used as a fuel or for synthesis of various chemicals.

Liquefaction consists of the reaction of wood wastes suspended in a liquid with H_2 or CO_2 under pressure, to produce an oil of high heating value. The presence of a catalyst is required. Although there is much research on this process, its commercial viability remains to be proved. One of the difficulties lies in the small particle size required and the cost of grinding lignocellulosic materials.

Hydrogasification is the conversion to methane by reaction with hydrogen. This process may be suitable for the upgrading of wood charcoal, but is less well adapted for direct conversion of wood waste.

Hydrolysis by acids or enzymes releases the sugars from the lignocellulose, which can then be fermented to produce ethanol by well established technologies. Anaerobic digestion of organic wastes by microorganisms, or biomethanation, produces a gas-containing methane and CO_2. Hydrolysis and biochemical methods are poorly adapted to the coversion of wood since hydrolysis does not attack lignin, which is also very difficult to b iodegrade. The conversion of agricultural wastes with higher cellulose content may be more attractive .

Although burning and gasifaction of wood include a pyrolysis step, pyrolysis may be considered as a process on its own right, when the subsequent steps of gasification and combustion are absent. Pyrolysis may offer advantages over these competing processes when:

1. There is a market for charcoal. This may be obtained by slow pyrolysis; gas and oil by products can be burned is close-coupled equipment to produce energy. Charcoal is a better fuel than wood, as it is more energy-intense, smokeless, and more convenient to use.

2. A medium heating value gas is desired. Here, flash pyrolysis is in direct competition with some established oxygen-blown gasifiers, producing a fuel gas of higher quality and with higher efficiency.

The success of a wood-waste carbonization process ultimately depends on its economics. Raw-material cost and availability are most important. Two extreme situations maybe envisaged: (a) industrial wastes, available on site at virtually no cost; or (b) logging wastes, more disperse and expensive due to collection and transportation costs. In the latter case, a promising approach is to use a mobile pyrolysis unit.

CHEMISTRY AND PHYSICS OF PYROLYSIS

Chemistry of Lignocellulosic Material

The average elemental composition of dry wood varies little with the source and can be expressed by the overall formula $C\,H_{1.44}O_{0.66}$. Only traces of sulfur and nitrogen can be found in wood. In terms of chemicals compostion, wood and other lignocellulosic materials consist essentially of three types of polymers:cellulose, hemicellulose, and lignin. In addition, there are extractives (terpenes, resins, fatty acids, tannins, pigments, and carbohydrates) and ash (oxides of calcium, potassium, sodium, magnesium, silicon, iron, and phosphorus). For most woods, the ash content is less than 1% by weight, while the extractives account for 2-5%.

Cellulose is the framework component of the cell walls in lignocellulosic materials. It is a linear polymer of glucose with degree of polymerization between 7,000 and 10,000. Bundles of cellulose molecules aggregate to form microfibrils, in which crystalline and amorphous zones can be found. These microfibrils are embedded in a matrix of hemicellluloses and lignin [9]. It is this structure, analogous to a fiber-reinforced plastic, that is responsible for the difficulty in separating the wood components prior to their degradation.

The hemicellulose are amorphous, branched polymers based on hexose (glucose, galactose, mannose), pentose (xylose, arabinose), and glucuronic acids (Figure 3). They have lower degrees of polymerization (50-300) and are easily hydrolyzed. Mannans are the main hemicellulose in softwoods. They have a linear chain of glucose and mannose, with side chains of galactose. Hardwoods contain mostly xylans, where the linear chain consists of xylose units, with side chains of arabinose or glucuronic acid.

Lignins are amorphous, cross-linked phenolic polymers, which are more abundant in softwoods than in hardwoods. Moreover, softwoods lignin contains mostly guaiacyl units (one methoxyl group in the aromatic ring), while in hardwood lignin, syringyl units are more frequent (two methoxyl groups). The higher contents of acetyl and methoxyl groups in hardwoods explain the higher yields in acetic acid and methanol obtained upon pyrolysis.

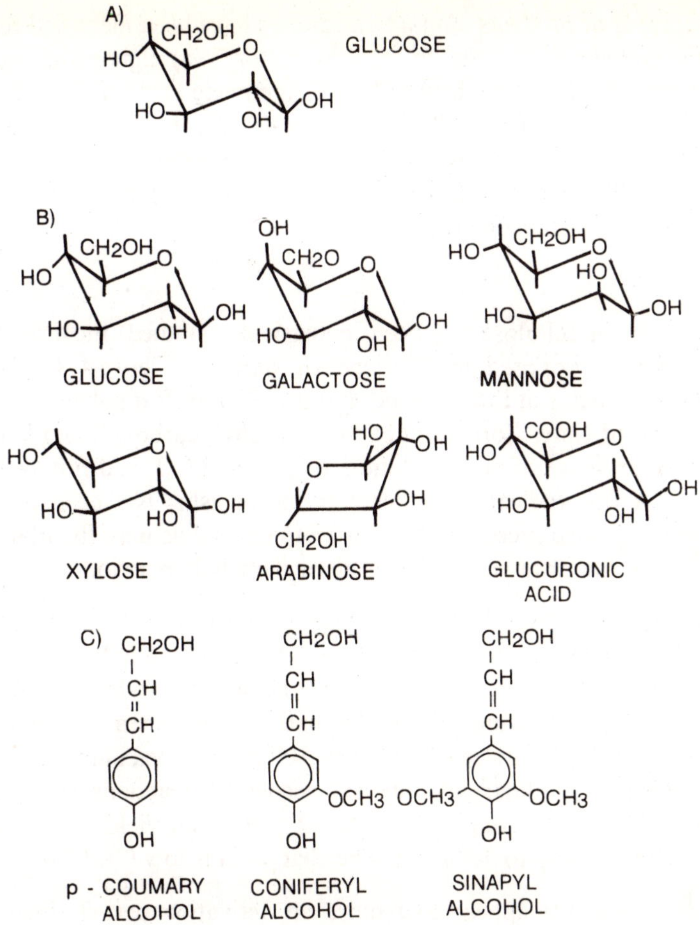

Fig. 14.3. Structural units of: (A) Cellulose, (B) Hemicellulose and (C) Lignin.

Figure 14.3 shows the structural units of the wood components, and Table 1 presents the average chemical composition of softwoods and hardwoods.

Table 14.1
Chemical Composition of Woods (% dry weight)

	Cellulose	Hemicellulose	Lignin	Extractives
Softwoods	42±2	27±2	28±3	3±2
Hardwoods	45±2	30±5	20±4	5±3

**Reprinted with permission from [90].*

Pyrolytic Behaviour of Lignocellulosic Materials

The pyrolysis of lignocellulosic materials is extremely complex. It is surely the result of hundreds of concurrent and consecutive reactions, which vary according to the pyrolysis conditions. Hundreds of products and intermediaries have been identified for these unidentified reactions. This complexity is due to various factors: (a) lignocellulosic materials are a mixture of components (cellulose, hemicellulose, lignin, etc.); (b) each of the major components is by itself a macromolecule which degrades in a very complex manner; and (c) the degradation products react among themselves in various ways (secondary reactions). Before dealing with the pyrolysis of wood, it is thus convenient to examine the thermal degradation of its major components.

Cellulose Pyrolysis

The pyrolysis of cellulose has been extensively studied most authors agree with the simplified mechanism proposed by Shafizadeyh, shown in Figure 4. Of the primary reactions (1)-(3), path (1) dominates at lower temperatures (<300°c). It involves reduction in molecular weight, elimination of water, formation of free carbonyl, carboxyl, and hydroperoxide groups (especially in the presence of air), evolution of CO_2 and CO, and finally the production of a charred residue. A free radical chain reaction mechanism has been proposed for this low temperature pathway. The presence of air or inorganic solids may facilitate the initiation step, thereby explaining the observed speeding up of the whole reaction.

At temperatures above 300°c, cellulose is decomposed by an alternative pathway (reaction (2). The liquid product contains levoglucosan (1,6-anhydro-B-D-glycopyransose), other anhydroglucose compounds , randomly linked oligosaccharides, and glucose decomposition products. The proposed mechanism for this pathway is shown in Figure 5. If cellulose is pyrolyzed at temperatures between 300°c and 500°c under vacuum to allow quick removal of the decomposition products and prevent secondary reactions (reactions (4) and (5) in Figure 4), tar yields are very high (up to 81%), with high yields of reducing sugars (up to 60%) and levoglucosan (up to 40%) [9]. The char yield is low (3-20%).

Reaction (1) and (2) appear to require an activation period. Indeed, to interpret the kinetics of pyrolysis between 250°c and 340°c, found it necessary to propose the mechanism shown in Figure 6. The activation step can be interpreted as a preliminary reduction in molecular weight, breaking of hydrogen bonds and a glass transition.

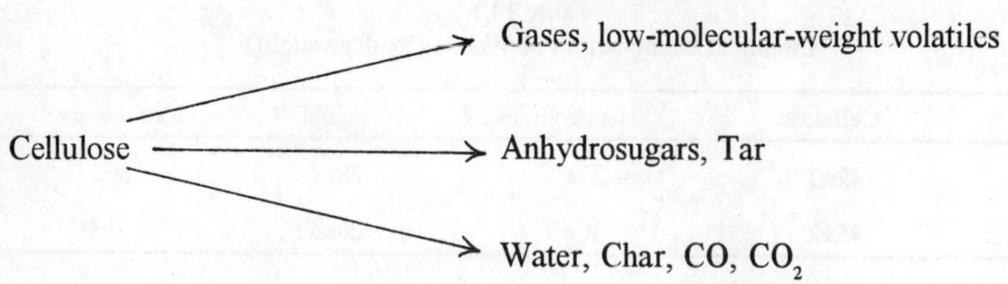

Fig. 14.4. Simplified mechanism of cellulose pyrolysis.

The 3rd parallel pathway, reaction in the overall mechanism (Figure 4), is of importance only above 500°C, hence with very high heating rates. It is attributed to fission, dehydration, disproportionation, decarboxylation, and decarbonylation of sugar units. It is difficult, however, to distinguish between the products of reaction and of the secondary reaction (5) of cracking

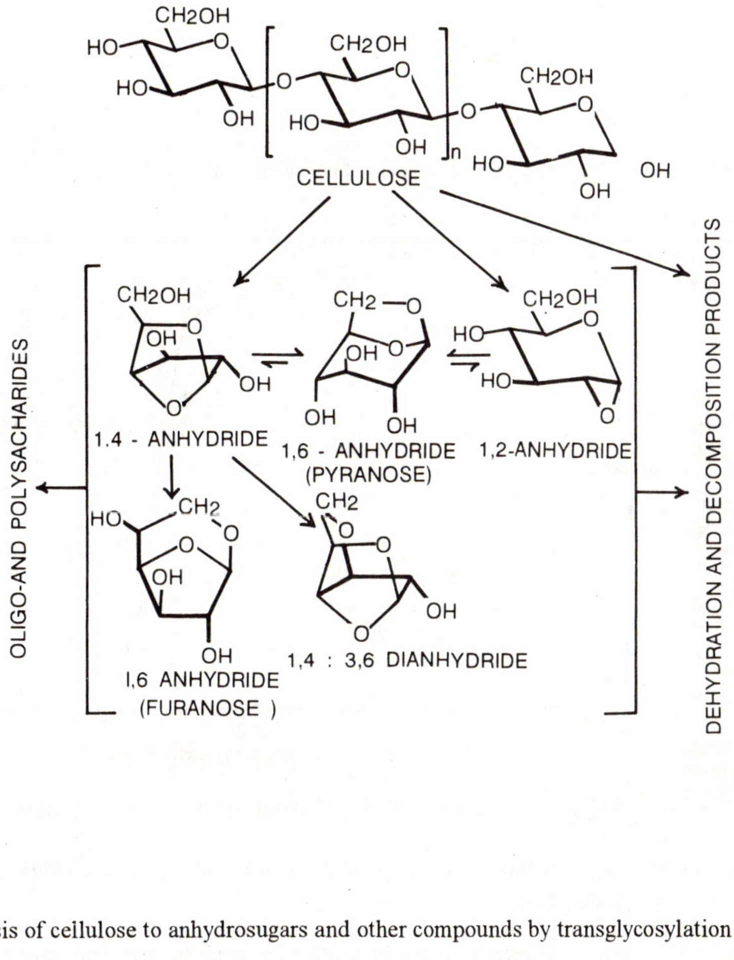

Fig. 14.5. Pyrolysis of cellulose to anhydrosugars and other compounds by transglycosylation reactions.

of sugar units aready in the vapour phase. Secondary reactions will be discussed again shortly.

Hemicellulose Pyrolysis

The hemicellulose component of wood is the least stable, decomposing in the temperature range 225-325°c. Although the pyrolysis of hemicelluloses has not been much studied, it may be expected to proceed through the intermediate formation of anhydrosugars, like cellulose. Shafizadeh and Chin reported the products obtained from xylan, and concluded that similar reaction to those of cellulose are involved.

Lignin pyrolysis

The products and kinetics of lignin pyrolysis reflect the complexity of its molecule, in which different monomeric units are linked together by several types of ether linkages as well

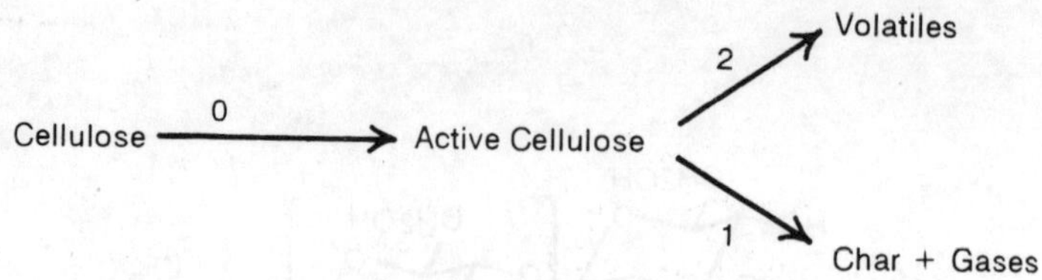

Fig. 14.6. Pyrolysis of cellulose with activatin step.

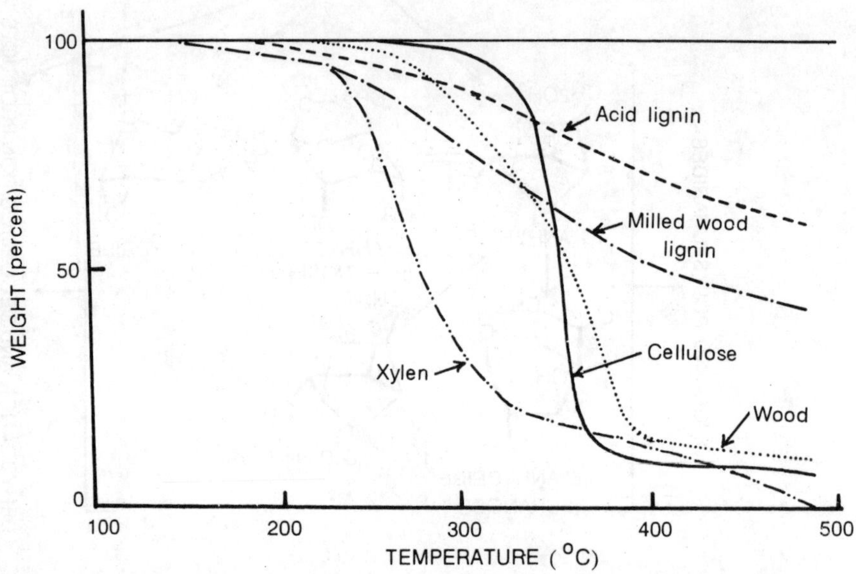

Fig. 14.7. Thermogravimetry of cottonwood and its components.

as C-C bonds. As a consequence, lignin pyrolysis never orignates a single product comparable to levoglucosan from cellulose.

Parts of the lignin molecule are less stable than cellulose, but most of it is more stable and reacts at higher temeratures, as sshown in Figure 7. Char is the major product (~55% yield), followed by an aqueous distillate (~20% yield) containing mostly water, with some mehtanol, acetone, and acetic acid. The remaining products are tar and gases. The tar contains mainly phenolic compounds, such as phenol, guaiacols, and cathechols.

Wood Pyrolysis

Primary reaction in the pyrolysis of wood are those ofthe pyrolysis of its components, which are expected to react independently. Hence, it is logical that the products, the kinetics, and the thermodynamics of the pyrolysis of wood and other lignocellulosic materials be the summation of those of their components. This was indeed shown to be so, as regards products (see Table 2 and references, kinetics (see figure 7), and heat of pyrolysis (see DTA results in references.

Secondary reactions between the volatiles escaping the solid matrix, both within it and in the gaseous environment, are extremely important in determining the final pyrolysis product distribution. Hundreds of such reactions occur, and are virtually impossible to determine individually. This is confirmed, e.g., by the number of products that have been identified. Soltes quotes 226 products identified by various researchers in the condensable products of pyrolysis. His lumped view of secondary reactions is shown in Figure 8.

Product Distribution

Table 2 shows the product distributions obtained in various laboratory, pilot, and industrial-scale pyrolysis processes. Additional data can be found in the literature [1, 29]. Different experimental conditions are responsible for the diversity of results shown, and the following parameters have been identified as playing a major role in determining the product distribution.

Solid Factor: Compostion
Surrounding temperature
Particle size
Impurities; catalysts
Moisture content

Gas Factors: Temperature
Residence time
Pressure
Composition

Solid Composition. The relative amounts of carbohydrates and lignin in the raw material have a major influence on the primary reactions and products of pyrolysis. Comparison between lignin and cellulose pyrolysis highlights this point: charcoal yields are high from ligin (~55%) and low from cellulose (~10%), while the corresponding tars are rich in phenolic compounds (from ligin) and levoglucosan (from cellulose).

Table 14.2a
Products of Pyrolysis of Lignocellulosic materials under various conditions (indirect Heating)

	High temperature Flash		Low Temperature Flash		
	Entrained Flow (Pipe)	Entrained Flow (cyclone)	Screen Heater	Fluidized Bed	Entrained Flow
Raw material	wheat straw	wood	cellulose	wood	grass straw
Particle size	0.16-0.27 mm	0.2-0.4 mm	0.010 mm thick	<0.595 mm	<1.15 mm
Temperature	950°C	1048°C	400°C	500° C	470°C
Gaseous environment	steam	He/steam	He, 5 psig	pyrolysis gas	pyrolysis gas
Residence time: solid	77 ms	1.4s	30s	—	—
gas	—	0.05 s	quick removal	0.55 s	—
Scale [ref.]	pilot [18]	lab [19]	lab [20]	pilot [21]	pilot [22] (Garrett flash)
Yields (wt% of dry feed)					

(Contd..)

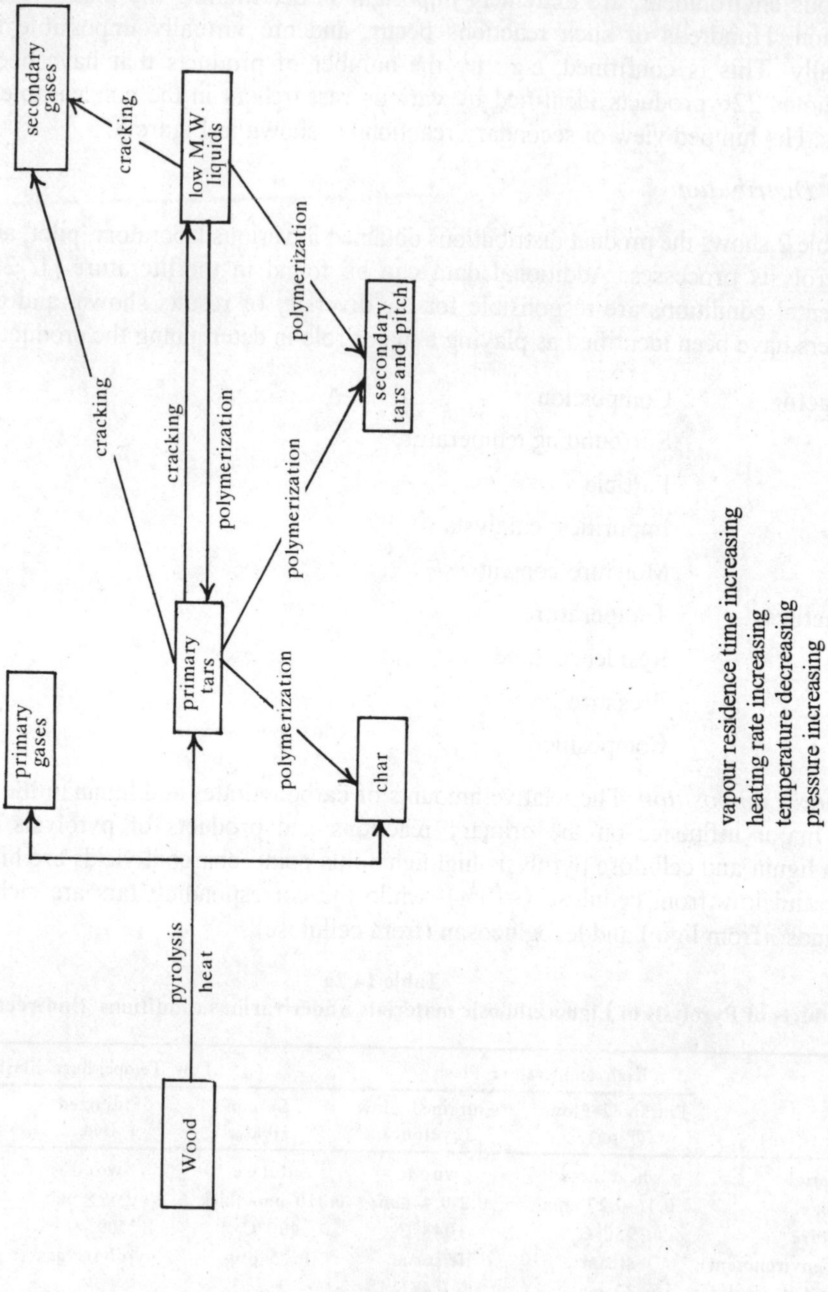

Fig. 14.8. Effect of vapor residence time, heating rate, temperture, and pressure on the formation of char, tars, and gases in the pyrolysis of wood.

Char		NA	4.0	6.2	12.2	*26.7
Condensate		NA	16.5	90.2	75.1	*67.4
	water	NA	NA	6.5	9.3	23.7
	organics	NA	NA	83.7	65.8	43.7
Gas		65.0	79.5	1.7	12.4	*6.2
	N_2	—	—	—	—	—
	H_2	8.6	2.1	—	0.02	0.04
	CO	43.4	44.7	0.25	5.3	3.3
	CO_2	10.2	17.4	1.45	6.3	2.6
	CH_4	4.7	6.4	—	0.5	0.1
	C_2H_2	—	1.1	—	—	—
	C_2H_4	4.4	6.6	—	0.2	0.05
	C_2H_6	—	0.9	—	0.04	0.2
	C_3	1.1	0.3	—	0.08	—

** Corrected to close mass balance at 100%.*

	Low Temperature slow pyrolysis +Gas-phase cracking (fixed bed)		Low Temperature Slow pyrolysis +Quick product Removal (single particle)
Raw material	cellulose	spruce lignin	wood
Particle size	~0.2 g sample	~0.2 g sample	5 mm depth
Temperature	500°C solid/ 700°C cracking	500°C solid 750°C cracking	800°C at surface
Gaseous environment	steam	steam	helium
Residence time: solid	minutes	minutes	>10 min
gas	3.8 s (cracking)	2.2 s (cracking)	quickly quenched
Scale [ref.]	lab [23]	lab [23]	lab [24]
Yields (wt% of dry feed)			
Char	9.6	*55.6	22.4
Condensate	2.4	*6.0	59.1
water	NA	NA	14.7
organics	NA	NA	44.4
Gas	86.3	*38.9	18.4
N_2	—	—	—
H_2	1.4	0.5	—
CO	55.4	18.7	8.9
CO_2	13.4	9.8	7.8
CH_4	8.5	7.5	1.1
C_2H_2	—	—	0.006
C_2H_4	6.4	1.9	0.4
C_2H_6	1.2	0.5	0.24
C_3	trace	—	—

**Corrected to close mass balance at 100%.*

Table 14.2b
Products of Pyrolysis of Lignocellulosic Materials under various conditions (direct heating)

	Sloping Hearth (air blast)	Gravitating Bed Counter current packed Bed (air blast)	Cocurrent Packed Bed (air blast)	Fluidized Bed (O_2 blast)
Raw material	pine cones (21% moisture)	wood waste	wood (3.6% moisture)	wood (10% moisture)
Particle size	coarsely milled	—	25 mm	—
Temperature	900°C (gas)	variable	variable	750 °c
Gaseous environment	pyrolysis gas + air	pyrolysis gas +air	pyrolysis gas +air	pyrolysis gas +O_2
Solid residence time	~1-2 h	long	long	—
Process [ref.]	Carbotecnica [25]	Moore-Canada [26]	Twente Univ. [27]	Mino [28]
Scale	Commercial, 1,000 kg wood/h	Commercial		pilot
Yield of char (wt% of dry feed)	~25		~0	~0
Dry gas composition (Volume %)				
N_2	45.2	45.8	54.7	0.3
H_2	6.1	18.0	15.0	32.5
CO	9.6	22.8	16.7	20.0
CO_2	28.2	9.2	12.6	30.0
CH_4	10.9	2.5	0.9	17.0
C_2^+	—	0.9	—	—

The slow pyrolysis of lignin and cellulose with subsequent gas-phase cracking of the volatiles was studied by Antal. Although yields of charcoal and gas were very different, similar gas compositions were obtained for lignin and cellulose. Thus, regardless of precursor, cracking of the tar in the gas phase tends to produce the same molecules.

Temperature surrounding the solid and particle size. These two factors together determine the reaction temperature at each point in the solid. With particle bigger than ~10 mm, there is always a considerable delay in temperature rise inside the solid, due to heat transfer limitations. As a result, the major reactions usually go to completion below 400°c (slow pyrolysis). For pyrolysis to occur at higher temperatures, the particles must be small (flash pyrolysis). The effect of temperature on the primary and secondary reaction pathways was discussed in the previous sections. Primary products tend to be mainly liquid organic compounds if the temperature is 400-600°C, and mainly permanent gases if the temperature is higher (~900°C) or if there is cracking in the gas phase. This becomes quite clear by comparing the product composition of the various experiments shown in Table 2.

Solid size also affects the yield of char. Smaller particles allow fast removal of primary products out of the solid matrix, preventing secondary reactions between the solid and the volatiles; this leads to higher yields of liquid and lower yields of gas and char. Figure 9 shows the effect of particle size on char yield from pine wood.

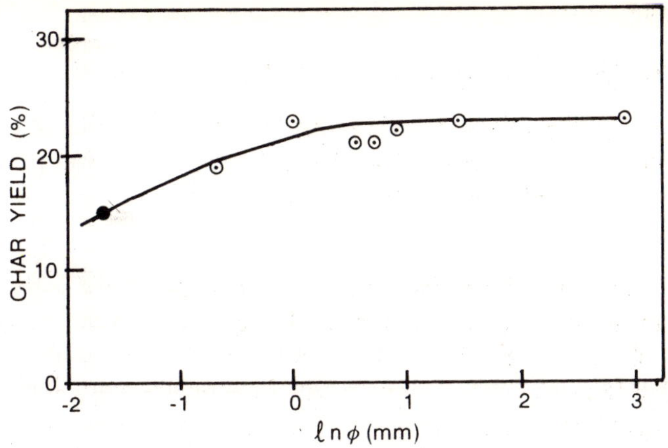

Fig. 14.9. Effect of particle size on char yield (slow pyrolysis, pine wood). Q= particle diameter, o= average of 13 points.

Impurities, catalysts, and moisture in the solid. It is well known that the ash compounds usually found in wood suppress tar formation. The same effect is obtained by addition of acidic or basic catalysts . Acidic catalysts promote dehydration and char formation, and basic catalysts increase both char and gas yields. The role of inorganic additives in the pyrolysis and combustion of cellulosic materials was investigated by Shafizadeh and coworkers. The general conclusion is that ash and other impurities present in the wood may be advantageous whenever minimum tar is desired.

Alkali compounds may also be added to wood in order to promote gasification. But perhaps more important is the use of secondary catalysts to adjust the product gas composition. This has been investigated, e.g., in the context of steam gasification of wood for methanol production.

Moisture in wood has a similar effect to ash, as it increases char and lowers far yields. This has been explained in terms of lower solid temperatures due to the heat requirements for moisture vaporization.

Gas factors. The nature of the sweeping gas affects the pyrolysis if it contains reactive components. Oxygen promotes combustion of the volatiles and gas-phase cracking reactions. Air has the same effect and additionally dilutes the resulting gas with nitrogen, thus lowering its heating value. Steam and carbon dioxide at sufficiently high temperatures gasify the char; in addition, they displace the water gas shift reaction equilibrium.

Pressure also affects the product distribution; pyrolysis at lower pressures implies faster removal of primary products from the solid matrix, origination more organic liquids, less char, and less gas (cf. Table 2).

Gas temperature and residence time are two of the main factors affecting the product distribution. Gas yields increase strongly with gas temperature and residence time. Cracking of the volatiles will be further promoted if the components are allowed to contact catalytic surfaces. It has been shown that charcoal is itself a good tar-cracking catalyst at temperatures of the order of 950°C.

Flash vs. Slow Pyrolysis, and a Recipe for Product Distribution

Flash and sow pyrolysis were defined in the introduction to this chapter in terms of high or low heating rates. The experimental evidence referred to in the previous section (and Table 2) indicates that whether pyrolysis is flash or slow is not very important to product distribution.

Shafizadeh's model for the pyrolysis of cellulose, presented earlier, was proposed on the basis of slow pyrolysis data. Yet, it appears to agree with flash pyrolysis data. Heating rate has also been shown to have little effect on the mechanism and product distribution of coal pyrolysis. In addition, Table 2 shows that high-temperature flash pyrolysis and low temperature slow pyrolysis with gas phase cracking yield similar product distributions (high-gas, low-tar yields). On the other hand, low temperature flash pyrolysis and low-temperature slow pyrolysis with quick product removal both give high tar yields and low yields of gas with similar composition. The main difference is that slower pyrolysis somewhat increases char yields.

Thus, the most important factor in product distribution is the volatile residence time. The recipe for high oil yield is therefore to remove and quench the volatiles as quickly as possible. The recipe for high gas yields is to allow the volatiles time to crack , preferably at higher temperatures. Lower char yields are obtained with smaller particles, higher temperatures, and the presence of steam.

Although possible in the laboratory, the manipulation of vapor residence time may find more costraints at an industrial scale. Thus, in a big reactor, very quick removal and quenching of volatiles may not be practical with big particles undergoing slow pyrolysis. As will be discussed later, flash pyrolysis in an entrained bed reactor is most flexible in this respect.

Kinetics of Pyrolysis

Knowledge of kinetics is important for the design of pyrolysis equipment. Two cases should be distinguished: (a) pyrolysis of small particles where kinetics are controlled by reaction rate; and (b) pyrolysis of big particles, where kinetics are affected and sometimes determined by heat transfer.

Kinetic of pyrolysis of Lignocellulosic materials are usually studied by thermogravimetry. Typical thermograms for the pyrolysis of wood and its major components are shown in Figure 14.7. In interpreting thermogravimetric data, the simplest possible reaction scheme is usually assumed. Most authors tend to lump pyrolysis as a single first-order reaction, whether dealing with the pyrolysis of a single component, like cellulose, or with that of a more complex material, like wood. Sometimes a set of competing reactions is considered, occasionally a more complex scheme. Kinetic parameters obtained by various authors are shown in Table 3.

Table 14.3
Kinetic Parameters of Pyrolysis of Lignocellulosic Materials

Material	Mechanism	Reaction no.	n	E (kj/mole)	K_0 (s^{-1})
Cellulose	Ce → C + G		1	224	3.8×10^{20}
Cellulose	Ce → C + G		0.46	153	2.2×10^{13}
Cellulose	Ce → C + G		1	133	7.2×10^{12}
	(2) V	1	1	243	1.0×10^{23}
Cellulose	Ce $\xrightarrow{1}$ Ca	2	1	198	1.1×10^{18}
	(3) C+G	3	1	151	4.7×10^{13}
Lignin	L → C+G		1	82	3.2×10^{5}
	1 G	1	1	89	5.2×10^{7}
Wood	W 2 → T	2	1	113	1.5×10^{10}
	3 C	3	1	107	2.6×10^{9}
	W → C + G	>330°C	1	18	1.9×10^{2}
Wood	W → C + G	<330°C	1	84	8.4×10^{7}
	W1 $\xrightarrow{1}$ G1	1	1	83	4.2×10^{6}
	W2 $\xrightarrow{2}$ G2	2	1	142	1.2×10^{12}
Pine wood	W3 $\xrightarrow{3}$ G3	3	1	77	2.6×10^{5}
	Wn $\xrightarrow{n}$ Gn	n	1		
	Wc → C				

n = reaction order
E = activation energy
K_0 = preexponential factor

C = char
Ca = active cellulose
Ce = cellulose
G = gas
L = Lignin
T = tar
V = volatiles
W = wood

W1 = TG-component 1 (W1/W=0.20)
W2 = TG-component 2 (W2/W=0.53)
W3—Wn = minor TG-comps. (W3 + ... + Wn)/W = 0.11)
Wc = fixed carbon (Wc/W = 0.18)

Too simple a reaction scheme prevents the use of the kinetic parameters over a wide range of temperatures. To try and account for the complexity of the phenomenon and extend the range of kinetic data, variable reaction order and variable activation energy have been proposed. More recently, results obtained by multistage isothermal thermogravimetry, show that pyrolysis kinetics may be described over a wide range of temperatures in a less arbitrary way by dividing wood into components which react in parallel. The major component is shown to be cellulose.

Particles with sizes greater than 1 mm pyrolyze with kinetics which may differ considerably from the intrinsic reaction kinetics. This is due to a significant heat transfer resistance, leading to temperature gradients inside the particle. Models proposed to describe the phenomenon are extensions and variations of the Bamford model. Equations have to be included to describe heat transfer to and through the particle, as well as the pyrolyis reaction ratc. These models require a fairly complicated numerical solution and knowledge of properties and parameters which are difficult to obtain. Pyrolysis times as a function of time and temperature are given in .

Enthalpy of pyrolysis

The heat of pyrolysis depends on the material and on the pyrolysis conditions, particularly on the heating rate, the extent of secondary reactions, and whether the gaseous atmosphere is inert or oxidizing. Values quoted for the enthalpy of pyrolysis as measured by various authors range from an endothermic value of +370 kj/kg (for pure cellulose) to an exothermic value of -1700 kj/ kg (for wood). Qualitatively, there is agreement between researchers on the following:

1. The pyrolysis of cellulose in an inert atmosphere is endothermic [30,31,54]. Beall [30] measured heats of pyrolysis under nitrogen of between + 61 and +230 kj/kg. The pyrolysis of cellulose in air, on the other hand, is exothermic [31].

2. The pyrolysis of lignin and of most hemicelluloses is usually exothermic [30,31]. Values measured by Beall [30] under nitrogen range from +21 to -510 kj/kg for lignin, and from +38 kj/kg to -1105 kj/kg for hemicelluloses.

3. The thermal degradation of wood is, overall, slightly exothermic. A plot of the enthalpy of pyrolysis measured by various authours versus temperature is given by Roberts [54]. It varies from -340 kj/kg at 180°c to -120 kj/kg at 450°c.

Gasification and Gas Phase Reactions

After pyrolysis of a lignocellulosic material, the remaining char is often gasified in the same reactor, Gasisfication requires higher temperatures (>600°C) and the presence of reactive (H_2O, CO_2, H_2). The main gas-solid reactions are:

$C+H_2O \rightarrow CO+H_2$ ΔH=131.4KJ/mole

$C+CO_2 \rightarrow 2CO$ ΔH=172.6KJ/mole

$C+2H_2 \rightarrow CH_4$ ΔH=-74.9 KJ/mole

If oxygen is present, combustion may occur:

$C+\frac{1}{2} O_2 \rightarrow CO$ ΔH=-110.5KJ/mole

$C+O_2 \rightarrow CO_2$ ΔH=-393.5 KJ/mole

The gas composition is affected by the following gas phase reactions:

$CO+H_2O \rightarrow CO_2+H_2$ ΔH=-41.2KJ/mole

$CH_4+H_2O \rightarrow CO+3H_2$ ΔH=206.1 KJ/mole

Enthalpies of reaction, ΔH, are at 25°. (Equilibrium constants and other data for these reactions can be found in various references. Kinetic data for charcoal gasification reactions by water vapor and CO_2 may be found in. Apparent gasification kinetics differ from intrinsic reaction kinetics due to diffusion limitations. A model for the gasification of big particles of charcoal is presented by Groeneveld.

Pyrolysis Technology

Table 4 classifies pyrolysis/gasification equipment and shows the typical products obtained with each type. Classification is primarily according to the way solids move inside the reactor. They may be moved batchwise or continuously, and in continuous reactors, they can move as gravitating ("fixed") beds, mechanically moved beds, fluidized beds, and entrained Gravitating and mechanically moved beds required big particle sizes, leading to slow pyrolysis. Flash pyrolysis may be achieved with fluidized and entrained beds, which operate with smaller particles.

Another important aspect in reactor design is the way energy is supplied to it. Pyrolysis at high temperatures and pyrolysis of wet materials may be considerably endothermic. Gasification reactions (if they occur) are very endothermic. Overall, energy is usually required, which may be supplied in two basic ways: directly or indirectly.

In directly heated reactors, part of the reacting material is subjected to an oxidation (usually a partial combustion) which gives energy to the rest of the process. Air or pure oxygen may be admitted somewhere in the reactor for this purpose. An alternative is the introduction of an oxygen-rich material, which is reduced inside the reactor, supplying the necessary oxygen. This is the case in the oxygen Donor process.

In *indirectly heated reactors*, heat is conveyed into the reactor through a wall or by solid particles which are heated elsewhere, as in a double fluidized bed.

Product gas quality from pyrolysis/gasification processes depends dramatically on the way the reactor is heated. Gas from directly heated reactors with air blast have a low heating value, due to dilution with nitrogen. Directly heated reactors with oxygen blast as well as indirectly heated reactors produce a gas with medium heating value.

Char and tar yields depend not only on pyrolysis conditions, as discussed earlier, but also on the extent of gasification, cracking, and reforming allowed in the reactor. In the next sections, each type of equipment is described in more detail.

Batch Carbonization

Since the bronze age (and still today, in many parts of the world) charcoal has been produced in pit kilns and earth kilns. These consist simply of blocks of wood stacked around a central air channel and then covered with earth or sod. Some air is allowed through the bottom of the pile, the pile is lit in the center, and combustion of small fraction of the wood provides heat for the carbonization, which lasts for days. Afterwards, days or even weeks are necessary for the cooling of the pile, to prevent ignition of the hot charcoal in presence of air. Experience is required even to achieve yields as low as 20% of charcoal. All gas and volatiles produced are wasted to the atmosphere.

An improvement on the earth kiln is the use of a permanent kiln to cover the wood pile. This may be made of metal, masonry, or cement blocks. capacity varies from 5 to 100 cords.

Table 14.4
Classification of Pyrolysis/Gasification Equipment

Reactor Type		Typical Particle size	Energy Supply		Products*
Batch	kilns	~300 mm	direct		charcoal
	retorts		direct or indirect		charcoal+[tar]
Continuous	gravitating beds (countercurrent or cocurrent)	10-200 mm	direct	air blast	charcoal +low HVgas+(tar) or low HV+(char+tar)
				O_2 blast	medium HV or synthesis gas + (tar (if countercurrent))
	mechanically moved beds	5-200 mm	direct (air blast)		low HVgas + (char +tar)
			indirect		medium HV gas +(char+tar)
	fluidized beds	0.5-10 mm	direct	air blast	low HV gas+(char +tar)
				O_2 blast	medium HV gas or synthesis gas (+tar +char)
			indirect		
	entrained beds	0.3-3 mm (pipe)	direct (air blast)		low HV gas+(char)
		1-10 mm (cyclone)	indirect		oil+[gas]+(char)-low temp. or medium HV gas+(tar+char)-high temp.

*[ ] = byproducts; ( )=usually undesirable byproducts.

doors are usually provided for the charging of the wood and removal of the charcoal. Holes and chimneys may be optimally located to regulate the air flow. Higher yields of charcoal(~30%[58]) can be obtained in these kilns than in earth kilns, with less dependence on operator skills. Metal kilns have the advantage of being transportable along the forest to the sites where the wood is available. Typically [8], for a metal kiln, loading takes 3 hours, carbonization takes 16-24 hours, and unloading 3 hours. With two experienced operators, two such kilns can produce 3 tons of charcoal per week. More details can be found in.

Disadvantages of the batch kilns to make charcoal are considerable, namely: low charcoal yield (maximum 30% with a modern design); low -quality charcoal; very low overall energy efficiency since gases and volatiles are lost; low capacity; and high labor and experienced labor requirements. Yet, this is still the most widely used technique for producing charcoal, and not only in developing countries. The main reasons are its simplicity, negligible capital costs, and sometimes ignorance of other options.

The efficiency of batch carbonization can be improved through the use of retort furnaces. A condensate may be obtained from the volatiles evolved during the pyrolysis. The rest of the gas may be burned to provide indirect heat to the process, making it unnecessary to let any air into the retort. This increases the charcoal yield, which can be as high as 30-40%. An example of a batch retortx is the constantine portable rector, developed in Australia to be used in the developing countries. It is a jacketed vessel located above a fire box, which uses wood wastes as well as pyrolysis process . Air-dried blocks 300 mm long are used. A complete cycle between loading and unloading takes about two days and produces 1-2 tons of charcoal. The system is simple and safe to operate, but still rather inefficient and laborintensive.

To make pyrolysis more efficient, continuous reactors are necessary.

Gravitating Beds

In gravitating bed reactors, the solids are fed at the top, move down the reactor solely by gravity, and are removed at the bottom either in the form of ash or charcoal, depending on the extent of gasification and combustion allowed. Gravitating beds are usually packed beds-simple, but with inherent problems such as bridging of solids and channeling of gases. Acceptable sizes range from 10 to 200 mm. Smaller sizes tend to fluidize at reasonable gas velocities, being better suited to processing in fluidzed beds or in entrained beds.

As heat transfer is slow in a packed bed, these reactors are usually directly heated, by allowing in either some air or pure oxygen. Gases inside the reactor the reactor may be led upwards or downwards, respectively counter-current and co-current to the sokids flow. This defines the two major types of gravitating bed pyrolyzers/ gasifiers, which are schematically shwon in Figure 10.

Counter-current reactors are simple and have low exit-gas temperature. There is little cracking of pyrolysis gases, which leads to a tar-loaded outlet gas. This is a disadvantage if the gas is to be further processed or transported cold, as there must be a tar removal step. A tar disposal problem is also created. This type of reactor is suitable for close-coupled gasification/combustion systems and retrofitting to existing oil-and gas-fired boilers. It has reached a high degree of refinement in the coal industry and can be scaled up to quite high capacities. The gas quality may be improved if a pure oxygen blast is used. Examples include the Union Carbide Purox process (up to 8 ton/h capacity, using O_2/H_2O blast) and the Georgia Institute of Technology/Tech Air process, which uses an air blast. Many other commercially available processes of this type are quoted by shand.

In the co-current reactor (see Figure 10), the pyrolysis gas is led through high temperature reduction and combustion zones, where tars are cracked. A hot clean gas is therefore produced. For decades this type of reactor was widely used close coupled with internal combustion engines for power or electricity generation. Charcoal was the most common raw material. Recent developments with wood wastes are reported by. Scale-up is more difficult than with counter-current reactors, limiting the acceptable size of gasifier. Oxygen blast and high pressure are therefore not attractive. Examples of available processes are quoted in.

As well as gas producers, gravitating beds have a long successful history in the continuous production of charcoal. Two example are stafford Badger and the Lambiotte retorts. The latter is a large-scale is (~20 ton charcoal/day) gravitating bed charcoal producer developed in Francc earlier in the century, but still very much in use. Pyrolysis gases flow upwards

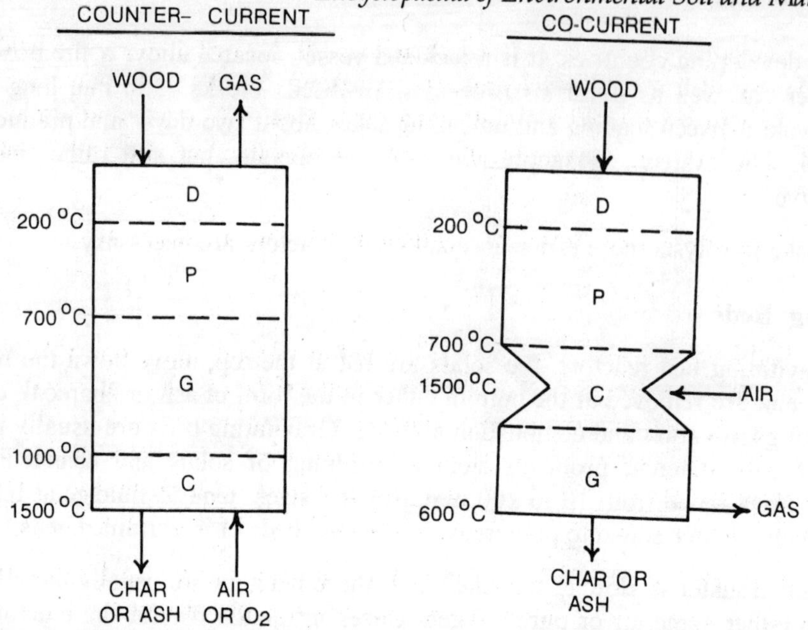

Fig. 14.10. Countercurrent and cocurrent gravitating bed gasifiers. D= heating and drying zone, P= pyrolysis zone, G= gasification zone, C=combustion zone.

counter-current to the solids, are withdrawn at the top at about 350ºC, conveyed to an external combustion chamber where they are partly oxidized, and then recycled to the lower part of the retort at ~700ºC, giving the necessary energy to the carbonization process. The gas purge is usable in a close-coupled burner.

A different design of a gravitating bed, which is not a packed bed, is the sloping hearth reactor of carbontecnica, Portugal, developed for the continuous production of charcoal and low heating value gas. As shown in Figure 11, wood waste is fed by variable-speed internal moving parts; instead, it is positioned at such an angle that the solids slide down while being exposed to high temperatures (~850ºc). Since the angle is greater than the angle of repose of the solids, these accumulate at the bottom for a residence time which varies with solids size and moisture content, and is typically 1-2 hours. The resulting charcoal is cooled by spraying with water on discharge. Since the gas does not flow through the solids and has little velocity above them, there is negligible carryover of fines, and raw-materials such as sawdust are acceptable. To ensure the high chamber temperature, some air is admitted through a register located at the bottom of the reactor, just above the solids level. Exothermic oxidation reactions between the air and the pyrolysis gas supply the necessary energy.

Table 14.5
Textural properties of charcoal from Pine cone obtained in the Carbotecnica Reactor

Real density	(g/cm^3)	1.86
Apparent density	(g/cm^3)	0.46
Surface area	(m^2/g)	202

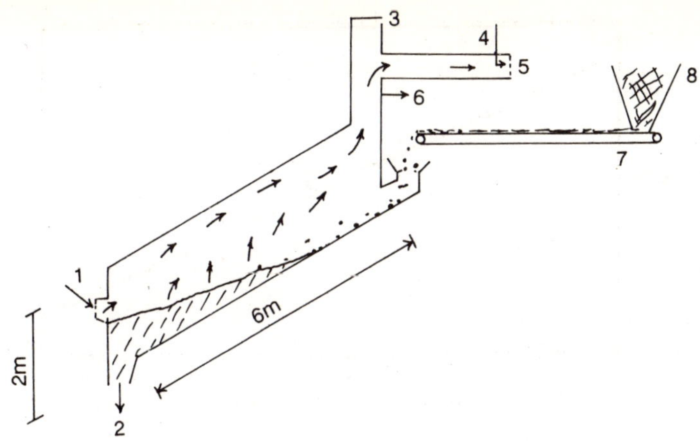

Fig. 14.11. Schematic cross section of the Carbotecnica sloping hearth carbonization reactor; (1) primary air inlet, (2) charcoal outlet, (3) stack, (4) secondary air inlet, (5) outlet to burner, (6) gas sampling point, (7) conveyor belt, (8) hopper. Reprinted with permission from.

The capacity is typically 1,000 kg/h of wood waste (20% moisture), but this depends on the type and size of the raw-material. Solid feed can vary widely in size, size distribution,shape, and ash content. This reactor versatility toward raw material is in fact one of its main advantages. The yield of charcoal is on average 20-25%, with typical properties shown in Table 5. The gas composition is given in Table 2, and its heating value is shown in Figure 12 as a function of feed moisture content (coarsely milled pine cones). The gas may be used in close-coupled boilers, kilns, or driers.

Mechanically Moved Beds

Some pyrolysis reactors include mechanical means to move the solids. They are often adapted from off-the-shelf equipment available for other gas/solid operations.

Based on a design intended for roasting operations in the minerals industry, the Nichols-Herreshof furnace is also used for charcoal production in the U.S. It is a vertical cylinder with alternating disc-and doughnut shaped trays. Fed at the top, solids fall from tray to tray moving radially in each tray, inwards through the doughnut-shaped trays and outwards through the disc-shaped trays. Solids movement through the trays is promoted by rotating arms attached to a central shaft. The shaft is hollow and cooled by air. The furnace is heated by controlled admission of air at each tray. A gas of low heating value is therefore produced. The charcoal yield is about 25% on a dry-wood basis.

Rotary kilns are also used as pyrolysis reactors. They can be directly or indirectly heated. A directly heated rotary kiln is used in Monsanto's Langard system for municipal solid waste disposal. It has a 35 ton/day capacity, is fired with fuel oil and air, and produces low heating value gas and char. An indirectly heated rotary kiln intended for medium heating value gas production by the Wright-Malta Co. was reported to operate at high pressure (300 psi) and at 600°C, with straw as the raw material. A catalyst (Na_2CO_3) and water vapor were used to reduce the amount of char.

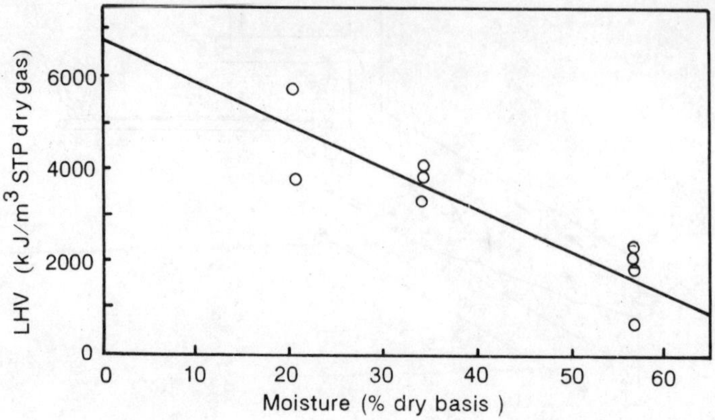

Fig. 14.12. Gas lower heating value (LHV) as a function of solid feed moisture (wt% d.b.).

Reactors with mechanical means to move the solids are costly, and maintenance problems may arise if there are moving parts subjected to high temperatures. They do, however, avoid the most frequent problem found in gravitating bed reactors: that of solids bridging.

Fluidized Beds

In Fluidized beds, the reacting solids are suspended and agitated in an upflowing gas stream, often together with a matrix of inert solids (e.g., silica or sand). To ensure fluidization, sufficient gas must be introduced. This may be recycled pyrolysis gas, air, oxygen, steam, or a mixture thereof. Several properties make fluidized beds advantageous for pyrolysis/gasification, particularly of forest and agricultural residues. Problems associated with gravitating beds such as bridging of solids and gas channeling do not occur. There is greater versatility toward the acceptable solids shape and size distribution. Typical particle sizes range from 0.5 to 10 mm. Fines in the feed are usually undesirable, as they get entrained in the gas stream. Fluidized beds may be intended as low HV gas, medium HV gas, or synthesis gas producers, always with medium to large throughputs. Complexity in design, control, and operation makes them expensive in a small scale. The small particle sizes required and consequent high heating rates make them unsuitable for charcoal production. Some char and tar are, however, inevitable in the gas stream, which complicates downstream processing.

Directly Heated Fluidized Beds

In directly heated fluidized-bed pyrolyzers /gasifiers, some air or pure oxygen is introduced with the fluidizing gases, most often together with water vapor. The vary good solids mixture achieved in a fluidized bed means that the reactor is not divided into drying, pyrolysis, gasification, and combustion zones, but particles undergoing each of these processes all coexist in close contact, thus lowering the maximum reactor temperature. Materials of construction are thus subjected to less severe conditions than in gravitating beds and ash sintering is unlikely. The extent of gasification achieved depends on the blast composition, the temperature, and the solids residence time.

Significant amounts of char are reported in the Thermax system (Alberta Industrial Development, Canada) which produces low HV gas using air and recycle gas as fluidized medium. The reported capacity is 75 ton/day. Pure oxygen as the oxidizing substance leads to medium HV gas or to synthesis gas production. The price of oxygen or of an oxygen plant makes this option interesting only for high gasifier capacities. Several systems are described in, mostly operating at high reassures (~10 bar) and bed temperatures of up to 1000°C. At pilot scale, the Creusot-Loire (2,500 kg/h), France, the Mino (1,000 kg/h), Sweden, and the Lurgi (500 kg/h), Germany, are processes intended for synthesis gas production, with methanol in view.

Indirectly Heated Fluidized Beds

If no air /O_2 is introduced in the reactor, heat must be supplied from elsewhere. This can be achieved in two basic ways, through a metallic wall or through the introduction of a hot substance, as in a double fluidized bed.

The concept of a double fluidized bed is illustrated in Figure 14.13. In one fluidized bed, the material is pyrolyzed at temperatures ~650-850°c in a fluidized matrix of inert solids, using water vapor and /or recycled pyrolysis gas as the fluidizing agent. The inert solids together with char particles are continuously transferred to a second fluidized bed, where the char is burned at temperature above 1,000 °c. The inert solids heated to this temperature are then returned to the pyrolysis bed, supplying their excess sensible heat. This system has been proven at commercial level with municipal solid waste. A 100 kg/h double fluidized bed gasifier has been run on wood at Batelle Columbus, U.S. Problems with double fluidized beds include the complexity of process control, attrition and erosion in the transfer lines and the choice of materials for the hot interface between the reducing atmosphere of the pyrolysis bed and the oxidizing atmosphere of the combustion bed.

Some of these problems are avoided in the AGIP Nuclear/Italenergie 800 kg/h fluidized bed. The fluidizing agent is recycled pyrolysis gas. The necessary energy is supplied by the combustion of some of this gas in a jacket around the reactor. This is possible because heat transfer is good in a fluidized bed. The pyrolysis bed produces medium HV gas at 730°c. The gas is reformed in a second fluidized bed at 1100°c by introduction of a little O_2, to crack tars and consume any entrained char.

Entrained Beds

In entrained flow reactors, the solids are carried by the gas either through a hot pipe or through a hot cyclone. Particle size must be very small, typically 0.3-3 mm. Solid and gas residence times are very small (~1 s), and very high heating rates are achieved. It is the realm of flash pyrolysis. The pyrolysis products depend mainly on the temperature at which the reactor is operated: the main product will be tar if the temperature is low (~500 °c), and it will be a gas if the temperature is high (~900°C).

Entrained flow pyrolysis /gasification has reached commercial scale on coal and peat (the Koppers-Totzek process). The Garrett Flash/Occidental Research process, intended mainly for the pyrolysis of municipal solid wastes, has also been run with lignocellulosic materials. At ~500°c, the yields were ~50% of oil, ~20% of char, and ~30 % of gas.

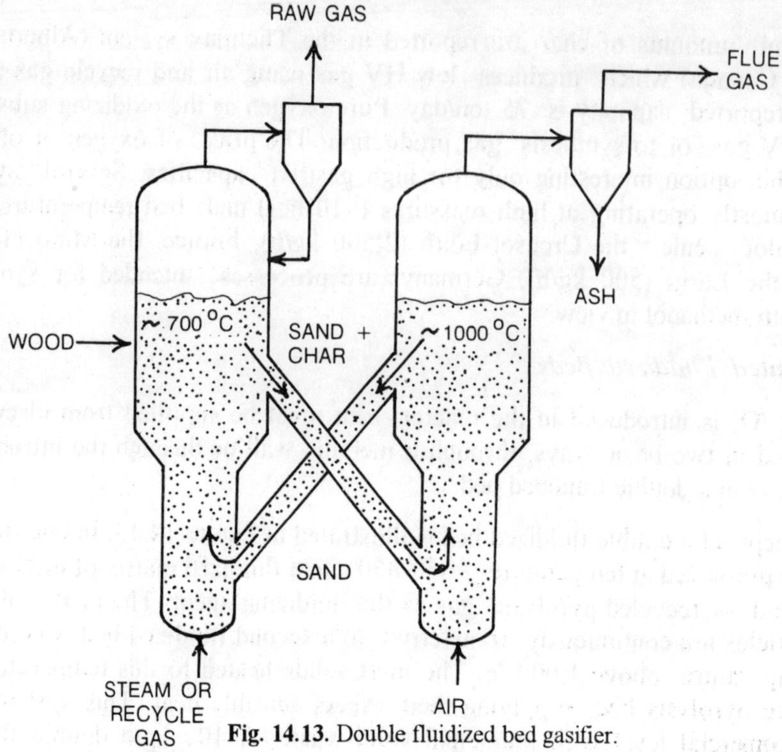

Fig. 14.13. Double fluidized bed gasifier.

Higher temperature entrained flow pyrolysis of lignocellulosic materials has been mostly studied at bench or pilot scale. Bohn and Benham at Seri, Colorado, used steam to convey wheat straw particles (~0.2 mm) through a pipe externally heated to ~950°c. With residence times of ~0.1 s, a 65% yield of a medium HV gas was obtained (gas compostion in Table 2).

Cyclonic reactors combine gas/solid, gas /wall, and solid /wall heat transfer. Pyrolysis of wood sawdust (~300 mm) in cyclonic reactors has been studied at Nancy University, France in a lab scale cyclone heated by solar radiation. Solid residence times of 1.4s were measured for a gas (helium+steam) residence time of 0.045 s. Char yields were always low (~4%). As the wall temperature was increased from 620°c to 1054°c, the gas yield increased from 16% to 93% (gas compostion in Table 2). The reactor may therefore be suitable for oil or for medium HV gas production, depending only on wall temperature. With a 32-mm diameter cyclone, a throughput of up to 35kg/h of solids was achieved! A bigger cyclone (400 mm in diameter) with a capacity of 200 kg/h is reported by for sawdust gasification with air to yield a low HV gas. The cyclone gasifier is simple to operate, reliable, and responds rapidly to changes in demand for gas.

As a conclusion, entrained flow reactors have large throughputs per unit volume, no great operational problems, and may yield gas or tar as their main product, depending only on the operating temperature. They are, however, expensive to operate on a small scale [28], while requiring very small feed particle size. Since size reduction of wood to sub-millimeter size is too costly, entrained flow beds are attractive for the pyrolysis of wood wastes only if large supplies of sawdust are available.

Use of Pyrolysis Products

Table 6 summarizes the products that can be obtained form the pyrolysis of wood or wood wastes in processes proven or under development. These can be divided into energy (direct), fuels (stored energy), and chemicals.

Energy from wood wastes may sometimes be more conveniently obtained through a intermediate pyrolysis/gasification step, rather than by direct burning. The production of electricity or power by coupling a gasifier to an internal combustion engine is viable in the small scale in forest/agricultural environments. Co-current gravitating bed gasifiers are favored for this application. Large-scale systems for electricity generation with gas turbines still await development.

Fuels (stored energy) from wood wastes are the next stage in flexibility. The ancient art of charcoal production has met modern challenges and requirements (i.e., low volatile content, Economics of charcoal production have been improved by burning the gas in close-coupled equipment.

Tar has been tried as a fuel oil with limited success due to low storage stability, corrosiveness, and high viscosity/Further processing (e.g., hydrotreating), has been proposed to improve its properties as a fuel.

Table 14.6
Energy, Fuels, and Chemicals from Wood Pyrolysis

Objective	Primary pyrolysis products	Conversion process	Final Product
		combustion	heat
Direct	low HV gas		
Energy	or	combustion in boilet	steam
	medium HV gas	internal comb. engine	power or electric.
		gas turbine	electricity
	charcoal		charcoal
Fuels		briquetting	briquets
(stored	tar		liquid fuel
energy)		methane synthesis	SNG
		Fisher-Tropsch synthesis	methanol
		ethylene synthesis	ethylene
			others
	charcoal		active carbon
Chemicals		separation	phenolics
	tar	separation +reaction	adhesives
			others
		separation	acetylene
	gas	separation	ethylene
			others

Other fuels, like *gasoline, substitute natural gas* (SNG), and mehtanol, can be obtained from wood wastes through synthesis gas conversions. The downstream technologies involved are well proven (methanol synthesis, Fischer-Tropsch, Mobil methanol to gasoline synthesis), but viable only on a large scale. The large scale production of synthesis gas from wood wastes is the subject of much current research in the United states and in the European Economic Community.

Although the production of fuels from wood wastes has received most of the attention in recent years, the production of chemicals may be more interesting, as they usually command higher prices. In this respect, the versatility of synthesis gas as an intermediate is again considerable, as ammonia, alcohols , and hydrocarbons can be obtained from it. The production of acetate chemicals, such as cellulose acetate (from synthesis gas and cellulose pulp) and vinyl acetate (from synthesis gas and ethanol), as well as a process to produce benzene and phenol from pyrolytic tar oil, have been evaluated by Klausmeier. Phenolics from tar have been reacted with formaldehyde to produce a phenol-formaldehyde adhesive suitable to be used in plywood laminates. The separation of higher value hydrocarbons from pyrolysis gas has also been proposed. High yields of acetylene, ethylene (e.g.,), and other light hydrocarbons have been reported in the pyrolysis gas at lab scale.

Char can also be considered as a chemical. Its properties are dependent on the ultimate carbonization temperature as shown in Table 7. the wide range of properties of charcoal results in a variety of metallurgical and chemical application.

Charcoal is one of the best raw materials for activated carbon production, since it retains the microscopic structure of the parent wood. Thus, the material presents some porosity, which is needed to provide access to the activating agents. Activation consists in the partial gasification of the carbon, in order to develop the porous texture of the material. Gasification is carried out by reaction with steam, air, or carbon dioxide, usually in the temperature range of 800°c-900°c. In this way, activated carbons are produced with very high specific surface areas (500-1,500 m²/g) and large adsorption capacities, which can be used as adsorbents in the purification of liquid and gaseous stream.

Table 14.7
Yields and properties of charcoal after pyrolysis (120 min) of Holm-oak wood

Temperature (°c)	Yield (%)	Proximate Analysis			Elementary analysis			Surface area* (m²/g)
		Volatile Matter %	Fixed carbon %	Ash %	C%	H%	N%	
100	89.5	80.00	17.81	2.19	53.26	5.50	0.21	—
200	89.3	79.33	18.48	2.22	50.75	5.42	0.19	—
300	53.7	50.87	43.79	5.35	62.63	4.29	0.33	—
400	31.7	27.11	67.01	5.88	74.97	3.82	0.34	11
500	29.2	16.91	76.34	6.74	83.42	2.33	0.37	438
600	25.5	10.45	84.06	5.49	86.92	1.67	0.33	468

(Contd..)

Temperature (°c)	Yield (%)	Proximate Analysis			Elementary analysis			Surface area* (m^2/g)
		Volatile Matter %	Fixed carbon %	Ash %	C%	H%	N%	
700	24.7	6.60	88.67	4.73	89.36	0.84	0.25	480
800	22.4	5.68	89.11	5.21	89.71	0.43	0.26	576
900	22.0	5.46	88.89	5.63	90.00	0.29	0.30	438
1000	22.0	5.28	88.68	5.04	89.92	0.29	0.43	—

Adapted with permission from.

$^*N_2$ adsorption data.

Table 14.8
Textural properties of Activated carbons from Almond shells (N_2 adsorption data)

Burn-off (%)	Micropore Volume (cm^3/g)	Surface Area (m^2/g)
8	0.26	687
19	0.30	795
34	0.36	936
52	0.41	1,068
70	0.48	1,245
80	0.51	1,327

Table 14.9
Activated Carbon from Various Lignocellulosic Material

Waste	Surface Area (m^2/g)
Almond shells	872
Olive stones	868
Sawdust (Pinewood)	1,084

The porous structure of activated carbons consists of a wide range of pore sizes, which can be classified into three groups according to their width : macropores (>50nm), mesopores (2-50 nm), and macropores (< 2nm). For the adsorption of gases and vapors, essentially microporous carbons are adequate, while for liquid-phase applications a well developed meso and macroporosity is additionally required.

The development of the porosity during activation is a function of the degree of burnoff (% of carbon gasified), and can be appreciated in Table 8, where the micropore volumes and apparent surface areas of activated carbons prepared form almond shells are presented. Table 9 presents the surface areas of activated carbons obtained from a variety of lignocellulosic wastes by carbonization at 850°c, followed by CO_2 activation at 825°c to about 20% burnoff.

[illegible]	[illegible]	Volatile Matter %	Fixed carbon %	[illegible]	Density [illegible]	[illegible]	Surface area (m^2/g)
[illegible]	[illegible]	[illegible]	[illegible]	[illegible]	[illegible]	0.25	480
[illegible]	[illegible]	[illegible]	[illegible]	[illegible]	0.43	0.25	[illegible]
[illegible]	[illegible]	[illegible]	[illegible]	[illegible]	0.38	[illegible]	[illegible]
[illegible]	[illegible]	[illegible]	[illegible]	[illegible]	[illegible]	[illegible]	[illegible]

[illegible]

Table 8

[illegible] properties of activated carbons from Almond shells [illegible]

[illegible]	Micropore volume [illegible]	[illegible]
[illegible]	[illegible]	[illegible]

Table 9

[illegible]

[illegible]	[illegible]
[illegible]	[illegible]

The pore structure of activated carbons consists of a wide range of pore sizes, which can be classified into three groups according to their width: micropores (< 2 nm), mesopores (2–50 nm) and macropores (> 50 nm). For the adsorption of gases and vapours [illegible] micropores [illegible] and macropores are [illegible] formed.

The development of the porosity during activation as a function of the degree of [illegible] (% of carbon [illegible]) and can be [illegible] in Table 8, while the micropore volumes and apparent surface areas of activated carbons [illegible] are [illegible]. Table 9 presents the surface areas of activated carbons obtained from a variety of [illegible] by CO_2 activation at 825°C [illegible].

15

Production of Biogas and Biomanure from Willlow-Dust, a Textile Mill Processing Residue

Introduction

An energy crisis is increasingly felt in all industrial activities, and biogas could be an alternative source of energy for some of these industrial operations. In India, considerable work has been done and technology developed for the production of biogas (Gobar Gas), which is primarily dependent on cattle waste. The solid content of cattle waste is comparatively low (15-16%) and it is dispersed in a slurry form. However, technology inivolving fermentation of other waste materials having higher solid content has not been developed to the same extent.

Textile mills in India consume about 7.85 million bales of cotton annually and generate an average 70,650 MT of cellulosic waste consisting of droppings, gutter fly, sweepings, etc., from blow rooms and other places. Many mills recover useful short fibres from blow room waste by passing it through willow machines. The residue is called willow-dust. It is estimated that the total amount of willow-dust generated in textile mills in India is 30,000-33,000 tonnes per year.

The possible ways of utilizing willow-dust are composting, direct burning, and anaerobic digestion. Composting is a biological process where in a considerable portion of organic mater is lost during aerobic fermentation. The work on composting of willow-dust indicates that about 35% of the material is lost during the process. Therefore, a better way would be to subject it to anaerobic digestion to get biogas as well as biomanure.

Work Done in India

In a pioneering attempt. Fowler and Joshi at the Indian Institute of Science, Bangalore, studied the anaerobic fermentation of materials, such as newspapers, filter papers, banana skins, etc. They found that materials rich in hemicelluloses produced a gas rich in methane. In 1939, the work was again started at the Indian Agricultural Research Institute (IARI), New Delhi. The effect of environmental factors such as pH and temperature on gas yield was studied. Desai and Biswas contributed significantly in the implementation of a small biogas plant which was improved later by Patel. The Khadi and Village Industries Commission (KVIC) adopted Patel's design and advocated its implementation.

The KVIC began to implement the Gobar gas scheme in a planned way in 1962. At first, the design was standardized for plant capacities varying from 2 m^3 to 140 m^3 of gas per day. In the meantime, it became obvious to the commission that a single type (Vertical gobar gas plant) might not be viable all over the country, and therefore work was begun to develop another type (horizontal) suitable for hilly areas and places where the water table is high.

Until the present, the effort in India has been directed at processing cattle waste in plug flow or semi-continuous digesters with 8-10% total solids. The Planning Research and Action Division (PRAD), Lucknow, U.P. established a Research Station at Ajitmal in 1960, and developed a fixed dome biogas plant known as the Janata Biogas Plant. The first community biogas plant in the country was established by the Division of Fateh Singh ka parwa in Etawah, U.P., through funds advanced by UNICEF.

The observed gas production per kg of fresh cattle waste from both KVIC and Janata Plants is 0.036 m^3 when the retention period is kept between 30-50 days, which total solids of 7-8%, and constant gas pressure of 10 cm water column.

A pilot plant for production of biogas and biomanure from agricultural wastes, utilizing the Hungarian semi-wet process, has existed since 1966 at the National Sugar Institute, Kanpur.

The pilot plant has a capacity of around 650 m^3/day of biogas production. The plant employs a battery of 12 digesters arranged in two rows. Each digesters has a capacity of 50 m^3. All the digesters are connected to a common gas holder of 300 m^3 capacity. The semi-wet process is a batch process and utilizes less water; the total solids content of the initial charge being 20-25% as compared to 8-10% total solids in the cattle waste based plants. This was constructed to process mainly sugarcane bagasse. Though considerable work has been done in India on the design aspects for processing cattle waste, there is little data regarding the microbiological aspects of boosting biogas production.

WORK DONE ABROAD

China is said to have seven million biogas plants. Scarcity of animal excreta forces the Chinese to use pretreated agricultural residues as feed material; pretreatment involved the use of lime water, and partial composting. Initial rectangular-type designs were changed to fixed types having a cylindrical shell with two hemispherical domes. Piggery waste and human excreta are also processed through anaerobic fermentation to obtain biogas and biomanure as a step in preventing environmental pollution. The amount of biogas produced from piggery waste is reported to be higher than cattle waste.

Biogas plants for processing piggery waste, agroindustrial waste, and cattle waste are also under operation in Fiji, the South Pacific, Jamaica, Latin America, Malaysia, Africa and South America.

In the Philippines, biogas work on Maya Farms, the agro-industrial division of Liberty Flour Mills, Inc., is very systematic. The farm spreads over 24 hectares in Manila, and has about 22,000 pigs, 6,000 ducks, and a number of cattle. The farm has developed a total waste recycling system which controls pollution and generates a fuel gas primarily from piggery waste. A cluster of 24 individual batch fed digesters with a total volume of 19,200 cu. ft. have been build above ground. Another set of digesters was constructed underground in the second phase. Continuously fed digesters were also constructed side by side to compare their performance. In all, the plants produce 66,000 cu.ft. of biogas per day. All operations such as cooking and generation of electricity to run various engines in the farm,flour mill, etc. are performed exclusively with biogas.

The global energy crisis and ever escalating cost of fossil fuels have triggered off research activities even in developed countries that seek to conserve energy and that are searching for alternate and renewable sources of energy.

One French program centers on biomass based on forest wastes. The work of Jean Pain on the utilization of compost to obtain biogas, biomanure, and hot water is worth mentioning. In the U.K., about 80% of the energy requirements could be met from biomass, according to the estimate of the Department of Energy if farm and industrial waste is used efficiently. Biogas plants of different sizes are in operation, processing various kinds of wastes in the U.K. A 227 m^3 below-ground digester with a floating top gas holder uses cattle waste in Bershire. Another 1,200 m^3 glass-lined tank digester with a water sealed gas holder is running on piggery waste. At Kent, a 3,000 m^3 digester is operating on wheat starch effluent. In the Netherlands there is an effort to recover methane gas from the anaerobic digestion of municipal waste in a sealed landfill. It has been estimated that 125 m^3 of gas can be produced per ton of waste.

In the USA, Biogas of Colorado, Inc. has developed a mobile methane digester. It has been reported that the degradable fraction of sorted household waste can be successfully digested in a fixed bed anaerobic system.

The potential for utilizing agricultural crop residues for biogas production is not new. Buswell and Hudfield speculated on the potential benefits of crop residue fermentation. Traditional sludge digestion practice would dictate that the solids content should not exceed 10% total solids. But, must agricultural crop residuees contain more than 90% solids. The potential benefits of fermentation at solids content in excess of 10% have been recognized.

In the Department of Engineering at Cornell University, several interesting variables in biogas production at minimal moisture content were investigated, and this gave a new impetus to processing crop residues at much lower cost. The process is called Dry Anaerobic Fermentation. It was shown that cattle waste can be fermented profitably at 33% initial total solids. A 5 m^3 pilot plant was built in 1980 to process wheat straw at 22% total solids. Much work has been done to show that dry fermentation is a good method for processing crop residues.

BIOGAS AND BIOMANURE FROM WILLOW-DUST; CHEMICAL ANALYSIS OF WILLOW-DUST

Willow-dust is rich in cellulose, hemicellulose and lignin. Cellulosic substrates with a carbon-nitrogen (C:N) ration of 25:1 are reported to be optimum for biogas production. As the C:N ration of willow-dust (28-30:1) comes close to this optimum level, it has proven suitable for biogas production from the investigations carried out at CTRL. The chemical composition of the willow-dust, which is akin to cattle waste used in the production of Gobar gas, is given in Table 1.

A number of experiments were undertaken to find out the optimum C:N ratios, solid-liquid ratios, pH conditions, and the effect of temperature on the production of biogas in a batch fermentation process. Trials were conducted for producing biogas in 10 liter bottles using 1 kg willow-dust soaked in 6 liters of water. This material was inoculated with cattle waste at the rate of 10%. Later trails were carried out by inoculating the material with spent slurry from previously fermented batches. It was observed that 1kg total solids (TS) of willow-dust yielded 176 liters of biogas in a period of 60 days.

To reduce the fermentation time, the willow-dust was treated aerobically with sodium hydroxide (100:1 W/W) and allowed to ferment in the open for 72 hours. This was followed by anaerobic fermentation in an airtight digester. This procedure reduced the time of anaerobic fermentation from 60 days to 30 days.

Table 15.1
Proximate Chemical Analysis of Willow-dust and Cattle Waste

Analysis*	Willow-dust %	Cattle waste** %
Ash (including sand)	10-12	15.0
Cellulose	26.0	28.5
Hemicellulose	16.0	26.0
Lignin	16.0	17.7
Nitrogen	1.5	1.3
Ether extractives	5.0	3.4
Water solubles	24.0	9.5
C/N	28-30	30.32

* On dry weight basis.
** Idnani and Varadarajan, *ICAR Tech. Bull.* 46 (1974).

The biogas produced from willow-dust had 55-60% methane similar to that obtained from many other organic materials (Table 2). The total gas produced from willow-dust (176 liters), and that reported for cattle waste (165-170 liters) as well as their calorific value, remained almost the same except perhaps for the fact that the generation of gas from willow-dust was completed in 30 days, compared to the 60 days required for cattle waste.

On the basis of the experience gained from the laboratory trials, an experimental plant for batch type fermentation was designed and fabricated (Figure 15.1). This plant has the capacity to handle 100 kg of willow-dust. The plant consisted of an anaerobic fermenter and a separate gas holder constructed from galvanized iron sheets. The anaerobic fermentation tank was totally closed except for one opening of 180 mm diameter at the top to feed the material and a similar opening at the bottom for removing the spent slurry. The tank also has a gas outlet, a pressure gauge, and two inspection windows. The biogas generated in the fermenter was led to the gas holder through flexible rubber hoses. The gas holder was of the floating type designed to store 500 liters of gas over water.

The studies carried out on this plant (Fig. 15.2) showed that during the first 6 days, after charging the digester with willow-dust, biogas rich in carbon dioxide was generated. In the until it was 55-60%. The production of biogas during the period of active gas generation

Table 15.2
Production and Analysis of Biogas from Willow-dust and Other Substrates

Substrate	Biogas (liters/Kg) Total Solids	CH_4 (%)	CO_2 (%)	Calorific values (see note) (MJ/m³)	Btu/ft³
Willow-dust	176*	55-60	38-40	18.9	500
Cattle waste	165-170**	54-56	30-40	18.3	492
	200-210†	55-60	40-45	18.9	500
Piggery waste	251	57	27	18.9	500
Bagasse	200-300	60-65	30	20.9	563
Water hyacinth	370	69	NA	22	580

Note: Calculated from the concentration of Methane. Pure methane = 33.3 MJ/m³ or 895 Btu/ft³.

* Batch fermentation, 30days.

** Batch fermentation, 60 days.

† Continuous fermentation with periodic addition of fresh material (retention period, 50 days).

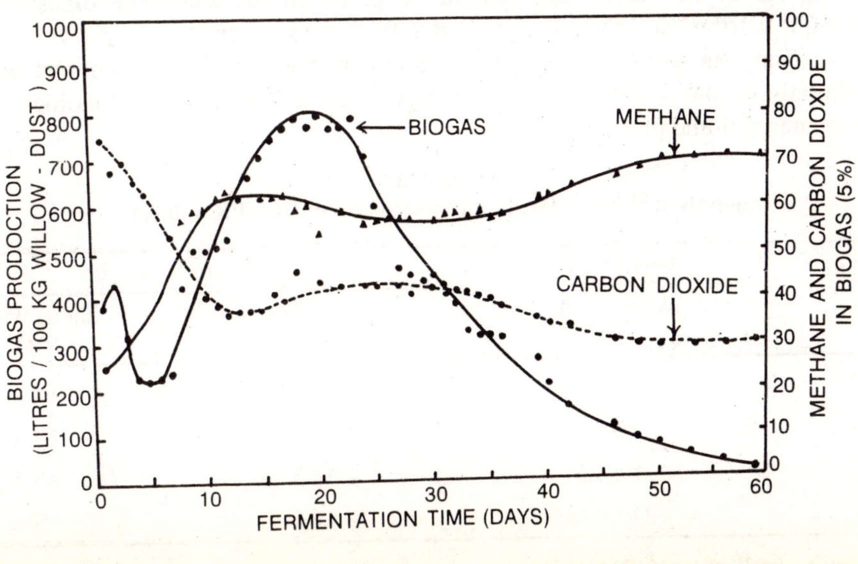

Fig. 15.2. Production of biogas from willow-dust (result from trials with 100 Kg experimental plant)

increased to a peak of 0.85-0.90 m^3 per day and then gradually decreased. On an average, 0.5-0.6 m^3 of biogas was obtained per day for 30 days from 100 kg willow-dust. Although about 20 m^3 of biogas was produced in 60 days, as much as 17 m^3 (85%) of it was confined to 30 days only.

The digested slurry was also tested for its manurial value. Analysis of the digested slurry for nitrogen, phosphorus potassium and sodium is given in Table 3. It was observed that the digested slurry has better NPK contents than the slurry obtained from cattle waste.

To assess the manurial value of the digested slurry. Pot culture trials were undertaken on a crop of Laxmi variety (Gossypium hirsutum Linn.) cotton using red sandy loam soil. Three sets of pots were taken. One set received 500g of the digested slurry, which was applied at the top and mixed with surface soil. Another set received a basal dose of 1.5g ammonium sulfate, 4.0g superphosphate and 1.5g potassium sulfate. This set again received the same dose of fertilizers after one month of sowing as top dressing. The third set served as control and did not receive either fertilizer or digested slurry. The difference in the stand of the crops after 45 days of growth is shown in Figure 3. The stand of the crop on digested slurry was comparable to the fertilized crop. This is attributed to the nutrients present in the slurry and also to the high lignin content. Lignin is known to have high water holding capacity and contributes to an overall improvement in the soil structure.

STUDIES ON DRY FERMENTATION

The technology developed in the above process for the production of biogas from willow-dust required only half the quantity of water (solid to liquid ratio, 1:6) as compared to the conventional cattle waste based biogas plants (solid to the ratio, 1:12). An interesting aspect of willow-dust fermentation is that the entire mass floats and remains in almost the same position during the 30-day active gas generation phase in the anaerobic digester. It was therefore thought worthwhile to reduce the water further to accommodate more fermentable material in the same unit space, and to obtain a biomanure with less water content that could be applied directly to plants. Also, in places where water is scarce, this technology becomes much more meaningful and practicable.

Table 15.3
Analysis of Willow-dust and Digested Slurry for NPK and Na

Nutrients	Undigested Willow-dust (%)	Digested slurry of Willow-dust (%)	Digested slurry of Cattle Waste (%)
Nitrogen	1.5-2.0	2.0-2.5	1.6-1.8
Phosphorus	0.5-0.7	0.8-1.0	1.1-1.2
Potassium	1.2-1.5	2.0-2.5	0.8-1.2
Sodium	0.04-0.045	0.3-0.4	NA

* Biswas, *Invention Intelligence*, 12, 71 (1977).

Attempts made to reduce the amount of water showed that it was possible to produce biogas as effectively as before with a substrate to liquid ratio of 1:1.5. In this process, conditions such as alkali treatment and aerobic fermentation remained unaltered. However, with this substrate to liquid ration of 1:1.5, biogas production started after 10 days, as compared to 7 days when substrate to liquid ratio was kept at 1:6. However, the total quantity of biogas (17 m^3) produced in the 30 days of active phase remained unaltered. By following this technology, it is now possible to charge the same 100 kg experimental biogas plant with 200 kg material, thereby getting twice the gas yield from the same unit space.

Since the sodium hydroxide added during the aerobic fermentation of willow-dust is not desirable from the point of view of obtaining agriculturally suitable biomanure, experiments were undertaken to replace sodium hydroxide with lime to the maximum extent possible. It was observed that the yield of biogas remained the same, in spite of reducing the use of NaOH to 0.1% with the addition of 1.5% lime and 15% inoculum. Further, about 8 m^3 of biogas was obtained in the subsequent 15 days of the normal gas generation phase, making the entire process viable for a total period of 45 days. The earlier process was economically viable for 30 days.

In the fermentation of cattle waste with different solids content, it has been reported that methane production proceeded only up to 32%, although hydrolytic bacteria functioned normally to 60% total solids. Jewell et al reported that in the case of wheat straw and corn stover with 28% initial volatile solids content, 90% of the biodegradable fraction was converted to methane in about 200 days. However, fermentation proceeded in the case of willow-dust at a volatile solids content of 53%. It has been reported that ammonia nitrogen at concentrations exceeding 3,000 mg/l^{-1} becomes toxic at a pH greater than 7.4. Degradation of cellulose is inherently slow, and in the present case, compaction of the dust due to a small amount of water present could have reduced the surface area of particles available to hydrolytic bacteria. Hobson and Shaw observed that the methane-producing bacterium Methanobacterium formaicicum was inhibited by high concentrations of ammonium. The smooth functioning of the methanogenesis in the case of willow-dust digestion even at a volatile solids content of 53% could be due to a steady pH of 7.0 and an ammoniacal nitrogen content in the range of 1,500-2,600 kg^{-1} of the fermenting material, which is below the reported toxic limit values.

The biomanure obtained after fermentation can be directly used on plants, unlike the spent slurry from Gobar gas plants which requires dewatering or sun drying for immediate use. During this process, nutrients are lost through either evaporation or percolation. The replacement of most of the sodium hydroxide with lime during the aerobic fermentation of willow-dust produced an agriculturally suitable biomanure. The bulk density of the material at the end of the fermentation was 0.846, compared to an initial value of 0.626. Since production of biogas from cellulosic waster in the batch fermentation process does not vary between small and large plants, and the shape of the plant does not matter, the technology already available for the production of biogas from willow-dust can be profitably exploited at the site of any textile mill.

PILOT PLANT TRIALS

The pilot plant to process willow-dust for biogas production was commissioned at Apollo Textile Mills, a unit of the National Textile Corporation, with financial assistance from the Department of Non-conventional Energy Sources (DNES), Govt. of India. Before setting up the plant, the mill was consuming about 65 kg of LPG daily in the laboratory, canteen, and primarily, for singeing operations. Based on the availability of the raw material, it was decided that the plant should process only the available willow-dust, because transportation of this bulky material from other mills is economically not feasible.

Description of the Pilot Plant

The general layout of the plant's design is given in Figure 4. The plant is constructed on 104 m³ land. There are six non-stirred digesters, each 2.4 x 2.40 x2.5 m in size, with a capacity to handle 2 tonnes of willow-dust when the substrate to liquid ratio is 1:6. The digesters are constructed of reinforced concrete and are partially underground. Provision has also been made for the initial aerobic fermentation in an open tank 8 x 2.4 x 2.5 m, and for a slurry tank 4.8 x 2.4 x 2.5 m for receiving the digested material after anaerobic fermentation. The same digesters can each handle 4 tonnes of willow-dust when the substrate to liquid ratio is 1:1.5. The digesters are separately connected through gas flow meters, which in turn are connected to a common gas holder through a header. Each digester has a separate bypass for the carbon dioxide generated during the initial stages of biogas production. The gas holder

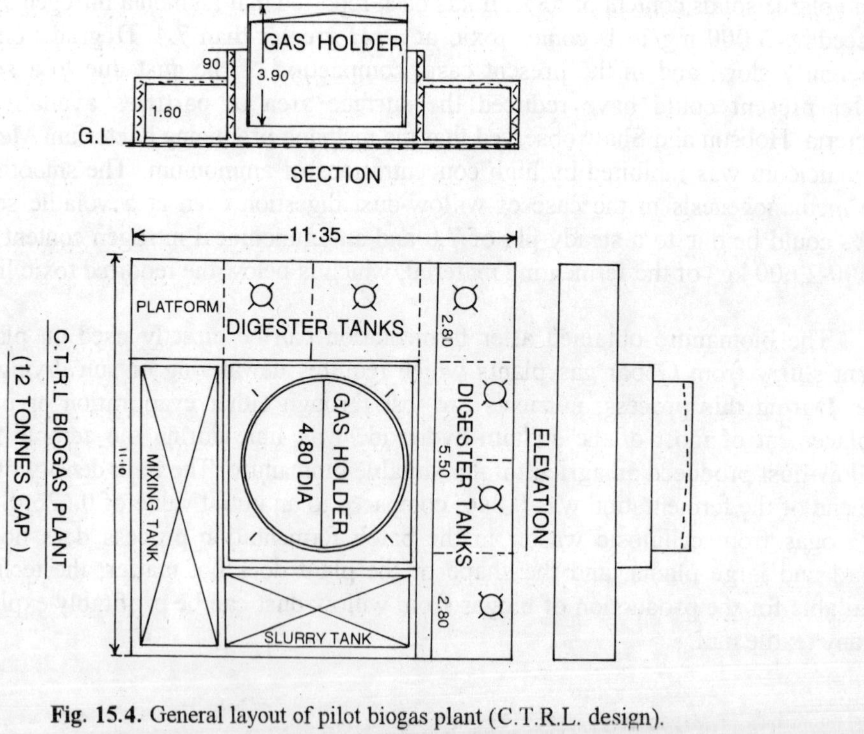

Fig. 15.4. General layout of pilot biogas plant (C.T.R.L. design).

is of floating type and constructed from cast iron sheets. The weight of the drum is adjusted to exert a constant pressure of 25 cm WG. The gas holder has a capacity of 70 m^3. A photograph of the plant is shown in Figure 5

Pretreatment and Aerobic Fermentation of Willow-dust

The mixing tank meant for pretreatment of willow-dust can accommodate a maximum of 6 tonnes at a time. To find out the effect of initial solid loadings on the subsequent biogas production, different quantities of willow-dust (2, 4 and 6 tonnes) were put into the tank.

Willow-dust (6 tonnes) was mixed in 9,000 liters of water containing 60 kg of sodium hydroxide. To achieve better mixing, 500 kg of the material was spread in a layer, the required quantity of alkaline water was sprinkled through a hose using a pump, and this layering process was repeated. The top of the material was covered with a polythene sheet to avoid excessive evaporation of moisture, and left for 72 hours. An inoculum was prepared from 600 kg of fresh cattle waste suspended in 26,400 liters of water. The pretreated material was divided into three parts and charged to three digesters by adding the cattle waste suspension simultaneously to achieve uniform mixing of the inoculum. The same procedure was followed when 2 and 4 tonnes of willow-dust were used. In all cases, the substrate to liquid ratio was maintained at 1:6. When this ratio was reduced to 1:1.5, willow-dust was mixed with alkaline water containing 10% of inoculum prepared as above, and left for aerobic pretreatment as before. The material was then inoculated with diluted cattle dung prior to anaerobic fermentation.

Since it was not possible to exactly apportion the pretreated material before charging, particularly when 4 and 6 tonnes were mixed, the biogas generated was totalled from either 2 or 3 digesters and the total quantity wad determined. The trials were run for two years in all seasons, as temperature was not a constraint limiting biogas production. Irrespective of the quantity of pretreated material, each digester received 2 tonnes of willow-dust when the substrate to liquid ratio was 1:6 and 4 tonnes when the ratio was 1:1.5. Data collected from four trails in the case of 2 and 4 tonnes of initial solid loadings and six trials with 6 tonnes during pretreatment were compiled and calculated to normal temperature and pressure. The pretreatment has been called aerobic fermentation as no special precautions were taken to exclude air from the pretreatment tank. However, as well as the chemical actions of the alkali, microbiological reactions could take place, leading to anaerobic conditions in the depths of the willow-dust, particularly in the dry material to which cow dung was added. The biogas productions from willow-dust are given in Figures 6 and 7. The biogas generation can be divided into four phases. The first phase is the generation of non-combustible gas and is spread over a period of one week. This period is the most critical stage, and decides the efficiency of the process with respect to the quality and quantity of biogas produced in the later period. Addition of inoculum during aerobic fermentation of willow-dust was resorted to in order to reduce the initial non-combustible gas phase to only one week instead of two weeks. Therefore, the entire gas generation can be divided into three phases of 30 days each. The percentage of methane remained between 55 and 60% throughout this gas generation period.

The initial solid loading during aerobic fermentation did significantly alter the pattern and quantity of biogas production during anaerobic fermentation when the substrate to liquid ratio was 1:6 (figure 6). The first 30 days accounted for 300 m^3 280 m^3 of biogas followed by 295 m^3, 240 m^3 and 230 m^3 in the subsequent 30 days against the initial solid loadings of 2,4 and 6 tonnes, respectively. However, in the last phase of 30 days, gas production was 145 m^3, 200 m^3, and 125 m^3. approximately 740 m^3 and 650 m^3 of biogas were produced per 2 tonnes of willow-dust in 90 days with 2, 4, and 6 tonnes of solid loadings during pretreatment.

In the case of dry fermentation, where the substrate to liquid ratio was 1:1.5, the solid loadings during the initial aerobic fermentation significantly altered the pattern and quantity of biogas production (Figure 7). When only 2 tonnes of willow-dust fermented aerobically about 500 m^3 of biogas was produced in the first 30 days of the active gas generation phase during anaerobic fermentation. When 4 tonnes were loaded for aerobic fermentation, the same quantity of 500 m^3 of biogas was distributed over 60 days. With 6 tonnes, a lower but steady gas production was maintained for more than 80 days.

No appreciable change in the daily temperature affected the biogas production. It is well known that a temperature of less than 20 C significantly lowers biomethanation. Bombay has an equable climate with temperatures ranging from 25 to 30°C (±5°C) depending on the seasons, excepting in the winter months when the night temperature occasionally drops below 20°C.

Complex chemical, physical and microbiological reactions took place which could be affected by the depth and consistency of material in the aerobic tank. The loading effect would probably differ with dimensions of the pretreatment tank.

The lignin, which is known to have a high moisture retention capacity, must be playing a vital role in the absorption and release of organic acids during willow-dust fermentation. Madamwar and Mithal have reported that activated charcoal increased biogas production and gave a higher methane content. This could be due to the absorption of toxic substances during biomethanation. A similar situation may take place during willow-dust fermentation.

The present study indicates that it is possible to obtain biogas from willow-dust in a largesized plant with results similar to small-scale batch digesters. It is possible to manipulate the biogas production by pretreatment. A constant supply of biogas can be ensured by having a series of digesters. The low-water digestion allows treatment of large weights of dust in a small digester plant, thus reducing the cost of the initial investment. Dry fermentation of willow-dust is an ideal method to obtain a constant supply of biogas from a minimum of two batch-fed digesters, apart from producing readily usable biomanure after fermentation.

After fermentation, about one tonne of biomanure was obtained from 2 tonnes of willow-dust from each digester, indicating that 50% of the material is converted to biogas. The biomanure is a stabilized organic matter, rich in plant nutrients.

The biogas generated from willow-dust in the textile mill is being used primarily for singeing. It is also used in the laboratory and the canteen.

In the textile mill processing house, a large quantity of alkaline water is first treated for recovery of as much alkali as possible and then drained. Since even this water contains 0.50

to 0.80% alkali, it is recommended that it be pumped to an overhead tank and then let into the gas storage tank by gravity. By keeping an overflow at the top of the water tank of the gas holder, the level can be maintained. This mild alkaline water will be useful for absorption of a major portion of the carbon dioxide, and thereby increase the methane content (and thus the calorific value) of the biogas stored. Some observations have indicated that the percentage of methane goes up to 85%. This helps further in reducing the cost of the plant.

BIOGAS PRODUCTION BY DRY ANAEROBIC FERMENTATION

After several successful trials at the Apollo Textile Mill Plant, an experimental plant to process willow-dust for biogas production by dry anaerobic fermentation has been fabricated at this laboratory. The plant consists of two batch-fed digesters, each with a capacity to process 500 kg material, and a common gas holder. A platform has been provided at the top of the assembly to mix willow-dust during aerobic fermentation, and outlets at the bottom to take out the digested slurry. The gas holder is a floating type and can store about 5 m^3 of biogas. The entire assembly is made out of mild iron sheets except the lids of the digesters, which are made out of G.I. sheets (Figure 8).

As much as 250 m^3 of biogas is being produced per tonne of willow-dust in 45 days, and 500 kg of readily usable biomanure is obtained after fermentation. Based on the successful trials undertaken at the experimental plant, a schematic diagram for scaling up the trials on a pilot plant is given in Figure 9. The proposed plant, has three digesters, each capable of handling 6 tonnes of willow-dust per batch, and two gas holders. Each gas holder can store about 17 m^3 of biogas. A top mixing platform has been provided for aerobic fermentation of willow-dust and there are bottom openings for taking out the digested slurry.

FIELD TRIALS FOR EVALUATING DIGESTED SLURRY

Field trials were laid out on a cotton crop under irrigated conditions at Arabhavi, Belgaum District, Karnataka for evaluating the manurial value of the digested slurry. The treatment details are given in Table 4. The trials were repeated for three seasons. The results indicate that slurry applied at the rate of 5 tonnes/ha is equivalent to 10 tonnes/ha of FYM and is also equivalent to the NPK applied at the rate of 80:40:40 kg/ha, based on the yield of seed cotton.

BIOGAS FROM CATTLE WASTE AND WILLOW-DUST

Two types of plant design (Khadi Village Industries Commission (KVIC) and Janata plant) are popular in India for the production of biogas from cattle waste. KVIC has a floating MS gas holder whereas the Janata plant is a modified fixed dome plant of Chinese design. Both have advantages and disadvantages, but the latter one seems to be most economical. Cattle waste can be fermented in these designs without problems. If the plants are not charged regularly, problems can arise in breaking up scum formation. Which can otherwise hinder the smooth flow of the slurry. The accumulation of sand creates a problem in the long run and at such times the only remedy is to empty out the entire digester and clean it.

Table 15.4
Effect of Digested Slurry on Cotton Crop

Center	Arabhavi, Karnataka (under irrigated conditions)
Variety	J K 119 (Gossypium hirsutum Linn.)
Design	R B D
Replications	3
Plot size	Gross 6 m × 6 m Net 4.5 m × 6m

Sl. No.	Treatments	Yield of Seed Cotton (kg/ha)
1.	Control	1,302
2.	FYM, 10 tons/ha	1,493
3.	Slurry, 5 tons/ha	1,479
4.	Slurry, 10 tons/ha	1,728
5.	Slurry, 15 tons/ha	1,593
6.	NPK, 80:40:40 kg/ha	1,549
7.	NPK, 160:80:80 kg/ha	1,821
8.	NPK, 80:40:40: kg/ha +Slurry, 10 tons/ha	1,759
	S.E. ± 157 C.V. 17%	

The improved plant design for processing cattle waste is shown in Figure 10. This is a semi-continuous plant into which a partition wall and scum breakers have been incorporated. The partition wall is Provided close to the inlet chamber, as opposed to in the center, to increase the retention time of the fermenting material. Another advantage of this wall is that the addition or fresh material will not be able to aerate the anaerobically stabilized slurry. Apart from this, the settled sand can be taken out from the inlet side without emptying out the entire digester. When gas accumulates in the gas storage space, its pressure compresses the slurry in the inlet and outlet ductings. In this process, the scum formed this way breaks away due to the presence of fixed scum breakers. When gas is utilized, the level of the slurry in the inlet and outlet ductings recedes and rises in the digester until it attains the original level. During this process too, the scum tends to pass through the scum breakers. As a result to this up and down movement of the slurry due to the gas collection and its usage, the scum breaks and the material will always be in a suspended state.

Trials conducted on a 2 m^3 plant indicated that the partition wall and fixed scum breakers have definite advantages. The improved plant design for processing willow-dust semi-continuously is shown in Figure 5.11. The inlet ducting has been given an inclined shape with a curvature at its bottom to facilitate easy flow. Although a partition wall is provided in the center of the digoster, the trials indicated that it hinders easy movement of the material.

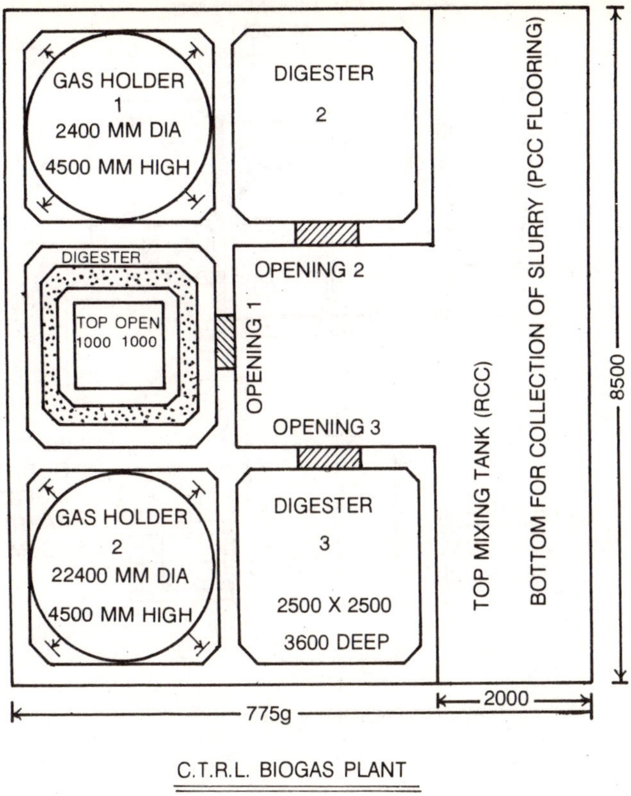

Fig. 15.9. Proposed biogas plant for willow-dust by dry fermentation (18 tons capacity).

Process

After loading, the willow-dust must be charged through the inlet chamber every week, as detailed below. The material, mixed with sodium hydroxide (1% W/w) in a substrate to liquid ratio of 1:1.5, is charged through the inlet at weekly intervals. Calculated quantities of water are to be added every day thereafter to maintain a substrate to liquid ratio of 1:12. The slurry is manually taken out of the outlet chamber every day. By following this method, it is possible to obtain 2 m^3 of biogas every day when the temperature is between 25°-30°C.

RELEVANCE OF BIOGAS RESEARCH TO INDIA AND OTHER TROPICAL COUNTRIES

India has an agricultural base, with more than 80% of the population engaged in agricultural activities, thus making biogas readily available. Moreover, the technology of biogas has been developed indigenously. This low cost fermentation technology needs only low energy input and is the only large-scale process available now for recycling cellulosic wastes to produce fuel and manure, and, at the same time, prevents pollution.

India is well situated for the production of biogas because it is a tropical country. Table 5 given an idea of the materials suitable for biogas production and Tables 6 and 7 list the

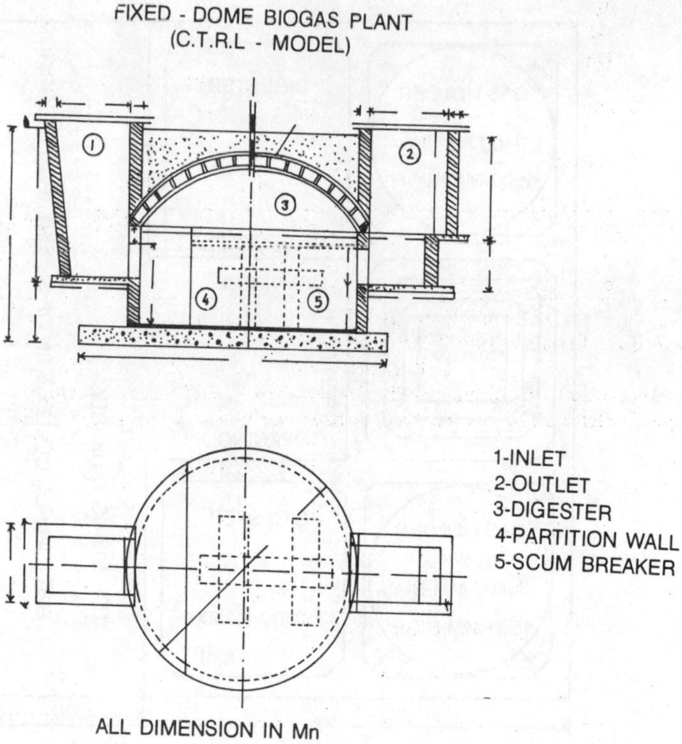

Fig. 15.10. Improved design for fermenting cattle waste for semicontinuous biogas production.

industrial wastes plentifully available in India. It is interesting to note that none of the materials, despite their potentials for biogas production, have been explored.

It is well known that biogas production falls at temperatures of 20°C and below, the optimum temperatures being 30-35°C with a maximum of 50-55°C. Temperature fluctuations are marked in India depending on the seasons. The extremely low temperatures in northern parts of the country significantly lower biogas production. In southern regions, temperature is not a problem, although production is below normal values during the winter months. At such times, dry anaerobic fermentation can come to the rescue to a great extent because the fermenting mass, due to a high solid load and its compaction, doesn't allow the heat to escape at a faster rate even during cooler months.

Under these circumstances, considering the huge quantities of cellulosic wastes available in this country and scarcity of water in most of the places, dry anaerobic fermentation seems to be the ideal method to process the material, either at an individual level or in community fermenters, and appears promising and practicable. The most important advantage of using dry anaerobic fermentation is the production of readily usable biomanure with the nutrients intact.

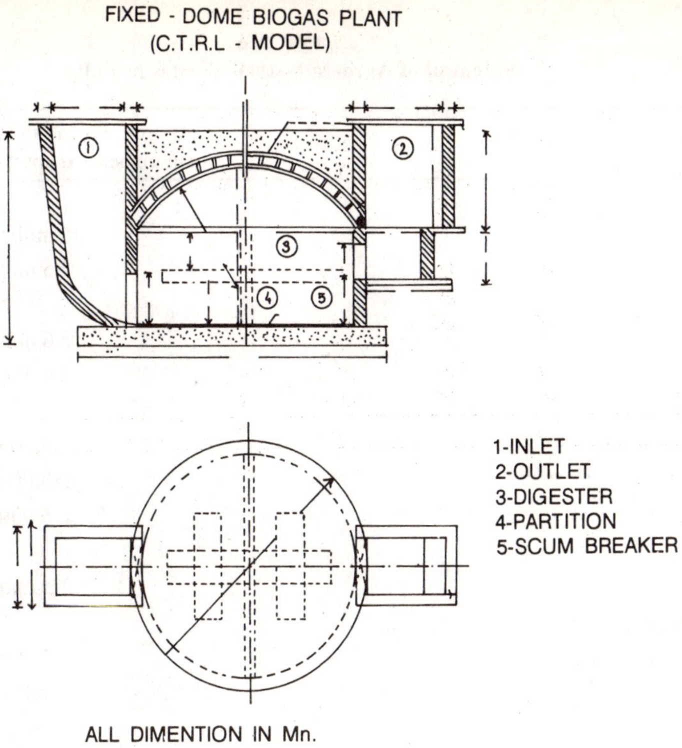

Fig. 15.11. Improved design for fermenting willow-dust for semicontinuous biogas production.

Table 15.5
Potential Sources of Biomass for Methane Generation

Item	Description
Animal wastes	Cattle waste, poultry litter, sheep and goat droppings, slaughterhouse wastes, fishery wastes, wool wastes
Crop residues	Sugarcane trash, weeds, corn and other crop stubble, straw, wasted fodder
Human wastes	Feces and urine, refuse
Forest residues	Twigs, bark, branches, leaves, undergrowth
Byproducts /wastes from agricultural-related industries	Bagasse, rice bran, tobacco wastes, seeds, wastes from fruit/vegetable processing, press mud from sugar manufacturing, tea waste, cotton dust, oil cakes
Aquatic based residue	Marine algae, seaweeds, water hyacinths
Solids wastes/urban	Paper wastes, household residues, trash

* Cheremisinoff, N. P., P. N. Cheremisinoff, and F. Ellerbusch, Biomass-Applications, Technology and Production, Marcel Decker, Inc., New York and Basel, 1980, p. 221.

Table 15.6
Potential of Agro-Industrial Wastes in India

Agro-industry	Approx. Potential/Year (in tonnes, except where noted)
Rice mill	
Rice husks	15 million
Rice bran	2.5 million
Sugar mill	
Bagasse	5.6 million
Press mud	2 million
Slaughterhouse waste	
Blood	55,000
Bones	450,000
Leather waste	5,000
Forest industry	
Sawdust	2 million
Fruit and vegetable industry	
Mango peels and stones	25,000
Citrus peels	4 million
Cotton mill	
Cotton stalks	12 million
Cotton dust	30,000-33,000
Marine industry	
Prawn shell waste	40,000
Fish waste	3,000
Frog legs	5,000
Tea industry	
Tea waste	10 million
Jute mill	
Jute sticks	2.5 million
Tobacco industry	
Tobacco leaves and stalks	34.5 million kg
Coffee industry	
Coffee husks	44,540
Major oilseeds industry	
Edible oil cakes	2 million
Minor oilseeds industry	
Neem seeds	100,000
Mahua seeds	70,000
Karanj seeds	26,000
Sal seeds	25,000

Source: Gaur, A.C. and K.V. Sadasivam, *Indian Fmg.* 31-37 (October, 1981).

Table 15.7
Potential of Crop Residue in India

Crop	Dry Matter Yield (000 tonnes)
Paddy	89,369.04
Jawar	14,287.80
Wheat	18,418.40
Bajra	8,589.00
Maize	10,554.00
Total cereals	—
Total pulses	9,851.25

Source: Gaur, A.C. and Sadasivam, K.V. Indian Fmg. 31-37 (October 1981).

Tolerance limits to different elements under anaerobic fermentation are very high. Although microorganisms can function normally even at very high concentrations of elements, a way has not yet been devised to exploit this to our benefit. In this regard, dry anaerobic fermentation offers great potential.

In conclusion, it can be said that production of biomass continues to increase under the present-day intensive agriculture. This can be further augmented by proper management of forests, development of fast-growing energy crops, intensive forestation, and widespread tree planting.

Hence, processing crop residues under dry anaerobic fermentation at individual or community levels appears appropriate when water is considered as a constraint to obtaining bioenergy and readily usable biomanure. Ultimately, this method takes care of environmental pollution also.

16

Economic Analysis of a Feedlot Waste Management Programme

Introduction

Recently, the U.S. Geological Survey emphasized concern about potential nitrate contamination of aquifers throughout the United States. Although nitrate itself is relatively nontoxic, it is reduced by intestinal bacterial to nitrite, a hazardous substance. Nitrate concentrations in drinking water above the Environmental Protection Agency's health standard of 10 parts per million (ppm) can cause infant mortality (methemoglobinemia). In addition, long exposure to nitrate is a suspected cause of cancer.

Stewart *et al.* evaluated ammonium and nitrate concentrations in groundwater under feedlots and adjacent irrigated fields in Colorado and concluded that the feedlots were a significant source of nitrate and ammonium in the groundwaters. Mink *et al.*, in a study of the land disposal of animal waste in the Boise Valley, Idaho, determined that due to denitrification, nitrate-nitrogen concentrations in the soil profile beneath two feedlots decreased rapidly from about 60 mg per L near the surface to 20 mg per L at the two meter depth. Where groundwater levels were less than 1.5 meters from the surface, the water was found to be affected by the feedlot. In a survey of high-nitrate groundwater in Missouri, Keller and Smith analyzed water from more than 5,000 wells and springs. About 42% of the samples contained more than 5 mg per L nitrate-nitrogen. They found the dominant source of nitrate in Missouri groundwater to be nitrogenous waste from livestock feedlots.

Farm wastes contribute a significant proportion of organic pollution to aquifers as well as the inorganic pollution causing eutrophication of lakes and rivers. Anaerobic digestion can reduce some of this load by lowering the organic content of these wastes being lost as gaseous methane and carbon dioxide. The inorganic salts, mainly nitrogen and phosphorus, are largely absorbed in the particulate material voided from the digester and can be settled out after digestion. Anaerobic digestion stabilizes the organic material, reduces odor, can produce a useful by-product (methane), and enhances the fertilizer value of the wastes.

While the cost of collection and distribution of feedlot waste coupled with that of conventional biological treatment is high, there is pressure on livestock producers to install such a system to minimize the pollution load on receiving waters. Much of the manure can go on the land, but agricultural runoff is a common form of "point source" pollution of rivers or lakes..

During the 1960s it became evident that with the increase in size and in numbers of intensive farming units, pollution from animal excreta, formerly negligible or ignored, could become a serious problem and should be controlled by legislation. Thus, it was the problem of aerobic digestion of animal waste products approximately 20 years ago. However, considering the relatively higher prices for petroleum fuels than were seen 20 years ago. However, considering the relatively higher prices for petroleum fuels than were seen 20 years ago. However, considering the relatively higher prices for petroleum fuels than were seen 20 years ago, the use of cattle feedlot manure for energy production may be coming into vogue.

In 1980 animal residue represented one of the major biomass feedstock sources for agricultural energy in Texas, USA. Bjornstad et al. predicted that biogas from animal residues will become competitive with natural gas by 1990. Van Dyne and Gilbertson estimated that 16 Mt of manure dry matter were voided by feeder cattle in the United States in 1974, of which approximately 97% was recoverable. Reviewed in previous studies are the physical characteristics of biomass technologies, the availability of biomass resources, the potential market for equipment to produce biomass fuels for on-farm use, and the institutional conditions that may ultimately govern biomass fuel penetration into national markets.

Another function proposed for the use of cattle manure is as a feedstuff. The utilization of animal wastes as feedstuffs is not a new phenomenon. In 1925 Evvard and Hennes reported that, on average, one pig following 1.9 steers recovered the equivalent of 142 kg of corn during the 120-day feeding period. In 1983 animal waste products were recognized as underutilized resources for animal feedstuffs by the National Research Council. The utilization of this resource offers a significant reduction in feed cost (as much as 40% or more) and also minimizes the requirements for expensive protein feeds and other competitive feed resources such as cereal grains and milling by-products. Processing livestock wastes for these purposes has the added benefit of effective disposal as well as alleviating a potential environmental problem.

There is a particular allure, in agricultural markets, for self-sufficiency. Cost-competitive biomass fuels provide some options for achieving at least part of this objective. While technologies exist that could allow the agricultural sector to attain energy self-sufficiency via biomass fuels, investment cost considerations will stand in the way of many biomass applications. High interest rates and depressed agricultural market prices reduce the willingness of farmers to purchase new capital equipment, including equipment to produce biomass fuels. Major uncertainties about the future of central fuel (diesel oil, electricity, fuel oil, gasoline, LPG, and natural gas) prices, interest rates, discount rates, and farm markets may also depress the growth of on-site digesters. This chapter analyzes the economic feasibility of investing in equipment for a biofuel production and solid recovery system for a feedlot on the Southern Great Plains of the United States.

Methods and Procedures

It was assumed that manure holds zero value for feedlot producers. The hypothetical biofuel production and solid recovery system (System) was established for a 40,000 head capacity feedlot. Figure 1 illustrates how the System involves placing animal waste into digesters which use anaerobic metabolism to produce a biogas consisting of approximately

60% methane (CH_4) and 40% carbon dioxide (CO_2). Only feedlots with concrete floors were considered in this analysis. Manure is washed off the feeding floor and collected for deposit directly into a tank where it is churned and held until ready for transfer to a digester. Within the digester manure ferments and produces biogas which is collected in an airtight hyphalon cover. Biogas is then employed to generate electricity used in part to operate the System while excess electricity is sold to a local public utility.

The fermenting process takes approximately 17 days to exhaust the gas producing properties. When exhausted, the manure floats, is drawn off the liquid surface, and piped to a solids recovery building. The still wet residue is then reduced to a solid state in a biogas-powered centrifuge dryer where 80 to 90% of the remaining solid mass is recovered and transformed into pellets. The remaining 10 to 20% of the effluent slurry can be used effectively on cropland as NPNK grade fertilizer. It was assumed in this analysis that the value of this material is equal to the expense of disposal.

Recovered pelleted residue is suitable for refeeding. Preston and Richardson reported that average daily gain for feedlot cattle could be increased slightly when substituting pelleted residue for alfalfa at 4% of the feedlot ration. Thus, it was assumed that in a ration consisting of 16% alfalfa, feed costs can be reduced by 25% of the cost of alfalfa when this amount of alfalfa is replaced with the pelleted residue.

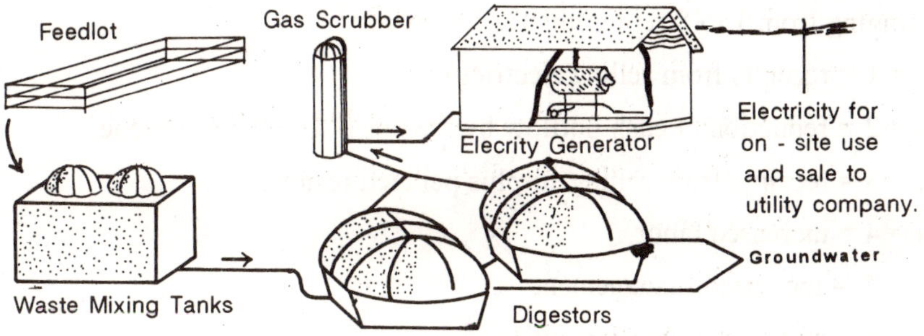

Fig. 16.1. Simplified diagram of a biofuel production and solid recovery system.

A conventional capital theory criterion for evaluating capital investments is the net present value (NPV) model used for evaluating the capital outlays required to implement the System. This model, along with related decision rules, theoretically ensures selection of investments that will contribute the largest gain to a firm's value. An NPV model evaluates an investment opportunity by considering aspects of the value of time with respect to expected costs and returns over its expected life. Assuming that feedlot cattle producers are profit maximizers, NPV analysis was used to determine the potential cost efficiency of the System. If the present value of all future cash outflows, resulting in a value greater than zero, the investment would be acceptable.

Returns were derived from:

1. Feeding the pelleted residue into the diets of feedlot cattle, thus reducing the cost of feed by the cost of the percentage of the alfalfa hay displaced by the pelleted residue (recognized as a reduction in cash outlays).

2. Selling excess pelleted residue to other feedlots.
3. Selling all excess electricity generated by the System to the local utility company Even though System-produced electricity should be less expensive than purchasing from the local utility, for the purpose of simplicity in the analysis it was not used in the feedlot. Thus, it was assumed that for its own use the feedlot continues buying electricity from the local utility company. Therefore, the electricity used by the feedlot is not considered in this study because it would be used whether or not the System were established. Manure transportation costs were assumed not to differ between disposal and inclusion into the System. Expenses were incurred in the capital outlay for implementing and continual operation of the System. The difference between the cash inflows are summarized in the following annual net cash flow (NCF) model:

$$NCF = \sum_{i=1}^{3} P_i Y_i - \left[\sum_{i=4}^{5} P_i X_i - \sum_{i=6}^{8} P_i\right]$$

where P_i = price of inputs and outputs

Y_i = quantity of output

X_i = quantity of input

subscript i ranging from 1 to 8 represents the following:

i = 1 = receipts from selling electricity

i = 2 = reduction in cash outflow by feeding 4% pelleted residue

i = 3 = receipts from selling surplus pelleted residue

i = 4 = increased labor

i = 5 = increased management

i = 6 = annual maintenance costs

i = 7 = annual loan payments

i = 8 = annual taxes paid on receipts

Capital Outlays

Capital outlays required to implement the System in the study feedlot were for the purchase of digestion equipment, electrical generation, and pelleted residue facilities. It was assumed that state-of-the-art equipment is used. Several federal agencies were consulted regarding a listing of the manufacturers of feedlot biomass fuel and waste meal systems. Few companies are involved in this activity on a commercial basis, resulting in a listing of companies involved in such enterprises not being available. A company which commercially produces the System provided cost estimates [R. W. Meyer, personal communication]. Capital outlays for the System consist of:

1. Twelve digesters with capacity to hold the manure of a 40,000-head feedlot, estimated at $150,000 each

2. System buildings, estimated at $200,000

3. Solid recovery equipment designed to separate and centrifuge the remaining solids, estimated at $500,000

4. Two 300-kW electrical generators and equipment estimated at $575,000 (the electrical generating facility utilize all biofuel not consumed within the power requirements of the System)

5. Annual costs of maintenance for biogas and electrical components of the System were estimated at $34,000 and $24,000, respectively, for the first year, which account for labor as well as other expenses required to maintain the System

6. Additional annual labor and management required to run the System, estimated at $6,240 and $3,000, respectively, for the first year.

Estimated System costs, generally consistent with those previously established in the U.S., are listed in Table 1. All annual variable expenses were arbitrarily assumed to increase by 5% per year due to inflation. Depreciation was computed under the 1986 U.S. Tax Reform Act using the 5-year 200% declining balance class of capital.

A reduction in cash outflow was realized by substitution of pelleted residue for 25% of the alfalfa used in a feed ration consisting of 16% alfalfa. The pelleted residue was valued at the cost of alfalfa when substituted. The market value was estimated at $92.4 t^{-1}.

Total value of consumed pelleted residue (VCPR) per steer per feeding cycle was developed through the following model:

$$VCPR = (TG) \times (FC) \times (PR) \times (PA)$$

where TG = expected total gain (estimated to be 226.799 kg)

FC = Feed conversion ratio between feed consumption and gain (estimated to be 3.515 kg of feed to 0.454 kg of gain)

PR = % of ration substituted by pelleted residue (estimated to be 4%)

PA = price of alfalfa per 0.454 (estimated at 92.4t^{-1}$)

Therefore, the total value of the pelleted residue per steer fed within the feedlot in a feeding cycle was $6.52, or total reduced feeding cost for 40,000 head per feeding cycle was $260,800. Assuming this 40,000-head feedlot operates 365 days per year and each feeding cycle consists of 161 days, there are 2.26 feeding cycles per year. Thus, the total annual reduced feeding costs for this feedlot operation was $589,408 for the first year. These values are recognized as a negative change in cash outflows for the purchase of alfalfa.

Revenues

Revenues are received from the sale of surplus pellets and selling electricity to the local public utility company. The output of biogas was estimated to be 1.012 m^3 per steer per day [R.W. Meyer, personal communication]. It was estimated that a 40,000-head feedlot producing this amount of biogas per steer per day requires two 300 kW generators to utilized the excess biogas not consumed within the System. Assuming both generators together produce 600 kWh^{-1}, 21 h per day (allowing 3 h per day for maintenance and repairs), 365 days per year, the estimated electricity revenues at \$0,0274 per kWh were \$126,000 annually [T. Buchanan, personal communication].

Table 16.1
Capital Outlays for Feedlot Biomass Fuel and Waste Meal System

Equipment	Cost($)
Digesters (12)	1,800,000
Buildings	200,000
Solids recovery equipment	500,000
Electrical generation equipment	575,000
Total capital outlay	3,075,000

* *Source:* R. W. Meyer, Energy Cycle of Leawood, Kansas, USA, personal communication, 1987.

R. W. Meyer [personal communication] estimated that 2.2722 kg of pelleted residue are produced per steer on a daily basis. Thus, it was conservatively assumed that 1.814 kg of pellets are produced per steer on a daily basis. Value of the pellets depends largely on consistency and nutritional value. It was assumed that surplus pellets could be marketed at a value less than alfalfa. Market information sensitive to the price of alfalfa hay for the Texas High Plains region indicated that a conservative estimate of the price a producer can expect to receive for the pelleted residue was \$33 t^{-1}. Therefore, given that 1.814 kg of pellets are produced and approximately 0.454 kg consumed daily (daily consumption of pelleted residue = (TG/161 days) x FC x Pr = (226.799 kg/161 days) x 7.75 x 0.04 = 0.437 kg), then approximately 1.361 kg of pellets per head are produced on a daily basis as surplus to be sold to other producers. Total revenue received from sales of surplus pellets per steer in a feeding cycle was \$7.25. Therefore, total revenue for this 40,000-head operation was \$289,800 per feeding cycle, or \$654,948 annually. These values are recognized as positive cash inflow. Revenue estimates were increased at a rate of 5% per year to reflect the effect of inflation.

Expenses

Operating expenses for the System include labor, management, and maintenance for the facility. Annual costs of maintenance for biogas and electrical components of the System were estimated at \$34,000 and \$24,000, respectively, for the first year. These estimates were much higher than those quoted by R. W. Meyer of Energy Cycle, but to remain conservative the values used represented cost proportions similar to those used by Bjornstad et al. These costs account for labor as well as other expenses required to maintain the System.

The operator is required to purchase additional labour inputs when implementing the system. The task of operation can be handled by existing feedlot personnel. For this study 20 additional hours per week were allotted for completing daily duties required to keep the System in operation. This labor was valued at \$6 h^{-1} for the Texas South Plains region and was sensitive to labor rates in a given region. Total annual labour expense was \$6,240 (i.e., 52 weeks × 20 h × \$6) for the first year.

A 5% increase to the cost of management was arbitrarily assumed with the implementation of the System. Therefore, assuming management for the study feedlot costs \$60,000 annually, \$3,000 was the expense of the annual management fee of the System for the first year. Expense estimates were increased at a rate of 5% per year to reflect the effect of inflation.

Loan, Depreciation, and Tax Structure

Funds required to finance a project of this nature vary according to the debt structure of the investing firm. It was assumed in this study that a debt structure of 60% borrowed and 40% equity funds were used. Debt (\$3,075,000 total investment x 0.6 = \$1,845,000) was amortized under Uniform Series Present Value tables over a 10-year period with 10 annual payments. To reflect the impact of debt cost on the profitability of the project, several interest rates were used in the analysis. Four worksheets were constructed varying only the interest rates (9, 12, 15 and 18%).

The valuation of equity capital affects the financial feasibility of the investment. Financial management textbooks suggest that the risk averse investors place higher values (discount rates) on equity capital. To illustrate the effect of different risk positions on the investment, worksheets were computed with discount rates of 0, 5, 10, and 15 percentage points above the interest rate. The discount is used to convert future values into present values accounting for inflation and opportunity costs.

Depreciation of the System was covered under the Accelerated Cost Recovery System Code and is eligible for the 5-year, 200% declining balance classification. Although the depreciation schedule does not affect the cash flow for the project, it does affect the tax structure. Taxes are computed on a marginal basis for the study because the investment is added to an existing operation. The marginal tax rate for a capital investment depends on the structure of the firm. A marginal tax rate of 34% is used in this study because the structure of the investing entity was assumed to be corporate.

Analysis

The NPV of the System's cash flows over its expected life are discounted and summed with the initial investment of the project to yield a present value of all inputs and outputs invested. The NPV model used in this study is:

$$NPV = \sum_{n=1}^{N}\{[P_n/(1 + d)^n] + [V/(1 + d)^N]\} - INV$$

where NPV = net present value

N = useful life of investment

n = time, in years

P = annual net cash flow (NCF)

V = terminal or salvage value

d= Discount rate

INV = initial investment

The useful life of investment in the System was assumed to be 20 years. The annual NCF was derived from Equation 1. The salvage value was set at zero because cost for dismantling was assumed to be equal to the value of scrap. The discount rate was used to translated future expenses and revenues into present values. The value of total initial investment was taken from Table 1. Defining the NPV requires that the value of the initial investment is recognized as an initial cash outflow consisting of borrowed funds and equity. The initial investment is added to the sum of the discounted annual cash flows over the useful life of the investment and the terminal value.

Interpretation of the NPV is based on the value obtained with the previous formula. If the value is greater than zero the investment is acceptable. If the value is less than zero the project is not acceptable. In the rare case of deriving a NPV of zero the investor must use personal preference in choosing whether to invest.

Results

The analysis was completed with the use of microcomputer worksheets. Using the date obtained from documented and current market information on the Texas South Plains, revenues and expenses were computed. Operating receipts and reduced cash outflows totaled $1,370,356 while total cash outflow (excluding capital outlay) was $323,131 for the first year. The NPV was found to be relatively responsive to interest and discount rates used. Intuitively, the higher the interest or discount rate, the lower the NPV. Analysis indicated the investment could result in an NPV of $7,517,325 if interest and discount rates are 9% for the time horizon. Table 2 presents an example of the worksheet consisting of some data used to derive the results of analysis while interest and discount rates are 9%.

Table 16.2

Example of Annual Present Values of Net Cash Flows for Feedlot Biomass Fuel and Waste Meal System: $1,230,000 Equity, $1,845,000 Debt, 34% Marginal Tax Rate, 9% Interest and Discount Rates

	Year			
	1 ($)	2 ($)	3 ($)	20 ($)
Operating receipts and reduced cash outflows				
Electricity receipts	126,000.00	132,300.00	138,915.00	318,395.72
Reduced feed expense	589,408.00	618,878.40	649,822.32	1,489,404.66
Surplus pelleted residue	654,948.00	687,695.40	722,080.17	1,655,020.98
Total receipts	1,370,356.00	1,438,873.80	1,510,817.49	3,462,821.36
Operating expenses				
Added labor	6,240.00	6,552.00	6,879.60	15,768.17
Added management	3,000.00	3,150.00	3,307.50	7,580.85
Maintenance	58,000.00	60,900.00	63,945.00	146,563.11
Total expenses	67,240.00	70,602.00	74,132.10	169,912.13

(Contd..)

	Year			
	1 ($)	2 ($)	3 ($)	20 ($)
Depreciation	1,230,000.00	738,000.00	516,600.00	
Loan interest payments	166,050.00	155,120.57	143,207.50	
Loan principal payments	121,438.07	132,367.49	144,280.57	
Total loan payments	287,488.07	287,488.07	287,488.07	
Principal remaining	1,723,561.93	1,591,194.44	1,446,913.88	
Taxable income	(92,934.00)	475,151.23	776,877.89	3,292.909.23
Marginal income tax	(31,597.56)	161,551.42	264,138.48	1,119,589.14
Total cash outflow	323,130.51	519,641.48	625,758.65	1,289,501.27
Net cash flow	1,047,225.49	919,232.32	885.058.84	2,173,320.09
Present value of NCF	960,757.33	773,699.45	683,427.82	387,787.44

Interest and discount rates are not expected to remain at 9%. One approach used to consider risk and uncertainty in conjunction with the NPV analysis method is to vary interest and discount rates. Therefore, scenario 1 of Table 3 presents the NPV analysis method is to vary interest and discount rates. Therefore, scenario 1 of Table 3 presents the NPVs resulting from analysis when interest rates are 9, 12, 15 and 18%, while discount rates are set at 0, 5, 10 and 15 percentage points above the interest rates.

An alternative criterion for investment decision making is the internal rate of return (IRR). The IRR is that discount rate that just equates discounted revenues to discounted expenses. Therefore, the IRR is that discount rate causing the NPV to equal zero. Analysis when interest rates on debt are 9, 12, 15 and 18%, indicated that IRRs for the investment were 31.7, 31.0, 30.3 and 29.6, respectively (scenario 1, Table 3). Thus, it is estimated from the study that interest rates ranging from 9 to 18% result in risk-adjusted discount rates of approximately 30% before the investment would provide no profit. Internal rates of return were also estimated for scenarios 2-6 of Table 3.

Although the authors chose to remain conservative, questions regarding the validity of the derived annual receipts and/or capital outlays and operating expenses may arise. Therefore, taking the previously mentioned approach of considering risk and uncertainty a step further, scenarios 2-5 of Table 3 mirror scenario 1 by presenting the NPVs resulting from analysis while decreasing receipts by 10 and 25% and increasing capital outlays and operating expenses by 10 and 25%, respectively.

The assumption that the study feedlot had existing concrete floors may limit the applicability of this analysis for feedlots in the Southern Great Plains. This assumption was made so that problems of soil buildup in the digesters could be omitted from the analysis. Assume that the materials and labor for a 4-inch thick concrete pad placed over wire mesh would cost the study feedlot an average price of $50 per head-capacity; $2 million in total. At $1.45 per 923.03 cm^2 this provides for 32,051.61 cm^2 per head. Scenario 6 of Table 3 lists the results of combining this added expense of investing in the System with the costs of Table 1.

Table 16.3
NPV ($000) of Investment, Varying Interest and Discount Rates*

Interest (%)	Discount Rate (%)				
	Interest	i+5	i+10	i+15	IRR
Scenario 1					
9	7,517	4,138	2,226	1,060	31.7
12	5,114	2,762	1,372	491	31.0
15	3,421	1,744	712	34	30.3
18	2,193	970	189	(341)	29.6
Scenario 2: Receipts are Decreased by 10%					
9	6,327	3,327	1,633	601	28.4
12	4,178	2,095	867	89	27.7
15	2,663	1,182	273	(323)	27.0
18	1,565	489	(197)	(661)	26.3
Scenario 3: Receipts are Decreased by 25%					
9	4,541	2,111	743	(87)	23.4
12	2,773	1,094	108	(514)	22.7
15	1,527	341	(384)	(858)	22.1
18	623	(233)	(775)	(1,142)	(21.4)
Scenario 4: Capital Outlays and Operating Expenses are Increased by 10%					
9	6,992	3,662	1,783	640	28.2
12	4,607	2,296	935	75	27.5
15	2,928	1,285	279	(381)	26.9
18	1,711	518	(214)	(754)	26.2
Scenario 5: Capital Outlays and Operating Expenses are Increased by 25%					
9	6,203	2,948	1,118	11	24.1
12	3,846	1,597	279	(551)	23.5
15	2,188	598	(371)	(1,003)	22.8
18	987	(160)	(885)	(1,374)	22.1
Scenario 6: Investment in Concrete Pads					
9	4,479	1,300	(446)	(1,524)	17.4
12	2,115	(56)	(1,311)	(2,091)	16.8
15	454	(1,057)	(1,965)	(2,549)	16.2
18	(746)	(1,817)	(2,483)	(2,925)	15.6

** Discount rates are equal to 0, 5, 10, and 15% points above the interest rates.*

Summary

With increased concern about potential nitrate contamination of aquifers throughout the

United States, feedlot producers may be interested in establishing a biomass fuel and waste meal system. This system returns to the producer pelleted waste meal, methane for the generation of electricity, and control of pollution resulting from the introduction of animal excreta to the nation's waterways and reservoirs. This study presents a financial and economic analysis of investment in a proposed feedlot biomass fuel and waste meal system. The high internal rate of return and net present values indicate apparent feasibility of the investment. Within the analysis it was implicitly assumed that the feedlot biomass fuel and waste meal system would function on a continuous basis without interruptions from major breakdowns. Results are dependent on this implicit assumption as well as the investor locating markets for excess pelleted residue and generated electricity. The accuracy of this analysis is dependent on the reliability of the assumptions and data used.

The robustness of the System's profitability is established by the consistency of results under the varying assumptions of Table 3. Interest and discount rates as high at 18% for scenarios 1-5, and as high as 15% for scenario 6 result in a profit being realized. Also, the IRRs remain consistently above a risk-adjusted discount rate of 21% for scenarios 1-5, and 15% for scenario 6 before the investment would provide no profit. Scenario 6, representing investment in concrete flooring as well as the System, is the most risky situation considered. This added expense may lead some cattle feedlot operators to terminate any further considerations of investing in the system.

The study was structured to be conservative on returns, and realistic in making assumptions where possible, but facility, feed, and energy costs may change. In addition, continuing research may increase efficiency of biofuel facilities as well as livestock feed ratios. Depending on current feedlot structural conditions and risk preference of the decision maker, a biomass fuel and waste meal system has been shown to be economically viable. Present value analysis indicated that under current economic conditions profits can be expected from an investment of this nature unless inflation and opportunity cost factors cause the discount rate to reach approximately 30%.

17

Marine Pollution—I

Introduction

The oceans cover nearly three-quarters of the surface of the planet. Their influence on the climatic and evolutionary development of the planet has been profound and, of course, they continue to be a major driving force for weather patterns. In addition, they are of great value as exploitable resources for food and minerals and as a means of transport. Over the past century the amount of wastes produced by human activities and subsequently entering the oceans has grown to the extent that significant changes from natural conditions can be seen. At first these were merely local aberrations but, increasingly, changes on the global scale are becoming observable. The extent to which the health of the oceans is being harmed by marine pollution is the subject of this article. The most important marine pollutants are given in Table 1.

What is pollution? The popular conception of pollution is "the wrong amount of the wrong thing in the wrong place at the wrong time." Fundamentally this definition is correct, but it is not a helpful one when attempts are made to control pollution because it is too vague for legislative purposes.

The most widely accepted definition of marine pollution is the one devised by the United Nations joint Group of Experts on the Scientific Aspects of Marine Pollution (GESAMP).[1] This states that marine pollution is the

> Introduction by man, directly or indirectly, of substances or energy into the marine environment (including estuaries) resulting in such deleterious effects as harm to living resources, hazards to human health, hindrance to marine activities including fishing, impairment of quality for use of sea water and reduction of amenities.

This definition is now widely incorporated into both natural and international marine pollution control legislation.

Implicit in this definition is the concept that there is a distinction between contamination, in which concentrations of a substance are elevated above natural levels without measurable effect, and pollution, in which concentrations are elevated to the point that harmful effects may be observed. This distinction is useful because it provides a criterion by which the seriousness of a problem may be reasonably assessed and by which a legal control may be implemented.

To illustrate this, consider the case of someone on a ship who throws an oily rag from the engine room into the ocean. The effects of this action on the marine environment are obviously trivial. If, on the other hand, we take the case of an oil tanker whose captain has released the washings from the cargo tanks close to a valuable shell fishery, the environmental consequences are likely to be much more severe. According to the above definition, the first case would be contamination and the second pollution. Neither action is desirable, but there are clearly major differences in scale in the ocean's ability to withstand the two actions.

The key question is, therefore, at what stage is the ability of the oceans to absorb waste without harm exceeded? The answer is clearly related to such factors as the toxicity of the waste to marine organisms, the quantity discharged, and the nature of the receiving environment. In practical terms the procedures for determining which, if any, wastes may be discharged to the oceans are determined by individual countries. Some have taken the view that no wastes at all should be discharged and wastes must be dealt with on land regardless of cost or environmental effects. Others derive some estimate of environmental capacity and license discharges to remain substantially below it. Marine disposal then becomes one option in a general waste management strategy.

Table 17.1
Marine Pollutants of Global Concern

Pollutant	Comment
Petroleum hydrocarbons	Crude oil and some of its refined products
Sewage	The effects of sewage pollution are generally local ones but sewage may contain some of the other chemicals listed in this table.
Halogenated hydrocarbons	Including compounds such as DDT and the PCB$_s$
Metals	Particularly mercury, cadmium, and lead
Radionuclides	Particularly cesium-137, strontium-90, and plutonium isotopes
Litter	Particularly persistent plastics and including lost fishing nets.

In both cases, two unavoidable features have to be recognized. The first is that, however much we try to minimize it, waste is inevitably created as a side product of human activities. There will always come a point at which the effort (in resource or energy terms) required to clean up a waste creates a greater problem than the waste itself. The second point is that waste has to be dealt with and that it is incumbent on every member of society to ensure that all wastes are disposed of in ways that cause the minimum environmental impact. It is therefore vital that rational arguments and not unreasoning prejudice should prevail in the formulation of waste management strategies.

II. The Major Marine Pollutants

A. Oil

Oil is not a single substance but a complex mixture of a great many individual chemicals. Crude oil contains thousands of compounds, and refined oils such as kerosene or gasoline consist of rather simpler mixtures. With such a complex mixture, many of the environmentally significant properties such as toxicity, solubility, and viscosity vary considerably from oil to oil, making the task of predicting effects very difficult. For the purposes of this article the discussions are limited to crude oil, which represents by far the greatest fraction of the oil likely to enter the oceans.

The total global production of crude oil is about 3 billion tonnes (metric tons) per year, and approximately half of this is transported by sea. This means that on any given day there are approximately 8.25 million tonnes afloat around the oceans. The major shipping routes for oil lie between the producers in the Arabian Gulf, the Gulf of Mexico, the North Sea, etc. and the major users in the United States, Western Europe and Japan (fig. 1). For the most part, the shipping lanes lie well out to sea in deep water, where the risks of accidents are small. However, on leaving and approaching ports ships are obviously close to land, and experience has shown that in these coastal zones the risks of serious incidents are greatest.

The most recent estimates indicate that the total amount of oil entering the oceans directly as a result of human activities is about 3 million tonnes per year, to which must be added another 250,000 tonnes deriving from natural seeps where oil is leaking from submarine oil-bearing strata. Table II gives a breakdown of the various sources of oil. From this it can be seen that, despite the widespread publicity they receive, tanker accidents are not the largest source of oil. From this it can be seen that, despite the widespread publicity they receive, tanker accidents are not the largest source of oil. Although they are probably the most important in terms of the environmental damage they cause.

Once spilled, crude oil spreads very rapidly, with the lighter fractions moving most quickly. From the moment of release the composition of the oil begins to change as the lighter, more volatile components tend to evaporate or dissolve, leading to a steady increase in the proportion of the heavier, tarry residues. The action of the winds and waves tends to mix the oil with water, leading (in much the same way that mayonnaise is made to the formation of water-in-oil emulsions that are often referred to as "chocolate mousse." These emulsions are very stable and may persist for long periods. The relatively toxic components that dissolve may have adverse effects on animals in the underlying water. These effects are normally fairly minor in extent because mobile creatures such as fish can avoid contaminated waters, although less mobile phytoplankton and zooplankton may be affected over limited areas.

Oil has entered the oceans from natural seeps over geological time, so is not a totally alien substance as some synthetic pollutants are. This means that a considerable variety of bacteria have evolved which are capable of feeding on at least some of the components of the oil, so that eventually all but the most resistant fractions (such as tars and asphaltenes) are degraded.

Degradation rates are highly temperature dependent, being much greater under warm conditions. It follows that the consequences of a major spill in cold waters (such as the Exxon Valdez in Alaska in 1989) are likely to be much more long-lived than had the spill been in, for example, the Gulf of Mexico. With the general trend toward oil exploration more and more remote regions, including polar seas, the risks of long-lasting damage from a large oil spill are increasing as time goes on.

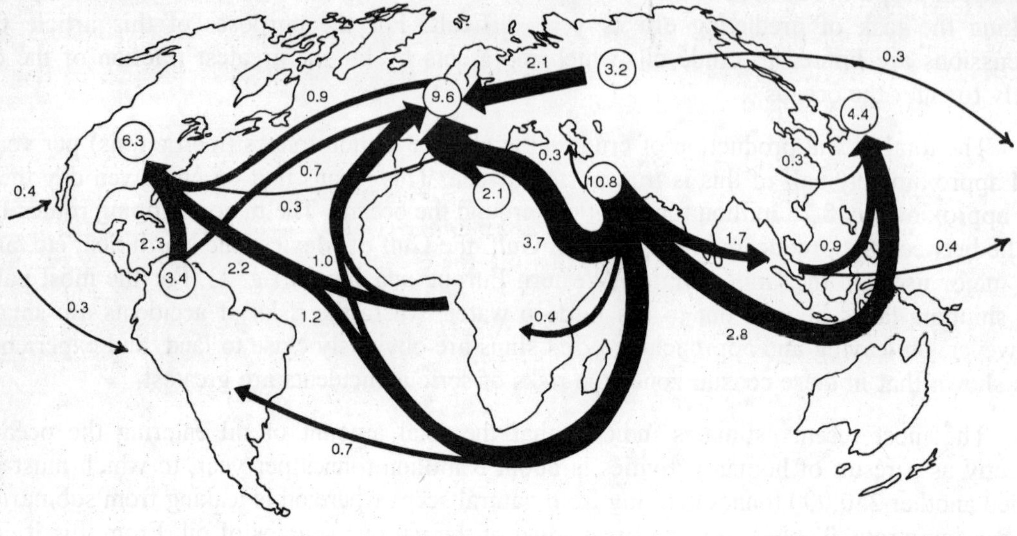

Fig. 17.1. Crude oil and products: seaborne trade, 1987. Units are in million barrels daily (1 barrel = 159 liters). [Reproduced with permission from Shell Briefing Service (1989). Shell Briefing Service Publication No. 1. Shell International Petroleum Company Ltd. (PAC/233), London.]

The most serious effects of oil pollution are also the most obvious. Gross fouling of beaches, plants, and animals can have devastating effects. Seabirds are particularly vulnerable because oil coats their feathers, causing loss of both thermal insulation and buoyancy. Oil that has been swallowed by birds trying to clean themselves is corrosive to stomach and gut walls, and few seriously affected birds can survive. Sea mammals are also vulnerable to spilled oil, which irritates sensitive membranes although it is not often directly fatal. An incident which resulted in fouling of breeding beaches could however, have very serious effects on breeding success. Fish are normally capable of avoiding contamination, but "tainting" of fish by small quantities of oil makes them unpalatable and therefore unsalable. As in the case of mammals, fouling of fish spawring areas or blockage of migration routes (for example, by a spill in the mouth of a salmon river) could have serious consequences for breeding patterns.

Table 17.2
Annual Inputs of Oil to the Marine Environment

Source	Annual input (million tonnes)
Offshore production	0.05
Routine ship operations	1.05
Ship accidents	0.42
Atmosphere	0.3
Land-based sources	1.23
Ocean dumping	0.02
Natural sources	0.25
Total	3.32

Source : Figures derived from National Research Council. (1985). "Oil in the Sea-Inputs, Fates and Effects." National Academy Press, Washington, D.C.

Shorelines affected by oil are prone to damage not only by the oil itself but also by inappropriate attempts to clean them. For example, the attempts to use the fire and then chemicals to treat the Torrey Canyon accident off the southwest peninsula of the United Kingdom in 1967 did considerably more damage than if the oil had been left alone. Guidelines to the best methods of cleaning different types of coastal environment are now available. In some cases these suggest that it is best to leave the situation alone and let it recover naturally. In others, mechanical or chemical treatments may be suitable. Chemical treatments have been considerably improved over the types used in the Torrey Canyon incident and can be usefully employed in appropriate circumstances and where legislation permits.

After extensive fouling has occurred, different components of the ecosystem recover at different rates. For example, seaweeds recover faster than the animals that feed on them, so a frequent consequence of a coastal oil spill is that the shoreline becomes covered in seaweeds. Only much (possibly several years) later do the grazing animals become reestablished so that something approaching a normal species balance returns.

Spilled oil is best treated at the source. It is much better to prevent the oil leaving the damaged vessel than to try to clean it up later. This was clearly demonstrated in the case of the Exxon Valdez, where quicker action on the part of those responsible could have greatly reduced the extent of the damage. Once oil has escaped, the primary line of defense is often a physical barrier such as a floating boom, which can be placed to encircle the oil and prevent further spreading. Such systems are most successful when weather conditions are calm and wave action cannot force the oil over or under the barrier. Mechanical systems designed to skim the oil from the waster surface are also very useful in calm conditions but are of little use when any kind of sea is running. Modern chemical treatments are designed to disperse the oil in the form of very fine droplets. This helps to increase the degradation rate and greatly reduce the coating properties of the oil. These treatments can be applied from ships or aircraft but are expensive and are usually effective only if applied when the oil is fresh. Once the oil has weathered for a few days, such treatments are largely ineffective.

1. Polynuclear Aromatic Hydrocarbons

Oil and its combustion products contain a variety of polynuclear aromatic hydrocarbons (PAHs), which are of particular concern because some of them are biologically active and are known to be carcinogens. In the 1970s there was concern that PAHs might bioaccumulate in marine systems in the same manner as chlorinated hydrocarbons. This fear has proved to be unfounded but PAHs are still considered to be of some environmental significance, although only 2-3% of the intake of PAHs by humans is normally likely to derive from the consumption of seafood. Most of the rest comes from inhalation of exhaust fumes and consumption of other foods (particularly smoked or barbequed foods). However, a relatively high incidence of human stomach cancers has been linked to diets dominated by smoked food, of which a significant proportion may be smoked fish.

Within the marine environment, there is some evidence that PAHs may have adverse, but sublethal, effects at sites where there are chronic, low-level contamination problems. Such effects may include changes in enzyme activity, reproductive failure, and reduced growth potential. The present state of knowledge is insufficient to make confident general statements about the extent of any problems, although regions affected by, for example, oil refinery effluents will clearly receive the greatest levels of exposure.

B. Sewage

The problem of how to dispose of human sewage has existed, either recognized or unrecognized, since stable communities began to be established around the world. The long-standing solution to the problem has been to place the sewage as close to the source as possible but where it can do no obvious harm. Sometimes this disposal has been on land, but frequently discharges have been made to rivers, to estuaries, or directly to the seas. Over the years the perception of what is an acceptable practice has altered, together with our understanding of what constitutes an environmentally safe option. To comprehend the major problems associated with sewage, it is necessary to understand what it is.

It used to be said that a particular brand of soap was the best for you because it was more than 99% pure. By the same logic, raw sewage should be harmless because it is >99% pure water. Unfortunately, it is the remaining fraction of a percent that causes all the problems. Dissolved in the water are a wide variety of inorganic (e.g., nutrients and trace metals) and organic (biodegradable and persistent) materials. Together with these there is a suspended solids loading consisting of human excrement, inorganic grit, and other organic or inorganic materials (rages, contraceptives, pieces of plastic, etc.). Finally, there are large quantities of pathogenic organisms (10^6-10^7 cells/ml). In some systems domestic wastes may be combined with industrial wastes, which creates additional problems.

The most important primary feature of raw sewage from a disposal point of view is its oxygen demand. The oxygen demand is the amount of oxygen that will be consumed by bacteria as they feed on and degrade the waste. It the amount of oxygen required is similar to or greater than the amount of oxygen required is similar to or greater than the amount of oxygen available, serious problems may arise. For example, if raw sewage (or any other waste with a high oxygen demand) is discharged to a river, the oxygen demand may be greater than the amount of dissolved oxygen available in the water. Under these circumstances all the oxygen may beconsumed, creating anoxic conditions which are extremely toxic to aquatic animals. Many rivers and estuaries around the world suffer from the problems of excessive oxygen-demanding wastes. These are frequently sewage wastes but may also be from manufacturing industry (e.g., silage liquor or slurry). In the case of occasional pollution incidents, the presence of such wastes is frequently marked by extensive fish kills. If the case of occasional pollution incidents, the presence of such wastes is frequently marked by extensive fish kills. If the problem is a chronic one (such as a piped outfall), the anoxic conditions that are created may persist and virtually no living creatures will survive in the water.

The various forms of sewage treatment that are available are all designed to reduce the oxygen demand of the effluent. In doing so they produce more highly mineralized wastewater and sewage sludge (the accumulated sewage solids). The mineral elements present in the wastewater are not necessarily benign. A proportion of them are nutrients such as nitrates and phosphates, which, if discharged in an inappropriate way, can stimulate excessive algal growth with a consequent high oxygen demand. This process, termed eutrophication, has recently created problems in a variety of regions. For instance, algal blooms in some Scandinavian and Mediterranean coastal waters have created large quantities of rotting organic matter on beaches, which has had serious consequences for some tourist regions. Eutrophication can also be caused by input of nutrients from agricultural fertilizers, and this was certainly a significant contribution to the eutrophication problems encountered in Scandinavia.

The problem is not simply that adding nutrient chemicals to the system stimulates growth. If the concentration ratios of the nutrients in the effluent are markedly different from those normally encountered in nature, significant changes in the species balance in the receiving environment may also be expected.

Sewage sludge must also be disposed of. Where the sludge is relatively uncontaminated by industrial wastes and where there is plenty of land, the sludge can be applied as a soil conditioner. Unfortunately, in many cases there is not enough land or the land is inaccessible because crops are growing on it, so other disposal options must be exploited. These may include landfill, incineration, or disposal at sea. Marine disposal has been used for many years by some countries and normally involves transporting the sludge to a designated disposal site in special ships.

The effects of sludge disposal at sea are the subject of considerable debate. Therefore is no doubt that levels of contaminants such as trace metals or persistent organic contaminants are higher at dump sites. There is also no doubt that animals living or feeding in such areas exhibit increased body burdens of these contaminants. Whether these changes are significant is the main point of debate. Suggested links between sludge disposal operations and outbreaks of fist hisease have generally been hard to substantiate, as have links between disease in humans and swimming in sewage-contaminated seawater.

One clear link between sewage disposal and environmental problems is related to the consumption of contaminated seafood. Fish, and to a much greater extent shellfish, are capable of accumulating a variety of pathogenic organisms from sewage-contaminated water. If eaten by humans, these contaminated organisms can have ill effects ranging from an upset stomach to serious diseases (see Table III). It is not uncommon for shellfisheries to be closed because of bacterial contamination. Most marine organisms can be restored to a nondangerous condition by keeping them in clean seawater for a period of time prior to consumption.

C. Pesticides and Other Persistent Organic Chemicals

Historically, the chemicals that have provoked the greatest concern in terms of their effects on the marine environment are the chlorinated hydrocarbons. These chemicals include such well-known substances as the pesticide DDT and the polychlorinated biphenyls (PCBs) (used largely in electrical apparatus). Although many hundreds of different chemicals come under the heading of chlorinated hydrocarbons (PCBs alone may consists of up to 209 distinct chemicals), many of them share a number of important properties. In particular, they are generally fairly toxic, are persistent in the environment, and are bioaccumulated. Because of these characteristics, very low concentrations in seawater can lead to much higher concentrations in fish, marine mammals,, and fish-eating birds. An estimate of the mass balance of U.S.-derived PCBs in the environment is given in Table IV.

Table 17.3
Diseases Associated with Sewage-Contaminated Fish or Shellfish

Typhoid
Paratyphoid
Salmonellosis
Septicemia
Gastroenteritis
Diarrhea
Infectious hepatitis
Skin infections

Source: Data from GESAMP (1976). "Principles for Developing Coastal Water Quality Criteria," United Nations Environment Programme (UNEP) Reports and Studies of GESAMP No. 5 (also published as UNEP Regional Seas Reports and Studies No. 42). UNEP, Nairobi, Kenya.

Table 17.4
Inventory of U.S.-Derived PCBs in the Environment

	Inventory (tonnes)
Atmosphere	18
Fresh waters	12-35
Freshwater sediment	1,400-7,100
Freshwater biota	30
Marine waters	6,000-66,000
Marine sediment	660-2,700
Marine biota	30
Soil	140-2,800
Sewage sludge	4,800
Total U.S. production	610,000
Total cumulative world production[a]	750,000[b]

Source: Data from Committee on the Assessment of Polychlorinated Biphenyls in the Environment. (1979). "Polychlorinated Biphenyls," National Academy of Sciences, Washington, D.C.

a Data from O. Hutzinger, ed. (1982). "Environmental Chemistry." Springer-Verlag, Berlin.

b Of which 300,000 tonnes is now widely dispersed in the environment and 450,000 tonnes is still in service or is in landfill sites.

Declines in the populations of predatory birds were first observed in the 1960s, when there was a steep fall in the reproductive success of such species as peregrine falcon (Falco porogrinus), osprey (Pandion haliaetus), American bald eagle (Haliaetus leucocephalus), and brown pelican (Pelicanus occidentalis). It was shown that chlorinated pesticides, particularly DDT and its degradation products, strongly influenced these declines by disrupting the metabolism of calcium in birds, causing them to lay eggs with abnormally thin shells. These eggs are prone to breakage in the nest and the young birds die. Chlorinated pesticides are not the only causes of declining populations; reduction in appropriate habitats is also a major factor. Nevertheless, since widespread bans on the use of these chemicals were introduced in most industrialized countries, many affected bird populations have staged recoveries. The bans often do not extend to developing countries, where DDT in particular is still a major weapon against insect-carried diseases such as malaria.

Marine mammals such as seals and sea lions are also top predators within the food web and accumulate relatively high concentrations of some chlorinated hydrocarbons in their bodies. The presence of PCBs in Baltic seal populations has often been cited as the cause of their high level of infertility. Although PCBs are known to interfere with ovulation in mammals and may well contribute to the problems of the Baltic seals, they have not been shown to cause the uterine occlusions which are an important cause of infertility in these populations. Other factors, including pollution from other sources, may also play a part. Similarly, there is no clear link between chemical contamination and the outbreak of disease (caused by phocine distemper virus) among common seals in the North Sea in 1988. It now seems most probable that the disease was an instance of a natural epidemic, to which all populations are vulnerable from time to time.

Despite much popular concern, neither DDT nor the PCBs pose major threats to human health under normal circumstances. DDT is a particularly safe pesticide with virtually no known instances of adverse effects on human health. Modern pesticides are generally much more dangerous to the user and are fairly frequently cited as a cause of death PCBs have been implicated in one or two outbreaks of disease where foodstuffs were severely contaminated and people consumed over 0.5 g of PCBs over a period of 3 or 4 months. In these cases symptoms include acne and responatory difficulties but fatalities are very rare. There are no known cases of illness arising from consumption of marine products contaminated with chlorinated hydrocarbons.

Disposal of waste chlorinated hydrocarbons (principally PCBs from decommissioned electrical apparatus) is proving to be difficult. At present, the best technology available is incineration at high temperature. There are considerable economic and, in my view, environmental advantages in carrying this out at sea, where the waste acid produced is readily absorbed by seawater. Marine incineration is, however, being phased out as a result of international pressure. The alternatives are either to incinerate on land, which is technically more difficult, or to stockpile the chemicals until some other strategy is developed, which is potentially a fairly high-risk option because of the risks of leakage or fire. The debate concerning this problem will undoubtedly continue.

D. Metals

All natural metals occur in seawater in greater or lesser amounts. Some, such as iron, copper, cobalt, and zinc, are essential in small quantities for the healthy growth of marine organisms. Others, such as mercury, lead, and cadmium, have no known biological role. All metals are toxic if present in excess, but the most important marine contaminants are generally considered to be among the non-essential elements.

1. Mercury

Mercury is probably the most serious metallic pollutant of the seas. In oxygenated marine sediments, bacteria may convert the less toxic, inorganic form of the metal to the more toxic, organic form of methyl mercury. This chemical form is relatively mobile in the environment and tends to accumulate in fish. Consumption of marine (or other) food containing more than 0.3 mg each day over an extended period may lead to the onset of Minimata disease. This disease, first observed in Minimata Bay, Japan, where the local fish-eating community received large doses of mercury that originated from a local factory, has symptoms which include mental disturbance, ataxia, and impairment of gait, speech, and hearing. The original incident resulted in a considerable number of fatalities and may more people who were irreversibly affected.

Mercury is an important industrial chemical used, for example, in the chlor-alkali industry and in the manufacture of small batteries. Many coastal waters adjacent to industrialized areas have elevated concentrations of mercury, which must be monitored regularly to ensure that health criteria are not exceeded. Some fish such as tuna have naturally high levels of mercury, which may approach or even exceed acceptable limits for human consumption in large, older specimens without apparent adverse health effects in the fish.

2. Cadmium

Cadmium, like mercury, is an industrially important metal which is used directly in plastics and electroplating and is also found in applications associated with zinc and phosphorus. Although cadmium has no known biological role, it is taken up by marine phytoplankton, apparently by the same mechanism employed for the uptake of the essential nutrient element phosphorus. Cadmium has been shown to stimulate phytoplankton growth and photosynthesis up to surprisingly high concentrations of the metal (i.e., cadmium). Even in heavily contaminated marine system, cadmium is unlikely to have acute toxic effects on marine animals, but in at least two instances cadmium pollution has apparently been responsible for adverse effects on humans.

The best-known example of this was the outbreak of itai-itai disease in a village on the Jintsu river in Japan. Irrigation of paddy fields with water contaminated with cadmium from a zinc processing plant was originally identified as the cause of the problem, but other factors such as malnutrition and vitamin deficiency have also been cited as major contributors. Itai-itai disease is characterized by brittle bones and considerable pain (the name means "ouch-ouch"), and the symptoms can be alleviated by the administration of large doses of vitamin D.

The other will-known instance of cadmium poisoning of humans was caused by the consumption of contaminated Tasmanian oysters, which led to nausea and vomiting.

3. Lead

Lead has been widely used for millenn and lead poisoning is a well-known phenomenon. Although use of lead, particularly as an additive to automobile fuel, has led to measurably higher concentrations in remote ocean waters, there is no suggestion that harmful effects are occurring. There have been one or two incidents in which ships carrying alkyl lead compounds (the chemical form used as fuel additive) have lost some of their cargoes, but these have not resulted in any obvious harm. The only known instance of marine lead pollution occurred in 1979 in the River Mersey estuary in the United Kingdom, where alkyl lead released from a nearby manufacturing plant caused deaths among local seabird populations (see Table V).

For humans the amount of lead received from marine food is trivial compared with that from other sources such as drinking water.

4. Tin

Tin, in the form of organotin, has recently been recognized as a serious marine pollutant. The use of organotin compounds in antifouling paints on boats has resulted in comparatively high concentrations of these chemicals in some enclosed waters (harbors, marinas, etc.). Although very effective as antifouling agents, organotins also affect nontarget organisms, causing shell deformations in oysters and imposex in the common dogwhelk (Nucella lapillus).

Table 17.5
Birds Found Dead or Sick on the Mersey Estuary, 1979

Birds	Number
Waders	
Dunlin (*Calidris aplina*)	1336
Redshank (*Tringa totanus*)	116
Curlew (*Numenius arquata*)	49
Others waders	26
Total	1527
Gulls	
Black-headed gull (*Larus ridibundus*)	368
Herring gull (*Larus argentatus*)	29
Common gull (*Larus canus*)	16
Other gulls	20
Total	433
Wildfowl	
Teal (*Anas crecca*)	48
Pintail (*Anas acuta*)	37
Mallard (*Anas platyrhynchos*)	34
Shelduck (*Tadorna tadorna*)	16
Other wildfowl	5
Total	140
Other identified and unidentified birds	242
Total all birds	2342

Source: Data from K.R. Bull, W.J. Every, P. Freestone, J.R. Hall, P. Osborn, A.S. Cooke, and T. Stowe (1983). Alkyl lead pollution and bird mortalities on the Mersey Estuary, U.K., 1979-1981. Environ. Pollut., Ser. A 31, 239-259.

The use of organotins in cages used for farming fish has also led to concern. Many countries have now banned or severely restricted the use of organotin antifoulants, although the chemicals are still used in others applications such as wood preservation.

E. Radioactivity

The oceans contain appreciable amounts of radioactive materials derived from natural sources in the atmosphere and the earth. Added to this natural radioactivity is a mixture of artificial radioactive substances derived from atomic weapons testing, other military activities, and peaceful uses of nuclear energy.

Since the Test Ban Treaty of 1963, which proscribed atmospheric and underwater tests of atomic weapons by signatory nations, the amount of radioactivity reaching the oceans from this source has dramatically decreased. Sources which are now more important include waste discharges from nuclear fuel reprocessing plants (such as that at Sellafield, U.K.) and nuclear power generating reactors, low-level radioactive waste dumped at sea, and accidental discharges resulting from the unplanned return to earth of nuclear-powered satellites, the loss of nuclear-powered submarines (about five to date), and the explosion at the reactor site at Chernobyl (USSR).

The fate of radionuclides in the oceans depends to a considerable extent on their chemistry. Isotopes such as cesium-137 are soluble in seawater and are not strongly reactive with particles. They therefore move with the currents and are gradually diluted and dispersed (Fig. 2). Other isotopes, such as those of plutonium, are largely insoluble in seawater and tend to adsorb onto sediment particles, which may eventually become buried.

The amount of radioactive material added as a result of human activities is small compared to the relatively high natural levels of radioactivity in the oceans. It is therefore not thought to be a threat to the health of marine organisms or the marine organisms or the marine environment in general. The principal concern with respect to the marine environment is the significance of any routes by which artificial radioactivity discharged to the oceans can find its way back to land and result in excessive exposure to individual humans (or populations). Potential routes include sea spray blown inland; the consumption of certain seaweeds, fish, or shellfish; and exposure to contaminated intertidal muds. Nevertheless, for nearly all people, the marine environment represents a very small (<< 1%) source of the radiation to which they are exposed, most of which comes from natural sources and the use of radioisotopes in medicine (about 80-85% and 15-20%, respectively, in the United Kingdom).

Dumping of low-level radioactive waste at sea has been stopped at the behest of the members of the London Dumping Convention. This decision was taken even though a scientific study indicated that no environmental problems would arise if the practice continued.

III. Other Marine Pollutants

Important pollutants which do not fit in the foregoing categories include solid wastes (industrial wastes and dredge spoil), garbage (particularly plastics and fishing nets, and waste heat.

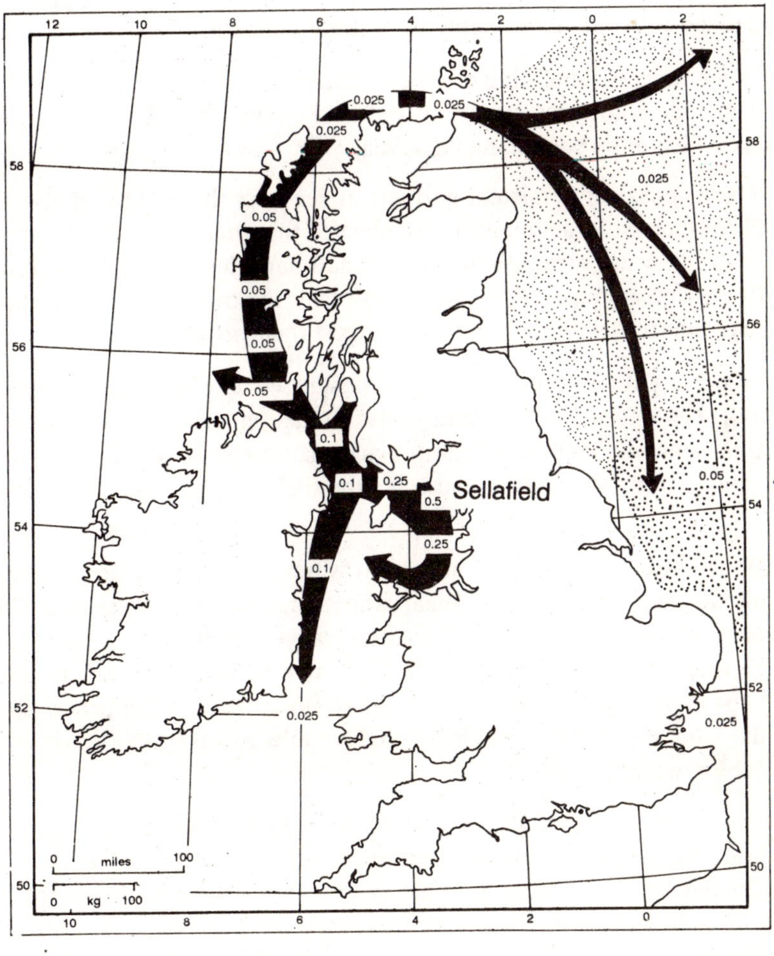

Fig. 17.2. Movements of the radioactive isotope cesium-137 from the Sellafield nuclear fuel reprocessing plant, 1987. Units are becquerels per kilogram (1 Bq is one radioactive disintegration per second and the natural radioactivity of sea water is about 14 Bq/kg). [Data from U.K. Ministry of Agriculture, Fisheries and Food. (1988). Aquatic Environment Monitoring Report No. 19].

1. Solid wastes from industry include "red mud" from the extraction of alumina from bauxite and "acid iron waste" from the production of titanium dioxide. Although iron is not generally considered to be a marine toxin when discharged to seawater, in large quantities it forms a fine precipitate which readily drifts around and may adhere to living organisms, clogging feeding or respiratory systems. The effects are normally localized, but in view of the large quantities discharged they are not insignificant.

Dredge spoil in enormous quantities is moved around the ocean floor as a result of efforts to keep harbors and navigable channels open. In one sense this does not involve the introduction of anything new to the marine environment, but it can be an important mechanism by which contaminants are moved from nearshore to coastal waters, and it is this aspect of dredge spoil disposal which is probably of greatest concern.

Other solid wastes which have been reported to have deleterious effects on the marine environment include china clay waste, fly ash and clinker from coal-fired power stations, and old munitions which have been dumped at sea.

2. The disposal of persistent garbage at sea is prohibited by international convention. Nevertheless, large quantities of plastics find their way into the marine environment, where they are generally found as small particles, some of which may be consumed by marine animals. Many marine birds are known to ingest plastic particles, which may interfere with their ability to digest their proper food.

Monofilament gill nets are increasingly recognized as a hazard to marine life, particularly in the Pacific. The enormous size and indiscrimination of these nets mean that they catch many nontarget animals, including seabirds and mammals. If, as fairly frequently occurs, the nets break away from the fishing vessels, they drift about and may cause considerable damage.

3. Waste heat is a pollution hazard to waters only where the ambient temperature is already high. In many tropical waters marine organisms live close to the upper limit of their temperature tolerance range. If the waste heat discharged from a factory or power station raises the water temperature still further, mortalities of marine organisms may occur. This is most likely to happen if the water receiving the heat is relatively enclosed and unable to dissipate the heat by mixing with cooler water.

Interesting extensions of the problems associated with energy generation are those associated with the renewable energy sources of waves and tides. Wave machines have not progressed far beyond the pilot study scale and the use of tidal barrages for power generation is also very limited. However, in view of the concerns over other power-generating methods, such schemes are increasingly being examined. Neither process is without environmental penalty. Wave machines would cause hazards to shipping and would be difficult and possible dangerous to maintain. By extracting energy they might also have an impact on the coastal zone behind them, which would be exposed to a lower-energy regime than it previously experienced, with possible changes in sedimentation patterns and plant and animal life. Tidal barrages have the effect of reducing the tidal range of an estuary, which in turn reduces the exposure of the intertidal sediments that are so important as feeding areas for birds.

IV. The Scale of Marine Pollution

In reading some of the more sensational literature it is difficult to escape the impression that if the oceans of the world are not already dead, their demise can be expected imminently. Conversely, some of the platitudes that are produced by (usually) official bodies are so all-embracing in their assurances that nothing at all is the matter that it is equally difficult to take them seriously. In this and the following sections I shall try to put marine pollution problems into perspective in terms of both the health of the oceans and, more generally, the health of the planet. It is necessary to have some understanding of the relative importance of the different pollution problems because any measures that may be required to alleviate them may be both extremely costly and not without consequences for other parts of the environment. It is also important to note that what is an appropriate course of action under one set of circumstances may be highly damaging in another.

Taking the definition of pollution as given in Section I and applying it to the oceans as a whole, it is difficult to find any evidence that marine pollution is yet having an adverse effect on a global scale. There is certainly evidence of widespread contamination by, for example, lead, certain radioisotopes, and chlorinated hydrocarbons, all of which can be detected in polar ice and in waters far removed from land. However, the fact that they are present does not mean that their presence is necessarily of great concern. Some modern analytical techniques can detect concentrations many thousands of times lower than those known to affect even the most sensitive organisms. There is certainly no evidence to suggest that the oceans in total are about to suffer a complete "ecocollapse." At present, the only activity with the capability of having a significant impact on the global scale is probably fishing, or more accurately overfishing, as the impact of vast fishing fleets is becoming increasingly significant.

Looking at the oceans from this point of view is, however, a little like stating that the only evidence of human activity observable from the moon is the Great Wall of China; it may be true, but it is not particularly helpful to an understanding of the complexities of planet Earth. The next scale down from the "world ocean" are what are commonly termed the regional seas. Examples of these include the North Sea, the Mediterranean Sea, the Baltic Sea, and the Caribbean. These environments are commonly characterized by being semienclosed with restricted circulations of water and having high population densities around their margins with the associated industrial activities. On this scale the evidence of human activities is more obvious, with anthropgenic inputs of certain chemicals sometimes being of a similar order of magnitude to the natural ones.

The Baltic Sea in particular suffers from the effects of both human activities (particularly sewage and paper/pulp industry wastes) and changes in the oceanography, which have led to a steady increase in the area of bottom water with depleted oxygen concentrations. Small areas of deep water are subject to total oxygen loss and the formation of highly toxic hydrogen sulfide.

The Mediterranean, like the Baltic, is almost completely enclosed, although subject to very different climatic conditions. The higher temperatures accentuate the risks associated with sewage-contaminated waters, and cases of bacterial, viral, and parasitic enteric diseases are fairly common. Significant amounts of oil are transported through the Mediterranean from the Suez Canal. Although there has not yet been a really serious oil pollution incident in this region, small oil spills are fairly common. Localized nutrient enrichment by sewage outfalls leading to algal blooms, as mentioned above, is sometimes a problem.

The North Sea also suffers from localized nutrient enrichment problems and excessive algal blooms in a few areas. Recent work has indicated that some types of bloom in the

North Sea can contribute to the acid rain problems of parts of Europe and Scandinavia. The mechanisms for this is discussed in Section VI.

Most of the other regional seas have similar problems of sewage, oil, and industrial waste contamination, but for many regions appropriate data are scarce and it is difficult to obtain a reliable picture of their state of pollution.

It is generally true that the evidence for significant disruption of these environments away from the immediate coasts is slight but closer to land the influence of human activities becomes more pronounced. Coastal zones are the most heavily exploited marine areas for many purposes, including fisheries and recreational activities. They are also among the areas most exposed to the adverse effects of human activities. Bacterial contamination of bathing beaches is a worldwide problem and, although frequently arising from poorly designed sewage disposal systems, can sometimes arise simply from the sheer numbers of people present. [See Coastal Pollution].

The juxtaposition of centers of population and major industries in many coastal areas can lead to conflicts of interest between recreational and industrial activities. For example, the construction of a new harbor or coastal defense system in one area may cause significant changes in the pattern of deposition of sediment such that an adjacent part of the coast suffers from erosion. In a number of cases, in order to increase the tourist capacity of a particular lagoon, local coral reefs have been destroyed to provide the raw materials for construction of a new hotel with the result that the environment that the tourists wanted to enjoy has been destroyed.

It is on beaches and within estuaries that the problems of marine pollution are generally at their worst. Visible pollution of beaches by oil, sewage solids, and litter detracts seriously from their amenity value. It must be said, however, that not all of the litter on beaches comes from the water, in many cases it is abandoned there by the people who use the beach. Pollution by pathogenic organisms is a small but not insignificant health risk. A further problem which arises occasionally on beaches adjacent to shipping lanes is the presence of materials and containers that have been lost overboard by passing ships. These items range from small pieces of rope to canisters containing dangerous chemicals or even, though rarely, full-size cargo containers. Munitions such as mines are also occasionally found on beaches. These may be of recent origin, as in the Arabian gulf, or from older sources such as World War II. Dangerous munitions are sometimes dredged up by fishing vessels, usually from areas close to old dump sites.

Estuaries around the world are very commonly sites of major population centers, harbors, and large industrial establishments. They are also subjected to the pollution loads that may be brought down by their source rivers. Coupled with these factors is the fact that the estuarine environment is extremely dynamic, both physically, as a result of the tides, and chemically because of the changing concentrations of salt in the water. From the biological point of view, estuaries are highly specialized environments and support a great diversity highly specialized environments and support a great diversity of aquatic plants and animals. For some migratory species estuaries are the gateway to the oceans or to the land, and these animals are particularly vulnerable to pollution-induced changes.

Virtually all estuaries with attendant human activities show signs of pollution-induced changes. One of the most common symptoms is the depletion of oxygen concentrations in the water induced by the presence of oxygen-demanding wastes. Chemical contaminant levels are also frequently high, with the particular range of contaminants reflecting the patterns of industrial and agricultural activity in the catchment area. Estuaries are very vulnerable to oil spills, which, because of the restricted nature of the environment, can do considerable damage even in small quantities.

V. Uses and Abuses of the Ocean

What is a legitimate use of the oceans? What are the acceptable limits of this use? The answers to these questions are not simple and are by no means universally agreed.

In trying to answer them it is probably easiest to start on the small scale and try to work upward. In Section I the example of an oily rag being thrown into the ocean was used, and we concluded that this could not be classified as pollution, although it would be hard to describe it as a legitimate use. If it is accepted that an event of this nature is of no environmental significance, but that if an oil tanker released some or all of its cargo in the same place there would be a serious problem of pollution, it follows that there is a limit to this kind of activity beyond which it is not possible to go without causing problems. Or, putting it another way, the oceans as an environment have a capacity to assimilate a limited amount of waste without harm.

Waste is a fact of life. By making all possible efforts to recycle it, the amount that requires disposal can be minimized. IT cannot, however, be reduced to zero because the law of diminishing returns (both financial and energetic) means that there comes a point at which the effort required to clean up a waste still further causes a greater problem through the use of the resources required. There is therefore an irreducible minimum amount of waste that will always require disposal, and in practice the amount will be much more than this theoretical limit.

So, given that there are wastes and that the oceans have a limited capacity to absorb wastes without harm, is it then legitimate to link these two concepts and to utilize the oceans as a receptacle for wastes up to some defined limits? Clearly, this in turn depends on whether it is possible to define appropriate limits. Science teaches us to look for the relationships between cause and effect, so it is not surprising to learn that many waste management strategies use the parameter of "observable harm" to set these limits. This is the logical implementation of the United Nations (GESAMP) definition of pollution, and this approach is often used as a guideline for making use of the oceans as a waste disposal resource. For example, in many situations the major environmental damage that might be caused by a particular amount of waste can be assessed. The waste management strategy is then to regulate the amount of waste discharged to (often) ten to one hundred times less than the load at which adverse effects might be expected.

It is possible to argue that the oceans are a special case and that, because of the dispersive nature of the aquatic environment and the consequent irreversibility of most forms of marine disposal, such practices should be prohibited. A decision along these lines has the consequence that the waste disposal options are reduced to (1) disposal or storage on land and (2) incineration or some other means of disposal to the atmosphere. These options too may prove to be environmentally or politically unacceptable even when all practicable steps have been taken to minimized the quantities involved. Sewages sludge, for example, is often contaminated with industrial wastes, but even that coming only from domestic sources has variable amounts of trace metals, chlorinated hydrocarbons, bacteria and viruses, and probably other contaminants such as dioxins. If it is decided that under no circumstances can it be placed in the sea, because it is not possible to know in every detail what happens to it after it has been released, the same argument also applies to land disposal, where inter actions with the biota (including edible crops) and leaching into ground waters, streams, and rivers are also highly unpredictable.

The policy of many countries is to phase out waste disposal at sea, but it is by no means certain that this attitude will prevail when all of the problems of land-based disposal become generally recognized. In particular, the NIMBY ("not in my back yard") syndrome is an increasingly important consideration for both environmentalists and politicians.

I have described some of the problems associated with waste disposal at sea without considering the effects of other uses such as shipping, fishing, naval activities, and recreation. All of these activities have attendant benefits and problems, although the issues are somewhat easier to resolve in most cases. There are no easy answers and one person's use is another's abuse, but when it comes to a consideration of the real problems of waste disposal it must surely be better to adopt a global strategy and to look for solutions that have the minimum harmful impact on the whole environment.

VI. The Role of the Oceans in Global Pollution Problems

A number of pollution issues are increasingly becoming of concern on a global scale. These include the greenhouse effect (and the possible changes in seal level the occurrence of acid deposition or acid rain, and the problems associated with the thinning of the ozone layer. Of these, the oceans have a definite role in the first two and are possible victims of the third.

Global warming caused by the increase in atmospheric carbon dioxide produced by the combustion of fossil fuels could have the effect of causing sea level to rise by two different mechanisms: thermal expansion of seawater and melting of the polar ice caps. One of the unanswered questions about the greenhouse effect is the rate at which fossil fuel carbon dioxide can be removed from the atmosphere by absorption in seawater. The oceans certainly have the capacity to remove the excess carbon dioxide, but it is not at all clear whether the process can occur quickly enough to mitigate at least some of the consequences. At present, considerable scientific effort is going into establishing this.

Some of the acidic emissions from industrial environments fall into the oceans as rain. Because the oceans are naturally slightly alkaline, the acid is neutralized and no harm should occur. Recent scientific work has, however, revealed an unexpected mechanism whereby the oceans can actually act as a source of acid rain. The mechanism was first noted in the North Sea, where, as noted previously, nutrient enrichment as a result of human activities can stimulate the growth of phytoplankton. Certain species of phytoplankton which bloom in the

North Sea produce the gas dimethyl sulfide (DMS) as a natural waste product. The DMS escapes to the atmosphere, where it reacts to form the sulfur oxides which are one of the components of acid depositions. The prevailing weather patterns carry this material over the main European landmass, where it may fall as acid rain. Much work is being done to try to estimate the significance of this problem.

The recently observed thinning of the ozone layer over polar regions has received a great deal of attention. Its overall effect at the earth's surface is an increase in the amount of incident ultraviolet radiation. The consequences of this increase for marine organisms on the global scale are not clear, but it is known that ultraviolet radiation has growth-inhibiting effects or marine phytoplankton. If the plankton are significantly harmed by the radiation, the consequences for the highly productive polar waters could be very serious. If the problem with the ozone layer continues to increase, the consequences for all surface-dwelling plants and animals, including humans, could be devastating.

VII. Summary and Conclusions

Is marine pollution a serious problems? As is tried to show in this article, the answer to this question is largely one of scale. Pollution of estuarine and nearshore waters is of concern in many parts of the world. Problems in more remote waters are much smaller and are often undetectable, but that is not to say that they do not exist at some level.

Humans have to live within the constraints that this planet imposes. Already there are many instances of human interference with natural systems leading to severe adverse consequences. The key to our survival must be an understanding of the nature of natural systems and the complex interactions between the natural environmental compartments. The oceans probably constitute the greatest of these compartments and are, in many ways, the least understood. It is of enormous importance to our ultimate survival as a species that we recognize what the oceans can do for us, not only as source of food, minerals, and energy but also as an aid to proper management of our waste products. It is equally important that we do all we can to understand the complexities of the oceans so that we do noting that causes irreversible damage.

18

Marine Pollution—II

Introduction

The International Oceanographic Commission (IOC) for the United Nations Educational and Scientific Commission (UNESCO) defines marine pollution as: "Introduction by man, directly or indirectly, of substances into the marine environment (including estuaries) resulting in such deleterious effects as harm to living resources, hazards to human health, or hindrance to marine activities and reduction of amenities." If we allow broad interpretation of the word "substances" this definition appears to be all-encompassing. Note however the strong contrast between the dictionary and UNESCO definitions. The connotation of the oceanographers is that unless man introduces the pollutant, there is no pollution. Apparently the inference is that if pollution is man-caused it can be man-cured.

Nearly 71% of the earth's surface is covered by oceans, which together comprise a total of approximately 1.37×10^{39} liters, weighing about 1.40×10^{36} metric or long tons. This is the tremendous amount of water and consequently for many years the oceans are regarded an ideal place to dump all of man's wastes. For example, the rivers of the world discharge close to 2.0×10^{14} metric tons of water into the ocean every year but this is much smaller than the water already into the oceon. Carried along with the water each year are 4.5×10^{11} metric tons of dissolved materials. Although this is a large amount of material compared to that already present it is literally a drop in he bucket. Unfortunately, the ocean does not receive all of its dissolved material input from the world's rivers. Some comes form the atmosphere as it is washed by falling rain. Most of the pollutants placed into the atmosphere and up in the oceans either directly or indirectly form precipitation. This also is a sizeable amount of material. It is to be noted that this is larger than the amount of sediment plus the amount of dissolved materials carried to the oceans by rivers.

The more important reason for being concerned about marine pollution stems from the fact that materials put into the oceans by man's activities is not evenly spread throughout the oceanic volume, pollution is generated where man lives and works and therefore finds its way for the most part into the oceanic areas the are closest to their activities. Consequently, the dilution factor is very small and the concentrations of these pollutants build up in coastal regions. They are increased to a much higher level than would be the case if the material were distributed throughout the entire oceanie volume.

Coastal are dumping grounds area having a much higher pollution concentration not only because the material is being put into these relatively shallow areas much more rapidly than it is being carried away by natural weve nations but also because the normal structure of the oceans tends to prevent the mixing of these inputs with the rest of the oceanic volume.

The oceans are not dead from pollution, not are they insensitive to the effects of pollutants. As long as we find it necessary to use the sea for a world dump, and there is no reason to believe we will ever stop, we must learn as much as we can about oceanic processes so that the assimilative capacity of the oceans is never exceeded.

Strictly speaking marine pollution may also be of natural origin. For example, it may be caused by underwater gas and oil eruptions, or by the intensive growth and subsequent death of certain micro-organisms, which are responsible for the familiar disastrous 'red tides'. However, such processes and effects are not regarded as 'pollution', even though they are often studied by methods similar or those used for investigating pollution effects caused by man.

Ecological studies on marine pollution fall into three principal categories:

(1) Biogeochemistry of pollutants.

(2) Marine ecotoxicology.

(3) Biological principles of anti-pollution measures.

The biogeochemistry of toxicants includes the investigation of the sources of pollutants, the pathways along which they enter the marine environment, patterns of accumulation in the biotic and abiotic components of ecosystems, mechanisms and rates of migration of pollutants, their transformations and other processes which determine the fate of toxicants in the sea. An important, though not decisive stage in such investigations is the determination of the content of toxicants in marine organisms and communities; this is a difficult analytical problem because the chemical composition of matter is complex, while the concentration levels of pollutants are low. The methods of chemical oceanography are used in order to obtain and analyse samples, the results being interpreted in a biogeochemical sense.

The second type of research work carried out on this problem, which is best described as 'marine ecotoxicology', involves the study of biological effects consequences of human interference with the composition of the marine environment. Finding answere to such question is not itself tantamount to solving the problem of pollution, but nevertheless yields an objective evaluation of present-day ecological anomalies in the world Ocean and prediction of those likely to occur in the future. These investigations provide a scientific foundation on which to base national or international efforts to preserve the seas from pollution.

Solutions of the immediate practical problems concerning pollution of the sea must be based on the results of closely inter-linked ecotoxicological and biogeochemical studies.

MARINE POLLUTANTS

In Table 1 a list of common pollutants associated with the marine environment have been given. These are not listed in any particular order of importance because it is found that in certain pollutants are more worrisome than others, while the opposite condition may be true in other areas. This difference in importance comes from local conditions like population pressure, flushing, and climate.

Theoretical calculations of the concentration of the principal toxicants in epipelagic waters of the World Ocean indicate that considerable pollution (lead pollution, in particular) of surface waters by atmospheric pollutants may have taken place. This is true of inland seas.

The global pollution zone of the marine environment may be characterised as follow: highest pollution levels in the euphotic layer; high pollution levels in the neritic zone and inland seas; latitudinal distribution of toxic pollutants; a mosaic distribution pattern of toxicant concentrations in water; localisation of toxicants in the hyponeuston and benthos biotope; and a coincidence of the maximum pollution zones with zones of high biomass and high productivity.

Table 18.1
Pollutants Associated with the Marine Environment.

Pathogens	Toxic organics
Sediments	Petroleum
Solid wastes	Nutrients
Heat	Radioactive materials
Fresh water	Oxygen demand materials
Brine	Acids and bases
Toxic inorganics	Aesthetically displeasing materials

Before we discuss the marine pollutants one by one, we shall give some general conclusions about these:

(1) The rate of biological accumulation and migration of micro-amounts of metals and radionuclides in the World Ocean is directly correlated with their physichcocmical properties and with the specific surface area of aquatic organisms.

(2) Concentrations of metals in sea water and in plankton tend to decrease with increasing atomic number of the metal.

(3) The relative content of artificial radionuclides in oceanic seston does not exceed 1 per cent of their total content in water; the corresponding figure for heavy metals and transition is 1-10 per cent, while in inland seas it may be as high as 25 per cent.

(4) The forms of microcomponents in sea water are largely determined by the supersaturation of surface waters by calcium carbonate, since the suspended and the colloidal forms of this compound are strong sorbents.

(5) Self-purification of surface waters by biosedimentation of impurities is mostly due to their accumulation in aquatic organisms at low tropic levels. If the accumulation factor in seston in 1000-10000, the rate of vertical biogeochemical migration of a given impurity is of more or less the same order of magnitude as the rate of its

hydrological transport (for biomasses and productivities considered as average for the World Ocean). This is true of many radionuclides (^{144}Ce, ^{95}Zr, ^{106}Ru, ^{65}Zn, ^{54}Mn, etc.), of heavy metals and transition metals present in microamounts (mercury, lead, cadmium, iron, zinc, etc.) chloro-organic compounds (DDT, PCB, aldrin, lindane, etc).

(6) The rate of input of petroleum and chloro-organic toxicants into the World Ocean is faster than their rate of decomposition under natural conditions. The degradation processes are particularly slow at great depths, and the latter must be considered as a depot of toxic substance and their metabolites.

(7) A typical feature of the geochemical behaviour of petroleum and chlorinated hydrocarbons in the World Ocean is that their distributions in biotic and abiotic components of the same ecosystem are altogether different.

(8) The contemporary radio-ecological status of the World Ocean is distinguished by the relatively stable contents of ^{90}Sr and ^{137}Cs in surface waters and in aquatic organisms. Another typical feature is the gradual increase of radionuclide contents in the ecosystems of inland seas.

(9) The concentration levels of artificial radionuclides (^{90}Sr and ^{137}Cs), heavy metals (mercury, lead and cadmium) and chloro-organic compounds (DDT, polychlorinated biphenyl, etc.) in commercially important marine products vary with the degree of pollution of marine biotopes, taxonomic identity of the organisms, and their ecological and physiological characteristics (trophic level in the community, diet, age, size, etc).

The most variable concentrations in industrial species are typical of substances with cumulative properties, DDT and DDT metabolites, polychlorinated biphenyls and other chloro-organic compounds, and methyl mercury.

(10) There is a generally valid relationship (a direct correlation in the case of radionuclides) between the concentration levels of toxicants in commercial animals and the pollution level of the marine environment as shown by the increasing sequence of pollutant concentrations: pelagic ocean water, neritic zone, inland sea, freshwater basins.

(11) Typical accumulation factors of ^{90}Sr and ^{137}Cs in commercial species are of the order of 10^1, - 10^2 for heavy metals 10^2 - 10^3, and for chloro organic compounds 10^3-10^4. The food chain effect - i.e., the dependence of the toxicant concentration in a given organism on the trophic level of the organism in the community-is definitely present only in the cases of mercury, polychlorinated biphenyls, DDT and other chloro-organic compounds. The contents of mercury and of chloro-organic compounds in fish show a regular increase with the size (weight, age) of the animal.

(12) The selective, irreversible accumulation of high concentrations of mercury in pelagic predatory fish and marine mammals is a natural process, and is not usually due to human factors.

(13) The distribution of taxicants in the organs and tissues of Commercial fish species is typically non-homogeneous. ^{90}Strontium, lead and cadmium become preferentially concentrated in ligaments and bones; chloro-organic substances and methyl mercury concentrate in the liver and in the lipid fractions; ^{137}caesium concentrates in muscles. Chloro-organic compounds concentrate in fatty tissues.

(14) Global concentration levels of radioactive and chemical toxicants in commercial marine products are usually much lower than the national and international norms for maximum permissible concentrations, and thus present no danger to human health. However, in view of the possibility of local pollution and human interference with the natural background of microcomponents in individual regions, concentrations of toxicants in commercial species and marine product must continue to be constantly monitored, especially for heavy metals and chloro-organic compounds.

(15) Generally speaking, monocellular algae my react to the presence of toxic components in the aqueous medium in two different ways:

(a) By inhibition of vital processes and production.

(b) By stimulation of cell development and cell division, which is usually followed by impaired vital activities.

The development of micro-algae is relatively independent of the presence of the toxicant in the medium.

(16) The inhibitory effect of the pollutants increases in the following sequence: oil, detergents, metals (arsenic, lead, cadmium, copper, mercury, methyl mercury), chloro-organic compounds. The most highly toxic compounds are DDT, polychlorinated biphenyls, mercury and methyl mercury, which inhibit photosynthesis in long-term experiments at concentrations as low as 1-10 ug/l. Arsenic, when present as arsenate ion in concentrations of up to 1-10 mg/l, had not effect on the rate of division of algae cells.

(17) Low concentrations of toxicants (0.1-10ug/l) acting for a short period of time tend to stimulate photosynthesis and the rate of cell division in algae. This may be caused by the overall activation of vital processes by small doses of poisons or by an alteration of the metabolic interrelationships with the cells of the microflora. In the case of petroleum products such a stimulation may be due to the presence of biologically active substanees in petroleum.

(18) Toxic (algicidal) and threshold concentration ranges are determined for various types and species of monocellular algae. Out of the several combinations of the effects of several factors studied, synergistic effects were observed for the combinations of petroleum with DDT, and combinations of certain metals with a detergent. A combined effect of several metals may be additive, or else may be manifested by a dominant inhibitory effect of the most toxic one (mercury, copper).

(19) The varying responses of different monocultured species to toxic impurities in short-term experiments (less than one day) are probably due to the differing rates

of absorption and penetration of toxicants into the cells. In long-term experiments the specific differences are usually much smaller and depend on molecular, cellular and tissue level mechanisms. A general protococcal algae to most pollutants, as compared with cultured marine phytoplankton forms.

(20) Of the species studied, Gyrodinium fissum is highly sensitive to mercury, the response being irreversible; the same applies to diatomous algae acted upon by a mixture of DDT and petroleum products.

(21) The inhibition of photosynthesis is a more sensitive reaction of monocellular algae to toxicity than is the decrease in the rate of cell division.

(22) The general picture of the effects and relative toxicities of the most common toxicants on natural communities of marine phytoplankton does not significantly differ from that obtained for cultures of single species.

(23) In a number of cases the intensity of photosynthesis of natural phytoplankton communities decrease in the presence of lower toxicant concentrations than in experiments carried out on monocultures. The higher sensitivity of natural phytoplankton to contaminants is probably due to interference with the interspecies relationships in the community.

The considerable geographic, seasonal and other fluctuations in the reaction of natural phytoplankton to the presence of toxicants are caused not only by the environmental factors, but also by the special features of its specific composition. This composition undergoes significant changes if the action of the contaminants on the phytoplankton is sufficiently prolonged. Nitzschia diatoms in marine phytocoenoses are more resistant than other species to certain heavy and transition metals.

PATHOGENS

The pathogenic materials are the living organisms that can produce sickness or biological unbalance in either plants or animals within the ocean itself or in humans who either come into contact with oceanic waters or eat the organisms caught in the water. These include a wide variety of bacteria, protozoa, viruses and fungi. The most common of these are normally found in sewage. However, other pathogens may occur in non-waste disposal areas where environmental conditions are such that the proper conditions exist for growth and reproduction.

SEDIMENTS

Sediments are always present to some extent in the marine environment. These sediments have a marked effect on plant growth because they block out a large portion of the light normally reaching greater depths and therefore decrease photosynthetic activity. In some cases rooted plants get completely destroyed by this process.

As sediments are deposited on the bottom they will cover up bottom dwelling organisms (benthos) such as oysters and may even smother them under extreme conditions. The deposited sediments tend to make the water shallower so that, in regions traversed by ships, periodic dredging is needed.

SOLID WASTE

The disposal of solid waste has been a critical urban problem because areas suitable for the dumping of these voluminous materials are becoming scarcer. Due to this reason the ocean has been used as a dumping ground for solid waste. The solid waste most frequently marine dumped is sludge material left over as a byproduct from domestic sewage treatment. However, the unused products of industry and the used-up products of society are also put into the sea. If the product discharged at sea contains materials which may be leached into the oceanic environment, a serious problem could arise especially when the materials are toxic. However, when the solid materials are inert, a little forethought may result in some benefit to the marine environment. Artificial fishing reefs, for example, have been found to be very successful in certain areas, serving to enhance the habitat area, especially for small fish.

EXCESS HEAT

Excess heat, if added to the marine environment, alters ambient conditions, and these changes may be detrimental to the organisms present. The amount of heat that is detrimental and the extent of the degradation is determined by a number of factors. The primary source of this heat has been of course, from electrical generating plants, whether they be fossil fueled or nuclear powered (see Thermal Pollution).

FRESH WATER AND BRINE

Although fresh water may be in great demand ashore, too much of it in the ocean obviously will produce a marked environmental change within a small area. This change may occur due to poorly designed storm drainage systems, water diversion networks associated with dams, and effluent of some industrial processes. Excessive fresh water usually is not a critical problem.

The introduction of brine into the marine environment is similar in its effect to that of fresh water except in the opposite direction. The organisms acclimated to a particular salinity now find themselves in a more saline environment which could make permanent damage. Brine is a byproduct of desalination plants.

TOXICS

Toxics inorganics have been materials commonly used in industry, many of them are relatively harmless; however, in larger quantities they may be quite destructive. There are perhaps, 35 to 40 commonly used toxic inorganics and these should be very closely controlled by the user.

The toxic organics have been the most disturbing of the modern day chemicals commonly discharged either purposefully or accidently into the marine environment. These include the biocides such as fungicides, herbicides, insecticides, rodenticides, and also the additional organics including halogenated hydrocarbons, petroleum and industrial chemicals. The most disturbing toxic organics have been the pesticides such as DDT and kepone which have the unfortunate characteristic of being more soluble in oil than in water so that they tend to collect within the fatty tissues of marine organisms. They have been also very stable compounds which do not deteriorate very easily over a long period of time.

These find their way into the ocean both as manufacturing effluent and as runoff after utilization. There long term effects are unknown upto this fine.

PETROLEUM

Although petroleum may be classed as a toxic organic, it is a naturally occurring material and is biodegradable, given enough time. Its effects are not too well known. Petroleum enter into the marine environment due to accidents such as tanker damage or transfer loss, natural seepage, off shore production losses, losses associated with refineries, from runoff originating as drippings or disposal of used automobile lubricants, and unburned hydrocarbons emitted into the atmosphere as internal combustion exhaust.

NUTRIENTS

Nutrients (commonly called fertilizers) are those chemicals which are required by plants. The activities of man have added to the total nutrient load of almost all coastal areas. When nutrient levels becomes out of hand, plants grow unchecked so that decaying plants exist in such great numbers that the oxygen supply becomes rapidly depleted. These nutrients are present in domestic sewage effluents, agricultural runoff, and the little understood but apparently important non-point source runoff from urban areas.

RADIOACTIVITY

Radioactive material have been not only discharged to the marine environment by nuclear power plants, nuclear power plant fuel production and reprocessing plants, and uranium activities of all sorts, but also result from more common activities such as the burning of coal. When coal is burned it gives out radioactive particles to the atmosphere which are the washed into the sea at a greater rate than any known nuclear power plant at this time. Other sources of radioactivity include the natural back-ground, weapons testing, mine drainage, accidental spillage, and a few isolated industries.

One of the partially unsolved problems associated with the use of nuclear energy for electrical power has been the long term storage or disposal of spent fuels. This storage must be in an area such that there have been no pathways back to man, and the deep oceans have been suggested as meeting this criterion.

OXYGEN DEMAND

Oxygen demand materials have been those which need oxygen for degeneration and therefore steal oxygen which would normally be utilized by marine animals. Thus if too many oxygen demand materials are kept in the marine environment, the animal population will be markedly decreased due to the lack of oxygen. Sewage sludge and any other organic waste material, even that resulting from excessive plant growth due to an over supply of nutrients have been examples of common oxygen demand materials.

ACIDS AND BASES

The discharge of acid and bases to the marine environment is quite disturbing to the natural ecological balance of the system. The normal pH of oceanic water has been somewhere around 8.0, slightly basic. This is maintained by the carbonate system. If a large amount of acid or base is introduced into the system, the carbonate reactions will be offset and an important element of the environment will get affected. In addition, there are large synergistic effects associated with pH. Most toxic materials, for example, increase their toxicity under conditions of low pH.

The sources of acidic or basic material have been primarily industrial with some of this material reaching the marine environment from accidental discharges and the rupturing of tankers.

AESTHETIC CONSIDERATIONS

Aesthetically displeasing materials include all the stuff one finds in the ocean which is unpleasant to look at or to smell. Tar balls, floatables, gas (often hydrogen sulphide-) producing materials, and coloring agents are some examples of pollutants offending the senses. Although in some cases these materials do not pose any real threat to the ecology of an area, when the area is being used for recreation, the quality of the surroundings becomes somewhat important.

MANAGEMENT PROBLEMS

Although the major pollutants are delineated, the management of marine pollution is quite complex. The reason for this is that the marine environment is used by man in many different ways and this multi-use is having a number of ramifications. In the first place each oceanic activity pursued by man may produce a whole range of pollutants. A moving ship, for example, may not only discharge pollutants in the form of waste products from the people living aboard, oil from pumping its bilges, and heat from the discharge of its condensers, it may also need deepening of a harbor channel. This may generate a salinity change in addition to sedimentation problems associated with the maintenance dredging required.

Each pollutant may come from many different users. The nutrients which primarily appear to come from the effluent of sewage disposal plants also may come from storm drainage of urban areas in the form of waste products of domestic animals, along with runoff from agricultural areas that have been fertilized by either commercial fertilizers or farm animals.

An example of the effect of one pollutant on another has been the attraction of heavy metals to suspended sediments. Most of the heavy metals in coastal waters are swept out of the water by sinking sedimentary particles and end up in the bottom deposits. If the heavy metals then enter their way into bottom dwelling organisms which, in turn, find their way into a food chain pathway to man, then this obviously is not desirable.

Another example of this synergistic effect of two or more pollutants has been the effect of oil on pesticides, combined with sediments. The pesticides are more soluble in oil than they are in water so that if there is oil present in oceanic water, it will tend to concentrate any pesticides that also may be present. Oil is also having an affinity for sediments and as it collects on the suspended particles, their weight gets increased to the point where they sink to the bottom. Portions of the inshore marine environment thus are covered with sediments very rich in both oil and pesticides.

The introduction of foreign material into the marine environment has been not always detrimental to man or to the environment either. In many areas the increased erosion produced by construction activities in large areas of siltation where these eroded sediments have been deposited in the shoreline areas. After stabilization, populations of marsh grass have been seen to spring up in many of these newly created areas. In the main there is no planning for this type of rehabilitation, but recent experimental work reveals that production of marshlands may very well be a common tool of the esturine planner of the future.

SOURCES OF MARINE POLLUTANTS

Now that the pollutants are delineated and described very briefly, we will concern ourselves with the activities involved in the production of these pollutants in an attempt to pinpoint the source of the problem. These activities are outlined in Table 2.

Table 18.2
Pollution Producing Activities

Marine Commerce
Industry
Electrical power generation
Sewage treatment
Other non-industrial wastes
Recreation
Construction

Let us discuss their one by one.

Marine Commerce

Marine commerce has been deeply involved with ships, and ships require a deep enough water environment so that they can pass to and from on their appointed rounds. This usually requires continuous dredging. Dredging is an expensive proposition requiring large expenditures of money. Once the sediments are extracted from the bottom and the channel is cut to the desired depth, the problem becomes one of disposing of the dredge spoils, the material dredged from the bottom. Various solutions have been suggested but the problem is still one that has not been adequately solved.

Another aspect of the ship problem has been oil. Whether these ships has been transporting oil or using it for fuel and lubrication, there always seems to be some lost to the sea.

Most harbors are having large storage areas open to the atmosphere so that a certain amount of leaching into the marine environment is inevitable. Lastly, there has been a whole family of pollutants associated with any ship. A ship is just a little city afloat and any problems found in urban areas will be found on ships also.

Industry

Whenever one mentions pollution to the average individual the first source of pollution that pops into mind has been industry. Most industries require the use of water. This water has either heat or some chemical added to it before being discharged into the environment. These chemicals may be anything from toxic substances to harmless coloring agents.

Electrical power generation is responsible for waste heat and certainly this is one of the by products of producing electrical power whether it be from the burning of fossil or nuclear fuels.

An additional problem involved with heat exchangers has been the fact that many of these are made from copper, a toxic material which is used in anti-fouling paints on boats for many years. Sometimes these heat exchangers are allowed to remain in an unused condition for extended periods of time with water sitting in the heat exchanger tubes. Some copper leaches into this water from the tubes and when the plant gets turned on again this copper-rich water is pumped into the environment where it may bring about some consternation among the local marine organisms.

Large amounts of water pumped at high speeds from a coastal regions are able to trap or entrain small or sometimes even large organisms. These creatures may suffer mechanical damage as they go through the system, or may get exposed to large amounts of heat for a short period of time.

Radioactive wastes are considered as one possible disadvantage of nuclear power plants, but the amount of radioactive waste actually transferred by cooling water to the surrounding environment has been measured as extremely low. The spent fuel problem appears to be of more concern at this time.

The pollution problems associated with mineral extraction could not be documented due to the small opportunity for data taking, but it is expected thy will be similar to marine construction. At this time the major oceanic mining effort has been directed toward the extraction of sand and gravel in the near shore areas. Within the imminent future it is expected that a concerted effort will be made to extract manganese nodules from the ocean floor, perhaps at depths as great as 3000 meters. This mining effort will be expected to have some effect on the environment but just how much has been not quite clear. The same can be said for the increase in activity connected with offshore oil drilling although it appears that there is less petroleum lost into the environment when the oil is obtained from an offshore well and transported to the continent by pipeline that when it is obtained from a land well and transported by means of tanker to its eventual user point. The spillage resulting from the use of tankers has been greater than that resulting from the use of pipe lines.

Sewage Treatment Systems

The process treating human waste products has been quite successful in removing suspended materials and pathogens, but it has been extremely difficult to extract all of the nitrogen and phosphorous compounds. Consequently, sewage treatment facility effluent is having reasonably high amounts of these nutrients which push plants toward uncontrolled growth. In addition, when the facility gets overloaded, as during a heavy rain, it is often completely bypassed, discharging raw sewage directly into the water. This needs a great deal of oxygen for decomposition so that the biological oxygen demand is increased.

One aspect of sewage treatment neglected up to now, primarily because the costs have been so overwhelming, has been the treatment of storm runoff. Unfortunately the solution is not as straightforward as simply increasing the capacity of existing plants to handle the heavy loads associated with storms, or even to add treatment to storm drainage systems when they exist separately. Much of the urban and rural runoff does not go through conventional pipe systems, so additional systems would have to be built if all runoff has to be treated. Storm drainage washes urban streets clean. All material deposited on the streets, including crankcase-drippings, rubber worn from auto tires, and bird, dog, and cat droppings are washed to sea. Modern studies have indicated that this runoff has been a very significant source of coliforms and petroleum, to name just two pollutants of some concern to man. How to handle these nonpoint sources has been a task presently facing marine environment managers and one that appears to have no inexpensive or simple resolution.

With most large urban areas disposal of the solid residue from sewage treatment plants has been a solid waste disposal problem of the first magnitude. Ocean dumping in the near-shore coastal regime is utilized for many years and, to a certain extent, will probably continue to be. Either the sludge must get treated more thoroughly before it is dumped or new deep sea dumping areas must be found. If new treatment or disposal techniques are not followed, there may occur major changes in the coastal environment.

Agricultural Sources

The largest single class of non-industrial polluters are farms. Agricultural pollutants have been commercial fertilizers, animal wastes, pesticides, herbicides, and sediments. In many cases the agricultural contribution of all these pollutants has been greater than any other single source. The pollution can be stopped, and there have been cases in the past where particularly bad problems have been solved.

Recreational Activities

Usually when we think of marine recreation we think of a non-polluted area since recreation is essentially limited by pollution. In shallow water motor boats stir up the sediment on the bottom, a condition which often persists for long periods of time. This increase in turbidity markedly influences the natural plant growth and probably has something to do with the decrease in the numbers of rooted plants reported in the last few years. The aspect of motor boating that has many investigators most concerned has been the discharge of petroleum products into the environment. There are some figures which indicate that recreational boats add as much petroleum and petroleum waste products to the environment as all other sources combined.

Construction

The last source of pollutants listed in Table 5.2 has been construction. By construction has been meant both that done at sea such as offshore oil rigs, and than done on land contiguous with the sea, such as port facilities. Implied here has been both the actual construction of the structure and the associated activities that go along with construction. There must be all sorts of support people. As in general, the amount of pollution gets directly related to numbers of people, any large construction effort will bring with it the usual people-waste products. The effect of a new physical structure where one had not previously been has been not too well known, but one thing is sure-the environment will never be the same as it was before. Some of these changes might be beneficial, but no matter how beneficial there have been always going to be some members of society who react to any change by claiming a degradation of the environment. Any marine construction project must get judged on the basis of the benefits to be derived by society as a whole rather than solely to those accruing to some small group.

EFFECTS OF MARINE POLLUTION

In this article, we shall attempt to get some idea as to what pollution does and why we consider it to be so undesirable. There have been three basic types of pollutants; the pathogenic, the aesthetic, and the ecomorphic. Pathogenic pollutants have been those which cause disease. This disease may be fatal if the pollutant is a lethal poison.

Aesthetic pollutants have been those pollutants causing a change in the environment displeasing to the eye, ear, or nose of man.

Ecomorphic pollutants, on the other hand, have been those pollutants which produce a change in the physical characteristics of the environment in such a way that there may be drastic changes in the structure or composition of the biosphere.

Obviously, these three types of pollutants have been not of equal importance. Pathogenic pollutants have been certainly much more serious than the other two; however, ecomorphic pollutants are often an indicator of more serious types of pollution to follow. Furthermore, since pollution has been man-produced and the effects have been suffered by man, aesthetic problems are certainly of interest.

Acute Effects

The pathogenic pollutants have been obviously the most important. The effects of pathogenic pollutants are either acute or chronic.

The acute effects have been the easiest to determine because of the short time between administration and affliction. The easiest way of measuring acute effect has been by feeding the pollutant in question to test organisms, such as mice or selected fish, and observing the dosage required to kill 50% of the organisms involved. This is usually carried out by feeding a population of test organisms increasing amounts of the pollutant, and for each dosage the number of individuals that succumb is noted. When these data are plotted, an S-shaped curve results.

Another way of examining acute effect is to observe the effect of a given pollutant concentration on organisms over a somewhat longer period of time, noting how long it takes for a given dosage to produce mortalities. This may be done by giving a predetermined concentration of the pollutant to the test animals and observing how long it takes to kill half of them. When data from an experiment such as this are plotted, a curve similar to vertical asymptote results.

Chronic Effects

There appears to be a dose size for each pathogen below which the pollutant does not have any effect. The concern with threshold values for various pollutants has been one aspect of the study of chronic effects. Chronic effects are the most difficult to measure because of the time involved; in some cases cancers have developed as long as 25 years after initial exposure to the carcinogenic agent. Nevertheless, measurements are made and for many materials a threshold value seems to describe the data reasonably well.

Most experiments require some measure of the effect on the organism other than death, so that many different measures are used. These are; deformity in the growth of organisms, damage to particular organs, change in heartbeat or breathing rate, genetic damage to individuals (rather than to the population as a whole), change in the rate of increase of a population, life expectancy of individuals within a particular population, and fecundity of females within the population. Unfortunately, an index may be very descriptive of a pollutant on a particular species, but it may be completely ineffective in describing what happens when this pollutant comes in contact with other species.

Synergism

Another complication in the examination and quantification of pathogenic pollutants has been the phenomenon of synergism. A synergistic effect takes place when the combined effect of two or more materials acting together has been greater than would be expected from the simple sum of the individual effects. Temperature, for example, has been a very strong synergistic parameter, generally enhancing most pollutant effects as temperature is increased. The same has been true of salinity and dissolved oxygen. Some pollutants, on the other hand, tend to decrease the effects of others, as might be in the case of a strong acid and a strong acid and a strong base present at the same time. Other pollutants may make strong pathogens to get precipitated out of the water column or to be collected on sediments, so that in either case they end up on the bottom. Thus it has been extremely important in the analysis of any pollutant to be aware of the other materials present and their synergistic tendencies.

Within the marine biosphere many organisms get affected by various pollutants. Whether or not a particular organism has been affected by any given pollutant will depend on many variables. For example, many pollutants occur only in the water column while others, such as some of the pesticides and heavy metals, are adsorbed into suspended sediments and usually find their way to the bottom within a relatively short period of time. Consequently, these latter types would tend to affect those organisms living on the bottom or feeding from bottom organisms more than they would influence the organisms living within the water column, like free swimming fish.

Also of importance in determining the effectiveness of a particular pollutant on living organisms has been its relative solubility in water and oil. Pollutants more soluble in oil will tend to find their way to organisms that have a larger content of oil in their body tissues. This has been particularly true of some of the pesticides which are more soluble in oil than water. The individual habits of marine organisms also influence their susceptibility to a particular pollutant. Feeding habits with respect to time of day, portion of water column foraged, type of material ingested, and the method of digestion utilized all determine to a great extent the types of pollutants the organism will be exposed to. Similarly spawning habits decide the type of area and time of year in which a species has been particularly sensitive to subtle changes in the environment. These changes influence not only the parents but also the offspring, and may be even their progeny.

Other habits not having to do with spawning also help to determine whether an organism will get exposed to particular pollutants. Whether the organism has been sensitive to light or sound, for example, will often determine whether it gets attracted to a particular out-fall. Thus we find that a pure and simple cause and effect relationship between a pollutant and an organism has been not enough to determine completely the actual response of the organism in the real world.

Pathways to Man

Once some of the pollutant gets ingested by marine organisms, the major concern of man has been the possibility of this pollutant appearing on his dinner table and causing a pathogenic response. Obviously if a fish ingests a pollutant and a human eats that fish, then that person will get ingested a dose of that pollutant and there will probably be some danger. The question is, "How much?" In order to arrive at an approximation of the actual amount of undesired material finding its way into the reader's stomach, one must be able to follow the pollutant through a number of steps.

Most organisms living in the sea tend to concentrate materials existing in the sea to much greater values than they have been in the ocean itself. For example, a diatom, a small form of marine algae having a silicate frustule, is having a body concentration of silicon about

40 thousand times as great as the oceanic waters from which the organism derives all of its material. Thus the diatom concentrates silicon very effectively so that if we ate diatoms, a large portion of our diet would be glassy. Similarly, other organisms, including fish, will concentrate other materials even if these do not naturally occur in the environment. But this amplification or concentration does not stop with one step. We must continually keep in mind the fact that an organism does not exist by itself in the marine environment. Any organism has been dependent upon many other organisms for its existence. An edible fish, for example, might very well subsist on smaller fish which, in turn, might subsist on small zooplankton (small floating forms of animal life), which, in turn, may subsist on phytoplankton (forms of floating plant life). There may be 5 to 10 individual links in a food chain leading to man, and there has been the distinct possibility that for each one of these links a concentration of the pollutant occurs. Hence, the total pathway to man for the pollutant might be a rather tortuous one, but it might result in a very high concentration of the pollutant material in the fish. This in itself may be of no concern to the individual unless he ingests the fish and if so, how often and how much he consumes. It has been not enough simply to know that there has been some pollutant present in marine organisms; one should also know how much is present, and in addition two other facts have been required. This first has been the accepted maximum level of ingestion of the pollutant before harm results, and the second has been the amount of fish that can be eaten before this level gets reached. In this way it might very well be possible to have a particular kind of fish one or two times a year with perfect safety, while if it has been eaten once a day, it might very well have a toxic effect.

Ecomorphic Pollutants

There are many different types of pollutants which may be classified as ecomorphic, while some pollutants fall into all three of the classes stipulated above and many fall into at least two. Sediment has been a pollutant whose effect has been primarily that of changing the environment since it is not pathogenic and very often is not aesthetically displeasing, although it can be. As it settles to the bottom, it will tend to bury the organisms that live on the bottom, such as oysters and clams, and eventually smother them. Sediment also tends to fill in marsh areas, killing the marsh grass, and completely changing the habitat. As marsh areas are breeding grounds for small fish and an important ecological link in most of the oceanic life, any destructive process such as this has been bound to markedly change the marine biological environment. On the other hand, in some cases where man's activities are able to produce large amounts of erosion and sediment has been brought down by rivers into estuaries, some estuarine areas have been filled enough to support marsh grasses. In same cases excess erosion can produce marsh lands rather than destroy them. Whenever there has been a depth change in an estuarine or coastal area, the effects have been many. They range from changing the biological environment so that larger fish will no longer live in the shoal areas to affecting the total circulation pattern of the area.

If the sediment has been in suspension it also has an effect on the ecosystem, the most obvious aspect being a change in transparency. By making the water more opaque, light will not penetrate as deeply as before so that the thickness of the layer of water in which plants grow (the euphotic zone) will get decreased. If the water has been fairly shallow to begin with, this decrease in transparency may very well result in the inability of rooted plants to grow and some marsh lands may be destroyed. The effect of sediment, then, has been two-fold: changing both bottom and water properties. Both of these effects tend to change the physical characteristics of the ecosystem.

Another ecomorphic pollutant has been oil, although oil sometimes produces toxic effects. The result of oil coating bird feathers has been a well known physical effect. The oil does not poison birds; it simply makes it impossible for them to fly, and consequently many of them die. A film of oil on the surface will also decrease the amount of sunlight entering the water and, similarly, will limit the exchange of oxygen from the atmosphere. In addition to the aesthetic and pathogenic effects of oil, there are thus also ecomorphic effects.

The toxic effects of various nitrogen compounds have been well known but fertilizers and other nutrients washed into the sea possess a nontoxic effect on marine waters. These materials will do just what they have been designed to do: make plants grow, whether these plants be terrestial or aquatic plants. Consequently, the first effect of a spike of fertilizer added to the sea has been an overstimulation of those plants normally there. Beyond that the undesirable plants (those that are not food for the edible types of marine animals) often get stimulated to grow and these marine weeds grow so rapidly that the overwhelm the other types of plants. As plants have been at the bottom of the food pyramid, the structure and characteristics of the ecosystem have been completely changed, even to the extent of markedly altering the species composition.

Once these plants have started to grow at greatly accelerated rates, they may completely cover the surface of the water, disallowing any light from penetrating more than a few centimetres. Those plants on the bottom of the blanket will die, due to lack of light, and those rotting plants will use oxygen in their decay processes. Even though there have been more plants than before, producing more oxygen in the process of photosynthesis, so many of them are dying and decaying the result has been to decrease the dissolved oxygen. This phenomenon it termed as eutrophication.

Solid waste has been another type of pollutant very often disposed of in oceanic areas. In some cases they tend to completely destroy the natural habitat and the result has been not only ecomorphic but also aesthetically displeasing. However, in some cases, the ocean dumping of waste products such as old automobile tires and even automobile bodies has produced an artificial habitat for many organisms that were not endemic to the region previously. However, there are cases where nontoxic solid waste materials have destroyed desirable habitats. Sometimes the dumping of solid wastes makes a change in water flow patterns producing changes in salinity and temperature which influence the biosphere markedly. This usually takes place when the water is reasonably shallow and the volume of solid waste products dumped is relatively large.

Another ecomorphic pollutant that seems to be receiving a lot of attention lately is heat which is supplied to the marine environment from electrical power plants or other industrial sources. This heat will increase the water temperature and may result in different kinds of damage to marine organisms. As most mutations have been caused by thermal activity under normal conditions, the possibility of genetic damage has been real. It is estimated that about 90% of the mutations occurring normally have been thermally caused, while only about 10%

have been caused by radiation. Thus with an increase in the temperature, one can expect greater mutation rates or more genetic damage.

In addition to genetic damage, heat can cause both acute and chronic problems. Animals can be killed by extreme thermal shock or they can suffer lingering effects and finally expire by being unable to acclimated to the new surroundings. More subtle changes are also notices as growth rates are changed with increase or decrease in temperature. Normally a small increase in temperature will make a small increase in growth rate; however, if the temperature change is extreme, this may be reversed. Changes in spawning patterns have also been noted in which fish that normally move upstream to spawn are unable or unwilling to cross a thermal barrier associated with a heated effluent. This may cause the cessation of spawning for a large portion of the species population. Another effect of temperature increase has been to cause certain organisms to grow more rapidly than others, upsetting the previous ecological balance and eventually causing a change in species composition.

A similar range of effects to that produced by heat has been caused by salinity changes. These might be produced from the input or rich brine solutions resulting from the effluent of water desalination efforts or a fresh water outfall such as a raw sewage treatment plant. A market change in salinity will bring about damage similar to that produced by heat, including genetic damage growth rate changes, spawning pattern changes, and changes in species composition. All organisms are used to living within some range of salinity and if this range gets exceeded, they will either die or move some place else.

Aesthetic Pollutants

The last type of pollutant mentioned above was aesthetic pollutants. Aesthetic pollutants have been those which offend the human senses and as such are regarded by many to be unimportant. In many environmental problem areas, though, the aesthetic considerations often receive more attention by environmentally oriented citizens than some of the pathogenic pollutants. This is simply because the aesthetic pollutants have been visible and what can be seen, smelled, or heard appears much more bothersome than that which is hidden, even though the hidden pollutant might be much more dangerous to life. Consequently, any scheme for controlling marine pollution must include some consideration of the aesthetic pollutants if it is to succeed. An excellent strategy seems to be to focus on the aesthetic pollutants and use them as a mechanism to gather momentum to clean up the rest of the pollution problems.

Solid waste material, when piled in shallow areas where it can be seen, has been a typical aesthetic pollutant. Building activities also may be regarded to be sources of aesthetic pollutants. Power transmission cable has caused more than one power company grief because of the environmental consideration given the local citizens who are concerned about the appearance of power cables running across the countryside. Oil wells, especially the offshore variety visible from resort areas, have been of great concern to most environmentalists. The same thing has been true of equipment required for dredging operations and the spoil areas resulting from these operations. To some people, especially those in living relatively pristine environments near the water, boats can also be an aesthetic pollutant, especially those that cause waves and noise. Effluents that make the color of the sea to change or increase the amount of suspended sediments also may be regarded aesthetic pollutants. It simply has been not as pleasant to go swimming in water that is cloudy or contains coloring material different from what one would expect in a normal, natural body of water.

In essence, any change from the pristine environment appears to many to be aesthetic pollution. Obviously, we are living in an unreal world if we think the marine environment can be maintained in an unspoiled form. It should be kept in mind that many complaints about aesthetic pollution come from individuals who would rather live in the primeval environment not realizing that it would not be available to them unless the commercial world about them supported it. There appears to be a dichotomy wherein we try to maintain the unspoiled environment in certain areas while letting others become overutilized. Nevertheless, there has been a growing movement to include in the cost of every project a relatively small sum to improve its aesthetic qualities. This is, of course, exactly the same philosophy that has been applied to the control of other forms of pollution. Additional money has been spent in a chemical process, for example, to control the pollution-producing properties of the effluent because society believes that this additional sum of money has been a desirable expenditure. It may be similarly be maintained that the expenditure for aesthetic pollution control is desirable as it also will tend to increase the quality of life.

CONTROL OF MARINE POLLUTION

Now that the nature of pollution and some of the sources of this undesirable product of society have been considered, we will now concern ourselves with some of the avenues open to us to accomplish some sort of control. The most obvious control strategy that can be used has been to some-how or other completely prevent the introduction of the pollutant into the environment. This can be carried out by not producing the pollutant or by restructuring the process so that the pollutant is no longer a by product. Let us consider a few major sources of pollutants and some possible methods of control in these particular cases.

Dredging

Maintenance dredging, that is, dredging required to maintain a particular depth, adds sediment to the water column and would not be needed if there were not sediment supply to shoal up major channel areas. This type of dredging has been not a one-shot affair, but it has been almost always necessary to periodically redredge the accumulated sediment from these channels and harbor areas. This continual sedimentation has been a normal process which occurs whether man is present or not. In some instances the activities of man tend to accelerate the sedimentation process but this has been not necessarily always true. For example, coastal plain estuaries, often the result of river basin flooding, tend to be a rather ephemeral event in geologic time. They get formed when water level rises, and they disappear with the continual process of deposition of material eroded from the higher reaches of the water shed.

Another example of natural erosion has been found on the west coast of the United States in the westernmost portion of California (that portion where more than 90% of the population lives). This area was formed by erosion of the Rocky Mountains and deposition of the eroded material at the base of the mountains.

The most obvious solution to the problem of maintenance dredging has been to cut off the sediment supply to the areas where sedimentation has been a major problem. This can be carried out in a number of different ways. The most effective of all has been to prevent upstream erosion, but at the same time this is perhaps the most difficult. The entire watershed must be policed very carefully to make sure that the natural landscape gets changed to such an extent that erosion no longer takes place. This implies extensive terracing and replanting of areas that have been denuded for one reason or another. Of course, this also implies that extreme measures must be taken in areas that do not have natural cover, such as agricultural areas or areas where construction is underway. A certain amount of this is being done.

Another method of preventing sediment from reaching the undesired area has been to provide settling ponds. The water is released after it has dropped the major portion of its load and is allowed to proceed downstream. The major disadvantage of this system has been that it is essentially a temporary solution since the ponds will fill up rather rapidly, needing either building of new ponds or dredging of some sort.

Sometimes the sedimentation that takes place in a channel area is a result of sediment loads that are introduced into the marine environment from nearby rather than those carried from far upstream. This takes place for example, when a channel gets dredged with unstable sides so that the material just tends to slough into the deeper regions. This needs continual dredging as the removed sediment has been almost immediately replaced by bottom material from the sides. One method of correcting this situation is by stabilization of the bottom. Adding larger, more dense material such as gravel to the bottom has been a technique utilized occasionally, although this has been usually more expensive than dredging so that it has been not done too often. Bottom stabilization, however, is a technique that is being used more and more to prevent the movement of bottom materials within relatively small contiguous areas.

Another possibility for preventing the accumulation of sediments in undesirable areas has been to change the natural circulation patterns so that sediments get deposited in an area where they do no damage. This may be carried out in various ways: by changing the physical dimensions of estuarine ares, by putting dams in selected areas, and even by changing the circulation pattern by planting grasses to increase bottom friction.

Another method to be considered in decreasing the amount of suspended sediments has been that involved in controlling the numbers and types or water craft allowed in particular areas. In regions that have been relatively shallow and narrow, high speed vessels tend to churn up the bottom, putting sediments back into the water column that had previously settled out. In addition, large wakes are generally produced that tend to erode the sides of the waterway by the mechanical action of these relatively high energy waves breaking on the shoreline. Both these effects can get decreased by the simple expedient of limiting the speed of boats in certain areas. It has been a fairly effective method; however, if the traffic is extremely heavy it may be necessary to limit the total number of vessels, in addition to their speed.

In addition to the general clarity aspect of pollution caused by dredging, there has been also a toxic one. Sediment retrieved from harbor bottoms has been especially likely to contain a relatively high concentration of industrial pollutants such as oil and heavy metals. During the dredging operation there has been a likelihood of some of this material's finding its way back into the water column, but of more concern has been the long-time leaching from dredge spoil dumping areas. Various methods, such as diking and the construction of artificial islands, are utilized to minimize this leaching but their performance under all conditions has been not too well known. Further research in the handling of dredge spoil gets needed at this time.

Oil Pollution

Another problem that, to a certain extent, appears amenable to prevention at the source is that involving oil. Unfortunately, the sources of oil in the marine environment have been extremely difficult to delineate.

All the experts agree that the greatest terrestial source of petroleum products in the ocean is used crankcase oil. Some go so far as to indicate that this source supplies more petroleum to the sea than any other source. This may be true because within the last decade or so the habits of American motorists are changed markedly to the extent that a large portion of them now change their own oil, discarding the used oil in a manner that probably makes it to end up in storm sewers. The major reason for this probably has been the almost complete disappearance of the used lubricating oil re-refining industry in recent years since it was not economically feasible to re-refine oil. However, with the modern continual increase in the price of new oil, the re-refining industry has been having a rebirth so that a large portion of the oil previously thrown away will now be reused.

Another entirely different source of oil in the marine environment has been that resulting from accidents that take place during shipping and transferring processes. Generally speaking, the more handling steps involved, the more accidents and associated losses can get expected. The figures as to how much has been actually lost in handling accidents compared to that lost in normal ship procedures, such as pumping bilges, has been not well known, but there has been no doubt that a very large portion of the oil reaching the marine environment has been the result of poor handling procedures in both the operation of vessels and the transfer of oil from one carrier to another. Better procedures are required, but perhaps even more than better procedures, stricter enforcement of accepted safe procedures would go a long way to decreasing the amount of oil that enters the ocean each year.

When a tanker empties its load of oil, the dynamic characteristics of this vessel get markedly changed. Its weight gets decreased about a thousand times and therefore it becomes extremely difficult to drive the ship, because it was designed to be driven with a full load. Therefore, when most tankers do not have a load of oil, they will fill their tanks with water for ballast. When they return for a fresh cargo of oil, they must pump this water from the tanks, releasing large amounts of oil as the residue left in the tanks gets washed out. There are many schemes which are suggested for decreasing the amount of oil that is wasted in the ballasting-unballasting process; however, it is not clear at this time how effective these processes have been or how effectively any requirements can be enforced.

Modern supertankers have been extremely large vessels and therefore have major structural weaknesses, so that occasionally they will break up in unusually heavy seas. They have been simply not designed to be utilized under unusual storm conditions, so that there will continue to be losses of this type in rough weather. In passing, it should be mentioned that once oil gets spilled at sea under large wave conditions (usually) the case when accidents occur), there has been no way for this oil to be retrieved. On the other hand, oil may be retriered in harbor areas where the sea surface is relatively clam, but only when there are waves no higher than approximately 1 metre. The obvious way-the only way-to prevent oil from getting to the marine environment when a ship founders has been to prevent the accident from occurring in the first place. Alternatively, once the accident has taken place the oil must be prevented from escaping from the ship. It has been somewhat doubtful whether these solutions will ever be possible.

Industrial Pollutants

Most industries are using large amounts of water for cooling purposes or as an integral part of the manufacturing process. Consequently, industrial effluents may be having waste products, heat, leached material from heat exchangers, or even incidental house-cleaning wastes. If the industry, for example, has been using water as a coolant, the heat exchangers must constantly be cleaned to keep the heat exchange efficiency high. These cleaners may be more toxic than process wastes. Consequently, it has been extremely desirable to treat all industrial effluents.

Some industries have tried to modify the entire process rather than just clean up the effluent. This often causes an increase in efficiency with a lower overall production cost. If this expenditure can be recouped in a relatively short period of time, this obviously might be expected to be a very cost-effective method of handling unwanted effluents. However, in many cases, there has been no increase in operating efficiency, with the net result being an increase in manufacturing costs passed on to the consumer. Society must then make a decision as to how much it has been willing to pay for a particular product vis a vis the amount of pollution introduced in the manufacture of this product.

Antifouling Paint

Sometimes pollution comes from the use of antifouling paint on boats. During the summer time in temperate waters the growth rate of marine organisms get increased markedly, especially those organisms setting on fixed surfaces. An untreated boat hull left in the water for a period of just a few weeks will need enough growth to cut the available speed for a given power input by fifty per cent. Consequently, all owners of boats left in the water have been very careful to apply some coating to the underside to prevent the accumulation of marine organisms. For large cargo ships or tankers, this could make savings of thousands of dollars in fuel costs. Common practice has been to paint the bottom of the boat with antifouling paint.

Antifouling paint has been so used to leach a toxic material into the water in just the proper amounts so that all organisms setting on the hull will get killed. However, the older types of antifouling paints, those that use copper as a toxin, released enough material to the environment not only to kill possible fouling organisms but also to be harmful to other marine creatures at some distance from the boat. Thus there has been a movement afoot to improve the quality of antifouling paints by drastically reducing this leaching to toxic material into the water. The solution that has been suggested has been to use new paint vehicles in conjunction with new toxic materials so that the toxic material has been almost completely retained within the vehicle, requiring organisms to come within a very short distance of the boat to be affected. One result of recent research is to develop a family of paints using acrylic resins as vehicles and tri-butal tin flouride (TBTF) as the primary antifouling ingredient. In the first few years of use this paint holds great promise because it is less toxic in the water, lasts longer, and does not allow bottom growth more effectively.

In their very effectiveness these new paints pose a problem, because the toxin gets retained within the paint for a longer period of time. Paint scrapings have been now more toxic than they were with the old copper bottom paint. This means that it has been somewhat more dangerous to strip the paint from the bottom of the boat when applying a new coat and precautions must be used to prevent the old paint from getting back into the water environment. When sandblasting large ships, for example, precautions have to be taken to retain the sandblasted material rather than allowing it to be simply washed away. Thus the solution of one problem presents another, but these problems have been solvable if some care has been taken to see that additional health hazards do not get created.

Pesticide and Herbicide Pollution

Some of the juicier pollution scandals that have taken place in recent years have involved various pesticides and the problem has been that many of the chemical compounds get introduced for public sale before the total effects of these materials have been known. It seems to be almost an insurmountable problem to determine what the total effects of a new compound will be before it has been put on the market because so many new products are introduced every year. However, a large portion of the problem has been undoubtedly due to basic carelessness. In both the manufacturing and use of these materials proper precautions are not being taken into consideration. Many users, for example, undoubtedly believe that if a certain amount per unit area will suffice, twice as much will produce twice as much control of either weeds or bugs. This excess toxin is washed into the streams and ends up killing marine organisms rather than protecting plants. It would be evident that more effective control of both the manufacture and use of pesticides and herbicides would go a long way toward eliminating undesirable side effects.

Solid Waste Pollution

The disposal of solid waste has been very necessary and there even seems to be a fear that if the solid waste problem is not solved, eventually there will be no room left for people. Because of this fear, large amounts of solid waste get dumped in the ocean. Some of these solid wastes take the form of sludge from sewage treatment plants whereas others consists of large pieces of discarded consumer products. In the latter case the problem might very well solve itself as the cost of scrap metal continues to increase.

One possible use for sewage treatment sludge has been in refurbishing land disfigured by strip mining. Often the overburden removed to get at the coal seam has been not capable of supporting life nor it has been even particularly stable. The addition of sludge from sewage treatment plants is considered as a method of soil treatment to make it possible to support a viable plant population. It is a concept that has not been tried yet but one which will be tried when it becomes economically feasible to transport the sludge to the strip mine. Another possibility for using the sewage sludge, if it contains only nutrient materials, has been suggested and has been considered by some of the Middle Eastern oil producing countries. The idea has been that empty tankers be filled with sewage sludge for the return voyage rather than with ballast water. This sludge will then be utilized to bring the desert regions back to a point where they have been capable of supporting a large agricultural industry.

Other forms of solid waste being dumped in nearshore areas have been dumped there primarily because there has been no other acceptable depository. One method of decreasing the severity of this produced by society. Packaging of consumer items, for example, seems to be a good place to start. From a disposal point of view the catsup bottle is probably the worst possible package. The package weighs more and takes up more space than the contents. The Navy has done excellent work in decreasing package size and weight because there are definite limitations aboard ship, but not too much has been done with the private sector of society simply because there hasn't been enough action on the part of consumers to insist upon small volume and weight packages. One possible exception to this has been the introduction of lightweight plastic bottles for large size drinks to replace the heavyweight glass variety. They are somewhat thinner than glass, weigh less, and may be incinerated, so the disposal problem has been somewhat alleviated. However, with the continuing increase in petroleum costs, plastics might not remain economically feasible in the future.

Recycling and Reclamation

It has been found that biological communities living in the sea need a certain amount of physical shelter. Some species just like to be alone, whereas in other cases the young of the species require places to hide to protect against predation. A marsh or wetlands area has been a good breeding place because both of these conditions prevail. It is found that in many cases these conditions can be created artificially by introducing large junk articles such as old automobiles or worn out ships to form effective artificial fishing reefs.

There has been disadvantages to using waste products for this sort of activity. One has been that in many cases these junk objects get transported to undesirable areas by dynamic oceanic forces. Pollution has been not controlled when automobile fenders and doors wash up on swimming beaches. When creating a finishing reef, steps must be taken to retain the integrity of these reefs for a long period of time. If proper attention has been paid to the magnitude and nature of oceanic forces, artificial night be made relatively permanent.

With the present states of the economy where things are some what more expensive, it appears that more items will find themselves in the near future on the list of materials to be recycled or reclaimed. Industrial processes will alter as it becomes more feasible to extract materials from the waste stream from a cost standpoint and, at the same time industrial, waste streams will become less and less polluted. Even in the area of multiple waste product use, like the combination of waste heat from either manufacturing or electrical power generation and waste nutrients from sewage treatment plants for the purposes of aquaculture, cost will always be the controlling factor.

It is known that elevated temperatures and the artificial addition of nutrients will increase growth rates of many organisms markedly. The choice of which organism to grow has been involved not only with which would be best to grow, but also which there has been a market for. Setting up an aquaculture system involves picking organisms which respond favourably to the environment, developing a system by which the optimum growth rate can get produced, developing a marketing system so that the end product will be sold at the maximum possible price, and assuring the continuing existence of a market for everything produced. Not only has been science included but also sociology and economics must be considered, and it is probably in these latter two areas that the major difficulties reside. Hence, no sophisticated aquaculture systems employing both waste heat and nutrients have been developed at this point, but it has been probably only a matter of time before we will have extensive aquaculture using man's waste products.

Storage of Pollutants

Another method of controlling pollution has been is to store them at some distance from man's activities so that they have been as inaccessible as possible to the major fraction of the population and pathways to man. Occasionally nuclear waste products are being dumped in deep oceanic areas in the hope that when they do break out of their canisters, they will not find an easy pathway back to man because the deeper life forms have been probably not in any food chain involving man. Perhaps not so well known has been the fact that deep ocean disposal is used for other materials such as overage ordnance and poison gases for which the military has no further use.

In the case of nuclear products, there has been a very vociferous school of thought that believes storage facilities should allow waste retrieval if a technological breakthrough is ever accomplished that would allow these nuclear waste products to be profitably utilized.

Asimov's suggestion has been to implant our hazarodus waste products in the oceanic trench areas so that within a few thousand years they will get carried into the interior of the earth and no longer be bothersome. The only problem with this suggestion has been that the exposure time before the canister having the waste products is effectively buried in the crust is relatively long. During this time the canister seal has to be absolute. Whether this can be accomplished with present technology is presently being debated within the scientific community.

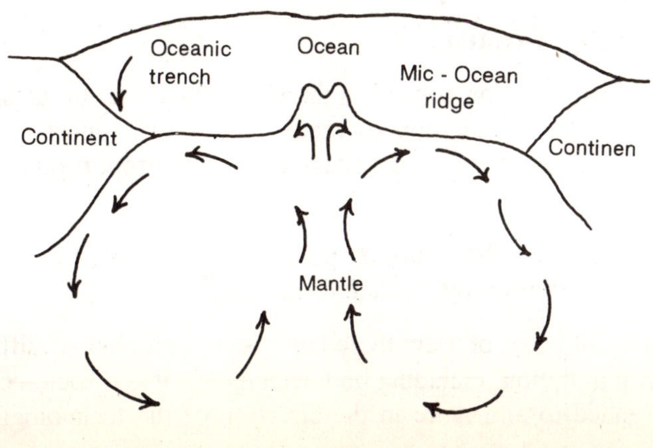

Fig. 18.2.1 Convection currents in the mantle resulting in mid-ocean ridges and oceanic trenches.

Nevertheless, storage has been still a viable method of controlling pollution, especially for the more toxic materials. At least theoretically, one knows where they are. The major concern with storage of toxic material has been accidental release, and the chances of this taking place always remain within the realm of possibility.

Controlling Pollution by Zoning

Another method of controlling pollution has been by specifying certain areas that can be utilized for effluent discharge and other areas that cannot. Although we are even starting to convert our city dumps into usable areas, such as golf courses or ski slopes by covering, planting, and land-scaping as rapidly as possible, but it will be some time before ocean dumping areas are reclaimed in this fashion.

Certain coastal regions would be reserved for industry, others for sewage treatment, while still others would be set aside for recreation. Some of these activities have been such that they may get pursued in conjunction with one another, whereas others cannot. The idea has been for every coastal area to try to get optimum utilization of marine resources available. In order to accomplish this type of zoning most effectively, a great deal more knowledge is needed than is presently available about the synergistic effects of various types of activities. One of the major advantages of optimum zoining has been that it would allow maximum avoidance of undesirable synergistic effects.

The biggest stumbling block to coastal zoining probably has been inovercoming entrenched interests in regions where the optimum activities have been found to be different from those presently underway: a socioeconomic problem rather than a technological one.

The concept is some what interesting one, because it essentially permits some pollution of certain areas while refusing to permit any pollution of other locales. This has been different from the way pollution has been presently controlled, where the effort has been made to limit pollution in all areas to the minimum possible amount. With the zoning concept pollution would still be limited as much as possible, but it might permissible to allow the water quality of ocean to degrade below the mean in some places, because in others it would be higher than the mean.

Controlling Pollution by Taxation

One last method of pollution control is taxation which can be utilized to encourage or discourage certain activities in addition to raising funds to support government programs. Pollution would be allowed, but it would be taxed at a rate proportional to the environmental insult.

Some economists suggest that a tax on pollution will simply be a license to pollute and will therefore encourage pollution rather than discourage it.

From a management point of view there have been a number of different directions that can be taken to control pollution, including both technological and socio-economic alternatives. So far society has tended to aim more in the direction of the technological alternatives, but there appears to be a definite swing toward other methods. In the years to come it might be possible that more and more emphasis will be kept on these non-technical methods, especially as the tradeoffs, costs, and benefits become better understood.

FINAL REMARK

It has been said that pollution is a biological problem that can be solved by the application of physical and chemical methods. Whether this is a completely true statement is often a moot point, however, because the opportunity to apply any solution is usually won only after the employment of politics. Thus the study of pollution is a truly interdisciplinary one, including as it does the three Ps-pollutants, processes, and people.

IMPORTANT CONCLUSIONS

(1) Effective control of marine pollution requires further research work aimed at evaluating the extent of pollution of marine ecosystems, determination of the biological effect of toxicants and prediction of biological consequences. However, such an approach alone is inadequate, since the problem is urgent and suitable measures must be taken as soon as feasible.

(2) Small marine organisms have a low resistance to toxicants. This fact was utilised to develop a system of rapid bioassays for immediate evaluation of the harmful effects produced by any given substance present in sea water.

(3) The more promising applications of the results and methods of toxicological research to the protection of the marine environment from pollution include: standardisation of the components of various toxicants in the sea based on biological criteria; ecotoxicological characterisation of sewage effluent and its components, for the determination of methods and norms of permissible dumping; recommendations on the priority and stepwise sequence of reducing the volume of sewage discharged by various industries; rapid tests of comparative toxicities of different substances, in order to select those which are ecologically safest; development and realisation of regional plans for protecting the marine environment from pollution.

19

Organic Matter Decomposition in the Marine Environment

Introduction

Organic matter is one of the most complex and least stable mixtures of largely undetermined components that exist in the earth system. During its passage through various segments of the earth system, organic matter is continuously degraded. This is illustrated with relevant examples from the marine environment. Decomposition of organic matter is responsible for the consumption of oxidants, for most early diagenetic processes, and for changes in the composition of sediments. The rate at which organic matter is decomposed varies over orders of magnitude and depends on the composition of the material being degraded and on the environmental conditions.

Almost all organic matter is formed by the process of photosynthesis. This process requires the presence of light and involves the uptake of carbon dioxide and nutrients and the production of oxygen. Estimates of the total marine photosynthetic production range from 20 to 150 Gt per year [1 gigaton (Gt) = 10^{15} g]. The marine biomass that is present at a certain times is estimated to vary from 2 to 4 Gt. According to these data, annual marine productivity exceeds the marine biomass by about an order of magnitude. This reflects the importance of recycling, hence decomposition, in the oceanic carbon budget.

Similarly, estimates of the global flux of particulate organic carbon leaving the surface ocean, i.e., net production, range from 2 to 20 Gt per year, which is 10 - 15% of the total photosynthetic production. This indicates that most of the organic matter produce is readily decomposed. Moreover, only a few percent of this new production (2-20 Gt yr^{-1}) becomes buried in marine sediments (0.126 Gt yr^{-1}). Most is decomposed during transport through the water column and during burial in the surface layers of the sediments. Accordingly, it seems that most of the organic matter produced by photosynthesis is degraded. [See Ocean Production Export: Hydrodynamics and Ecosystems].

In parallel with the decomposition of organic matter, there is liberation of carbon dioxide, phosphate, and other nutrients. This recycling of nutrients is a prerequisite for the continuation of primary production because productivity is limited by the availability of nutrients. If there were no liberation of nutrients, all nutrients available would be consumed rapidly and productivity would consequently cease. [See Carbon Dioxide Transport in Oceans].

At present there is some uncertainty about the importance of marine dissolved organic matter in global cycles. Although dissolved organic matter constitutes by far the largest reservoir of organic carbon (500-1600 Gt), it was conventionally considered to the inert because of its homogeneous distribution, its condensed structure, and its average[14] C age of 6000 yr. However, more recent data suggest that this pool of dissolved organic matter is actively mineralized. Active degradation of dissolved organic matter is consistent with the fact that dissolved organic matter provides the main substrate for bacterial growth. More data are needed to reveal the reactivity toward decomposition of dissolved organic matter.

II. Organic Matter

A. Its Composition

Organic matter is normally considered to be a composite of residual materials derived from marine and terrestrial sources. Differentiation of marine and terrestrial organic matter input can be based on microscopic, elemental, isotopic, and molecular biomarker parameters. As a rule, the following characteristics can be used for discrimination between organic mater sources.

Table 19.1
Preservation Potential

Compound	Preservation Potential[a]	Compound	Preservation Potential[a]
Starch	1	Chitin	2
Glycogen	1	Gums	2
Fructans	1	Glycolipids	2-3
DNA, RNA	1	Resins, amber	2-3
Cellulose	1-2	Sporopollenin	4
Pectins	1-2	Tannin	4-5
Proteins	1-2	Lignins	5
Dextrans	2	Cutan	5

a The preservation potential ranges from 1 (extensive degradation under any depositional conditions) to 5 (no degradation).

Marine organisms have C/N ratios ranging from 6 to 8, while land plants generally have ratios higher than 12. Similarly, marine organic carbon is isotopically heavier (g^{13C}_{org} -18 to -23%) than most terrigenous organic carbon ($\delta^{13}C_{org}$, —24 to —29%). The co-occurrence of longchain alkanes (C_{23}-C_{30})m with a strong odd over even predominance and fatty acids and alcohols (C_{20}-C_{30}) with a strong even over odd predominance is indicative of terrigenous organic matter. Marine organic particulate matter has a characteristic composition given by the Redfield ratio: $(CH_2\ O)_{106}\ (NH_3)_{16}\ (H_3\ PO_4)$.

The use of these characteristics to discriminate organic matter may be restricted because of factors such as the choice of the number and composition of end members (C/N ratios and $g^{13}C$ values) and the representativity of the small fractions of total organic matter that are actually analyzed in the molecular biomarker approach. Another potential problem is the presence of organic matter from bacterial sources (e.g., benthic biomass), which is normally assumed to be unimportant but which might influence interpretations considerably.

B. Changes in the Composition during Decomposition

The organic compounds that make up organic matter vary in their susceptibility to decomposition because of differences in chemical structure. For a few groups of organic compounds the potential to survive decomposition (i.e., preservation potential) is presented in Table I. It seems that components with nonhydrolyzable macromolecular structures have the highest preservation potential.

It has been shown that the microbial community tend to utilize organic substrates sequentially, the more reactive substrates being consumed first. As a consequence of the preferential utilization of specific components, the bulk composition of organic matter changes considerably during decomposition. There is preferential loss of N and P during the mineralization of organic matter. For example, in suspended matter the C/N ratio changes progressively with depth from the initial value of 6.6 to 10 at 1000 m and to 15 at depths greater than 3000 m. A similar increase in C/N ratios with depth has been observed in organic-rich sediments. Accordingly, nitrogen-rich material is preferentially consumed. Because the isotopic composition varies among the different components, changes in the isotopic composition of total organic matter are expected due to preferential decomposition. There is, however, little evidence for a significant change in the isotopic composition during organic matter decomposition. Finally, the most abundant components in extant biomass. polysaccharides and polypeptides, are readily degraded and the resistant compounds, i.e., those with a high preservation potential, are consequently relatively enriched. [See Nitrogen Cycle of the Sea].

The organic matter actually preserved is an intensely modified residue of that originally supplied as a consequence of the preferential utilization and preservation processes. The organic matter preserved is enriched in resistant biomacromolecules and depleted in nitrogen. Any partitioning into organic matter sources based on this modified residue will therefore be biased toward more terrestrial organic matter. There is no simple way to correct for these changes.

The mineralization of organic matter involves not only the decomposition of extant biomass but also the formation of high-molecular-weight organic matter (e.g., geopolymer formation or natural vulcanization). This new organic matter is formed in an environment usually rich in ammonium, phosphate, and hydrogen sulfide and may therefore incorporate these components. These natural vulcanization processes are responsible for most of the organic sulfur compounds found in marine sediments.

III. Decomposition Reactions

A. Oxidants Involved

Decomposition of organic matter to carbon dioxide involves an oxidation reaction because zero-valent carbon in organic matter is oxidized to tetravalent carbon in carbon dioxide.

Suitable electron acceptors for organic matter oxidation are oxygen, nitrate, nitrite, manganese oxides, iron oxides, sulfate, and oxygen bound in organic matter itself. These oxidants are utilized not simultaneously, but sequentially. the sequence of oxidant consumption corresponds to the order of decreasing energy yield per mole or organic carbon oxidized and is given in Table 2.

Table 19.2
Oxidation Reactions of Organic Matter[a]

Oxygen reduction:

$138O_2 + (CH_2O)_{106}(NH_3)_{16}(H_3PO_4) + 18HCO_3^- \rightarrow 124CO_2 + 16NO_3^- + HPO^{2-}_4 + 140OH_2O$

Nitrate reduction:

$94.4NO_3^- + (CH_2O)_{106}(NH_3)_{16}(H_3PO_4) \rightarrow 13.6CO_2 + 92.4HCO_3^- + 55.2N_2 + HPO^{2-}_4 + 84.8H_2O$

Manganese (IV) reduction:

$236MnO_2 + (CH_2O)_{106}(NH_3)_{16}(H_3PO_4) + 364CO_2 + 104H_2O \rightarrow 470HCO_3^- + 8N_2 + 236Mn^{2+} + HPO^{2-}_4$

Iron (III) reduction:

$212Fe_2O_3 + (CH_2O)_{106}(NH_3)_{16}(H_3PO_4) + 740CO_2 + 316H_2O \rightarrow 846HCO_3^- + 424Fe^{2+} + 16NH_3 + HPO^{2-}_4$

Sulfate reduction:

$53SO^{2-}_4 + (CH_2O)_{106}(NH_3)_{16}(H_3PO_4) \rightarrow 39CO_2 + 67HCO_3^- + 16NH^+_4 + 53HS^- + 39H_2O + HPO^{2-}_4$

Methane formation:

$(CH_2O)_{106}(NH_3)_{16}(H_3PO_4) \rightarrow 53CO_2 + 53CH_4 + 16NH_3 + H_3PO_4$

a Stoichiometry of the organic matter oxidation reactions listed in the order in which they should occur (based on free energy released) in the environment. Manganese reduction and nitrate reduction are interchangeable depending on conditions. For organic matter the ideal Redfield ratio is assumed: $(CH_2O)_{106}(NH_3)_{16}(H_3Po_4)$. The organic matter involved does not necessarily have this ratio.

The sequence is determined not only by thermodynamics but also by other factors related to process kinetics and bacterial physiology. Because these degradation reactions are catalyzed microbially, they are subject to limitation of the physiological potential of the living organisms. For instance, oxygen may kill anaerobic bacteria and so inhibit anoxic metabolism. Similarly, nitrate may inhibit the bacterial reduction of ferric iron. As a consequence of the thermodynamic and physiological constraints, there is a distinct zonation in marine sediments in which these reactions occur (Fig. 19.1).

The six major oxidants that are sequentially consumed by decomposing organic carbon are oxygen, nitrate, manganese oxides, iron oxides, sulfate, and the oxygen of organic matter itself (methane generation) (Fig. 19.1, Table 2). On the basis of these oxidations, three types of diagenesis are recognized: (1) oxic diagenesis, which involves oxygenation: (2) suboxic diagenesis, which comprises all processes having nitrate, nitrite, manganese, or iron oxides as electron acceptors; and (3) anoxic diagenesis, which comprises all processes ;related to sulfate reduction and methane generation.

B. Importance of Various Oxidants in Sediment

The relative importance of various oxidants on a global scale can be calculated from estimates of the amount of organic carbon oxidized in certain areas, the areal coverages of these areas, and the relative contribution of the oxidants in these areas. Oxygen and sulfate appear to be quantitatively the most important electron acceptors in marine sediments, accounting for 0.86-1.9 Gt and 0.15-1.3 Gt carbon per year, respectively. The amounts of organic carbon mineralized by nitrate and methanogenesis account only for 0.02-0.08 Gt and 0.02-0.05 Gt carbon per year, respectively.

The relative importance of various oxidants may differ considerably depending on the sedimentary setting (Table III). In the deep-sea environment, nearly all organic carbon is oxidized by oxygen. Secondary oxidants such a nitrate, manganese, and iron oxides never oxidize more than 5-10% of the total amount of organic matter that is degraded. In sediments from the continental slope and rise, oxygen is still the major oxidant but nitrate, manganese and iron oxides, and sulfate are also quantitatively important (Table III). In near-shore marine sediments, oxygen and sulfate are about equally important in oxidizing organic matter. Finally, all organic matter in anoxic sediments is decomposed by sulfate reduction and methane generation.

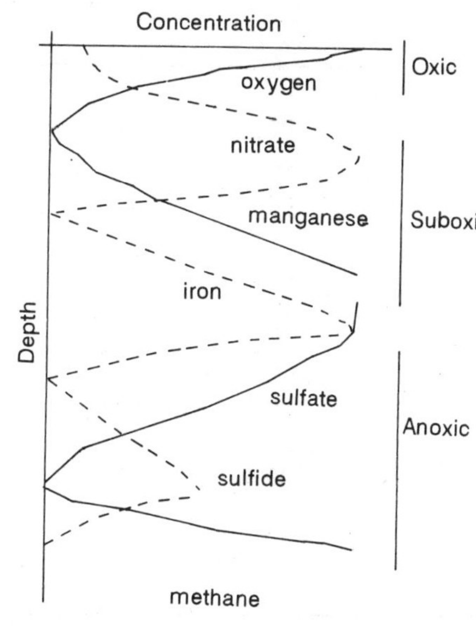

Fig. 19.1. Schematic representation of depth trends in concentrations of dissolved species involved in organic matter decomposition reactions. The succession of processes is: oxygen reduction, nitrate formation, nitrate reduction, manganese oxide reduction, iron oxide reduction, sulfate reduction, and methane generation. Absolute values of depths and concentrations will depend on the sediment. The three types of diagenesis (oxic, suboxic, and anoxic) are indicated on the right.

Although coastal sediments (i.e., water depths less than 200 m) constitute only 8.6% of the total area of the ocean, they are responsible for 83% of the sedimentary mineralization of organic matter. Sediments from the continental slope and rise (water depth 200-4000 m) cover 33.2% of the ocean floor and contribute 15% to the sedimentary mineralization of organic matter. The deep sea below a water depth of 4000 m covers about 58% of the ocean floor, but only 2% of the organic matter is oxidized here. In pelagic regions most of the organic matter is oxidized in the water column.

The flux of organic matter to the sediment depends on the primary productivity at the ocean surface and on the water depth. When the organic carbon flux is considered in relation to local productivity, the fraction that reaches the r\sediment is mainly a function of water depth (Fig. 19.2). As a consequence, pelagic sediments, because of their water depth, are unimportant for the cycling of organic carbon. On the other hand, coastal sediments are very important because they not only are responsible for 83% of the decomposition of organic matter in sediments but also accumulate more than 90% of the total organic matter buried in sediments. These numbers show that relatively small, but active, areas of the ocean are responsible for most of mineralization in sediments. As a further example, although anoxic sediments cover only 0.03% of the ocean floor, about 10% of the total anoxic decomposition occurs in these sediments.

Table 19.3
Importance of Various Oxidants at Some Representative Sites

Site	Percentage of total				
	O_2	NO_3	Mn/FeO_x	SO_4	CH_4
Deep sea					
Manop H	99.2	0.8	0.4	—	—
Manop M	90.7	6.9	0.9	1.1	—
Hatteras abyssal plain	96	4	—	—	—
Continental slope and rise					
Indian Ocean	86.6	12.5	0.3	0-6	—
Hatteras continental rise	76	8.3	1.7	13.8	—
Bermuda rise	78.2	11.7	1.4	8.7	—
Savu basin	60.9	25.4	6.5	7.2	—
Coastal sediments					
Danish coastal sediments	39.9	3.1	—	57	—
Anoxic sediments					
Cape Lookout Bight	—	—	—	68	32
Saanich Inlet	—	—	—	76	24

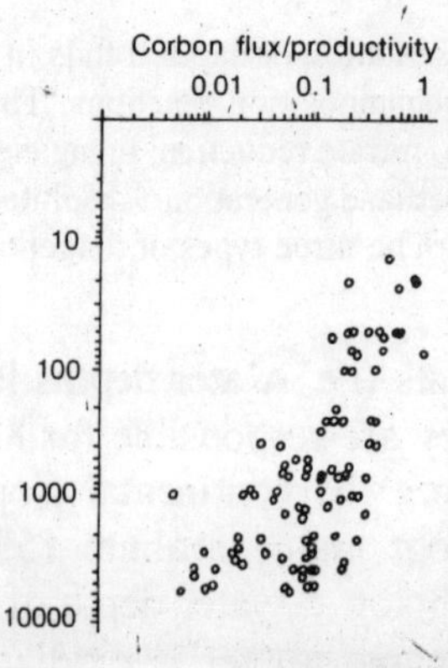

Fig. 19.2. Relationship between the carbon flux, normalized to local primary productivity rate, and water depth in the ocean.

IV. Decomposition Rates

A. Determination of Decomposition Rates

Our knowledge of the rate of decomposition of organic carbon is based on information from two sources. One is the geochemical approach, which combines pore water and solid phase dat with mass balance calculations. This approach involves the application of kinetic models in which decomposition rates are calculated from observed chemical gradients in combination with estimates of rates of diffusion and burial.

The other source of information is based on experimental rate measurements. The experimental determination of organic matter decomposition rates involves techniques such as direct injection of radiotracers or laboratory simulations using sediment samples. There is, however, one major uncertainty when applying this method: how well do experimental rate measurements reflect actual in situ process rates? Changes in temperature and, in particular, pressure (i.e., depth) are known to affect microbial activity and thermodynamic equilibria.

Information on decomposition rates in pelagic sediments is mainly based on geochemical evidence because experimental rate determinations are hampered by the pressure decrease during sampling of deep-sea sediments. Moreover, chemical gradients in deep-sea sediment can be sampled in sufficient detail for modeling purposes. Information on decomposition rates in coastal sediments has been obtained by using either the geochemical or the experimental approach. There are only a few studies in which the two approaches have been combined. These studies have shown that the organic matter decomposition rates based on the geochemical approach are strongly dependent on the kinetic model applied.

B. Rate Models

The overall process of organic matter decomposition is generally described by simple models.

1. One-G Model

In this model it is assumed that reactive organic matter decomposes at an overall rate directly proportional to its own concentration (i.e., by first-order kinetics):

$$dG_m/dt = -kG_m \tag{1}$$

where G_m is the concentration of metaboliziable organic carbon, k is the first-order decay constant, and *t* is time. According to this model, the concentraiton of reactive organic matter and the rate of organic matter decomposition both decrease exponentially with time.

2. Multi-G Model

On the basis of actual rate measurements it has been proposed that organic matter can be divided into various groups of compound of different reactivities, each of which undergoes first-order decomposition. This multi-G model can be expressed as

$$G_{Tm} = \Sigma G_i \quad (2)$$

$$dG_i/dt = -k_i G_i \quad (3)$$

$$dG_{Tm}/dt = -\Sigma k_i G_i \quad (4)$$

where G_i is the concentration of organic carbon is group i, G_{Tm} is the concentration of total metabolizable organic carbon, k, is the first-order decay constant for group i, and *t* is time. Selective removal according to the reactivity of each group accounts for the change in the reactivity and the amount of organic matter with time. This multi-G model is consistent with the well-established sequential utilization principle mentationed earlier.

3. Power Model

In the G-type models the organic matter is considered to be composed of a refractory fraction and one or more labile fractions. This partitioning between labile and refractory compounds is highly subjective and dependent on the time scale and environment under consideration. The power model does not have these limitations. In this model, total organic matter is considered to decompose according to first-order kinetics

$$\frac{dG_T}{dt} = -k(t)\, G_T \quad (5)$$

where G_T is the concentration of total organic carbon and k is the time-dependent first-order rate parameter. Assuming that k is constant over an interval Δt, k can be calculated from

$$k = \frac{1}{\Delta t} \ln \frac{G_t}{G_{(t+\Delta t)}} \quad (6)$$

where G_t is the organic matter concentration at a certain time.

Using result of decomposition experiments, profiles of organic carbon concentration versus depth in cores with a known sedimentation rate, and profiles of particulate organic

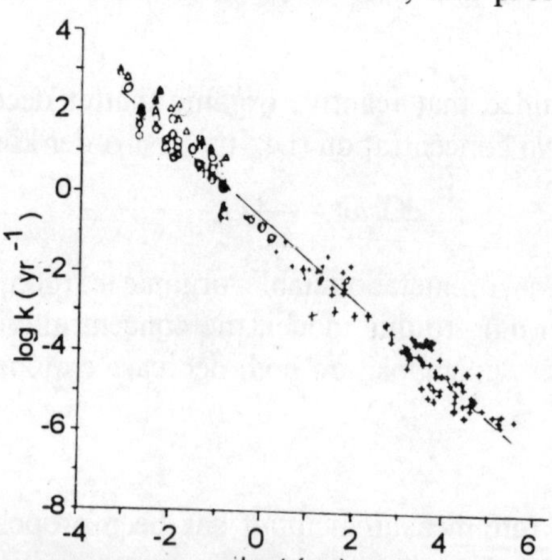

Fig. 19.3. Reactivity or organic carbon (k) versus time (t) in years. First-order decay parameters have been calculated using Eq. (6). (O) Experiments; (Δ) particulate matter; (+) sedimentary organic matter.

carbon concentration in the water column versus water depth, rate parameters were derived according to Eq. (6). Rate parameters based on experimental and field data versus time are described by a single straight line with a correlation coefficient of - 0.99 (Fig. 3).

This result implies that there is a continuous decrease in the reactivity or organic matter over more than eight orders of magnitude. This strong relationship also suggests that there is no significant discrepancy between rate parameters derived from laboratory experiments and field data follow the same reactivity decrease with time, their starting point (i.e., apparent initial age) is different. This starting point provides a quantitative measure of the initial reactivity, i.e., type of organic matter.

V. Implications for the Sedimentary Record

A. Organic Matter in Marine Sediments

The concentration of organic carbon in marine sediments is controlled by a variety of processes that regulate supply, degradation, and preservation of various components. Most of the organic matter supplied to sediments is readily degraded and recycled and only a few percent becomes permanently buried. A useful measure of the degree of recycling is the carbon burial efficiency, i.e., the ratio of the carbon burial rate and the carbon flux to the sediment surface. The carbon burial efficiency is mainly a function of the sediment accumulation rate.

This empirical relationship between the carbon burial efficiency and sediments accumulation should ;be used with some caution because there are also correlations between the organic carbon flux to the sediment and the sediment accumulation rate, between the organic carbon flux and the reactivity of organic matter, and between the reactivity of organic matter and the sedimentation rate. Clearly the single influence of one parameter on the preservation of organic carbon is difficult to determine because of these dependent varibles. Other important variables that may affect the burial of organic carbon are the degree of bioturbation and the bottomwater oxygen content. There is considerable disagreement and confusion in the literature concerning the influence of the redox state on benthic rates of organic matter decomposition (and conversely burial). Some scientists believe that organic matter is degraded faster and more completely under oxic rather than suboxic or anoxic conditions, whereas others believe that rates of organic matter decomposition by aerobic and anaerobic microorganisms are very similar.

Despite the complexity of the processes involved in organic matter preservation, an empirical transfer equation has been derived:

$$PaP = C_{org}\, p\, (1-\phi)/(0.0030s^{0.3}) \tag{7}$$

In this transfer function, paleoproductivity (PaP in grams of carbon per square meter per year) is estimated as the product of organic carbon percentage (C_{org}), dry sediment density (p), and sediment volume (1-) divided by a function of sedimentation rate (S). On the basis of Eq. (7), it has been suggested that oceanic productivity was approximately the same as today during past inter-glacial periods but was higher by a factor of 2-3 during glacial stages.

B. Formation and Dissolution of Mineral Phases

The decomposition of organic matter results in (1) consumption of oxidants, (2) production of carbon dioxide, and (3) production of nutrients and hydrogen sulfied. These chemical changes may result in the dissolution or precipitation of mineral phases and may so affect the inorganic chemical composition of sediments.

The consumption of oxidants by decomposing organic matter result in the establishment of reducing conditions in which iron (III) and manganese (IV/III) oxides become unstable and dissolve. This reduction from iron (III) to iron (II) can easily be recognized in recent marine sediments because it is accompanied by a color transition from brown to green. The depth from the sediment-watter interface to the brown-green transition can therefore be used as a measure of the rate at which oxidants are consumed within the sediment during organic carbon oxidation. Sediments underlying highly productive areas have a thin brown Fe(III) — bearing layer, whereas sediments from low-productivity waters have a thick brown surface layer.

The production of carbon dioxide and the consumption of alkalinity during oxic decomposition of organic matter (see Table II) is balanced by the dissolution of calcium carbonate according to

$$Co_2 + H_2O + CaCO_3 \rightarrow Ca^2 + 2HCO_3^-$$

The oxic decomposition of each mole of organic carbon results in the dissolution of about I mol of carbonate C. This process of metabolically driven $CaCO_3$ dissolution occurs at all depths and is relatively independent of the water column saturation state (lysocline depth).

The production of phosphate, ammonium, hydrogen sulfide, and bicarbonate during organic matter decomposition may result in supersaturation of pore ;waters with respect of minerals such as dolomite [$CaMg(CO_3)_2$], pyrite (FeS_2), and carbonate-fluorapatite [$Ca_5 (PO_4)_3 (OH,F)$]. These authigenic minerals are often referred to as organogenic because their precipitation is a direct result of organic matter diagnesis. In addition to pyrite, dolomite, and apatite, the organogenic mineral assemblage may include minerals such as rhodochrosite ($MnCO_3$), reddingite [$Mn_3(PO_4)_2\ 3H_2O$], alabandite (mnS), siderite ($FeCO_3$), vivianite [$Fe_3(PO_4)_2\ 8H_2\ O$]. amorphous FeS, and struvite ($MgNH_4PO_4.6H_2O$). The presence of some of these minerals may be characteristic of certain environments.

GLOSSARY

Authlgenic minerals. Minerals formed in situ, often as a consequence of diagenetic reactions.

Diagenesis. Processes that occur during and after deposition of sediments.

Nutrients. Elements or compounds that are essential for living organisms, e.g., nitrogen and phosphorus.

Organic matter. Material derived from living organisms.

Organic matter decomposition/degradation/mineralization. Decay/ breakdown/ transformation of organic material.

Oxidants. Compounds that are able to accept electrons, e.g., oxygen nitrate, and sulfate.

Preferential Preservation/decomposition. Utilization of the most labile substrate by bacteria which results in the preservation/breakdown of specific compounds.

20

Coastal Pollution

Introduction

Coastal pollution is the introduction by human activities of harmful amounts of substances into water along our coasts from the upper limit of penetration of salty water into rivers out to the edge of the continental shelf. The most severely affected areas are semi-enclosed with poor water exchange with the sea, such as estuaries and harbors. Pollutants are both natural compounds, which our activities have caused to become damagingly concentrated in coastal waters, and newly created or unnatural materials with which ecosystems have had no previous experience. Cleanup is difficult and expensive. Prevention is the most important remedial activity.

Pollutants are substances present in and damaging to environment. Contaminants are substances foreign to a system but not necessarily damaging, DDT, PCBs, mercury, oils and fertilizers are substances that can be pollutants. Some are harmful in almost any amount, such as PCBs. Others such as mercury or fertilizers occur naturally in coastal waters and sediments are become pollutant only when they are present in excessive amounts. Thus the definition is relative.(See Marine Pollution)

We can all think of instances in which one person's pollutant is considered by someone else to be a harmless contaminant. DDT, before its effects on shell thickness in bird eggs was known was considered by many only as a contaminant. The change in attitude came as the result of research into the causes do decline in bird populations, and came slowly as the data on its effects were accumulated. In this example, DDT moved from a contaminant to a pollutant not because its relative presence in the environment changed but rather because our understanding of its effects on the environment caused it to be reclassified.

To understand the importance of pollutants in coastal areas we need to be able to answer a series of questions about any a pollutant:

1. Where does it come from?
2. What is its fate in the coastal area? Does it get washed out? Does it degrade? How long does it remain a pollutant in any particular area?
3. What effects does it have on the coastal system in which it is found? How and why does it damage the coastal system?

4. To what extent do the damages that result from a pollutant affect humanity, either directly or indirectly?

5. What can be done to improve the polluted system? Is just stopping the input enough or will more have to be done to restore the coastal area to good health?

A. The Coastal Zone

"Coastal" means not only the place where the land meets the sea, but also the inland areas where the water in rivers begins to become salty and the offshore areas to the edge of the continental shelf where water depths rapidly increase to those of the deep sea, usually taken as the 200-m depth contour. The coastal zone includes small and large estuaries such as the Chesapeake, banks like the Grand Banks of Newfoundland, and inland seas like the Baltic. The coastal zone can be hundreds of kilometers in width, as it is off the west coast of florida, or very narrow,as it is off the west coast of Africa. Salinities can be comparable to oceanic values (35%) as on the west coasts of Africa or South America or greatly diluted by Fresh water as by the mouths of great rivers or in estuaries.

B. Pollution Pressure

In the United States, the coastal counties are the most rapidly growing areas of the country. In 1980, 110 million people of a total of 228 million, or nearly half of the U.S. population, lived in the coastal counties that account for only 11% of the total land area (the calculation excludes Alaska). Populations in coastal areas are expect to grow most' rapidly in the future. With such a concentration of people living close to the sea, the pollution problems of today can only get worse unless better remedial and preventive measures are taken.

II. Natural Pollutants

Substances with the potential to be pollutants are present in coastal areas whether or not there is any human influence on their origin. They become pollutants only when their concentrations have been increased by human activity to a level at which they are damaging to the coastal ecosystems. The materials are nutrients (inorganic compounds that support plant growth such as nitrates and phosphates), sediments, metals, natural organic compounds including petroleum, and pathogenic organism.

A. Nutrients

Nitrogen and phosphorus compounds are the nutrients that most often limit plant production in the coastal zones. Potassium, another important nutrient, is abundant in seawater. The micronutrients are present in about the necessary ratio with regard to the macronutrients. Phosphate in readily bound to sediments, so that in waters shallow enough to support rooted plants that can take up nutrients from sediments, nitrogen may often be more limiting than phosphorus. Eutrophication is a widespread problem in shallow, enclosed coastal areas where exchange with the open ocean is restricted

Nitrogen and phosphorus enter the coastal areas in large amounts in river runoff. The rivers are fed by nonpoint sources-runoff from agricultural use of fertilizers and soil erosion-and from point sources such as sewage and industrial waste outfalls. Nearly 600 municipalities discharge wastes directly into marine coastal waters in the United States. Signficant amounts of nitrogen can also come from the air in the form of nitrate in acid rain. The source can contribute over 10% of load of nitrogen to coastal waters in the northeast U. S. coastal waters. Typically, the concentrations of nutrients in coastal waters are low, and plants respond to increases in concentratiory with increases in production. In moderate amounts this merely increases the primary production, increasing the food for consumers such as fish and shellfish. In high concentrations, it stimulates production to the extent that, in areas with restricted circulation, night time oxygen consumption by plants and bacteria may produce anoxia, bad odors, and fish and shellfish kills. The areas affected can be large: The deep parts of the central Baltic and of Chesapeake Bay, and the Louisiana coast west of the Mississippi mouth all experience anoxic events thought to be connected with eutrophication. [See Acid Rain; Oxygen, Biogeochemical Cycle]

But there are more subtle effects at intermediate concentrations, less easy to understand and not so readily traced to eutrophication. Sea grass beds have declined in estuaries as a result of stimulation of the growth of epiphytes on their leaf blades. The epiphytes shade the grass leaves and limit their growth at the deeper parts of their depth distribution. An overabundance of phytoplankton in the water as the result of a nutrient-stimulated bloom can produce a similar effect. In both cases, the algae intercept photosynthetically active light before it can reach the seagrass leaves. Excess nutrients also stimulate the growth of green algae, such as Ulva (sea lettuce) Cladophora and Enteromorpha, more than that of sea grasses and lead to a replacement of the latter by the former. On a large scale of where important fisheries (e.g., bay scallops) are dependent on sea grasses, the results are often deleterious. In another setting, the green algae may be contributing to an important food chain.

Both phosphorus and nitrogen tend to accumulate in estuaries. Phosphorus is strongly attached to clay particles and precipitated by iron under oxic conditions. Since estuarine circulation tends to trap sediments in the estuary, phosphorus is similarly trapped. Ammonium is also sorbed onto sediments. But nitrate, the oxidized form, is not and moves with the water. Both nutrients are built into plankton and sink with organisms that settle out or in fecal pellets of grazers, moving to the bottom and accumulating there. Sediments on the shelf also tend to be brought into the estuarine trap with bottom currents. Toward the shelf edge or at the heads of submarine canyons, sediments may be transported off the shelf into deeper waters and be lost from coastal waters. [See Phosphorus, Biogeochemistry.]

Sediments are not a final sink for nutrients, however. Microbial activity mineralizes organic matter in sediments, releasing nutrients that can diffuse up into the overlying water and enter the productive cycle again. Thus nutrients can contribute to eutrophication many times before they may be lost by deep burial, harvest of fisheries, or sedimentation off the shelf edge.

There is one significant difference between phsophorus and nitrogen in their contribution to coastal pollution. Nitrogen in the form of nitrate can take the place of oxygen in the respiration of certain bacteria in anoxic environment, supporting microbial degradation of organic matter that is nearly as efficient as that supported by oxygen. The end product of this denitrification is dinitrogen gas (N_2), which passes to the atmosphere and is lost from the coastal pollution cycle. [See Nitrogen Cycle of the Sea.]

Groundwater entering the ocean is another source of nutrients, but because of the strong sorption of phsophate by soil particles and the oxidizing condition in groundwater, the only significant nutrient transferred by this route is nitrate. A properly functioning traditional septic system (one without a denitrifying step built in) directly transfers its nitrate to groundwater. Nitrate transfer does not occur when the drainfield is clogged and the effluent comes out on the surface of the ground; there it gets into the soil and is partially denitrified and partly taken up by plants. In areas with dense housing close to the coast, served by working on-site septic systems, groundwater transport of nitrate is an important source of pollution.

The effect is exacerbated by the fact that the polluted water added to groundwater closest to the ocean's edge will enter the sea in the shallow, nearshore waters where the shellfish beds and fishery nursery areas are usually most important. Since the nitrogen is transported as nitrate, this transport may be partially offset by denitrification if the groundwater passes through an anoxic, organic-rich layer at the bottom of the coastal water. [See Groundwater Transport of Contaminants.]

Another locally important source of nutrient pollution can be mariculture. Feeding of confined fish or rafted shellfish results in unconsumed feed and feces falling go the bottom. Here, nutrients can be regenerated into the water column resulting in problems associated with eutrophication such as anoxia. This is parallel to problems with land-based agriculture except that here the nutrients are placed directly into the coastal waters rather than arriving in runoff or groundwater. While mariculture is not a large source of nutrient pollution on a global basis, it can be a problem in a local, enclosed area with poor circulation, as in some waters of Hong Kong.

B. Sediments

Sediments are not usually thought of as pollutants, although they are responsible for transport of sorbed pollutants and for their distribution in coastal waters. But sediments themselves act as pollutants, damaging coastal systems where and when their abundance and/or type has been drastically changed by human activity. Rivers such as the Nile, Amazon, Yangtze, and Mississippi carry enormous loads of sediments to their deltas. The coastal waters in such areas are highly turbid; benthic faunas are subject to burial and to anoxic events. To the extent that these are natural conditions, and the faunas and floras are adapted to them, we do not class the sediments as pollutants. But when sediments are introduced into otherwise clear or relatively clear waters, the effects can be significant. [See Rivers.]

Increased deposition can bury benthic organisms and start the creation of intertidal flats and marshes. Land clearing for farming in Virginia during the seventeenth and eighteenth centuries greatly increased erosion, loading previously clear rivers with sediments, filling in harbors, and creating new marshes beyond the previous shoreline. [See Erosion and Weathering.]

The reduction of sediment loads can be equally important and should be mentioned here. Damming of rivers causes sediment to be held in the reservoirs and reduces the supply to the coasts. Leveeing of the river-mouth regions can prevent floods from depositing sediments on deltas. In either case, as a result of sediment compaction and/or sea level, rise, the delta may be progressively submerged without new sediment supplies to maintain its level. This is happening dramatically in the Mississippi delta, with annual land losses of many square kilometers. The reduction in sediment transport also reduces the nutrients carried sorbed to sediments and can cause a reduction in coastal productivity. This appears to be happening in the eastern Mediterranean since the construction of the Aswan Dam on the Nile. Eventually the sediments trapped in the river basin will be released as the reservoirs fill or river processes remobilize sediments stored on floodplains with consequences for coastal waters mentioned above. [See Sediment Transport Processes.]

Damming of rivers and diversion of water for consumption by cities or agriculture can significantly increase salinities in estuaries and deltas and change the biota. For example, oysters are most productive in brackish waters that are too fresh for their predators to survive in. Freshwater diversions may increase salinity and allow predators to move upstream and destroy the oyster beds.

C. Pathogens

Human pathogens, bacteria, protozoa, and viruses are what many people first think of in connection with coastal pollution. Measurements of coliform bacteria in the water are used to regulate closing of shellfish beds for harvest and beaches to swimming. Both fecal coliform and total coliform are typically used to indicate the presence of pathogens because they are easy to measure. They are not the best indicators of probable presence of human pathogens. In some coastal areas, fecal coliforms from dogs and waterfowl are abundant enough to cause shellfish beds to be closed. Coliform bacteria themselves are generally not disease causing, although they may be responsible for mild dysentery.

Public health officers may close areas to shellfishing or swimming because of high coliform counts without evidence of disease-causing organisms. While this action may result in a financial loss to local industry dependent on those areas, at least the error is in the direction of caution. A real danger can result from the presence of pathogenic organisms in the absence of enough coliforms (which do not survive particularly well in seawater) to signal danger. Authorities are beginning to look at fecal streptococcus and Clostridium perfringens as additional indicators of human fecal pollution.

We have little information on the fates of viruses in coastal waters. It is clear from numerous examples that some viruses such as enteric hepatitis A (which causes the potentially fatal disease infectious hepatitis) can survive long enough in coastal waters to be transmitted to people, most frequently through raw or insufficiently cooked shellfish. Viruses are also responsible for the most prevalent gastroenteritis acquired by swimming. It only requires a few viable bacteria, and perhaps only a single viral particle to transmit a disease in appropriate conditions. Bacteria and viruses tend to accumulate on particles. This improves their survival in water, concentrates them in packets, and thus increases the chances that they will be eaten and accumulated by filter-feeders such as shellfish.

It may be possible for people to be infected by pathogens in seawater without going into the water or eating raw seafood. Bubbles from waves breaking at the surface shoot tiny droplets up into the air. These can carry bacteria (and presumably viruses) into a land breeze, which can carry them up onto the coast. Bacteria from the sea have been captured and grown on bacterial culture media exposed to seaside air. Perhaps they could also infect people breathing that air.

D. Heavy Metals

Heavy metals or trace metals are present in relatively small amounts in ocean waters or sediments. Iron and calcium are relatively abundant and used by many organisms. Cobalt, copper, and zinc are used in low concentrations; at larger concentrations they are toxic to organisms. Some metals, such as lead or mercury, have no known biological function. A town in Japan is an example of what can happen with metal pollution concentrated in a small coastal area. A chemical plant discharged waste containing mercury into Minimata Bay for a number of years. In the monomethyl form, mercury is soluble in water and readily taken up by fish and shellfish. When these were eaten by the local residents, the mercury caused a wide range of problems from death and paralysis to birth defects.

One way to avoid metal pollution in coastal waters is to ensure that metals in runoff or point-source discharge are diluted. In oxygenated seawater, metals are all soluble at low concentrations as chloride, carbonate, or hydroxide complexes so the "dilution solution," if achievable, would result in metal concentrations too low to be damaging.

The difficulty for all the pollutant metals lies in local, excessive concentrations. Even so there are not many examples of concentrations of metals in coastal waters that present problems. In areas contaminated with metals, they are stored in the sediments often as highly insoluble sulfides, or oxides. The metals tend to remain where they come to lie because of this insolubility and because they are frequently buried by subsequent sedimentation once the inputs have ceased. Potential problems from metal pollution are also reduced by the fact that many animals have biochemical systems for metal sequestering and excretion.

In the late 1980s alarm was generated by the use of toxic tributyl tin as an ingredient in antifouling paints for boats. It was subsequently banned in the United States for civilian use. But in this example, Minimata Bay, and other similar cases, the problems are local. If the metals do escape from the sediments they tend to be diluted rapidly by currents and not reconcentrated locally. [See Butyltins, Effect on Marine Chemistry.]

The two metals of most concern are mercury (because it can be methylated in sediments, in which form it is readily taken up and accumulated by organisms) and cadmium (because it is also accumulated in organisms and is more readily available from sediments than most heavy metals, although it is not methylated). Cadmium, widely used in metal plating, is soluble in the metallic form, and so it is present in municipal sewage discharges. It has, like mercury, been responsible for poisoning of people who have eaten polluted shellfish.

E. Natural Organics

Petroleum hydrocarbons (PHC), of which there are hundreds of thousands of different chemical compounds, are the natural organic compounds commonly involved in coastal pollution events that get wide publicity. Other compounds that could fall into the category of natural organic pollutants-antibiotics and allelopathic compounds (by which organisms reduce competition with other species)—are not ordinarily though of as pollutants, although they have effects similar to small amounts of pollutants.

Petroleum can enter coastal ocean in huge amounts from oil well blowouts, tanker accidents, and other human mishaps. There are other sources of hydrocarbon inputs, less obvious but no less significant. High-temperature combustion from forest fires, automobiles, and industrial sources produce hydrocarbons that are discharged into the air and enter coastal waters directly as atmospheric fallout or fall on the land and are washed into rivers and coastal waters. Oil spilled on streets washes into water bodies and eventually finds it way into the ocean. Leaks from natural oil reservoirs, oil seeps, contribute to coastal pollution and are large enough in some areas (e.g., off the California coast at Santa Barbara) to be smelled from shore.

In massive oil spills the quantities of oil can be sufficient to smother organisms, kill them by simple entrapment, or, in the cases of mammals and birds, kill by destroying the insulating value of their fur or feathers. But in most spills the toxic properties of the compounds are the most damaging. Many of the smaller molecules are anesthetics and disrupt normal functions. The aromatic compounds and some of the simpler nonaromatic ring compounds (cyclic alkanes) are directly toxic. The larger aromatic or their metabolic products are also carcinogenic and teratogenic (causing birth defects). Aromatic compounds with more than one benzene ring in their structure are called polycyclic aromatic hydrocarbons (PAH) and are lumped together in this discussion.

Straight-chain hydrocarbons are also found in plant and animal waxes and are readily used as energy sources by many microbes in oxygenated waters. The lighter PHC are volatile and are lost to the air. The cyclic alkanes and PAH are most resistant to decomposition, especially in anoxic situations such as marine sediments, where they can last for decades. They are degraded slowly by microorganisms and often only as the microbes are metabolizing other, more readily used organic molecules. Since PHC are preserved for millions of years in the oxygen-free sediments and rocks from which we obtain them, their persistence is not surprising.

The result of this long residence time in marine sediments is that, in highly polluted areas such as major harbors, there can be a high level of PAH in the sediments and slow degradation balanced by continual inputs of new pollutants. This assures that organisms living in and on such sediments will be exposed to the PAH throughout their lifetimes. Hydrocarbons are only slightly soluble in water but very soluble in lipids, including the lipids that compose cellular membranes. PHC are therefore concentrated in organisms and passed along the food web, often concentrated as they move to higher trophic levels.

Organisms have biochemical enzymes systems for the degradation of organic pollutants such as PAH. These work by introducing oxygen into the molecules, which can then be joined to more soluble molecules and excreted or further degraded. Unfortunately these initial products of the metabolic process are much more reactive than the parent compounds and will combine with cellular constituents such as DNA to disrupt normal gene function; thus they are carcinogenic and teratogenic. These partial breakdown products are also persistent in the environment.

In areas of considerable PAH pollution, such as Boston Harbor, fish have a high incidence of liver tumors of types shown to be caused by PAH exposure. Whether or not be causative agents of these tumors can be transmitted through the food chain has not been demonstrated.

III. Unnatural Pollutants

The pollutants discussed above are all materials found in nature, which enter the coastal oceans in some amounts regardless of human presence: The main influence of human activity has been to accelerate the rate at which these materials enter the oceans and to concentrate them in certain areas. The net result is that we have produced harmful concentrations of materials the were previously if not harmful concentrtions or materials that were previously, if not harmless, at least manageable by the organisms and ecosystems which had evolved in the presence of these materials. The following categories are of compounds or elements completely new to life, created by human activity.

A. Radioisotopes

Radioisotopes were always present in the coastal oceans as they were everywhere on earth, but at very low concentrations. Now we are confronted with much higher concentrations, made by humans, and with the transuranic elements that have not been present on earth since life evolved. Our concern with these elements is not their direct toxicity, but the fact that they are radioactive and therefore damaging to life at much lower concentrations than other elements. We measure their presence in terms of the effect of radio decay on humans rather than absolute concentrations. [The present limit is 5 mSv (milliSieverts) or 500 mrem (Millirem) per year.]

The United Kingdom is one nation that has evolved strict regulations for coastal discharge of radioisotopes. A nuclear fuel processing facility on the coast of the Irish Sea discharges radioactive waste into the sea. The Permissible limits are set by finding the individual(s) who would be most at risk,"the critical pathway." The discharge limit is then set to protect that individual. When they system was set up in the 1950s the individual(s) at highest risk used a red algae called laver (Porphyra) to make laverbread and their risk was from shortlived insotopes accumulated by the algae. The short-lived insotopes were discharged only through the early 1970s, after which the cesium plutonium, and americium insotopes provided the limit through a critical pathway involving fish, crustacea and mollusks. It is estimated that 80% of the plutonium discharged in locked up in sediments. The rest in mostly flushed away in the water. But the remainder, less than 0.01% of the discharge, is accumulated by alge and passes through the food chain to the human foods listed.

Note that this manner of regulation is concerned only with the human effects, not any effects on other organisms or ecosystems. Since human are the top predators and stand in a position to accumulate the most radiation, this may be a reasonable approach. It is certainly the easiest.

B. Synthetic Organic Compounds

PCBs and DDTs are among the best known of the compounds created by humans that subsequently caused pollution problems in coastal waters and elsewhere. On the order of 100,000 compounds have been created, with half that many in common use. About 1000 of these are made in sufficient quantity to pollute the world if released. The U. S. EPA lists over 120 compounds as priority pollutants. Most of the latter list are halogenated compounds, organic molecules that have chlorine or bromine as a part of their structure. The halogens are responsible for both the biological activity and long environmental life of these compounds. Most of the compounds are also more lipid-soluble than water soluble, and so they concentrate in organisms and sorb to particles. Few of these compounds are problems only in coastal waters, but since all runoff from the land and a considerable fraction of airborne pollutants eventually end up in the ocean, the problems are often exacerbated at the coast.

The greatest concentrations of organic pollutants are found in harbors and river estuaries (which tend to be the same places). The compounds are carried down the rivers attached to sediments , which settle out in the estuaries or, if they are relatively soluble get precipitated as the water becomes saltier(are salted out). The worldwide "Mussel watch" program, which used the concentrations of organic pollutants in mussels as a monitoring tool in coastal waters, found much higher concentrations of a wide variety of natural and synthetic organic pollutants in busy estuaries than along open coasts. Since the synthetic organic are even more resistant to decomposition, they are even more resistant to decomposition than ? AH, because their "halogen coats" inhibit degradation, they are even more long lived in marine sediments than PAH. They are also more readily concentrated in organisms because the enzyme systems for ridding the body of pollutants are also less effective against the heavily halogenated compounds.

New Bedford harbor in Massachusetts was the site of two plants that manufactured electrical capacitors for about 30 years beginning in 1947. Capacitors were filled with a PCB mixture because it was a good insulator and fireproof. Moreover, PCBs were considered to be biologically inert and harmless. The factories discharged their wastes, which contained a large amount of spilled PCB mixture, directly into the harbor. Spoiled capacitors were dumped on land but groundwater continued to carry the PCBs to the coast.

It is difficult to separate out the effect of the high concentrations of PCBs now found in the harbor sediments from that of all the other pollutants, but it has been shown that reproductive failures in clams are due to those pollutants. A more serious result for people directly is the closing of a large area in the adjacent bay and all of the harbor to fishing. clamming,and lobstering, all this in one of the most important fishing ports in the United States. But PCBs are accumulated in animals, especially in fatty tissues; they cause nerve, skin, and liver damage,and promote tumor formation-so they should be kept out of human diet. Many people no longer eat the "green" in lobster as they used to do, because this is the site of greatest concentration of PCBs and other pollutants.

Another class of synthetic organic compounds that pollute coastal waters in large amounts are plasticizers (phthalates). These compounds are used in the manufacture of many plastics and plastic objects. Although their solubility in water is similar to that of petroleum hydrocarbons, they are so widely used they become ubiquitous in polluted waters. They are not particularly toxic, but some are teratogenic.

Plastic objects themselves have become another important pollutant in coastal waters. Objects such as six-pack holders can catch around bird necks and strangle them. Plastic bags, eaten by sea turtles mistaking them for jellyfish, may clog digestive tracts and lead to starvation. Tiny plastic particles, lost in manufacturing processes or from breakup of objects such as cups, are eaten by small marine animals, larval fish, and affect them the same way that the larger objects do the turtles.

Chlorinated pesticides are the last large category of synthetic pollutant organic compounds important in the sea.The story of DDT and its degradation products DDE and DDD well known. These compounds are gradually disappearied from U. S. coastal waters since their use in this country we banned in the early 1970s. But DDT is still widely used other part of the world, especially in the tropics and waste use other chlorinated compounds with long-lasting environmental effects despite increasing regulation. These conpounds are of concern both for their potenial for transfer to man through marine food sources. They have a range of toxicities, and many at carcinogens.

Maricultural use of pesticides, like the use of nutrients introduces the chemicals directly into coastal waters. Most of the use is of antibiotics, but an anticholesterinase,, nerve poison(Duvan) that will kill a variety of other marine invertebrates, is used against fish lice in salmon culture. The eclogcal consequences of their use is not understood.

C. Complex Mixtures

Most of what we dump directly into coastal waters is a broad mixture of pollutants and contaminants. Sewage sludge and dredge spoils are good examples. Even when the materials are relatively simple at their sources, they are mixed in the receiving coastal waters. The result is that the ecosystems are subject to insults from a variety of pollutants at the same time. The effect of one pollutant is frequently enhanced in the negative sense by the effect of another; this synergism results in more severe effects than would be produced by just the sum of the separate effects.

Not surprisingly, these synergisms are not well understood. Considerable study in required to understand the effects of just one group of pollutants. Study of the effects of coupling of that first group with other groups increases the research effort exponentially. Some of the more impressive sites where we have found damage to sea life-Boston, New York, and Seattle-are undoubtedly due to synergisms between the complex mixtures of pollutants present in those areas.

IV. Remedies

A. Prevention

In theory, much of the input of coastal pollution could be reduced by stronger regulations, interstate compacts, and incentives or alternatives to disposing of contaminants in the streams, road sides, groundwater, and coastal waters. We could modify our practices to stop putting these materials into the sea. We could stop manufacturing and using many dangerous chemicals just as we have stopped the use of the most persistent chlorinated pesticides in the United states. We could greatly improve the recovery of metals from industry and sewage treatment systems. Sediment erosion, water overuse and misuse can be stopped by better conservation measures. The wealthiest countries are in the best position to do this, but the records of North America, Europe, and Japan are not impressive. The poorer countries are unlikely to do much better. It is more likely that they will do worse as they invest their limited resources in trying to improve their national economies rather than in preventing pollution.

Even in the best of circumstances, some sorts of pollution will be hard to prevent. The occasional oil spill seems inevitable, though we can certainly do much better than we have at reducing its occurrence. The occasional accident on land that results in pollution of a river will continue to cause coastal pollution. Probably the hardest to prevent will be pollution by nutrients. Runoff of nutrients from farms can be reduced be careful application at rate that the plans can use effectively, and by preserving and increasing natural vegetation and wetlands bordering rivers and coasts. The latter both intercept nutrients before they reach open water and stimulate denitrification. Of course, nutrients in groundwater not intercepted will continue to show up for many decades to come even if we stop all inputs now.

B. Cleanup

Cleanup of coastal ocean areas means cleaning up the ocean bottom, not the water column. Waters are exchanged rapidly enough so that if inputs cease, existing pollutants will be diluted to low concentrations. Cleaning the bottom of a harbor is a tremendous undertaking. Dredging polluted sediments and burying them beneath a layer of clean sediment deep enough that animals will not burrow through it is a potenial, costly solution for limited areas. Even the process of dredging will stir up a lot of material, putting it back into the water, where it can have a deleterious by currents. Preventing this makes the process much more difficult and costly. In restricted areas this may be a viable solution. However, in most areas, preventing additional pollution, combined with natural degradation and burial processes, is the most likely cleanup scenario. Nature in capable of remedying many of our mistakes, if we will just refrain from making things worse.

GLOSSARY

Anoxla. Complete absence of dissolved oxygen in water.

Coliforms. Bacteria that normally live in intestines; fecal coliforms normally live at body temperature of warm-blooded animals, 37°C.

Denitrification. Production of nitrogen gas by bacteria from nitrate (NO_3) in anoxic environments.

Eutrophication. Enrichment by plant nutrients resulting in higher productivity; at highest levels can result in anoxia.

PAH. Polynuclear aromatic hydrocarbons, compounds composed of two or more benzene rings.

PCB. Polychlorinated biphenyls, a family of toxic, oily, non-combustible compounds manufactured for insulating fluids.

21

Soil Analysis

Soil/Sediment Sampling

Soil is nothing but a weathered layer of the earth's crust which is having living organisms and products of their decay intermingled.

When the soil is submerged in water, it is known as an aquatic sediment.

Collection of Soil Sample

Whenever a soil sample is to be collected, one should keep in mind the purpose of study and parameter(s) in question while selecting sampling sites, depth, and method to be used for collection of soil sample. In order to study general characterisation of soil, it becomes desirable to collect a few random samples from the study area to the depth of about 15 cm.

It is possible to collect surface soil samples by employing tools like shade or shovel. One should not collect soil sample from just near the root of a large tree. Further one should also avoid the soil samples near the foundations of civil constructions because these do not represent their natural make up. Whenever soil samples from deeper profiles have to be used, one should use special borer samplers.

Soil samples collected by the above methods should be stored in thick quality polythene bags and immediately transferred to the laboratory. In order to determine parameters such as pH, redox potential, nitrogen, and phosphorus etc., analysis should be undertaken as early as possible. For many other edaphic factors the sample could be stored after drying it at about 40°C. The dried soil could be ground using mortar and pestle and sieved through a 2 mm meshsized sieve.

Collection and Handling of Aquatic Sediments

It is possible to collect the aquatic sediments by using special instruments, like scoop, grab, and core samplers.

PHYSICAL PARAMETERS

Bulk Density

One may define the bulk density of the soil as the dry weight of a unit volume of it. It may be expressed as g/cm^3. Generally the bulk density has been found to range from 1.1 to 1.5 g/cm^3 for medium to fine textured soil and from 1.2 to 1.65 g/cm^3 for coarse textured soil. However, it has been slightly higher in case of alkaline saline soils. The soils having high slightly higher in case of alkaline saline soils. The soils having high bulk density have been found to be inhibitive to root penetration, and have low permeability and infiltration. The bulk density has been inversely related to pore space of soil.

Material

Oven; measuring cylinder; and chemical balance.

Method

Collect the sample and dry it in an oven at 105°C until we get its constant weight. Now put a little dried soil to a measuring cylinder and record the volume. Now find out the weight of this volume of soil on a balance.

Calculation, It is found out be the following equation:

$$\text{Bulk density} = \frac{\text{Weight of soil (g)}}{\text{Volume of soil (cm}^3\text{)}}$$

$$= \ldots \text{ g/cm}^3$$

Specific Gravity

The specific gravity of the soil has been directly related to its bulk density. It has been used as an index of some aspects of soil quality as in case of bulk density.

Material

Oven; glass bottles; and chemical balance.

Method

Collect the soil sample and dry it in an oven at 105°C until we get a constant weight. Fill a pre-weighed glass bottle of known volume with dried soil and note down its weight. Fill it with distilled water and record its weigh.

Calculation

$$\text{Specific gravity} = \frac{x_2 - x_1}{x_3 - x_1}$$

where x_1 = weight of empty bottle

x_2 = weight of empty bottle + soil

x_3 = weight of empty bottle + water

Moisture Content

Water present in the soil serves the following purposes:

(i) It acts as a solvent.

(ii) It acts as a transporting agent.

(iii) It maintains texture and compactness of soil, thereby making it habitable for plants and animals.

Soil gets moisture from infiltration of precipitated water and irrigation. Its constent in the soil at any time is more or less dependent upon the water holding capacity of soil. However, it drains through percolation, evaporation, and uptake by plants.

Material

Oven; and chemical balance.

Method

Collect a fresh homogenized sample of soil. Now weigh it. Then dry it in an oven at 105°C until we get a constant weight. Cool it in a desicator. Finally record the final weight of sample.

Calculation

$$\text{Moisture content (\%)} = \frac{x_1 - x_2}{x_1} \times 100$$

where x_1 = Initial weight of sample (g)

x_2 = Final weight of dried sample.

Water Holding Capacity

When we saparate the soil absolutely with water, water fills all the proes between the particles of soil and no air space exists as in the case of aquatic sediments. Such a soil is said to be at its maximum water holding capacity or saturation.

Material

Perforated circular soil boxes filter paper Whatman NO. (1); Petri dish; chemical balances and oven.

Method

Collect the soil sample and crush it. Dry this sample in an oven at 105°C. Keep a filter paper (Whatman No. 1) inside the perforated bottom of the circular soil box. Then weigh the box. Now fill it with dried soil sample. Then, find out the weight of box filled with dried soil. Keep the box in Petri dish of 10 cm diameter, having water, for about 12 hours, so that water enters the box and gets saturated the soil. Remove the box out of water, wipe it dry on the outside, and find out its weight.

Calculations

$$\text{WHC}(5) = \frac{(W_3 - W_1) - (W_2 - W_1)}{(W_2 - W_1)}$$

where W_1 = Weight of empty box

W_2 = Weight of box with dried soil

W_3 = Weight of box with water saturated soil

WHC = Water holding capacity.

Chemical Parameters

The major inorganic constituents of soil are the compounds of Al, si, ca, Mg, Fe, k and K. However, it also contains minor quantities of B, Mn, Cu, Zn, Mo, Co, I and F.

The main organic constituent of soil is humus. Other organic constituents of present in the soil are proteins, amino acids, purines, pyrimidines, sugars, fats, aromatic compounds, alcohold, waxes and resins.

Hydrogen Ion Concentration (pH)

Weight 10 g af air-dry soil/sediment. To this add 100 ml of distilled water for making a suspension of 1 : 10 w/v dilution. Find out the pH of suspension by using the method discusses for water. Give the dilution while reporting the result.

Electrical Conductivity (EC)

Weigh 10 g of air-dry soil/ sediment. To this add 100 ml of distilled water a for making suspension (1 : 10 w/v). Determine the EC as described earlier. Give the dilution while expressing the result.

Redox Potential

Weigh 10 g of air-dry soil/sediment. To this add 100 ml of distilled water for making a suspension (1 : 10 w/v). Record the redox potential of suspension as that of water. Give the dilution while expressing the result.

Alkalinity

Weigh 10 g of air-dry soil/sediment. To this add 100 ml of distilled water to get a suspension (1 : 10 w/v). Now filter it through a filter paper (Whatman No. 44). Measure the alkalinity of filterate by using the method described for water. In a fraction of same soil/ sediment sample the moisture content is found out be the method as given earlier.

Calculation

$$\text{Alkalinity (mg/g)} = \frac{A}{10} \times \frac{V}{x} \times \frac{1}{(100 - M)}$$

where, A = alkalinity of filtrate (mg/1);

V = total volume of suspension (ml);

X = weight of soil/sediment used in suspension (g); and

M = moisture content of soil/sediment (%).

Chloride

Weigh 10 g of soil/sediment. To this add 100 ml of distilled water for making a 1 : 10 w/v suspension. Filter it through a filter paper (Whatman No. 44). Now determine the chlorides in filtrate by employing the method given for water. From a fraction of same soil/sediment sample find out moisture content as described earlier.

Calculation

$$\text{Chloride (mg/g)} = \frac{A}{10} \times \frac{V}{x} \times \frac{1}{(100 - M)}$$

where, A = chloride estimated in filtrate (mg/1);

V = total volume of suspension (ml);

x = weight of soil/sediment used in suspension (g); and

M = moisture content of soil/sediment (%).

Sulphate

Weigh 10 g of air-dry soil/sediment. To this add 100 ml of distilled water for making a suspension (1 : 10 w/v). Filter it through a filter paper (Whatman No. 44). Now find out the sulphate in it by using the method described for water. From a fraction of same soil/sediment sample the moisture content is estimated by following the method given earlier.

Calculation

$$\text{Sulphate (mg/g)} = \frac{A}{10} \times \frac{V}{x} \times \frac{1}{(100 - M)}$$

where, A = sulphate estimated in filtrate (mg/1);

V = total volume of suspension (ml);

x = Weight of soil/sediment used in suspension (g);

M = moisture content of soil/sediment (%).

Nitrogen (Kjeldahl nitrogen)

Kjedahl nitrogen is including organic and ammonia forms of nitrogen but is not including the nitrate-nitrogen.

Material

1. Kjeldahl distillation assembly (Fig. 8.1).

(a) *Catalyst mixture*. Now grind a mixture of 20 g of copper sulphate, 3 g of murcuric oxide, and 1 g of selenium powder. Mix 2 g of this mixture with 40 g of sodium sulphate to get the catalyst mixture.

(b) *Sulphuric acid* (concentrated).

(c) *Sodium hydroxide solution* (40%). Weigh 40 g of sodium hydroxide. Dissolve in 100 ml of distilled water.

(d) *Zinc granules*.

(e) *Boric acid cum indicator solution*.

(f) *Hydrochloric acid* (0.1 N). Dilute 8.34 ml of concentrated hydrochloric acid (12N) with distilled water to get 100 ml of 1.0N HCL. Dilute this 100 ml of 1.0N HCL to 1 *l* by adding distilled water to obtain 0.1N HCL.

Method

Weigh 8 g of air-dry soil/sediment. Put it in a 300 ml round bottom flast. To this add 20 g of catalyst mixture (reagent A) and 35 ml of sulphuric acid (reagent B). Now heat the contents from bottom of the flask for nearly 2 hours. After cooling the contents (digest) add about 100 ml of distilled water. After waiting for about 5 minutes, deliver the supernatant into 1 *l* distillation flask A of Kjeldahl distillation assembly (Fig. 8.1). Wash the residue with a little of distilled water many times and transfer the supernatant each time to the same distillation flask. To this add 100 ml of sodium hydroxide solution (reagent C) and a few granules of zinc (regent D). Now put 25 ml of boric acid cum indicator solution (regent E) in a 500 ml Erlemeyer flask B and keep it below the condensor C of distillation assembly so that the lower open end of the condensor gets dipped in solution. Now heat the distillation flask of a hot plate and collect about 150 ml of distillation flask on a hot plate and collect about 150 ml of distillate in flask B. Remove the flask with distillate. Then titrate the distillate (which has become blue due to dissolution of ammonia) against 0.1 N hydrochloric acid (reagent F). Changing of blue colour to light brown-pink reveals the end point. Also run a distilled water by using the same method.

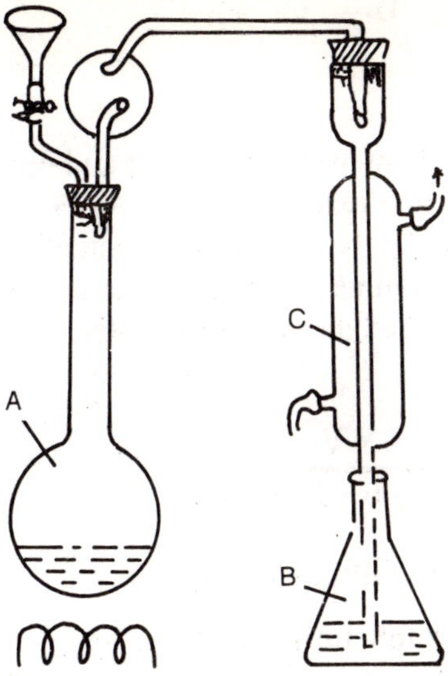

Fig. 21.1. Kjeldahl disttilation assembly

Calculation

$$\text{Kjeldahl Nitrogen (mg/g)} = \frac{(V_1 - V_2) \times N \times 14}{X}$$

$$\text{Kjeldahl Nitrogen (\%)} = \frac{(V_1 - V_2) \times N \times 14}{X}$$

where, V_2 = volume of titrant which is used against sample (ml);

V_2 = volume of titrant which is used against distilled water blank (ml);

N = normality of titrant (o.1); and

X = weight of soil/sediment which is used (g).

NITRATE

Material

Laboratory reagents (in addition to those used for nitrate estimation in water as given below:

(a) *Extraction reagent.* (a) Weigh 12.5 g of copper sulphate. Dissolve if in distilled water to make 100 ml of solution. (b) Weigh 0.6 g of silver sulphate. Dissove it in distilled water to make 100 ml of solution. Mix 20 ml and 100 ml of solutions (a) and (b) respectively. To this add distilled water to make 1 *l* of nitrate extraction reagent.

(b) *Calcium hydroxide*. Dry, powdered.

(c) *Magnesium carbonate*. Dry, powdered.

Method

Dry the soil/sediment in air. Now weigh 50 g of it and put it in an Erlenmeyer flask (500 ml). To this add 250 ml of extraction reagent (reagent A). Shake for 5 minutes. Then add 1 g of magnesium carbonate (reagent C). Now filter the contents through filter paper (Whatman No. 50). Then measure the total volume of filtrate. Find out the nitrate content in filtrate by usiung phenol disulphonic acid method discussed for water.

Calculation

$$NO_3\text{-N (mg/g)} = \frac{Y \times V}{1000 \times X}$$

where, Y = NO_3-N determined in filtrate (mg l);

V = total volume of filterate (ml); and

X = weight of dried soil/sediment used (g).

TOTAL PHOSPHORUS

Material

(i) All reagents which are used in estimation of inorganic phosphorus in water; (ii) filter paper (Whatman No. 44); (iii) other reagents which are given below :

(a) *Nitric acid* (concentrated).

(b) *Perchloric acid* (concentrated).

(c) *Sulphuric acid* (diluted). Dilute 5 ml of concentrated sulphuric acid with distilled water to 100 ml.

Method

Dry the soil/sediment in air. Then grind to fine powder. Now put 0.5 g of it in a round bottom flask. To this add a few drops of distilled water, 2 ml of nitric acid (reagent A), and 2 ml of perchloric acid (reagent B). Heat gently on a hot plate to dryness stage.

Cool it. To this add 20 ml of sulphuric acid (reagent C). Then boil for 15 minutes. Cool the flask. Then filter the digest through a filter paper (Whatman No. 44). Now make up the volume of it to 250 ml by adding distilled water in a volumetric flask. Measure out the phosphate content in solution by using the method discussed for estimation of inorganic phosphorus in water.

Calculation

$$\text{Total phosphorus (mg/g)} = \frac{Pd \times V}{1000 \times X}$$

where, Pd=PO_4 - P in digest (mg/1); V = total volume of solution (ml); and X = weight of air-dry soil/sediment taken (g).

PHOSPHATE

Material

(i) All regents, which are needed for the estimation of inorganic phosphorus in water : (ii) filter paper (Whatman No. 50); and (iii) other reagent, such as given below:

(a) *Sulphuric acid* (0.102 N). 2.8 ml of concentrated sulphuric acid is diluted to 1 *l* by adding distilled water. Measure out 20 ml it. Then, dilute to 1 *l* with distilled water to obtain 0.002 N sulphuric acid.

Method

Dry the soil/sediment sample in air. Now put 1 g of it in a 500 ml flask. To this add 200 ml of sulphuric acid (0.002N). Now shake for about half an hour. Now filter the suspension through a filter paper (Whatman No. 50). Find out the phosphate content in the filtrate by using the method discussed for inorganic phosphorus in water.

Calculation

$$PO_4\text{—P (mg/1)} = \frac{Ps \times V}{1000 \times X}$$

where, Ps = PO_4 - P estimated in suspension (mg/1); V = total volume of suspension (ml); and X = weight of air-dry soil/sediment taken (g).

CALCIUM

Material

(i) All reagents, which are needed for estimation of calcium in water; (ii) filter paper (Whatman No. 50); and (iii) other reagents such as discribed below :

(a) *Ethyl alcohol*. Absolute and 40%.

(b) *Ammonium acetate solution* (1N). Measure out 57 ml of glacial acetic acid. Now dilute it to 800 ml with distilled water. Add gradually the concentrated ammonium hydroxide solution till pH 7.0 has been attained. Now add distilled water to make the final volume 1 *l*.

Method

Preparation of soil extract. Dry the soil/sediment in air. Now put 50 g of it in a flask. To this add 100 ml of 40% of ethyl alcohol (reagent A). Then shake well, wait for about 10 minutes, and filter the suspension through filter paper (Whatman No. 50). Then wash the soil residue on filter paper with 40% ethyl alcohol and finally with absolute ethyl alcohol. Now put the residue into a breaker. To this add 100 ml of ammonium acetate solution (reagent B).

.Now stir, and keep it over-night. Filter the supernatant through filter paper (What man No. 50). Then collect the filtrate (soil extract).

Determination of calcium. Measure out the calcium in soil extract by using the method described for water.

Calculation

$$\text{Calcium (mg/g)} = \frac{V_3 \times 400.4 \times V_1}{V_2 \times X \times 10000}$$

where, V_3 = volume of EDTA titrant used (ml); V_1 = total volume of soil extract (ml); V_2 = volume of soil extract titrated (ml); and X = weight of air-dry soil/sediment which is taken for extaction (g).

MAGNESIUM

Material

(i) All reagents, which are needed for determination of magnesium in water; (ii) filter paper (Whatman No. 50); and (iii) other reagents, such as given below :

(a) *Ethyl alcohol*. Absolute and 40%.

(b) Ammonium acetate solution.

Method

The soil extract is prepared as described in case of determination of calcium. Determine magnesium content in the extact by using the method for determination of magnesium in water.

Calculation

$$\text{Magnesium (mg/g)} = \frac{(V_3 - V_2) \times 440.4 \times V_1}{V \times 1.645 \times X \times 10000}$$

where, V_3 = volume of EDTA titrant which is used for determination of calcium and magnesium (ml); V_2 = volume EDTA titrant which is used for determination of calcium (ml): V_1 = total volume of soil extract (ml); V_2 = volume of soil extract titrated (ml); and X = weight of air-dry soil/sediment which is taken for extraction (g).

SODIUM

Material

(i) All reagents which are needed for determination of sodium in water; (ii) filter paper (whatman No. 50); and (iii) other reagents, such as given below :

(a) *Ethyl alcohol*. Absolute and 40%.

(b) Ammonium acetate solution.

Method

The soil extract is prepared by the method as described for calcium. Find out the sodium content in extract by using the method of sodium determination in water.

Calculation

$$\text{Sodium (mg/g)} = \frac{Y \times V}{W \times 10000}$$

where, X = sodium content of soil extract (mg/1); V = total volume of soil extract (ml); and W = weight of air-dry soil/ sediment which is taken for extraction (g).

POTASSIUM

Material

(i) All reagents, needed for the estimation of potassium in water; (ii) filter paper (Whatman No. 50); and (iii) other reagents, such as given below :

(a) *Ethyl alcohol.* Absolute and 40%.

(b) Ammonium acetate solution.

Method

The soil extract is prepared by the method as described for determination calcium. Determine the potassium in extract as it has been determined in water.

Calculation

$$\text{Potassium (mg/g)} = \frac{Y \times V}{X \times 10000}$$

Here, Y = potassium content of soil extract (mg/1); V = total volume of soil extract (ml); and X = weight of air-dry soil/sediment which is taken for extraction (g).

IRON

Material

All reagents which are needed for the estimation of iron in water.

Method

Weigh 10 g of air-dry soil/sediment. To this add 100 ml of distilled water for making 1 : 10 w/v suspension. Filter it through a filter paper (Whatman No. 44). Now find out the iron in filtrate by using the method for iron estimation in water. From a fraction of soil/sediment sample find out moisture content as described earlier.

Calculation

$$\text{Iron (mg/g)} = \frac{A}{10} \times \frac{V}{X} \times \frac{1}{(100 - M)}$$

where, A = iron determined of filtrate (mg/1); V = total volume of suspension (ml); X = weight of soil/sediment which is used in suspension (g); and M = moisture content of soil/ sediment (%).

ORGANIC MATTER

Material

(a) *Potassium dichromate solution* (IN). It is prepared by dissolving 49.04 of potassium dichromate in distilled water to get 1 *l* of solution.

(b) *Sulphuric acid* (concentrated).

(c) *Phosphoric acid* (concentrated).

(d) *Sodium fluoride*. Dry, powdered.

(e) *Diphenylamine indicator*. It is prepared by dissolving 0.25 g of diphenylamine in 10 ml of distilled water and 50 ml of concentrated sulphuric acid is gradually added to it.

(f) *Ferrous ammonium sulphate solution* (0.5 N). 20 ml of concentrated sulphuric acid, is added to 800 ml of distilled water. Then dissolve in it 196.1 g of ferrous ammonium sulphate. Further add more distilled water to make the volume 1 *l*.

Method

Dry the soil/sediment in air. Now weigh 0.5 g of it. Then put it in 500 ml Erlenmeyer flask. To this add 10 ml of potassium dichromate solution (reagent A) and gradually 20 ml of sulphuric acid (reagent B). Keep it as such for about half an hour. To this add 200 ml of distilled water, 10 ml of phosphoric acid (reagent C), 0.2 g of sodium flouride (Reagent D), and 1 ml of diphenylamine indicator) reagent E). Then contents are tirated against ferrous ammonium sulphate solution (reagent F). At the end point the dull green colour gets changed through turbid blue to thc brilliant green.

Run a distilled water blank simultaneously.

Calculation

$$\text{OM (mg/g)} = \frac{6.79}{X}\left[1 - \frac{V_1}{V_2}\right] \times 10$$

$$\text{OM (\%)} = \frac{6.79}{X}\left[1 - \frac{V_1}{V_2}\right]$$

$$\text{Carbon (\%)} = \frac{6.791}{X \times 1.724}\left[1 - \frac{V_1}{V_2}\right]$$

In the above relation, OM = organic matter; X = weight of soil/sediment taken (g); V_1 = volume of titrant used against sample (ml); and V_2 = volume of titrant used againt distilled water blank (Ml).

22

Infrared and Raman Spectroscopy in the Earth Science

Introduction

Infrared and Raman spectroscopies are the two major techniques used to study the vibrational properties of materials. Vibrational spectroscopy is applied in a variety of ways in the earth sciences; as a versatile analytical tool, in detailed studies of structure and bonding of minerals and other earth materials, and as a technique for remote sensing of atmospheres and planetary surfaces.

A. Vibrations of Molecules and Solids

The atoms in a molecule or condensed phase are held together by chemical bonds, which determine the equilibrium structure. If the atoms are displaced from their equilibrium positions, the chemical forces tend to restore the structure to its original state. At any temperature, the atoms are in constant motion within the molecule, undergoing slight excursions from their equilibrium state. The interatomic force field determines that these excursions take the form of small periodic displacements about the equilibrium position of each atom, known as the normal modes of vibration of the molecule or material. Because each atom can move in three independent directions in space, there are 3N - 6 normal modes of vibration for a system with N atoms. The remaining 6 degrees of freedom correspond to translation (3) and rotation (3) of the system as a whole. Each normal mode is associated with a characteristic vibrational frequency, determined by the strength of the interatomic forces, the masses of the atoms involved, and their relative positions in the molecule. In a crystalline solid, the atoms are arranged in a regular periodic array, and the vibrational modes take the form of vibrational waves traveling through the crystal. These are characterized not only by the oscillation frequency of the atoms about their equilibrium positions but also by the wave-length of the lattice vibrational modes observed in infrared or Raman spectroscopy are determined by the symmetry of the molecule or crystal.

B. Infrared Spectroscopy

Vibrational frequencies lie in the range 10^{12} - 10^{14} s^{-1}, which corresponds to light in the infrared region of the electromagnetic spectrum (vibrational frequencies are often expressed in wave number units, with 1 cm^{-1} = 2.998 x 10^{10} s^{-1}). In infrared spectroscopy, a beam of infrared light is passed through the sample, and the absorption spectrum gives the vibrational frequencies of the material. For solid or liquid samples, the infrared beam can be reflected from the surface to give an infrared reflectance spectrum. Materials at a given temperature also emit infrared radiation, with an emission spectrum which reflects the thermally excited vibrations of the material.

C. Raman Spectroscopy

This technique is named after Sir C. V. Raman and K. S. Krishnan in Calcutta and G. Landsberg and L. (Moscow, both in 1928). An intense beam of light in the visible region of the spectrum is passed through the material, and small frequency shifts due to interaction with vibrational modes of the sample are observed. In earlier work, sunlight or light from a mercury lamp was used as the source of the visible radiation; now lasers are used to provide the incident beam. Raman scattering is complementary to infrared spectroscopy for studying molecular vibrations. In general, not all vibrational modes of a given sample are active in the infrared or the Raman spectrum (this is determined by the molecular symmetry), and obtaining both gives a more complete record of the vibrational spectrum.

II. Applications

A. Phase Identification and Analysis

Because the vibrational spectrum of a substance is determined by the details of its structure and bonding, each material has a characteristic vibrational spectrum, which can be used as a "fingerprint" for identification of that phase in an unknown sample. This type of qualitative analysis is complementary to other laboratory techniques in common use, especially x-ray diffraction, but vibrational spectroscopy can have certain advantages over these. For example, some minerals, such as the Al_2SiO_5 polymorphs andalusite and sillimanite, or the chain (a) and ring (b) polymorphs as $CaSiO_3$ wollastonite, have almost identical x-ray patterns but very different infrared and Raman spectra. In addition, the vibrational spectra are often extremely sensitive to changes in structural order, even on a distance scale of several tens of angstrom units (10^{-10} m), and can be used to distinguish minerals in slightly different disordered states. Because vibrational spectroscopy does not depend on the existence of long-range structural order, it is also extremely useful for studying highly disordered systems such as glasses, liquids, and dense fluids. Finally, x-ray diffraction is relatively insensitive to light elements such as hydrogen, lithium, or boron, which are important in the earth sciences. Vibrations involving these atoms can usually be easily observed in infrared or Raman spectra. [See Powder Diffraction Techniques.]

With suitable calibration, both infrared and Raman spectroscopic analyses can be made highly quantitative. In general, the absolute intensity of a particular infrared-or Raman-active vibrational mode cannot be calculated from first principles with any degree of certainty, so empirical correlations based on series of standards containing known concentrations of the species or phase of interest must be established. Quantitative infrared and Raman studies which have had a major impact in the earth sciences include the determination of total water content and H_2O/OH speciation in silicate glasses via infrared adsorption spectroscopy and the in situ nondestructive analysis of fluid inclusions by Raman spectroscopy.

B. Structural Studies

Most of the vibrational spectroscopic studies to date in mineralogy have been carried out to investigate some structural feature of the mineral in question. This generally requires an interpretation of the vibrational spectra observed in quires an interpretation of the vibrational spectra observed in terms of a structural model, hence a correlation between the dynamic (*vibrational) structure and the static structure of the mineral concerned. This is often obtained by carrying out a vibrational calculation, constrained by the experimental spectra, which gives the atomic displacements corresponding to each vibrational mode. These calculations can be extremely useful, but the results must be treated with some care because they can be highly model dependent. Major progress is being made in determining vibrational properties from first principles via solid-state ab initio calculations.

Useful structural information can be gained from partial interpretation of the vibrational spectra. For example, some types of vibrational fall in characteristic frequency ranges, such as O-H stretching (normally 3000-3700 cm^{-1}) or Si-O stretching (800-1200 cm^{-1}) vibrations. The Si-O stretching frequency depends on the polymerization state of the tetrahedral SiO_4 groups present (i.e., the number of SiOSi linkages per tetrahedron), or the coordination state (tetrahedral SiO_4 or octahedral SiO_6) of the silicate units, so infrared and Raman spectroscopy can be a useful complement to other methods of structure determination. The O-H stretching frequency is highly dependent on the immediate OH environment and the presence or absence of hydrogen bonding. Vibrational studies can be invaluable in helping locate the hydrogen atoms in OH-containing minerals. Both infrared and Raman spectroscopy can be carried out with plane-polarized light, and the resulting polarized spectra as a function of crystal orientation give valuable information on the symmetry of vibrational modes, hence the symmetry of the structural groups involved (for example, distortion of CO^{2-}_3 groups in carbonate minerals from trigonal planar symmetry), and their orientation (for example, the orientation of H_2O molecules within channels in natural cordierites).

C. Thermodynamic Studies

Most rock-forming minerals have vibrational frequencies in the range 50-1500 cm^{-1}, which corresponds to vibrational energies of approximately 0.5 - 15 kj/mol or to thermodynamic temperatures up to 2000 K (1727°C). This means that over the temperature range of the earth's crust and much of the mantle, changes in internal energy of minerals with temperature are mainly due to excitation of vibrational modes. It follows that the heat capacity of minerals over the temperature range of the crust and mantle can be calculated from a knowledge of their vibrational spectra. Methods for carrying out these calculations have been developed and applied to a wide range of rock-forming minerals. Despite some ambiguities in the heat capacity calculations, due mainly to the lack of complete vibrational data, they form a useful complement to experimental determinations of mineral heat capacities. This is especially true for phases stable only at high pressure such as $MgSiO_3$ garnet, ilmenite, and perovskite, which are probably major constituents of the earth's mantle and for the direct determinations of high-temperature heat capacities difficult to carry out in the laboratory. Heat capacity calculation of this type have been useful in constraining high pressure-high temperature phase diagrams for minerals in the system (Mg, Fe)O-SiO_2, resulting in a better understanding of the likely mineralogy of the earth's mantle. [See Mantle.]

D. Phase Transitions

A major fraction of vibrational spectroscopic studies in mineralogy has been devoted to in situ studies of temperature-and pressure-induced phase transitions. One of the classical studies was carried out by Raman and T. M. K. Nedungadi in 1940, when they followed the Raman spectrum of quartz as a function of temperature through the transition between the a and b forms at 573°C. They observed that one vibrational mode, at 207 cm^{-1} at room temperature, decreased markedly in frequency with increasing temperature and disappeared at the a-b phase transition temperature. This type of behavior is known as soft mode behavior. A "soft" vibrational mode is one whose atomic displacement pattern mimics the structural changes occurring during the phase transition, so that exciting the vibrational mode "drivers" the phase change. Many such thermal phase transitions driven by soft modes have now been studied by infrared and Raman spectroscopy, including a number relevant to mineralogy, such as the a-b quartz transition and phase transitions in minerals with the rutile and perovskite structures.

With the advent of the diamond anvil cell coupled with infrared and Raman spectroscopic techniques, a growing number of pressure-induced soft mode phase transitions have been studied. In the diamond cell, the sample is compressed between two opposed diamond planar faces. Because of the small surface area of the diamond faces, enormous pressures (in the kilobar to megabar range) can be generated by application of even modest forces to the back plates of the diamonds. Diamond is an ideal material for this type of cell because of its great strength, enough to withstand the large strains generated in the anvil. In addition, it is transparent to much of the visible and infrared spectrum, permitting a wide range of spectroscopic studies to be carried out in situ at high pressure. Some studies are already being carried out under combined high pressure and high temperature conditions within the diamond anvil cell, and soon it will be possible to investigate a wide range of phase transitions under the pressure-temperature conditions of the de p earth. Such studies will be important for understanding both the detailed mineralogy of the mantle and core and the thermal budget of the earth, since displacive phase transitions at high pressure and high temperature in minerals such as $(Mg,Fe)SiO_3$ perovskite and SiO_2 stishovite could be important in transporting energy within the earth.

There is currently also much interest in the structures and vibrational properties of molecular gases such as hydrogen, nitrogen, and methane solidified in the diamond cell. These gases form the major constituents of the giant outer planets, and this work is particularly important for gaining an understanding of the structure of the planetary interiors. Intense interest has focused on the possibility of metallization in nitrogen, oxygen, and even hydrogen at high pressures, associated with a transition from the molecular solid to an atomic solid state. In this case the deep interiors of the Jovian planets could be composed of metallic condensed gases, giving rise to a planetary magnetic field.

III. Applications to Geochemistry

A. Structure of Silicate Glasses and Melts

Vibrational spectroscopy has been sued extensively to study the structures of silicate glasses and melts in an effort to understand the nature and properties of natural magmas. In most of these studies, systematic changes in the spectra with composition (or some other variable) are noted, and the glass spectra are interpreted by comparison with the spectra of corresponding crystals or by carrying out vibrational calculations. For simple alkali and alkaline earth silicate glass series, the variations of the spectra with composition and temperature have given valuable insights into the number and distribution of different silicate polymer species, which have been correlated with variations in the melt and glass viscosity density, and immiscibility behavior. The spectra of alumino-silicate glasses have been much more complex to interpret, but a better understanding of the structural role of aluminum in aluminosilicate glasses and melts is beginning to emerge. These studies have been particularly powerful when coupled with other structural techniques, such as nuclear magnetic resonance spectroscopy, Mossbauer spectroscopy, or x-ray absorption spectroscopy and diffraction. Of considerable interest are recent studies of melts in situ at high temperatures (which give insight into not only structural properties but also dynamic transport and relaxation processes) and at high pressures in the diamond anvil cell. It is already possible to carry out vibrational and other spectroscopic studies on silicate melts under the pressure and temperature conditions of magma generation in the earth's crust and mantle. These studies are certain to lead to a better understanding of melt processes and crystallization phenomena within the earth and terrestrial planets in terms of the physics and structural chemistry of the melt phase.

B. Dissolution of Water and Carbon Dioxide in Molten Silicates

Water and carbon dioxide are the two most important volatile species which dissolve in natural aluminosilicate melts, in terms of both their abundance and their effect on melt behavior. Both infrared and Raman spectroscopies have been used to study the dissolution mechanisms of these volatile species. Quantitative infrared absorption studies have shown that water dissolves in silicate glasses and melts as both molecular H_2O and bound hydroxyl (OH) species, and the relative proportions of these species vary with temperature, pressure, and total water content. The OH/H_2O speciation reaction can be correlated with the effect of dissolved water on melt properties such as viscosity and electrical conductivity and aids in understanding the relationship between hydrated aluminosilicate melts and aqueous fluids containing dissolved aluminate and silicate species. Infrared and Raman studies have shown that carbon dioxide also dissolves in silicate melts both as molecular CO_2 and via reaction with the silicate species to form carbonate (CO^{2-}_3) groups. These investigations have led to an understanding at the molecular level of how dissolution of carbon dioxide should increase the viscosity of silicate melts at depth, and they are beginning to yield insights into the mechanisms for separation of carbonate from silicate magmas at high pressure within the earth.

C. Speciation in Aqueous Fluids

Water is the most abundant volatile species in the earth system, and an area of intense interest in geochemistry concerns the structure of aqueous fluids and liquids and the speciation of metals and gases dissolved in them. Because the infrared absorption of water itself is so strong, it is difficult to carry out infrared spectroscopic studies on species dissolved in water, and Raman spectroscopy has been the technique of choice in most studies of this type. The structure of water and the interpretation of its vibrational spectrum remain topics of current debate. Some features of the spectra can be understood in terms of the gas-phase molecule H_2O_2 but others cannot, and the roles of hydrogen bonding and rapid proton exchange are subjects of continuing interest and discussion. The results of Raman spectroscopic experiments with increasing temperature have been interpreted as indicating a reduction in the number of hydrogen bonds. At low temperatures, application of pressure appears to have little effect on either the Raman spectrum or the structure, but at temperatures in the supercritical regime increasing pressure further decreases the importance of hydrogen-bonded structures and water begins to behave as a normal polar fluid. There have been several studies of the effects of dissolved gases and electrolytes on the Raman spectrum of water, and Raman spectroscopy has been used to investigate the structures and equilibria of species dissolved in water as a function of pressure and temperature, but much work remains before a systematic pattern of the behavior of geochemically important species can begin to emerge. [See Crustal Fluids; Water, Physical Properties and Structure.]

D. Studies of Fluid Inclusions

Fluid inclusions are microscopic bubbles of gas or liquid trapped during formation of a mineral or glassy material, and they provide a record of the fluid phases with which the mineral was in contact within the earth. Many techniques have been developed for qualitative and quantitative analysis of the fluid and any dissolved species, including infrared spectroscopy of bulk samples, but one of the most successful has been micro-Raman spectroscopy. Because the incident light beam for Raman spectroscopy is in the visible region of the spectrum, it can easily be focused to a small spot by an ordinary microscope. In the Raman microprobe, developed independently by P. Dhamelincourt in France and G. Rosasco in the United States, the laser light is focused to a small spot (1-5 m) by a microscope objective and the Raman scattered light is also collected by the same objective before being sent into the spectrometer. The result is that a Raman spectrum can be obtained from a region as small as 1 m in diameter at the focus of the incident and scattered beams. If the sample containing the inclusions is transparent to visible radiation, it is easy to focus the beam inside micrometer-sized inclusions and obtain the characteristic spectrum of any molecular species inside. This technique provides a rapid, in situ; nondestructive means of obtaining qualitative and quantitative analyses of the major species in natural fluid inclusions, and it has had a major influence on our understanding of the nature of fluids in the earth's crust and mantle. In addition, micro-Raman spectroscopy combined with thermodynamic observations of fluid inclusions in minerals from ore deposits has led to a better understanding of the role of fluids ins deposition and transport processes in ore-forming bodies. Micro infrared instruments are now also being developed, using reflection (Cassegrain) objective of focus the infrared reduction and are beginning to provide information complementary to results of micro-Raman studies of fluid inclusions.

IV. Other Applications in Earth and Planetary Science

A. Infrared Emission Spectroscopy: Remote Sensing of Planetary Surface

In infrared absorption or reflection spectroscopy, the infrared beam incident on the sample is generated by a source and then passed through a spectrometer or interferometer to the detector. In infrared emission spectroscopy, the sample itself acts as the source, and the thermal emission spectrum in the infrared region is characteristic of the vibrational spectrum and temperature of the material. This technique is useful in the laboratory for studying samples which are opaque to infrared radiation and hence not amenable to absorption studies, such as materials adsorbed on metals and samples with rough, irregular surfaces with low reflectivity. This is the case of soils, and infrared emission studies in the laboratory have been useful in determining the mineral constituents and absorbate species in natural soil samples. In addition, aircraft-mounted infrared emission spectrometers have been used both to measure thermal profiles of terrains and to identify the water content and mineralogy of soils from their characteristic vibrational signatures. Along with other teledetection techniques, airborne infrared emission spectroscopy is a valuable tool in terrestrial prospecting and resource evaluation. In an exciting extension of this technique, infrared emission spectrometers mounted on spacecraft in the near future will provide valuable new information about the surface mineralogy of the moon, Mars, and perhaps other terrestrial planets. [See Remote Sensing of Earth Resources From Air and Space.]

B. Raman Lidar: Remote Sensing of Atmospheric Gases

Lidar stands for light detection and ranging and refers to a family of laser-based remote sensing techniques which involve sending a powerful laser beam into the atmosphere and collecting and analyzing the backscattered signal. In Raman lidar techniques, the characteristic vibrational or rotational Raman lines of gas-phase molecules in the atmosphere are monitored to provide a three-dimensional map of their concentration profiles. In addition, the relative intensities of the rotational Raman lines of O_2 and N_2 molecules give a useful measure of the temperature profile in the atmosphere. Vibrational Raman lidar is particularly useful for monitoring SO_2, CO_2 and water vapor and has been used to study atmospheric pollution by power plants and aircraft and automobile emissions. The technique has begun to be used to measure the concentrations and spatial distribution of gases in volcanic gas plumes. Two of the major advantages of Raman lidar are that it can be carried out remotely throughout all phases of an eruption, eliminating the need to fly aircraft through the plume, and that several gas species can be monitored simultaneously, which is not possible with other remote sensing methods in current use. [See Remote Sensing of the Upper Atmosphere, Instrumentation and Observational Techniques.]

GLOSSARY

Fluid inclusion. Bubble of liquid or gas trapped inside a solid mineral-phase, usually formed during crystallization of the mineral in the presence of a fluid phase.

Infrared emission. Light in the infrared region of the spectrum emitted by a heated sample.

Infrared radiation. Light with wavelength between approximately 0.001 and 1 μm, in the infrared region of the spectrum.

Infrared spectroscopy. Interaction of infrared radiation with a sample, usually to excite vibrational modes of the material.

Normal modes of variation. Natural resonances of the coupled periodic displacements of atoms in a material about their equilibrium positions.

Raman lidar. Light detection and ranging technique, base on the Raman effect.

Raman scattering. Scattering of visible light accompanied by a shift in wavelength due to interaction with molecular vibrations.

Raman spectroscopy. Analysis of Raman scattered light to give information on vibrational modes of the sample.

Soft mode. Vibrational mode for which the atomic displacements track the structural changes followed during a displacive phase transition. The frequency of the soft mode goes to zero at the transition temperature or pressure.

23

NMR Spectroscopy in the Land Pollution

Introduction

NMR has recently begun to be used to characterize local order in both minerals and amorphous materials and viscous liquids. The detailed information obtained can be used to estimate the configurational contribution to the entropy of materials and better constrain their thermodynamic properties. Dynamic NMR techniques are revealing specific atomic motions which appear to control viscosity and therefore flow in silicate liquids.

In common with other spectrosdcopies, nuclear magnetic resonance (NMR) involves the interaction of electromagnetic radiation with matter. In particular, NMR relies on the absorption and emission of radio-frequency (rf) radiation by the nuclei of atoms in a sample placed in a large magnetic field. Most common elements have an isotope which possesses a property called nuclear spin; this means that the nucleus has angular momentum and an associated magnetic field the many nuclear magnetic moments have no preferred orientation. When placed in a magnetic field, they align either parallel or antiparallel to the field and the previously degenerate angular momentum states become separated by an energy difference E. The difference in energy between the two states in proportional to the magnetic field and the magnetogyric ratio (ratio of magnetic moment to angular momentum) of the magnetic nucleus. The spectroscopy involves inducing transitions between these two states of the "spin system" by rf irradiation (satisfying the Bohr condition, E=hv) and subsequently detecting the emitted radiation. One of the great advantages of NMR spectroscopy is that the different magnetogyric ratios of the nuclei in a constant magnetic field, and if the field is large enough they may be excited separately. Therefore, by choosing the correct resonance frequency it is possible to observe just the interaction of the rf radiation with the silicon atoms, say, in a sample. However, this is not the limit of specificity of the technique because the resonance frequency, which in determined by the magnetic field at the nucleus, is modified by the local magnetic field due to the electrons surrounding the nucleus. This is the most powerful aspect of NMR because is provides information about the local environments of atoms, and detailed work (described later) or known structures has made it possible to relate these differences in resonance frequency to differences in nearest and next-nearest neighbor atoms and differences in bond lengths and angles.

The information obtained from NMR spectroscopy is oflocal structure in the region of ~4-5 A(4-5 X 10^{-10}m) from the resonant nucleus. This information may appear to by very microscopic compared with earth processes occurring over the scale of 10^{0} -10^{3} M. However,it is inevitable in the study of the earth that experiments performed in investigate important mechanisms and processes can never simulate nature exactly because of the large physical or long time scales involved. A sound basic understanding of the physics and chemistry of earth materials is essential for extrapolation of laboratory data to real systems with any confidence.

One of the fundamental questions in geochemistry is why a particular mineral or assemblage of minerals exists under certain conditions of composition, temperature, and pressure. At present, much of the accumulated knowledge of the phase equilibria of minerals comes from studies from studies which reproduce experimentally the interaction of many components and the effects of temperature and pressure. However, to cover every combination of variables would require an impossibly large number of detailed experiments. Advances in computing power and techniques have meant that eventually the phase equilibria for any combination of oxide components and temperature and pressure may be calculable. There is a need to provide accurate thermodynamic data, such as heat capacities, enthalpies of formation, entropies, and molar volumes, for input into these computer programs and a need for detailed structural information to allow these properties to be calculated. One way in which this information can be obtained in via NMR spectroscopy. X-ray diffraction continues to be the method of choice for determining the structure of earth materials, with spectroscopic methods providing of earth materials, with spectroscopic methods providing complementary local information about atomic environments. NMR has made some important contributions to the understanding of atomic ordering in minerals, which can be related to the configurational estropy of a material. The other advantage of NMR is that the time scale motions and can monitor the dynamic processes which control transport properties such as viscosity and diffusivity. These are important in the description of the thermodynamic properties of liquids.

II. Principles of NMR Spectroscopy

An NMR spectrum in obtained by exciting the magnetic nuclei in a sample by a pulse of rf radiation and recording the subsequent signal induced in the transmitter-receiver coil of the NMR probe as function of time. It is usual practice to repeat the process as rapidly as possible, adding the signal digitally to improve the signal-to-noise ratio of the experiment. To render to data in an interpretable form, the experiment. To render the data in an interpretable form, the accumulated signal in Fourier transformed and displayed as a frequency spectrum in which the absorption peaks are characteristic of the transitions of the nuclear spin system of the sample. The time between pulse repetitions depends on the ability of the spin system to return to an equilibrium spin population(i.e.,to relax) via interaction with lattice motions or other motions of nearby magnetic nuclei. These motions cause fluctuating magnetic fields which stimulate the transitions necessary to return to spin populations to equilibrium. This process is characterized by the relaxation time, t_1, and it can affect the experimental sensitivity of a nucleus in very rigid materials or materials with very dilute magnetic nuclei, because the pulse repetition rate is limited by T_1 and insufficient signal accumulation can be achieved in a reasonable time.

Although the rf excitation and detection of the transitions of the spin system are the same, sample mobility divides NMR spectroscopy more or less into two disciplines:liquid and solid-state spectroscopy. The first category is important because it identifies molecular species n solution or in the liquid state and has developed closely with modern chemistry, in which most reactions that are studied take place in solution. Liquid state spectroscopy is characterized by such narrow spectral lines that in general resonances from structurally distinct nuclei do not overlap; thus it is called high-resolution NMR. Traditionally,NMR in the solid state consisted of broad overlapping signals due to the influence of anisotropic (orientation-dependent) magnetic or electric interactions and fell into the realm of solid-state physics. These interactions result in different signals from equivalent nuclei at different orientations to the applied magnetic field. In the spectra of powders or particle or molecular orientations, resonances from structurally distinct nuclei and usually obscured by these interactions. Because of the rapid tumbling and diffusive motions of molecules in solution, these orientation effects are averaged out to leave a scalar (orientation-independent) contribution from the interaction and narrow line is observed. Most of the effort in developing the high-resolution solid-state NMR being used in the earth sciences has gone into removing the anisotropic part of these orientation-dependent interactions to leave just the scalar part, which is the equivalent of the solution state measurements if there are no intrinsic solid-state effects.

The lifting of the degeneracy of the angular momentum states of nuclei in a magnetic field is called the Zeeman interaction. In the magnetic fields commonly used in NMR spectroscopy this results in an energy gap of ~10-600 MHz. The remaining magnetic interactions can generally be considered as a perturbation of the Zeeman energy levels and are of the order of a hertz to a few kilohertz. Consequently, they are expressed in part per million (ppm) and in the case of the chemical shift are referenced to a standard material, usually a solution. One important perturbation of the magnetic field at a nucleus is the influence of another adjacent nucleus which possesses a magnetic moment. The local field at one spin due to the other depends on the magnitude of the magnetic moment and the internuclear distance; the perturbation to the Zeeman level depends on the angle the internuclear vector makes with the magnetic field. One of the first demonstrations of the nature of the dipolar interactions between nuclear spins was in gypsum ($CaSO_4$ $2H_2O$). In a single crystal there are two water molecules in which the proton-proton internuclear vector in fairly well isolated well isolated and at a well-defined angle to the magnetic field. Two ^1H resonances are observed for these chemically equivalent protons, each corresponding to the parallel or antiparallel polarization of its proton neighbor. The splitting of the lines varies with the orientation of the crystal to the magnetic field; see Fig.1A. When a powder sample is used the split lines form many spins in different orientations are superposed, resulting in a powder spectrum which is known as s Pake doublet (Fig.1B).When more than two spins are involved in the mutual dipolar interaction, the calculation of the powder line shape rapidly becomes intractable ands in any case a broad featureless experimental line is generally observed.

When the dipolar interaction is between two unlike spins the splitting or broadening effect may be removed from the spectrum of one spin by irradiation at the resonance frequency of the other; since this rapidly changes the second spin's polarization, the first spin no longer experiences the splitting effect. This is known as dipolar decoupling. In a solution the direct dipolar interaction, which is orientation dependent (direct, through space), is removed by its rapid variation as the molecules tumble. However, an additional weaker coupling to the other nuclear spin is mediated through the chemical bond (indirect or J coupling) and the spin

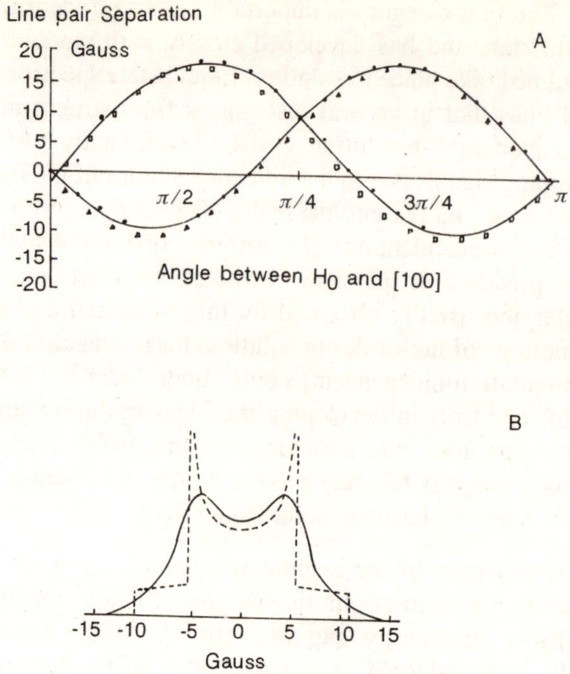

Fig. 23.1. (A) Plot of the dipolar splitting between the two pairs of lines in the ^1H NMR spectrum of a single crystal of gypsum ($CaSO_4 2H_2O$) as the crystal orientation is changed relative to the field. (B) ^1H NMR spectrum of gypsum powder, all possible orientations of the crystal are present and so a broad doublet is observed. [Adapted from G. E. Pake(1948).J Chem. Phys.16,327, with permission.]

is still sensitive to the polarization of the adjacent spin. This scalar interaction results in a binomial splitting pattern for each chemically equivalent magnetic neighbor. Decoupling is used in the liquid-state NMR of coal-derived liquids and extracted of humic materials described in Section III. These are typical hydrocarbon systems and ^{13}C spins are observed while the attached or adjacent ^1H spins are irradiated at their resonance frequency. This has two effects. First, it removes the splitting from the spectrum,,since each ^{13}C sees only an average ^1H polarization (zero). Second, the irradiation disturbs the populations of the coupled spin system ^1H-^{13}C so that the ^{13}C signal enhanced. This is the nuclear Overhauser effect (NOE). Unfortunately,carbons in different environments, especially those not directly bound to hydrogen, are subject to different enhancements and the spectrum is no longer equantitative. NOE effects can be overcome be gating the decoupler off after anquisition to allow the spin system populations to rețurn to equilibrium without decoupling before proceeding with next pulse.

Solid earth materials which originate from biological material and contain many C-H bonds require intense rf irradiation at the ^1H-^{13}C direct dipolar coupling. Apart from these materials, most earth materials examined by NMR contain nuclear spins which are distant from each other because of the low natural abundance of the isotopes or simply because of the presence of oxygen at the network anion;since ^{1717} is 0.037%0.037% abundant in naturally occurring oxygen, most magnetic nuclei are at least two bond lengths away from the next magnetic nucleus in the structure. As the dipolar interaction falls off rapidly with internuclear distance, dipolar coupling is not a great problem in the spectroscopy of these materials.

The main obstacles to high-resolution solid-state NMR of inorganic earth materialized are anisotropic interactions caused by the nonspherical symmetry of the bonding environment around the nucleus of interest. Third manifests itself as chemical shift anisotropy (CSA) and is the predominant broadening interaction for I=1/2 nuclei (e.g.,^{29}Si). Nuclei with I>1/2 are subject to quadrupolar broadening which is usually dominant, in addition to CSA. The effect for ^{29}Si NMR can be seen in the single-crystal spectrum of quartz in Fig.2A. The tetrahedral coordination of Si in quartz is not perfect, and the electronic environment around the Si atom does not exhibit spherical symmetry. The three crystallographically equivalent SiO_4 units are

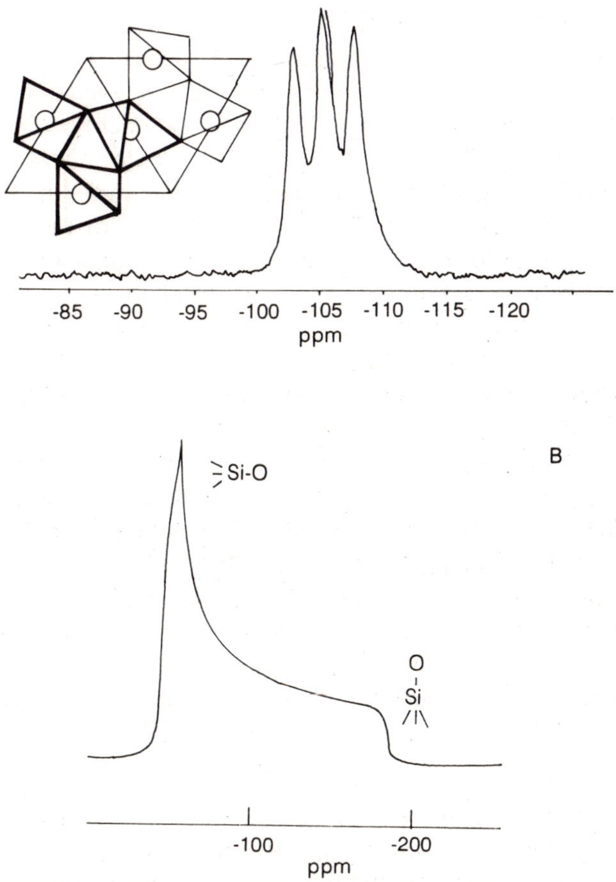

Fig. 23. 2. (A) ^{29}Si NMR spectrum of a single crystal of quartz. Shown inset are the three orientations of the SiO_4 tetrahedra in the unit cell; these give rise to the three lines in the spectrum. (B) Simulated ^{29}Si powder pattern for a tetrahedral Si with uniaxial symmetry; the two orientations of the unit shown correspond to the extrema of the powder pattern

related by a threefold screw axis; each Si in the unit cell has the same electron density distribution,but it is oriented differently with respect to an external axis such as a magnetic field. Hence, the shielding for each Si atom is different and the resonance for each SiO_4 tetrahedron in the unit cell may be observed. If the symmetry of electron distribution about an SiO_4 tetrahedron is reduced, compared with quartz, by the presence of one nonbridging oxygen, say, then the effect is much more dramatic. Fig. 2B showed as idealized spectrum of powder containing such units. The extrema of the "powder pattern" represent the chemical shifted for the orientation of the unique axis of the SiO_4 unit eitherperpendicular or parallel to the external magnetic field. In a randomly oriented powder there are more ways to have the SiO_4 unit perpendicular to the field than parallel to it, so the resonance pattern has the greatest intensity to the perpendicular (less shielded) side of the spectrum. The scale of Fig.2B shows that the anisotropy is of the order of 130 ppm; since the isotropic shift range of silicon in tetrahedral coordination with oxygen in much less than this, when many slightly different silicon environments are present they will be obscured by the CSA effect.

To remove the CSA effect in the solid state the orientations of the silicate tetrahedra must be made to very in time, just as the orientations of species in solution do. If the sample could be macroscopically rotated during the experiment, the orientations of the tetrahedra with respect to the magnetic field would very with time and each would have some average orientation during the acquisition of the NMR signal. The average orientation during spinning for each separate tetrahedral orientation in a random powder varies if the inclination of the spinning axis to the magnetic field is changed. The inclination of the spinning axis can be chosen so that the average orientation for all tetrahedral orientation is the same. The resonance frequency associated with this average orientation is equal to the isotropic chemical shift. This angle of inclination of the spinning axis to the magnetic field is known as the magic angle(54.7^0) and the technique as magic angle spinning (MAS). The technology of MAS has progressed so that modern double air bearing spinning systems can routinely spin at speeds in excess of 10 kHz (600,000 rpm). Sample rotors are machined from some of the strongest ceramics (e.g. silicon nitride and stabilized zirconia) in order to withstand the enormous forces on the rotor during spinning. Such high speeds are needed because unless the spinning speed exceeds the total anisotropy of the static line, spinning sidebands will appear in the spectrum at multiples of the spinning frequency at either side of the isotropic peak. These complicate the spectrum and remove intensity peak. These complicate the spectrum and remove intensity from thc isotropic peak. These complicate the spectrum and remove intensity from the isotropic peak. As the magnetic fields available for NMR increase in strength, the CSA,which is proportional to the magnetic field, increases and proportionally higher spinning speeds are required.

The progress of NMR as a major spectroscopic tool was accelerated by a technique which enhanced the sensitivity of the spectra of low-abundance nuclei with lesser magnetic moments by transferring polarization from an abundant nucleus with a large magnetic moment, almost always ^1H. This cross-polarization (CP) technique is most important in the spectroscopy of hydrocarbon systems such as coal and humic material, where the polarization transfer is from ^1H to ^{13}C, but has also been employed in silicate systems using ^1H-^{29}Si CP where there are Si-OH groups. The transfer of polarization occurs when transitions in both unclear spin systems are isoenergetic; this is achieved by applying rf field simultaneously to both spin

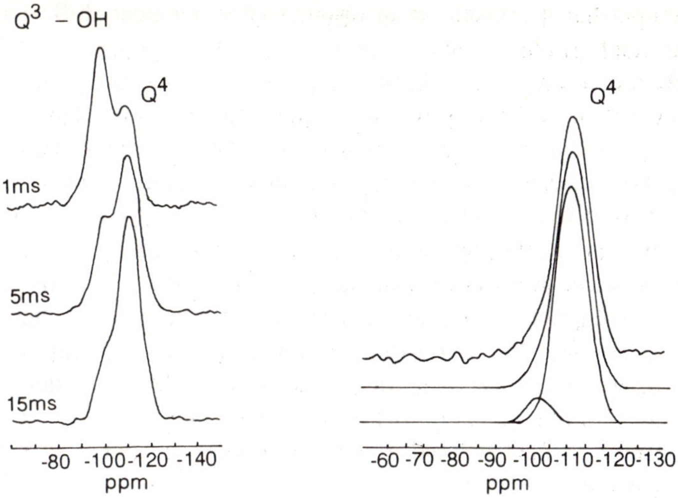

Fig. 23.3 (Left) ^{1}H-^{29}Si cross polarization spectra of a silica glass with 2.5 wt% dissolved water using contact times as shown. (Right) The true abundance of the Q^3-OH group is shown in the deconvolution of the single-pulse spectrum. [Adapted from I. Farnan, S.C. Kohn, and R. Dupree (1987). Geochim. Cosmochim. Acta 51,2869-2873, with permission.]

systems with amplitudes which match their energy gaps. The CP occurs at a certain rate (R_{xh}, where X is the dilute nucleus),so the enhancement of an Si nucleus,say, depends on how close it is an ^{1}H, how long that ^{1}H can maintain its polarization [$T_{1p}(^{1}H)$], and how much time is allowed for the CP to take place (contract time). Although this greatly improves the signal-to-noise ratio, the spectra are nonquantitative because nuclei close to ^{1}H are greatly enhanced for short contacts times. Figure.3 shows ^{1}H-^{29}Si CP for different contract times and the quantitative single-pulse spectrum of a hydrous silica glass;the great enchancement of the SiO_3OH group at short contact times is very clear. The selectivity of CP for species near ^{1}H nuclei may be used to observe the surface sites in a material, even when that material nominally contains no $_1H$ nuclei,by transferring polarization from surface surface hydroxyl groups.

Nuclei with electric quadrupole moments(I>1/2), in particular those with half-integer spin (3/2,5/2,...), are the most common in the periodic table. They can yield high-resolution spectra under MAS conditions provided they are in fairly symmetric sites with modest electric field gradients. In general, the spectra will be less resolved than for 1=1/2 nuclei. This is because in the absence of a magnetic field the angular momentum of a nucleus with an electric quadrupole moment will be quantized along any electric field gradient at the nucleus. In the presence of a large magnetic field the angular momentum will be quantized along the magnetic field provided the Zeeman interaction is much greater than the electric quadrupole interaction. In this situation the quadrupolar effect is a perturbation which introduces an orientation-dependent shift in the nuclear spin energy levels and hence in the resonance frequencies. For half-integer nuclei the central (1/2-1/2) transition is not affected to first order, and this makes these nuclei favorable for NMR because we can observe the central transition without first-order broadening. However, the secord—order quadrupole effect is often large enough that it introduces an orientation-dependent shift. This effect is greater for larger electric field gradients but is reduced as the magnetic field is increased,so high magnetic fields are desirable for these nuclei. The second-order quadrupolar effect can be reduced only by a factor of 3 to 4 by magic angle spinning and cannot be removed entirely by spinning about any single axis. Consequently, the MAS spectrum will reveal a powder pattern for a well-ordered material. This can be simulated and they value of the insotropic chemical shift and the electric field gradient at the site determined together with its asymmetry. It is often the case that the resolution gained by MAS will be enough to resolve the chemical shift of nuclei with different numbers of nearest neighbors,since there will be additional narrowing effects such as reduction of dipolar coupling and removal of chemical shift anisotropy. However, if two or more MAS powder pattern overlap,deconvolution of the spectrum may be difficult. Figure 4 shows a theoretical quadrupolar MAS powder pattern narrowed from the static pattern. Note the position of the isotropic chemical shift, which does not lie under the absorption curve, a fact has led to some erroneous chemical shifts being reported in the literature for quadrupolar nuclei.

If the electric field gradient at a site is very large, it will produce a very broad NMR line which may be difficult or impossible to excite completely or to observe, since its signal will decay rapidly and be lost in the instrument "dead time" at the start of signal acquisition. This has implications for the quantitative nature of the NMR signal because nuclei in these sites will not be represented in the spectrum. Another factor which can affect quantitative is the presence of paramagnetic species in the sample. An unpaired electron has a very large local field, and this can perturb the local magnetic field at adjacent nuclei so that the resonance frequency is shifted by a large amount or the signal from nuclei in these sites is broadened beyond detection. Obviously,an effect like this is a problem for sample of natural earth

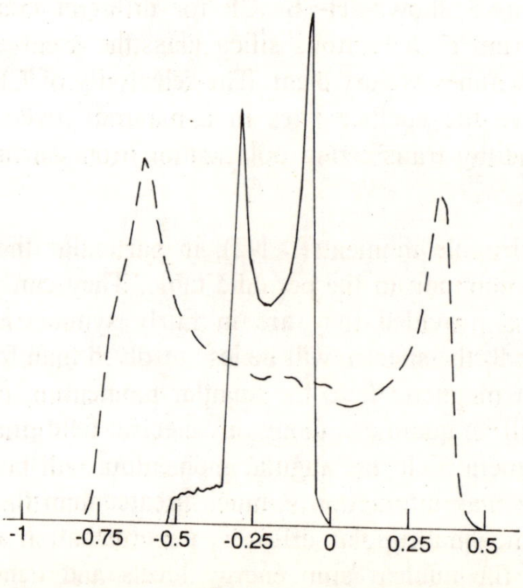

Fig. 23. 4. Static (dashed line) and narrowed MAS (full line) powder patterns for a half-integer quadrupolar nucleus. F is a width factor calculated from the magnitude of the electric field gradient, the magnetic strength, and the spin quantum number I.

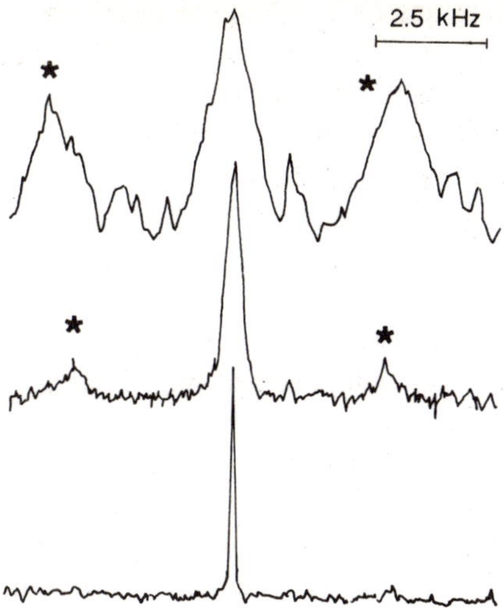

Fig. 23.5. ^{29}Si MAS spectra of forsterite; from bottom to top: Mg_2SiO_4, $Mg_{1.98}Fe_{0.02}SiO_4$, $Mg_{1.90}Fe_{0.10}SiO_4$. The increasing amounts of Fe^{2+} cause not only broadening of the isotropic line but also an increase in the static magnetic susceptibility causing sidebands (*). [Adapted from A. R. Grimmer, F. von Lammpe, M. Magi, and E. Lipmaa (1983). Z. Chem. 23, 342-343, with permission.]

materials, which generally contain iron or some other paramagnetic species. This is demonstrated in Fig.5 where foresterite (Mg_2SiO_4) with various amounts of Fe^{2+} substituted for Mg^{2+} shows increased broadening with Fe^{2+} content; in.fact, no signal at all could be obtained from a sample containing 10 mol% Fe^{2+} Small amounts (a few hundred parts pet million) of paramagnetic impurities can be advantageous because they reduce the T_1 of the sample and allow rapid data acquisition.

III. Organic Geochemistry And Solution NMR

One of the first applications of NMR is geochemistry, as in chemistry, was to analyze organic materials; in geochemistry these were coals, oils, and related humic substances, i.e.,geoorganic materials derived from plant debris. These are all complex materials,and the task of identifying the structural units which compose them and determining how those units are connected together is a daunting one. Most studies were simplified by working on solutions of derivative liquids. For example, coal is a heterogeneous mineral, both compositionally and phyysicallt,which is composed of carbon,hydrogen, and oxygen plus some inorganic realized as ash when the coal is burned. Various stages of transformation are termed peat,lignite, bituminous coal and anthracite. The coal phases of the organic material have been classified by their reflectivity and range from bright vitrain (anthracite) to dull durain (brown coal). The different stages of transformation of plant debris into coal can be correlated by elemental analysis to the gradual removal of oxygen and then hydrogen from the original plant material until the highly carbonaceous (~97 wt %) anthracite is formed. Because of the high carbon content of anthracite coals it was inferred that they must be highly aromatic, since this is the most likely molecular arrangement for a low hydrogen-to-carbon ratio. One of the first successful applications of NMR to the problem of coal structure was an ^{1}H NMR study of coal-derived liquids by J.K. Brown and W.S. Ladner in 1960. The aromatic and alihatic hydrogens in the liquid could be distinguished by their chemical shift. This meant that the previously descriptive classifications of coal could be put on a quantitative basis by measuring the ratio of aromatic to aliphatic hydrogen, fa; this would then give the maturity of the coal in terms of how much aromatic material is contained. There are problems with this would then method in that some resonances form nonaromatic hydrogen overlapped with the broad aromatic peak and caused some overestimation of fa, but this could be corrected for by using evidence from chemical analysis or chromatography. Once pulsed NMR spectrometers with high-fields superconducting magnets became avilable, the ability to observe different carbon environments in coal-derived liquids provided direct evidence of the carbon environments in coal-derived liquids provided direct evidence of the carbon environments existing in the liquids. The resolution compared with ^{1}H NMR is enhanced because of the greater chemical shift range of carbon (~300 ppm) and the higher magnetic field, which provides a separation of lines which overlap at low fields. The amount of aromatic carbon can be read directly from the integration of the spectrum, and different aliphatic functional groups may be distinguished as well as quaternary (not bonded to H) carbons in the aromatic band. The latter are important because they are formed when aromatic groups bond directly together (sharing a carbon atom)and not when they are linked by aliphatic bridges, which is an indicator of maturity. The combination of high-field ^{1}GH and ^{13}C NMR provides a method of very precisely defining the aromatic-to-aliphatic carbon and hydrogen ratios. Together with confident assignments of the many resolved peaks in these spectra and a knowledge of the average molecular weight,this yields a picture of the chemical structure of the coal which is termed the average structure and can be used successfully to design processes for production of economic products from specific coal samples.

The most desirable information about the chemical structure of coal or humic substances would be that obtained directly from the solid. When the application of magic angle spinning, cross-polarization, and high-power ^{1}H decoupling made ^{13}C NMR possible in the solid state, it was quickly applied to solid coals. For high-rank coals, broad relatively featureless bands were observed for aromatic and aliphatic carbons, and lower-rank coals showed similar bands along with signals from C-O linkages and more volatile aliphatic carbons. Thus, the result obtained from NMR of coals in the solid state tended to reinforce the picture of increasing aromaticity and decreasing oxygen and hydrogen content with maturing which had been established by indirect measurements of derivative liquids. Typical solid-state ^{13}C spectra of coal at various stage of maturity are shown in Fig. 6. One problem in going from solution-state to solid-state ^{13}C data is quantitation. In the solution state there has been much careful work on the quantitative nature of the technique, with the development of relation reagents and gated decoupling pulse sequences which eliminate the inequivalent NOE enhancement factors for different carbons in the sample and produce quantitative spectra. In solid-state experiments,because of the need to enbance sensitivity by cross-polarization from ^{1}H, not all of the carbons are enhanced equally and the experiment in inherently is inherently nonquantitative. Considerable efforts have been made to evaluate R_{ch} and $T_{1p}(^{1}H)$ for coals in order to establish a relationship between the intersity observed for each carbon in a coal and it true abundance. One problem is the very short components of $t_{1p}(^{1})$ for ^{1}H nuclei close to

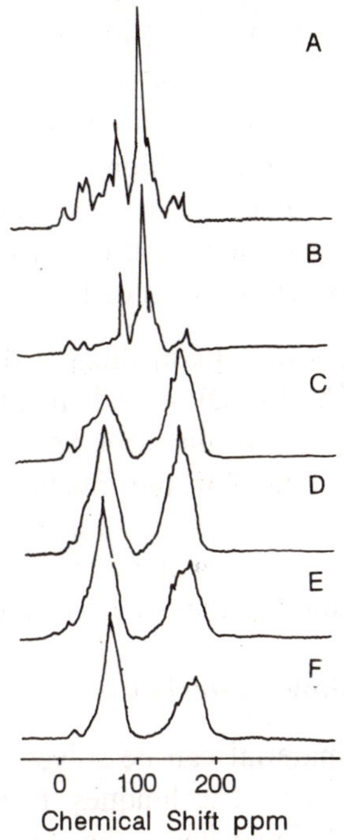

Fig. 23.6 ^{13}C CP MAS spectra of peats (a,b), lignites (C,D,E),and bituminous coal (F). The aromatic peak between 50 and 80ppm grows in intensity with the maturity of the material. [Adapted from J.M. Dereppe, J.P. Boudou, C. Morreaux, and B Durand (1983). Fuel 62,575-580, with permission.]

paramagnetic, which decay too rapidly for the cross-polarization process to occur. The unpaired elections associated with the organic free radicals or inorganictions present in coal produce a strong dipolar field, which either shifts the frequencies of adjacent nuclei or broadens them beyond detection. Nuclei a little farther away suffer from line broadening and contribute to the broad featureless background of the aliphatic and aromatic bands in coal spectra. Improvement in resolution can be obtained if the ignorganic ions are washed out of the coal before running the NMR spectrum.

In common with coal, ^{13}C and ^{1}H NMR of liquid extracts from humic substances such as peats and soils has provided insight into the chemical structure of these matertials,and average chemical structure may be inferred by using a combination of average molecular weights, clemental analyses, chromatography, and mass sepectrometry. the acquisition of solid-state spectra of these materials in fairly routine and serves as a useful check on the inferences made from the indirect solution-state data.

Dissolution of silicates and subsequent precipitation are important in the hydrothermal alternation of minerals. Knowledge of the structure of the silicate anions in solution and how they aggregate and eventually precipitate would provide essential information in the elucidation of these processes. The nature of silicates in solution has been quite extensively studied.

However, silicates are poorly soluble in neutral solutions and ^{29}Si studies of these solutions have been hampered by low sensitivity. Solubility in much greater in very basic solutions, and an array of silicate anion structures have been identified. These range from a simple monomer to arrangements which include up to 10 linked SiO_4 tetrahedra. Silicate polymers can be identified by the different chemical shifts for tetrahedra in dimers with one bridging oxygen(Q^1) and those with increasing degrees of condensation. In the Q^n nomenclature, Q refers to a quaternary silicon, bonded to four oxygens, and n is the number of oxygens shared with other SiO_4 tetrahedra(bridging oxygen). This nomenclature was later adopted in the description of solid silicates. [See Hydrothermal Systems; Silicates.]

Aluminosilicates are even less soluble than silicates, but mixed aluminosilicate anions in solution have been detected by ^{29}AI NMR. Dissolved silicates and aluminosilicates both form gels acid solution, and the time scale of the process is such that ^{29}Si NMR can detect intermediate oligomeric species and elucidate the kinetics of gel formation. These reaction are of considerable interest in the production of synthetic aluminosilicates such as zeolites. The ability to obtain similar kinetic information about mineral dissolution or precipitation would be very useful in determining alteration mechanisms for minerals in contact with water.

IV. Structural Studies of Crystalline Materials

The structure of a crystalline material can be solved by single-crystal x-ray diffraction or powder x-ray diffraction with refinement techniques. In general, this means the long-range average structure, i.e., the unit cell dimensions, the number and type of sites within the materials, and which atoms occupy which sites. The details of the distribution of atoms with similar structural roles among similar sites may be obscured because of the requirement of the diffraction technique to average over a large number of the repeat units of the structure and the similarity of the scattering factors of the elements. The great sensitivity of NMR to the local structure around a nucleus derives from the fact that very small changes in electron density brought about by extremely subtle structural changes can be accurately measured as relative changes in NMR frequency, which for some high-resolution liquid spectroscopy can be as small as $4x10^{-10}$. For ^{29}Si in a crystalline silicate, resolution of ~0.1 ppm ($1x10^{-7}$) is routinely achievable. Hence, ^{29}Si NMR has rapidly evolved as the method complementary to x-ray diffraction for determining site occupancies and ordering schemes in silicates and aluminosilicates. The low natural abundance of ^{29}Si (4.7%) was not an obstacle to routine observation in crystalline silicates once relatively high field (>4 T) superconducting magnets became available for NMR spectroscopy. The magic angle spinning technique provided spectra with narrow linewidths so that acceptable signal-to-noise ratios could be achieved in experiments lasting 4-12 h. [See Powder Diffraction Techniques.]

The ability to obtain ^{29}Si NMR spectra goes only a part of the way to developing a characterization technique for crystalline silicates. The information collected from an NMR experiment is rather indirect, consisting of peak frequencies, intensities, and peak widths; the relationship between this information and the structure of the material under investigation needs to be determined. One way to do this would be to calculate chemical shifts for the different structural possibilities available to the material, compare them with the measured values, and thus identify the structure. At present, it is not possible to calculate the electronic structure of a material and predict its isotropic chemical shifts to anywhere near the degree of accuracy with which the shifts can be measured. Consequently, all structural identifications must be made by comparison with the chemical shifts of compounds or materials with a known structure. A large range of silicates and aluminosilicates have been studied by ^{29}Si NMR, and this has provided a body of information which can be used to identify the silicon environments present in a material. Together with x-ray diffraction data, this information can be used to provide a complete description of the structure of a material.

The primary factor in determining the silicon chemical shifts is the number of a oxygen neighbors. Silicon in stishovite (-911.3 ppm), where it has six oxygen near neighbors, is about 84 ppm more shielded than in quartz (-107.4 ppm), where it has four oxygen neighbors. However, the great advantage in using ^{29}Si NMR in silicate structural studies is its sensitivity to the next-nearest-neighbor atom, since in silicates the nearest neighbors are predominantly four oxygen atoms. The secondary effect on the chemical shift depends on whether the next nearest neighbor is another silicon atom, i.e., whether the nearest oxygen atom is shared between two tetrahedra as a bridging oxygen. This effect is systematic and results in deshielding the silicon (making its chemical shift more positive) as the number of bridging oxygens in the tetrahedron decreases. For example, tale [$Mg_3Sio_{10}(OH)_2$], which contains only Q^3 silicon environments, has a chemical shift of -98.1 ppm and is deshielded by ~10 ppm from quartz, which has a Q^4 silicon environment. The trend is exemplified by the shifts of diopside ($CaMgSi_2O_6$, 84.7 ppm), akermanite ($Ca_2MgSi_2O_7$, -73.7 ppm), and forsterite (Mg_2SiO_4, 61.9 ppm), which have single Q^2, Q^1, and Q^0 silicon environments, respectively. The preceding trend ignores more subtle effects on the chemical shift, such as the effect of the field strength of the cation coordinated to the nonbridging oxygen(s) and the effect of the average Si-O-Si bond angle with chemical shift for a number of SiO_2 polymorphs in which, obviously, cation effects are absent.

The presence of aluminum in the framework of an aluminosilicate introduces another systematic effect on the chemical shift, and that is the deshielding of silicon caused by having an aluminum next nearest neighbor. The effect is approximately 5 ppm for each Al next nearest neighbor with all other influences being equal. Figure 8 shows the ^{29}Si spectrum of a plutonic nepheline (ideally, $Na_{75}K_{25}AlSiO_4$) from Bancroft, Ontario; the spectrum consists of four resonances from four different silicon environments in the material. The structure determined by x-ray diffraction shows only two tetrahedral framework sites. There are four resonances because of the distribution of Si and Al in the structure. Lowenstein's rule excludes the possibility of Al-O-Al linkages in the structure, so for an Si/Al ratio of 1:1 the Si atoms form Q^4(4Al) units and the Al atoms Q^4(4Si) units in an ordered distribution; any random distribution would require Al-O-Al linkages. Nephelines produced under low-temperature condition were proposed to be ordered and those formed at higher temperatures disordered, based on x-ray diffraction refinements. This situation illustrates the unique ability of ^{29}Si NMR to elucidate Si-Al ordering in minerals and other aluminosilicates. The two peaks on the deshielded side of the spectrum are resonances for silicon in the two tetrahedral sites, each with 4 Al neighbors, and are distinguished by the different average Si-O-Al angle for each site. The extra peaks are due to excess SiO_2 in this natural material. The Si/Al ratio calculated directly from the intensities in the ^{29}Si NMR spectrum was found to be 1.11 and was later confirmed by electron microprobe analysis. The extra peaks are accounted for

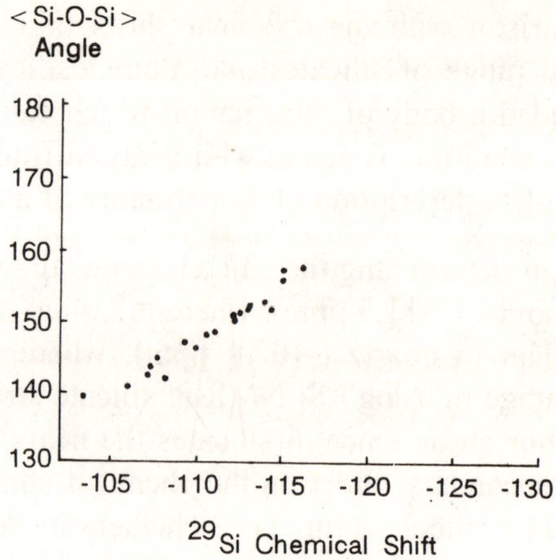

Fig. 23.7. Plot of the average intertetrahedral angle (Si-O-Si) versus chemical shift for SiO_2 polymorphs and dealuminated zeolites, where it is the dominant effect on the silicon chemical shift. [Adapted from R. Oestrike, W.-H. Wong, R.J. Kirpatrick, R. L. Hervig, A. Navrotsky, and B. Montez (1987). *Geochim. Cosmochim. Acta* 51, 2199-2209, with permission.]

extremely well by assuming that the extra Si is substituted randomly into a completely ordered arrangement of Si and Al, which creates one Q^4(OAI) and three Q^4(3AI). The fact that nephelines produced at higher temperatures also exhibit complete local ordering has implications for the entropy of fusion of nepheline and hence for the prediction of phase equilibria at high temperatures. A similar pattern of excess Si substitution in an ordered arrangement of silicon and aluminum was observed by ^{29}Si NMR in plagioclase feldspars. A general consensus derived from ^{29}Si MAS NMR work on Si-Al ordering in minerals is that the aluminum avoidance principle holds when Si/Al is 1 or more with a few well-known exceptions. Synthetic cordierite and synthetic ultramarine both show abundant Si-Al disorder. Both may have been formed far from equilibrium, as subsequent annealing restored the order in cordierite and natural ultramarines show local ordering and no disorder. The preponderance of AlO^{5-}_4 as tetrahedral framework species in minerals, oxide ceramics, and catalysts makes the ordering information obtained from NMR essential to constrain the thermodynamics of these materials and hence calculate phase equilibria.

The chemical shift of aluminum is sensitive to the number of oxygen nearest neighbors, and the shifts for 4-, 5-, and 6-fold coordinated aluminum in many minerals have been recorded and fall in the ranges 50 to 80 ppm, 25 to 40 ppm, and -20 to +15 ppm, respectively. Therefore, it is generally possible to distinguish between tetrahedrally coordinated and octahedrally coordinated aluminum in minerals by ^{27}Al NMR, 5-fold coordination being very rare. This means that framework (4-fold) and nonframework (6-fold) aluminums may be distinguished and allow framework Si/Al ratios to be determined. This information can be important if the Si/Al ratios to be determined. This information can be important if the Si/Al ratio is known only from elemental analysis and questions of framework ordering need to be resolved. In addition, disorder in minerals such as spinel ($MgAl_2O_4$) may be determined. The

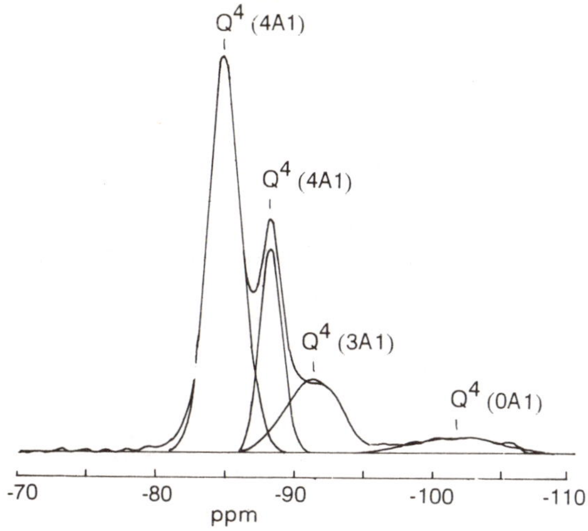

Fig. 23.8. ^{29}Si MAS spectrum of nepheline from Bancroft, Ontario. The peaks are assigned to Q^4 units as shown and correspond to different numbers of aluminum neighbors which agree with a completely ordered arrangement of Si and Al in the structure. [Adapted from J. F. Stebbins, J. B. Murdoch, I.S.E. Carmichael, and A. Pines (1986). *Phys. Chem. Miner.* 13, 371-381, with permission.]

occupation of tetrahedral and octahedral sites by aluminum is an important contribution to the entropy of spinel, which is often used in the calculation of the pressures and temperatures of coexisting mineral phases.

The second-order quadrupolar broadening (see Section II) of ^{27}Al MAS NMR spectra usually prevents resolution of different tetrahedral Al sites in the same material. Different shifts for different numbers of Si next nearest neighbors or numbers of bridging oxygen neighbors are not usually resolved in framework aluminosilicates except where the ample is very well crystallized and the electric field gradient of each site fairly small. However, by correlating the shifts of minerals containing single sites or well-resolved multiple sites and correcting for the second-order quadrupolar shift of the NMR line, empirical relationships between T-O-T angle, condensation, and chemical shift which were established for ^{29}Si can be demonstrated for ^{27}Al.

The most fundamental information about oxides and silicates may potentially be available from ^{17}O NMR. Because it is only 0.037% abundant, ^{17}O is extremely difficult to observe in natural minerals and synthetic samples enriched in ^{17}O must be prepared. These have been used to investigate the relationship between chemical shifts and structure and electric field gradients and bond ionicity. In static and MAS NMR of powders the spectra are dominated by the second-order quadrupolar effect (^{17}O has $I = 5/2$) and powder patterns can usually be observed when there are only two or three inequivalent oxygens in the material. The difference between the chemical shifts of Si-O-Si and Si-O-Al is not very large, but the electric field gradients are significantly different and correlate well with the average ionicities of the two bonds calculated from Pauling electronegativities. An example in Fig. 9 shows how this allows the Si/Al ratio for a sample of synthetic zeolite Na-Y to be calculated by matching the

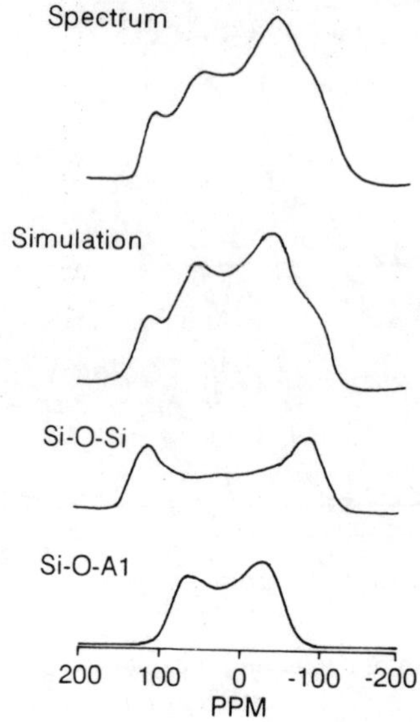

Fig. 23.9. ^{17}O NMR spectra of zeolite Na-Y, which contains Si-O-Al and Si-O-Si linkages. These can be identified in the static spectrum by their characteristic electric field gradients in the ^{17}O spectra. The experimental spectrum may be simulated by superposition of the two individual powder patterns. [Adapted from H.K.C. Tinken, G.L. Turner, J.-P. Wilson, L. B. Welch, and E. Oldfield (1986). *J. Am. Chem. Soc.* 108, 7231-7235, with permission. Copyright 1986 American Chemical Society.]

intensities of the two well-known powder patterns to the overlapped line shape. The result agrees with that obtained from ^{29}Si NMR. ^{17}O NMR can distinguish between bridging and nonbridging oxygens in a similar way, in that the electric field gradients for nonbridging oxygens fall in the range 1-3 MHz and those for bridging oxygens in the range 3-5.5 Mhz. There is some systematic difference in the chemical shifts of bridging and nonbridging oxygens in silicates, with nonbridging oxygens falling in the range 40-70 ppm, depending on the nonnetwork cation, and bridging oxygens (Al, Si) in the range 31-52 ppm. In crystal structures in which there are many crystallographically inequivalent oxygen sites, MAS can usually resolve only bridging and nonbridging oxygens unless there are large differences in electric field gradient between two oxygens of the same type. A new sample reorientation technique has been developed which involves spinning the sample about two axes during the experiment. One version of the technique involves spinning about two axes sequentially and performing a two-dimensional experiment to correlate the signal from the two angles (dynamic angle spinning, DAS). The second technique spins the sample about two axes simultaneously (double rotation, DOR). These reorientation schemes remove the anisotropic contribution of the second-order quadrupolar nuclei such as ^{17}O. Figure 10 shows the MAS, DAS, and DOR spectra of wollastonite ($CaSiO_3$); three oxygen sites are resolved in the MAS spectrum but in the DAS and DOR spectra of wollastonite ($CaSiO_3$) three oxygen sites are resolved in the MAS spectrum but in the DAS and DOR spectra eight of the nine crystallographically inequivalent oxygen atoms can be resolved, in the latter spectrum among many spinning sidebands. This should provide a very powerful tool with which to approach structural problems in mineralogy, although detailed work to establish the empirical relationships between oxygen isotropic chemical and quadrupolar shifts will be needed.

V. Amorphous Materials and Viscous Liquids

Most rocks which make up the earth's crust and probably the mantle originally crystallized from a silicate liquid, so the details of the microscopic behavior of such liquids are of great importance in developing an understanding of the evolution of the planer. Natural magmas are complex liquids in terms of both composition and structure. They often contain up to 10 oxide components, and their chemistry and physics are thought to lie somewhere between those of molten salts and organic polymers. Structural information has been limited because they are liquids and lack the long-range translational symmetry required for diffraction methods and because of the technological difficulties of spectroscopic work at high temperatures. One way to understand the structure of silicate liquids is to quench the liquid to a glass and perform the spectroscopy on the glass at room temperature. Two potential problems with this method are that there may be structural features of the liquid which cannot be quenched into

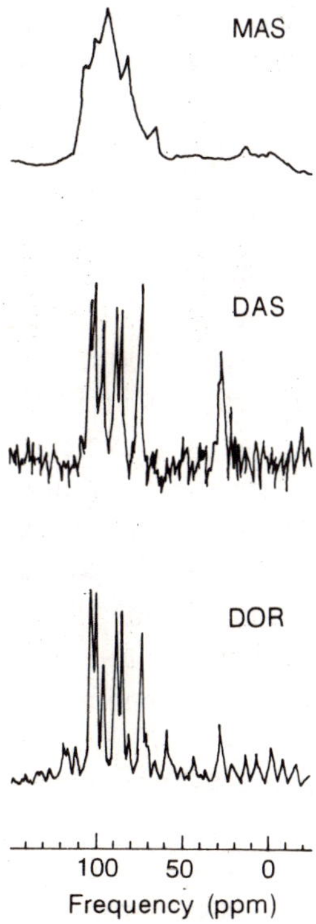

Fig. 23.10. Magic angle spinning, dynamic angle spinning, and double rotation ^{17}O NMR spectra of wollastonite ($CaSiO_3$). The latter two techniques involve sample reorientation about two axes, and narrow lines are observed for half-integer quadrupolar nuclei. Here, eight of the nine crystallographically inequivalent oxygens can be distinguished in the DAS; they are present in the DOR among numerous sidebands. (Spectra courtesy of B. Chmelka, University of California, Berkeley.)

the glass and that the structure of a glass represents the structure of the liquid only at the glass transition. Even very rapid quenching produces a glass transition temperature several hundred degrees below "geologically relevant" temperatures for most melt compositions. Despite these possible problems, most geological melt compositions are good glass structure and the original liquid structure has been made by the spectroscopy of glasses. [See Magma Chamber Dynamics.]

The models of glass structure which have developed over more than 50 years all have as a starting point the tetrahedral coordinations of silicon by four oxygen atoms. In the prototype glass, vitreous SiO_2, these tetrahedra are linked by shared corners (bridging oxygen atoms) to four other tetrahedra and the considerable variation in the intertetrahedral (Si-O-Si) bond angles is responsible for the lack of long-range order. More generally, the structure of glasses with geochemical compositions can be thought of in terms of how the different oxide components disrupt the continuous network formers by forming strong, partially covalent bonds with oxygen (in tetrahedral coordination), and cations such as Na^+ and Ca^{2+} act as network modifiers by forming weaker, more ionic bonds to oxygen with coordination numbers of six or more. The local structure of the glass can be characterized by the distribution of nonbridging oxygens in the network, i.e., the distribution of Q^n species, where n denotes the number of bridging oxygens per SiO_4 tetrahedron. ^{29}Si MAS NMR can distinguish these sites by reference to the chemical shifts for similar units in minerals. The resolution is poorer in glass spectra due to the breadth of the lines for silicons with the same Q speciation but different average intertetrahedral bond angles, which results in a distribution of chemical shifts. Vibrational spectroscopy can distinguish the presence of different Q species in a glass with greater resolution, but ^{29}Si NMR can give quantitative information about the abundances of the Q species. This allows models of structure to be tested. For a random distribution of bridging oxygens a ^{29}Si NMR can give quantitative information about the abundances of the Q species. This allows models of structure to be tested. For a random distribution of bridging oxygen a ^{29}Si NMR spectrum can be predicted and compared with experiment, as can a constrained distribution of nonbridging oxygens where only two Q species. This allows models of structure to be tested. For a random distribution of bridging oxygens a ^{20}Si NMR spectrum can be predicted and compared with experiment, as can a constrained distribution of nonbridging oxygens where only two Q species can coexist at any general composition or a single Q species at a stoichio-metric composition. Some of the early ^{29}Si MAS results on the binary system Na_2O. SiO_2 indicated that the constrained distribution was operating and only Q^3 and Q^4 species were observed for compositions of less than 33.3 mol% Na_2O (see Fig. 11A). However, the signal-to-noise ratios of these early spectra were not good enough for minor deviations from the constrained distribution to the observed. Non-MAS (i.e., static) experiments on samples enriched in ^{29}Si showed the presence of a narrow symmetric Q^4 peak when according to the constrained model none should exist (see Fig. 11B). This can be explained by a disproportionation reaction of the form $2Q^n = Q^{n-1} + Q^{n+1}$ (n = 1, 2, 3). IT was demonstrated that the nonbridging oxygen distribution was close to constrained but deviated from it depending on the field strength of the network-modifying cation; this deviation appeared to increase in glasses with a higher fictive temperature (more rapidly quenched). Therefore a substantial contribution to the entropy of silicate liquids may come from the increasingly random distribution of nonbridging oxygens with temperature.

The distinction between a liquid and a glass is the mobility of the liquid structure. IT is possible to carry out NMR experiments on silicates at high temperatures and take advantage of the time scale of the technique to see how the liquid structure changes with time. Figure 12 shows the ^{29}Si spectrum of $K_2Si_4O_9$ at 700 and 997°C. At 700°C the spectrum consists of two resonances for Q^3 and Q^4 silicons in the sample and is nearly identical to the spectrum at room temperature. The demonstrates the very slow structural rearrangements occurring in a liquid with a viscosity of ~10^9 Pa s, which are not rapid enough to cause the line narrowing

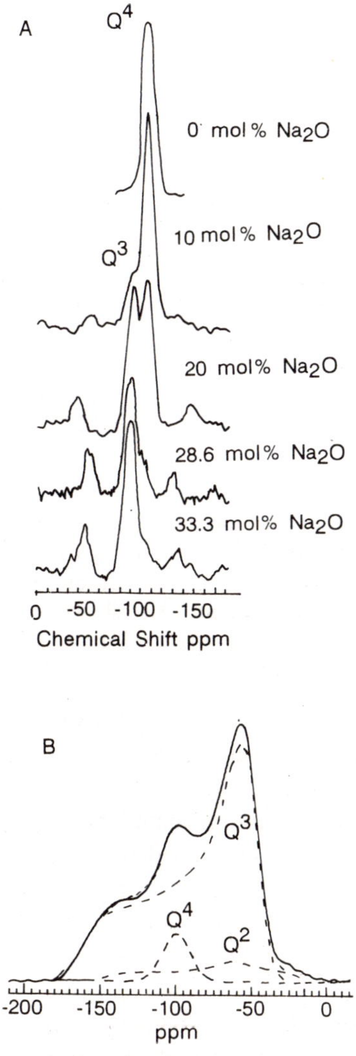

Fig. 23.11. (A) Series of ^{29}Si spectra for Na_2O. SiO_2 compositions showing the gradual conversion of Q^4 units to Q^3. (B) Static spectrum of 34 mol % Na_2O. SiO_2 which clearly shows the presence of Q^4 silicons although by a constrained distribution of nonbridging Q^4 Silicons although by a constrained distribution of nonbridging oxygens they should not exist. [Adapted from (A) R. Dupree, D. Holland, P. W. McMillan, and R. F. Pettifer (1984). J. Non-Cryst. Solids 68, 399-410, and (B) J. F. Stebbins (1988). *J. Non-Cryst. Solids* 106, 359-369, both with permission.]

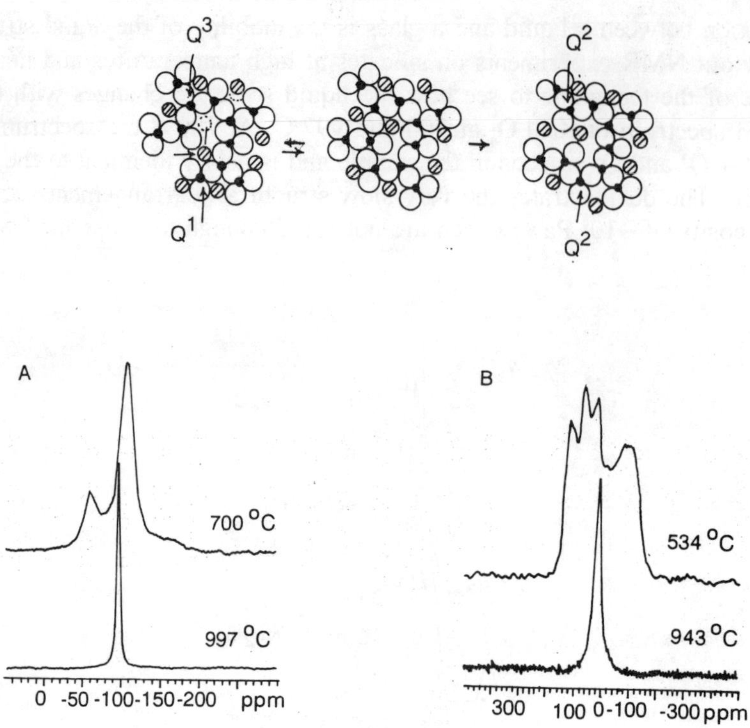

Fig. 23.12. (A) ^{29}Si spectra at high temperature; two resonances for Q^3 and Q^4 collapse to a single line at high temperature. (B) ^{17}O spectra at high temperature; powder patterns for bridging oxygens (broad) and nonbridging oxygens (narrow) exist above the glass transition but collapse to a single line at high temperature. Also shown is an exchange mechanism consistent with the high-temperature spectra in which bridging and nonbridging oxygens and Q species are exchanging rapidly (~500 kHz).

common in NMR experiments on liquids. As the temperature of the sample is raised, the lines do begin to narrow and eventually a single narrow line is observed at 997°C. This is a very informative result because the existence of a single NMR line when two Q species are known to exist means that silicons are not only undergoing rapid motion but also exchanging rapidly between Q sites in the liquid. The immediate conclusion from this result is that large silicate anionic species do not exist in the liquid for more than a microsecond or so. In fact, the exchange process causing the line narrowing can be modeled and exchange rates at different temperatures used to evaluate an activation energy for the process. This is within 10% of the activation energy for viscous flow derived from viscosity measurements. Therefore NMR appears to be observing one of the fundamental steps in viscous flow. ^{17}O experiments at high temperature also show rapid exchange among bridging and nonbridging oxygens with the average lifetime of a bridging oxygen being ~1 s, which would be the lifetime of any large silicate anion. A possible mechanism for the Q species exchange is shown in Fig. 12. In two dimensions it is easier to picture $Q^1 + Q^3 = 2Q^2$ exchange in which a nonbridging oxygen from a Q^1 unit approaches a Q^3 unit as a modifier cation moves away, an intermediate state in which silicon is over-coordinated is formed, a former bridging oxygen then dissociates from the intermediate, and a Q species exchange is complete. A process of this sort is consistent with both the ^{29}Si and 17^{0} high-temperature spectra. These results present a picture of atomic-scale dynamics in silicate melts which are dominated by short-range processes and not longer-range effects such as the entanglement of large polymeric anionic species.

As the temperature of a silicate liquid is increased, the atomic structure appears to become more random with local rearrangements of structural units occurring, which strongly suggest the formation of additional intermediate units with silicon in higher coordination. When pressure is applied to a crystalline silicate, the structure rearranges to reduce the volume; this is done by increasing the coordination of silicon from four to six. The lower mantle of the earth is probably dominated by silicon in sixfold coordination with oxygen. The effect of pressure on a silicate liquid might be expected to produce a similar coordination change. ^{29}Si NMR is very sensitive to these coordination changes, and spectra of silicate glasses quenched from liquids at bout 1500°C and pressures of 1.9-8 GPa have shown small (~1-10%) peaks for not only six-coordination changes, and spectra of silicate glasses quenched from liquids at about 1500°C and pressures of 1.9-8 GPa have shown small (~1-10%) peaks for not only six-coordinated silicon but also five-coordinated silicon. The presence of these coordinations in the glass implies that they existed in the equilibrium liquid and strengthens the case for the existence of a five-coordinated intermediate in the mechanism for viscous flow. Similar coordination changes from four to six have been observed for aluminum in albite glass quenched from above 6 GPa. These coordination changes have implications not only for the density of silicate liquids but also for their viscosity. Silicon in sixfold coordination is more like a network modifier than a network former, which should reduce the viscosity of a silicate liquid at high pressure.

GLOSSARY

Chemical shift. Small difference in frequency of a nuclear resonance due to the chemical environment of the nucleus. Measured in parts per million (ppm) from a reference material.

Electric field gradient. Second derivative of the potential at a crystal site.

MAS. Magic angle spinning, rapid rotation of the sample at an angle of 54.7° to the magnetic field during the experiment.

NMR. Nuclear magnetic resonance.

Nuclear quadrupole moment. Property of a nucleus with a less than symmetrical charge distribution.

24

Powder Diffraction Techniques

Introduction

Most of the materials that comprise the crust of the earth are crystalline, and one of the most powerful and versatile methods for characterizing and utilizing crystalline properties is powder diffraction. When a small powdered mineral sample is placed in a collimated beam of X rays or other equivalent radiation, many diffracted beams are generated with characteristic angles and intensities. Careful measurement of these beam will yield information or each of the minerals that make up the sample independent of the other minerals in the mixture. Modern diffraction equipment is safe and easy to use, and only a small amount of training is required for a user to obtain some very useful information on a sample under study. Highly skilled diffractionists are able to determine many properties that are related to the structure of the mineral and its state of crystallinity.

Modern diffraction equipment range from very, sophisticated computer-controlled devices to simple cameras. The expefimental measurements are usually obtained in short times from minutes to a few hours, and the resulting data are relatively straightforward to interpret for determining the identification of the minerals present in a sample. Frequently, such identifications are sufficiently unique to confirm the chemistry of the mineral without required an independent elemental analysis. These identifications are valid even where the different compounds have identical chemical compositions. The diffraction data may also be used to determine the values of other crystal properties and to estimate the amount of each mineral present in a mixture. More advanced analyses lead to information on the crystal structure and the state of perfection of the crystal.

Most common rock-forming minerals are quite complex crystals, and subtle variations in the crystal structures of many mineral species reveal considerable information on their history of formation. Once researchers are able to interpret these differences by correlations with other measurements, the differences become useful tools in determining the geologic history in other samples containing the same minerals. Powder diffraction can then be used to look for sensitive crystal properties, such as the size and shape of the crystal cell of some variations in intensities, which indicate the same conditions as in the characterized sample. Many samples may be examined in a short time, allowing the mapping of the desired property with respect to the field locations of the samples.

Powder diffraction includes both neutron and electron diffraction as well as X-ray diffraction. Each technique has its special attributes. X-ray diffraction is by far the most common, because the cost of the equipment is the lowest and the ease of use is the greatest. Electron diffraction in rarely used for powder experiments because the grains in a sample usually yield single-crystal data,which are easier in interpret in their own right, and dimensional measurements are not as accurate as for X-ray diffraction. Sample preparation is also more difficult.

Neutron diffraction facilities are only located at a few sites in the world are used primarily for research projects. Experimental times are also considerably longer than for x-ray diffraction. For these reasons, this article will concentration on X-ray diffraction theory and applications. The modern synchrotron, which is also restricted to a few sites, is essentially an X-ray generator. It is primarily a research tool.

II. The Powder Diffraction Experiment

The diffraction process requires two conditions. The radiation must have wave properties and the sample must act like a periodic grating with grating spacing of the same order of magnitude as the wavelength of the radiation. X rays have wavelengths in the range of 0.1-10.0 A. Crystals are characterized by the periodic arrangement of the atoms to form a three-dimensional grating with periodicities in the range of 3-20 A for most minerals. The combination is ideal for producing diffraction effects. It was in 1911 that M. vonLaue postulated and proved that such effects could be detected. In 1913, W. L. Bragg showed how in interpret diffraction intensities in terms of the crystal structure. Since that time,the technique has been perfected and the theory advanced such that modern mineralogy and materials research rely heavily on the method for many crystal measurements.

A. The Periodic Nature of the Crystalline State

The crystal structure of a mineral may be described as consisting of a basic unit cell that contains each of the elements of the compound in a characteristic arrangement. The unit cell is a parallelepiped. The crystal is composed of many unit cells packed in an array like bricks in a well, resulting in a periodic aggrangement of all the atoms in the crystal. It is this periodicity that gives rise to the diffraction effects. The crystal may be characterized by the size and shape of the unit cell. In fact, all a crystals are classified by the cell shape into six crystal system. These systems are shown in fig.1. The dimensions of the lengths of the unit-cell edges and the inter-edge angles are readily determined by diffraction measurements, and they form the basic of one means for the identification of crystalline materials. Using the Crystal Data Database produced by the Office of Standard Reference Data of the National Institutes for Standards and Technology, a diffractionists may compare cell values from an unknown to the cell data for over 140,000 inorganic and organic compounds.

The periodicity of a crystal structure is usually characterized by selecting a selecting a specific point in each unit cell and examining the distribution of this point in space. Such an array of points. Known as a lattice, is shown in Fig2. The unit cell of this lattice in identical to the unit cell of the crystal. All geometry of diffraction may be explained by the interaction of this lattice with the wave properties of the radiation. The most general concept imagines that there are planes of the lattice points that reflect the radiation. Because every lattice point has an identical plane passing through it, there are many set of equally spaced planes in the crystal space. Figure 3 illustrates the concept in two dimensions. Each set of planes reflects the incident radiation, and the reflected rays insteract. In general, these reflected rays cancel each other because of the phase shifts. But when the angular settings are just right, all the reflected rays will add to produce a very strong beam. This geometry,first described by W. L. Bragg, is Fig.4. The perpendicular spacing, and angel of incidence is known as the Bragg angle, 0. For a specific wavelength of incident radiation the relation for strong reflection is =2d sin0. This relation is generally called the Bragg law. In reality, the diffraction process is not one of reflection in the crystal structure. A detailed derivation of the geometrical principles leads to the same bragg relation.

Because there are many possible sets of Bragg planes in a crystal, there are many possible positions in which a crystal will generate a measurable diffracted ray. In a diffraction experiment, the incident beam in culminated and rendered monochromatic. A single crystal in such a culminated beam will, in general, not be in the proper position to create a diffracted analysis are used. In the first, a single crystal may be moved by rotation and/or tilting to achieve the diffracting conditions. In the second, the sample is made up of many crystals in random orientation, so that least some are in position to diffract. This latter method is known as the powder method.

Because there are so many possible Bragg planes in a crystal, some means of specifying the various sets in needed. The system used in Miller indices. Figure 5 indicates how these integers are assigned. The crystal coordinate system in set with its origin on any of the lattice points. One Bragg plane of the set must pass through this point also. Starting at the origin, one counts the number of Bragg planes that interest each crystal axis. Miller indices are usually form. Miller indices appear in most diffraction calculations and in tabulations of diffraction data.

B. The Diffraction Process

Because the incident beam is collimated, the diffracted rays in the powder method form a family of concentric circular cones around the beam path,as illustrated in Fig.6.These diffracted rays have a characteristic angle and intensity that can be measured by placing a film the crystal or by using an angular tracking device with an X-ray detector that sends a signal proportional to the beam intensity that can be measured by placing a film behind the crystal or by using an angular tracking device with an X-ray detector that sends a signal proportional to the beam intensity to electronic recording equipment. The collection of all the allowed diffracted beams from a specific material is the diffraction spectrum or pattern of the substance. This spectrum is usually recorded as a table of the diffraction angles converted to d-spacings and the corresponding intensity I on a relative scale. Using the d's and I's obtained from an unknown, the diffractionist can identify minerals in a mixture and determine other crystal properties.

Most diffraction experiments are performed with X-rays, but beams of neutrons or electrons will also be diffracted by cystalline materials. Electron beams are easily absorbed in materials, so the samples must be very thin and contained in the high vacuum chamber of the

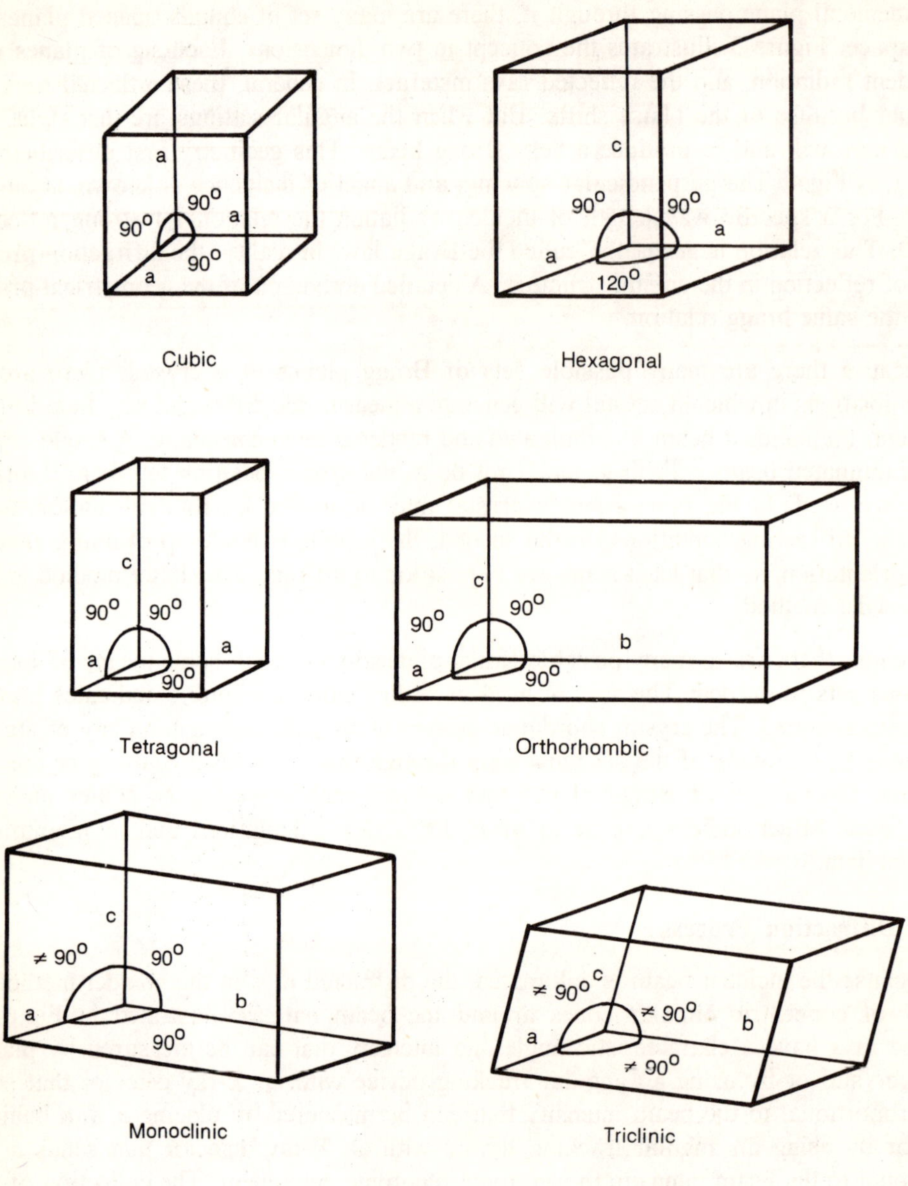

Fig. 24.1. The shapes of the unit cells of the six crystal systems.

electron microscope. Electron diffraction originates as scattering from the electrons in the sample, and the diffraction results are similar to those obtained with X-rays. Neutron beams are generated by nuclear reactors, and beam tunnels collimate the neutrons to diffractions apparatus similar in design to devices used with X-rays. Neutron scattering is by the nucleus of the atoms;thus, the diffracted beams may show considerably different relative intensities when compared to the data obtained from X rays. Neutron diffraction is used to study

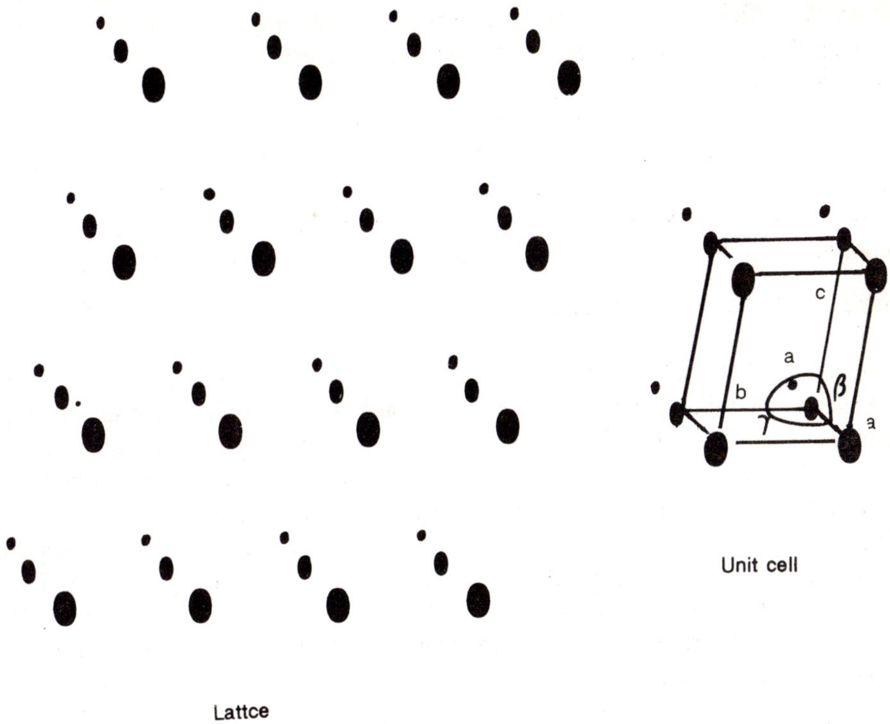

Fig. 24.2 A lattice and its unit cell.

magnetic materials, and it will often reveal information on elements closely spaced in the periodic table that are indistinguishable in X-ray diffraction. It is particularly useful for studying the studying the role of hydrogen in a structure.

C. Modern Diffraction Equipment

1. X-ray Sources

X ray are generated whenever a high-voltage electron beam is rapidly decelerated by hitting a target block. Some of the energy is also absorbed in the atomic structure of the target elements by exciting the electronic state, which can then decay and re-radiate the energy. The first process emits X rays of various wavelengths, but the second process emits X rays of various wavelengths, but the second process emit high intensities of only a few specific wavelengths characteristic of the target material. It is this second type that is so important for diffraction applicationed, because the strong characteristic radiation can be selected by filtering through a metal foil or diffracted by a special crystal to yield an essentially monochromatic X-ray beam.

Conventional generators use a sealed X-ray tube that contains an election gun and a metal target. The electrons are accelerateds by an applied high voltage and are focused on the target, causing the X rays to be generated from a very small volume of the target. The X rays may then leave the tube through thin beryllium windows. The limit of beam power depends on the efficiency of cooling on the target. Modern generators can produce 2kW of beam power. Copper is used most commonly as the target material for two reasons. In can be cooled effectively, and the characteristic wavelength is 1.54060 A-a value which is convenient for the study of most crystal intensity by maintaining the applied voltage at a constant potential and the electron flow at a constant current. Tubes need to be replaced after a few years of use because the target gets pitted and the beam distribution is affect. These types of generators may bc sct up wherever there is electricity and cooling water, but their operation is governed by state laws because of the inherent dangers of the X rays.

Some diffraction experiments are limited by the low beam power of a conventional source. There are two ways for achieving higher power. Generators have been designed which use a target that in rotating, thus distributing the heat over a larger volume and allowing for effective cooling while employing higher electron-beam currents. Rotating anode generators have been designed for producing beam power up to 35kW. These generators are expensive to purchase and to operate, but they can be set up in any laboratory. Alternatively, the synchrotron facilities around the world have ports that yield beam power in excess of 100 kW. These facilities are available to requstors who submit an acceptable proposal to the supervising committee and cover their own expenses while collecting the data. Time will be scheduled when the system in operating with the appropriate apparatus in place. Obviously, these specialized generators are used for nonconventional experiments. [See Synchrotron Radiation In Earth Sciences.]

Conventional generators usually have the X-ray tube mounted in a shielded tube tower that allows up to four tube housings are remote and connected to the generator by a flexible cable. If space permits,one tube can serve as the source for several experiments. Cameras and diffractometers are usually mounted on the generator and become and integral package.

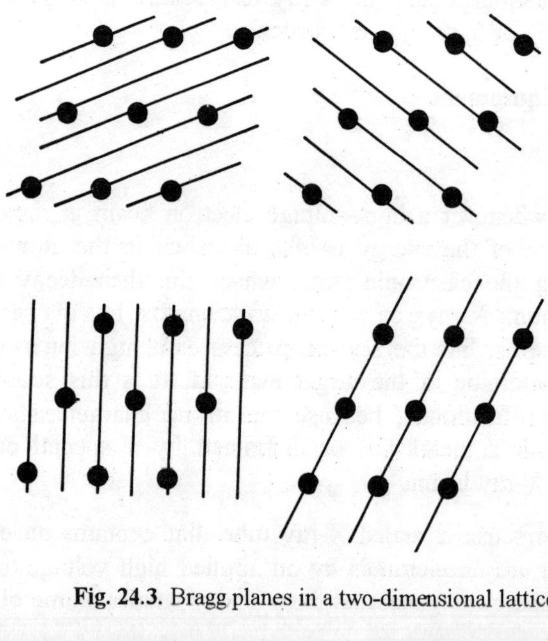

Fig. 24.3. Bragg planes in a two-dimensional lattice.

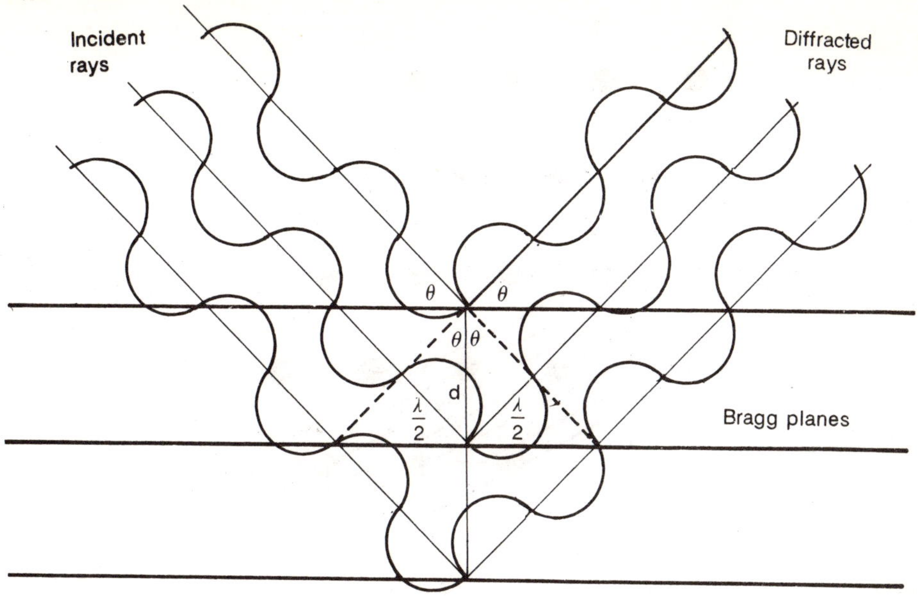

Fig. 24.4. Conditions for Bragg reflection; λ =2d sin θ.

2. Cameras

The simplest recording devices employ film as the recording medium. Usually, there are very few moving parts other than some means for rotating the sample during the exposure. The most commonly used camera is the Debye - Scherrer, in which the sample is placed in a thin glass tube and rotated on an axis perpendicular to the X-ray beam. The film is a 35X360-mm strip forming a circular cylinder about the rotation axis. Using this film, it is possible to measure the diffraction angle and the intensity of every diffracted beam from the sample. The resolution and accuracy of this camera is limited by the size of the sample, the divergence of the beam, the radius of the camera, and the graininess of the film. These radius of the camera, where used routinely up to the 1960s as the major recording device. More recently they are used only when limited amounts of sample are available or when the sample must be confined because of reactivity or toxicity. Good diffraction patterns may be obtained from nanogram for rotating a carefully centered single particle about two axes simultaneously allowed powder patterns to be produced from single crystals. This device is known as a Gandolfi camera and can be used for particles down to 1 um in sixe.

Another family of cameras is the facusing camera. One type of focusing camera employs a curved sample support, which does not gain any resolution but does gain intersity because it can use a large sample. This type is rarely used today. The other type of focusing camera uses a carefully cut and bent single crystal to diffract and reconverge a diverging beam to a focus equivalent to the target image. This device is known as the guinier camera and achieves very high resolution with a good gain in intersity. It does record dthe diffracteds beams on film which is held in a cylindrical cassette. The diffracting sample is a thin layer on a stickly tape mounted on the film circle. The sample is rotated in its own plane. When a camera is used with a fine-force X-ray tube and is optimally aligned, the diffraction line widths are around 0.02°. The diffraction angles and intensities are easily measured from the film.

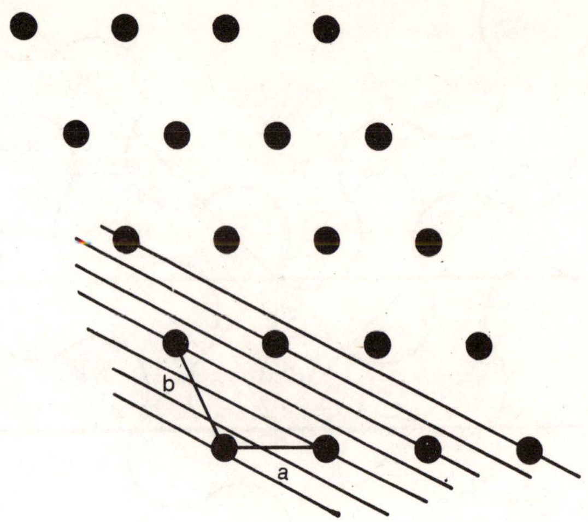

Fig. 24.5. Miller indices for Bragg planes in two dimensions. There are two planes on the a axis and three planes on the B axis. The Miller index is then.

3 Diffractometers

Diffractometer is the name given to a device that records the diffracted X-ray beam by moving an electronic detector on a rotating arm to intersect the diffracted beams and determine their diffraction angle and intensity. The earliest diffractometers did not use focusing were too weak, but in the 1940s devices were developed that used the first type of focusing optics employed in cameras with considerable success. The optics are illustrated in Fig. 7. The curved sample was approximated by a flat sample, which was successful with the large used for focusing circle.

The most successful diffractometer employs symmetrical Bragg-Brentano optics, which allows the detector arm to be a constant radius. The X rays are detected by quantum counting device that convert the beam intensity into voltage pulses whose energy is proportional to the X-rays are detected by quantum counting devices that convert the beam intensity into voltage pulses whose energy is proportional to the X-ray energy and whose frequency is

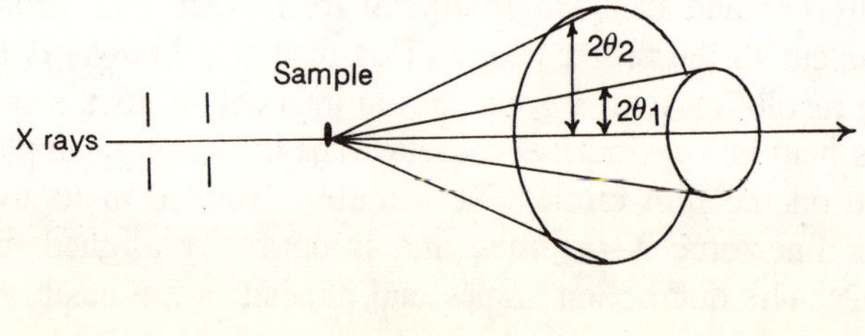

Fig. 24.6. Cones of diffracted rays around the incident beam

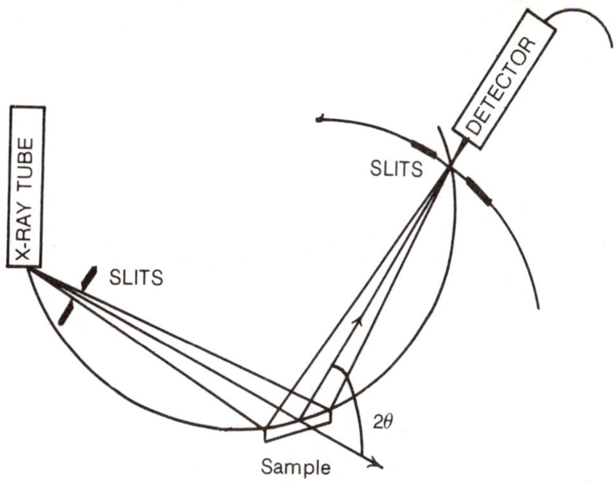

Fig. 24.7. The geometry of the powder diffractometer.

proportional to the beam intensity. Angular scanning of the detector arm allows a plot of the intensity of the diffracted beams versus the angle from which the diffraction information is derived. The angular resolution of the modern diffractometer is 0.005°, but the width of the diffraction peaks is usually of the order of 0.07° 20. Advantages of the diffractometer over cameras include the ability to determine the detailed shape of the diffraction peak and to measure the intensity values more accurately. This type of diffractometer is the most used diffraction device today. Diffraction patterns are usually displayed in graphical form, as illustrated in Fig.8.

Other types of diffractometers have been developed for special applications. Some employ unsymmetrical focusing optics. Others use Guinier optics with a focusing monochromator. These diffractometers are effective to study very thin films on a supporting substrate. Attachments to either type of films on a supporting substrate. Attachments to either type of diffractometer may be used to confine the sample in special atompheres and to heat the material to temperatures up to 2500° or cool it down to near absolute zero. One device which rotates both the detector and the X-ray tube about a stationary sample allows liquids to be studied. New types of detectors, usually called position-sensitive detectors, are collecting data over large angular ranges, simultaneously allowing time-resolved studies that can follow reactions while they are occurring.

4. Role of Computers in Data Collection and Analysis

The computer now plays many roles in diffraction analysis. The diffractometer is very amenable to control by computer, and all companies manufacture an automated powder diffractometer, APD. The main advantage of computer control is the ability to step the profiles to be collected as digitized intensity-versus-angle traces for mathematical analysis techniques. Computers can also monitor the voltage and current settings, carry out peak integrations, and change samples for 24-hour operations. [See Computer Applications In Geology.]

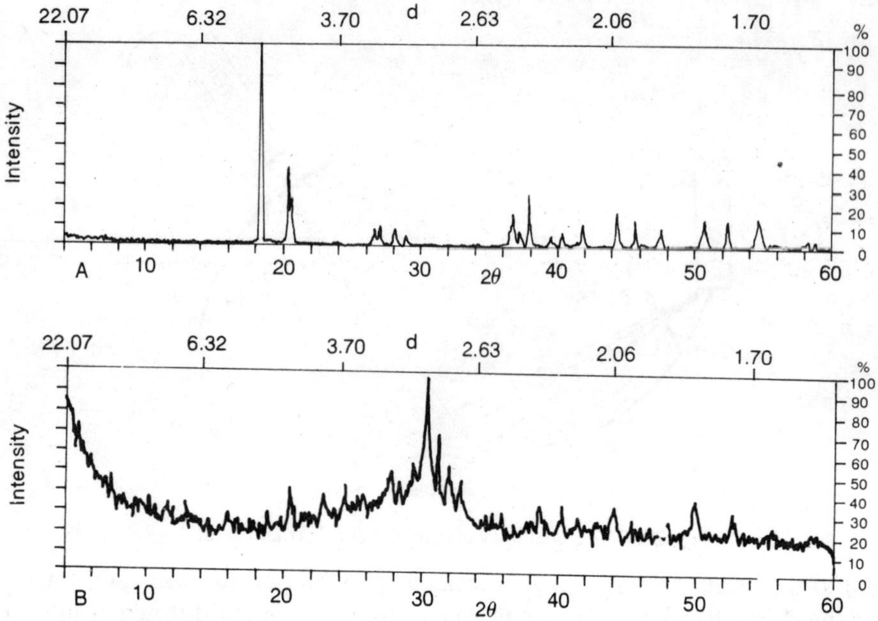

Fig. 24.8. Diffractometer pattern from (A) a well-crystallized sample and (B) a poorly-crystallized sample.

Of equal importance is the analysis of the digitized data.Computers easily reduce the data to d's and I's for identification, unit cell determination, and the refinement of lattice parameters. There are many methods to determine the accurate positions of the diffraction time - which is now feasible. The most elaborate methods use profile-fitting elaborate; these methods usually produce the most accurate results. Good profile fitting allows the unraveling of multiple peaks and results in resolutions comparable to the Guinier method. Identification methods that were initially developed using tables and books now use computer assistance to test databases of up to 50,000 diffraction patterns for possible candidate matches. The short list presented to the user saves considerable time in selecting the best candidates for detailed comparisons.

Digitized data opens up many new possibilities for data analysis where theoretical diffraction patterns may be calculated form structure models and compared directly with the raw data. These "whole pattern" methods are being used for quantification, refining crystal structures, determining crystalline size and strain, and revealing orientational textures. It is also being developed for identification and cell refinement. More than 280 computer programs are now in use for the analysis of powder diffraction data; in addition, software is provided by the APD manufactures.

III. Information in a Diffraction Pattern

Compared to most other spectroscopic methods, powder diffraction patterns are unique in that all the peaks of the pattern of a pure single phase are related to the same unit cell and crystal structure. The positions of all the peaks are determined by size and shape of the unit cell. The intensities are related to the types of atoms in the unit cell and their arrangement. A third feature is the shape of the profile, which can be related to be perfection of the crystal.

Each of these measurements has its particular applications. Another important feature of diffraction patterns is that when the sample is a mixture, the diffraction pattern of the composite is a simple addition of the patterns of each of the individual minerals, where the strengths of the patterns are related to the amount present in the mixture.

A. Diffraction Angles and Crystal Geometry

The diffraction angles are determined by the sets of Bragg planes allowed by the lattice. The number of possible planes is actually unlimited, but the size of the unit cell and the wavelength of the X rays used limit those that satisfy the Bragg equation to a small number whose d-spacings are greater than /2. When locating the peaks even with computer algorithms, many corrections are necessary to achieve the accuracy that is possible. Both the instrument and the sample cause problems, w2which can mostly be minimized by adding a certified standards to the sample and referencing all measurements to the peaks of this standard.

Accurate d-spacings lead to accurate values for the unit cell parameters. Accurate cells are requisite for the characterization of new materials and reveal the subtle differences between similar materials. Small chemical substitutions will cause measurable differences in the lattice dimensions, and many times the cell parameters will be the most sensitive indicator of the substitutions. Accurate d-spacings are also required for comparisons to references databases and are very important when applying computer methods to find the unit cell geometry when it is now known a priori.

B. Diffraction Intensities and Crystal Structures

While the positions of the peaks are determined by the size and shape of the unit cell, the intensities are determined by the type and arrangement of the atoms in the unit cell. Accurate measurement of the intensities are beset with many experimental problems, the most difficult of which is the lack of random orientation of the crystallities in the sample. Theoretical modeling of diffraction intensities must assume that there is an equal probability for all possible diffraction planes to be in a position to diffract. When the particle shapes are flattened on elongated and the sizes are larger than 10 um, it is very difficult to prepare to truly random sample. Another problem with powder patterns is the overlap of many peaks, which makes it very difficult to decide how much intensity should be assigned to each component of a complex peak. With APD systems, the actual measurement of the intensity values is no longer a problem.

Diffraction intensities have many uses besides determining the crystal structures. In combination with the measured d-spacings,they provide the best for identification using a reference database. When the minerals are known, the intensities can used to estimate to amounts present in a mixture. Differences in intensities in a pattern when compared to a reference pattern may indicate chemical substitutions, different states of structural order, differences in the crystal structures, and preferred orientation in the sample. In time-resolved studies, the increase or decrease in intensity of a specific peak is an indication of reactions that are taking place.

C. Diffraction Profiles and Crystal Perfection

The shapes of diffraction profiles are the most difficult to analyze because of sensitivity to the spectral distribution of the X rays and instrumental aberrations. Advanced mathematical techniques are required to separate out true sample profile; but where it can be done, there is considerable information to be obtained. Very small crystallities cause diffraction profiles to be broadened, and the peak width at half intensity can be used as a measure of the average crystallite size. Anything that affects the lattice dimensions nonuniformly will cause asymmetric broadening. Applied or residual stresses will distort the lattice as well as inhomogeneous chemical substitutions. Concentrations of defects in crystals will usually result distortions in the diffraction profiles also.

Crystal distortions are usually modeled, and the calculated profiles are compared with measured profiles. The combination of the spectral and instrumental profiles can be determined by measuring the profiles obtained from a sample with no defects. Several such standards are available including Si, Ge, and LaB_6. The modeling in calculationally intense, but there are several programs available for such analyses.

IV. Databases of Diffraction Information

Although diffraction data can be used to characterize new materials without referring to any previous information, it is usually desirable in most experiments to compare measured data with data reported on other samples. There are considerable crystal data in the literature, but locating the desired data could be very time consuming. Fortunately,there are four major databases that have collected this literature information and are readily available to the diffractionist. All these databases are available in computer-readable form.

A. The Powder Diffraction File

The Powder Diffraction File (PDF) is published by the JCPDS- International Centre for Diffraction Data (ICDD). This database diffraction information recorded as d-I sets for over 50,000 compounds along with supported crystal, chemical patterns in the PDF. In addition to being a rich source of diffraction and crystal data, the references to the literature often lead to the diffractionist to additional information. An example of a PDF data set in card image form is illustrated in Fig 9.

The PDF has been a source of information for nearly 50 years;the first published in 11941941, was actually as reprinting of data that appears in the PDF has been collected from several sources, including the open literature, data solicited directly from diffractionists, and data collect or projects supported by the ICDD. All the data receive extensive review by a board of experienced editors, and only the best data for a particular compound are usually accepted for publication. New data may be used to replace older data when the editors consider the new data superior. Thus, the average quality of the data in the PDF improves with time; but because it depends on the literature data for much of the information, considerable data are still included that were not collected under adequate control.

The PDF is the basis for many computer programs that attempt to identify compounds using d's and I's. Because of variations in compounds and inaccuracies in both the experimental and the reference data, the search procedures do not always find an unequivocal solution. Where the sample is a single phase, the solution is usually apparent; but where there is a mixture and some minerals are present in small amounts, the solution is considerably more difficult, and the computer usually presents the diffractionist with a list of several candidates based on figures-of-merit determined by the possibility that the known pattern is in the unknown pattern. The diffractionist must then select the best solution by making detailed comparisons through matching the entire d-I set.

B. NIST Crystal Data File

The Office of Standard Reference Data of the National Institute for Standards and Technology produce the Crystal Data File (CDF) contains a tabulation of crystal unit-cell information along with chemical and crystal symmetry information, ands the reference from which the data were collected. The CDF is published and distributed by the ICDD and is available by lease only.

The CDF relies on the open literature exclusively and includes references to all reported data. Careful editing assures that the information is faithfully reproduced, but there is no replacement of various data with new data. There are now over 146,000 inorganic and organic entries in this database. The CDF was fist considered as a reference for crystal identification when it was introduced in 1954, but it has become considerably more than that. In addition to being a database for identification through matching of crystal cell values,it serves as an reference for locating compounds with specific crystal properties.

Identification using a set of unit-cell parameters has some interesting ramifications. Finding a direct match with a previous decryption of the same compound is usually straightforward if the crystal is cubic, hexagonal,or tetragonal. When the crystal in orthorhombic, there are six different ways the a,b and c axes could be defined; this situation could lead to difficulties if there were not rules for nomenclature. If the user follows the database rules, there is no ambiguity. When the crystal system in monoclinic and especially when it is triclinic, the cell setting is very critical. Programs supplied with the CDF allow the user to redefine cells to follow the rules.

When the crystal is a new compound,the exact match will no be in the database, but there is a high probability that a related compound might be. Chemical substitutions that do not rearrange the crystal structure generally result in the magnitude of the cell parameters changing proportionally. Parameter ratios usually do not change much, and the similaritities can often be recognized by comparing these ratios with ratios listed in CDF.

Because the CDF contains considerable symmetry and chemical information, it may also be used as a source to locate potential compounds that may have certain desired properties. By using computerized search procedures, the CDF can be examined for all compounds without a center of symmetry or whose symmetry is polar. Likewise, compounds with specific chemistry or elements may be located. Because every entry contains the literature reference, the CDF is an excellent way to locate references on any compound in the file.

C. FIZ-4 Inorganic Crystal Structure Database

The Inorganic Crystal Structure Database (ICSD) is maintained as a computer-based service by the Fachsinformation-zentrum in Karlsruhe and the Gmelin Institute in Frankfurt, Germany . It contains coordinates of the atoms in the crystal structures for over 32,000 crystals whose structures have been reported in the literature. The supporting information includes the crystal cell and symmetry, chemistry, references, and some physical property data. The information has been edited by calculating crystal densities and bond distances, and comments were added when difficulties are encountered.

The ICSD is valuable to the diffractionist as a source of reference information. Also, computer programs are available that will use the crystal description and calculate the theoretical powder diffraction pattern. When there is no powder pattern recorded, this calculation offers a quick method to generate a good reference. Calculated patterns are useful for checking experimental patterns where preferred orientation may be present. They are also useful in modeling structure variations such as different orderings of atomic substitutions.

D. NRCC Metals Structure Database

The Metals Structure Database (MSD) is maintained as a computer-based database by the National Research Council of Canada and contains the same kind of data as the ICSD except for metals and alloys. At present, this database is available as an on-line service and in computer-readable form through CYSTI of Canada. ITs uses are the same as for the ICSD.

V. Diffraction Analyses

A. Identification of Phases

The most common use of powder diffraction is for the confirmation and identification of phases. Powder patterns are essentially a fingerprint of a crystalline compound and are usually sufficiently unique to allow positive identification when the reference pattern matches the experimental pattern in every detail. Even when there are small differences, the experienced diffractionist can usually guess what the differences indicate with respect to the unknown material.

The identification of an unknown using only d's and I's measured experimentally may be successful when there is a pattern for the compound or a similar one in the reference database. When there is no appropriate reference, no identification should be expected. The procedure takes two steps. The first step, which is called the search step, locates candidate phases in the database by using characteristic parts of the diffraction pattern, such as a few strong or high d-spacing peaks in a comparison with the same information in the complete diffraction pattern of best candidates from the search step by locating the corresponding peaks and comparing the d-values and intensities. If the match shows a complete one-for-one correspondence, the identification is complete. Missing peaks or incorrect intensities must be explained. Unaccounted-for peaks in the unknown pattern usually imply another phase is present and other candidates must be tested.

$Ca_3(V,Fe,Al)(SiO_4)_3$
Calcium Vanadium Iron Aluminum Silicate Grossular

Rad. CuKα λ 1.54178 Filter Mono. **d-sp** Diff.
Cut off 17.7 **Int.** Diffractometer **$I/I_{cor.}$**
Ref. Keller, L., Rask. J., Buseck, P., Arozona State University,
Temple, Arizona, USA, JCPDS Grant-in-Aid Report, (1987)

Sys. Cubic **S.G.** Ia3d (230)
a 11.8493(8) b c A C
α β γ **Z 8** **mp**
Ref. Ibid.
$\mathbf{D_x}$ 3.64 $\mathbf{D_m}$ 3.65 **SS/FOM** F_{30} = 184 (.005.31)

εα nωβ 1.753 εγ Sign 2V
Ref. Deer, W., Howie. R. Zussman, J., Rock Forming Minerals, 1A
604 (1982)

Color Orange
Peak height intensities. Speciemen from Eden Mills, Vermont, USA. Microprobe analysis, average of ten (wt.%): SiO_2 39.37. Al_2O_3 21.06. CaO 35.08, Fe_2O_3 1.92. FeO 2.08. MnO 0.13. This material was examined by single-crystal X-ray diffraction. HRTEM and IR spectroscopy (Allen et al., 1986). This garnet is oprically anisotropic and is ver slightly distorted from cubic symmetry. Validated by calculated pattern 33-260. Opptical data from specimen from Hunting See following card

d A	Int	hkl	dA	Int	hkl
4.837	1	211	1.3417	1	752
4.189	1	220	1.3249	6	840
3.167	1	321	1.2929	9	842
2.962	35	400	1.2777	1	921
2.650	100	420	1.2631	6	664
2.256	7	332	1.2490	1	851
2.418	18	422	1.2222	1	932
2.323	16	431	1.1971	2	941
2.163	12	521	1.1733	1	1011
2.094	3	440	1.1618	2	1020
1.922	18	611	1.1509	1	943
1.874	2	620	1.1298	1	1031
1.7471	1	631	1.1098	1	871
1.7099	12	444	1.1003	6	1040
1.6757	1	543	1.0817	5	1042
1.6432	18	640	1.0727	1	873
1.6123	1	721	1.0556	1	1121
1.5834	30	642	1.0473	4	880
1.5050	1	651	1.0235	1	1132
1.4810	6	800	1.0160	1	1060
1.4587	2	741	1.0087	1	1141
1.4369	1	820	0.9873	1	1200
1.4162	1	653	0.9741	2	1220
1.3964	1	660	0.9611	4	1222
1.3774	1	831	0.9548	1	1231

$Ca_3(V,Fe,Al)\ (SiO_4)_3$
Calcium Vanadium Iron Aluminum Silicate Grossular

Rad. λ **Filter** **d-sp**
Cut off **Int.** **I/I_{cor}**
Ref.

Sys. **S.G.**
a **b** **c** **A** **C**
α β γ **Z** **mp**
Ref.

$\mathbf{D_x}$ $\mathbf{D_m}$ **SS/FOM**

Hill quarry. Montgomery. Maryland, USA. Garnet gourp, garnet subgroup. Silicon used as internal standard. PSC: c1152. To replace 3-826, 24-177 and 26-292.

See preceding card

d A	Int	hkl	dA	Int	hkl
0.9368	1	1240			
0.9310	1	1154			
0.9253	1	1080			
0.9197	1	992			
0.9088	1	1251			
0.8983	1	1231			
0.8931	1	1244			
0.8881	1	1253			
0.8833	4	1260			
0.8736	1	1262			
0.8688	1	1341			
0.8596	1	1093			
0.8552	1	888			
0.8507	1	1271			
0.8421	1	1411			
0.8379	1	1420			
0.8337	1	1273			
0.8256	1	1431			
0.8216	1	1280			

Fig. 24.9. A card image of X-ray powder diffraction data from the Powder Diffraction File. (Courtesy of JCPDS-International Centre for Diffraction Data, Publishers of the Powder Diffraction File, 1601 Park Lane, Swarthmore, Pennsylvania.)

Search methods may be done by using specifically prepared tables using the strongest peaks (the Hanawalt method), or the largest d-spacing (the Fink method), or by employing one of several computer search programs. Book searching can be slow and tedious, but when there is only one phase involved and the patterns are accurate, results are usually obtained easily. Mixtures require more time, and the computer is especially useful in this situation. Phases present in small quantities are difficult to locate and confirm because much of its diffraction pattern is not recorded-the intensities are too weak.

The procedure for identifying a compound may vary depending on how much is known about the material a priori. The experienced diffractionist uses every clue available. When a compound is expected, the pattern should be called up from the PDF and compared directly with the unknown data. When some specific family of compounds is expected from a knowledge of the elements involved and the treatment, the patterns may be collected into a mini-database for quick checking. Chemistry often is the key to correct identification. Either it leads the diffractionist to a small set of compounds to consider or it distinguishes which of several similar compounds is the most probable.

The PDF is classified into subfiles where possible. The best-defined and most useful subfile is the Mineral File. All patterns made from mineral specimens or from synthesized materials that correspond to known minerals comprise the Mineral File. These patterns have been published in hard-copy form by the ICDD, and all patterns in the computer-readable database are flagged so the search need only consider them when asked. Thus, to identify any specimen that comes from a natural source, the answer can be expected in this subfile. The smaller number of patterns (3800) that need to be considered in the search and match usually result in faster and more accurate identifications. For mineral identification, the powder diffraction method is very powerful.

B. Determination of Crystal Geometry

The classification of minerals is very important for understanding their crystal properties. The first step is to determine the correct crystal system. There are several ways to obtain evidence suggesting the crystal system, such as crystal morphology and optics in polarized light, but diffraction methods are best, but powder diffraction methods are very useful also [See Optical Mineralogy.]

All d-spacings are directly related to the dimensions of the unit cell, so it follows that unit cells should be determinable from the full set of d-spacings measured on a compound. In practice, this determination is easier said than done; but it can be done. Where the crystal system is cubic, experienced diffractionists can usually recognize that fact by simple inspection of the pattern. Other systems require considerable effort to solve. Computer assistance is very useful for this problem because it usually requires many steps of testing data consistency with the possible unit cell suggested by using the largest d-spacings. Where the experimental data are very accurate, the probability of locating a satisfactory solution is very high. Low-quality experimental data are useless for cell determination.

Usually, the compounds is not a total unknown. Identification methods may locate a related phase whose cell is similar. This information may be transferred to the new material, and the cell values adjusted for a better fit. Where the morphology or optics suggest a specific crystal system, that knowledge may be useful in directing the computer to examine the right test cell.

If a cell is determined for an unknown material and the powder pattern is not in the PDF, the cell information may be used to search the CDF. Many times the cell dimensions will suggest isostructural relations with some known compound, and this information will even suggest the crystal structure on the new material. Unit-cell information is important in the characterization of any new material.

C. Precision Determination of Cell Parameters

Once the crystal system is determined and a trial set of cell parameters are obtained, or when a data set is obtained on a known material, it is often necessary to calculate the best-fit cell parameters. This process is known as cell refinement and is computer assisted. Many programs for cell refinement are available, most of which have means to consider systematic errors that might be present in the data set. Some programs start with a probable cell and assign Miller indices and refine as an interative procedure. The test of the results is to examine the differences between the calculated and the measured d-spacings.

Accurate cell parameters have many uses. Deviations from cells determined on other specimens of the same mineral indicate subtle differences in the crystal structures that suggest compositional differences or structural differences. A very important example is the behavior of feldspars. Composition differences in the solid solutions causes measurable changes in the cell parameters. Even when the composition is fixed, as for the K-spars, the ordering the Al/Si has a marked effect on the cell. Now that much of the behavior has been described and related to thermal history, the cell data from other feldspars can also be interpreted in terms of thermal history.

D. Quantification of Phase Composition

In the diffraction pattern of a mixture, the intensity of each component pattern is proportional to the amount of the phase in the mixture, thus the diffraction intensities can be used to estimate quantitatively the amount present. The proportionality is not linear, because of absorption effects of the elements involved; and it is difficult to calculate directly without knowing the results beforehand. Fortunately, several methods can be used to get around this problem, and quantification by X-ray powder diffraction (QXRPD) is a useful technique.

The biggest problem affecting QXRPD is the preparation of the sample. To have statistically significant intensities when the diffraction data are collected, the sample must have a truly random particle orientation with a sufficient number of particles in the irradiated volume to allow all orientations to be achieved with equal probability. These conditions are not as easy to meet as first seem possible; but there are ways, especially if the particle size is smaller than 10 m. With careful work, accuracies of +2% absolute can be attained.

There are three basic techniques for intensity data. If the mixture is composed of cubic, hexagonal, and/or tetragonal compounds, each phase may be characterized by a single isolated diffraction peak, which can be integrated by a programmed diffractometer. For more complex phases, several peaks can be used to correct for interference. The peaks may be integrated or fit by profile matching to get the desired intensity values. Alternatively, the whole diffraction trace may be used, and the fitting procedure would then involve either measured or calculated traces for each of the component phases.

The interpretation of the measurements also follows several different procedures. If the adsorption effects can be measured or calculated, the direct equations can be solved. Alternative methods employ an added standard to which every measurement is ratioed or premeasurement of pure phase intensities ratioed to a standard. These two approaches compensate for the absorption effects. When setting up a quantitative analysis scheme, it is usually possible to select the best-suited method, depending on the phases involved and the accuracy needed.

E. Crystal Structure Modeling and Refinement

Because the intensities of a diffraction pattern are controlled by the crystal structure, it should be possible to determine crystal structures from powder data. Although theoretically possible, in practice it is very difficult to achieve. The difficulty arises because of the need to separate all the diffraction peaks so that all the possible observable peaks can be assigned unique intensity values. Single-crystal techniques allow peak separation, but powder methods in general do not, due to the overlap of peaks with similar d-spacings.

In spite of the difficulties, several approaches provide crystal structure information from the powder data, all of which employ analyzing the digitized diffraction trace. If an unknown can be matched with an isostructural compound by cell matching or d-I matching, the structure of the known compound can be used as a model for the new compound. By making appropriate elemental substitutions and adjusting position coordinates for some atoms to compensate for size effects, the theoretical powder pattern of the new compound can be calculated and compared to the experimental data. If there are differences, the model can be adjusted and a new pattern calculated. By recognizing which changes improve the pattern fit, structure may be refined. Solid-solution effects and order-disorder may be studied by this approach.

Another method is to use modern peak-fitting to decompose complex profiles into their component peaks and to assign unique intensities to each component. If accurate intensities can be assigned, the values may be used in standard singly-crystal structure-solving techniques. This method is most successful when the profile fitting is coupled with calculations that lock the peak positions to the dimensions of the unit cell. There are still problems when two speaks are exactly superimposed and it is not possible to decide how much of the total intensity should be assigned to each part. Usually, there are still enough determinable intensities to be able to solve the structure. This technique is most useful where the structure is not known and it is impossible to obtain a single crystal of the material large enough to yield measurable intensities.

A third method is useful where a model structure is available. This approach, known as the Rieveld method, uses the differences between a calculated digitized diffraction trace and the experimental data to correct parameters involved in the calculation. The parameters in this refinement include the positions and shapes of the diffraction peaks as well as the positions of the atoms. Careful selection of the order of the parameters varied during the refinement usually leads to a best-fit structure that is as accurate as could be achieved using discrete intensity values. This method is very powerful; it can even be used where the sample is a mixture of several compounds. For mixtures, not only are the structures of the component minerals determined simultaneously, the amounts of each mineral may also be determined. Considerable time is required to collect data of sufficient accuracy for this method to produce its results, but the results, but the results are well worth the effort.

F. Crystallite Size and Strain

The shape of the diffraction peaks are controlled by several aspects of the sample in addition to the spectral distribution and the instrumental aberrations. When the crystallites are smaller than 0.2 m, the peaks develop a perceptible symmetrical broadening, which increases significantly as the size decreases. When there are no other sample parameters contributing to the peak shape, the width of the peak may be used as a measure of the crystallite size. The usual measure used is the full width of the profile at half the peak intensity (FWHM). It leads to an estimate of the average crystallite size. More elaborate profile-fitting methods yield little more information but sometimes can be used to estimate the size distribution.

When the sample is under stress, either from external forces or forces developed in a crystallite due to dislocations, or at the grain boundaries due to residual boundary forces, the crystal lattices are deformed. The amount of deformation can be determined by measuring the d-spacings in the stressed and unstressed states. This difference divided by the unstressed states. This difference divided by the unstressed value is known as the strain. If the strains were totally elastic and the lattices of all the grains in the sample were deformed identically, the diffraction peaks would shift from the unstrained position, and the strain could be easily determined. In general, the strains are not elastic, and different parts of the sample are affected differently; so the peaks are smeared over the range from the unstrained position to the position corresponding to the greatest strain. The resulting peak is very asymmetric and represents the distribution of strain values. Complete analysis is possible if there is no broadening from small crystallites; however, the deformation almost always causes the crystal grains to segregate into very small domains. Elaborate mathematical analysis can yield both size and strain data from these samples. The most common analysis is known as the Warren-Averbach method. This method requires deconvolution of the profile of a set of peaks.

Using modern profile-fitting methods, these deformed shapes can be analyzed by assuming a reasonable model of size and strain and refining the profile parameters. The refinement may be done by adjusting the parameters by trial and error or by using an approach similar to the Reitveld method.

Another factor that may contribute to the broadening of a peak is any inhomogeneous variation in the chemical composition of the sample. This situation is very common when a compound is a solid solution. When solid solutions crystallize from the melt, the melt always has a different composition from the solid except at the pure end-member composition. It is very difficult for the atoms to diffuse in the solid state to reequilibrate as a homogeneous solid. Thus different parts of the crystal will have different lattice parameters due to the different composition. Because all parts of the sample contribute to the diffraction peak, the experimental peak will be broadened. This type of broadening is difficult to analyze.

G. Percent Crystallinity and Crystallinity Indexes

Crystallinity has several meanings in diffraction analysis. The most common use of the term is to imply the fraction of crystalline materials in an X-ray amorphous matrix. Although amorphous materials do not create classical diffraction pattern, they do scatter Z rays and contribute to the general background of the diffraction pattern. The amorphous component of the background is usually one or more broad humps added to the other noise. Measuring percent crystallinity with the diffraction pattern usually one or more broad humps added tot he other noise. Measuring percent crystallinity with the diffraction pattern usually consists of taking the ratios of the intensities of the diffraction peaks to the intensity of the amorphous hump. These measurements may be based on area ratios, or on some measure of peak heights and the height of the amorphous peak. In all cases, there must be some sample whose crystallinity was determined by independent determination to calibrate the X-ray measurements.

Crystallinity may also imply the perfection of the crystallites. A sample may contain no truly amorphous fraction, but the peaks are broadened because of imperfections in the way the crystallites are formed; for example, in kaolinite, the layers may be perfect, but they are imperfectly stacked. The diffraction pattern contains both sharp and broad lines. Crystallinity indexes are defined that are some measure of this imperfection. Usually the heights of several peaks are ratioed to the heights of the valleys between the peaks. Well-crystallized samples have well-resolved peaks; but low-crystallinity samples have broad peaks that do not resolve well, and the adjacent peaks interfere. Again the interpretation requires independent measurements for calibration, but usually only the relative indexes to other samples is really important.

H. Crystal Changes with Temperature and Pressure

By using special attachments on diffractometers and cameras, it is possible to measure diffraction patterns when the sample is confined in special atmospheres-at high or low temperatures and/or at high or low pressures. Changes in the crystal structure commonly occur with changes in temperature and pressure. These phase changes are easily revealed by diffraction methods through differences in the diffraction patterns. Lattice parameter measurements indicate the thermal expansion or compressibility of the sample. Temperatures down to near absolute zero and up to 2800°C can be achieved with relatively simple devices. Pressures from 10^{-10} torr to 5 Mbar can also be attained. The very high pressures require tiny diamond anvil cells. This range of conditions allows the study of mineral behavior in almost any part of the earth above the core.

Special atmospheres are sometimes necessary to prevent oxidation and hydration or to control the partial pressure of some component in the gas. Examination of very pure metals may require ultra-high vacuums. Samples may have to be prepared in enclosures and then transferred to the diffraction system. Elaborate devices have been devised to accomplish these tasks, and almost any situation can now be simulated in a diffraction system.

1. Reaction Kinetics

Not only is it important to determine that a reaction does occur, it is also desirable to determine the rate of the reaction under different conditions. Reactions that are slow may be studied by the methods described in the previous section. For more rapid reactions, high-intensity sources are used with diffraction devices that use special detectors that can record simultaneously all the diffracted rays in a wide angular range. When these data are plotted as a function of time, usually in fractions of a second, the changes are easily to follow. Peaks due to the reacting phase disappear together, and the new peaks belong to the reacting phase disappear together, and the new peaks belong to the product phase. Sometimes intermediate phases may be detected that are so transient that they would not be seen in normal experiments.

VI. Applications in Earth System Science

Although all of the analyses described above have applications in earth system science, several important uses will be illustrated in more detail.

A. Mineral Identification

Most common minerals found in the rocks of the earth's crust are complex both chemically and structurally. Powder diffraction is one of the best tools both for identification and for the measurement of quantities like the lattice parameters that are sensitive to the history of formation of the mineral.

The PDF is organized to assist the diffractionist in the identification of a mineral with the designation of the Mineral File. This file contains reference patterns for over 95% of the accepted 3000+ mineral species; the ones that are not represented are very rare species or ill-defined species. Most of the common minerals that show considerable chemical of variation are represented by several patterns. Great care has been used in selecting the patterns to be included in the file, because many more patterns in the literature have been judged less qualified. Even with this selectivity, the diffractionist must use considerable judgment when making the matches because of the potential for chemical variability.

Usually, the mineral family can be easily determined. The particular specie in the family may be distinguished by the intensities or by the lattice dimensions. If the match is not obvious, the unit cell can be refined; and the cell can be compared to determinative curves that can be found in the literature. Most important solid solutions have been studied extensively, and such determinative curves may be found in the original references and in many mineral compendia. Amphibole minerals are especially hard to speciate because of the wide variation of chemical substitution that may occur. Even the end members are not well defined for many of the possible substitutions. Determinative curves are only available for some of the better-studied series. Usually just an identification that the mineral is an amphibole is sufficient for most purposes. When a more definitive identification is necessary, a chemical analysis will be required.

A special case of identification is when a row mineral is encountered. In this situation, the match will not be in the Mineral File. Several approaches may be taken to characterize this new phase. Searching the Mineral File may produce a match with an isostructural mineral; with the new chemistry, this information is enough to confirm the new mineral status. If a match is not found in the Mineral File, the whole PDF should be searched. Many compounds have been described that may occur in nature but have not yet been observed. If this step fails, the diffraction pattern may be "indexed" to determine the until cell, and the other databases searched for related compounds. If all else fails and a single crystal is available, then single-crystal diffraction methods should be used.

B. Clay Mineral Analysis

The identification of clay minerals is a special case of the identification of minerals. All clay minerals are by nature extremely fine grained, and powder diffraction is the best method for their identification. Because clay minerals have a layer-like crystal structure, the crystals may be highly disordered in the way the layers are superimposed, and grains are usually tabular in shape. Consequently, the best approach to obtaining good diffraction patterns from clays is to deliberately cause the grains to orient so that the layers are all parallel. The result is to have all the 001 peaks appear in the diffraction patterns, the clay minerals can be classified into several groups depending on the d-spacing of the first peak and the behavior of these d-spacings with several treatments. If the d-spacing is 10 A, the sample is a mica clay. If the d-spacing is 7 A, the sample is a kaolin or serpentine. A 14-A d-spacing that is unaffected by treatment is a chlorite, and one which changes with dehydration or glycol saturation is a smectite or a vermiculite. More complicated forms can also be recognized. These distinctions can be made even in patterns of mixtures. In most studies, this level of identification is sufficient. Where the identification of specific clay species is required, considerable additional effort is necessary. [See Soil Physics.]

Further, clay mineral analysis includes examining the diffraction intensities of the 001 to determine which elements are in the clay minerals, and examining the d-spacings of the non-001 peaks in a random sample to determine some details of the structures of the layers.

Clay minerals are quite reactive in geologic environments, and accurate identifications provide considerable information on the history of the sample. Most sedimentary rocks contain significant quantities of clay minerals, some of which was among the original particulate materials to comprise the rock and some developed after the rock was formed. It is this latter clay that is indicative of the rock reactions. The clay components can be separated from the rest of the rock by differential settling in water, and then the diffraction methods can be applied. [See Clay Sedimentology.]

Soils also contain abundant clay among their components, which needs diffraction analysis to confirm any identifications. Soil clays, although related to other clay minerals, are usually distinct because they arc in the process of formation and are of low crystallinity. Positive identification involves pretreatments such as controlling the humidity, exchanging the cation, and heating to delineate the species involved. The oriented sample mount is also very useful in these studies.

C. The Characterization of Zeolite

Zeolites are a class of aluminosilicate minerals that have very open crystal structures. They are surprisingly common minerals; and they have economic value as catalysts, as molecular sieves, and in agricultural applications. In addition to being found in cavities in basalts, they form as late-stage alteration products of igneous rocks and volcanic ashes, in hydrothermal ore deposits, and in some sedimentary and metamorphic environments. Diffraction methods are the best way to identify and study these minerals. The aluminosilicate framework is stable, but the other elements and water can easily be exchanged to cause distinct changes in the overall structure that modify the diffraction patterns. An experienced diffractionsit can recognize these differences and identify the mineral variety.

Many zeolites crystallize as spectacular crystals of museum quality, but those samples that form as alterations of other minerals and the industrial forms of zeolites are extremely fine grained. Most of the general crystal structure information was obtained on the single crystals. The fine-grained forms are never identical to these well-crystallized samples. Their study requires starting with the determined structure and modeling changes, then calculating theoretical powder patterns for comparison with the experimental diffraction patterns. The alternate approach is to use the Rietveld refinement to adjust the crystal parameters to better fit the powder data. Once the framework portion of the structure is known, models for the intra-framework cations and water can be devised.

D. Nonmetallic Industrial Raw Materials

Many earth materials are used for industrial products. In addition to rocks like granite and limestone being used directly for building stones and fill materials, there are products like ceramics, cement, and glass, which use special rocks as their source material. The source material must be very pure in must cases, and diffraction methods are used to confirm the purity. Limestone used for cement and quick lime must be low-magnesium calcite with limited amounts of clay or quartz. Glass sand must be a high-quartz sandstone with no refractory contaminants in the trace grains. Industrial clays must have no quartz. In most cases the supply of raw material must be monitored continuously as the products are processed. Diffraction methods are very useful to monitor the supply bins and the feed lines because the analytical results can be obtained in minutes and changes can be made without interrupting the processing.

The determination of ore grades in some mining depends on knowing the minerals and their abundances in the source rocks. The mining of phosphates, halides, sulfates, zeolites, bauxites, and clays requires a knowledge of the dominant mineral or minerals to determine which layers are economic and which are not. Bauxites may contain the aluminum in oxyhydroxides, hydroxides, clays, or oxides. Because the clays and oxides are not amenable to processing, they must be avoided. Clay usefulness depends on the purity and dominant mineral type. Kaolin clays are used for ceramics, and montmorillonites are used for lubricants, fillers, catalysts, and construction materials. Some feldspars may be used for special ceramics like whiteware and false teeth. All of these materials require crystallographic characterization. [See Bauxite and Laterite Soil Ores.]

E. Metallic Mineral Deposits

Deposits of metallic minerals may have the element(s) of interest present in quantities as high as 55% to as low as parts-per-million. Although the element abundances are usually determined by chemical analysis, it is very important to know the mineral containing the elements because this knowledge affects the interpretation of the geology for exploration purposes as well as the method to be used to extract the element from the mineral. Copper is a good example of an element that may occur in many different minerals. Common minerals include chalcopyrite, $CuFeS_2$; bornite, Cu_5FeS_4; chalcocite, Cu_2S; and cubanite, $CuFe_2S_3$; these vary considerably in Cu/Fe ratio. Excessive Fe can complicate the ore processing; the mill operator must know which minerals are being processed, because even small changes in the Cu/Fe ratio can upset the mill procedures. The mineral identification is accomplished by diffraction methods. Knowledge of the specific ore minerals and the associated minerals can indicate the temperatures of formation of the ore deposit and can be used to map spatial zones, which helps the exploration for more ore.

Where the element concentration is very low, the elements are usually substituted in some other host mineral. Rare-earth elements are commonly substituted in appetites; Fe, Mn, and Cd may be substituted in sphalerites; Hf substitutes for Zr in several minerals; Cs, Rb, Ba, and Sr substitute in feldspars; and Sc and Ga substitute for Al in some aluminosilicate Diffraction methods can be used to reveal the lattice distortions due to the substitutions and provide information on the role of the host phase. Where the abundance of the host mineral is small, preconcentration methods may be required to obtain enough sample for a good analysis.

F. Phase Transformations in the Earth

Direct sampling of the earth's crust is limited to the surface and some deep mines and drill holes. Mineral reactions occurring deeper in the crust and in the mantle must be modeled or simulated in the laboratory to be studied. Where the experimental reactions can be quenched to trap the phases formed, diffraction analysis can easily identify the phases. But, most of the time, the reactions cannot be quenched, and it is necessary to examine the sample in situ. Temperature-driven reactions can be examined in resistance-heated furnaces that have windows for the X-ray beam. Alternatively, devices using laser beams, induction coupling, and focused are images for heating have been employed. Diffractometers are the most versatile for temperatures above 1000°C, and the samples can be large enough to yield good data in short periods of time. Pressure devices usually employ camera geometries. Many of the discontinuities in the mantle are interpreted to be caused by density changes associated with phase changes or reactions involving the prominent minerals caused by the high pressures. Tiny samples between diamonds in X-ray cells at pressures up to 5 Mbar can be used to simulate the conditions in the lower mantle. Much can be used to simulate the conditions in the lower mantle. Much has been learned by these studies. [See Mantle Metasomatism.]

G. Environmental Hazards

Modern society is becoming more aware of hazards in the environment. The primary hazards of concern are toxic and carcinogenic materials and undesirable components in the air. Two hazards that are monitored by diffraction analyses are asbestos types and silica forms.

Asbestos has attracted considerable attention in recent years as a potential carcinogen. Federal regulations require its removal in public buildings, so its proper identification is critical. Regulations define asbestos by shape where the length is over four times the other dimensions. Many different minerals meet this definition, but they differ considerably in their potential as a carcinogen. The amphibole asbestos types are well documented as highly dangerous, but the serpentine types have not been confirmed to be a hazard. Other fibrous minerals are not considered hazardous. Scanning electron microscopy may determine the shape, but diffraction is necessary to determine the mineral identification. Diffraction techniques also determine the other minerals when the asbestos in incorporated in plasters and wall boards.

All the crystalline forms of silica, SiO_2, are considered hazardous. The most common form, quartz, is a very common mineral and a major component of dust in the household and industrial facilities. Freshly broken particles in the respirable size range, 1-10 m, are confirmed as toxic. Other silica minerals in this range are also considered toxic, so all forms of silica must be monitored in industrial plants and in mines. The standard method established by OSHA (Occupational Safety and Health Administration) includes an X-ray diffraction procedure to identify and quantify the quartz, cristobalite, and tridymite content of the airborne dust in plant atmospheres. Also, any product containing free silica minerals must be so labeled if the concentration is greater than 0.1 weight percent. Most industries with this requirement use X-ray diffraction methods to determine the quartz.

Airborne particulates can also be identified as individual particles. Using the Gandolfi modification of the Debye-Scherrer camera with a high-intensity X-ray source, particles down to 2 m can be made to produce identifiable diffraction patterns with exposure times around 24 hours. Particles are collected on filters and isolated under a microscope. Successful results require careful centering in the camera. This technique also is applicable to analyzing point-by-point reactions in samples that contain so many phases that a bulk analysis fails to confirm which phase belongs to which spot.

H. Solid-Gas Reactions and Corrosion

Oxidation, carbonation, and hydration are reactions that plague the lifetime and usefulness of products. The corrosion products are always very fine grained, usually occur as thin surface films, and are often poorly crystallized. Diffraction methods are very useful to identify the phases that make up these coatings. Many times there are several phases that cannot be separated for individual identification. Such mixtures can be unraveled using the diffraction pattern and some help from the scanning electron microscope, micro-Raman spectroscopy, and Fourier transform infrared spectroscopy for additional information.

Where the corrosion products defy complete characterization, reactions may be induced in diffraction chambers in which appropriate atmospheres surround the reacting samples. The surface of the sample is monitored with time, and it may be heated to induce the reaction. Special diffractometers with the incident beam at a very low angle to the surface limits the beam penetration and enhances the diffracted intensities. Films as thin as 50 nm can be studied with this setup.

VII. Summary

Powder diffraction techniques have many uses and the method is very versatile. Suitable equipment can be set up in almost any laboratory. Many types of analysis are quite routine, and operators do not need extensive training to perform these analyses. In the hands of experienced diffractionists, the technique can provide many useful results on a wide variety of materials. Like all materials analytical techniques, there are unique advantages and definite limitations.

A. Sampling and Sample Representation

Diffraction techniques use relatively little sample, which can be both an advantage and a disadvantage. The chief advantage in the small size means that it is usually easier to obtain a phase-pure sample, which is easier to interpret. Frequently, the extra effort made to purify the sample prior to obtaining the diffraction data saves many hours doing the analysis. The small sample also allows the diffractionist to select specific parts of a complex piece for determining the spatial distribution of the phases. The disadvantages appear whenever accurate intensities are required, because it is difficult to randomize the crystal orientations in the small sample volume. Many times it is desirable to make up more than one sample and compare the measurements from each by statistical methods. When the analysis performed is for quantifying the phase composition, there is the question as to how well the small volume of sample used represents the bulk sample being analyzed.

B. Significance of Data and Measurements

Like other techniques that involve the measurement of numerical data, results may be evaluated by statistical methods. However, the statistical measures indicate only the precision of the data where only random errors are involved. Careful alignment of the instruments and the use of both internal and external reference standards can minimize the systematic errors, and mathematical methods can compensate in part for the remaining systematic errors. Using modern diffraction equipment, the level of accuracy is often limited by the crystallinity of the sample and how it is prepared. Under ideal conditions, d-spacings and cell parameters may be measured to 1 part in 10,000 with routine methods and to 1 part in 50,000 with more care. Intensity measurements have more problems and 1 part in 100 may be difficult to achieve.

Because many diffraction peaks are usually measured for each experiment, there is an over-determination of the data required for the calculations involved. Consequently, it is possible to use mathematical fitting procedures to calculate the best fit results and their corresponding statistics. The best way to test procedures is to measure "round-robin" samples for comparisons to expected values determined by other techniques and in other laboratories. These tests reveal the differences between accuracy and precision, and usually indicate that the statistics underestimate the real errors. Such tests also reveal significant differences in values obtained from instruments of different manufacture because of different philosophies of collecting and analyzing the raw data. Experienced diffractionists are usually conservative in the results they report.

Some diffraction techniques are successful even with low-quality data. Identification by pattern matching may often be successful even with partial data. An experienced diffractionist may be able to guess possibilities using all the supporting information, such as probable chemistry and conditions of formation. Intuition in diffraction analysis is often the key to success.

C. Diffraction Service Laboratories

Many industries, research centers, and universities maintain diffraction facilities. In general, these facilities are only available to employed personnel. For the user who needs less frequent analyses, there are many private laboratories that offer diffraction services along with other types of analyses. These laboratories may be located by referring to directories of laboratory services or the ads in professional journals like Powder Diffraction. Equipment manufacturers will analyze a few samples for prospective customers.

Very specialized facilities are maintained by several of the national laboratories and some universities. Because these facilities are sponsored at least in part by federal funds, their use is open to anyone with a justifiable project. The procedures to acquire time on these systems require presenting a proposal to the governing committee and covering the expenses of preparing and running the experiment.

D. Using Corroborating Information

Although powder diffraction techniques provide considerable information on crystalline materials, additional data from other types of measurements often assist in achieving more complete identifications. Chemical analyses help confirm the identifications. Microscopic observations, either optical or electron, can indicate the number of phases and often suggest the crystal systems of the grains. Infrared and Raman spectroscopies provide measurements that indicate the nature of the bonding and the presence of order/disorder effects. [See Infrared and Raman Spectroscopy in the Geosciences.]

As has been presented above, powder diffraction is an important tool when used to characterize crystals in support of interpretations of other physical property measurements. The interest in superconductivity, for example, illustrates the need to know the nature of the active phases because of their similarities to inactive phases. Physicists need to know the structure of their crystals before performing time-consuming measurements of crystal properties. The chemist needs to confirm the starting materials and products of reactions. The materials scientist needs to know the crystalline state of the components of complex materials. The geologist must identify the minerals in the rocks and in experiments. There is hardly any field of research that cannot benefit from diffraction studies.

GLOSSARY

Bragg Plane. Imaginary planes of lattice points.

Diffraction Intensities. Strength of the diffracted beam measured in counts per second.

Diffraction Pattern. Collection of all observable diffraction peaks characteristic of a compound.